Plant
Pathology

SECOND EDITION

Plant Pathology

GEORGE N. AGRIOS

Professor
Department of Plant Pathology
University of Massachusetts

ACADEMIC PRESS

New York   San Francisco   London
A Subsidiary of Harcourt Brace Jovanovich, Publishers

To Nick, Tony, and Alexy

ACADEMIC PRESS, INC.
111 Fifth Avenue, New York, New York 10003

United Kingdom Edition published by
ACADEMIC PRESS, INC. (LONDON) LTD.
24/28 Oval Road, London NW1

ISBN: 0-12-044560-3
Library of Congress Catalog Card Number: 78-202

PRINTED IN THE UNITED STATES OF AMERICA

preface to second edition

Since the first edition of PLANT PATHOLOGY appeared, several important developments have emerged, both in the field of plant pathology and in public attitudes toward food production. More people have become aware that all the food and much of the beauty around us come from plants. These people now want to know more about how to grow plants and keep them healthy. Many people also have become more aware of, and concerned about, chemical pollutants in the environment and in our food. The number of students majoring or taking courses in the various agriculture-related disciplines has increased dramatically. These students come from varied backgrounds, and many are not aiming for comprehension of the biochemical intricacies that lead to the development of disease. In fact, most students and practitioners of plant pathology are concerned primarily with the practical aspects of plant pathology: what causes the disease and what can be done about it. Most of these students are likely to take only one course in plant pathology. This course will form the basis for their attempts to diagnose and control diseases in the field, as well as to develop a general understanding of plant diseases that they can apply in their disciplines and their jobs.

The second edition of PLANT PATHOLOGY incorporates many of the new developments in the field. It also provides information and visual aids that should prove useful in illustrating concepts, as well as in identifying pathogens and diagnosing diseases. Thus, new chapters or sections have been added on diseases caused by mycoplasmalike organisms, rickettsialike bacteria, viroids, and protozoa. New information has been added on the genetics of plant diseases, the development of resistant varieties, and their vulnerability to new pathogen races; on the development of epidemics; on systemic fungicides and biological control of diseases; on postharvest diseases of plant products and on mycotoxins and mycotoxicoses; on techniques of isolation, culturing, indexing, and identification of pathogens; and on mycorrhizae and root-nodule bacteria.

More diseases have been described in each chapter to give the instructor a greater selection in the diseases he or she chooses to cover. The diseases caused by fungi and those caused by bacteria have been organized in logical, cohesive groups according to their most important symptoms; thus, more of them are available to the instructor and student while the actual number of groups that must be covered is relatively small. In these and in the other chapters on diseases caused by specific pathogens, *selection* is the key word. It is not intended that the instructor cover all, or even half, of the diseases described. Rather, after discussion of the general properties of each group of diseases, the instructor can select one or a few of the diseases within each group for more or less detailed study.

Many more diagrams of disease cycles, of groups of pathogens and of symptoms, and of techniques and concepts of plant pathology have been added. The diagrams should help students visualize many of the concepts and will provide a handy reference for the identification of most common pathogens and the diagnosis of most important plant diseases. Moreover, numerous new photographs (macroscopic, microscopic, electron micrographs, and scanning electron micrographs) that illustrate concepts, pathogens, and symptoms have been added to make the book more stimulating and more effective.

Many people helped make the second edition possible. Of these, Ms. Joan Weeks, who prepared most of the new drawings, and Mrs. Joyce Mieg, who typed the entire manuscript with cheerfulness, efficiency, and excellence, have my most sincere appreciation and thanks. I am also thankful to Ms. Janet Murray, who prepared some of the drawings.

I wish to acknowledge the support which R. A. Rohde, Head of the Department of Plant Pathology, gave to my work on the book by making the resources of the Department available to me. I also want to thank A. J. Browning, R. J. Campana, R. E. Davis, T. O. Diener, C. A. Martinson, M. S. Mount, R. A. Rohde, D. Sands, S. H. Smith, T. A. Tattar, and R. F. Whitcomb for reviewing parts of the manuscript. Also, I am indebted to M. F. Brown, D. J. Politis, and R. N. Goodman, all of the Department of Plant Pathology, University of Missouri, for their help with numerous excellent electron micrographs; E. P. DuCharme, M. C. Shurtleff, and G. Nyland for their help in locating several photos for the book; the USDA photo library and numerous other colleagues who loaned me photographs and whose names are listed in the legends of these photographs. Finally, I want to express my thanks to the editors and the staff of Academic Press for their unexcelled cooperation in the production of this text.

GEORGE N. AGRIOS

contents

part
1
general
aspects

1
introduction

The welfare of plants is of particular interest to those most directly concerned with the growth of plants and the manufacture and distribution of plant products. It is of concern not only to farmers and workers in industries that process agricultural products, but also to innumerable workers in supporting industries whose livelihood depends on making equipment or products used in processing plant products—for example, machinery for textile and canning industries—or on distributing the raw or manufactured agricultural products. Most importantly, however, the welfare of plants should be of concern to every one of us as growers of plants for food or pleasure, as individuals concerned with the beauty and safety of our natural environment and, particularly, as consumers of plants and of the endless series of products derived from plants.

The growth and yield of plants depend on the availability of nutrients and water in the soil where they grow and on the maintenance within certain ranges of such environmental factors as light, temperature, and moisture. Plant growth and yield depend also on protecting the plants from parasites. Anything that affects the health of plants is likely to affect their growth and yield and may seriously reduce their usefulness to themselves and to mankind. Plant pathogens, unfavorable weather, weeds, and insect pests are the most common causes of reduction or destruction of plant growth and production. Plants suffer from diseases whose causes are similar to those affecting animals and man. Although there is no evidence that plants feel pain and discomfort, the development of disease follows the same steps and is usually as complex in plants as it is in animals and man.

Plant pathology is the study of (1) the living entities and the environmental conditions that cause disease in plants; (2) the mechanisms by which these factors produce disease in plants; (3) the interactions be-

3

tween the disease-causing agents and the diseased plant; and (4) the methods of preventing disease, alleviating the damage it causes, or controlling a disease either before or after it develops in a plant.

Plant pathologists study the diseases caused by fungi, bacteria, mycoplasmas, parasitic higher plants, viruses, viroids, nematodes, and protozoa. They also study plant disorders caused by the excess, imbalance, or lack of certain physical or chemical factors, such as moisture, temperature, and nutrients. Plant damages caused by insects, man, or other animals are not ordinarily included in the study of plant pathology.

Plant pathology utilizes the basic techniques and knowledge of botany, mycology, bacteriology, virology, nematology, plant anatomy, plant physiology, genetics, biochemistry, horticulture, soil science, forestry, chemistry, physics, meteorology, and many other branches of science. Plant pathology profits from advances in any one of these sciences, and many advances in other sciences have been made in the attempt to solve phytopathological problems. A good knowledge of at least the basic facts of the related sciences is indispensable for efficient performance by any plant pathologist.

Although plant pathology as a science attempts to increase our knowledge of the causes and the development of plant diseases, it is also a science with a more practical goal. The purpose is to develop controls for all plant diseases. The goal is to save the produce which today is destroyed by plant diseases and to make it available to the growers who toil to produce it and to the hungry and ill-clothed millions of our increasingly overpopulated world.

the concept of
disease in plants

A plant is healthy or normal when it can carry out its physiological functions to the best of its genetic potential. These functions include normal cell division, differentiation, and development; absorption of water and minerals from the soil and translocation of these throughout the plant; photosynthesis and translocation of the photosynthetic products to areas of utilization or storage; metabolism of synthesized compounds; reproduction; and storage of food supplies for overwintering or reproduction.

Whenever plants are disturbed by pathogens or by certain environmental conditions and one or more of these functions are interfered with beyond a certain deviation from the normal, then the plants become diseased. The primary causes of disease are either pathogens or factors in the physical environment. The specific mechanisms by which diseases are produced vary considerably with the causal agent and sometimes with the plant. At first the reaction of the plant to the disease-causing agent is at the site of affliction, is of a chemical nature, and is invisible. Soon, however, the reaction becomes more widespread and histological changes take place that manifest themselves macroscopically and constitute the symptoms of the disease.

Affected cells and tissues of diseased plants are usually weakened or destroyed by the disease-causing agents. The ability of such cells and tissues to perform their normal physiological functions is reduced or completely eliminated; as a result, plant growth is reduced or the plant dies. The kinds of cells and tissues that become infected determine the type of physiological function of the plant that will be interfered with first. Thus, infection of the root (e.g., root rots) interferes with absorption of water and nutrients from the soil; infection of the xylem vessels (vascular wilts, certain cankers) interferes with translocation of water and minerals to the crown of the plant; infection of the foliage (leaf spots, blights, mosaics) interferes with photosynthesis; infection of the cortex (cortical canker, viral infections of phloem) interferes with the downward translocation of photosynthetic products; flower infections (bacterial and fungal blights, viral, mycoplasmal, and fungal infections of flowers) interfere with reproduction; and infections of fruit (fruit rots) interfere with reproduction and/or storage of reserve foods for the new plant (Fig. 1).

In contrast to the above, there is another group of diseases in which the affected cells, instead of being weakened or destroyed, are stimulated to divide much faster (hyperplasia) or to enlarge a great deal more (hypertrophy) than normal cells. Such hyperplastic or hypertrophied cells result in the development of usually nonfunctioning, abnormally large, or abnormally proliferating organs or in the production of amorphous overgrowths on normal-looking organs. Overstimulated cells and tissues not only divert much of the available food stuffs to themselves and away from the normal tissues, but frequently, by their excessive growth, crush adjacent normal tissues and interfere with the physiological functions of the plant.

Disease in plants, then, can be defined as any disturbance brought about by a pathogen or an environmental factor which interferes with manufacture, translocation, or utilization of food, mineral nutrients, and water in such a way that the affected plant changes in appearance and/or yields less than a normal, healthy plant of the same variety. Pathogens may cause disease in plants by (1) consuming the contents of the host cells upon contact; (2) killing or disturbing the metabolism of host cells through toxins, enzymes, or growth-regulating substances they secrete; (3) weakening the host by continually absorbing food from the host cells for their own use; and (4) blocking the transportation of food, mineral nutrients, and water through the conductive tissues. Diseases caused by environmental factors result from extremes in the conditions supporting life (temperature, light, etc.) and in amounts of chemicals absorbed or required by plants.

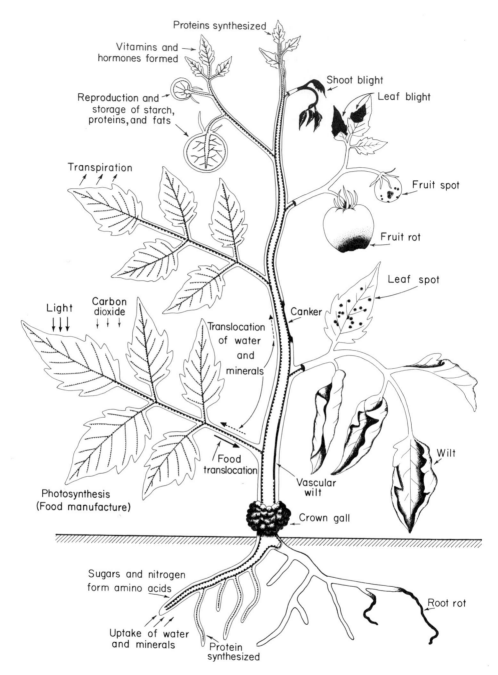

Proteins synthesized

Vitamins and
hormones formed

Shoot blight

Leaf blight

Reproduction and
storage of starch,
proteins, and fats

Fruit spot

Transpiration

Fruit rot

Leaf spot

Light   Carbon
dioxide

Canker

Translocation
of water
and
minerals

Food
translocation

Wilt

Photosynthesis
(Food manufacture)

Vascular
wilt

Crown gall

Sugars and nitrogen
form amino acids

Root rot

Uptake of water
and minerals

Protein
synthesized

FIGURE 1.
Schematic representation of the basic functions in a plant and of the interference
with these functions caused by some common types of plant diseases.

classification of plant diseases

There are tens of thousands of diseases that affect cultivated plants. On the average, each kind of crop plant can be affected by one hundred or more plant diseases. Each kind of pathogen may affect anywhere from one variety to several dozen or even hundreds of species of plants. To facilitate the study of plant diseases, they must be grouped in some orderly fashion. This is necessary also for the identification and subsequent control of any given plant disease. Any one of several criteria may be used as a basis for classification of plant diseases. Plant diseases are sometimes classified according to symptoms they cause (root rots, cankers, wilts, leaf spots, scabs, blights, anthracnoses, rusts, smuts, mosaics, yellows, ring spots), according to the plant organ they affect (root diseases, stem diseases, foliage diseases, fruit diseases), or according to the types of plants affected (field crops diseases, vegetable diseases, fruit tree diseases, forest diseases, turf diseases, diseases of ornamental plants). However, the most useful criterion for classification of a disease is the type of pathogen that causes the disease (Figs. 2 and 3). Such a classification has the advantage that it indicates the cause of the disease, knowledge of which suggests the probable development and spread of the disease and also possible control measures for the disease. On this basis plant diseases are classified as follows:

   I. Infectious Plant Diseases
      1. Diseases caused by fungi
      2. Diseases caused by bacteria
      3. Diseases caused by mycoplasmas
      4. Diseases caused by parasitic higher plants
      5. Diseases caused by viruses and viroids
      6. Diseases caused by nematodes
      7. Diseases caused by protozoa
  II. Noninfectious or Physiological Disorders

1. Too low or too high temperature	6. Nutrient deficiencies
2. Lack or excess of soil moisture	7. Mineral toxicities
3. Lack or excess of light	8. Soil acidity or alkalinity (pH)
4. Lack of oxygen	9. Toxicity of pesticides
5. Air pollution	10. Improper agricultural practices

history of plant pathology

Man became painfully aware of plant diseases in the early times of antiquity. This is evidenced by the inclusion in the Old Testament of blasting and mildew, along with human diseases and war, among the

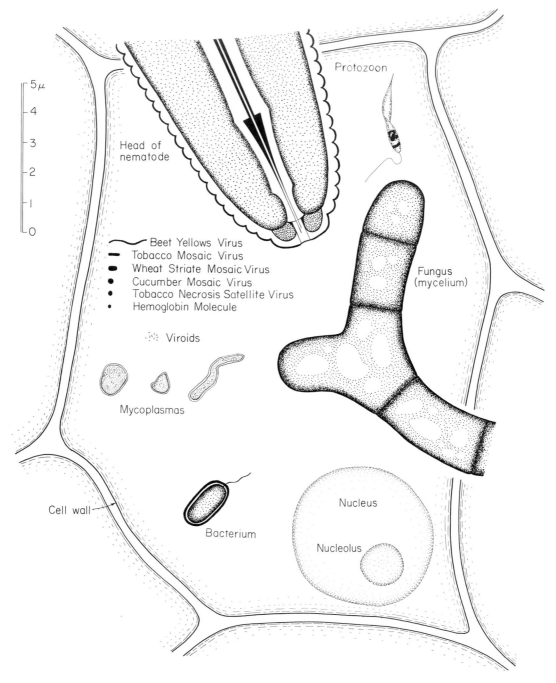

5μ
4
3
2
1
0

Protozoon

Head of
nematode

Beet Yellows Virus
Tobacco Mosaic Virus
Wheat Striate Mosaic Virus
Cucumber Mosaic Virus
Tobacco Necrosis Satellite Virus
Hemoglobin Molecule

Viroids

Fungus
(mycelium)

Mycoplasmas

Cell wall

Bacterium

Nucleus

Nucleolus

FIGURE 2.
Schematic diagram of the shapes and sizes of certain plant pathogens in relation
to a plant cell.

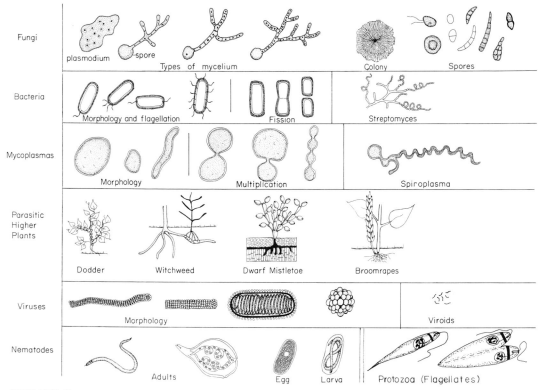

FIGURE 3.
Morphology and multiplication of some of the groups of plant pathogens.

great scourges of mankind. The Greek philosopher Theophrastus (370–286 B.C.) was the first actually to study and write about diseases of trees, cereals, and legumes, although his approach was observational and speculative rather than experimental. During the following 2000 years, little was added to the knowledge of plant pathology, although references to the ravages of plant diseases appeared in the writings of several contemporary historians.

The discovery of the compound microscope around the middle of the 17th century opened a new era in the life sciences. The anatomy of plants was studied and described, and the fungi, bacteria, and many other microorganisms were discovered.

Fungi. In 1755, Tillet added the black dust from bunted wheat to seed from healthy wheat and observed that bunt was much more prevalent in plants produced from such seed than from nondusted seed. He thus showed that bunt, or stinking smut, of wheat is a contagious plant disease. He also showed that its occurrence can be reduced by seed treatments. Tillet, however, believed that it was a poisonous substance contained in the dust, rather than living microorganisms, that caused the disease.

In 1807, Prevost proved conclusively that bunt is caused by a fungus; he studied the spores, their production and germination. He could control the disease by dipping the seed in a copper sulfate solution, and he pointed out the importance of the environment in induction and development of the disease. Prevost's findings, however, were ahead of his time and were rejected by almost all his contemporaries, who believed in spontaneous generation.

The devastating epidemics of late blight of potato in Northern Europe, particularly Ireland, in the 1840s tragically dramatized the importance of plant diseases and greatly stimulated interest in their causes. The destruction of the potato crop in Ireland in 1845 and 1846 caused widespread famine which resulted in the death of hundreds of thousands of people and the immigration of more than one and a half million Irish to the U.S. Several investigators described various aspects of the disease and of the pathogen, but it was DeBary (1861) who finally proved experimentally that the fungus *Phytophthora infestans* is the cause of the disease.

DeBary (1853), working at first with smut and rust fungi, established conclusively that fungi are causes, not results, of plant disease. He described the microscopical structure and development of many smut and rust fungi and the relationships of these fungi to the tissues of the diseased plants. DeBary also made great contributions with his studies of the Peronosporaceae and the diseases they incite (downy mildews), especially the late blight of potato, his discovery of the occurrence of two alternate hosts in the rusts, and his studies of the physiology of the *Sclerotinia* rot diseases of carrots and other vegetables. In the *Sclerotinia* diseases, DeBary noted that host cells were killed in advance of the invading hyphae of the fungus and that juice from rotted tissue could break down healthy host tissue. Boiled juice from rotted tissue had no effect on healthy tissue. DeBary concluded that the pathogen produces enzymes that degrade and kill plant cells from which the fungus can then obtain its nutrients.

Brefeld (1875, 1883, 1912) contributed greatly to plant pathology by introducing and developing modern techniques for growing microorganisms in pure culture. In this he was assisted a great deal by the methods and refinements developed by Koch, Petri, and others. Brefeld also studied and illustrated the complete life cycles of the smut fungi and diseases of cereal crops.

In 1878, a new disease, the downy mildew of grape, was introduced into Europe from the United States, spread rapidly, and threatened to ruin the vineyards of Europe. In 1882, Millardet noticed that vines which had been sprayed with the bluish-white mixture of copper sulfate and lime to deter pilferers retained their leaves through the season, whereas the leaves of untreated vines had been killed by the disease and had fallen to the ground. After numerous spraying experiments, Millardet concluded in 1885 that a mixture of copper sulfate and hydrated lime could effectively control the downy mildew of grape. This mixture became known as "Bordeaux mixture," and its success in controlling downy mildews and many other foliage diseases was spectacular. Even today Bordeaux mixture is one of the most widely used fungicides all over the world. The discovery of Bordeaux mixture gave great encouragement and stimulus to the study of the nature and control of plant diseases.

In the early 1900s, studies of the genetics of disease resistance in the cereal rusts by Biffen (1905) and in the *Fusarium* wilts of cotton, watermelon, and cowpea by Orton (1900) led to the selection and breeding of resistant varieties in these and in other crops.

Bacteria. In the meantime, Pasteur and Koch had proved in 1876 that the animal disease anthrax is incited by a bacterium. In 1878, Burrill showed that fire blight of pear and apple is also caused by a bacterium. Soon after that, several other plant diseases were shown to be caused by bacteria; E. F. Smith's numerous and excellent contributions from 1895 on to the study of bacterial diseases of plants established beyond any doubt the importance of bacteria as phytopathogens.

Nematodes. The first plant parasitic nematodes were reported by Needham in 1743 within wheat galls (kernels). It was in the 1850s, however, before other nematodes such as the root knot, the bulb and stem, and the cyst nematodes were observed. A series of studies on plant parasitic nematodes were made by Cobb from 1913 to 1932 and these contributed greatly to nematode taxonomy, morphology, and methodology.

Viruses. In 1886, Mayer reproduced the "tobacco mosaic" disease by injecting juice from infected tobacco plants into healthy plants. The juice of diseased plants remained infective even after continual heating at 60°C, although it lost its infectivity after several hours of heating at 80°C. Since no fungi were present on the diseased plant or the filtered juice, he concluded that tobacco mosaic was probably caused by a bacterium. In 1892, Ivanowski showed that the causal agent of tobacco mosaic could even go through a filter that retains bacteria. This led him to believe that the disease was caused by a toxin secreted by bacteria or by small bacteria that passed through the pores of the filter. Beijerinck (1898) finally concluded that tobacco mosaic was caused not by a microorganism but by a *contagium vivum fluidum*, which he called a virus.

It was not until 1935, however, that Stanley obtained an infectious crystalline protein by treating juice from infected tobacco plants with ammonium sulfate and concluded that the virus could be considered as an autocatalytic protein which could multiply within living cells. In 1936, Bawden and his colleagues demonstrated that the crystalline preparations of the virus actually consisted of protein and nucleic acid. The first virus particles were viewed with the electron microscope by Kausche and his colleagues in 1939. In 1956, Gierer and Schramm showed that the protein could be removed from the virus and that the nucleic acid alone could infect a plant and could reproduce the complete virus.

Protozoa. In 1909, Lafont observed flagellate protozoa in the latex-bearing cells of laticiferous plants of the family Euphorbiaceae, but protozoa found in laticiferous plants were thought to be parasitizing the latex without causing disease to the host plants. In 1931, Stahel found flagellates infecting the phloem of coffee trees and causing abnormal phloem formation and wilting of the trees. In 1963, Vermeulen presented additional and more convincing evidence of the pathogenicity of flagellates to coffee trees, and in 1976 flagellates were also reported from the phloem of coconut palm trees infected with the "hartrot" disease.

Mycoplasmas. In 1967, Doi and his colleagues in Japan observed mycoplasmalike bodies in the phloem of plants infected with several leafhopper-transmitted diseases. The same year Ishiie and his colleagues showed that the mycoplasmalike bodies and the symptoms disappeared temporarily when the plants were treated with tetracycline antibiotics. Similar bodies have since been found in many other yellows- or witches'-broom-type diseases previously thought to be caused by viruses. In 1972, Davis and his colleagues observed a motile, helical microorganism associated with corn stunt disease; they called it spiroplasma. It has since been shown that spiroplasmas are the cause of corn stunt and other plant diseases. Spiroplasmas resemble mycoplasmas and bacteria in some respects but their relationship to mycoplasmas, bacteria or any other microorganisms is still unknown.

Viroids. In 1971, Diener determined that the potato spindle tuber disease was caused by a small molecule of infectious ribonucleic acid (RNA) which he called a "viroid." Viroids are too small to multiply themselves and comprise the smallest known agents of infectious diseases. Several other plant diseases are now known to be caused by viroids.

Rickettsialike bacteria. In 1972, Windsor and Black observed rickettsialike organisms in the phloem of clover plants infected with the club leaf disease. The following year similar organisms were observed in grape infected with Pierce's disease, in peach infected with phony peach, and others. These pathogens are transmitted by leafhoppers and are present exclusively or primarily in the phloem or xylem elements of plants. They are apparently a new kind of bacteria but so far little is known about their nature and properties.

During the 20th century, plant pathology has matured as a science. Thousands of diseases have been described, pathogens have been identified, new kinds of plant pathogens have been discovered, and control measures have been developed. The studies of genetics and of the physiology of diseases have been expanded greatly, and new chemical compounds are being developed continually to combat plant diseases. Still, this is probably just the beginning of plant pathology and of the hope that it holds for the future. The huge losses in plants and plant products that occur annually are the single best reminder of how much is yet to be learned about plant diseases and their control. There are thousands of plant diseases that we know little or nothing about; there are probably new types of pathogens that cause plant diseases and are awaiting discovery; our knowledge of the physiology of plant diseases is dreadfully incomplete; and there must surely be better materials and methods for controlling plant diseases that are waiting to be produced and developed. Progress in any and all of these areas is the goal of plant pathology. And a hungry, overpopulated world is anxiously awaiting the results.

*importance
of plant diseases*

KINDS AND AMOUNTS OF LOSSES

Plant diseases are important to man because they cause damage to plants and plant products. For millions of people all over the world who still depend on their own plant produce for their existence, plant diseases can make the difference between a happy life and a life haunted by hunger or can even result in death from starvation. The death from starvation of a quarter million Irish people in 1845 and much of the hunger of the underfed millions living in the underdeveloped, rural countries today are morbid examples of the consequences of plant diseases. For countries where food is plentiful, plant diseases are important because they cause economic losses to growers, they result in increased prices of products to consumers, and they destroy the beauty of the environment by damaging plants around homes, along streets, in parks, and in forests.

Plant diseases may limit the kinds of plants that can grow in a large geographical area by destroying all plants of certain species that are extremely susceptible to a particular disease; this is exemplified by the American chestnut, which was annihilated in North America as a timber tree by the chestnut blight disease, and by the American elm, which is being eliminated as a shade tree by the Dutch elm disease. Plant diseases may also determine the kinds of agricultural industries and the level of employment in an area by affecting the amount and kind of produce available for canning or processing by the industries in the area. On the other hand, plant diseases are responsible also for the creation of new industries which develop chemicals, machinery, and methods to control plant diseases; the annual expenditures to this end amount to billions of dollars in the U.S. alone.

The kinds and amounts of losses caused by plant diseases vary with the plant or plant product, the pathogen, locality, environment, control measures practiced, etc., or combinations of these factors. The amount of losses may range from slight loss to 100 percent loss. Plants or plant products may be reduced in quantity by disease in the field, as indeed is the case with most plant diseases, or by disease during storage, as is the case of the rots of stored fruits, vegetables, grains, and fibers. Frequently, severe losses are caused by reduction in the quality of plant products. For instance, spots, scabs, blemishes, and blotches on fruit, vegetables, or ornamental plants may have little effect on the quantity produced, but the inferior quality of the product may reduce the market value so much that production is unprofitable or a total loss. Some diseases, e.g., ergot of rye, make plant products unfit for human or animal consumption by making them poisonous.

Financial losses resulting from plant diseases may be incurred indirectly by the farmer having to plant varieties or species of plants that are resistant to disease but are less productive, or more costly, or commercially less profitable; by having to spray or otherwise control a disease, thus incurring expenses for chemicals, machinery, storage space, and labor; by having to provide refrigerated warehouses and transportation vehicles, thereby increasing expenses; by limiting the time during which products can be kept fresh and healthy, thus forcing growers to sell during a short period when products are abundant and prices are low; by necessitating the sorting of healthy from diseased products, and thus increasing costs of handling plant products.

Some plant diseases can be controlled almost entirely by one or another method, thus resulting in financial losses only to the amount of the cost of the control. Sometimes, however, this cost may be almost as high as, or even higher than, the return expected from the crop, as in the case of some diseases of small grains. For other diseases no effective control measures are known as yet, and only a combination of cultural practices and somewhat resistant varieties make it possible to raise a crop. For most plant diseases, however, practical controls are available although some losses may be incurred in spite of the control measures taken. In these cases, though, the benefits from the control applied are generally much greater than the combined direct losses from the disease and the indirect losses due to expenses for control.

SOME HISTORICAL AND PRESENT EXAMPLES OF LOSSES CAUSED BY PLANT DISEASES

For thousands of years, mankind has depended on a few crop plants for its sustenance and survival. Wheat, rice, corn, a few other cereals, potatoes, and some legumes have provided the staple food for man in different parts of the world. The same or related plants are used as feed for all domesticated animals, which are then used by humans for food, as energy sources, or for pleasure. As human societies developed, the needs of people for fiber plants for more and better clothing kept on increasing. Cotton was and still is the main fiber crop, but flax, hemp, jute, and sisal have been important in some parts of the world. Wood and wood products filled, at first, the needs for tools, shelter, and furniture, but recently industrial uses for paper, plastics, etc., have increased their demand tremendously. Industry has also been reaching more and more for plants as raw materials, e.g., rubber, synthetic fibers, drugs, and a large variety of organic compounds. Improved living conditions also created and increased the need for more and better fruits, vegetables, sugar, and oil crops, which are part of a normal, healthful diet, as well as the need for luxury or pleasure crops such as tobacco, coffee, tea, and cacao. Finally, plants have always been a necessary part of man's environment for aesthetic reasons, but also because they provide a moderating force in balancing the concentration of carbon dioxide in the atmosphere and in preventing floods and erosion of the soil and because they improve the physical properties and fertility of the soil by adding organic matter to it.

Plant diseases have affected the existence, adequate growth, or productivity of each of the above kinds of plants and thereby one or more of the basic prerequisites for a healthy, safe life for humans since the time they gave up their dependence on wild game and wild fruit for their existence, became more stationary and domesticated and began to practice agriculture more than 6000 years ago. Destruction of food and feed production and supplies by diseases has been an all too common occurrence in the past and has resulted in malnutrition, starvation, migration, or death of people and animals in numerous occasions, several of which are well documented in history. Similar effects are observed annually in underdeveloped, agrarian societies, in which the families and nations are dependent for their sustenance on their own produce. In more developed societies, losses from diseases in food and feed produce result primarily in financial losses and higher prices with much less direct local malnutrition and starvation. But it should be kept in mind that lost food or feed produce means less such available in the world economy and, considering the perennially inadequate amounts of food available, a lot of poor people somewhere in the world will be the worse off for these losses and will go hungry.

Some examples of plant diseases that have caused severe losses in the distant and/or recent past are listed below.

EXAMPLES OF SEVERE LOSSES CAUSED BY PLANT DISEASES

Fungal Diseases
1. Cereal rusts—Worldwide—Frequent severe epidemics. Huge annual losses.
2. Cereal smuts—Worldwide—Continuous losses on all grains.
3. Ergot of rye and wheat—Worldwide—Poisonous to humans and animals.
4. Late blight of potato—Cool, humid climates—Epidemics—Irish famine (1845–1846).
5. Brown spot of rice—Asia—Epidemics—The great Bengal famine (1943).
6. Southern corn leaf blight—U.S.—Epidemic 1970—$1 billion lost.
7. Powdery mildew of grapes—Worldwide—European epidemics (1840–1850s).
8. Downy mildew of grapes—U.S.—Europe—European epidemic (1870s–1880s).
9. Downy mildew of tobacco—U.S.—Europe—European epidemic (1950s–1960s).
10. Chestnut blight—U.S.—Destroyed all American chestnut trees (1904–1940).
11. Dutch elm disease—U.S.—Europe—Destroying all American elm trees (1930 to date).
12. Coffee rust—Destroyed all coffee in S. E. Asia (1870s–1880s). Since 1970 present in Brazil.
13. Banana leaf spot or Sigatoka disease—Worldwide—Great annual losses.
14. Rubber leaf blight—S. America—Destroys rubber tree plantations.

Viral Diseases

15. Sugarcane mosaic—Worldwide—Great losses on sugarcane and corn.
16. Sugarbeet yellows—Worldwide—Great losses every year.
17. Citrus quick decline (tristeza)—Africa, Americas—Millions of trees being killed.
18. Swollen shoot of cacao—Africa—Continuous heavy losses.
19. Plum pox or sharka—Europe—Spreading severe epidemic on plums, peaches, apricots.

Bacterial Diseases

20. Citrus canker—Asia, Africa, Brazil. Killed millions of trees in Florida 1910s.
21. Fire blight of pome fruits—North America, Europe. Kills numerous trees annually.

Mycoplasmal Diseases

22. Peach yellows—Eastern United States, Russia. 10 million peach trees.
23. Pear decline—Pacific coast states and Canada (1960s), Europe. Millions of pear trees killed.

Nematode Diseases

24. Root knot—Worldwide—Continuous losses on vegetables and most other plants.
25. Sugarbeet cyst nematode—Severe in northern Europe and the western U.S.

Additional Diseases Likely to Cause Severe Losses in the Future

Fungal

1. Downy mildew of corn and sorghum—Just spreading out of S.E. Asia.
2. Soybean rust—Also spreading from S.E. Asia and from Russia.
3. Monilia pod rot of cocoa—Very destructive in S. America—Spreading elsewhere.

Viral

4. African cassava mosaic—Destructive in Africa—Threatening Asia and Americas.
5. Streak disease of maize (corn)—Spread throughout Africa on sugarcane, corn, wheat, etc.
6. *Hoja blanca* (white tip) of rice—Destructive in the Americas so far.
7. Bunchy top of banana—Destructive in Asia, Australia, Egypt, Pacific islands.

Bacterial

8. Bacterial leaf blight of rice—Destructive in Japan and India—Spreading.
9. Bacterial wilt of banana—Destructive in the Americas—Spreading elsewhere.

Mycoplasmal

10. Lethal yellowing of coconut palms—Destructive in Central America. Spreading into U.S.

Viroid

11. Cadang-cadang disease of coconut—Killed more than 15 million trees in the Philippines to date.

Nematode

12. Burrowing nematode—Severe on citrus in Florida and on banana in many areas.

PLANT DISEASES AND
WORLD CROP PRODUCTION

World population today is about 4.0 billion and, at the present rate of growth, it is expected to be 4.4 billion by 1980 and 6.4 billion by the year 2000. It is paradoxical that countries in which a high percentage of the population is engaged in agriculture have the lowest agricultural output, their people are living on a substandard diet and have the highest population growth rates. As a result of the current distribution of usable land and population, educational, and technical levels for food production, and of general world economics, it is estimated that even today some 600 million are undernourished and 1.8 billion suffer from hunger or malnutrition or both. To meet the food needs of these people and of the additional millions to come in the next few years, all possible methods to increase the world food supply must and are presently being pursued, including: (1) expansion of crop acreages, (2) improved methods of cultivation, (3) increased fertilization, (4) use of improved varieties, (5) increased irrigation, and (6) improved crop protection.

There is no doubt that the first five of the above measures must provide the larger amounts of food. Crop protection from pests and diseases can only reduce the amount lost after the potential for increased food production has been attained by proper utilization of the other parameters. Crop protection, of course, has been important in the past and is important now. For example, it was estimated by the United States Department of Agriculture that in the United States alone, in 1965, crops worth $3.7 billion were lost to diseases, 3.3 billion to insects, and 2.5 billion to weeds. But crop protection becomes even more important in an intensive agriculture where increased fertilization, genetically uniform high-yielding varieties, increased irrigation, etc., are used. Crop losses to diseases and pests not only affect national and world food supplies and economies but affect even more the individual farmer whether he grows the crop for direct consumption or for sale. Since operating expenditures for the production of the crop remain the same, harvest losses due to disease and pests directly lower the crop and the net return.

It is estimated that 506 million tons (M.T.) of cereals (35 percent of the potential world production), 129 M.T. of potatoes (32 percent), 636 M.T. of sugar beets and sugar cane (45 percent), 78 M.T. of vegetables (28 percent), 56 M.T. of fruits (29 percent), 6 M.T. of coffee, cocoa, tobacco, etc. (37 percent), 42 M.T. of oil crops (32 percent), and 8 M.T. of fiber crops and natural rubber (32 percent) are lost to diseases, insects, and weeds. The total value of these losses in 1978 prices amounts to about 200 billion dollars.

Plant diseases alone are responsible for losses of 135 M.T. of cereals, 89 M.T. of potatoes, 232 M.T. of sugar beets and sugar cane, 31 M.T. of vegetables, 33 M.T. of fruits, 2.6 M.T. of coffee, cocoa, tobacco, etc., 14 M.T. of oil crops, and more than 3 M.T. of fiber crops and natural rubber. The total value of the losses caused by plant diseases, in 1978 prices, amounts to about 70 billion dollars, i.e., a third of all losses to crops. This does not include losses caused by reduced quality of the harvested products or by the cost of control measures applied to keep the losses at the above levels.

When one considers the potential and actual value of the crops produced in the various continents, the value of the losses caused to crops by diseases, insects and weeds, and the total regional percentage losses differ considerably with the continent. Thus, 25 percent of all produce in Europe is lost, 28 percent in Oceania, 29 percent in North and Central America, 30 percent in the U.S.S.R. and China, 33 percent in South America, 42 percent in Africa, and 43 percent in Asia. It is apparent that losses are much greater in underdeveloped areas than they are in the more developed ones. Another point that can be made is that, although diseases and insects cause approximately equal losses to world crops overall, insects are much easier controlled in developed countries than in the underdeveloped ones, especially Asia, while losses caused by diseases seem to be as great in developed as they are in underdeveloped countries.

EFFECTS OF CHANGES IN AGRICULTURAL METHODS AND IN HUMAN SOCIETY ON THE DEVELOPMENT AND SPREAD OF PLANT DISEASES

Among the many changes in human society in the last several decades have been the rapid increase in population with the resultant food deficits, the greater mobility of people and products over the earth, the rapid increase in knowledge in every field of endeavor, industrialization, and the increased cooperation of scientists and governments in solving problems common to several parts of the world. As a result of all these, new agricultural methods have been developed to meet the food and economic needs of the growers, the nations, and the world. However, all these changes in human society and agricultural methods have been having an effect on the kinds, severity of development, and rates of spread of the diseases that attack crop plants.

Improvement of crop plants by breeding high-yielding varieties has been and continues to be one of the better and cheaper ways of increasing crop production. This is being done with every single kind of cultivated crop plant. However, it has achieved its greatest success and was responsible for a tremendous upsurge in food production, the so-called "green revolution," in the case of the high-yielding dwarf and semidwarf wheat varieties that were also resistant, at least for some years, to the stem rust disease. These varieties, produced and distributed at first by the International Maize and Wheat Improvement Center in Mexico, not only increased wheat production in Mexico in the mid-1960s 6.5 times that in 1945, thus changing Mexico from a wheat-importing to a wheat-exporting country, they also behaved very similarly and were just as productive in Africa and Asia. To produce high yields with these varieties, many agronomic practices had to be altered drastically. Plant density per acre was increased, the date of planting had to be changed, higher levels of fertilizers and heavier and more frequent irrigation had to be employed. Soon, enormous areas were sown with a few high-yielding, genetically uniform dwarf varieties and in many areas local pathogens or local strains of common pathogens attacked the dwarf wheats. For example, monocultures of these wheats in the area of India, West Pakistan, Afghanistan, and Turkey increased from about 23,000 acres in 1966 to 30 million acres in 1971, replacing hundreds of local varieties and coming in contact with numerous new pathogens or new pathogen races. When virulent pathogens or new virulent races that may arise come in contact with such huge expanses of genetically uniform crops, devastating epidemics may develop within a short time. Already, new races and biotypes of stem rust (*Puccinia graminis* f. sp. *tritici*), leaf rust (*P. recondita*), and stripe rust (*P. striiformis*) have been identified and have caused severe epidemics in some areas that reduced yields of dwarf varieties by as much as 55 percent. In the same or other areas *Septoria* leaf blotch and glume blotch also caused severe losses on some dwarf varieties. For example, in the Mediterranean countries, *Septoria* almost completely destroyed one dwarf variety causing yield losses between 80 and 87 percent. Many of the dwarf wheats are susceptible to powdery mildew, while others are more susceptible than the older local varieties to seedling blights, to ergot, to smuts, or to certain local bacterial, viral, or nematode diseases.

A similar "green revolution" with respect to improvement of rice varieties has been carried out by the International Rice Research Institute in the Philippines. New nonlodging dwarf rice varieties that respond favorably to high nitrogen fertilization and produce high yields were developed and distributed widely in southeast Asia and elsewhere. Soon, however, many of these varieties became susceptible to diseases, such as bacterial blight caused by *Xanthomonas oryzae* and bacterial leaf streak caused by *X. oryzicola*, that were either unknown or unimportant when old local varieties were planted, but which now, due to high nitrogen fertilization and double cropping of large expanses of genetically homogeneous varieties, reached catastrophic proportions. In some countries, rice blast, caused by the fungus *Pyricularia oryzae* also became severe on the new high-nitrogen fertilized rice varieties.

The need to reduce costs in the production of high-yielding hybrid corn seed led to the search for and development of the male-sterile plants that would not need detasseling. This, however, led to hybrids that were genetically uniform in carrying the trait for male sterility which also made them susceptible to a previously unimportant race of the fungus *Helminthosporium maydis*, and as a result the southern corn leaf blight destroyed more than a billion dollars worth of corn in the U.S. in just one year.

Expansion of irrigation in Venezuela made possible the production of two rice crops per year where only one was grown before. As a result, a serious outbreak of the virus disease *hoja blanca* occurred because the new conditions favored the multiplication and spread of the insect vector of the virus from the one rice crop to the other. Irrigation also increases the population and distribution of many fungal, bacterial, and nematode pathogens that affect the roots and lower parts of the stem.

The grafting of varieties on different rootstocks, either to secure trueness of the variety or to provide resistance to a factor to which the variety is susceptible, often leads to complications and heavy losses. In addition to the cases of true horticultural incompatibility between rootstock and scion, rootstocks often carry latent viruses or mycoplasmas that may be injurious to the scion, e.g., East Malling clonal stocks used in apple tree propagation. In some cases, new pathogens attack the rootstocks through the soil, e.g., *Fusarium javanicum* var. *ensiforme* attacking the cucurbit rootstock on which greenhouse cucumbers were grafted because the rootstock was resistant to *F. oxysporum*. Finally, the rootstocks may be susceptible to viruses and mycoplasmas that are transmitted by insect vectors to resistant scions, as it happened with the citrus tristeza virus causing the decline of sweet orange trees grafted on sour orange rootstocks and with the mycoplasma causing the decline of pear varieties grafted on oriental pear.

Mechanization of agricultural practices often results in a number of plant disease problems. This is usually the result of unnoticed and more indiscriminate contamination of cultivators, harvesters, conveyors, and farm equipment with pathogens upon contact with diseased plants or infested soil, and of the more widespread dissemination of these pathogens to other products, other fields or other parts of the same, large field.

The increased use and amounts of fertilizers, particularly nitrogen, for production of greater yields is generally considered to increase the severity of diseases such as powdery mildew, rusts, fire blight, etc., caused by pathogens that prefer young succulent tissues, and to decrease the diseases caused by pathogens that attack primarily mature or senescent tissues. However, it is now known that it is generally the form of nitrogen (nitrate or ammonium) available to the host or pathogen that affects disease severity or resistance rather than the amount of nitrogen. In either case, increased fertilization does affect the susceptibility of plants to diseases and this must be taken into account in the efforts to increase productivity through fertilization.

The weed killers which are increasingly used in cultivated fields not only cause injury to cultivated crop plants directly sometimes but they also influence several soil pathogens and soil microorganisms antagonistic to pathogens. Other chemicals, too, such as fertilizers, insecticides, fungicides, etc., alter the types of microorganisms that survive and thrive in the soil and this sometimes leads to reduction in the numbers of useful predators and antagonistic microorganisms of pathogens or their vectors. The use of fungicides and other pesticides specific against a particular pathogen often leads to increased populations and disease severity caused by other pathogens not affected by the specific pesticide. This occurs even with some rather broad spectrum systemic fungicides, e.g., benomyl, which control most but not all pathogens. Where such fungicides are used regularly and widely, those of the fungi, such as *Pythium* and *Alternaria*, that are not affected by them, soon become more important as pathogens than when other more general fungicides were used.

The use of pesticides to control plant diseases and other pests has been increasing steadily at an annual rate of about 14 percent since the mid-

1950s. In 1970 approximately 1 billion pounds of pesticides, including fungicides, insecticides and herbicides, were produced by United States companies for domestic and foreign use. There is little doubt that the pesticides increased yields of crops in most cases in which they were applied. The cost of production, distribution, and application of pesticides is, of course, another form of economic loss caused by plant diseases and pests. Furthermore, such huge amounts of poisonous substances do damage to our environment and food as they are spread over our crop plants several times each year.

The public awareness of the direct, indirect, and cumulative effects of pesticides on organisms other than the pests they are intended to control has led to increased emphasis on the protection of the environment. As a result, many pesticides had to be abandoned or restricted in their use and their functions had to be taken over by other less effective or more specific pesticides or by other more costly or less efficient methods of control. The interest and effort to control diseases and other pests by biological and cultural methods is still growing while at the same time more restrictions are being imposed in the testing, licensing, and application of pesticides for disease control. Not only is it necessary for the pesticide producers to provide more detailed data on the effectiveness, toxicity, and persistence of each pesticide, but the application of each pesticide must be licensed for each crop on which it is going to be applied and, furthermore, in some countries each prospective commercial applicator of pesticides must pass an examination and be licensed to legally apply pesticides on crop plants.

The desirability of using fewer and safer pesticides, however, is counteracted by the increasing demand of consumers over the last several decades for high-quality produce, especially fruits and vegetables free of any kind of blemishes caused by diseases or insects. A change in the attitude of consumers to demand less extravagant esthetic quality of produce could reduce use of pesticides and waste of perfectly wholesome foodstuffs, but such a change in attitude may not come for some time yet.

The economics of agricultural production continue to lead to consolidation of smaller farms into increasingly larger ones often devoted to monoculture of a single profitable crop or a single stage of it. Monoculture is made more imperative by mechanization, since different crops would require additional expenditures for the specialized equipment needed for sowing, cultivation, spraying, harvesting, storage, and handling of various crops. The concentration into a continuous area of many fields and many plants of the same species and variety, however, carries many special risks, particularly the appearance or introduction and rapid spread of a destructive pathogen.

The tendency in recent years of farm enterprises to specialize in the year-round production of young seedlings or cuttings, e.g., tomato, chrysanthemum, which they subsequently sell to commercial growers throughout the world, carries with it the danger not only of a destructive disease spreading rapidly within that farm, but, much more importantly, of a destructive disease being carried on the propagative material to the rest of the country and the world. This has already happened with a chrysanthemum rust which spread from Japan to South Africa and from there throughout the globe.

The increased mobility of all kinds of nursery stock and produce throughout the world has been a major factor in the spread or appearance of many new diseases in many parts of the world. In agriculturally advanced countries, plant quarantine inspectors at each port of entry intercept most of the pathogens and other pests. But many pathogens do get through nevertheless and, if they happen to be carried in an area where the environment is favorable and there are susceptible host plants, a new disease may appear. The chances for new diseases to appear are much greater in underdeveloped countries where new improved varieties are constantly imported from other, developed countries. Many times the imported propagative material carries pathogens that may be serious not only to this same variety but, more importantly, to some or all of the local varieties of the same and related species. Moreover, even when the imported propagative material is disease free, once it has been planted extensively in the new area or country, it may be attacked by one of the locally existing pathogens or races of pathogens and this may lead to an unexpected epidemic and the failure of the new variety.

The increased travel for tourism and business has undoubtedly contributed to the introduction of some plant pathogens to new areas, but no specific cases are known.

Industrialization and increased travel harm plants in more direct ways. The production of air pollutants by factories, automobiles, airplanes, etc., causes direct injury to most plants and reduces their growth and productivity. Also, much productive land is constantly turned into residential areas, huge industrial complexes, shopping centers, parking lots, highways, and lesser roads. It is estimated that in the U.S. and Canada highway building alone takes a quarter of a million acres of arable land and that much more pasture land out of production in a single year! In the U.S., two million acres of land each year are converted from agricultural to nonagricultural uses, including 420,000 acres for urban development, an equal amount for reservoirs and flood control and nearly one million acres for parks, wilderness, and wildlife areas. The amount of cropland is decreasing at an annual rate of 3 percent. How long can this continue before we run out of food-producing land?

diagnosis of plant diseases

INTRODUCTION: PATHOGEN OR ENVIRONMENT?

For diagnosis of a plant disease it is prudent to first determine whether the disease is caused by a pathogen or an environmental factor. In some cases, in which typical symptoms of a disease are present, it is fairly easy for a somewhat experienced person to determine not only whether the disease is caused by a pathogen or an environmental factor but by which one of them. In most cases, however, a detailed examination of the symptoms and an inquiry into characteristics beyond the obvious symptoms are necessary for a correct diagnosis.

INFECTIOUS DISEASES

Diseases caused by pathogens (fungi, bacteria, parasitic higher plants, nematodes, viruses, mycoplasmas, and protozoa) are characterized by the presence of these pathogens on the surface of these plants (some fungi, bacteria, parasitic higher plants, and nematodes) or inside the plants (most pathogens). The presence of such pathogens at an active state on the surface of a plant would indicate that they are probably the cause of the disease. Their detection and identification can, in some cases, be determined with the experienced naked eye or with a magnifying lens (some fungi, all parasitic higher plants, some nematodes) or, more frequently, by microscopic examination (fungi, bacteria, and nematodes). If no such pathogens are present on the surface of the diseased plants, then it will be necessary to look for additional symptoms and, especially, for pathogens inside the diseased plant. These are usually at the margins of the affected tissues, at the vascular tissues, or at the base of the plant, and on or in its roots.

DISEASES CAUSED BY PARASITIC HIGHER PLANTS   The presence of a parasitic higher plant (e.g., dodder, mistletoe, witchweed, broomrape, etc.) growing on a plant is sufficient for the diagnosis of the disease.

DISEASES CAUSED BY NEMATODES   The presence on or in a plant of a species of plant parasitic nematodes, which can be distinguished from the nonparasitic ones by the stylet (spear) they possess, indicates that the nematode is probably the pathogen that causes the disease, or at least involved in the production of the disease. If the nematode can be identified as belonging to a species or genus known to cause such a disease, then the diagnosis of the disease can be made with a degree of certainty.

DISEASES CAUSED BY FUNGI AND BACTERIA   When fungal mycelium and spores or bacteria are present on the affected area of a diseased plant, two possibilities must be considered: (1) The fungus or bacterium may be the actual cause of the disease; or (2) they can be one of the many saprophytic fungi or bacteria that can grow on dead plant tissue once the latter has been killed by some other cause—even other fungi or bacteria.

Fungi. Determination of whether the observed fungus is a pathogen or a saprophyte is initiated by microscopically studying the morphology of its mycelium, fruiting structures, and spores. From these, the fungus can be identified and can be checked in appropriate books of mycology or plant pathology to see whether it has been reported to be pathogenic or not, especially on the plant on which it was found. If the symptoms caused on the plant correspond to those listed in the books as caused by that particular fungus, then the diagnosis of the disease is in most cases considered complete. If no such fungus is known to cause a disease on plants, especially one with symptoms similar to the ones under study, then the fungus found should be considered a saprophyte and the search for the cause of the disease must continue. In many cases, neither fruiting structures nor spores are initially present on the diseased plant tissue and therefore no identification of the fungus is possible. With most fungi, however, fruiting structures and spores are produced in the diseased tissue if the latter is placed in a glass, plastic, etc., "moisture chamber," i.e., a container in which wet paper towels, etc., are added to increase the humidity in the air of the container.

Bacteria. The diagnosis of a bacterial disease and the identification of the causal bacterium is based primarily on the symptoms of the disease, the constant presence of large numbers of bacteria in the affected area and the absence of any other pathogens from it. However, bacteria are small ($0.8 \times 1-2$ μm) and, although they can be seen with the compound microscope, they are all tiny rods and have no distinguishing morphological characteristics for their identification. Care must be taken, therefore, to exclude the possibility that the observed bacteria are saprophytic, growing in the dead tissue that was killed by some other cause. The easiest and surest way of proving that the observed bacterium is the pathogen is through isolation and growth of the bacterium in pure culture and, using a single colony for reinoculation of a susceptible host plant, reproduce the symptoms of the disease. This is usually, but not always, the fastest and most accurate method for identification of the bacterium by comparison of the resultant symptoms with those produced by known species of bacteria.

DISEASES CAUSED BY VIRUSES, VIROIDS, MYCOPLASMAS, RICKETTSIALIKE BACTERIA, AND PROTOZOA   The diagnosis of diseases caused by the other pathogens, i.e., viruses, mycoplasmalike organisms (MLO), rickettsialike bacteria, and protozoa, is much more difficult because it is complicated by two very important factors: (1) Because of their small size, transparent bodies, small numbers, etc., most of these pathogens cannot be seen with the regular compound microscope and, due to their distribution in the diseased plant, frequently they cannot be found and observed even with the electron microscope; (2) the symptoms of many of the diseases they cause are nonspecific and resemble each other and those caused on plants by many environmental factors, by insect damage or by other pathogens of the root system. Of course, several diseases caused by these pathogens develop very distinct symptoms and these diseases can be diagnosed and the pathogen identified easily, quickly and with a great degree of accuracy.

The diagnosis of diseases caused by the above pathogens without production of such diagnostic symptoms proceeds by first proving that such a disease is caused by a pathogen and not an environmental factor. This is accomplished by transmitting the pathogen from a diseased to a healthy plant and reproducing the symptoms on the inoculated plant.

The most common methods of such transmission are by budding or grafting part of a diseased plant onto a healthy plant; by rubbing sap from a diseased plant onto a healthy plant; or by allowing certain insects, nematodes, or other potential vectors of the pathogen to feed on the diseased plant and then transferring them onto a healthy plant. If by any of these methods the inoculated healthy plant develops symptoms identical to those of the diseased plant, then the disease is certain to be caused by one of these pathogens and not by an environmental factor.

Because at the present state of our knowledge the known MLO, rickettsialike bacteria, and protozoa are transmitted only by budding or grafting and by certain insect vectors, transmission of the pathogen through sap, nematodes, or through certain other ways would be taken to indicate that the pathogen is a virus or a viroid.

Further diagnosis of a disease caused by either a virus, viroid, MLO, rickettsialike bacteria, or protozoa may involve a series of tests, the most common of which are: (1) inoculation of several host plants with the pathogen and comparison of the symptoms on these hosts with those produced on the same hosts by known pathogens; (2) electron microscopy of infected tissues and comparison of the morphology of the pathogen in them—if found—with that of other known pathogens; (3) application of certain antibiotics on the diseased plant to determine whether the pathogen is susceptible to any of them or not as expressed by recovery of the plant from the disease. For example, susceptibility to tetracyclines would tend to indicate a possible MLO etiology, susceptibility to penicillin would indicate a possible rickettsialike bacterium etiology, while no effect would indicate possible viral etiology; (4) thermotherapy of the disease by exposing diseased plants or parts of them to hot water or high air temperatures for different periods of time. Recovery from symptoms at lower temperatures or in shorter periods would tend to indicate MLO or rickettsialike etiology, the opposite would suggest viral etiology; (5) if the pathogen can be isolated and purified, e.g., virus, *Spiroplasma citri* cultures, antisera may be produced and subsequently used for diagnostic serological tests. Sometimes sap from the diseased plant mixed with available antisera to known pathogens may give a quick identification of the unknown pathogen—if one of the antisera had been produced by use of the same species of the pathogen as the antigen.

DISEASES CAUSED BY MORE THAN ONE PATHOGEN   Quite frequently a plant may be attacked by two or more pathogens of the same or different kinds and may develop one or more types of disease symptoms. The most important aspect of such a situation is that the presence of the additional pathogen(s) be recognized. Once this is ascertained, the diagnosis of the disease(s) and the identification of the pathogen(s) proceeds as described above for each kind of pathogen.

NONINFECTIOUS DISEASES

If no pathogen can be found, cultured from or transmitted from a diseased plant, then it would have to be assumed that the disease is caused by a nonliving, environmental factor. The number of environmental factors that can cause disease in plants is almost unlimited, but most of them affect plants by interfering with the normal physiological processes either by causing an excess of a toxic substance in the soil or in the air or lack of an essential substance such as water, oxygen or mineral nutrients, or by causing an extreme in the conditions supporting plant life, such as temperature, humidity, oxygen, CO_2, or light. Some of these effects are the result of normal conditions, e.g., low temperatures, occurring at the wrong time, or of abnormal conditons brought about naturally, e.g., flooding, drought, or by the activities of people and their machines, e.g., pollutants, soil compaction, or weed killers.

Diagnosis of the specific environmental factor that causes or has caused a disease is sometimes made easy by the apparent change in the environment, e.g., a flood or a late or an early frost. Some environmental factors cause specific symptoms on the plants that help diagnose the cause of the malady, but most of them cause nonspecific symptoms that, unless the prehistory of the environmental conditons, applied treatments, etc., in the area are known, make it very difficult to arrive at an accurate diagnosis of the cause.

identification of a previously unknown disease— Koch's postulates

When a pathogen is found on a diseased plant, the pathogen is identified by reference to special manuals; if the pathogen is known to cause such a disease, then the diagnosis of the disease may be considered completed. If, however, the pathogen found seems to be the cause of the disease, but no previous reports exist to support this, then the following steps are taken to verify the hypothesis that the isolated pathogen is the cause of the disease:

1. The pathogen must be found associated with the disease in all the diseased plants examined.
2. The pathogen must be isolated and grown in pure culture on nutrient media, and its characteristics described (nonobligate parasites), or on a susceptible host plant (obligate parasites), and its appearance and effects recorded.
3. The pathogen from pure culture must be inoculated on healthy plants of the same species or variety on which the disease appears, and it must produce the same disease on the inoculated plants.
4. The pathogen must be isolated in pure culture again and its characteristics must be exactly like those observed in step 2.

If all the above steps (usually known as Koch's postulates) have been followed and proved true, then the isolated pathogen is identified as the organism responsible for the disease.

Koch's postulates are possible, although not always easy to carry out, with pathogens such as fungi, bacteria, parasitic higher plants, nematodes, some viruses, some viroids, and the spiroplasmas. These organisms can be isolated and cultured, or can be purified, and can then be introduced into the plant and cause the disease. With the other pathogens, however, such as most viruses, some viroids, mycoplasmas, rickettsialike bacteria, and protozoa, culture or purification of the pathogen is not yet possible, and the pathogen often cannot be reintroduced into the plant to reproduce the disease. Thus, with these pathogens, Koch's postulates cannot be carried out and their acceptance as "the" pathogens of the diseases with which they are associated is more or less tentative. However, in most cases, the circumstantial evidence that shows that these are "the" pathogens of these diseases is overwhelming, and it is assumed that further improvement of techniques of isolation, culture, and inoculation of pathogens will someday prove that today's assumptions are justified.

SELECTED REFERENCES

Cramer, H. H. 1967. Plant Protection and Crop Production (translated from German by J. H. Edwards). *Pflantzenschutz Nachrichten* 20/1967, 1. Farbenfabriken Bayer AG, Leverkusen. 524 p.

FAO (1961 to 1965). Production Yearbook. Rome, 15–19.

Horsfall, J. G., and E. B. Cowling (eds). 1977. "Plant Disease." Vol. 1. Academic Press, New York. 465 p.

Keitt, G. W. 1959. History of plant pathology. *In* "Plant Pathology" (J. G. Horsfall and A. E. Dimond, eds.), Vol. 1, pp. 61–97. Academic Press, New York.

Klinkowski, M. 1970. Catastrophic plant diseases. *Ann. Rev. Phytopathol.* **8**:37–60.

Paddock, W. C. 1967. Phytopathology in a hungry world. *Ann. Rev. Phytopathol.* **5**:375–390.

Parris, G. K. 1968. "A Chronology of Plant Pathology." Johnson & Sons, Starkville, Mississippi.

Stakman, E. C., and J. G. Harrar. 1957. "Principles of Plant Pathology." The Ronald Press Co., New York. 581 p.

Ten Houten, J. G. 1974. Plant pathology: Changing agricultural methods and human society. *Ann. Rev. Phytopathol.* **12**:1–11.

Thurston, H. D. 1973. Threatening plant diseases. *Ann. Rev. Phytopathol.* **11**:27–52.

U.S. Dept. of Agriculture. 1965. Losses in Agriculture. Agric. Handbook No. 291, 120 p.

Walker, J. C. 1969. "Plant Pathology." 3rd edition. McGraw-Hill Book Co., New York. 819 p.

2
parasitism and disease development

The pathogens that attack plants belong to the same groups of organisms that cause diseases in humans and animals. With the exception of some insect-transmitted plant pathogens, however, which cause diseases in both their host plants and their insect vectors, none of the pathogen species that attack plants are known to affect humans or animals. Plants are moreover attacked by a number of other plants.

Infectious diseases are those that result from infection of a plant by a pathogen. They are characterized by the ability of the pathogen to grow and multiply rapidly on diseased plants and also by its ability to spread from diseased to healthy plants and, thereby, to cause new diseases.

parasitism and pathogenicity

An organism that lives on or in some other organism and obtains its food from the latter is called a parasite. The relationship between a parasite and its host is called parasitism. A plant parasite is an organism that becomes intimately associated with a plant and multiplies or grows at the expense of the plant. The removal by the parasite of nutrients and water from the host plant usually leads to reduced efficiency in the normal growth of the plant and becomes detrimental to its further development and reproduction. Thus, in many cases, parasitism is intimately associated with pathogenicity, since the ability of the parasite to invade and become established in the host generally results in the development of a diseased condition in the host.

28

In some cases of parasitism, as with the root nodule bacteria of legume plants and the mycorrhizal infection of feeder roots of most flowering plants, both the plant and the microorganism are beneficial to the other's development, and this phenomenon is known as symbiosis.

In most plant diseases, however, the amount of damage caused to plants is often much greater than would be expected from the mere removal of nutrients by the parasite. This additional damage results from substances secreted by the parasite or produced by the host in response to stimuli originating in the parasite. Tissues affected by such substances may show increased respiration, disintegration or collapse of cells, wilting, abscission, abnormal cell division and enlargement, and degeneration of specific components such as chlorophyll. These conditions in themselves do not seem directly to improve the welfare of the parasite. It would appear therefore that the degree of pathogenicity exhibited by a parasite is not always proportional to the nutritional affiliation of the parasite and its host. Pathogenicity then may be defined as the interference of the parasite with one or more of the essential functions of the plant, with parasitism playing, frequently, an important, but not always the most important, role.

Of the large number of groups of living organisms, only a few members of a few groups can parasitize plants: fungi, bacteria, mycoplasmas, rickettsialike bacteria, and parasitic higher plants (all belonging to the plant kingdom), nematodes and protozoa (of the animal kingdom), viruses, and viroids. These parasites to be successful must be able to invade a host plant, feed and proliferate in it, and withstand the conditions in which the host lives. Some parasites, including viruses, viroids, mycoplasmas, rickettsialike bacteria, nematodes, and protozoa, and, of the fungi, those causing downy mildews, powdery mildews, and rusts, can grow and reproduce in nature only on living hosts, and they are called obligate parasites. Other parasites (most fungi and bacteria) can live on either living or dead hosts and on various nutrient media and are, therefore, called nonobligate parasites. Some nonobligate parasites live most of the time or most of their life cycles as parasites but, under certain conditions, may grow saprophytically on dead organic matter, whereas others live most of the time and thrive well on dead organic matter but, under certain circumstances, may attack living plants and become parasitic. There is usually no correlation between the degree of parasitism of a pathogen and the severity of disease it can cause, since many diseases caused by weakly parasitic pathogens are much more damaging to the plant than others caused even by obligate parasites. Moreover, certain fungi, e.g., the slime molds and those causing sooty molds, can cause disease by just covering the surface of the plant without feeding at all or by feeding on insect excretions rather than by parasitizing the plant.

Obligate and nonobligate parasites generally differ in the ways by which they attack their host plants and procure their nutrients from the host. Many nonobligate parasites secrete enzymes which bring about the disintegration of the cell components of plants and which alone or with the toxins secreted by the pathogen result in the death and degradation of the cells. The invading pathogen then utilizes the contents of the cells for its growth. Many fungi and all bacteria act in this fashion, growing on a nonliving substrate within a living plant. This mode of nutrition is like that of saprophytes. On the other hand, all obligate (and some nonobligate) parasites do not kill cells in advance but get their nutrients either by penetrating living cells or by establishing close contact with them. The association of these pathogens with their host cells is a very intimate one and results in continuous absorption or diversion of nutrients, which would normally be utilized by the host, into the body of the parasite. The depletion of nutrients, however, although it restricts the growth of the host and results in symptoms, does not always kill the host. In the case of obligate parasites death of the host cells restricts the further development of the parasite and may result in its death.

Parasitism of cultivated crops is a common phenomenon. In North America, for example, some 8000 species of fungi cause approximately 80,000 diseases, and at least 200 species of bacteria, about 75 mycoplasmas, more than 500 different viruses, and over 500 species of nematodes attack crops. Although there are about 2500 species of higher plants parasitic on other plants, only a few of them are serious parasites of crop plants. Recently several rickettsialike organisms and viroids have also been shown to cause diseases in plants. A single crop, tomato, is attacked by more than 80 species of fungi, 11 bacteria, 16 viruses, several mycoplasmas, and several nematodes. This is an average number of diseases since corn has 100, wheat 80, and apple and potato each have about 200 diseases.

host range
of pathogens

Pathogens differ with respect to the kinds of plants that they can attack, with respect to the organs and tissues that they can infect, and with respect to the age of the same organ or tissue of the same plant on which they can grow. Some pathogens are restricted to a single species, others to one genus of plants, while others have a wide host range, including many taxonomic groups of higher plants. Some pathogens grow especially on roots, others on stems, some mainly on the leaves or on fleshy fruit or vegetables. Some pathogens attack specifically certain kinds of tissues, e.g., vascular parasites. Others may produce different effects on different parts of the same plant. In regard to age of plants, some pathogens attack seedlings or tender parts of plants, while others attack only mature tissues.

Most obligate parasites are usually very specific as to the kind of host they attack, possibly because they have evolved in parallel with their host and require certain nutrients that are produced or become available to the pathogen only in these hosts. Nonobligate parasites usually attack many different plants and plant parts of varying age, possibly because they depend for their attack on nonspecific toxins or enzymes that affect substances or processes found commonly among plants. Some nonobligate parasites, however, produce disease on only one or a few plant species. In any case, the number of plant species presently known to be susceptible to a single pathogen is surely smaller than the actual number in nature since only a few species out of thousands have been studied for their susceptibility to each pathogen. Furthermore, because of genetic changes, a pathogen may be able to attack hosts previously immune to it.

stages in the development of disease

In every infectious disease there is a series of more or less distinct events that occur in succession of one another and lead to the development and perpetuation of the disease and the pathogen. This chain of events is called a disease cycle. A disease cycle sometimes corresponds fairly closely to the life cycle of the pathogen but it refers primarily to the appearance, development, and perpetuation of the disease rather than the pathogen. The disease cycle involves the changes in the plant and the plant symptoms as well as those in the pathogen and spans periods within a growing season and from one growing season to the next. The main events in a disease cycle include inoculation, penetration, infection, growth and reproduction of the pathogen, dissemination of the pathogen, and overwintering or oversummering of the pathogen.

INOCULATION

Inoculation is the coming in contact of a pathogen with a plant. The pathogen or pathogens that land on, or are otherwise brought into contact with the plant are called inoculum.

INOCULUM

Inoculum is any part of the pathogen that can cause infection. Thus, in fungi inoculum may be fragments of mycelium, spores, or sclerotia (compact mass of mycelium); in bacteria, mycoplasmas, rickettsialike bacteria, viruses, and viroids, inoculum is always whole individuals of bacteria, mycoplasmas, rickettsialike bacteria, viruses, and viroids, respectively; in nematodes, inoculum may be adult nematodes, nematode larvae, or eggs; in parasitic higher plants inoculum may be plant fragments or seeds. Inoculum may consist of a single pathogen, e.g., one spore, or of millions of pathogens, e.g., bacteria carried in a drop of water.

TYPES OF INOCULUM Inoculum that survives the winter and causes the original infections in the spring or early summer is called primary inoculum and the infections it causes are called primary infections. Inoculum produced from primary infections is called secondary inoculum and that in turn causes secondary infections. Generally, the more abundant the primary inoculum and the closer it is to the crop, the more severe the disease and the losses that result. In some diseases, e.g., smuts, there is only primary inoculum and, therefore, one disease cycle per season. Such diseases produce spores at the end of the season and these spores serve as primary inoculum for the following year. In most diseases, however, there are many disease cycles per season and with each cycle the amount of inoculum is multiplied manyfold. These are the kinds of diseases that cause most of the explosive epidemics on most crops.

SOURCES OF INOCULUM   The inoculum sometimes is present right in the plant debris or soil of the field where the crop is grown, other times it comes into the field with the seed, transplants, tubers, or other propagative organs, or it may come from sources outside the field. Outside sources of inoculum may be nearby plants or fields, or fields many miles away. In many plant diseases, especially those of annual crops, the inoculum survives in perennial weeds or alternate hosts and every season it is carried from them to the annual and other plants. Fungi, bacteria, parasitic higher plants, and nematodes either produce their inoculum on the surface of infected plants or their inoculum reaches the plant surface when the infected tissue breaks down. Viruses, viroids, mycoplasmas, and rickettsialike bacteria produce their inoculum within the plants; such inoculum almost never reaches the plant surface in nature and, therefore, cannot by itself escape from one plant and spread to another.

STEPS IN INOCULATION

LANDING OR ARRIVAL OF INOCULUM   The inoculum of most pathogens is carried to host plants passively by wind, water, insects, etc., and only a tiny fraction of the inoculum produced actually lands on susceptible host plants; the bulk of the inoculum is wasted because it lands on things that cannot become infected. Some types of inoculum in the soil, e.g., zoospores and nematodes, may be attracted to the host plant by substances diffusing out of the plant roots. Vector-transmitted pathogens are usually carried to their host plants with an extremely high efficiency.

GERMINATION OF SPORES AND SEEDS—HATCHING OF EGGS   All pathogens in their vegetative state are capable of initiating infection immediately. However, fungal spores and seeds of parasitic higher plants must first germinate; for that they require favorable temperature and also moisture in the form of rain, dew, or a film of water on the plant surface or at least high relative humidity. The moist conditions must last long enough for the pathogen to penetrate; otherwise it desiccates and dies. Most spores can germinate immediately after their maturation and release, but others, the so-called resting spores, require a dormancy period of varying duration before they can germinate. When a spore germinates it produces a germ tube, i.e., the first part of the mycelium, that can penetrate the host plant. Some fungal spores germinate by producing other spores, such as zoospores or basidiospores. Seeds germinate by producing a radicle which either penetrates the host plant or produces a small plant that penetrates the host plant by means of haustoria.

Nematode eggs also require favorable temperature and moisture conditions to become activated and hatch. In most nematodes, the egg contains the first larval stage before or soon after the egg is laid. This immediately undergoes a molt and gives rise to the second larval stage that may remain dormant in the egg for various periods of time. Thus, when the egg finally hatches it is the second stage larva that emerges, and it either finds and penetrates a host plant or undergoes additional molts that produce the other larval stages and the adults.

CONDITIONS FAVORING INOCULATION

The frequency of successful inoculations depends on the amount of primary and/or secondary inoculum available, the occurrence of long periods of temperature and moisture that favor release of inoculum or activity of its vectors, the direction of air currents or windblown rain, the distance of the inoculum from the host plants, the density of host plants, and the number and size of host plants.

Once the pathogen reaches the host surface, successful inoculation still depends on several factors such as the nature of the plant surface, favorable temperature and moisture for spore germination, the presence of stimulatory or inhibitory substances secreted by the host plant, and the presence of other, often antagonistic microorganisms on the plant surface. Plant exudates and surface microflora seem to have little effect on pathogens attacking the aboveground parts of plants, but they may be quite important in root diseases in which they may influence the ability of the pathogen to germinate and its ability to survive.

PENETRATION

Pathogens penetrate plant surfaces by direct penetration, through natural openings, or through wounds (Fig. 4). Some fungi penetrate tissues in one way only, others in more than one. Bacteria enter plants mostly through wounds, less frequently through natural openings, and never directly (Fig. 5). Viruses, viroids, mycoplasmas, and rickettsialike bacteria enter through wounds made by vectors, although some viruses and viroids may also enter through wounds made by tools and other means. Parasitic higher plants enter their hosts by direct penetration. Nematodes enter plants by direct penetration and, sometimes, through natural openings.

Penetration does not always lead to infection. Many organisms actually penetrate cells of plants which are not susceptible to these organisms and which do not become diseased; these organisms cannot proceed beyond the stage of penetration and die without producing disease.

DIRECT PENETRATION
THROUGH INTACT PLANT
SURFACES

It is probably the most common type of penetration in fungi and nematodes and the only type of penetration in parasitic higher plants. None of the other pathogens can enter plants by direct penetration.

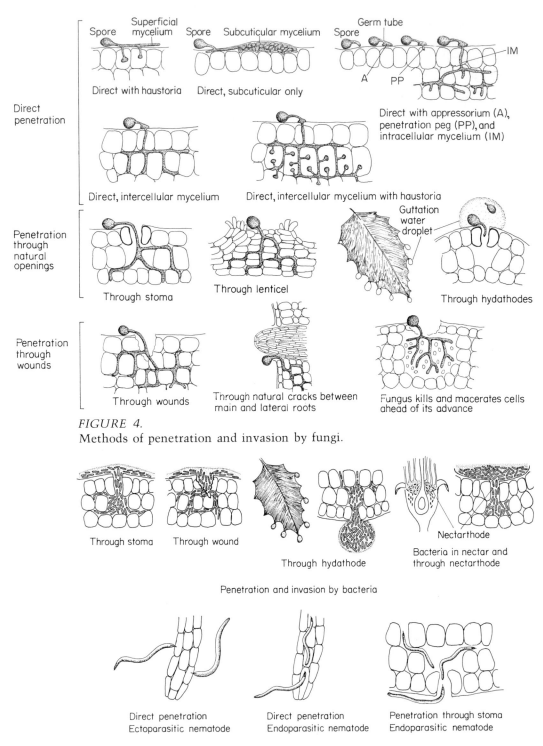

FIGURE 4.
Methods of penetration and invasion by fungi.

FIGURE 5.
Methods of penetration and invasion by bacteria and nematodes.

Fungi that penetrate their host plants directly do so through a fine hypha or an appressorium (Figs. 4, 6, and 7). These are formed at the point of contact of the germ tube or mycelium with a plant surface. The fine hypha grows toward the plant surface and pierces the cuticle and the cell wall through mechanical force and enzymatic softening of the cell wall substances. Most fungi, however, form an appressorium at the end of the germ tube, the appressorium being usually bulbous or cylindrical with a flat surface in contact with the host plant's surface. Then, a fine hypha, usually called a penetration peg, grows from the flat surface of the appressorium toward the host and pierces the cuticle and the cell wall. The penetration peg is generally of much smaller diameter than a normal hypha of the fungus but it resumes its normal diameter once inside the cell lumen. In most fungal diseases the fungus penetrates the plant cuticle and the cell wall but in some, e.g., apple scab, the fungus penetrates only the cuticle and stays between the cuticle and the cell wall.

Parasitic higher plants also form an appressorium and penetration peg at the point of contact of the radicle with the host plant, and penetration is similar to that in fungi.

Direct penetration in nematodes is by means of repeated back-and-forth thrusts of their stylets. This finally creates a small opening in the

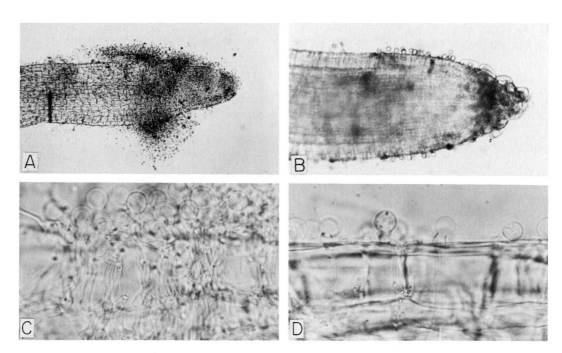

FIGURE 6.
Attraction of zoospores of *Phytophthora cinnamomi* to roots of two types of blueberry (A and B) and infection of the roots by the zoospores (C and D). Attraction of zoospores to roots one hour after inoculation (A and B) and infection and colonization of the root after 24 hours (C and D) are greater in the susceptible highbush blueberry (A, C) than in the more resistant rabbiteye blueberry (B, D). (Photos courtesy R. D. Milholland, from *Phytopathology* **65**:789–793.)

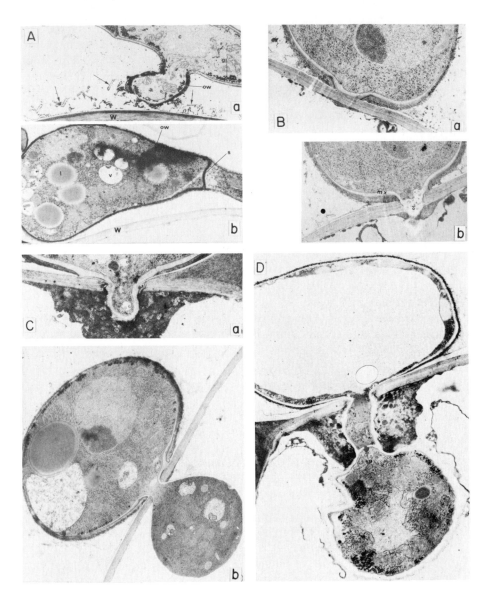

FIGURE 7.
Electron micrographs of direct penetration of a fungus (*Colletotrichum graminicola*) into an epidermal leaf cell. (A) (a) Developing appressorium from a conidium. Note wax rods (arrows) on leaf surface. (b) A mature appressorium separated by a septum from the germination tube. (B) (a) Formation of a penetration peg at central point of contact of appressorium with cell wall. (b) Lomasomelike structures in the infection peg which has already penetrated the cell wall and the papilla produced by the invaded cell. (C) Development of infection hyphae. (a) Infection peg penetrating the papilla. (b) Appressorium and swollen infection hypha after penetration. (D) Upon completion of penetration and establishment of infection, the appressorium consists mostly of a large vacuole and is cut off from the infection hypha by a septum. (Photos courtesy D. J. Politis and H. Wheeler, from *Physiol. Plant Pathol.* **3:**465–471.)

cell wall and the nematode sends its stylet into the cell or the entire nematode enters the cell (Fig. 5).

PENETRATION THROUGH WOUNDS

All bacteria and most fungi can enter plants through various types of wounds (Figs. 4 and 5) and, in nature, all viruses, viroids, mycoplasmas, and rickettsialike bacteria enter plants through wounds made by their vectors. The wounds utilized by bacteria and fungi may be fresh or old and may consist of lacerated or killed tissue. These pathogens may grow briefly on such tissue prior to their advance into healthy tissue. Laceration or death of tissues may be the result of: environmental factors, e.g., wind breakage or rubbing, sand blasting, hail, frost, heat scorching, fire; animal feeding, e.g., insects, nematodes, worms, large animals; cultural practices of man, e.g., cultivation, weeding, pruning, grafting, transplanting, spraying, harvesting; self-inflicted injuries, e.g., leaf scars, root cracks, etc.; and, finally, wounds or lesions caused by other pathogens. Bacteria and fungi penetrating through wounds apparently germinate or multiply in the sap present in fresh wounds or in a film of rain or dew water present on the wound. Subsequently the pathogen invades adjacent plant cells directly or through haustoria, or it secretes enzymes and toxins which kill and macerate the nearby cells.

Penetration of viruses, viroids, mycoplasmas, and rickettsialike bacteria through wounds depends on the deposition of these pathogens by their vectors (insects for all four pathogens, and also nematodes, mites, and fungi for viruses, and human hands and tools for some viruses and viroids) in fresh wounds created at the time of inoculation. In most cases these pathogens are carried by one or a few kinds of specific vectors and can be inoculated successfully only when they are brought to the plant by these vectors.

PENETRATION
THROUGH NATURAL OPENINGS

Many fungi and bacteria enter plants through stomata and some enter through hydathodes, nectarthodes, and lenticels (Figs. 4 and 5). Most stomata are present in large numbers on the lower side of leaves, measure about 10–20 × 5–8 μm, and are open in the daytime but more or less closed at night. Bacteria present in a film of water over a stoma can easily swim through the stoma and into the substomatal cavity where they can multiply and start infection. Fungal spores generally germinate on the plant surface and the germ tube may then grow through the stoma; frequently, however, the germ tube forms an appressorium that fits tightly over the stoma and usually one fine hypha grows from it into the stoma (Fig. 8). In the substomatal cavity the hypha enlarges and from it grow one or several small hyphae which actually invade the cells of the host plant directly or through haustoria. Although some fungi can apparently penetrate even closed stomata, others penetrate stomata only while they are open and some, e.g., the powdery mildew fungi, may grow over open stomata without entering them.

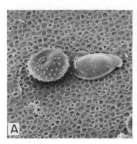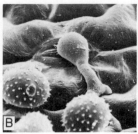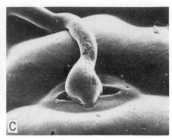

FIGURE 8.
Scanning electron micrographs of appressorium formation and penetration
through a stoma by the bean rust fungus *Uromyces phaseoli.* (A) Uredospore,
short germ tube, and large, flattened appressorium forming on a membrane. (B)
Uredospore, germ tube, and appressorium formed after 6-hour germination over
closed stoma on bean leaf. (C) Young appressorium over open stoma of bean leaf.
(Photos courtesy W. K. Wynn, from *Phytopathol.* **66**:136–146).

    Hydathodes are more or less permanently open pores at the margin and
tip of leaves; they are connected to the veins and secrete droplets of liquid
containing various nutrients. Some bacteria utilize these pores as a
means of entry into leaves but few fungi seem to enter plants through
hydathodes. Some bacteria also enter blossoms through the nectarthodes
or nectaries, which are similar to hydathodes.

    Lenticels are openings on fruit, stem, tubers, etc., which are filled with
loosely connected cells to allow passage of air. During the growing season, lenticels are open but even so relatively few fungi and bacteria
penetrate tissues through them, growing and advancing mostly between
the cells. Most pathogens that penetrate through lenticels can also enter
through wounds, lenticel penetration being apparently a less efficient,
secondary pathway.

INFECTION

Infection is the process by which pathogens establish contact with the
susceptible cells or tissues of the host and procure nutrients from them.
During infection pathogens grow and/or multiply within the plant tissues
and invade the plant to a lesser or greater extent. Thus invasion of the
plant tissues by the pathogen, and growth and reproduction of the pathogen in or on infected tissues are actually two concurrent substages of
disease development within the stage of infection.

    Successful infections result in the appearance of discolored, mal-
formed, or necrotic areas on the host plant, which are called symptoms.
Some infections, however, remain latent, i.e., they do not produce symptoms right away but at a later time when the environmental conditions
may be more favorable, or at a different stage of maturity of the plant.

    All the visible and invisible changes in the appearance and functions of
infected plants comprise the symptoms of the disease. Symptoms may

change continuously from the moment of their appearance until the entire plant dies, or they may develop up to a point and then remain more or less unchanged for the rest of the growing season. Symptoms may appear as soon as 2 to 4 days after inoculation, as happens in some localized virus diseases of herbaceous plants, or as late as 2 to 3 years, as in the case of some viral, mycoplasmal, and other diseases of trees, following inoculation. In most plant diseases, however, symptoms appear from a few days to a few weeks after inoculation.

The time interval between inoculation and appearance of disease symptoms is called incubation period. The length of the incubation period of various diseases varies with the particular pathogen–host combination, with the stage of development of the host, and with the temperature in the environment of the infected plant.

During infection, some pathogens obtain nutrients from living cells, often without killing the cells, or at least not for a long time; others kill cells and utilize their contents as they invade them; still others kill cells and disorganize tissues ahead of the invading pathogen. During infection pathogens release in the host a number of biologically active substances (e.g., enzymes, toxins, growth regulators) which may affect the structural integrity of the host cells or their physiological processes. In response to these, the host reacts with a variety of defense mechanisms which result in various degrees of protection of the plant from the pathogen.

For a successful infection to occur it is not sufficient that a pathogen comes in contact with its host but several other conditions must also be satisfied. First of all, the plant variety must be susceptible to the particular race of the pathogen—in which case that race of the pathogen is said to be virulent on that variety of the host plant; the host plant must be in a susceptible stage, since some pathogens attack only young seedlings, others only mature or senescing plants, some only the leaves, others only the flowers or the fruit, or only ripe fruit, etc.; the pathogen must be in a pathogenic stage, e.g., fungal mycelium or spores and seeds that can germinate and infect immediately without requiring a resting (dormancy) period first, or infective larval stages or adults of nematodes; and finally, the temperature and moisture conditions in the environment of the plant must favor the growth and multiplication of the pathogen. When these conditions occur at an optimum, the pathogen can invade the host plant up to the maximum of its potential even in the presence of plant defenses, and as a consequence disease develops.

INVASION

Various pathogens invade hosts in different ways and to different extents (Figs. 4 and 5). Some fungi, e.g., those causing apple scab and black spot of rose, produce mycelium which grows only in the area between the cuticle and the epidermis (subcuticular); others, e.g., those causing the powdery mildews, produce mycelium only on the surface of the plant but send haustoria into the epidermal cells. Most fungi spread into all the tissues of the plant organs (leaves, stems, roots, etc.) they infect, either by growing directly through the cells (intracellular mycelium), or by growing between the cells (intercellular mycelium). The fungi that cause vascular wilts invade the xylem vessels of plants.

Bacteria invade tissues intercellularly although, when parts of the cell walls dissolve, bacteria also grow intracellularly. Bacteria causing vascular wilts, like the vascular wilt fungi, invade the xylem vessels. Most nematodes invade tissues intercellularly but some can invade intracellularly as well. Many nematodes do not invade cells or tissues at all but feed by piercing epidermal cells with their stylet.

Viruses, viroids, mycoplasmas, and rickettsialike bacteria invade tissues by moving from cell to cell intracellularly. Viruses and viroids invade all types of living plant cells, mycoplasmas invade phloem sieve tubes and perhaps a few adjacent phloem parenchyma cells, while rickettsialike organisms invade either xylem vessels or phloem sieve tubes.

Many infections caused by fungi, bacteria, nematodes, viruses, and parasitic higher plants are local, i.e., they involve a single cell, a few cells, or a small area of the plant. These infections may remain localized throughout the growing season or they may enlarge slightly or very slowly. Other infections enlarge more or less rapidly and may involve an entire plant organ, e.g., flower, fruit, leaf, a large part of the plant, e.g., a branch, or the entire plant.

All infections caused by mycoplasmas and rickettsialike bacteria and all natural infections caused by viruses and viroids are systemic, i.e., the pathogen, from one initial point, spreads and invades most or all susceptible cells and tissues throughout the plant. Vascular wilt fungi and bacteria invade xylem vessels internally but they are usually confined to a few vessels in the roots, the stem or the top of infected plants and only in the final stages of the disease they invade most or all xylem vessels of the plant. Some fungi, primarily among those causing downy mildews, smuts, and rusts, also invade their hosts systemically although in most cases the older mycelium degenerates and disappears, and only younger mycelium survives in actively growing plant tissues.

GROWTH AND
REPRODUCTION
OF THE PATHOGEN

Individual fungi and parasitic higher plants generally invade and infect tissues by growing into them from one initial point of inoculation. Most of these pathogens, whether producing a small spot, a large infected area, or a general necrosis of the plant, continue to grow and branch out within the infected host indefinitely so that the same pathogen individual spreads into more and more plant tissues until the spread of the infection stops or the plant is dead. In some fungal infections, however, while the younger hyphae continue to grow into new healthy tissues, the older ones in the already infected areas die out and disappear, so that an infected plant may have several points where separate units of mycelium are active. Also, fungi causing vascular wilts often invade plants by producing and releasing spores within the vessels and, as the spores are carried in the sap stream, they invade vessels far away from the mycelium, germinate there and produce mycelium which invades more vessels.

All other pathogens, i.e., bacteria, mycoplasmas, viruses, viroids, nematodes, and protozoa, do not increase much, if at all, in size with time, since their size and shape remain relatively unchanged throughout their existence. These pathogens invade and infect new tissues within the plant by reproducing at a rapid rate and increasing their numbers tremendously in the infected tissues; the progeny then are either carried passively into new cells and tissues through plasmodesmata (viruses and viroids only), phloem (viruses, viroids, mycoplasmas, rickettsialike bacteria, protozoa), xylem (some bacteria), etc., or, as happens with bacteria, protozoa and nematodes, they may move through cells by swimming on their own power.

Plant pathogens reproduce in a variety of ways (Fig. 3). Fungi reproduce by means of spores which may be either asexual (equivalent to the buds on a twig or the tubers of a potato plant), or sexual (equivalent to the seeds of plants). Parasitic higher plants reproduce just like all plants, i.e., by seeds. Bacteria, mycoplasmas, and protozoa reproduce by fission, i.e., one mature individual splits into two equal, smaller individuals. Viruses and viroids are replicated by the cell, just as a page placed on a photocopying machine is replicated by the machine as long as the machine is operating and paper supplies last. Nematodes reproduce by means of eggs.

The great majority of plant pathogenic fungi produce mycelium only within the plants they infect. Relatively few fungi produce mycelium on the surface of their host plants, and only the powdery mildew fungi produce mycelium on the surface of, and not within, their hosts. The great majority of fungi produce spores on, or just below, the surface of the infected area of the host and the spores are released outward into the environment. Some of the lower fungi, however, e.g., the clubroot pathogen, and the fungi causing vascular wilts produce spores within the host tissues and these spores are not released outward until the host dies and disintegrates. Parasitic higher plants produce their seeds on aerial branches, and some nematodes lay their eggs at or near the surface of the host plant. Bacteria reproduce between or within host cells, generally inside the host plant, and come to the host surface only through wounds, cracks, etc. Viruses, viroids, plant mycoplasmas, and rickettsialike bacteria reproduce only inside cells and apparently do not reach or exist on the surface of the host plant.

The rate of reproduction varies considerably among the various kinds of pathogens but, in all of them, one or a few pathogens can produce tremendous numbers of individuals within one growing season. Some fungi produce spores more or less continuously while others produce them in successive crops. In either case several thousand to several hundreds of thousands of spores may be produced per square centimeter of infected tissue. Even small specialized sporophores can produce millions of spores and the number of spores produced per infected plant are often in the billions or trillions. The numbers of spores produced in an acre of heavily infected plants, therefore, are generally astronomical and, as they are released, there are enough spores to land on and inoculate every conceivable surface in the field and the surrounding areas.

Bacteria reproduce very rapidly within infected tissues. Under optimum nutritional and environmental conditions, e.g., in culture, bacteria divide, i.e., double their numbers, every 20 to 30 minutes and, presumably, bacteria multiply just as fast in a susceptible plant as long as the temperature is favorable. Millions of bacteria may be present in a single drop of infected plant sap and the number of bacteria per plant must be astronomical. Rickettsialike organisms and mycoplasmas appear to reproduce slower than bacteria and although they spread systemically throughout the vascular system of the plant they are present in relatively few xylem or phloem vessels and the total number of these pathogens in infected plants is relatively small.

Viruses and viroids reproduce within living host cells, the first new virus particles being detected several hours after infection. Soon after that, however, virus particles accumulate within the infected living cell until as many as 100,000 to 10,000,000 virus particles may be present in a single cell. Viruses and viroids infect and multiply in most or all living cells of their hosts and it is apparent that each plant may contain innumerable particles of these pathogens.

Nematode females lay about 300 to 600 eggs, about half of which produce females and these again lay 300 to 600 eggs each. Depending on the climate, the availability of hosts, and the duration of each life cycle of the particular nematode, the nematode may have from two to more than a dozen generations per year. If even half of the females survived and reproduced, each generation time would increase the number of nematodes in the soil by more than 100-fold; thus the buildup of nematode populations within a growing season and in successive seasons is often quite dramatic.

DISSEMINATION OF THE PATHOGEN

A few pathogens, e.g., nematodes, fungal zoospores, and bacteria can move very short distances on their own power and thus can move from one host to another one very close to it. Fungal hyphae and rhizomorphs can grow between tissues in contact and sometimes through the soil toward nearby roots. Both of these means of dissemination, however, are very limited, especially in the cases of zoospores and bacteria.

The spores of some fungi are expelled forcibly from the sporophore or sporocarp by a squirting or puffing action that results in successive or simultaneous discharge of spores up to a centimeter or so above the sporophore. The seeds of some parasitic plants are also expelled forcibly and may arch over distances of several meters.

Almost all dissemination of pathogens that is responsible for plant disease outbreaks, and even for disease occurrences of minor economic importance, is carried out passively by agents such as air, water, insects, certain other animals, and man (Fig. 9).

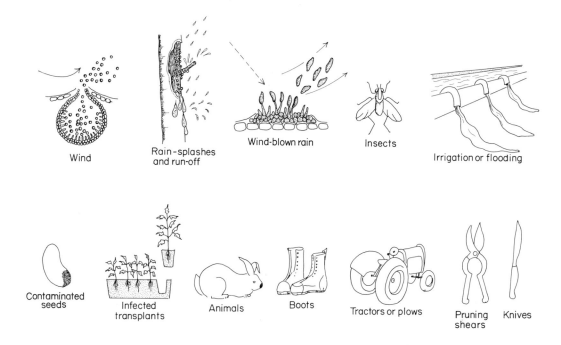

FIGURE 9.
Means of dissemination of fungi and bacteria.

DISSEMINATION BY AIR

Most fungal spores and the seeds of most parasitic plants are disseminated by air currents that carry them as inert particles to various distances. Air currents pick up spores and seeds off the sporophores, or while they are being forcibly expelled or are falling at maturity, and, depending on the air turbulence and velocity, may carry the spores upward or horizontally in a way similar to that of particles contained in smoke. While airborne, some of the spores may touch wet surfaces and get trapped, and when air movement stops or when it rains the rest of the spores land or are brought down by the raindrops. Most of the spores, of course, land on anything but a susceptible host plant and are, therefore, wasted. The spores of many fungi are actually too delicate to survive a long trip through the air and are therefore successfully disseminated for only a few hundred or a few thousand meters. The spores of other fungi, however, particularly those of the cereal rusts, are very hardy and occur commonly at all levels and at high altitudes (several thousand meters) above infected fields. Spores of these fungi are often carried over distances of several kilometers, even hundreds of kilometers, and in favorable weather may cause widespread epidemics.

Air dissemination of other pathogens occurs rather infrequently and only under special conditions, or indirectly. Thus, the bacteria causing fire blight of apple and pear produce fine strands of dried bacterial exudate containing bacteria and these strands may be broken off and disseminated by wind. Bacteria and nematodes present in the soil may be blown away along with soil particles in the dust. Wind also helps in the dissemination of bacteria, fungal spores, and nematodes by blowing away rain splash droplets containing these pathogens, and wind carries away insects that may contain or are smeared with viruses, bacteria or fungal spores. Finally, wind causes adjacent plants or plant parts to rub against each other and this may help the spread by contact of bacteria, fungi, some viruses and viroids, and possibly of some nematodes.

DISSEMINATION BY WATER

Water is important in disseminating pathogens in three ways. (1) Bacteria, nematodes, and spores, sclerotia and mycelial fragments of fungi present in the soil are disseminated by rain or irrigation water that moves on the surface or through the soil. (2) All bacteria and the spores of many fungi are exuded in a sticky liquid and depend for their dissemination on rain or (overhead) irrigation water which either washes them downward or splashes them in all directions. (3) Raindrops or drops from overhead irrigation pick up the fungal spores and any bacteria present in the air and wash them downward where some of them may land on susceptible plants. Although water is less important than air in long-distance transport of pathogens, water dissemination of pathogens is more efficient in that the pathogens land on an already wet surface and can move or germinate immediately.

DISSEMINATION BY INSECTS, MITES, NEMATODES, AND OTHER VECTORS

Insects, particularly aphids and leafhoppers, are by far the most important vectors of viruses, mycoplasmas, and rickettsialike bacteria. Each one of these pathogens is transmitted internally by only one or a few species of insects during feeding and movement of the insect vectors from plant to plant. Specific insects also transmit certain fungal and bacterial pathogens, such as those causing Dutch elm disease and the bacterial wilt of cucurbits. In addition, many insects become smeared with any kind of bacteria or sticky fungal spores as they move among plants and carry these externally from plant to plant where they deposit the pathogens on the plant surface or in the wounds the insects make on the plants during feeding. Insects may disseminate pathogens over short or long distances depending on the kind of insect, the insect–pathogen association, and the prevailing weather conditions, particularly wind.

A few species of mites and of nematodes can transmit internally several viruses from plant to plant. In addition, mites and nematodes probably carry externally bacteria and sticky fungal spores with which they become smeared as they move on infected plant surfaces.

Almost all animals, small and large, that move among plants and touch the plants along the way, can disseminate pathogens such as fungal spores, bacteria, seeds of parasitic plants, nematodes, and perhaps some viruses and viroids. Most of these pathogens adhere to the feet or the body of the animals, but some may be carried in contaminated mouthparts.

Finally, some plant pathogens, e.g., the zoospores of some fungi and certain parasitic plants, can transmit viruses as they move from one plant to another (zoospores), or as they grow and form a bridge between two plants (dodder).

DISSEMINATION BY MAN

Man disseminates all kinds of pathogens over short and long distances in a variety of ways. Within a field, man disseminates pathogens through his successive handling of diseased and healthy plants, through tools contaminated when used on diseased plants and then carrying the pathogen to healthy plants, through transport of contaminated soil on feet or equipment, etc. Man also disseminates pathogens on infected transplants, seed, nursery stock, budwood, etc., and by using contaminated containers. Finally, man disseminates pathogens by importing into his area new varieties that may carry pathogens that go undetected for a while, and by his travels through the world and importation of food or other items that may carry harmful plant pathogens.

OVERWINTERING AND/OR OVERSUMMERING OF THE PATHOGEN

Pathogens that infect perennial plants can survive in them during the low winter temperatures and/or during the hot, dry weather of the summer, regardless of whether the host plants are actively growing or are dormant at the time.

Annual plants, however, die at the end of the growing season, as do the leaves and fruits of deciduous perennial plants and even the stems of some perennial plants. In colder climates, annual plants and tops of some perennial plants die with the advent of the low winter temperatures and their pathogens are left without a host for the several months of cold weather. On the other hand, in hot, dry climates, annual plants die during the summer and their pathogens must be able to survive such periods in the absence of their hosts. Thus, pathogens that attack annual plants and renewable parts of perennial plants have evolved mechanisms by which they can survive the cold winters or dry summers that may intervene between crops or growing seasons (Fig. 10).

Fungi have evolved a great variety of mechanisms of overwintering or oversummering. On perennial plants, fungi overwinter as mycelium in infected tissues, e.g., cankers, and as spores at or near the infected surface of the plant or on the bud scales. Fungi affecting leaves or fruits of deciduous trees usually overwinter as mycelium or spores on fallen, infected leaves or fruits, or on the bud scales. Fungi affecting annual plants usually survive the winter or summer as mycelium, as resting or other spores and as sclerotia in infected plant debris, in the soil, and in or on seeds and other propagative organs, e.g., tubers. In some areas fungi survive by continuous infection of host plants grown outdoors through-

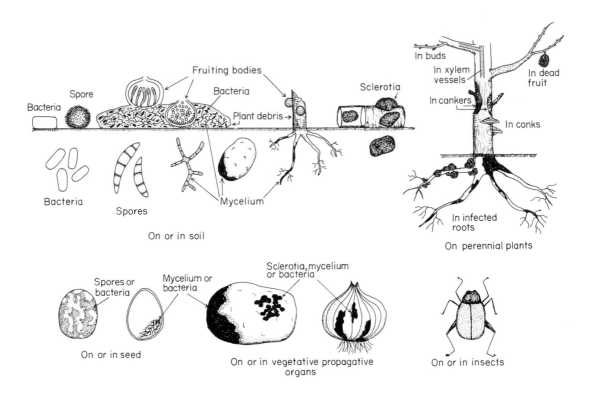

FIGURE 10.
Sources of inoculum of fungi and bacteria.

out the year, e.g., cabbage, or of plants grown in the greenhouse in the winter and outdoors in the summer. Similarly, some rust and other fungi overwinter on winter crops grown in warmer climates and move from them to the same hosts grown as spring crops in colder climates. Also, some fungi infect cultivated or wild perennial, as well as annual, plants and move from the perennial to the annual ones each growth season. Some rust fungi infect alternately an annual and a perennial host and the fungus goes from the one to the other host and overwinters, of course, in the perennial host.

Bacteria overwinter and oversummer as bacteria in essentially the same ways described for fungi, i.e., in infected plants, seeds, tubers, etc., in infected plant debris and, some of them, in the soil. Bacteria survive poorly when present in small numbers and free in the soil but survive well when masses of them are embedded in the hardened, slimy polysaccharides that usually surround them. Some bacteria also overwinter within the bodies of their insect vectors.

Viruses, viroids, mycoplasmas, rickettsialike bacteria, and protozoa survive only in living plant tissues such as the tops and roots of perennial plants, the roots of perennial plants that die to the soil line in the winter or summer, in vegetative propagating organs, and in the seeds of some hosts. A few viruses survive within their insect vectors and some viruses and viroids may survive on contaminated tools and in infected plant debris.

Nematodes usually overwinter or oversummer as eggs in the soil, and as eggs or nematodes in plant roots or in plant debris. Some nematodes produce larval stages or adults that can remain dormant in seeds, on bulbs, etc. for many months or years.

Parasitic higher plants survive either as seeds, usually in the soil, or as their infective vegetative form on their host.

SELECTED REFERENCES

Ellingboe, A. H. 1968. Inoculum production and infection by foliage pathogens. *Ann. Rev. Phytopathol.* **6**:317–330.

Emmett, R. W., and D. G. Parbery. 1975. Appressoria. *Ann. Rev. Phytopathol.* **13**:147–167.

Horsfall, J. G., and A. E. Dimond (eds.). 1959, 1960. "Plant Pathology." Vols. 1, 2, 3. Academic Press, New York.

Meredith, D. S. 1973. Significance of spore release and dispersal mechanisms in plant disease epidemiology. *Ann. Rev. Phytopathol.* **11**:313–342.

Rotem, J., and J. Palti. 1969. Irrigation and plant diseases. *Ann. Rev. Phytopathol.* **7**:267–288.

Schuster, M. L., and D. P. Coyne. 1974. Survival mechanisms of phytopathogenic bacteria. *Ann. Rev. Phytopathol.* **12**:199–221.

Stakman, E. C., and J. G. Harrar. 1957. "Principles of Plant Pathology." The Ronald Press, New York, 581 p.

Tarr, S. A. J. 1972. "The Principles of Plant Pathology." Winchester Press, New York, 632 p.

Van der Plank, J. E. 1963. "Plant Diseases: Epidemics and Control." Academic Press, New York, 349 p.

Van der Plank, J. E. 1975. "Principles of Plant Infection." Academic Press, New York, 216 p.

3
how pathogens attack plants

The intact, healthy plant is a community of cells built in a fortresslike fashion. The plant surfaces that come in contact with the environment either consist of cellulose as in the epidermal cells of roots and in the intercellular spaces of leaf parenchyma cells, or they consist of a layer of cuticle which covers the epidermal cell walls, as is the case in the aerial parts of plants. Often an additional layer, consisting of waxes, is deposited outside the cuticle, especially on younger parts of plants (Fig. 11).

Pathogens attack plants because during their evolutionary development they have acquired the ability to live off the substances manufactured by the host plants, and some of the pathogens depend on these substances for survival. Such substances, however, are contained in the protoplast of the plant cells, and, if pathogens are to gain access to them, they must first penetrate the outer barriers formed by the cuticle and/or cell walls. Even after the outer cell walls have been penetrated, further invasion of the plant by the pathogen necessitates penetration of more cell walls. Furthermore, the plant cell contents are not always found in forms immediately utilizable by the pathogen and must be transformed to units which the pathogen can absorb and assimilate. Moreover, the plant, reacting to the presence and activities of the pathogen, produces structures and chemical substances that interfere with the advance or the existence of the pathogen; if the pathogen is to survive and to continue living off the plant, it must be able to overcome such obstacles.

Therefore, for a pathogen to infect a plant it must be able to make its way into and through the plant, obtain nutrients from the plant, and neutralize the defense reactions of the plant. Pathogens accomplish these activities mostly through secretions of chemical substances that affect certain components or metabolic mechanisms of their hosts. Penetration and invasion, however, seem to be aided by, or in some cases be entirely

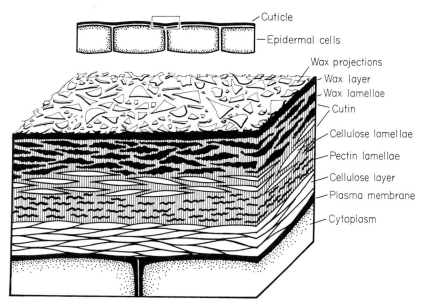

FIGURE 11.
Schematic representation of the structure and composition of the cuticle and cell wall of foliar epidermal cells. [Adapted from Goodman, Király, and Zaitlin (1967). "The Biochemistry and Physiology of Infectious Plant Disease." Van Nostrand, Princeton, New Jersey.]

the result of, mechanical force exerted by certain pathogens on the cell walls of the plant.

mechanical forces
exerted by pathogens
on host tissues

Plant pathogens are, generally, tiny microorganisms which cannot develop a "voluntary" force and apply it on a plant surface. Only some fungi, parasitic higher plants, and the nematodes appear to apply mechanical pressure to the plant surface they are about to penetrate. The amount of pressure, however, may vary greatly with the degree of "presoftening" of the plant surface by the enzymatic secretions of the pathogen.

For fungi and parasitic higher plants to penetrate a plant surface, they must, generally, first adhere to it. Although hyphae and radicles are usually surrounded by mucilaginous substances, their adhesion to the plant seems to be brought about primarily by the intermoleculer forces developing between the surfaces of plant and pathogen upon close contact with each other. After contact is established, the diameter of the part of hypha or radicle in contact with the host increases and forms a flattened, bulblike structure called the "appressorium." This increases the adherent area between the two organisms and securely fastens the pathogen to the plant. From the appressorium, a fine growing point, called the "penetration peg," arises and advances into and through the cuticle and cell wall. If the underlying host wall is soft, penetration occurs easily. When the underlying wall is hard, however, the force of the growing point may be greater than the adhesion force of the two surfaces and may cause the separation of the appressorial and host walls, thus averting infection. Penetration of plant barriers by fungi and parasitic higher plants is almost always assisted by the presence of enzymes secreted by the pathogen at the penetration site, resulting in the softening or dissolution of the barrier through enzymatic action.

While the penetration tube is passing through the cuticle, it usually attains its smallest diameter and appears threadlike. Following penetration of the cuticle, the hyphal tube diameter often increases considerably. The penetration tube attains the diameter normal for the hyphae of the particular fungus only after it has passed through the cell wall (Figs. 4 and 7).

Nematodes penetrate plant surfaces by means of the stylet, which is thrust back and forth and exerts mechanical pressure on the cell wall. The nematode first adheres to the plant surface by suction which it develops by bringing its fused lips in contact with the plant. After adhesion is accomplished, the nematode brings its body, or at least its forward portion, to a position vertical to the cell wall. The nematode then, with its head stationary and fixed to the cell wall, thrusts its stylet forward while the rear part of its body sways or rotates slowly round and round. After several consecutive thrusts of the stylet, the cell wall is pierced and the stylet or the entire nematode enters the cell.

Once a fungus or nematode has entered a cell it generally secretes increased amounts of enzymes which, presumably, soften or dissolve the opposite cell wall and make its penetration easier. Mechanical force, however, probably is brought to bear in most such penetrations, although to a lesser extent. It should be noted that in many fungal infections the diameter of the hypha becomes much smaller than the normal whenever it penetrates a cell wall and resumes its normal size once the wall has been penetrated.

Considerable mechanical force is also exerted on host tissues by some pathogenic fungi upon formation of their fructifications in the tissues beneath the plant surface. Through increased pressure the sporophore hyphae as well as fruiting bodies, such as pycnidia and perithecia, push outward and cause the cell walls and the cuticle to expand, become raised in the form of blisterlike protuberances, and finally break.

chemical
weapons of pathogens

Although some pathogens may use mechanical force to penetrate plant tissues, the activities of pathogens in plants are largely chemical in nature. Therefore, the effects caused by pathogens on plants are almost entirely the result of biochemical reactions taking place between substances secreted by the pathogen and those present in, or produced by, the plant.

The main groups of substances secreted by pathogens in plants, and which seem to be involved in production of disease, either directly or indirectly, include enzymes, toxins, growth regulators, and polysaccharides. These groups vary greatly as to their importance in pathogenicity, and their relative importance may be different from one disease to another. Thus in some diseases, e.g., soft rots, enzymes seem to be by far the most important, whereas in diseases like crown gall, growth regulators are apparently the main substances involved, and in the *Helminthosporium* blight of Victoria oats, the disease is primarily the result of a toxin secreted in the plant by the pathogen. Enzymes, toxins, and growth regulators, probably in that order, are considerably more common and more important in plant disease development than are polysaccharides.

Among the plant pathogens, all except viruses and viroids can probably produce enzymes, growth regulators, polysaccharides, and possibly toxins. Plant viruses and viroids are not known to produce any substances themselves, but they may induce the host cell to produce either excessive amounts of certain substances already found in healthy host cells or substances completely new to the host, some of which may belong to the groups mentioned above.

Pathogens produce these enzymes, etc., either in the normal course of their activities or upon growth on certain substrates. Undoubtedly, natural selection has favored the survival of pathogens that are assisted in their parasitism through the production of such substances. The presence or the amount of any such substance produced, however, is not always a measure of the ability of the pathogen to cause disease. As a matter of fact, many substances, identical to those produced by pathogens, are also produced by the healthy host plant.

In general, plant pathogenic enzymes disintegrate the structural components of host cells, break down inert food substances in the cell, or affect the protoplast directly and interfere with its functioning systems. Toxins seem to act directly on the protoplast and interfere with the permeability of its membranes and with its function. Growth regulators exert a hormonal effect on the cells and either increase or decrease their ability to divide and enlarge. Polysaccharides seem to play a role only in the vascular diseases in which they passively interfere with the translocation of water in the plants, or they may also be toxic.

ENZYMES

Enzymes are large protein molecules which catalyze all the interrelated reactions in a living cell. For each kind of chemical reaction that occurs in a cell there is a different enzyme which catalyzes that reaction.

ENZYMATIC DEGRADATION
OF CELL WALL SUBSTANCES

Most plant pathogens secrete enzymes throughout their existence or upon contact with a substrate. Usually, the first contact of pathogens with their host plants occurs at a plant surface. Such a surface may consist primarily of cellulose which makes up the epidermal cell walls or, on the aerial plant parts, of cellulose plus cuticle. Cuticle, which consists of cutin, is frequently covered with a layer of wax. Protein and lignin may also be found in epidermal cell walls. Penetration of pathogens into, and collapse of, parenchymatous tissues is brought about by the breakdown of the cell walls, consisting of cellulose, pectins, and hemicelluloses, and of the middle lamella, consisting primarily of pectins. Complete plant tissue disintegration involves, in addition, breakdown of lignin. The degradation of each of these substances is brought about by the action of one or more sets of enzymes secreted by the pathogen.

CUTICULAR WAX   Plant waxes are found as granular or rodlike projections, or as continuous layers outside or within the cuticle of many aerial plant parts. No pathogens are known to date to produce enzymes that can degrade waxes. Wax layers on plant surfaces are apparently penetrated by fungi and parasitic higher plants by means of mechanical force alone.

CUTIN   Cutin is the main component of the cuticular layer. The upper part of the layer is admixed with waxes, while its lower part, in the region where it merges into the outer walls of epidermal cells, cutin is admixed with pectin and cellulose (Fig. 11). There is evidence that at least some phytopathogenic fungi produce cutinases, i.e., enzymes that catalyze the breakdown and dissolution of cutin.

PECTIC SUBSTANCES   Pectic substances constitute the main components of the middle lamella, i.e., the intercellular cement which holds in place the cells of plant tissues (Fig. 12) and also a large portion of the primary cell wall, in which they form an amorphous gel filling the spaces between the cellulose microfibrils (Fig. 13).

Pectic substances are polysaccharides containing a very high percentage of galacturonic acid molecules. Several enzymes, known as pectinases, degrade pectic substances. Some remove small branches off the pectin chains and have no effect on the length of the pectin chains but they alter their solubility and affect the rate at which they can be attacked by the chain-splitting pectinases. The latter cleave the pectic chain and release shorter chain portions containing one or a few molecules of galacturonic acid.

The pectin-degrading enzymes have been shown to be involved in the production of many diseases. Pectic enzymes are produced by germinating spores and apparently, acting together with other pathogen metabolites, assist in the penetration of the host. Pectin degradation results in weakening of cell walls or tissue maceration which undoubtedly facilitates the inter- or intracellular invasion of the tissues by a pathogen. Pectic enzymes also provide nutrients for the pathogen in infected tis-

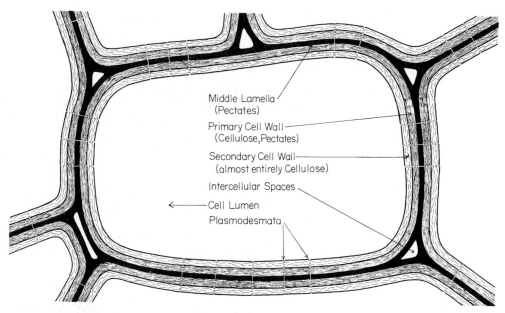

Middle Lamella
(Pectates)

Primary Cell Wall
(Cellulose, Pectates)

Secondary Cell Wall
(almost entirely Cellulose)

Intercellular Spaces

← Cell Lumen

Plasmodesmata

FIGURE 12.
Schematic representation of structure and composition of plant cell walls.

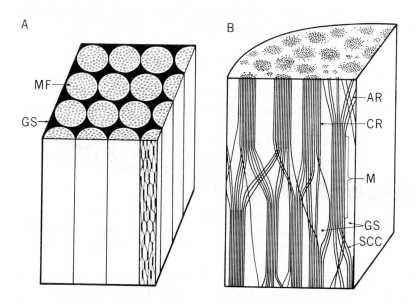

A

MF

GS

B

AR

CR

M

GS
SCC

FIGURE 13.
Schematic diagram of the gross structure of cellulose and microfibrils (A), and of the arrangement of cellulose molecules within a microfibril (B). MF = microfibril; GS = ground substance (pectin, hemicelluloses, or lignin); AR = amorphous region of cellulose; CR = crystalline region; M = micelle; SCC = single cellulose chain (molecule). [Adapted from H. P. Brown, A. J. Panshing, and C. C. Forsaith (1949). "Textbook of Wood Technology," Vol. 1. McGraw-Hill, New York.]

sues. Pectic enzymes, by the debris they create, seem to be involved in the induction of vascular plugs and occlusions in the vascular wilt diseases. Cells are usually quickly killed in tissues macerated by pectic enzymes, but how these enzymes kill cells is not clear yet.

CELLULOSE   Cellulose is also a polysaccharide, but it consists of chains of glucose molecules. It occurs in all higher plants as the skeletal substance of cell walls in the form of microfibrils (Figs. 12 and 13). Microfibrils, which can be perceived as bundles of iron bars in a reinforced concrete building, are the basic structural unit of the wall even though they account for less than 20 percent of the wall volume in most meristematic cells. The cellulose content of tissues varies from about 12 percent in the nonwoody tissues of grasses to about 50 percent in mature wood tissues to more than 90 percent in the cotton fibers. The spaces between microfibrils and between micelles or cellulose chains within the microfibrils may be filled with pectins and hemicellulose and probably some lignin at maturation.

Cellulose-degrading enzymes have been shown to be produced by several phytopathogenic fungi, bacteria, and nematodes, and they are undoubtedly produced by parasitic higher plants. Saprophytic fungi, mainly certain groups of basidiomycetes, and to a smaller degree saprophytic bacteria cause the breakdown of most of the cellulose decomposed in nature. In living plant tissues, however, cellulolytic enzymes secreted by pathogens play a role in the softening and/or disintegration of cell-wall material; they facilitate the penetration and spread of the pathogen in the host and cause the collapse and disintegration of the cellular structure, thereby aiding the pathogen in the production of disease. Cellulolytic enzymes may further participate indirectly in disease development by releasing, from cellulose chains, soluble sugars which serve as food for the pathogen, and, in the vascular diseases, by liberating into the transpiration stream large molecules from cellulose which interfere with the normal movement of water.

The enzymatic breakdown of cellulose results in the final production of glucose molecules. This is brought about by a series of enzymatic reactions carried out by several enzymes called cellulases.

LIGNIN   Lignin is found in the middle lamella and the cell wall of xylem vessels and the fibers that strengthen plants. It is also found in epidermal and occasionally hypodermal cell walls of some plants. The lignin content of mature woody plants varies from 15 to 38 percent and is second only to cellulose.

Lignin is different from both carbohydrates and proteins in composition and properties. The most common basic structural unit of lignin is a phenylpropanoid. The lignin polymer is perhaps more resistant to enzymatic degradation than any other plant substance.

It is obvious that enormous amounts of lignin are degraded by microorganisms in nature, as is evidenced by the decomposition of all annual plants and a large portion of perennial plants that disintegrate annually. It is generally accepted, however, that only a small group of microorganisms is capable of degrading lignin. Actually, only about 500 fungus species, almost all of them basidiomycetes, have been reported, so far, as being capable of decomposing wood. About one-fourth of these (the brown-rot fungi) seem to cause some degradation of lignin but cannot utilize it. Most of the lignin in the world is degraded and utilized by a group of fungi called white-rot fungi. It appears that the white-rot fungi secrete one or more enzymes (ligninases) which enable them to utilize lignin.

It appears that, with the exception of the wood-rotting fungi, the other pathogens produce few or no lignin-degrading enzymes and that the diseases they cause are not dependent on the presence of such enzymes.

ENZYMATIC DEGRADATION OF SUBSTANCES CONTAINED IN PLANT CELLS

Most kinds of pathogens live all or part of their lives in association with or inside the living protoplast. These pathogens obviously derive nutrients from the protoplast. All the other pathogens, i.e., the great majority of fungi and bacteria, obtain nutrients from protoplasts after the latter have been killed. Some of the nutrients, e.g., sugars and amino acids, are probably sufficiently small molecules to be absorbed by the pathogen directly. Some of the other plant cell constituents, however, such as starch, proteins, and fats, can be utilized only after degradation by the enzymes secreted by the pathogen.

PROTEINS   Plant cells contain innumerable different proteins which play diverse roles as catalysts of cellular reactions (enzymes) or as structural material (membranes). Proteins are formed by the joining together of numerous molecules of about twenty different kinds of amino acids.

All pathogens seem to be capable of degrading many kinds of protein molecules. The plant pathogenic enzymes involved in protein degradation are similar to those present in higher plants and animals and are called proteinases.

Considering the paramount importance of proteins as enzymes, constituents of cell membranes, and structural components of plant cells, degradation of host proteins by proteolytic enzymes secreted by pathogens can profoundly affect the organization and function of the host cells. The nature and extent of such effects, however, has been investigated little so far, and the significance in disease development is not known.

STARCH   Starch is the main reserve polysaccharide found in plant cells. Starch is synthesized in the chloroplasts and, in nonphotosynthetic organs, in the amyloplasts. Starch is a glucose polymer and exists in two forms: amylose, an essentially linear molecule, and amylopectin, a highly branched molecule of various chain lengths.

Most pathogens utilize starch, and other reserve polysaccharides, in their metabolic activities. The degradation of starch is brought about by the action of enzymes called amylases. The end product of starch breakdown is glucose and it is used by the pathogens directly.

LIPIDS   Various types of lipids occur in all plant cells, the most important being: the oils and fats found in many cells, especially in seeds where they function as energy storage compounds; the wax lipids, found on most aerial epidermal cells; and the phospholipids and the glycolipids, both of which, along with protein, are the main constituents of all plant cell membranes. The common characteristic of all lipids is that they contain fatty acids that may be saturated or unsaturated.

Several fungi, bacteria, and nematodes are known to be capable of degrading lipids. Lipolytic enzymes, called lipases, phospholipidases, etc., hydrolyze the liberation of the fatty acids from the lipid molecule. The fatty acids are presumably utilized by the pathogen directly.

MICROBIAL TOXINS IN PLANT DISEASE

Living plant cells are complex systems in which many interdependent biochemical reactions are taking place concurrently or in a well-defined succession. These reactions result in the intricate and well-organized processes essential for life. Disturbance of any of these metabolic reactions causes disruption of the physiological processes that sustain the plant and leads to development of disease. Among the factors inducing such disturbances are substances that are produced by plant pathogenic microorganisms and are called toxins. Toxins act directly on living host protoplasts, seriously damaging or killing the cells of the plant. Some toxins act as general protoplasmic poisons and affect many species of plants representing different families. Others are toxic to only a few plant species or varieties and completely harmless to others.

Fungi and bacteria may produce toxins in infected plants as well as in culture medium. Toxins, however, are extremely poisonous substances and are effective in very low concentrations. Some are unstable or react quickly and are tightly bound to specific sites within the plant cell.

Toxins injure host cells either by affecting the permeability of the cell membrane or by inactivating or inhibiting enzymes and subsequently interrupting the corresponding enzymatic reactions. Certain toxins act as antimetabolites inducing a deficiency for an essential growth factor.

TOXINS THAT AFFECT A WIDE RANGE OF HOST PLANTS

Several toxic substances produced by phytopathogenic microorganisms produce all or part of the disease syndrome not only on the host plant but also on other species of plants which are not normally attacked by the pathogen in nature.

THE WILDFIRE TOXIN OR TABTOXIN   It is produced by the bacterium *Pseudomonas tabaci* which causes the wildfire disease of tobacco. The disease is characterized by necrotic spots on leaves, each surrounded by a yellow halo. Sterile culture filtrates of the organism produce symptoms identical to those characteristic of wildfire of tobacco not only on tobacco, but in a large number of plant species belonging to many different families. Although several mechanisms of action of tabtoxin have been proposed at different times, the mechanism of action is still uncertain.

FUSARIAL TOXINS   Many species of *Fusarium* cause wilt diseases on a number of plants. Symptoms consist of epinasty, plugging and browning of xylem vessels, necrosis, wilting, and finally death of the plant. One compound, called lycomarasmin, was isolated from culture filtrates of the *Fusarium* causing the tomato wilt disease. Lycomarasmin causes wilting and necrosis between the veins of excised tomato leaves but seems to be of little or no importance in disease development. A second toxin produced by *Fusarium* and called fusaric acid produces, in addition to wilt, water-soaked spots on the leaves and browning of the vascular tissue rather than necrosis between the leaf veins. Both toxins bind heavy metals such as Fe^{3+} and Cu^{2+}. This in turn affects the permeability of cell membranes and the enzymatic reactions in the cells by inhibition of enzymes.

PYRICULARIN   The fungus *Pyricularia oryzae* is the cause of the blast disease of rice. Rice blast appears as yellowing, striping, and stunting of seedlings; and as leaf spots and rot of the culm at the base of the leaf in mature plants. Culture filtrates of the fungus contain the toxin pyricularin and can reproduce the disease symptoms in seedlings and in mature plants. Pyricularin is a fairly potent toxin and affects a number of species of higher plants. Low concentrations of pyricularin stimulate growth and respiration of the host, while higher concentrations inhibit both.

Several other toxic substances have been isolated from cultures of pathogenic fungi and bacteria and have been implicated as contributing factors to the development of the disease caused by the pathogen. Among such fungi are species of *Alternaria, Ascochyta, Botrytis, Ceratocystis, Cercospora, Colletotrichum, Endothia, Fusicoccum,* and *Phytophthora.* Several species of bacteria of *Pseudomonas, Xanthomonas, Corynebacterium,* and *Erwinia* also produce toxins and, with one possible exception, apparently all of them are nonspecific.

HOST-SPECIFIC TOXINS

A host-specific toxin is a substance produced by a pathogenic microorganism which at a certain concentration is toxic only to the host of that pathogen. Such toxins are produced by only a few pathogens.

VICTORIN

Victorin is produced by the fungus *Helminthosporium victoriae.* This pathogen occurs as a saprophyte or weak parasite on many grasses. It is a soil- and seed-borne organism, and when it infects susceptible oat plants it remains in the basal portions of the plant, where it causes a necrosis of the root and stem. It also produces a powerful toxin that acts at a distance from the site of infection, disrupts the permeability of cell membranes and causes a leaf blight, and rapidly destroys the entire plant. Infection usually occurs near the soil line and, within 4 to 5 days from inoculation, host cells begin to collapse and the area of damage spreads from the point of infection. The first observable symptoms, however, appear as yellow to orange-red stripes on the leaves. When the fungus enters the cell of a resistant host, the host cell responds to the disturbance quickly and both host cell and fungus die immediately without further growth of the fungus. In susceptible hosts this response seems to be prevented by the action of the fungus toxin.

The toxicity of the toxin is limited to plants of the oat variety Victoria (*Avena sativa* var. *Victoria*) and to those derived from crosses of Victoria with other oat varieties. All other plant species tested were immune to the toxin, since only concentrated culture filtrates had any effect on them.

PERICONIA CIRCINATA TOXIN

It is produced by a fungus that invades the roots and the lower internodes of susceptible sorghum plants. It causes a scalded appearance in the foliage of infected young seedlings, stunting, early blooming, and premature death. Leaves of older plants, although free from the pathogen, roll, wilt, turn yellow, and show the usual blight symptoms. In older roots the cortex decays, and the central cylinder turns red and dies.

ALTERNARIA KIKUCHIANA TOXIN

It is produced in the black spot disease of Japanese pears (*Pyrus serotina*). Pears of susceptible varieties sprayed with culture filtrates of the fungus are damaged while those of resistant varieties are unharmed.

Several other cases of fungal plant pathogens producing host-specific toxins have been reported. Among such fungi are several more species of *Helminthosporium* and species of *Hypoxylon*, *Ophiobolus*, and *Phyllosticta*. There is little or no information, however, as to how any of these toxins bring about their toxic effects on the plant. The bacterium *Erwinia amylovora* has also been reported to produce a host-specific toxin called amylovorin.

growth regulators
in plant disease

Plant growth is regulated by a small number of groups of naturally occurring compounds which act as hormones and are generally called growth regulators. The most important growth regulators are auxins, gibberellins, and cytokinins, but other compounds, such as ethylene and growth inhibitors, play important regulatory roles in the life of the plant. Growth regulators act in very small concentrations, and even slight deviations from the normal concentration may bring about strikingly different plant growth patterns. The concentration of a specific growth regulator in the plant is not constant, but it usually rises quickly to a peak and then quickly declines as a result of the action of hormone-inhibitory systems present in the plant. Growth regulators appear to act, at least in some cases, by promoting synthesis of messenger-RNA molecules which leads to the formation of specific enzymes and which, in turn, control the biochemistry and the physiology of the plant.

Plant pathogens may produce more of the same growth regulators as those produced by the plant or more of the same inhibitors of the growth regulators as those produced by the plant; they may produce new and different growth regulators or inhibitors of growth regulators; or they may produce substances that stimulate or retard the production of growth regulators or growth inhibitors by the plant.

It is obvious that, whatever the mechanism of action involved, pathogens often cause an imbalance in the hormonal system of the plant and bring about abnormal growth responses incompatible with the healthy development of a plant. That pathogens can cause disease through secretion of growth regulators in the infected plant or through their effects on the growth-regulatory systems of the infected plant is made evident by the variety of abnormal plant growth responses they cause, such as stunting, overgrowths, rosetting, excessive root branching, stem malformation, leaf epinasty, defoliation, suppression of bud growth, etc. The most important groups of plant growth regulators, their function in the plant and their role in disease development, where known, are discussed below.

AUXINS

The auxin naturally occurring in plants is indole-3-acetic acid (IAA). Continually produced in growing plant tissues, IAA moves rapidly from the young green tissues to older tissues, but is constantly being destroyed by the enzyme indole-3-acetic acid oxidase, which explains the low concentration of the auxin.

The effects of IAA on the plant are numerous. Required for cell elongation and differentiation, absorption of IAA to the cell membrane also affects the permeability of the membrane; IAA causes a general increase in respiration of plant tissues and promotes the synthesis of messenger RNA and, subsequently, of proteins—enzymes as well as structural proteins.

Increased auxin (IAA) levels occur in many plants infected by fungi, bacteria, viruses, mycoplasmas, and nematodes although some pathogens seem to lower the auxin level of the host. Thus, the fungi causing late blight of potato (*Phytophthora infestans*), corn smut (*Ustilago maydis*), cedar apple rust (*Gymnosporangium juniperi-virginianae*), banana wilt (*Fusarium oxysporum* f. *cubense*), the root-knot nematode (*Meloidogyne* sp.), and others, not only induce increased levels of IAA in their respective hosts but are themselves capable of producing IAA. In some diseases, however, increased levels of IAA are wholly or partly due to the decreased degradation of IAA through inhibition of IAA oxidase, as has been shown to be the case in several diseases, including corn smut and stem rust of wheat.

The production and role of auxin in plant disease have been studied more extensively in some bacterial diseases of plants. *Pseudomonas solanacearum,* the cause of bacterial wilt of solanaceous plants, induces a 100-fold increase in the IAA level of diseased plants compared to that of healthy plants. How the increased levels of IAA contribute to the development of wilt of plants is not yet clear, but the increased plasticity of cell walls as a result of high IAA levels renders the pectin, cellulose, and protein components of the cell wall more accessible to, and may facilitate their degradation by, the respective enzymes secreted by the pathogen. Increase in IAA levels seems to inhibit lignification of tissues and may thus prolong the period of exposure of the nonlignified tissues to the cell wall degrading enzymes of the pathogen. Increased respiratory rates in the infected tissues may also be due to high IAA levels and, since auxin affects cell permeability, it may be responsible for the increased transpiration of the infected plants.

In crown gall, a disease caused by the bacterium *Agrobacterium tumefaciens* on more than one hundred plant species, galls or tumors develop on the roots, stems, petioles, etc., of the host plants. Crown gall tumors develop when crown gall bacteria enter fresh wounds of a susceptible host. Immediately after wounding, cells around the wound are activated to divide. During the intense cell division of the second and third days after wounding, the cells are somehow conditioned and made receptive to a stimulus produced by the bacteria or by the host cells in response to the bacteria. This stimulus, known as the tumor-inducing principle (TIP) transforms normal plant cells into tumor cells. Tumor cells subsequently grow and divide independently of the bacteria and their organization, rate of growth, and rate of division can no longer be controlled by the host plant.

Tumor cells contain higher than normal amounts of IAA and also of cytokinin. The crown gall bacteria, of course, produce IAA, but since even tumors free of bacteria contain increased levels of IAA, it is certain that the tumor cells themselves are capable of generating the abnormal levels of IAA they contain. However, although the increased levels of IAA of tumor cells are sufficient to cause the autonomous enlargement and division of these cells once they have been transformed to tumor cells, high IAA levels alone cannot cause the transformation of healthy cells into tumor cells. What other substances are involved in the tumor-inducing principle is not known.

Many plant viruses, viroids, and mycoplasmas cause stunting of plant growth, stimulation of axillary buds, and various morphological abnormalities on organs of infected plants. Such manifestations are very similar to the symptoms produced by imbalance in the growth substances of the plant. The mechanisms by which these pathogens bring about changes in the auxin levels of their hosts are presently unknown and in some of these diseases there is not even correlation between auxin content and symptoms exhibited by the infected plants.

GIBBERELLINS

Gibberellins are normal constituents of green plants and are also produced by several microorganisms. Gibberellins were first isolated from the fungus *Gibberella fujikuroi*, the cause of the "foolish seedling disease" of rice. The best known gibberellin is gibberellic acid. Compounds such as vitamin E and helminthosporol also have gibberellinlike activity.

Gibberellins have striking growth-promoting effects. They speed elongation of dwarf varieties to normal sizes, promote flowering, stem and root elongation, and growth of fruit. These types of elongation resemble in some respects that caused by IAA, and gibberellin also induces IAA formation. Auxin and gibberellin may also act synergistically. Gibberellins seem to activate genes that had been previously "turned off."

The foolish seedling disease of rice, in which rice seedlings, infected with the fungus *Gibberella fujikuroi*, grow rapidly and become much taller than healthy plants, is apparently the result, to a considerable extent at least, of the gibberellin secreted by the pathogen.

Although no difference has been reported so far in the gibberellin content of healthy and virus- or mycoplasma-infected plants, spraying of diseased plants with gibberellin overcomes some of the symptoms caused by these pathogens. Thus, stunting of corn plants infected with corn stunt mycoplasma and of tobacco plants infected with severe etch virus was reversed after treatment with gibberellin. Axillary bud suppression, caused by sour cherry yellows virus (SCYV) on cherry and by leaf curl virus on tobacco, was also overcome by gibberellin sprays. The same treatment also increased fruit production in SCYV-infected cherries. In most of these treatments the pathogen itself does not seem to be affected and the symptoms reappear on the plants after gibberellin applications are stopped. It is not known, however, whether the pathogen-caused stunting of plants is actually due to reduced gibberellin concentration in the diseased plant, especially since the growth of even healthy plants is equally increased after gibberellin treatments.

CYTOKININS

Cytokinins are potent growth factors necessary for cell growth and differentiation. In addition, they inhibit the breakdown of proteins and nucleic acids, thereby causing inhibition of senescence, and they have the capacity to direct the flow of amino acids and other nutrients through the plant, toward the point of high cytokinin concentration. Cytokinins occur in very small concentrations in green plants, their seeds, and in the sap stream.

The first compound with cytokinin activity to be identified was kinetin which, however, was isolated from herring sperm DNA and does not occur naturally in plants. Several cytokinins, e.g., zeatin and isopentenyl adenosine (IPA), have since been isolated from plants.

Cytokinins act by preventing genes from being "turned off" and by activating genes that had been previously "turned off."

The role of cytokinins in plant disease is just beginning to be studied. Cytokinin activity increases in clubroot galls, in crown galls, rust galls, and in rust-infected bean and broad bean leaves. In the latter, cytokinin activity seems to be related to both the juvenile feature of the green islands around the infection centers and to the senescence outside the green island. On the other hand, cytokinin activity is lower in the sap and in tissue extracts of cotton plants infected with Verticillium wilt and in plants suffering from drought. In the *Helminthosporium* blight disease of Victoria oats, cytokinins increase the quantity of toxin absorbed by the cells but tobacco leaves injected with the wildfire toxin and treated with kinetin fail to develop the typical toxin-induced chlorosis. A cytokinin is partly responsible for the "leafy" gall disease caused by the bacterium *Corynebacterium fascians,* and it has been suggested that cytokinins may be responsible for the witches'-broom diseases caused by fungi and mycoplasmas.

Treating plants with kinetin before or shortly after inoculation with a virus seems to reduce the number of infections in local-lesion hosts and to reduce virus multiplication in systemically infected hosts.

ETHYLENE

Ethylene is naturally produced by plants and exerts a variety of effects on plants, including leaf abscission, epinasty, and fruit ripening. Ethylene is produced by several plant pathogenic bacteria of the genera *Pseudomonas, Xanthomonas,* and *Erwinia.* In the fruit of banana infected with *Pseudomonas solanacearum,* the ethylene content increases proportionately with the (premature) yellowing of the fruit, while no ethylene can be detected in healthy fruits. Ethylene has also been implicated in the leaf epinasty symptom of the vascular wilt syndromes, and in the premature defoliation observed in several types of plant diseases.

POLYSACCHARIDES

Fungi, bacteria, nematodes, and possibly other pathogens, constantly release varying amounts of mucilaginous substances which coat their bodies and provide the interface between the outer surface of the micro-organism and its environment.

The role of slimy polysaccharides in plant disease appears to be limited primarily in the wilt diseases caused by pathogens that invade the vascular system of the plant. In the vascular wilts, large polysaccharide molecules released by the pathogen in the xylem may be sufficient to cause a mechanical blockage of vascular bundles and thus initiate wilting. Although such an effect by the polysaccharides alone may occur rarely in nature, when it is considered together with the effect caused by the macromolecular substances released in the vessels through the breakdown of host substances by pathogen enzymes, the possibility of polysaccharide involvement in blockage of vessels during vascular wilts becomes obvious.

SELECTED REFERENCES

Albersheim, P., T. M. Jones, and P. D. English. 1969. Biochemistry of the cell wall in relation to infective processes. *Ann. Rev. Phytopathol.* **7:**171–194.

Bateman, D. F., and R. L. Millar. 1966. Pectic enzymes in tissue degradation. *Ann. Rev. Phytopathol.* **4**:119–146.

Brown, W. 1965. Toxins and cell-wall dissolving enzymes in relation to plant disease. *Ann. Rev. Phytopathol.* **3**:1–18.

Dekhuijzen, H. M., and J. C. Overeem. 1971. The role of cytokinins in clubroot formation. *Physiol. Plant Pathol.* **1**:151–161.

Dickinson, S. 1959. The mechanical ability to breach the host barriers, *in* "Plant Pathology" (J. G. Horsfall, and A. E. Dimond, eds.), Vol. 2, pp. 203–232. Academic Press, New York.

Durbin, R. D. 1971. Chlorosis-inducing pseudomonad toxins: Their mechanism of action and structure, *in* "Morphological and biochemical events in plant–parasite interaction" (S. Akai and S. Ouchi, eds.), pp. 369–385. The Phytopathol. Soc. of Japan, Tokyo.

Goodman, R. N., J. S. Huang, and P. Y. Huang. 1974. Host-specific phytotoxic polysaccharide from apple tissue infected by *Erwinia amylovora. Science* **183**:1081–1082.

Heitefuss, R., and P. H. Williams (eds.). 1976. "Physiological Plant Pathology." Encyclopedia of Plant Physiology, New Series, Vol. 4. Springer-Verlag, New York.

Király, Z., M. El Hammady, and B. I. Pozsár. 1967. Increased cytokinin activity of rust-infected bean and broad bean leaves. *Phytopathology* **57**:93–94.

Kirk, T. K. 1971. Effects of microorganisms on lignin. *Ann. Rev. Phytopathol.* **9**:185–210.

Kuriger, W. E., and G. N. Agrios. 1977. Cytokinin levels and kinetin-virus interactions in tobacco ringspot virus-infected cowpea plants. *Phytopathology* **67**:604–609.

Maramorosch, K. 1957. Reversal of virus-caused stunting in plants by gibberellic acid. *Science* **126**:651–652.

Misaghi, I., J. E. DeVay, and T. Kosuge. 1972. Changes in cytokinin activity associated with the development of Verticillium wilt and water stress in cotton plants. *Physiol. Plant Pathol.* **2**:187–196.

Norkrans, Birgitta. 1963. Degradation of cellulose. *Ann. Rev. Phytopathol.* **1**:325–350.

Patil, S. S. 1974. Toxins produced by phytopathogenic bacteria. *Ann. Rev. Phytopathol.* **12**:259–279.

Pozsár, B. I., and Z. Király. 1966. Phloem-transport in rust-infected plants and the cytokinin directed long-distance movement of nutrients. *Phytopathol. Z.* **56**:297–309.

Pringle, R. B., and R. P. Scheffer. 1964. Host-specific plant toxins. *Ann. Rev. Phytopathol.* **2**:133–156.

Rowan, S. J. 1970. Fusiform rust gall formation and cytokinin of Loblolly pine. *Phytopathology* **60**:1225–1226.

Scheffer, R. P., and R. B. Pringle. 1967. Pathogen-produced determinants of disease and their effects on host plants, *in* "The Dynamic Role of Molecular Constituents in Plant–Parasite Interaction" (C. J. Mirocha and I. Uritani, eds.), pp. 217–236. Bruce, St. Paul, Minnesota.

Sequeira, L. 1973. Hormone metabolism in diseased plants. *Ann. Rev. Plant Physiol.* **24**:353–380.

Strobel, G. A. 1974. Phytotoxins produced by plant parasites. *Ann. Rev. Microbiol.* **25**:541–566.

Van den Ende, G., and H. F. Linskens. 1974. Cutinolytic enzymes in relation to pathogenesis. *Ann. Rev. Phytopathol.* **12**:247–258.

Wheeler, H. 1975. "Plant Pathogenesis." Springer-Verlag, New York. 106 p.

4
pathogen effects on plant physiological functions

effect of pathogens on photosynthesis

Photosynthesis is a basic function of green plants that enables them to transform light energy into chemical energy which they can utilize in their cell activities. Photosynthesis is the ultimate source of all energy used in plant or animal cells, since, in a living cell, all activities except photosynthesis expend the energy provided by photosynthesis.

In photosynthesis, carbon dioxide from the atmosphere and water from the soil are brought together in the chloroplasts of the green parts of plants and, in the presence of light, react to form glucose with concurrent release of oxygen:

$$6\ CO_2 + 6\ H_2O \xrightarrow[\text{chlorophyll}]{\text{light}} C_6H_{12}O_6 + 6\ O_2$$

In view of the fundamental position of photosynthesis in the life of plants, it is apparent that any interference of pathogens with photosynthesis results in a diseased condition in the plant. That pathogens do interfere with photosynthesis is obvious from the chlorosis they cause on many infected plants, from the necrotic lesions or large necrotic areas they produce on green plant parts, and from the reduced amounts of growth, fruits, etc., produced by many infected plants.

In leaf spot, blight, and other kinds of diseases in which there is destruction of leaf tissue, photosynthesis is obviously reduced because of the reduction, through death, of the photosynthetic surface of the plant. Even in other diseases, however, plant pathogens reduce photosynthesis, especially in the late stages of diseases, by affecting the chloroplasts and causing their degeneration. The overall chlorophyll content of leaves in

64

many fungal and bacterial diseases is reduced, but the photosynthetic activity of the remaining chlorophyll seems to remain unaffected. In plants infected by vascular pathogens, chlorophyll is reduced and photosynthesis stops even before the eventual wilting of the plant. Most virus, mycoplasma, and nematode diseases induce varying degrees of chlorosis. In the majority of such diseases photosynthesis of infected plants is reduced qreatly, in advanced stages of the disease the rate of photosynthesis being no more than one-fourth the normal rate.

effect of pathogens on translocation of water and nutrients in the host plant

All living plant cells require an abundance of water and an adequate amount of organic and inorganic nutrients in order to live and to carry out their respective physiological functions. Plants absorb water and inorganic (mineral) nutrients from the soil through their root system. These are generally translocated upward through the xylem vessels of the stem and into the vascular bundles of the petioles and leaf veins, from which they enter the leaf cells. The minerals and part of the water are utilized by the leaf and other cells for synthesis of the various plant substances, but most of the water evaporates out of the leaf cells into the intercellular spaces and from there diffuses into the atmosphere through the stomata. On the other hand, nearly all organic nutrients of plants are produced in the leaf cells, following photosynthesis, and are translocated downward and distributed to all the living plant cells by passing for the most part through the phloem tissues. It is apparent that interference by the pathogen with the upward movement of inorganic nutrients and water or with the downward movement of organic substances will result in diseased conditions in the parts of the plant denied these materials. These diseased parts, in turn, will be unable to carry out their own functions and will deny the rest of the plant their services or their products, thus resulting in disease of the entire plant. For example, if water movement to the leaves is inhibited, the leaves cannot function properly, photosynthesis is reduced or stopped, and few or no nutrients are available to move to the roots, which, in turn, become starved, diseased, and may die.

INTERFERENCE WITH UPWARD TRANSLOCATION OF WATER AND INORGANIC NUTRIENTS

Many plant pathogens interfere in one or more ways with the translocation of water and inorganic nutrients through the plants. Some pathogens affect the integrity or function of the roots and cause decreased absorption of water by them; other pathogens, by growing in the xylem vessels or by other means, interfere with the translocation of water through the stem; and, in some diseases, pathogens also interfere with the water economy of the plant by causing excessive transpiration through their effects on leaves and stomata.

EFFECT ON ABSORPTION OF WATER BY ROOTS

Many pathogens, such as the damping-off fungi, the root-rotting fungi and bacteria, most nematodes, and some viruses cause an extensive destruction of the roots before any symptoms appear on the aboveground parts of the plant. Root injury affects directly the amount of functioning roots and decreases proportionately the amount of water absorbed by the roots. Some vascular parasites, along with their other effects, seem to inhibit root hair production, which reduces water absorption. These and other pathogens also alter the permeability of root cells, an effect that further interferes with the normal absorption of water by roots.

EFFECT ON TRANSLOCATION OF WATER THROUGH THE XYLEM

Fungal and bacterial pathogens that cause damping-off, stem rots, and cankers may reach the xylem vessels in the area of the infection and, if the affected plants are young, may cause their destruction and collapse. Affected vessels may also be filled with the bodies of the pathogen and with substances secreted by the pathogen or by the host in response to the pathogen, and may become clogged. Whether destroyed or clogged the affected vessels cease to function properly and allow little or no water to pass through them. Certain pathogens, such as the crown gall bacterium (*Agrobacterium tumefaciens*), the clubroot fungus (*Plasmodiophora brassicae*), and the root-knot nematode (*Meloidogyne* sp.) induce gall formation in the stem and/or the roots. The enlarged and proliferating cells near or around the xylem exert pressure on the xylem vessels, which may be crushed and dislocated and, thereby, become less efficient in transporting water.

The most typical and complete dysfunction of xylem in translocating water, however, is observed in the vascular wilts caused by the fungi *Ceratocystis*, *Fusarium*, and *Verticillium*, and bacteria like *Pseudomonas* and *Erwinia*. These pathogens invade the xylem of roots and stems and produce diseases primarily by interfering with the upward movement of water through the xylem. In many plants infected by these pathogens the water flow through the stem xylem is reduced to a mere 2 to 4 percent of that flowing through the stems of healthy plants. In general, the rate of flow through infected stems seems to be inversely proportional to the number of vessels blocked by the pathogen and by the substances resulting from the infection. Evidently, more than one factor is usually responsible for vascular dysfunction in the wilt diseases. Although the pathogen is the single cause of the disease, some of the factors responsible for the disease syndrome originate directly from the pathogen, while others originate from the host in response to the pathogen. The pathogen can reduce the flow of water through its physical presence in the xylem as mycelium, spores, or bacterial cells and by production of large molecules (polysaccharides) in the vessels. The infected host may reduce the flow of water through reduction in the size or collapse of vessels due to infection, development of tyloses in the vessels, release of large-molecule compounds in the vessels as a result of cell wall breakdown by pathogenic enzymes, and reduced water tension in the vessels due to pathogen-induced alterations in foliar transpiration.

EFFECT ON TRANSPIRATION

In plant diseases in which the pathogen infects the leaves, transpiration is usually increased. This is the result of destruction of at least part of the protection afforded the leaf by the cuticle, increase in permeability of leaf cells, and dysfunction of stomata. Diseases like the rusts, mildews, and apple scab destroy a considerable portion of the cuticle and epidermis and this results in uncontrolled loss of water from the affected areas. If water absorption and translocation cannot keep up with the excessive loss of water, loss of turgor and wilting of leaves follows. The suction force of excessively transpiring leaves is abnormally increased and may lead to collapse and/or dysfunction of underlying vessels through production of tyloses and gums.

INTERFERENCE WITH THE TRANSLOCATION OF ORGANIC NUTRIENTS THROUGH THE PHLOEM

Organic nutrients produced in leaf cells through photosynthesis move through plasmodesmata into adjoining phloem elements. From there they move down the phloem sieve tubes and eventually, again through plasmodesmata, into the protoplasm of living nonphotosynthetic cells, where they are utilized, or into storage organs, where they are stored. Thus, in both cases, they are removed from "circulation." Plant pathogens may interfere with the movement of organic nutrients from the leaf cells to the phloem or with their translocation through the phloem elements and, possibly, with their movement from the phloem into the cells that will utilize them.

Obligate fungal parasites, such as the rust and mildew fungi, cause an accumulation of photosynthetic products, as well as inorganic nutrients, in the areas invaded by the pathogen. In these diseases, the infected areas are characterized by reduced photosynthesis and increased respiration. However, synthesis of starch and of other compounds as well as dry weight are temporarily increased in the infected areas, indicating translocation of organic nutrients from uninfected areas of the leaves or from healthy leaves toward the infected areas.

In some virus diseases, particularly the leaf-curling type and some

yellows diseases, starch accumulation in the leaves is a common phenomenon. In most of these diseases, starch accumulation in the leaves is
mainly the result of degeneration (necrosis) of the phloem of infected
plants which is one of the first symptoms of these diseases. It is also
possible, however, at least in some virus diseases, that the interference
with translocation of starch stems from inhibition by the virus of the
enzymes that break down starch into smaller translocatable molecules.
This is suggested by the observation that in some mosaic diseases, in
which there is no phloem necrosis, infected, discolored areas of leaves
contain less starch than "healthy," greener areas at the end of the day, a
period favorable for photosynthesis; but the same leaf areas contain more
starch than the "healthy" areas after a period in the dark, which favors
starch hydrolysis and translocation. This suggests that virus-infected
areas not only synthesize less starch than healthy ones, but also that
starch is not easily degraded and translocated from virus-infected areas,
although no damage to the phloem is present.

effect of pathogens
on host plant respiration

Respiration is the process by which cells, through enzymatically controlled
oxidation (burning) of the energy-rich carbohydrates and fatty acids, liberate energy in a form that can be utilized for the performance of various
cellular processes. Plant cells carry out respiration in, basically, two
steps. The first step involves the degradation of glucose to pyruvate and is
carried out, either in the presence or in the absence of oxygen, by enzymes
found in the ground cytoplasm of the cells. The production of pyruvate
from glucose follows either the glycolytic pathway, otherwise known as
glycolysis, or, to a lesser extent, the pentose pathway. The second step
involves the degradation of pyruvate, however produced, to CO_2 and
water. This is accomplished by a series of reactions known as the Krebs
cycle which is accompanied by the so-called terminal oxidation and is
carried out in the mitochondria only in the presence of oxygen. Under
normal (aerobic) conditions, that is, in the presence of oxygen, both steps
are carried out and one molecule of glucose yields, as final products, six
molecules of CO_2 and six molecules of water,

$$C_6H_{12}O_6 + 6 O_2 \longrightarrow 6 CO_2 + 6 H_2O$$

with concomitant release of energy (678,000 calories). Some of this energy
is lost, but almost half is converted to 20–30 reusable high-energy bonds of
adenosine triphosphate (ATP). The first step of respiration contributes two
ATP molecules per mole of glucose, and the second step contributes the
rest. Under anaerobic conditions, however—that is, in the absence of
oxygen—pyruvate cannot be oxidized but it instead undergoes fermentation and yields lactic acid or alcohol. Since the main process of energy
generation is cut off, for the cell to secure the necessary energy a much

greater rate of glucose utilization by glycolysis is required in the absence of oxygen than is in its presence.

The energy-storing bonds of ATP are formed by the attachment of a phosphate (PO_4) group to adenosine diphosphate (ADP), at the expense of energy released from the oxidation of sugars. The coupling of oxidation of glucose with the addition of phosphate to ADP to produce ATP is called oxidative phosphorylation. Any cell activity that requires energy utilizes the energy stored in ATP by simultaneously breaking down ATP to ADP and inorganic phosphate. The presence of ADP and phosphate in the cell, in turn, stimulate the rate of respiration. If, on the other hand, ATP is not utilized sufficiently by the cell for some reason, there is little or no regeneration of ADP and respiration is slowed down. The amount of ADP (and phosphate) in the cell is determined, therefore, by the rate of energy utilization; this, in turn, determines the rate of respiration in plant tissues.

The energy produced through respiration is utilized by the plant for all types of cellular work, such as accumulation and mobilization of compounds, synthesis of proteins, activation of enzymes, cell growth and division, defense reactions, and a host of other processes. The complexity of respiration, the number of enzymes involved in respiration, its occurrence in every single cell, and its far-reaching effects on the functions and existence of the cell, make it easy to understand why respiration of plant tissues is one of the first functions to be affected during infection by plant pathogens.

RESPIRATION
OF DISEASED PLANTS

When plants are infected by pathogens, the rate of respiration generally increases. The increase in respiration appears shortly after inoculation, certainly by the time of appearance of visible symptoms, and continues to rise during the multiplication and sporulation of the pathogen. After that, respiration declines to normal levels or to levels even lower than those of healthy plants. Respiration increases more rapidly in infections of resistant varieties, but it also declines quickly after it reaches its maximum. In susceptible varieties, respiration increases slowly after inoculation, but it continues to rise and it remains at a high level for much longer periods.

Several changes in the metabolism of the diseased plant accompany the increase in respiration following infection. Thus, the activity or concentration of several enzymes of, or related to, the respiratory pathways seem to be increased. The accumulation and oxidation of phenolic compounds are also greater during increased respiration. Increased respiration in diseased plants is also accompanied by an increased activation of the pentose pathway and, sometimes, by considerably more fermentation than observed in healthy plants.

The increased respiration in diseased plants is apparently brought about, at least in part, by uncoupling of the oxidative phosphorylation. In that case no utilizable energy (ATP) is produced through normal respiration in spite of use of the existing ATP and accumulation of ADP which stimulates respiration. The required energy by the cell for its vital processes is then produced through other, less efficient ways including the pentose pathway and fermentation.

The increased respiration of diseased plants can also be explained as the result of increased metabolism in the plant. In many plant diseases, growth is first stimulated, protoplasmic streaming increases, and materials are synthesized, translocated, and accumulated in the diseased area. The energy required for these activities derives from ATP produced through respiration. The more ATP is utilized, the more ADP is produced and further stimulates respiration. It is also possible that the plant, because of the infection, utilizes ATP energy less efficiently than a healthy plant. Because of the waste of part of the energy, an increase in respiration is induced and the resulting greater amount of energy enables the plant cells to utilize sufficient energy to carry out their accelerated processes.

Although oxidation of glucose via the glycolytic pathway is by far the most common way through which plant cells obtain their energy, part of the energy is produced via the pentose pathway. The latter seems to be an alternate pathway of energy production to which plants resort under conditions of stress. Thus, the pentose pathway tends to replace the glycolytic pathway as the plants grow older and differentiate, and to increase upon treatment of the plants with hormones, toxins, wounding, starvation, etc. Infection of plants with pathogens also tends, in general, to activate the pentose pathway over the level at which it operates in the healthy plant. Since the pentose pathway is not directly linked to ATP production, the increased respiration through this pathway fails to produce as much utilizable energy as the glycolytic pathway and is, therefore, a less efficient source of energy for the functions of the diseased plant. On the other hand, the pentose pathway is the main source of phenolic compounds which play important roles in the defense mechanisms of the plant against infection.

SELECTED REFERENCES

Allen, R. J. 1942. Changes in the metabolism of wheat leaves induced by infection with powdery mildew. *Am. J. Bot.* **29**:425–435.

Daly, J. M. 1967. Some metabolic consequences of infection by obligate parasites, *in* "The Dynamic Role of Molecular Constituents in Plant–Parasite Interaction" (C. J. Mirocha and I. Uritani, eds.), pp. 144–164. Bruce, St. Paul, Minnesota.

Dimond, A. E. 1970. Biophysics and biochemistry of the vascular wilt syndrome. *Ann. Rev. Phytopathol.* **8**:301–322.

Duniway, J. M. 1971. Water relations in *Fusarium* wilt in tomato. *Physiol. Plant Pathol.* **1**:537–546.

Hampson, M. C., and W. A. Sinclair. 1973. Xylem dysfunction in peach caused by *Cytospora leucostoma. Phytopathology* **63**:676–681.

Kuć, J. 1967. Shifts in oxidative metabolism during pathogenesis, *in* "The Dynamic Role of Molecular Constituents in Plant–Parasite Interaction" (C. J. Mirocha and I. Uritani, eds.), pp. 183–202. Bruce, St. Paul, Minnesota.

Livne, A. 1964. Photosynthesis in healthy and rust-affected plants. *Plant Physiol.* **39**:614–621.

Livne, A., and J. M. Daly. 1966. Translocation in healthy and rust-infected beans. *Phytopathology* **56**:170–175.

Millerd, A., and K. J. Scott. 1962. Respiration of the diseased plant. *Ann. Rev. Plant Physiol.* **13**:559–574.

Nelson, P. E., and R. S. Dickey. 1970. Histopathology of plants infected with vascular bacterial pathogens. *Ann. Rev. Phytopathol.* **8**:259–280.

Roberts, D. A., and M. K. Corbett. 1965. Reduced photosynthesis in tobacco plants infected with tobacco ringspot virus. *Phytopathology* **55**:370–371.

Sempio, C. 1959. The host is starved, *in* "Plant Pathology" (J. G. Horsfall and A. E. Dimond, eds.), Vol. 1, pp. 278–312. Academic Press, New York.

Uritani, I., and T. Akazawa. 1959. Alteration of the respiratory pattern in infected plants, *in* "Plant Pathology" (J. G. Horsfall and A. E. Dimond, eds.), Vol. 1, pp. 349–390. Academic Press, New York.

Wheeler, H. 1975. "Plant Pathogenesis." Springer-Verlag, New York, 106 p.

5
how plants defend themselves against pathogens

Each plant species is affected by approximately one hundred different kinds of fungi, bacteria, mycoplasmas, viruses, nematodes, etc. Frequently, a single plant is attacked by hundreds, thousands, and, in the leafspot diseases of large trees, probably by hundreds of thousands of individuals of a single kind of pathogen. Yet, although such plants may suffer damage to a lesser or greater extent, many survive all these attacks and, not uncommonly, manage to grow well and to produce appreciable yields.

In general, plants defend themselves against pathogens either by means of structural characteristics that act as physical barriers and inhibit the pathogen from gaining entrance and spreading through the plant, or by means of biochemical reactions that take place in cells and tissues of the plant and produce substances that are toxic to the pathogen or create conditions that inhibit the growth of the pathogen in the plant.

structural defense

PREEXISTING DEFENSE STRUCTURES

The first line of defense of plants against pathogens is their surface which the pathogen must penetrate if it is to cause infection. Some structural defenses are present in the plant even before the pathogen comes in contact with the plant. Such structures include the amount and quality of wax and cuticle that cover the epidermal cells, the structure of the epidermal cell walls, the size, location, and shapes of stomata and lenticels, and the presence on the plant of tissues made of thick-walled cells that hinder the advance of the pathogen.

Waxes on leaf and fruit surfaces form a water-repellent surface and thereby prevent the formation on the tissue of a film of water on which pathogens might be deposited and germinate (fungi) or multiply (bacteria). A thick mat of hairs on a plant surface may also, conceivably, exert a similar water-repelling effect and may reduce infection.

Cuticle thickness may increase resistance to infection in diseases in which the pathogen enters its host only through direct penetration. Cuticle thickness, however, is not always correlated with resistance and many plant varieties with cuticle of considerable thickness are easily invaded by directly penetrating pathogens.

The thickness and toughness of the outer wall of epidermal cells are apparently important factors in the resistance of some plants to certain pathogens. Thick, tough walls of epidermal cells make direct penetration by fungal pathogens difficult or impossible. Plants with such walls are often resistant, although, if the pathogen is introduced beyond the epidermis of the same plants by means of a wound, the inner tissues of the plant are easily invaded by the pathogen.

Many pathogenic fungi and bacteria enter plants only through stomata. Although the majority of them can force their way through closed stomata, some, like the stem rust of wheat, can enter only when stomata are open. Thus, some wheat varieties, in which the stomata open late in the day, are resistant because the germ tubes of spores germinating in the night dew desiccate owing to evaporation of the dew before the stomata begin to open. The kind of structure of stomata, e.g., a very narrow entrance and broad, elevated guard cells, may also confer resistance to some varieties against certain of their pathogens.

The thickness and toughness of the cell walls of the tissues being invaded vary and may sometimes make the advance of the pathogen quite difficult. The presence, in particular, of bundles or extended areas of sclerenchyma cells, such as are found in the stems of many cereal crops, may stop the further spread of pathogens like the stem rust fungi. Also, the xylem, bundle sheath, and sclerenchyma cells of the leaf veins effectively block the spread of some fungal, bacterial, and nematode pathogens which thus cause the various "angular" leaf spots because of their spread only into areas between, but not across, veins.

DEFENSE STRUCTURES FORMED IN RESPONSE TO INFECTION BY THE PATHOGEN

Although some pathogens may be blocked from entering or from invading their host plants by the preformed superficial or internal defense structures, most pathogens manage to penetrate their hosts and to produce various degrees of infection. Even after the pathogen has penetrated the preformed defense structures, however, plants exhibiting various degrees of resistance usually respond by forming one or more types of structures that are more or less successful in defending the plant from further invasion by the pathogen. Some of the defense structures formed involve tissues ahead of the pathogen and are called histological defense structures; others involve the walls of invaded cells and are called cellular defense structures; still others involve the cytoplasm of the cells under attack and the process is called cytoplasmic defense reaction. Finally, death of the invaded cell may protect the plant from further invasion and this is called necrotic or hypersensitive defense reaction.

HISTOLOGICAL DEFENSE STRUCTURES

FORMATION OF CORK LAYERS   Infection of plants by fungi or bacteria and even by some viruses and nematodes frequently induces formation of several layers of cork cells beyond the point of infection (Figs. 14 and 15), apparently as a result of stimulation of the host cells by substances secreted by the pathogen. The cork layers not only inhibit the further invasion by the pathogen beyond the initial lesion but also block the spread of any toxic substances that the pathogen may secrete. Furthermore, cork layers stop the flow of nutrients and water from the healthy to the infected area and deprive the pathogen of nourishment. The dead tissues, including the pathogen, are thus delimited by the cork layers and either remain in place forming a necrotic lesion (spot) or are pushed outward by the underlying healthy tissues and form scabs that may further be sloughed off and thus remove the pathogen from the host completely.

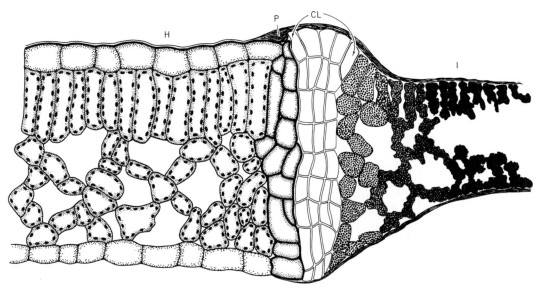

FIGURE 14.
Formation of cork layer between infected and healthy areas of leaf. CL = cork layer; H = healthy leaf area; I = infected; P = phellogen. (After Cunningham, 1928.)

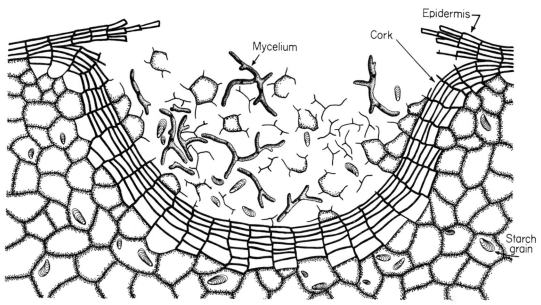

FIGURE 15.
Formation of cork layer on potato tuber following infection with *Rhizoctonia*.
[After G. E. Ramsey (1917). *J. Agr. Res.* **9:**421–426.]

FORMATION OF ABSCISSION LAYERS   Abscission layers are formed on
young, active leaves of stone fruit trees following infection by any of
several fungi, bacteria, or viruses. An abscission layer consists of a gap
between two circular layers of cells of a leaf surrounding the locus of
infection. Upon infection, the middle lamella between these two layers of
cells is dissolved throughout the thickness of the leaf completely cutting
off the central area from the rest of the leaf (Fig. 16). Gradually this area
shrivels, dies, and sloughs off, carrying with it the pathogen. Thus, the
plant, by discarding the infected area along with a few yet uninfected
cells, protects the rest of the leaf tissue from becoming invaded by the
pathogen and from becoming affected by the toxic secretions of the
pathogen.

FORMATION OF TYLOSES   Tyloses form in xylem vessels of most plants
under various conditions of stress and during invasion by most of the
vascular pathogens. Tyloses are overgrowths of the protoplast of adjacent
living parenchymatous cells which protrude into xylem vessels through
pits (Fig. 17). Tyloses have cellulosic walls and may, by their size and
numbers, clog the vessel completely. In some varieties, tyloses form
abundantly and quickly ahead of the pathogen while the pathogen is still
in the young roots, block the further advance of the pathogen, and the
plants of these varieties remain free of, and, therefore, resistant to this
pathogen. Varieties in which few, if any, tyloses form ahead of the pathogen are susceptible to the disease.

DEPOSITION OF GUMS   Various types of gums are produced by many
plants around lesions following infection by pathogens or injury. Gum

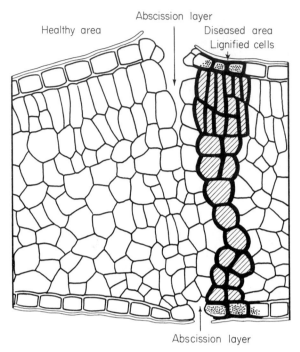

FIGURE 16.
Formation of abscission layer around a diseased spot of a *Prunus* leaf. (After Samuel, 1927.)

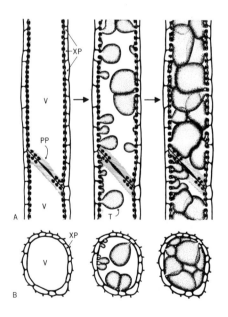

FIGURE 17.
Development of tyloses in xylem vessels. Longitudinal (A) and cross-section (B) views of healthy vessels (left), and of vessels with tyloses. Vessels on right are completely clogged with tyloses. PP = perforation plate; V = xylem vessel; XP = xylem parenchyma cell; T = tylosis.

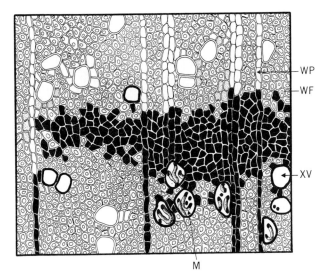

FIGURE 18.
Gum barrier in apple twig infected with *Physalospora cydoniae*. M = mycelium in vessels; XV = xylem vessel; WF = wood fiber; WP = wood parenchyma. (After Hesler, 1916.)

secretion is most common in stone fruit trees but occurs in most plants. The defensive role of gum stems from the fact that they are quickly deposited in the intercellular spaces and within the cells surrounding the locus of infection, thus forming an impenetrable barrier which completely encloses the pathogen (Fig. 18). The pathogen then becomes isolated, starved, and sooner or later dies.

CELLULAR DEFENSE STRUCTURES

The cellular defense structures involve morphological changes in the cell wall, or derived from the cell wall, of the cell being invaded. The effectiveness of these structures as defense mechanisms seems to be rather limited, however. Two main types of such structures have been observed in fungal diseases: (a) swelling of the cell wall of epidermal and subepidermal cells during direct penetration which may inhibit host penetration and establishment of infection by the pathogen, and (b) sheathing of hyphae penetrating a cell wall by enveloping them in a sheath formed by the extension of the cell wall inward in a way that surrounds and precedes the invading hypha (Fig. 19).

CYTOPLASMIC DEFENSE REACTION

In a few cases of slowly growing, weakly pathogenic fungi that induce chronic diseases or nearly symbiotic conditions, the cytoplasm invests the clump of hyphae and the nucleus is stretched to the point where it breaks in two. In some cells, the cytoplasmic reaction is overcome and the protoplast disappears while fungal growth increases. In some of the invaded cells, however, the cytoplasm and nucleus enlarge. The cyto-

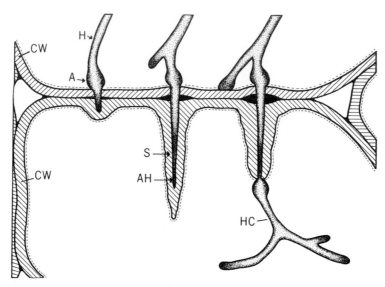

FIGURE 19.
Formation of sheath around hypha penetrating a cell wall. CW = cell wall; H = hypha; A = appressorium; AH = advancing hypha still enclosed in sheath; HC = hypha in cytoplasm; S = sheath.

plasm becomes granular and dense, and various particles or structures appear in it. Finally, the mycelium of the pathogen disintegrates and the advance of the invasion stops.

NECROTIC DEFENSE REACTION: DEFENSE THROUGH HYPERSENSITIVITY

In many host–pathogen combinations, the pathogen may penetrate the cell wall, but as soon as it establishes contact with the protoplast of the cell, the nucleus moves toward the intruding pathogen and soon disintegrates, and brown, resinlike granules form in the cytoplasm, first around the pathogen and then throughout the cytoplasm. As the browning discoloration of the cytoplasm of the plant cell continues and death sets in, the invading hypha begins to degenerate (Fig. 20). In most cases the hypha does not grow out of such cells and further invasion is stopped.

The necrotic or hypersensitive type of defense is very common, particularly in diseases caused by obligate fungal parasites and by viruses and nematodes. Apparently, the necrotic tissue isolates the obligate parasite from the living substance, on which it depends absolutely for its nutrition for growth and multiplication, and, therefore, results in its starvation and death. The faster the host cell dies following invasion the more resistant to infection the plant seems to be.

SELECTED REFERENCES

Akai, S. 1959. Histology of defense in plants, *in* "Plant Pathology" (J. G. Horsfall and A. E. Dimond, eds.), Vol. 1, pp. 391–434. Academic Press, New York.

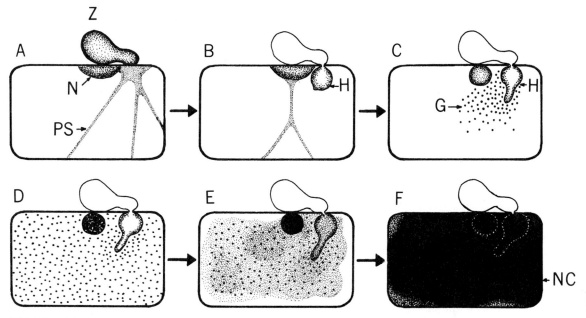

FIGURE 20.
Stages in the development of necrotic defense reaction in cell of a very resistant
potato variety infected by *Phytophthora infestans*. N = nucleus; PS =
protoplasmic strands; Z = zoospore; H = hypha; G = granular material; NC =
necrotic cell. [After K. Tomiyama (1956). *Ann. Phytopathol. Soc. Japan*
21:54–62.]

Cunningham, H. S. 1928. A study of the histologic changes induced in leaves by
    certain leaf-spotting fungi. *Phytopathology* **18**:717–751.
Hart, Helen. 1929. Relation of stomatal behaviour to stem-rust resistance in
    wheat. *J. Agr. Res.* **39**:929–948.
Hart, Helen, 1931. Morphologic and physiologic studies on stem-rust resistance
    in cereals. *U. S. Dept. Agr. Minn. Agr. Expt. Sta. Tech. Bull.* **266**:76 pp.
Hesler, L. R. 1916. Black rot, leaf spot, and canker of pomaceous fruits. *N. Y.
    (Cornell) Agr. Expt. Sta. Bull.* **379**:53–148.
Martin, J. T. 1964. Role of cuticle in the defense against plant disease. *Ann. Rev.
    Phytopathol.* **2**:81–100.
Muller, K. O. 1959. Hypersensitivity, *in* "Plant Pathology" (J. G. Horsfall and A. E.
    Dimond, eds.), Vol. 1, pp. 469–519. Academic Press, New York.
Samuel, G. 1927. On the shot-hole disease caused by *Cladosporium carpophilum*
    and on the "shothole" effect. *Ann. Botany (London)* **41**:375–404.
Weimer, J. L., and L. L. Harter. 1921. Wound-cork formation in the sweet potato. *J.
    Agr. Res.* **21**:637–647.

biochemical defense

Although structural characteristics may provide a plant with various degrees of defense against attacking pathogens, it is becoming increasingly
clear that the resistance of a plant against pathogen attacks depends not
so much on its structural barriers as on the substances produced in its cells
preceding or following infection. This becomes apparent from the fact

that a particular pathogen will not infect certain plant varieties although no structural barriers of any kind seem to be present or to form in these varieties. Similarly, in resistant varieties, the rate of disease development soon slows down and, finally, in the absence of structural defenses, the disease is completely checked. Moreover, many pathogens which enter nonhost plants naturally, or which are introduced into nonhost plants artificially, fail to cause infection although no apparent visible host structures inhibit them from doing so. These examples suggest that defense mechanisms of a chemical rather than a structural nature are responsible for the resistance to infection exhibited by plants against certain pathogens.

PREEXISTING BIOCHEMICAL DEFENSE

INHIBITORS RELEASED BY THE PLANT IN ITS ENVIRONMENT

Plants, generally, exude a variety of substances through the surface of their aboveground parts as well as through the surface of their roots. Some of the compounds released by certain kinds of plants, however, seem to have an inhibitory action against certain pathogens. Fungitoxic exudates on leaves of some plants seem to be present in sufficient concentrations to inhibit germination of fungal spores present in dew or rain droplets on these leaves. Similarly, in the presence of water drops or soil moisture containing conidia of the onion smudge fungus on the surface of red onions, fungitoxic substances diffuse into the liquid, inhibit the germination of the conidia and cause them to burst, thus protecting the plant from infection. Both the fungitoxic exudates and the inhibition of infection are missing in the white-scaled, susceptible onion varieties.

INHIBITORS PRESENT IN PLANT CELLS BEFORE INFECTION

It is uncertain whether any plant is resistant to a disease because of an inhibitory compound present in the cell before infection. Chlorogenic acid, a phenolic compound toxic to many microorganisms, may be such an inhibitor in some diseases. Thus, in the case of the potato scab, caused by *Streptomyces scabies*, tubers of resistant varieties contain higher concentrations of chlorogenic acid than do tubers of susceptible varieties. The concentration of chlorogenic acid in resistant varieties is especially high in tissues through which the pathogen enters (lenticels) and in which it normally grows (outer layers of tuber). Also, roots of certain potato varieties resistant to *Verticillium* contain more chlorogenic acid than do roots of susceptible varieties. Moreover, even susceptible varieties are not attacked while young, when their roots contain high concentrations of chlorogenic acid, but become susceptible later, when their content in chlorogenic acid declines.

DEFENSE THROUGH
DEFICIENCY IN NUTRIENTS
ESSENTIAL FOR THE
PATHOGEN

Species or varieties of plants that for some reason do not produce one of the substances essential for the survival of an obligate parasite or for development of infection by any parasite would be resistant to the pathogen that requires it. Thus, for *Rhizoctonia* to infect a plant, the plant must have a substance necessary for formation of a hyphal cushion from which the fungus sends into the plant its penetration hyphae. In plants in which this substance is apparently lacking, cushions do not form, infection does not occur, and the plants are resistant. The fungus does not normally form hyphal cushions in pure cultures, but forms them when extracts from a susceptible but not a resistant plant are added to the culture. Also, certain mutants of *Venturia inaequalis,* the cause of apple scab, which had lost the ability to synthesize a certain growth factor, also lost the ability to cause infection. When, however, the particular growth factor is sprayed on the apple leaves during inoculation with the mutant, the mutant not only survives, but it also causes infection. The advance of the infection though continues only as long as the growth factor is supplied to the mutant externally.

DEFENSE THROUGH
ABSENCE OF COMMON ANTIGENS

Plants do not produce antibodies against invading pathogens such as fungi, bacteria, or viruses, but some kind of immunological response may also be operating in plants. When the antigens of a given number of races of a pathogen are compared with the antigens of the same number of plant varieties, each of which is infected by one or more of the pathogen races, it can be shown that a specific antigen in each of the pathogen races is commonly shared by only those plant varieties that are susceptible to a particular race. When a given variety does not have an antigen that is present in a particular pathogen race, the variety is resistant to that race, suggesting that susceptibility and resistance are due to the presence or absence of the specific pathogen antigens in the plant varieties.

BIOCHEMICAL DEFENSE
INDUCED BY THE ATTACKING
PATHOGEN

BIOCHEMICAL INHIBITORS
PRODUCED IN PLANTS IN RESPONSE
TO INJURY BY THE PATHOGEN

Plant cells and tissues respond to injury, whether caused by a pathogen, or mechanical or chemical agent, through a series of biochemical reactions which seem to be aimed at isolating the irritant and at healing the wound. This reaction is often associated with the production of fungitoxic substances around the site of injury as well as formation of layers of protective tissue such as callus and cork. Some of the compounds thus produced are present in concentrations high enough to inhibit growth of most fungi and bacteria that cannot infect that host. These compounds include mostly phenolic compounds such as chlorogenic and caffeic acids, oxidation products of phenolic compounds, and also the phyto-alexins, most of which are also phenolic compounds.

ROLE OF PHENOLIC COMPOUNDS   Some of the phenolics implicated in disease resistance occur widely in plants and are found in healthy as well as diseased plants, but their synthesis or accumulation seems to be accelerated following infection. Such compounds may be called "common" phenolic compounds. Certain other phenolics, however, are not present in healthy plants but are produced upon stimulation of a plant by a pathogen or by a mechanical or chemical injury. Such compounds are known as phytoalexins.

"Common" phenolics.   It has often been observed that certain "common" phenolic compounds that are toxic to pathogens are produced and accumulate at a faster rate after infection in a resistant variety than in a susceptible variety. Examples of such phenolic compounds are chlorogenic acid, caffeic acid, scopoletin, etc. Although some of the common phenolics may each reach concentrations that could be toxic to the pathogen, it should be noted that several of them appear concurrently in the same diseased tissue and it is possible that the combined toxic effect of all fungitoxic phenolics present, rather than that of each one separately, is responsible for the inhibition of infection in resistant varieties.

Phytoalexins.   Phytoalexins are fungitoxic substances produced in appreciable amounts in plants only after stimulation by microorganisms, or chemical and mechanical injury, and inhibit the growth of microorganisms pathogenic to plants. They include several compounds such as ipomeamarone, orchinol, pisatin, phaseolin, and rishitin.

Phytoalexins in general are not produced by healthy plants but are produced by plants following infection, injury, or at least stimulation by certain fungal, but not bacterial, secretions. Fungi pathogenic to a particular plant species seem to stimulate production of generally lower concentrations of phytoalexins than nonpathogens and, besides, pathogenic fungi seem to be less sensitive to the toxicity of the phytoalexin produced by their host plant than are nonpathogenic fungi. For example, in the case of pisatin production by pea pods inoculated with the pathogen *Ascochyta pisi,* different varieties of pea produce different concentrations of pisatin which approximately parallel the resistance of the variety to the pathogen. When the same pea variety is inoculated with different strains of the fungus, the concentration of pisatin produced varies with the fungus strain used for inoculation and it is, approximately, inversely proportional to the virulence of each particular strain on the pea variety.

Fungitoxic phenolics released from nontoxic phenolic complexes. Several fungi are known to produce or to liberate from plant tissues an enzyme that can hydrolyze complex phenolic molecules and release the phenolic compound from the complex. Some of these phenolics are quite toxic to the pathogen and appear to play a role in the defense of the plant against infection.

Role of phenol-oxidizing enzymes in disease resistance.   The activity of many phenol-oxidizing enzymes is generally higher in the infected tissue of resistant varieties than in the infected susceptible ones or the uninfected healthy plants. The importance of polyphenoloxidase activity in disease resistance probably stems from its property to oxidize phenolic compounds to quinones which are often more toxic to microorganisms than the original phenols. It is reasonable to assume that an increased activity of polyphenoloxidases will result in higher concentrations of toxic products of oxidation and therefore in greater degrees of resistance to infection.

DEFENSE THROUGH INDUCED SYNTHESIS OF PROTEINS AND ENZYMES Pathogen attacks on plants appear to induce alterations in protein synthesis in the plant that can lead to the development of a local resistant or immune layer around infection sites. The resistance or immunity of plants to a pathogen may depend on the speed and extent of protein synthesis induced in the host by the pathogen or closely related nonpathogens. This type of defense seems to be related to that afforded by phytoalexins, the additional proteins or enzymes being those required for synthesis of phytoalexins, although it is possible that the two mechanisms operate separately.

DEFENSE THROUGH FORMATION OF SUBSTRATES RESISTING THE ENZYMES OF THE PATHOGEN   Plant resistance to some pathogens is apparently due to the presence or appearance of compounds which are not easily degraded by the enzymes of pathogens attempting to invade the plant. These compounds are usually complexes between pectins, proteins, and polyvalent cations such as calcium or magnesium. The availability or accumulation of either cation near the infection results in formation of pectic salts or other complexes that resist degradation by the pathogen enzymes. Thus they inhibit tissue maceration and confine the pathogen to lesions of limited size.

DEFENSE THROUGH INACTIVATION OF PATHOGEN ENZYMES Several phenolic compounds or their oxidation products seem to induce resistance to disease through their inhibitory action on pathogenic enzymes rather than on the pathogen itself. In some diseases the more resistant the varieties the higher is their content in polyphenols and although these phenols do not inhibit the growth of the pathogen, they do inhibit the activity of its pectinolytic enzymes and apparently contribute to the resistance of the plant.

DEFENSE THROUGH DETOXIFICATION OF PATHOGEN TOXINS   In at least some of the diseases in which the pathogen produces a toxin, resistance to disease is apparently the same as resistance to toxin. However, no satisfactory explanation of the resistance to toxin is yet available.

Detoxification of at least some toxins, e.g., fusaric acid, pyricularin, etc., is known to occur in plants and to play a role in disease resistance. These toxins are rapidly metabolized by resistant varieties or are combined with other substances and form nontoxic compounds. The amount of the nontoxic compound formed is often proportional to the disease resistance of the variety.

Resistant plants and nonhosts are not affected by the specific toxins produced by *Helminthosporium, Periconia,* and *Alternaria,* but it is not yet known whether the selective action of these toxins is dependent upon the presence of receptor sites in susceptible but not in resistant varieties or on the detoxification of the toxins in resistant plants.

DEFENSE THROUGH ALTERED RESPIRATION   Following infection, resistant varieties often show a greater initial increase in respiration than do susceptible varieties, but they also show a decline in respiration within a few days after the infection, whereas the susceptible varieties do not. The increased respiration in infected tissues indicates a general acceleration of the metabolism of the host which is apparently a necessary condition for the development of the defense reaction in the host.

DEFENSE THROUGH ALTERED BIOSYNTHETIC PATHWAYS   Injury or infection of plants triggers a physiological condition of stress, during which respiration is often increased and several enzymes are activated. Under some stress conditions, new enzyme proteins are produced and compounds peculiar to stress physiology are rapidly synthesized and accumulate in concentrations that are toxic to many microorganisms. Infection or wounding also causes a shift from the glycolytic to the pentose pathway, which, in turn, provides a substance necessary for the production of most phenolic compounds that are toxic to pathogens.

DEFENSE THROUGH THE HYPERSENSITIVE REACTION   The hypersensitive reaction is one of the most important defense mechanisms in plants. It occurs only in incompatible combinations of host plants with fungi, bacteria, viruses, and nematodes. In such combinations, no difference is observable in the manner of penetration of epidermis in susceptible and in resistant plants. Following infection, however, loss of turgor, browning, and death of infected cells occur rapidly in resistant varieties while infected cells of susceptible varieties can survive considerably longer. In resistant varieties, a number of physiological changes occur in the infected cells and in the cells surrounding them, while in susceptible varieties such changes either do not occur or they occur at a much slower rate. Such changes in hypersensitive reactions include loss of permeability of cell membranes, increased respiration, accumulation and oxidation of phenolic compounds, production of phytoalexins, and others. The end result of all these intermediate stages is always death and collapse of the infected and, perhaps, a few surrounding cells. Fungal and bacterial pathogens within the area of operation of the hypersensitive reaction are isolated by necrotic tissue and quickly die. In virus diseases, the hypersensitive reaction always results in formation of the so-called local lesions in which the virus may survive for considerable time but is, generally, found in low concentrations and its spread beyond the lesion is, as a rule, checked.

SELECTED REFERENCES

Akai, S., and S. Ouchi (eds.). 1971. "Morphological and Biochemical Events in Plant–Parasite Interaction." The Phytopathol. Soc. of Japan, Tokyo. 415 p.

Bateman, D. F. 1967. Alteration of cell wall components during pathogenesis by *Rhizoctonia solani*, *in* "The Dynamic Role of Molecular Constituents in Plant–Parasite Interaction" (C. J. Mirocha and I. Uritani, eds.), pp. 58–75. Bruce, St. Paul, Minnesota.

DeVay, J. E., W. C. Schnathorst, and M. S. Foda. 1967. Common antigens and host-parasite interactions, *in* "The Dynamic Role of Molecular Constituents in Plant–Parasite Interaction" (C. J. Mirocha and I. Uritani, eds.), pp. 313–325. Bruce, St. Paul, Minnesota.

Deverall, B. J. 1977. "Defense Mechanisms of Plants." Cambridge Univ. Press, Cambridge. 110 p.

Farkas, G. L., and Z. Király. 1962. Role of phenolic compounds in the physiology of plant disease and disease resistance. *Phytopathol. Z.* **44**:105–150.

Heitefuss, R. and P. H. Williams. (eds.). 1976. "Physiological Plant Pathology". Encyclopedia of Plant Physiology, New Series, Vol. 4. Springer-Verlag, New York. 890 p.

Klement, Z., and R. N. Goodman. 1967. The hypersensitive reaction to infection by bacterial plant pathogens. *Ann. Rev. Phytopathol.* **5**:17–44.

Kosuge, T. 1969. The role of phenolics in host response to infection. *Ann. Rev. Phytopathol.* **7**:195–222.

Kuć, J. 1966. Resistance of plants to infectious agents. *Ann. Rev. Microbiol.* **20**:337–370.

Kuć, J. 1972. Phytoalexins. *Ann. Rev. Phytopathol.* **10**:207–232.

Muller, K. O. 1959. Hypersensitivity. *In* "Plant Pathology" (J. G. Horsfall and A. E. Dimond, eds.), Vol. 1, pp. 469–519. Academic Press, New York.

Scheffer, R. P., and R. B. Pringle. 1967. Pathogen-produced determinants of disease and their effects on host plants, *in* "The Dynamic Role of Molecular Constituents in Plant–Parasite Interaction" (C. J. Mirocha and I. Uritani, eds.), pp. 217–234. Bruce, St. Paul, Minnesota.

Schoeneweiss, D. F. 1975. Predisposition, stress and plant disease. *Ann. Rev. Phytopathol.* **13**:193–211.

Stahmann, M. A. 1967. Influence of host–parasite interactions on proteins, enzymes, and resistance, *in* "The Dynamic Role of Molecular Constituents in Plant–Parasite Interaction" (C. J. Mirocha and I. Uritani, eds.), pp. 357–369. Bruce, St. Paul, Minnesota.

Tomiyama, D. 1963. Physiology and biochemistry of disease resistance of plants. *Ann. Rev. Phytopathol.* **1**:295–324.

Uritani, I. 1971. Protein changes in diseased plants. *Ann. Rev. Phytopathol.* **9**:211–234.

Wheeler, H. 1975. "Plant Pathogenesis." Springer-Verlag, New York. 106 p.

Wilson, C. L. 1973. A lysosomal concept for plant pathology. *Ann. Rev. Phytopathol.* **11**:247–272.

6
genetics
and
plant
disease

introduction

One of the most dynamic and significant aspects of biology is that characteristics of individuals within a species are not "fixed" in their morphology and physiology but vary from one individual to another. As a matter of fact, all individuals produced as a result of a sexual process are expected to be different from each other and from their parents in a number of characteristics, although they retain most similarities with them and belong to the same species. This is true of fungi produced from sexual spores such as oospores, ascospores, and basidiospores, of parasitic higher plants produced from seeds and of nematodes produced from fertilized eggs, as well as of cultivated plants produced from seeds. When individuals are produced asexually, the frequency and degree of variability among the progeny are reduced greatly but even then certain individuals among the progeny will show different characteristics. This is the case in the overwhelmingly asexual reproduction of fungi by means of conidia, zoospores, sclerotia, uredospores, etc., and in bacteria, mycoplasmas and viruses, as well as in the asexual propagation of plants by means of buds, cuttings, tubers, etc.

mechanisms of variability

In host plants and in pathogens, such as most fungi, parasitic higher plants, and nematodes, which can, and usually do, reproduce by means of a sexual process, variation in the progeny is introduced primarily through

86

segregation and recombination of genes during the meiotic division of the zygote. Bacteria too, however, and even viruses, exhibit variation which seems to be the result of a sexuallike process. In many fungi certain parasexual processes lead to variation. On the other hand, all plants and all pathogens, especially bacteria, viruses, and fungi, and probably mycoplasmas, can and do produce variants in the absence of any sexual process by means of mutations and, perhaps, by means of cytoplasmic adaptation.

GENERAL
MECHANISMS OF VARIABILITY

Three mechanisms of variability—hybridization, mutation, and cytoplasmic inheritance—occur in both plants and pathogens.

HYBRIDIZATION

Hybridization occurs during sexual reproduction of plants, fungi, and nematodes whenever two haploid (1N) nuclei, containing slightly different genetic material, unite to form a diploid (2N) nucleus, called a zygote. In hybridization, a recombination of genetic factors occurs during the meiotic division of the zygote as a result of genetic crossovers in which parts of chromatids (and the genes they carry) of the one chromosome of the pair are exchanged with parts of chromatids of the other chromosome of the pair. In this way a recombination of the genes of the two parental nuclei takes place in the zygote, and the haploid nuclei or gametes resulting after meiosis are different both from gametes that produced the zygote and from each other. In the fungi, the haploid nuclei or gametes often divide mitotically to produce mycelium and spores which result in genetically different groups of homogeneous individuals that may produce large populations asexually until the next sexual cycle.

MUTATION

Mutation is a more or less abrupt change in the genetic material of an organism, which is then transmitted in a hereditary fashion to the progeny. Mutations occur spontaneously in nature in all living organisms, those that reproduce only sexually or only asexually and those that reproduce both sexually and asexually. Mutations in single-celled organisms, such as bacteria, in fungi with haploid mycelium, and in viruses, are expressed immediately after their occurrence. Most mutant factors, however, are usually recessive; therefore, in diploid or dikaryotic organisms mutations can remain unexpressed until they are brought together in a hybrid.

Mutations for virulence probably occur no more frequently than for any other inherited characteristic but, given the great number of progeny produced by pathogens, it is probable that large numbers of mutants differing in virulence from their parent appear in nature every year. Besides, considering that only a few genetically homogeneous varieties of each crop plant are planted continuously over enormous land expanses for a number of years, and considering the difficulties involved in shifting from one variety to another on short notice, the threat of new, more virulent, mutants appearing and attacking a previously resistant variety is a real one. Moreover, once a new factor for virulence appears in a mutant, this factor will take part in the sexual or parasexual processes of the pathogen and may produce recombinants possessing virulence quite different in degree or nature from that existing in the parental strains.

CYTOPLASMIC INHERITANCE

Cytoplasmic inheritance is the acquisition by a plant or a pathogen, through extrachromosomal inheritance, of the ability to carry out a physiological process which it could not before. Cytoplasmic inheritance presumably occurs in all organisms except viruses and viroids, which lack cytoplasm. Three types of adaptations brought about by changes in the genetic material of the cytoplasm have been shown in pathogens. Pathogens may acquire the ability to tolerate previously toxic substances, to utilize new substances for growth, and to change their virulence toward host plants. Several characteristics of plants are also inherited through the cytoplasm, including the resistance to infection by certain pathogens.

SPECIALIZED MECHANISMS OF VARIABILITY IN PATHOGENS

Certain mechanisms of variability appear to be operating only in certain kinds of organisms or to be operating in a rather different manner than those described as general mechanisms of variability. These specialized mechanisms of variability are sexuallike or parasexual processes and include heterokaryosis and parasexualism in fungi; conjugation, transformation and transduction in bacteria; and genetic recombination in viruses.

SEXUALLIKE PROCESSES IN FUNGI

HETEROKARYOSIS   Heterokaryosis is the condition in which, as a result of fertilization or anastomosis, cells of fungus hyphae or parts of hyphae contain nuclei that are genetically different. In the Basidiomycetes, the dikaryotic state may differ drastically from the haploid mycelium and spores of the fungus. Thus in *Puccinia graminis tritici*, the fungus causing stem rust of wheat, the haploid basidiospores can infect barberry but not wheat, and the haploid mycelium can grow only in barberry, while the dikaryotic aeciospores and uredospores can infect wheat but not barberry and the dikaryotic mycelium can grow in both barberry and wheat. Heterokaryosis also occurs in other fungi but its importance in plant disease development in nature is not known.

PARASEXUALISM   Parasexualism is the process by which genetic recombinations can occur within fungal heterokaryons. This comes about by the occasional fusion of the two nuclei and the formation of a diploid nucleus. During multiplication, crossing over occurs in a few mitotic divisions and results in the appearance of genetic recombinants by the occasional separation of the diploid nucleus into its haploid components.

SEXUALLIKE PROCESSES IN BACTERIA

New biotypes of bacteria seem to arise with varying frequency by means of at least three sexuallike processes (Fig. 21). It is probable that similar processes occur in mycoplasmas and rickettsialike organisms. (1) *Conjugation,* in which two compatible bacteria come in contact with each other and a small portion of the chromosomal or nonchromosomal genetic material of the one bacteriun is transferred to the genetic material of the other. (2) *Transformation,* in which bacterial cells are transformed genetically by absorbing and incorporating in their own cells genetic material secreted by, or released during rupture of, other compatible bacteria. (3) *Transduction,* in which a bacterial virus (phage) transfers genetic material from the bacterium in which the phage was produced to the bacterium it infects next.

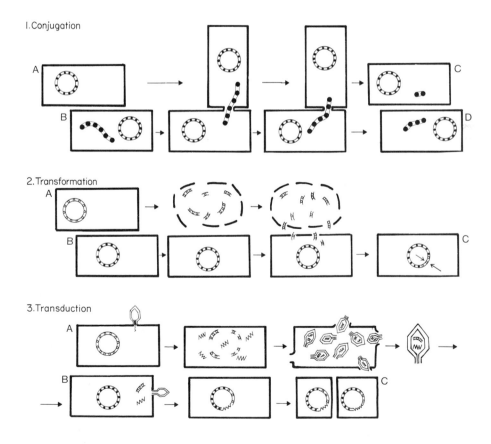

FIGURE 21.
Mechanisms of variability in bacteria through sexuallike processes.

GENETIC
RECOMBINATION IN VIRUSES

When two strains of the same virus are inoculated into the same host plant, one or more new virus strains are recovered with properties (virulence, symptomatology, etc.) different from those of either of the original strains introduced into the host. The new strains probably are hybrids (recombinants) although their appearance through mutation, not hybridization, cannot always be ruled out.

In multicomponent viruses consisting of 2 to 4 components, new virus strains may also arise in host plants or vectors from recombination of the appropriate components of two or more strains of such viruses.

stages of variation in pathogens

The entire population of a particular organism on earth, e.g., a fungal pathogen, has certain morphological characteristics in common and comprises the species of the pathogen, e.g., *Puccinia graminis*, the cause of stem rust of cereals. Some individuals of this species, however, attack only wheat or only barley, or oats, etc., and these individuals comprise groups that are called varieties or special forms (*formae specialis*) such as *P. graminis tritici*, *P. g. hordei*, *P. g. avenae*, etc. But even within each special form, some individuals attack some of the varieties of the host plant but not the others, some attack another set of host plant varieties, and so on, each group of such individuals comprising a race. Thus, there are more than 200 races of *Puccinia graminis tritici* (race 1, race 15, race 59, etc.). Occasionally, one of the offspring of a race can suddenly attack a new variety or can cause severe symptoms on a variety that it could barely infect before. This individual is called a variant. The identical individuals produced asexually by the variant comprise a biotype. Each race consists of one or of several biotypes (race 15A, 15B, etc.).

The appearance of new pathogen biotypes may be very dramatic when the change involves the host range of the pathogen. If the variant has lost the ability to infect a plant variety that is widely cultivated, this pathogen simply loses its ability to procure a livelihood for itself and will die without even making its existence known to us. If, on the other hand, the change in the variant pathogen enables it to infect a plant variety cultivated because of its resistance to the parental strain, the variant individual, being the only one that can survive on this plant variety, grows and multiplies on the new variety without any competition and soon produces large populations that spread and destroy the heretofore resistant variety. This is the way the resistance of a plant variety is said to be "broken down," although it was the change in the pathogen, not the host plant, that brought it about.

types of plant resistance to pathogens

Different plants are resistant to certain pathogens for various reasons. Some plants, of course, are immune to a particular pathogen even under the most favorable conditions for disease development. Others exhibit certain degrees of resistance to a pathogen under most environmental conditions. Still others are actually susceptible to the pathogen but, under the conditions they are normally grown, may appear resistant.

Some very susceptible varieties exhibit apparent resistance. Such varieties can escape disease because of rapid growth or early maturity and of some inherent quality which makes them resistant for a period of their life (earliness or lateness) and which, with proper planting, can be made to coincide with the period of abundance of inoculum. Other varieties show tolerance to a disease and can produce a good crop in spite of infection either because of exceptional vigor or because of a hardy structure. Still other varieties are not infected by certain pathogens because their stomata are too few, closed, or plugged with masses of cells, or because the waxy coating on their leaves, the thick skin of their fruit, etc., do not allow the pathogen to enter the host. In all these cases, however, once the pathogen has established infection in the host it can develop freely and can produce symptoms as though the host is susceptible.

Truly resistant varieties, on the other hand, are those in which the pathogen and the host are incompatible with each other, or the host plant can defend itself against the pathogen by the various defense mechanisms activated in response to infection by the pathogen. If resistance of a plant to a pathogen is provided by one or a few defense mechanisms controlled by one or a few genes, respectively, such resistance is called specific or vertical and is either monogenic (one gene) or oligogenic (a few genes), and the genes responsible for it are called major genes. If resistance is provided by a combination of lesser defense mechanisms, each of which alone is rather ineffective against the pathogen, and such mechanisms are controlled by a group or groups of complementary genes, such resistance is called general or horizontal resistance, it is almost always polygenic and the genes are called minor genes. In addition, resistance is sometimes controlled by genetic determinants contained in the cytoplasm of the cell and is called cytoplasmic resistance; the two best known cases of cytoplasmic resistance occur in corn in which resistance to two leaf blights, the southern corn leaf blight caused by *Helminthosporium maydis* and the yellow leaf blight caused by *Phyllosticta maydis,* is conferred by characteristics present in normal cytoplasm of various types of corn but absent or suppressed in Texas male-sterile cytoplasm.

Varieties with specific (monogenic or oligogenic) resistance generally show complete resistance to a specific pathogen under most environmental conditions, but a single or a few mutations in the pathogen may produce a new race that may infect the previously resistant variety. On the contrary, varieties with general (polygenic) resistance are less stable and may vary in their reaction to the pathogen under different environmental conditions, but a pathogen will have to undergo many more mutations to completely break down the resistance of the host. As a rule, a combination of major and minor genes for resistance against a pathogen is the most desirable makeup for any plant variety.

genetics of virulence in pathogens and of resistance in host plants

Infectious plant diseases are the result of the interaction of at least two organisms, the host plant and the pathogen. The properties of each of these two organisms are governed by their genetic material, the DNA, which is organized in numerous segments comprising the genes.

The inheritance of host reaction—degree of susceptibility or resistance—to various pathogens has been known for a long time and has been used quite effectively in breeding and distributing varieties resistant to pathogens causing particular diseases. The inheritance of infection type—degree of virulence or avirulence—however, has been overlooked until relatively recently. It has now become clear that pathogens consist of a multitude of races, each differing from others in its ability to attack certain varieties of a plant species but not other varieties. Thus when a variety is inoculated with two appropriately chosen races of a pathogen, the variety is susceptible to one race but resistant to the other. Conversely, when the same race of a pathogen is inoculated on two appropriately chosen varieties of a host plant, one variety is susceptible while the other is resistant to the same pathogen. This clearly indicates that, in the first case, one race possesses a genetic characteristic that enables it to attack the plant, while the other race does not, and in the second case, that the one variety possesses a genetic characteristic that enables it to defend itself against the pathogen, so that it remains resistant, while the other variety does not. When several varieties are inoculated separately with one of several races of the pathogen, it is again noted that one pathogen race can infect a certain group of varieties, another race can infect another group of varieties, including some that can and some that cannot be infected by the previous race, and so on. Thus, varieties possessing certain genes of resistance or susceptibility react differently against the various pathogen races and their genes of virulence or avirulence. The progeny of these varieties react to the same pathogens in exactly the same manner as did the parent plants, indicating that the property of resistance or susceptibility against a pathogen is genetically controlled (inherited). Similarly, the progeny of each pathogen causes on each variety the same effect that was caused by the parent pathogens, indicating that the property of virulence or avirulence of the pathogen on a particular variety is genetically controlled (inherited).

It appears from the above that, under favorable environmental conditions, the outcome—infection or noninfection—in each host–pathogen combination is predetermined by the genetic material of the host and of the pathogen. The number of genes determining resistance or susceptibility varies from plant to plant, as the number of genes determining virulence or avirulence varies from pathogen to pathogen. In most host–pathogen combinations the numbers of genes involved and what they control are not yet known. In some diseases, however, particularly those caused by fungi, e.g., potato late blight, apple scab, powdery mildews, tomato leaf mold, the cereal smuts and rusts, and also in tobacco mosaic, considerable information regarding the genetics of host–pathogen interactions is available.

THE GENE-FOR-GENE CONCEPT

The coexistence of host plants and their pathogens side by side in nature indicates that the two have been evolving together. Changes in the virulence of the pathogens must be continually balanced by changes in the resistance of the host, and vice versa, so that a dynamic equilibrium of resistance and virulence is maintained and both host and pathogen survive. If either the virulence of the pathogen or the resistance of the host increased unopposed, it would have led to the elimination of either the host or the pathogen, respectively, which obviously has not happened. Such a stepwise evolution of resistance and virulence can be explained by the gene-for-gene concept, according to which for each gene that confers resistance in the host there is a corresponding gene in the pathogen that confers virulence to the pathogen, and vice versa.

The gene-for-gene concept was first proven in the case of flax and flax rust, but it has since been shown to operate in many other rusts, in the smuts, powdery mildews, apple scab, late blight of potato, and other diseases caused by fungi, as well as some diseases caused by bacteria, viruses, parasitic higher plants, and nematodes. In all these diseases it was shown that whenever a variety is resistant to a pathogen as a result of 1, 2, or 3 resistance genes, the pathogen also contains 1, 2, or 3 virulence genes, respectively. Each gene in the host can be detected and identified only by its counterpart gene in the pathogen, and vice versa. Generally, genes for resistance are dominant while genes for virulence are recessive. Whenever a new gene for virulence appears, the resistance of the host breaks down and plant breeders introduce another gene for resistance in the plant which counteracts the new gene for virulence in the pathogen. This produces a resistant variety—until another gene for virulence appears in the pathogen.

The gene-for-gene concept has been demonstrated only in plants with monogenic and oligogenic types of resistance to a certain disease. Plant breeders apply the gene-for-gene concept every time they incorporate a new resistance gene into a desirable variety that becomes susceptible to a new strain of the pathogen. With the diseases of some crops new resistance genes must be found and introduced into old varieties at relatively frequent intervals, while in others a single gene confers resistance to the varieties for many years. The gene-for-gene concept presumably applies to polygenic or general resistance as well, although so far proof for this and for polygenic control of virulence in pathogens is lacking.

THE NATURE OF RESISTANCE TO DISEASE

A plant is either immune to a pathogen—that is, it is not attacked by the pathogen even under the most favorable conditions—or it may show various degrees of resistance ranging from almost immunity to complete susceptibility. Resistance may be conditioned by a number of internal and external factors which operate to reduce the chance and degree of infection. Any heritable characteristic of the plant that contributes toward localization and isolation of the pathogen at the points of entry, toward reduction of the harmful effects of toxic substances produced by the pathogen, or toward inhibition of the reproduction and, thereby, of the further spread of the pathogen contributes toward the resistance of the plant to disease. Furthermore, any heritable characteristic that enables a particular variety to complete its development and maturation under conditions that do not favor the development of the pathogen, also contributes to resistance (disease escape).

The contribution of the genes conditioning resistance in the host seems to consist of, primarily, providing the genetic potential in the plant for development of one or more of the morphological or physiological characters—including those described in the chapters on structural and biochemical defense—that contribute toward disease resistance. With the exception of virus and viroid diseases of plants, in which the genes of the host could conceivably come into "face-to-face" confrontation with the "genes" of the viral nucleic acid, the genes of plants infected by other types of pathogens seem to never come in contact with the genes of the pathogen. In general, in all host–pathogen combinations, viruses and viroids included, the interactions between genes of host and genes of pathogen are believed to be brought about indirectly through the physiological processes controlled by the respective genes.

The mechanisms by which genes control the physiological processes that lead to disease resistance or susceptibility are not yet clear but they are, presumably, no different than are the mechanisms controlling any other physiological process in living organisms. Thus, it is known that the genes are each carried by the genetic material (DNA) as successive groups of nucleotide triplets (triplet code) which are first read and transcribed on messenger RNA as the latter is synthesized. The messenger RNA then becomes associated with clusters of ribosomes (polyribosomes) and leads to the production of a specific protein which is either an enzyme or a structural protein. The produced enzyme participates or initiates biochemical reactions related to one or another of the cellular processes, and may result in the production of a certain morphological characteristic or accumulation of a certain chemical substance.

The genes responsible for determining the kind and properties of a protein are called structural genes. The timing of activation of the structural genes, the rate of their activity—protein synthesis—and the timing of their inactivation are controlled by other genes called regulatory genes. Furthermore, messenger-RNA formation seems to be initiated only at certain points of the DNA strand, and these points are called operators. A single operator may control the transcription into a messenger-RNA of only one structural gene, or of a series of structural genes concerned with the different steps of one particular metabolic function, for example the biosynthesis of a fungitoxic phenolic compound or a compound that reacts with and detoxifies a pathogen toxin. The group of genes controlled by such an operator is called an operon.

Thus, it is possible that for the production of an inducible enzyme or a fungitoxic substance, a stimulant (inducer) secreted by the pathogen inactivates a repressor molecule, which is the product of a regulatory gene. The function of the repressor was to combine with a specific operator locus and prevent the transcription of that operon, thereby blocking the synthesis and action of the relevant proteins in the absence of infection. Following infection, however, and after inactivation of the repressor by the pathogenic stimulant, transcription of the operon can take place, the particular substance is produced and, if this substance is toxic enough to the pathogen, the infection stops and the variety is resistant. On the other hand, if a pathogen mutant appears that does not secrete the particular stimulant (inducer) that inactivates the repressor molecule, the defense reaction does not take place, the pathogen infects the host without opposition and so it causes disease. In that case, the resistance of the host is said to have broken down but it is actually bypassed by the pathogen rather than broken down. Other possible ways by which a pathogen could "break down" the resistance of a host would be through a mutation in the pathogen, which enables it to produce a substance that can react with and neutralize the defensive toxic substance of the host that is directed against the pathogen; or through a mutation in the pathogen that would eliminate or block its receptor site on which the host defensive substance becomes attached, and the pathogen then can operate in the presence of that substance and of the defense mechanism that produces it.

breeding of resistant varieties

The value of resistance in controlling plant diseases was recognized in the early 1900s. Advances in the science of genetics and the obvious advantages of avoiding losses from plant diseases by simply planting a resistant instead of a susceptible variety made the breeding of resistant varieties possible, and very desirable. The more recent realization of the dangers of polluting the environment through chemical control of plant diseases gave additional impetus and importance to the breeding of resistant varieties. Thus the breeding of resistant varieties, which is but one part of broader plant breeding programs, is more popular and more intensive today than it ever was in the past. Its usefulness and importance are paramount in the production of food and fiber. Yet, some aspects of plant breeding, and of breeding of resistant varieties in particular, have shown certain weaknesses and have allowed the occurrence of some plant disease epidemics that could not have developed if it were not for the uniformity created in crops through plant breeding.

NATURAL VARIABILITY IN PLANTS

Today's cultivated crop plants are the result of selection, or selection and breeding, of plant lines that evolved naturally in one or many geographical areas over millions of years. The evolution of plants, from their ancient ancestors to the present day crop plants has occurred slowly and has, in the meantime, produced countless genetically diverse forms of these plants. Many such plants still exist as wild types at the point(s) of origin or in the areas of natural spread of the plant. Although these plants may appear as useless remnants of evolution that are not likely to play a role in any future advances in agriculture, their diversity and survival in the face of the various pathogens that affect that crop indicates that they carry numerous genes for resistance against these pathogens.

Since the beginning of agriculture some of the wild plants in each locality have been selected and cultivated and thus produced numerous cultivated lines or varieties. The most productive of these varieties were perpetuated in each locality from year to year and those that survived the local climate and the pathogens continued to be cultivated. Nature and pathogens eliminated the weak and susceptible ones while the farmers selected the best yielders among the survivors. Surviving varieties had different sets of major and minor genes for resistance. In this fashion, selection of crop plants continued wherever they were grown, each locality independently selecting varieties adapted to its own environment and resistant to its own pathogens. Thus, numerous varieties of each crop plant were cultivated throughout the world and, by their own genetic diversity, contributed to making the crop locally adapted but overall genetically nonuniform and, thereby, safe from any sudden outbreak of a single pathogen over a large area.

EFFECTS OF PLANT BREEDING ON VARIABILITY IN PLANTS

During the present century, widespread, intensive, and systematic efforts have been made and continue to be made by plant breeders throughout the world toward breeding plants that combine the most useful genes for higher yields, better quality, uniform size of plants and fruit, uniform ripening, cold hardiness, disease resistance, etc. In searching for new useful genes, plant breeders cross existing, local, cultivated varieties with each other and with those of other localities, both here and abroad, and with wild species of crop plants from wherever they can be obtained. Furthermore, plant breeders often attempt to generate additional genetic variation by treating their plant material with mutagenic agents.

The initial steps in plant breeding generally increase the variability of genetic characteristics of plants in a certain locality by combining in such plants genes that were more or less widely separated before. As breeding programs advance, however, and as several of the most useful genes are identified, subsequent steps in breeding tend to eliminate variability by combining the best genes in a few cultivated varieties and leaving behind or discarding plant lines that seem to have no usefulness at the time. In a short time a few "improved" varieties replace most or all others over large expanses of land. The most successful improved varieties are also adopted abroad and, before too long, some of them replace the numerous but commercially inferior local varieties. Occasionally, even the wild types themselves may be replaced by such a variety. Thus, Red Delicious apples, Elberta peaches, certain dwarf wheat and rice varieties, certain genetic lines of corn and potatoes, one or two types of bananas and sugarcane, etc. are grown in huge acreages throughout the world. In almost every crop, relatively few varieties make up the great bulk of the cultivated acreage of the crop throughout a country or throughout the world. The genetic base of these varieties is often narrow, especially since many of them have been derived from crosses of the same or related ancestors. These few varieties are used so widely because they are the best available, they are stable and uniform, and therefore everybody wants to grow them. At the same time, however, because they are so widely cultivated, they carry with them not only the blessings but also the dangers of uniformity. The most serious of these dangers is the vulnerability of large uniform plantings to sudden outbreaks of catastrophic plant disease epidemics.

PLANT BREEDING
FOR DISEASE RESISTANCE

Most plant breeding is done for development of varieties that produce greater yields of better quality. When such varieties become available they are then tested for resistance against some of the most important pathogens present in the area where the variety is developed and where it is expected to be cultivated. If the variety is resistant to these pathogens, it may be released to growers for immediate production. If, however, it is susceptible to one or more of these pathogens, the variety is usually shelved or discarded; or sometimes it is released for production if the pathogen can be controlled by other means, e.g., chemical; but more often it is subjected to further breeding in an attempt to incorporate into the variety genes that would make it resistant to the pathogens without changing any of its desirable characteristics.

The sources of genes for resistance are the same gene pool of the crop that provides genes for every other inherited characteristic, i.e., other native or foreign commercial varieties, older varieties abandoned earlier or discarded breeders' stock, wild plant relatives and, occasionally, induced mutations. Resistance may be obtained by incorporating one, a few or many, major or minor resistance genes to the variety. Specific or vertical resistance is easy to manipulate in a breeding program and, therefore, it is often preferred to general or horizontal resistance but both, specific and general, have their advantages and limitations. Specific resistance is aimed against specific pathogens or pathogen races. It is most effective when incorporated in annual crops that are easy to breed, e.g., small grains; when it is directed against pathogens that do not reproduce and spread very rapidly, e.g., *Fusarium*; or pathogens that do not mutate very frequently, e.g., *Puccinia graminis*; when it consists of "strong" genes that confer complete protection to the plant that carries it; and when the host population does not consist of a single genetically uniform variety grown over large acreages. If one or more of these, and of several other, conditions are not met, specific resistance becomes short lived, i.e., it breaks down as a result of appearance of new pathogen mutants and hybrids that can bypass it or overcome it.

On the other hand, general or horizontal resistance confers incomplete but permanent protection—it does not break down. General resistance involves mechanisms of defense which are beyond the limits of the capacity of the pathogen to change, i.e., beyond the probable limits of its variability. General resistance is universally present in wild and in domesticated plants but is at its highest in wild plants and at its lowest in greatly "improved" varieties. General resistance operates against all races of a pathogen, including the most pathogenic ones. Actually, the more pathogenic the races present the greater the selection for general resistance in the host. General resistance is eroded in the absence of the pathogen because there is no selection pressure for resistance. In many cases, general (polygenic) resistance is an important part of the resistance exhibited by plants possessing specific resistance but it is overshadowed by the latter. Considerable general resistance can be incorporated into cultivated crops by cross-breeding of existing genetically different varieties. The wider the genetic base of a crop the greater its general resistance and the smaller the need for the usually temporary specific resistance.

Varieties with general (polygenic, horizontal, or nonspecific) resistance remain resistant much longer than do varieties with specific (oligogenic or vertical) resistance, but the resistance of the second group is much more effective, in the short run, than that of the first group. Also, varieties with specific resistance are often attacked suddenly and rapidly by a new virulent race and lead to severe epidemics. These disadvantages can be avoided in some crops by the use of multilines, which are either mixtures of individual varieties (lines or cultivars) that are agronomically similar but differ in their resistance genes, or varieties that are derived from crossing several to many varieties that contain different resistance genes and then selecting from those that contain the mixtures of genes.

Multilines have been developed mostly in small grains against the rust fungi but their use is likely to increase in these and in other crops as the control of plant diseases with specific resistance and with chemicals becomes more risky or less acceptable.

Often genes of resistance are present in the varieties or species normally grown in the area where the disease is severe and in which the need for resistant varieties is most pressing. With most diseases, a few plants remain virtually unaffected by the pathogen although most or all other plants in the area may be severely diseased. Such survivor plants are likely to have remained healthy because of resistant characters present in them. Such plants may be cross-pollinated with the new one to fortify the latter with their genes for resistance. On the other hand, if these plants are propagated asexually and continue to be resistant to the pathogen in subsequent years they may become the stock plants for the development of one or more resistant varieties.

If no resistant plants can be found within the local population of the species, other species, cultivated or wild, are checked for resistance, and, if resistant, are crossed with the cultivated varieties in efforts to incorporate the resistance genes of the other species into the cultivated varieties. With some diseases, e.g., late blight of potatoes, it has been necessary to look for resistance genes in species growing in the area where the disease originated and where, presumably, existing plants managed to survive the long, continuous presence of the pathogen because of their resistance to it. Finally, it is possible to increase or make apparent resistance in plants by the use of chemicals such as colchicine, which induce polyploidy in plants and result in creation of a homozygous condition by doubling heterozygous alleles, or by the use of mutagenic chemicals and radiations resulting in the occasional appearance of mutants which exhibit greater resistance to the pathogen than did the parent plant.

Incorporating genes for resistance from wild or unsatisfactory plants into susceptible, but agronomically desirable, varieties is a difficult and painstaking process involving a series of crossings, testing, backcrossing to the desirable varieties, and so forth. The feasibility of the method in most cases, however, has been proved repeatedly. Through breeding, varieties of some crops, e.g., tobacco, have been developed in which genes for resistance against several different diseases have been incorporated.

VULNERABILITY OF GENETICALLY UNIFORM CROPS TO PLANT DISEASE EPIDEMICS

Even completely resistant varieties do not remain so forever. The continuous production of mutants and hybrids in pathogens sooner or later leads to the appearance of races that can infect previously resistant varieties. Sometimes, races may exist in an area in small populations and avoid detection until after the introduction of a new variety, or virulent races of the pathogen existing elsewhere may be brought in after introduction of the resistant variety. In all cases, widespread cultivation of a single, previously resistant variety provides an excellent substrate for rapid development and spread of the new race of the pathogen, and usually leads to an epidemic. Thus, genetic uniformity in crops, although very desirable when it concerns horticultural characteristics, is undesirable and often catastrophic when it occurs in the genes of resistance to diseases.

The cultivation of varieties with genetically uniform disease resistance is possible and quite safe if other means of plant disease control, e.g., chemical, are possible. Thus, a few fruit tree varieties, e.g., Delicious apples, Bartlett pears, Elberta peaches, Navel oranges, etc., are cultivated throughout the world in the face of numerous virulent fungal and bacterial pathogens that would destory them in a short time were it not for the fact that the trees are protected from the pathogens by numerous chemical sprays annually. Even such varieties, however, suffer tremendous losses when affected by pathogens that cannot be controlled with chemicals as is the case of fire blight of pears and pear decline, of tristeza disease of citrus, etc.

Another case in which varieties with genetically uniform disease resistance are not likely to suffer from severe disease epidemics is when the resistance is aimed against slow-moving soil pathogens such as *Fusarium* and *Verticillium*. Aside from the fact that some pathogens normally produce fewer races than others, even if new races are produced at the same rate, soilborne pathogens lack the dispersal potential of airborne ones. As a result, a new race of a soilborne pathogen would be limited to a relatively small area for a long time and, although it could cause a locally severe disease, it would not spread rapidly and widely to cause an epidemic. The slow spread of such virulent new races of soilborne pathogens allows time for the control of the disease by other means or the replacement of the variety with another one resistant to the new race.

Genetic uniformity in plant varieties becomes a serious disadvantage in the production of major crops because of the potential danger of sudden and widespread disease epidemics caused by airborne or insect-borne pathogens in the vast acreages in which each of these varieties are often grown. Several examples of epidemics that resulted from genetic uniformity are known and some of them have already been mentioned. Southern corn leaf blight was the result of the widespread use of corn hybrids containing the Texas male-sterile cytoplasm; the destruction of the "Ceres" spring wheat by race 56 of *Puccinia graminis* and of the "Hope" and its relative bread wheats by race 15B of *P. graminis* was the result of replacement of numerous genetically diverse varieties by a few uniform ones. The Helminthosporium blight of Victoria oats was the result of replacing many varieties with the rust-resistant Victoria oats; coffee rust destroyed all coffee trees in Ceylon because all of them originated from uniform susceptible stock of *Coffea arabica*; tristeza destroyed millions of orange trees in South and North America because they were all propagated on susceptible sour orange rootstocks; and pear decline destroyed millions of pear trees in the Pacific coast states because they were propagated on susceptible oriental rootstocks. In spite of these and many other well-known examples of plant disease epidemics which occurred because of the concentrated cultivation of genetically uniform crops over large areas, crop production continues to depend on genetic uniformity. A few varieties of each crop make up the bulk of the cultivated crop over as vast an area as the U. S. (Table I). As can be seen in Table I, although a relatively large number of varieties are available for each crop, only a few, often 2 or 3, varieties are grown in more than half the acreage of each crop, and in some they make up more than three-fourths of the crop. It is easy to see that two pea varieties make up almost the entire (96 percent) pea crop of the country, i.e., about 400,000 acres, that two varieties account for 42 percent of the sugarbeet crop, i.e., about 600,000 acres. But the figures become even more spectacular when one considers the most popular varieties of the truly large acreage crops. Thus, although six corn varieties account for 71 percent or 47 million acres, one of them alone accounts for 26 percent or 17 million acres. Furthermore, most of the varieties shared the same male-sterile cytoplasm. Similarly, six varieties of soybean account for about 24 million acres of that crop and most of these varieties share common ancestors.

It is apparent that several hundreds of thousands or several million acres planted to one variety present a huge opportunity for the development of an epidemic. The variety, of course, is planted so widely because it is resistant to existing pathogens. But this resistance puts extreme survival pressure on the pathogens over that area. It takes one "right" change in one of the zillions of pathogen individuals in the area to produce a new virulent race that can attack the variety. When that happens it is a matter of time—and, usually, of favorable weather—before

TABLE I.
ACREAGE AND FARM VALUE OF MAJOR U.S. CROPS IN 1969 AND EXTENT TO WHICH SMALL NUMBERS OF VARIETIES DOMINATE CROP ACREAGE (Reproduced with permission of the National Academy of Sciences.)

Crop	Acreage (millions)	Value (millions of dollars)	Total Varieties	Major Varieties	Acreage (percent)
Bean, dry	1.4	143	25	2	60
Bean, snap	0.3	99	70	3	76
Cotton	11.2	1200	50	3	53
Corn[a]	66.3	5200	197	6	71
Millet	2.0	?		3	100
Peanut	1.4	312	15	9	95
Peas	0.4	80	50	2	96
Potato	1.4	616	82	4	72
Rice	1.8	449	14	4	65
Sorghum	16.8	795	?	?	?
Soybean	42.4	2500	62	6	56
Sugar beet	1.4	367	16	2	42
Sweet potato	0.13	63	48	1	69
Wheat	44.3	1800	269	9	50

[a] Corn includes only released Agricultural Experimental Station inbreds for seed, forage, and silage.

the race breaks loose, the epidemic develops and the variety is wiped out! In most cases the appearance of the new race is detected early and the variety is replaced with another one, resistant to the new race, before a widespread epidemic occurs; this of course requires that varieties of a crop with different genetic base are available at all times. For this reason, most varieties must usually be replaced within about 3 to 5 years from the time of their widespread distribution.

In addition to the genetic uniformity within one variety, plant breeding often introduces genetic uniformity to several or all cultivated varieties of a crop by introducing one or several genes in all of these varieties or by replacing the cytoplasm of the varieties with a single type of cytoplasm. Induced uniformity through introduced genes includes, for example, the dwarfism gene in the dwarf wheats and rice varieties, the monogerm gene in sugar beet varieties, the determinate gene in tomato varieties, and the stringless gene in bean varieties. Uniformity through replacement of the cytoplasm occurred, of course, in most corn hybrids in the later 1960s when the Texas male-sterile cytoplasm replaced the normal cytoplasm; and cytoplasmic uniformity is commercially employed in several varieties of sorghum, sugar beet and onions, it is studied in wheat, and is also present in cotton and in cantaloupe. Neither the introduced genes nor the replacement cytoplasm, of course, make the plant less resistant to diseases, but if a pathogen appears that is favored by or can take advantage of the characters controlled by that gene or that cytoplasm, then the stage is set for a major epidemic. That this can happen was proven by the southern corn leaf blight epidemic of 1970, the susceptibility of dwarf wheats to new races of *Septoria* and *Puccinia*, of tomatoes with the determinate gene to *Alternaria*, and others.

SELECTED REFERENCES

Boone, D. M. 1971. Genetics of *Venturia inaequalis. Ann. Rev. Phytopathol.* 9:297–318.

Browder, L. E. 1971. Pathogenic specialization in cereal rust fungi, especially *Puccinia recondita* f. sp. *tritici:* Concepts, methods of study and application. *USDA and Kansas Agr. Exp. Sta. Tech. Bull.* 1432:51 p.

Browning, J. A. and K. J. Frey. 1969. Multiline cultivars as a means of disease control. *Ann. Rev. Phytopathol.* 7:355–382.

Day, P. R. 1973. Genetic variability of crops. *Ann. Rev. Phytopathol.* **11**:293–312.

Day, P. R. 1974. "Genetics of Host–Parasite Interaction." W. H. Freeman Co., San Francisco. 238 p.

Day, P. R. (ed.) 1977. "The Genetic Basis of Epidemics in Agriculture." *Ann. N.Y. Acad. Sci.* **287,** 400 p.

Fincham, J. R. S., and P. R. Day. 1971. "Fungal Genetics" (3rd ed.). Oxford: Blackwell; Philadelphia: Davis. 402 p.

Flor, H. H. 1956. The complementary genic systems in flax and flax rust. *Advan. Genet.* **8**:29–54.

Flor, H. H. 1971. Current status of the gene-for-gene concept. *Ann. Rev. Phytopathol.* 9:275–296.

Gallegly, M. E. 1970. Genetics of *Phytophthora. Phytopathology* **60**:1135–1141.

Halisky, P. M. 1965. Physiologic specialization and genetics of the smut fungi. III. *Bot. Rev.* **31**:114–150.

Holmes, F. O. 1965. Genetics of pathogenicity in viruses and of resistance in host plants. *Advan. Virus Res.* **11**:139–162.

Holton, C. S., J. A. Hoffman, and R. Duran. 1968. Variation in the smut fungi. *Ann. Rev. Phytopathol.* **6**:213–242.

Hooker, A. L. 1967. The genetics and expression of resistance in plants to rusts of the genus *Puccinia. Ann. Rev. Phytopathol.* **5**:163–182.

Hooker, A. L. 1974. Cytoplasmic susceptibility in plant disease. *Ann. Rev. Phytopathol.* **12**:167–179.

Lincoln, R. E. 1940. Bacterial wilt resistance and genetic host–parasite interactions in maize. *J. Agr. Res.* **60**:217–240.

Moseman, J. C. 1966. Genetics of powdery mildews. *Ann. Rev. Phytopathol.* **4**:269–290.

National Academy of Sciences. 1972. Genetic vulnerability of major crops. Washington, D.C. 307 p.

Nelson, R. R. (ed.). 1973. Breeding Plants for Disease Resistance. Concepts and Applications. The Pennsylvania State University Press, University Park, 401 p.

Robinson, R. A. 1971. Vertical resistance. *Rev. Plant Pathol.* **50**:233–239.

Robinson, R. A. 1973. Horizontal resistance. *Rev. Plant Pathol.* **52**:483–501.

Schafer, J. F. 1971. Tolerance to plant disease. *Ann. Rev. Phytopathol.* **9**:235–252.

Shepherd, K. W., and G. M. E. Mayo. 1972. Genes conferring specific plant disease resistance. *Science* **175**:375–380.

Sidhu, G. S. 1975. Gene-for-gene relationships in plant parasitic systems. *Sci. Prog., Oxford* **62**:467–485.

Stewart, D. M., and B. J. Roberts. 1970. Identifying races of *Puccinia graminis* f. sp. *avenae.* A modified international system. *USDA and Minn. Agr. Exp. Sta. Tech. Bull.* **1416**:23 p. plus 5 plates.

Van der Plank, J. E. 1968. "Disease Resistance in Plants." Academic Press, New York.

Van der Plank, J. E. 1963. "Plant Diseases; Epidemics and Control." Academic Press, New York.

Walter, J. M. 1967. Hereditary resistance to disease in tomato. *Ann. Rev. Phytopathol.* **5**:131–160.

Watson, I. A. 1970. Changes in virulence and population shifts in plant pathogens. *Ann. Rev. Phytopathol.* **8**:209–230.

Watson, I. A., and N. H. Luig. 1968. Progressive increase in virulence in *Puccinia graminis* f. sp. *tritici. Phytopathology* **58**:70–73.

Webster, R. K. 1974. Recent advances in the genetics of plant pathogenic fungi. *Ann. Rev. Phytopathol.* **12**:331–353.

Wellhausen, E. J. 1937. Effect of genetic constitution of the host on the virulence of *Phytomonas stewartii. Phytopathology* **27**:1070–1089.

7
effect of environment on development of infectious plant diseases

Although all pathogens, all perennial, and, in warmer climates, many annual plants are present in the field throughout the year, almost all diseases occur only, or develop best, during the warmer part of the year. Also, it is common knowledge that most diseases appear and develop best during wet, warm days, or that plants heavily fertilized with nitrogen usually are much more severely attacked by some pathogens than are less fertilized plants. These general examples clearly indicate that the environmental conditions prevailing in both air and soil, after contact of a pathogen with its host, may greatly affect the development of the disease, and frequently they determine whether a disease will occur or not. The environmental factors that most seriously affect the initiation and development of infectious plant diseases are temperature, moisture, light, soil nutrients, and soil pH. Their effects on disease may be brought about through their influence on the growth and/or susceptibility of the host, on the multiplication and activity of the pathogen or on the interaction of host and pathogen and its effect on the severity of symptom development.

It is obvious then that, for a disease to occur and to develop optimally, a combination of three factors must be present: susceptible plant, infective pathogen, and favorable environment. However, although plant susceptibility and pathogen infectivity remain essentially unchanged in the same plant for at least several days, and sometimes for weeks or months, the environmental conditions may change more or less suddenly and in various degrees. Such changes influence the development of diseases in progress, or the initiation of new ones, more or less drastically. Of course, a change in any environmental factor may favor the host or the pathogen or both, or it may be more favorable to the one than it is to the other, and the expression of disease will be affected accordingly. Plant diseases generally occur over a fairly wide range of the various environmental

104

conditions. Nevertheless, the extent and frequency of their occurrence, as well as the severity of the disease on individual plants, are influenced by the degree of deviation of each environmental condition from the point at which disease development is optimal.

effect of temperature

Plants as well as pathogens require certain minimum temperatures in order to grow and carry out their activities. The low temperatures of late fall, winter, and early spring are below the minimum required by most pathogens. Therefore, diseases are not, as a rule, initiated during that time and those in progress generally come to a halt. With the advent of higher temperatures, however, pathogens become active and, when other conditions are favorable, they can infect plants and cause disease. Pathogens differ in their preference for higher or lower temperatures, and many diseases develop best in areas, seasons, or years with cooler temperatures, while others develop best where and when relatively high temperatures prevail. Thus, some species of the fungi *Typhula* and *Fusarium*, which cause snow mold of cereals and turf grasses, thrive only in cool seasons or cold regions. Also, the late blight pathogen *Phytophthora infestans* is most serious in the northern latitudes, whereas in the subtropics it is serious only during the winter. On the other hand, most diseases are favored by high temperatures and are limited to within areas and during seasons in which such temperatures are prevalent. Such diseases include the fusarial wilts of plants, the *Phymatotrichum* root rots of plants, the brown rot of stone fruits caused by *Monilinia fructicola*, the southern bacterial wilt of solanaceous plants caused by *Pseudomonas solanacearum*, etc.

The effect of temperature on the development of a particular disease after infection depends on the particular host–pathogen combination. The most rapid disease development, i.e., the shortest time required for the completion of a disease cycle, usually occurs when the temperature is optimum for the development of the pathogen but is above or below the optimum for the development of the host. At temperatures much below or above the optimum for the pathogen, or at temperatures near the optimum for the host, disease development is slower. Thus, for stem rust of wheat, caused by *Puccinia graminis tritici*, the time required for a disease cycle (from inoculation with uredospores to new uredospore formation) is 22 days at 5°C, 15 days at 10°C, and 5 to 6 days at 23°C. Similar time periods for the completion of a disease cycle are required in many other diseases caused by fungi, bacteria, and nematodes. Since the duration of a disease cycle determines the number of disease cycles and, approximately, the number of new infections in one season, it is clear that the effect of temperature on the prevalence of a disease in a given season may be very great.

If the minimum, optimum, and maximum temperatures for the pathogen, the host, and the disease are about the same, the effect of temperature in disease development is apparently through its influence on the
pathogen which becomes so activated at the optimum temperature that
the host, even at its optimum growth, cannot contain it.

In many diseases, the optimum temperature for disease development
seems to be different from those of both the pathogen and the host. Thus,
in the black root rot of tobacco, caused by the fungus *Thielaviopsis
basicola*, the optimum for disease is at 17 to 23°C, while that for tobacco
is 28 to 29°C and for the pathogen is 22 to 28°C. Evidently, neither the
pathogen nor the host grow well at 17 to 23°C, but the host grows so
much more poorly, and is so much weaker, than the pathogen that even
the weakened pathogen can cause maximum disease development. In the
root rots of wheat and corn, caused by the fungus *Gibberella zeae*, the
maximum disease development on wheat occurs at temperatures above
the optimum for development of both the pathogen and wheat, but on
corn, it occurs at temperatures below the optimum for the pathogen and
for corn. Considering that wheat grows best at low temperatures while
corn grows best at high temperatures, it would appear that the more
severe damage to wheat at high temperatures and to corn at low temperatures is due to disproportionate weakening of the plants, in relation to the
weakening of the pathogen, at the unfavorable temperatures.

The effect of temperature on virus diseases of plants is a great deal
more unpredictable. Temperature determines not only the ease with
which plants can become infected with a virus but also whether or not a
virus multiplies in the plant and, if it does, whether the symptoms
produced will be of one kind or another. The severity of the disease may
vary greatly in various virus–host combinations depending on the temperature during certain stages of the disease. Temperature, probably in
combination with sunlight, seems to determine the seasonal appearance
of symptoms in the various virus diseases of plants. Viruses producing
yellows or leaf-roll symptoms are most severe in the summer, while
those causing mosaic or ringspot symptoms are most pronounced in the
spring. New growth produced during the summer on mosaic- or ringspot-
infected plants usually shows only mild symptoms or is completely free
from symptoms.

effect of moisture

Moisture, like temperature, influences the initiation and development of
infectious plant diseases in many interrelated ways. It may exist as rain or
irrigation water on the plant surface or around the roots, as relative
humidity in the air and as dew. The most important influence of moisture seems to be on the germination of fungal spores and on the penetration of the host by the germ tube. Moisture also activates the bacterial,
fungal, and nematode pathogens, which may then infect the plant. Moisture, such as splashing rain and running water, also plays an important

role on the distribution and spread of many of these pathogens on the same plant or from one plant to another. Finally, moisture affects disease by increasing the succulence of host plants, thus considerably increasing their susceptibility to certain pathogens.

The occurrence of many diseases in a particular region is closely correlated with the amount and distribution of rainfall within the year. Thus, late blight of potato, apple scab, downy mildew of grapes, and fire blight, are found or are severe only in areas with high rainfall or high relative humidity during the growing season. As a matter of fact, in all these, and other diseases, the rainfall determines not only the severity of the disease, but also whether the disease will occur at all in a given season. In the cases of the fungal diseases, the effect of moisture is on the germination of spores of fungi, which require a film of water on the tissues in order to germinate and, also, on the liberation of spores from the sporophores which, as in apple scab, can occur only in the presence of moisture. The number of disease cycles per season of many of these diseases is closely correlated with the number of rainfalls per season, particularly of rainfalls that are of sufficient duration to allow establishment of new infections. Thus in apple scab, for example, continuous wetting of the leaves, fruit, etc., for at least 9 hours is required for any infection to take place even at the optimum range (18 to 23°C) of temperature for the pathogen. At lower or higher temperatures the minimum wetting period required is higher, i.e., 14 hours at 10°C, 28 hours at 6°C, and so on. If the wetting period is less than the minimum for the particular temperature, the pathogen fails to establish itself in the host and to produce disease.

Most fungal pathogens are dependent on the presence of free moisture on the host or of high relative humidity in the atmosphere only during germination of their spores and become independent once they can obtain nutrients and water from the host. Some pathogens, however, such as those causing late blight of potato and the downy mildews, require at least high relative humidity in the environment throughout their development. In these diseases, the growth and sporulation of the pathogen, and also the production of symptoms, come to a halt as soon as dry, hot weather sets in and resume only after a rain or after the return of humid weather.

Though most fungal and bacterial pathogens of aboveground parts of plants require a film of water in order to produce successful infections, the spores of the powdery mildew fungi can germinate, penetrate, and cause infection even when there is only high relative humidity in the atmosphere surrounding the plant. In the powdery mildews, spore germination and infection are actually lower in the presence of free moisture on the plant surface than they are in its absence and, in some of them, the most severe infections take place when the relative humidity is rather low (50 to 70 percent). In these diseases, the amount of disease is limited rather than increased by wet weather. This is also indicated by the fact that the powdery mildews are more common and more severe in the drier areas of the world and their relative importance decreases as rainfall increases.

In many diseases affecting underground parts of plants, such as roots, tubers, and young seedlings—for example in the *Pythium* damping off of seedlings and seed decays—the severity of the disease is proportional to the soil moisture and is greatest near the saturation point. The increased moisture seems to affect primarily the pathogen, which multiplies, and moves (zoospores in the case of *Pythium*) best in wet soils, but it may also decrease the ability of the host to defend itself through reduced availability of oxygen in water-logged soil and by lowering the temperature of such soils. Many other soil fungi, e.g., *Phytophthora*, *Rhizoctonia*, *Sclerotinia*, and *Sclerotium*, some bacteria, e.g., *Erwinia* and *Pseudomonas*, and most nematodes usually cause their most severe symptoms on plants when the soil is wet but not flooded, while yet others such as *Streptomyces scabies*, causing the common scab of potato, are most severe in rather dry soils.

Most bacterial diseases, and also many fungal diseases of young tender tissues, are particularly favored by high moisture or high relative humidity. Bacterial pathogens and fungal spores are usually disseminated in water drops splashed by rain, in rain water moving from the surfaces of infected tissues to those of healthy ones, or in free water in the soil. Bacteria penetrate plants through wounds or natural openings and cause severe disease when present in large numbers. Once inside the plant tissues, the bacteria multiply faster and are more active during wet weather, probably because the plants, through increased water absorption and resulting succulence, can provide the high concentrations of water that favor bacteria. The increased bacterial activity produces greater damage to tissues, and this damage, in turn, helps release greater numbers of bacteria on the plant surface where they are available to start more infections if wet weather continues.

effect of wind

Wind affects infectious plant diseases primarily through its importance in the spread of plant pathogens and, to a smaller extent, through its speeding up of the drying of wet plant surfaces. Most plant diseases that spread rapidly and are likely to assume epidemic proportions are caused by pathogens, e.g., fungi, bacteria, viruses, that are either spread directly by the wind or are spread by insect vectors which themselves can be carried over long distances by the wind. Some spores, e.g., zoosporangia, basidiospores, and some conidia, are usually quite delicate and do not survive long-distance transport in the wind. Others, e.g., uredospores and many kinds of conidia, can be transported by the wind for many miles. Wind is even more important in disease development when it is accompanied by rain. Wind-blown rain helps release spores and bacteria from infected tissue and then carries them through the air and deposits them on wet surfaces which, if susceptible, can be infected immediately. Wind also injures plant surfaces while they are blown about and rubbing against each other; this facilitates infection by many fungi and bacteria and also by some mechanically transmitted viruses. Wind sometimes helps prevent infection by speeding up the drying of wet plant surfaces on which fungal spores or bacteria may have landed. If the plant surfaces dry before penetration has taken place, any germinating spores or bacteria present on the plant are likely to desiccate and die and no infection will occur.

effect of light

The effect of light on disease development, especially under natural conditions, is far less than that of temperature or moisture although several diseases are known in which the intensity and/or the duration of light may either increase or decrease the susceptibility of plants to infection and also the severity of the disease. In nature, however, the effect of light is limited to the production of more or less etiolated plants as a result of reduced light intensity. This usually increases the susceptibility of plants to nonobligate parasites, e.g., of lettuce and tomato plants to *Botrytis* or of tomato to *Fusarium,* but decreases their susceptibility to obligate parasites, e.g., of wheat to the stem rust fungus *Puccinia.*

Reduced light intensity generally increases the susceptibility of plants to virus infections. Holding plants in the dark for one to two days before inoculation increases the number of lesions (i.e., infections) appearing after inoculation and this has become a routine procedure in many laboratories. Generally, darkening affects the sensitivity of plants to virus infection if it precedes inoculation with the virus, but seems to have little or no effect on symptom development if it occurs after inoculation. On the other hand, low light intensities following inoculation tend to mask the symptoms of some diseases, which are much more severe when the plants grow in normal light than when they are shaded.

effect of soil pH

The pH of the soil is important in the occurrence and severity of plant diseases caused by certain soil-borne pathogens. For example, the club-root of crucifers, caused by *Plasmodiophora brassicae,* is most prevalent and severe at about pH 5.7, while its development drops sharply between 5.7 and 6.2, and is completely checked at pH 7.8. On the other hand, the common scab of potato, caused by *Streptomyces scabies,* can be severe at a pH range from 5.2 to 8.0 or above, but its development drops sharply at pH below 5.2. It is obvious that such diseases are most serious in areas whose soil pH favors the particular pathogen. In these, and in many other diseases, the effect of soil acidity (pH) seems to be principally on the pathogen, although in some, a weakening of the host through altered nutrition induced by the soil acidity may affect the incidence and severity of the disease.

effect of
host–plant nutrition

Nutrition affects the rate of growth and the state of readiness of plants to defend themselves against pathogenic attack. Abundance of certain nutrients, e.g., nitrogen, results in the production of young, succulent growth and may prolong the vegetative period and delay maturity of the plant, making it more susceptible to pathogens that prefer to attack such tissues—and for longer periods. Conversely, lack of nitrogen would make plants weaker, slower growing, and faster aging and would make them susceptible to pathogens that are best able to attack weak, slow growing plants. Thus, it is known that high nitrogen fertilization increases the susceptibility of pear to fire blight (*Erwinia amylovora*), of wheat to rust (*Puccinia*) and to powdery mildew (*Erysiphe*), etc. Reduced nitrogen may also increase the susceptibility of some plants to certain diseases, e.g., of tomato to *Fusarium* wilt, of many solanaceous plants to the *Pseudomonas solanacearum* wilt, of sugar beets to *Sclerotium rolfsii*, of most seedlings to *Pythium* damping off. It is possible, however, that it is the form of nitrogen (ammonium or nitrate) available to the host or pathogen that affects disease severity or resistance rather than the amount of nitrogen. Of numerous root rots, wilts, foliar diseases, etc., treated with either form of nitrogen, almost as many decreased or increased in severity when treated with ammonium nitrogen as did with nitrate nitrogen, but each form of nitrogen had exactly the opposite effect on a disease than did the other form of nitrogen.

Although nitrogen nutrition, because of its profound effects on growth, has been studied the most extensively in relation to disease development, studies with other elements such as phosphorus, potassium, and calcium, and also with micronutrients, have revealed similar relationships between levels of the particular nutrients and susceptibility or resistance to certain diseases. In general, plants receiving a balanced nutrition, in which all required elements are supplied in appropriate amounts, are more capable of protecting themselves from new infections and of limiting existing infections than when one or more nutrients are supplied in excessive or deficient amounts. Even a balanced nutrition, however, may affect the development of a disease when the concentration of all the nutrients is increased or decreased beyond a certain range.

the role of
environmental factors in
plant disease epidemics

For infection to occur at least once, the plant and the pathogen must be brought together and the range of temperature and moisture must be such that the pathogen will be able to grow and cause infection. Predisposition of the plant for disease by improper nutrition, lighting, soil pH, etc. may also be of some importance but these factors seldom determine whether disease will occur or not.

For a disease to become important in a field, however, and particularly if it is to spread over a large area and develop into an epidemic, the right combinations of environmental factors must occur and spread either constantly, or repeatedly and at frequent intervals, over a large area. Even in a single, small field, which contains the pathogen, plant diseases almost never become severe from just one set of favorable environmental conditions. It takes repeated disease cycles and considerable time before the pathogen produces large enough numbers of individuals that can cause an economically severe disease in the field. However, once large populations of the pathogens are available, they can then attack, spread to nearby fields, and cause a severe disease in a very short time—just a few days.

Plant disease epidemics are caused primarily by fungi, bacteria, viruses, and mycoplasmas which can either be carried directly by the wind (fungi, some bacteria) or by insect vectors (viruses, mycoplasmas, some fungi, and bacteria). In some cases minor epidemics occur when the pathogen is carried in or on the seed or any other propagative organ, but in this case the epidemic is limited to the plants propagated from such contaminated seed, etc., unless again the pathogen can become airborne and spreads to other plants.

A plant disease epidemic implies the development and rapid spread of a disease on a particular kind of crop plant cultivated over a large area—be that a large field, a valley, a section of a country, the entire country or even part of a continent. So, the first component of a plant disease epidemic is a large area planted to one, more or less genetically uniform crop plant, with the plants and the fields being close together. The second component of an epidemic is the presence or appearance of a virulent pathogen at some point among or near the cultivated host plants. Such cohabitations of host plants and pathogens occur, of course, daily in countless locations and cause local diseases of varying degrees of severity. But most of these destroy crop plants to a limited extent and do not develop into epidemics. Epidemics develop only when the combinations and progression of the right sets of environmental conditions, i.e., moisture, temperature, and wind or insect vector, coincide with the susceptible stage(s) of the plant and with the production, spread, inoculation, penetration, infection, and reproduction of the pathogen.

Thus, for an epidemic to develop the small original or primary inoculum of the pathogen must be carried by wind or vector to some of the crop plants as soon as they begin to become susceptible to that pathogen. The moisture and temperature must then be right for germination or infection to take place. Following infection the temperature must be favorable for rapid growth and reproduction of the pathogen (short incubation period, short disease cycle) so that numerous new spores, etc., will appear as quickly as possible. The moisture (rain, fog, dew) then must be sufficient and should last long enough for abundant release of spores, etc. Winds of proper humidity and velocity and blowing toward the susceptible crop plants must then pick up the spores and carry them to the plants while the latter are still susceptible. It so happens, of course, that most plant disease epidemics are effectively spread from south to north in the northern hemisphere (and in the opposite direction in the southern hemisphere). This occurs because the warmer spring and early summer weather, and the growth seasons also move in the same direction and so the pathogens constantly find plants in their susceptible stage as everything progresses northward.

In each new location, however, the same set of favorable moisture, temperature and wind or vectors must be repeated so that infection, reproduction and dispersal of the pathogen must occur as quickly as possible. Furthermore, these conditions must be repeated several times within each location so that the pathogen can multiply itself and the number of infections it causes on the host plants. It is these repeated infections that usually result in the more or less complete destruction of almost every plant within the area of an epidemic, although the uniformity and area of cultivation of the plant along with the prevailing weather (i.e., moisture, temperature, winds) determine the final spread of the epidemic.

Fortunately, the most favorable combinations of conditions for disease development do not occur very often over very large areas and, therefore, spectacular plant disease epidemics that destroy crops over large areas are relatively rare. However, small epidemics involving the plants in a field, a valley, etc. occur quite frequently. With many diseases, e.g., potato late blight, apple scab, and cereal rusts, the environmental conditions seem to be usually favorable and disease epidemics occur or would occur every year were it not for the control measures (chemical sprays, resistant varieties, etc.) employed annually to avoid such epidemics.

weather and forecasting of plant disease epidemics

The appearance and development of many plant diseases depend on the kinds of weather conditions that preceded, prevail during or are likely to follow certain stages in the cultivation of crop plants. Therefore, it is often possible to predict the likelihood and severity of a particular disease if the association of the weather conditions and of the development of the disease are known and if the changes in the weather are followed closely. Ability to forecast the appearance and occurrence of diseases, especially diseases that fluctuate widely from season to season, can help farmers reduce or avoid losses from them either by applying control measures or by planting crops not susceptible to these diseases.

The occurrence of some diseases during a growing season can be predicted quite acurately by analyzing the weather conditions, i.e., temperature, during the previous winter. For example, the incidence of bacterial wilt of corn depends on the amount of inoculum that survived the previous winter in the bodies of its vector, corn flea beetles. When the sum of the mean temperatures for the three winter months at a given location is less than −1°C, most of the beetle vectors are killed and so there is little or no bacterial wilt! Warmer winters allow greater survival of the beetle vectors and proportionately more severe wilt outbreaks. Similarly, when January temperatures are above normal, the downy mildew (blue mold) of tobacco appears early in seedbeds in the following season and causes severe losses, while when January temperatures are below normal, blue mold appears in seedbeds late and causes little damage. If the disease is expected and can be controlled in seedbeds, subsequent control in the field is made much easier.

With most diseases, however, the pathogen responds quickly to changes in temperature and moisture during the growing season and in these cases disease forecasting depends on much closer observation and correlation of the weather and the pathogen. For example, the occurrence of an infection period in apple scab is a function of temperature and duration of wetness in the area. Thus, if it is 6°C and the spores are wet for 28 hours, infection will occur, and similarly with 14 hours wetness at 10°C, 9 hours at 18 to 24°C, 12 hours at 26°C, etc. A grower need spray only when these combinations occur and he can get good control with less expense.

Because downy mildews of plants and late blight of potatoes are among the most destructive diseases where they occur, several systems of forecasting have been developed for these diseases. Weather forecasting for late blight of potato, caused by *Phytophthora infestans*, has been extensively studied in several countries in Europe and in the U. S. and effective systems for disease forecasting have been developed for various areas. The main conditions for late blight epidemics appear to be constant, cool temperatures between 10 and 24°C while the relative humidity is 75 percent or more for at least 48 consecutive hours. When this combination occurs, a late blight outbreak can be expected from 2 to 3 weeks later. Several hours of rainfall, dew, or relative humidity close to saturation point within that period and afterward serve to increase the severity of the disease and the likelihood of a major epidemic.

SELECTED REFERENCES

Chupp, C. 1928. Club root in relation to soil alkalinity. *Phytopathology* **18**:301–306.

Colhoun, J. 1973. Effects of environmental factors on plant disease. *Ann. Rev. Phytopathol.* **11**:343–364.

Dickson, J. G. 1923. Influence of soil temperature and moisture on the development of seedling blight of wheat and corn caused by *Gibberella saubinetii. J. Agr. Res.* **23**:837–870.

Gallegly, M. E., Jr., and J. C. Walker. 1949. Plant nutrition in relation to disease development. V. *Am. J. Bot.* **36**:613–623.

Hepting, G. H. 1963. Climate and forest diseases. *Ann. Rev. Phytopathol.* **1**:31–50.

Huber, D. M., and R. D. Watson. 1974. Nitrogen form and plant disease. *Ann. Rev. Phytopathol.* **12**:139–165.

Jones, L. R., J. Johnson, and J. G. Dickson. 1926. Wisconsin studies upon the relation of soil temperature to plant diseases. *Wisc. Agr. Expt. Sta. Res. Bull.* **71**.

Kassanis, B. 1957. Effect of changing temperature on plant virus diseases. *Advan. Virus Res.* **4**:169–186.

Keitt, G. W., and K. L. Jones. 1926. Studies of the epidemiology and control of apple scab. *Wisc. Agr. Expt. Sta. Res. Bull.* **73**.

Miller, P. R. 1953. The effect of weather on diseases. *Yearbook Agr.* (U. S. Dept. Agr.), pp. 83–93.

Miller, P. R., and M. J. O'Brien. 1957. Prediction of plant disease epidemics. *Ann. Rev. Microbiol.* **11**:77–110.

Schnathorst, W. C. 1965. Environmental relationships in the powdery mildews. *Ann. Rev. Phytopathol.* **3**:343–366.

8
control
of plant
diseases

Information on symptoms, causes, and mechanisms of development of plant diseases is intellectually interesting and scientifically justified, but, most important of all, it is useful because it makes feasible the development of methods to combat plant diseases and, thus, increase the quantity and improve the quality of plant products.

Methods of control vary considerably from one disease to another depending on the kind of pathogen, the host, and the interaction of the two. In controlling diseases, plants are generally treated as populations rather than individuals, although certain hosts, especially trees, ornamentals, and, sometimes, virus-infected plants, often are treated individually. With the exception of trees, however, damage or loss of one or a few plants is usually considered insignificant and control measures are generally aimed at saving the populations rather than a few individual plants.

Considering the regularity with which most serious diseases of crop plants appear in an area year after year, the rapidity of spread of most plant diseases, and the difficulties, when at all possible, in curing a disease after it has begun to develop, it is easy to understand why almost all control methods are aimed at protecting plants from becoming diseased rather than at curing them after they have become diseased. As a matter of fact, there are few infectious plant diseases that can be satisfactorily controlled in the field by therapeutic means, although certain diseases can be cured under experimental conditions.

The various control methods could be generally classified as regulatory, cultural, biological, physical, and chemical, depending on the nature of the agents employed to control the disease.

regulatory methods

In order to prevent the import and spread of plant pathogens into the country or individual states, certain federal and state laws regulate the conditions under which certain crops may be grown and distributed between states and countries. Such regulatory control is applied by means of quarantines, inspections of plants in the field or warehouse, and occasionally by voluntary or compulsory eradication of certain host plants.

QUARANTINES
AND INSPECTIONS

Plant pathogens introduced into an area in which they did not exist before are likely to cause much more catastrophic epidemics than do existing pathogens, because plants developing in the absence of a pathogen have no opportunity to select resistance factors specific against this pathogen and are, therefore, extremely vulnerable to attack by such a pathogen. Some of the worst plant disease epidemics that have occurred throughout the world, e.g., the downy mildew of grapes in Europe, the bacterial canker of citrus, the chestnut blight, the Dutch elm disease, and the soybean cyst nematode, in the U.S., are all diseases caused by pathogens introduced from abroad.

In order to keep out foreign plant pathogens and to protect the nation's farms, gardens, and forests, plant quarantine regulations prohibit or restrict entry into or passage through the U.S. from foreign countries of plant pathogens not known to be widely established in this country, and of plants, plant products, soil, or other materials carrying or likely to carry such pathogens. Similar quarantine regulations also exist in most other countries.

Plant quarantines are carried out by experienced inspectors stationed in all points of entry into the country of persons or produce likely to introduce new pathogens. Plant quarantines are already credited for interception of numerous foreign plant pathogens and, thereby, saving the country's plant world from potentially catastrophic diseases. Yet, the introduction of pathogens in the form of spores, eggs, etc. on unsuspected carriers, the existence of latent infections of seeds and other plant propagative organs with viruses, fungi, bacteria, or nematodes, even after treatment, make plant quarantines considerably less than foolproof. Various steps taken by plant quarantine stations, such as growing plants under observation for certain periods of time before they are released to the importer, tend to reduce the chances of introduction of harmful pathogens. In specific cases, for example with annual imports of flower bulbs from Holland, U.S. quarantine inspectors may, following previous agreement between the parties involved, visit and inspect for diseases the flower fields in Holland; if they find the fields to be disease free, they issue inspection certificates allowing the import of such bulbs into the U.S. without further tests.

Similar quarantine regulations govern the interstate, and even the intrastate, sale of nursery stock, tubers, bulbs, seeds, and other propagative organs, especially of certain crops, such as potatoes and fruit trees. The movement and sale of such materials within and between states, however, is controlled by the regulatory agencies of each state by mutual agreement and arrangement.

Several voluntary inspection systems are also in effect in various states in which appreciable amounts of nursery stock, potato seed tubers, etc., are produced. Growers interested in producing and selling disease-free seed potatoes, woody ornamentals, etc., submit to a voluntary inspection and/or indexing of their crop in the field and in storage by the state regulatory agency, by experiment station personnel, or others. If, following certain procedures recommended by the inspecting agency, the plant material is found to be free of certain, usually virus, diseases, the inspecting agency issues a certificate indicating the freedom of the plants from these specific diseases, and the grower may then advertise his produce as disease free, thus securing a better and higher-priced market.

cultural methods

They include the activities of man aimed at controlling disease through the cultural manipulation of plants. Some of these methods are aimed at eliminating the pathogen from the plant or from the area in which the plants are growing (eradication), others at increasing the resistance of the host to the pathogen or creating conditions unfavorable to the pathogen, and still others at obtaining pathogen-free propagative material from infected plants.

HOST ERADICATION

When a pathogen has been introduced into a new area in spite of quarantine, a plant disease epidemic frequently follows. If the epidemic is to be prevented, all the host plants heretofore infected by or suspected to harbor the pathogen may have to be removed and burned. This results in elimination of the pathogen and prevention of greater losses from the spread of the pathogen to more plants. Such host eradication has controlled, for example, the bacterial canker of citrus in Florida and other southern states where more than three million trees have had to be destroyed for that reason. Host eradication is also carried out routinely in many nurseries, greenhouses, and fields to prevent the spread of numerous diseases through elimination of infected plants that provide a ready source of inoculum within the crop.

Certain pathogens of annual crops, for example cucumber mosaic virus and potato yellow dwarf virus, overwinter only or mainly in other perennial, usually wild, plants. Eradication of the host in which the pathogen overwinters sometimes suffices to eliminate completely or to reduce drastically the amount of inoculum that can cause infections the follow-ing season. Similarly, some pathogens require two alternate hosts to complete their life cycles: for example, *Puccinia graminis tritici* requires wheat and barberry; *Cronartium ribicola* requires pine and currant (*Ribes*); and *Gymnosporangium juniperi-virginianae* requires cedar and apple. Eradication of the wild or economically less important alternate host interrupts the life cycle of the pathogen and leads to control of the disease. This has been carried out somewhat successfully with stem rust of wheat and white pine blister rust through eradication of barberry and currant, respectively, although, owing to other factors, both diseases are still widespread and catastrophic. In cases like the cedar-apple rust, how-ever, in which both hosts may be important, control through eradication of the alternate host is impractical.

CROP ROTATION

Soil pathogens, which can attack plants of one or a few species or even families, can sometimes be eliminated from the soil by planting, for 3 or 4 years, crops belonging to species or families not attacked by the pathogen. Complete control through crop rotation is possible with pathogens that survive only on living plants or only so long as the host residue persists as a substrate for their saprophytic existence. When the pathogen, however, produces long-lived spores or can live as saprophytes for more than 5 or 6 years, crop rotation becomes ineffective or impractical. In the latter cases, crop rotation can still be useful by reducing, although not eliminating, the pathogen populations in the soil so that appreciable yields from the susceptible crop, which otherwise would be impossible, can be obtained every third or fourth year of the rotation.

SANITATION

Sanitation includes all activities aimed at eliminating or reducing the amount of inoculum present in a plant, field, or warehouse and at pre-venting the spread of the pathogen to other healthy plants and plant products. Thus, plowing under or removal and proper disposal of infected leaves, branches, or other plant debris that may harbor the pathogen reduce the amount and the spread of the pathogen and the amount of disease that will develop later on. Workers who smoke, by washing their hands before handling certain kinds of plants, e.g., tomato, may reduce the spread of tobacco mosaic virus. Washing the soil off farm equipment before moving it from one field to another may also help to avoid spread-ing any pathogens present in the soil. Similarly, by washing the produce, its containers, and the walls of storage houses, the amount of inoculum and subsequent infections may be reduced considerably.

IMPROVEMENT
OF GROWING CONDITIONS
OF PLANTS

Cultural practices aiming at improving the vigor of the plant often help increase its resistance to pathogen attack. Thus, proper fertilization, drainage of fields, irrigation, proper spacing of plants, and weed control improve the growth of plants and may have a direct or indirect effect on the control of a particular disease. For example, the most important measures for controlling *Valsa* canker of fruit and other trees are adequate irrigation and proper fertilization of the trees.

CREATING
CONDITIONS UNFAVORABLE
TO THE PATHOGEN

Proper aeration of stored products hastens drying of their surface and inhibits germination and infection by any fungal or bacterial pathogens present on them. Similarly, proper spacing of plants in the field or greenhouse prevents creation of high humidity conditions on the plant surfaces and inhibits infection by certain pathogens, such as *Botrytis*. Good soil drainage also reduces the number and activity of certain fungal pathogens (e.g., *Pythium*) and nematodes and may result in significant disease control. Appropriate choice of fertilizers or soil amendments may also lead to changes in the soil pH which may influence unfavorably the development of the pathogen. Flooding of fields for long periods of time or dry fallowing may also reduce the number of certain pathogens in the soil by starvation, by lack of oxygen, or by desiccation.

TISSUE CULTURE

With certain plants, such as carnation and chrysanthemum, that are generally propagated by cuttings, control of the vascular diseases caused by *Fusarium*, *Verticillium*, etc. may be obtained through tissue culture of the meristem tips. Since these pathogens do not reach the apical meristems until the very late stages of the disease, the culture of meristem tips provides pathogen-free cuttings for starting new healthy plants. Similarly, most viruses do not invade the uppermost millimeter or so of the growing meristem and, by tissue culture of the meristematic tip, healthy plants may be produced. Tissue culture is, however, a rather specialized technique and usually only a few healthy plants are produced by tissue culture and are then used for further asexual propagation.

biological methods

Biological control of plant diseases can be achieved by selecting and breeding plants for resistance to particular pathogens or by using other microorganisms that are either antagonistic to the pathogen or parasitize the pathogen itself. Although the use and breeding of resistant varieties is the oldest, cheapest, and overall best means of controlling plant diseases, the use of hyperparasites or antagonistic microorganisms has been attracting considerable interest in recent years.

RESISTANT VARIETIES

The use of resistant varieties is the cheapest, easiest, safest, and most effective means of controlling plant diseases in crops for which such varieties are available. Cultivation of resistant varieties not only eliminates losses from disease but also eliminates expenses for sprays and for other ways of disease control, and makes unnecessary the contamination of the environment with toxic chemicals that would otherwise be used to control plant diseases. Moreover, for many diseases, e.g., those caused by vascular pathogens and viruses, which cannot be adequately controlled by any available means, and for others, e.g., cereal rusts and root rots, which are economically impractical to control in other ways, the use of resistant varieties provides the only means of producing acceptable yields.

Varieties of various crops resistant to some of the most important or most difficult to control diseases are made available to growers by federal and state experiment stations and by commercial seed companies. More than 75 percent of the total agricultural acreage in the United States is planted with varieties that are resistant to one or more diseases, and with some crops, such as small grains and alfalfa, varieties planted because they are resistant to certain disease(s) make up 95 to 98 percent of the crop. Growers and consumers alike have gained the most from the use of varieties resistant to the fungi causing rusts, smuts, powdery mildews, and vascular wilts, but several other kinds of fungal diseases, and many diseases caused by viruses, bacteria and nematodes are controlled through resistant varieties.

Resistant varieties have been used in only a few cases, e.g., blister rust and fusiform rust of pine, for disease control in fruit and forest trees. This is due to the difficulty in replacing susceptible varieties with resistant ones and in keeping the resistant ones from being attacked by new races of the pathogen that are likely to develop over the long life span of trees.

CROSS PROTECTION
AND INTERFERENCE

The term cross protection is used specifically for the protection of a plant by a mild strain of a virus from infection by a strain of the same virus which causes much more severe symptoms. This appears to be a general phenomenon among virus strains. Its application, however, in controlling virus diseases has met with little success because of the laboriousness of the method for field crops and because of the dangers of mutations, double infections and the danger of spread to, and higher virulence in, other crops.

Two cases resembling cross protection but probably due to interference of one pathogen with the other have been reported recently. Certain plants, e.g., bean and sugarbeet, inoculated with virus exhibit a greater resistance to infection by certain obligate fungal pathogens causing rusts and powdery mildews than do virus-free plants. However, in other host–virus–fungus systems, virus-infected plants are less resistant to fungus infections than healthy ones. A less documented case of "cross protection" involves the inhibition of infection of pear with the fire blight bacterium by inoculation with a nonpathogenic bacterium. Similarly, it has been reported that cucurbit plants can be protected from infection by the anthracnose fungus *Colletotrichum lagenarium* by inoculating the plants while young with the same fungus.

A somewhat different form of interference seems to occur in roots infected with mycorrhizae. It appears that the presence of the mycorrhizal fungi on and in the roots acts as a protective barrier to infection by the highly pathogenic fungi *Pythium*, *Phytophthora*, *Fusarium*, and others.

Excellent control of crown gall was recently obtained in the greenhouse and in the field by treating seeds, germinated seeds, or the roots of nursery stock with a suspension of a strain of *Agrobacterium radiobacter*. The presence of this bacterium on the plants prevented infection by the virulent strains of *Agrobacterium tumefaciens*.

HYPERPARASITISM

Control of pathogenic microorganisms with other microorganisms or viruses which parasitize or antagonize the pathogens has not yet met with appreciable practical success, but recent experimental results and the increased interest in and information about such possibilities hold great promise for future developments. The best known cases of hyperparasitism include the bacteriophages, mycoparasites, and nematophagous fungi.

BACTERIOIOPHAGES

Bacteriophages or phage (=bacteria-destroying viruses) are known to exist in nature for most plant pathogenic bacteria. Successful experimental control of several bacterial diseases was obtained when the bacteriophages were mixed with the inoculated bacteria, when the plants were first treated with bacteriophages and then inoculated with bacteria, and when the seed was treated with the phage. However, not one bacterial disease is controlled effectively by bacteriophage in practice. Also, no plant disease caused by a bacterium has been cured yet by treatment with phage after the disease has developed.

MYCOPARASITISM

The mycelium and spores of several phytopathogenic soil fungi such as *Pythium*, *Fusarium*, and *Helminthosporium*, are attacked and parasitized in culture and, probably, in the soil by one or more fungi which, as a rule, are not pathogenic to plants. The growth of some of these and of other fungi in the soil is also inhibited by the presence in their environment of certain other fungi and bacteria. No bacteria have been shown yet to attack and parasitize fungi directly but some bacteria, e.g., *Bacillus subtilis*, as well as some fungi appear to be antagonistic to certain soil-inhabiting phytopathogenic fungi and through their enzymatic or toxic secretions cause lysis and death of the fungus. Attempts to control soil-inhabiting fungal phytopathogens through addition of the parasitic or antagonistic fungi and bacteria in the soil have given rather disappointing results. Some fungal and bacterial diseases of plants, however, such as *Fusarium* root rot of corn and crown gall of fruit trees and ornamentals, can be controlled by treating the seeds or dipping the seedlings in preparations containing fungi and bacteria antagonistic to these plant pathogens. Also, inoculating freshly cut pine stumps with spores of *Peniophora gigantea* effectively inhibits infection by *Fomes annosus*.

Addition of soil amendments favoring antagonistic microorganisms has in several cases induced an increase in the populations of the hyperparasites with a concomitant reduction of the populations of the phytopathogenic fungi and a parallel reduction in disease severity. Similar results can be obtained by partial sterilization of the soil or by soil treatment with selective fungicides. These treatments usually affect the pathogen by encouraging the growth and activity of fungi such as *Trichoderma*, of several Actinomycetes, and of other microorganisms which either produce antibiotics toxic to the pathogen or are otherwise antagonistic to it and inhibit its growth and activity.

Hyperparasitic fungi that attack other fungi are known for several plant pathogens, including some rusts, powdery mildews, downy mildews, *Pythium*, *Helminthosporium*, and *Sclerotinia*. None of the hyperparasitic fungi, however, have given satisfactory results in field trials.

PARASITES OF NEMATODES

Many plant-parasitic nematodes are parasitized by soil fungi, protozoa, and possibly by bacteria and viruses. Several predators, including protozoa, fungi, other nematodes and microarthropods also seem to attack phytopathogenic nematodes in the soil. The use of hyperparasites or predators to control plant-parasitic nematodes has been little investigated yet. The greatest emphasis has been placed upon the use of predacious fungi but, while their effectiveness *in vitro* and in pot tests has been encouraging, their application on a field scale has been disappointing.

CONTROL THROUGH TRAP CROPS AND ANTAGONISTIC PLANTS

Some plants that are not actually susceptible to certain sedentary plant-parasitic nematodes produce exudates that stimulate hatching of eggs of these nematodes. The larvae enter these plants but are not able to develop into adults and lay eggs, and finally they die. Such plants are called trap crops. By using trap crops in a crop rotation program, the nematode population in the soil is reduced considerably. For example, *Crotalaria* plants trap the larvae of the root-knot nematode *Meloidogyne* sp. and black nightshade plants (*Solanum nigrum*) reduce the populations of the golden nematode, *Heterodera rostochiensis*. Similar results can be obtained by planting highly susceptible plants which, after infection by the nematodes, are destroyed before the nematodes reach maturity and begin to reproduce.

A few kinds of plants, e.g., asparagus and marigolds, are antagonistic to nematodes because they release certain substances in the soil which are toxic to several plant-parasitic nematodes and, when interplanted with nematode-susceptible crops, they decrease the number of nematodes in the soil and in the roots of the susceptible crops.

Unfortunately, neither trap nor antagonistic plants give a sufficient degree of control to offset the expense involved and, therefore, they have been little used in practical control of nematode diseases of plants.

physical methods

The physical agents most commonly used in controlling plant diseases are temperature (high or low) and various types of radiation.

CONTROL BY HEAT TREATMENT

Heat treatments have been used for soil sterilization, for disinfection of propagative organs, for freeing plants from viruses, and for healing plant products before storage.

SOIL STERILIZATION BY HEAT

Soil sterilization in greenhouses, and sometimes in seed beds and cold frames, is usually achieved by the heat carried in live or aerated steam or hot water. The soil is steam sterilized either in special containers (soil sterilizers) into which steam is supplied under pressure, or on the greenhouse benches, in which case steam is piped into, and is allowed to diffuse through, the soil. Soil sterilization is completed when the temperature in the coldest part of the soil has remained for at least 30 minutes at

82°C or above, at which temperature all plant pathogens in the soil are killed. Heat sterilization of soil is frequently achieved by heat produced electrically rather than supplied by steam or hot water.

HOT-WATER TREATMENT
OF PROPAGATIVE ORGANS

Hot-water treatment of certain seeds, bulbs, and nursery stock is commonly used to kill any pathogens with which they are infected or which may be present inside seed coats, bulb scales, etc. In some diseases, seed treatment with hot water was for many years the only means of control, as in the loose smut of cereals, in which the fungus overwinters as mycelium inside the seed where it could not be reached by chemicals. Similarly, treatment of bulbs and nursery stock with hot water frees them from nematodes that may be present within these organs, e.g., *Ditylenchus dipsaci* in bulbs of various ornamentals, *Radopholus similis* in citrus rootstocks.

The effectiveness of the method is based on the fact that these dormant plant organs can withstand higher temperatures than those in which their respective pathogens can survive for a given period of time. The temperature of the hot water used and the duration of the treatment varies with the different host–pathogen combinations. Thus, in the loose smut of wheat the seed is kept in hot water at 52°C for 11 minutes, whereas bulbs treated for *Ditylenchus dipsaci* are kept at 43°C for 3 hours.

ELIMINATION OF
PATHOGENS FROM PLANTS
BY HEAT

Heat treatment has been the most successful and widely used therapeutic method against virus, mycoplasma, and rickettsialike diseases of plants. Dormant plant material, such as budwood, dormant trees, and tubers, is usually treated with hot water at temperatures ranging from 35 to 54°C, and treatment times from a few minutes to several hours. Actively growing plants are sometimes treated with hot water, but much more frequently they are treated with hot air, which gives both better survival of the plant and better elimination of the pathogen than does hot water. Temperatures of 35 to 40°C seem to be optimal for air treatment of growing plants. For hot-air treatment, the infected plants are usually grown in the greenhouse or in growth chambers for periods varying for different host–pathogen combinations, but generally lasting 2 to 4 weeks, although some viruses require treatment for 2 to 8 months and others may be eliminated in just one week. Although mycoplasmas, rickettsialike bacteria and many viruses can be eliminated from their hosts by heat treatment, for several viruses such treatment has been unsuccessful.

HOT-AIR TREATMENT
OF STORAGE ORGANS

Treatment of certain storage organs with hot air removes the excess moisture from their surfaces and hastens healing of wounds and thus prevents their infection by certain weak pathogens. For example, keeping sweet potatoes at 28 to 32°C for 2 weeks helps the wounds to heal and prevents infection by *Rhizopus* and by soft-rotting bacteria. Also, hot-air "curing" of harvested tobacco leaves removes most moisture from them and protects them from attack by fungal and bacterial saprophytes.

DISEASE CONTROL
BY REFRIGERATION

Refrigeration is probably the most widely used method of controlling postharvest diseases of fleshy plant products. Low temperatures at or slightly above the freezing point do not, of course, kill any of the pathogens that may be on or in the plant tissues but they inhibit or greatly retard the growth and activities of all such pathogens and thereby prevent the spread of existing infections and the initiation of new ones. Most perishable fruits and vegetables are usually refrigerated immediately after harvest, transported in refrigerated vehicles, and kept refrigerated until they are used by the consumer. Regular refrigeration of especially succulent fruits and vegetables is sometimes preceded by a quick hydrocooling or air cooling of these products, aiming at removing the excess heat, carried in them from the field, as quickly as possible to prevent any new infections that might start otherwise. The magnitude of disease control through refrigeration and its value to the growers and the consumers can hardly be exaggerated.

DISEASE CONTROL
BY RADIATIONS

Various types of electromagnetic radiations, such as ultraviolet (UV) light, X-rays, and γ-rays, as well as particulate radiation, such as α-particles and β-particles have been studied for their ability to control postharvest diseases of fruits and vegetables by killing the pathogens present on them. Some satisfactory results were obtained in experimental studies using γ-rays to control postharvest infections of peaches, strawberries, tomatoes, etc., by some of their fungal pathogens. Unfortunately, with many of these diseases the dosage of radiation required to kill the pathogen also injures the plant tissues on which the pathogens exist. So far, no plant diseases are commercially controlled by radiations.

chemical control

The most commonly known means of controlling plant diseases in the field and in the greenhouse and, sometimes, in storage, is through the use of chemical compounds that are toxic to the pathogens. Such chemicals either inhibit germination, growth, and multiplication of the pathogen or are outright lethal to the pathogen. Depending on the kind of pathogens they affect, the chemicals are called fungicides, bactericides, nematicides, viricides or, for the parasitic higher plants, herbicides. Some chemicals are toxic to all or most kinds of pathogens, others affect only one kind of pathogen, and certain compounds are toxic to only a few or a single specific pathogen.

Most of the chemicals are used to control diseases of the foliage and of other aboveground parts of plants. Others are used to disinfest and protect from infection seeds, tubers, and bulbs. Some are used to disinfest the soil, others to disinfest warehouses, to treat wounds, or to protect stored fruit and vegetables from infection. Still others (insecticides) are used to control the insect vectors of some pathogens.

The great majority of the chemicals applied on plants or plant organs can only protect them from subsequent infection and cannot stop or cure a disease after it has started. Also, the great majority of these chemicals are effective only in the plant area to which they have been applied (local action) and are not absorbed or translocated by the plants. Some chemicals, however, do have a therapeutic (eradicant) action, and several new chemicals are absorbed and systemically translocated by the plant (systemic fungicides and antibiotics).

METHODS OF
PLANT DISEASE CONTROL
WITH CHEMICALS

FOLIAGE SPRAYS AND DUSTS

Chemicals applied as sprays or dusts on the foliage of plants are usually aimed at control of fungus diseases and to a lesser extent of bacterial diseases. Most fungicides and bactericides are protectants and must be present on the surface of the plant in advance of the pathogen in order to prevent infection. Their presence usually does not allow fungus spores to germinate or they may kill spores upon germination. Contact of bacteria with bactericides may inhibit their multiplication or cause their death.

Some fungicides may also have a direct effect on pathogens which have already invaded the leaves, fruit, and stem, and in this case they act as eradicants by killing the fungus inside the host, or they may suppress the sporulation of the fungus without killing it. Some fungicides, e.g., dodine, have a partial systemic action because they can be absorbed by a part of the leaf tissues and be translocated internally into the whole leaf area. Several fungicides, e.g., the benzimidazoles, benomyl and thiabendazole, and the oxanthiins, carboxin and oxycarboxin, are clearly systemics and can be translocated internally throughout the host plant. Some bactericides, e.g., streptomycin, are also systemics, as are most antibiotics.

Fungicides and bactericides applied as sprays appear to be more efficient than when applied as dusts. Dusts may be preferable to sprays if application is to be made during a rain because they adhere better to wet plant tissues. Sometimes other compounds, e.g., lime, may be added to the active chemical in order to reduce its phytotoxicity and make it safer for the plant. Compounds with a low surface tension, such as detergents, are often added to fungicides in order to increase their spreading and thereby the contact area between fungicide and the sprayed surface. Some compounds, finally, are added to increase the adherence of the fungicide to the plant surface, e.g., starch and oils.

Since most fungicides and bactericides are protectant in their action, it is important that they be at the plant surface before the pathogen arrives there or at least before it has time to germinate, enter, and establish itself in the host. Because spores require a film of water on the leaf surface or at least atmospheric humidity near saturation before they can germinate, sprays or dusts seem to be most effective when they are applied before, during, or immediately after every rain. Considering that most fungicides and bactericides are effective only upon contact with the pathogen, it is important that the whole surface of the plant be covered completely with the chemical in order to be protected. For this reason, young, expanding leaves, twigs, and fruits must be sprayed more often than mature tissues, since small, growing leaves may outgrow protection after 3 to 5 days from spraying. The interval between sprays of mature tissue may vary from 7 to 14 days or longer, depending on the particular disease, the frequency and duration of rains, and the season of the year. The same factors also determine the number of sprays per season which may vary from 2 or 3 to 15 or more. Figure 22 shows some types of equipment used for spraying and dusting plants and for injecting chemicals into plants or into the soil.

The number and variety of chemicals used for foliar sprays and dusts is quite large. Some of these compounds are specific against certain diseases, others are effective against a wide spectrum of pathogens. Sprays with these materials usually contain 0.5 to 2 pounds of the compound per hundred gallons of water, although some, e.g., sulfur, are applied at 4 to 6 pounds per 100 gallons of water. Some of the fungicides used for foliar sprays or dusts are also used for seed treatments.

SEED TREATMENT

Seeds, tubers, bulbs, and roots are usually treated with chemicals to prevent their decay after planting by controlling pathogens carried on them or existing in the soil where they will be planted. Chemicals can be applied on the seed as dusts, as thick water suspensions mixed with the seed, or the seed can be soaked in a water solution of the chemical and then be allowed to dry. Tubers, bulbs, corms, and roots can be treated in similar ways.

In treating seeds or any other propagative organs with chemicals, precautions must be taken so that their viability is not lowered or de-

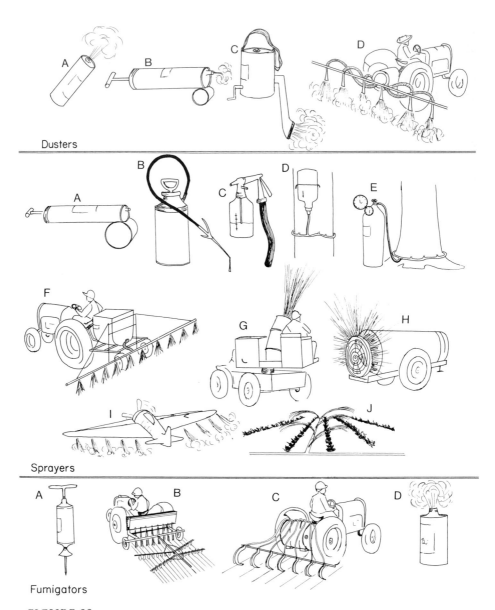

Dusters

Sprayers

Fumigators

FIGURE 22.
Various types of equipment used for dusting, spraying, injection, or fumigation for control of plant diseases. Dusters: (A–C) Portable dusters, (D) Tractor mounted. Sprayers: (A–C) Portable sprayers, (D) Tree injection gravity flow apparatus, (E) Tree injection under pressure, (F–H) Tractor-mounted sprayers for annuals (F) and for trees (G,H), (I) Airplane spraying (or dusting), (J) Spraying through the irrigation system. Fumigators: (A) Handgun fumigator, (B,C) Tractor mounted gravity-flow or pump-driven injectors, (D) Fumigation can for greenhouse or warehouse.

stroyed. At the same time enough chemical must stick to the seed to protect it from attacks of pathogens and, when the seed is planted, to diffuse into, and disinfest a sphere of soil around the seed in which the new plant will grow without being attacked at this particularly vulnerable period of growth.

Chemicals used in treating seeds, bulbs, corms, tubers, and roots include some inorganic copper and zinc compounds but mostly organic compounds such as captan, carboxin, oxycarboxin, chloroneb, chloranil, dichlone, hexachlorobenzene, maneb, zineb, thiram, pentachloronitrobenzene (PCNB), and streptomycin. Some chemicals may control specific diseases of some plants while others are more general in their action and may control many diseases of a number of plants.

SOIL TREATMENT

Soil to be planted with vegetables, ornamentals, or trees is frequently treated (fumigated) with volatile chemicals (fumigants) for control of nematodes, fungi, and bacteria. Treatment is usually done several days, weeks or months before planting. The chemicals are applied either with tractor-mounted, chisel-tooth injection shanks or disks, or, for small areas, with hand applicators (Fig. 22). The chemical is injected 10 to 15 cm deep in the soil and is applied either throughout the field or along the rows in which the plants will be planted. Some of the fumigants are so volatile that the treated soil must be covered immediately with a plastic or other covering to retain the fumes. Certain fumigants move through the soil slowly so that no covering other than the soil is needed. The most common fumigants are chloropicrin, methyl bromide, ethylene dibromide (EDB), dichloropropene–dichloropropane (D–D), Mylone, Nemagon, Vapam, Vorlex, and Zinophos.

Certain fungicides are applied to the soil as dusts, drenches, or granules to control damping off, seedling blights, crown and root rots, and other diseases. Such fungicides include captan, diazoben, PCNB, and chloroneb.

TREATMENT OF TREE WOUNDS

Large pruning cuts and wounds made on the bark of branches and trunks accidentally or in the process of removing infections by fungi and bacteria, need to be protected from drying and from becoming ports of entry of new pathogens. Drying of the margins of large tree wounds is usually prevented by painting them with shellac or any commercial wound dressing. The exposed wood is then sterilized by swabbing it with a solution of either 0.5 to 1.0 percent sodium hypochlorite (10 to 20 percent Clorox), or with 70 percent ethyl alcohol. Finally, the entire wound is painted with a permanent-type, tree wound dressing, such as a 10:2:2 mixture of lanolin, rosin, and gum, or Cerano, or Bordeaux paint, or an asphalt-varnish tree paint. Some wound dressings, e.g., Cerano and Bordeaux paint, are themselves disinfectants, while most others require the addition of a disinfectant, such as 0.25 percent phenyl mercuric nitrate or 6 percent phenol.

CONTROL OF
POSTHARVEST DISEASES

The use of chemicals for the control of postharvest diseases of fruits and vegetables is complicated enormously by the fact that most compounds effective against storage diseases leave on the produce high concentrations of residues that are toxic to consumers. Many chemicals also cause injury to the products under storage conditions and give off undesirable odors.

A number of fungitoxic chemicals, however, most of them used specifically for control of postharvest diseases, have been developed. Most of these are used as dilute solutions into which the fruits or vegetables are dipped before storage, or as solutions used for washing or hydrocooling of fruits and vegetables immediately after harvest. Some chemicals, e.g., elemental sulfur, are used as dusts or crystals that undergo sublimation in storage, and others, e.g., SO_2, as gases. Finally, some chemicals are impregnated in the boxes or wrappers containing the fruit. Among the compounds used for commercial control of postharvest diseases of, primarily, citrus fruits but also of other fruits are borax, biphenyl, sodium o-phenylphenate, and thiabendazole. Certain other chemicals, such as elemental sulfur, sulfur dioxide, dichloran, captan, and benzoic acid, have been used mostly for the control of storage rots of stone and pome fruits, bananas, grapes, strawberries, melons, potatoes, etc.

DISINFESTATION
OF WAREHOUSES

To avoid infection of stored products by pathogens left over in the warehouse from previous years, the storage rooms are first cleaned thoroughly and the debris is removed and burned. This is usually followed by washing the walls and floors with a copper sulfate solution (1 pound in 5 gallons of water), or by spraying with a 1:240 solution of formaldehyde. Warehouses that can be closed airtight and in which the relative humidity can be kept at nearly 100 percent while the temperature is between 25 and 30°C can be effectively fumigated with chloropicrin (tear gas) used at 1 pound per each 1000 cubic feet. Fumigation of warehouses can also be carried out by burning sulfur in the warehouse at the rate of 1 pound per 1000 cubic feet of space, or with formaldehyde gas generated by adding 23 ounces of potassium permanganate to 3 pints of formaldehyde per 1000 cubic feet of space. In all cases the fumigants should be allowed to act for at least 24 hours before the warehouse doors are opened for aeration.

CONTROL OF INSECT VECTORS

When the pathogen is introduced or disseminated by an insect vector, control of the vector is as important as, and sometimes easier than, the control of the pathogen itself. Application of insecticides for the control of insect carriers of fungus spores and bacteria has been fairly successful and is a recommended procedure in the control of several such insect-carried pathogens.

In the case of viruses, mycoplasmas, and rickettsialike bacteria, however, of which insects are the most important disseminating agents, insect control has been helpful in controlling the spread of their diseases only when it has been carried out at the area and on the plants on which the insects overwinter, or on which the insects feed before they enter the crop. Control of such diseases by killing the insect vectors with insecticides after they have arrived at the crop has seldom proved adequate. This is probably because, even with good insect control, enough insects survive for sufficiently long periods to spread the pathogen. Nevertheless, appreciable reduction in losses from certain such diseases has been obtained by controlling their insect vectors, and the practice of good insect control is always desirable.

TYPES OF CHEMICALS USED FOR PLANT DISEASE CONTROL

Many hundreds of chemicals have been advanced to date for crop protection as fumigants, soil treatments, sprays, dusts, paints, pastes, and systemics. The most important of these chemicals and some of their properties and uses are described below.

COPPER COMPOUNDS

Bordeaux mixture, the product of reaction of copper sulfate and calcium hydroxide (lime), is the most widely used copper fungicide all over the world. It controls many fungus and bacterial leaf spots, blights, anthracnoses, downy mildews, and cankers, but causes burning of leaves or russeting of fruit such as apples when applied in cool, wet weather. The phytotoxicity of Bordeaux is reduced by increasing the ratio of lime to copper sulfate, since copper is the only ingredient in the Bordeaux mixture that is toxic to pathogens and, sometimes, to plants, while lime's role is primarily that of a "safener." For dormant sprays, concentrated Bordeaux is made by mixing 10 pounds of copper sulfate, 10 pounds of lime, and 100 gallons of water; it has the formula 10:10:100. The most commonly used formula for Bordeaux is 8:8:100. For spraying young, actively growing plants the amounts of copper sulfate and lime are reduced, and the formulas used may be 2:2:100, 2:6:100, etc. For plants known to be sensitive to Bordeaux, a much greater concentration of lime may be used, as in the formula 8:24:100.

Fixed copper. In the "fixed" or "insoluble" copper compounds the copper ion is only slightly soluble, and these compounds are, therefore, less phytotoxic than Bordeaux, but also less effective as fungicides. The "fixed" coppers are used for control of the same diseases as Bordeaux and they can also be used as dusts. The "fixed" coppers contain either basic copper sulfate (e.g., Basicop, Tribasic), or basic copper chlorides (e.g.,

C–O–C–S), or copper oxides (e.g., Cuprocide, Perenox), or miscellaneous other formulations. Most of them are recommended as sprays at the rate of 4 pounds per 100 gallons of water or as 7 percent copper dusts.

Kocide is a copper formulation fungicide and bactericide that contains cupric hydroxide and dissolves readily in water. It controls the same diseases as the other copper compounds but, because it is water soluble, it does not clog spray nozzles as often.

SULFUR COMPOUNDS

Several inorganic sulfur formulations and numerous organic sulfur compounds have proven to be excellent fungicides and are used to control a variety of diseases.

INORGANIC SULFUR COMPOUNDS *Sulfur.* The element sulfur as a dust, wettable powder, paste, or liquid is used primarily to control powdery mildews on many plants, but it is also effective against certain rusts, leaf blights, and fruit rots. Sulfur, in its different forms, is available under a variety of trade names, such as Kolodust, Microfine sulfur, Magnetic spray wettable sulfur, Micronized sulfur, Kolofog, etc. Most sulfur formulations are applied at the rate of 1 to 6 pounds per 100 gallons of water and may cause injury in hot (temperatures above 30°C), dry weather, especially to sulfur-sensitive plants such as tomato, melons, and grape.

Lime-sulfur. By boiling lime and sulfur together Lime-Sulfur, Self-Boiled Lime-Sulfur, and Dry Lime-Sulfur are produced which are used as sprays for dormant fruit trees to control blight or anthracnose, powdery mildew, apple scab, brown rot of stone fruits, peach leaf curl, etc., and are sometimes used for summer control of the same diseases. The various lime-sulfurs are applied at the rate of 2 to 10 gallons per 100 gallons of water.

ORGANIC SULFUR COMPOUNDS—CARBAMATES The organic sulfur compounds comprise unquestionably the most important, most versatile, and most widely used group of modern fungicides. They include thiram, ferbam, ziram, nabam, maneb, and zineb and are all derivatives of dithiocarbamic acid.

Thiram consists of two molecules of dithiocarbamic acid joined together. It is used mostly for seed and bulb treatment for vegetables, flowers, and grasses, but also for the control of certain foliage diseases, e.g., rusts of lawn, fruits, and vegetables. Thiram is also good as soil drench for control of damping off and seedling blights. Thiram, in various formulations, is sold under many trade names: Thiram, Arasan, Tersan, Spottrete, Thylate, Ortho Lawn and Turf Fungicide, etc.

Ferbam consists of three molecules of dithiocarbamic acid reacted to one atom of iron. Ferbam is used to control many foliage diseases of fruit trees and ornamentals. It is sold as Fermate, Ferbam, Karbam Black, Coromate, Carbamate, etc.

Ziram contains two molecules of dithiocarbamic acid joined to zinc. It is sold as Zerlate, Karbam White, Corozate, etc., and controls many foliage diseases of vegetables and ornamentals; especially safe for tender seedlings.

Another group of dithiocarbamic acid derivates with different molecular configurations contains the fungicides nabam (Na), zineb (Zn) and maneb (Mn). *Nabam* is sold as Dithane D-14, Parzate Liquid, etc., and gives fair control of some foliage diseases of flowers and vegetables. Nabam, ziram, and even ferbam have been largely replaced by newer organic fungicides.

Zineb is sold as Dithane Z-78, Parzate, etc.; it is an excellent, safe, multipurpose foliar and soil fungicide for the control of leaf spots, blights, fruit rots, etc. of vegetables, flowers, fruit trees, and shrubs.

Maneb contains manganese, it is sold as Manzate, Dithane M-22, Tersan LSR, etc., and is an excellent, broad-spectrum fungicide for the control of foliage and fruit diseases of many vegetables, especially tomato, potato, and vine crops, and of flowers, trees, turf, and some fruits. Maneb is one of the most frequently used fungicides for control of vegetable diseases. Maneb is often mixed with zinc or with zinc ion and results in the formulations known as maneb–zinc (sold as Manzate D or Dithane M-22 Special) and as zinc ion maneb called mancozeb (sold as Manzate 200, Dithane M-45, and Fore). The addition of zinc reduces the phytotoxicity of maneb and improves its fungicidal properties.

QUINONES

Quinones, which occur naturally in many plants and are also produced upon oxidation of plant phenolic compounds, often show antimicrobial activity and are often considered to be associated with the innate resistance of plants to disease. Only two quinone compounds, chloranil and dichlone, however, have been developed and are used commercially as fungicides.

Chloranil is sold as Spergon; it is used mainly as seed and bulb treatment for flowers, vegetables and some grasses. It is also used as a soil drench, as a dip for flower corms and bulbs, and as sprays and dusts for certain foliage diseases, e.g., downy mildews of melons, damping off.

Dichlone is sold as Phygon, Phygon XL, etc., and is used mainly as a seed treatment for certain vegetables and grasses. Dichlone is also used as a protectant or eradicant spray for certain blights, fruit rots, and cankers of vegetables and fruit.

BENZENE COMPOUNDS

Many rather unrelated compounds that have a benzene ring are toxic to microorganisms, and several have been developed into fungicides and are used commercially.

Dinitro-o-cresol is contained in the formulations called Elgetol, Krenite, etc., and is used as a dormant spray for control of certain diseases of fruit and ornamental trees, as a preemergence ground spray, and as a tree wound treatment.

Hexachlorobenzene, or HCB, is used as seed treatment for the control of seed- and soil-borne bunt in wheat and other grains.

Pentachloronitrobenzene, sold as PCNB, Terrachlor, etc., is a long-lasting soil fungicide. It controls various soil-borne diseases of vegetables, turf, and ornamentals and is applied as a dip or in the furrow at planting time. It is used primarily against *Rhizoctonia*, *Sclerotinia*, and *Plasmodiophora* but has no effect on *Pythium*.

Dichloran, sold as Botran, DCNA, etc., is used as a foliar, fruit and soil fungicide for diseases of vegetables and flowers caused mostly by sclerotia-producing fungi, and as postharvest dip or spray for fruits, vegetables and flowers affected by the same fungi or by *Rhizopus* and *Penicillium*.

Dinocap, sold as Karathane, Mildex, etc., is specific against powdery mildews. It also suppresses mites.

Diazoben, sold as Dexon, is used as a seed and soil fungicide against damping-off and root rots of many ornamentals, vegetables, and fruits caused by *Pythium*, *Aphanomyces* and *Phytophthora*.

Chlorothalonil, available as Bravo, is an excellent broad-spectrum fungicide against many leaf spots, blights, downy mildews, rusts, anthracnoses, scabs, fruit rots of many vegetables, field crops and ornamentals and even trees. Another formulation of chlorothalonil is sold as Daconil 2787 and is used primarily against foliage diseases of turf grasses and of some ornamentals. A tablet formulation called Termil is thermally dispersed in greenhouses for control of *Botrytis* on many ornamentals and for several leaf molds and blights of tomato.

HETEROCYCLIC COMPOUNDS

This is a rather heterogeneous group but includes some of the best fungicides, e.g., captan.

Captan is sold as Captan, Orthocide, etc.; it is an excellent, safe fungicide for control of leaf spots, blights, fruit rots, etc. on fruits, vegetables, ornamentals and turf. It is also used as a seed protectant for vegetables, flowers and grasses and as a postharvest dip for certain fruits and vegetables.

Folpet is sold as Folpet, Phaltan, Orthophaltan, etc.; it is similar to captan in spectrum and effectiveness. In addition, it controls many powdery mildews.

Captafol is sold as Difolatan, Ortho Difolatan, etc., and has properties similar to those of captan and folpet. Moreover, Difolatan exhibits unusual resistance to weathering which provides extended redistribution and residual activity. These properties, combined with its low phytotoxicity, also allow the use of up to three times the regular amount of Difolatan as a single application treatment (SAT) on apples against apple scab, cherry leaf spot, citrus melanose and scab, and against several foliage diseases of tomato. Such concentrated sprays may provide protection for longer periods and reduce the number of sprays needed.

Glyodin is a liquid fungicide with excellent wetting and sticking properties. It is sold as Glyodin, Crag Glyodin, etc., and is effective against apple scab and certain other foliar diseases of fruit trees and ornamentals. It is often combined with dodine (Glyodex).

Dyrene is sold as Dyrene, Turftox, etc., it is used for spraying ornamentals, turf, and vegetables.

SYSTEMIC FUNGICIDES

Systemic fungicides are absorbed through the foliage or roots and are translocated upward internally by the plant through the xylem. Systemic fungicides generally move upward in the transpiration stream and may accumulate at the leaf margins, while downward translocation in the phloem is rare or does not occur at all. They are not reexported to new growth. Some of them become systemically translocated when sprayed on herbaceous plants but most are only locally systemic on the sprayed leaves. Many systemics are most effective when applied as seed treatments, root-dip, in-furrow treatment or soil drench, and in trees when injected into the trunks.

Several systemic fungicides are presently available in the market and many more are in the experimental stage. Systemic fungicides belong to at least three groups of compounds although some of them are not related to these groups.

OXANTHIINS   They include primarily carboxin and oxycarboxin and are effective against some smut and rust fungi and against *Rhizoctonia.*

Carboxin is sold as Vitavax. It is used as a seed treatment and is effective against damping-off diseases caused by *Rhizoctonia* and the various smuts of grain crops.

Oxycarboxin is sold as Plantvax. It is used as a seed or foliar treatment and is effective in controlling a wide variety of rust diseases.

BENZIMIDAZOLES   They include the most important, up to now, systemic fungicides such as benomyl, thiabendazole, and thiophanate. They are effective against numerous types of diseases caused by a wide variety of fungi.

Benomyl is sold as Benlate, Tersan 1991, etc. It is a safe, broad-spectrum fungicide against a large number of important fungus pathogens and it also suppresses mites. It controls a wide range of leaf spots and blotches, blights, rots, scabs, plus seed- and soil-borne diseases. Benomyl is particularly effective for powdery mildew of all crops; scab of apples, peaches, and pecans; brown rot of stone fruits; fruit rots in general; *Cercospora* leaf spots; cherry leaf spot; black spot of roses; blast of rice; various *Sclerotinia* and *Botrytis* diseases; loose and cover smuts of wheat. It is highly active against and suppresses infections by *Rhizoctonia, Thielaviopsis, Ceratocystis, Fusarium,* and *Verticillium.* It has no effect on Phycomycetes, on some dark-spored Imperfects such as *Helminthosporium* and *Alternaria,* on some Basidiomycetes, nor on bacteria. Benomyl may be applied as seed treatment, foliar spray, trunk injection, root dip or row treatment and as a fruit dip. Benomyl seems to be mutagenic and to hasten the appearance of pathogen races resistant to it.

Thiabendazole is sold as Mertect, Tobaz, etc. It is also a safe, broad-spectrum fungicide and effective against many Imperfect fungi causing leaf spot diseases of turf and ornamentals, diseases of bulbs and corms. It is commonly used as a postharvest treatment for the control of storage rots of citrus, apples, pears and banana, potato and squash.

Thiophanate ethyl, under the trade names Topsin, Cercobin, and Cleary, is effective against several root and foliage fungi affecting turfgrasses.

Thiophanate methyl, under the trade names Fungo, Topsin M, Cerco-bin M, Zyban, Chipco Spot Klean, etc., is a broad-spectrum preventive and curative fungicide for use on turf and as a foliar spray to control powdery and downy mildews, *Botrytis* diseases, numerous leaf and fruit spots, scabs and rots. Also used as a soil drench or dry soil mix to control soil-borne fungi attacking bedding plants, foliage plants, and container-grown plants.

PYRIMIDINES   They include primarily the compounds dimethirimol, ethirimol and triarimol which are effective against powdery mildews and certain other fungi. They are used primarily as seed and soil treatments.

MISCELLANEOUS SYSTEMICS

Chloroneb, sold as Demosan, Tersan SP, or Chloroneb, is a seed and soil fungicide for turf and ornamentals.

Ethazole, sold as Truban, Terrazole, Koban, is a soil, seed, and turf fungicide that controls *Pythium* and *Phytophthora* damping off and seed-ling blights of ornamental and nursery crops. Often sold combined with PCNB (Terrachlor-Super-X, Terra-Coat), or with thiophanate methyl (Banrot) for broader spectrum.

Triforine, effective against many foliar diseases of fruit trees and or-namentals.

MISCELLANEOUS
ORGANIC FUNGICIDES

A number of other, chemically diverse compounds are excellent protec-tant fungicides for certain diseases or groups of diseases.

Dodine is sold as Cyprex. It is an excellent fungicide against apple scab, and it also controls certain foliage diseases of cherry, strawberry, pecan, and roses. It gives long-lasting protection and is also a good eradi-cant. It appears to have local systemic action in leaves. Strains of the apple scab fungus resistant to dodine have appeared and predominate in some areas.

Fentin hydroxide, sold as Du-Ter, is a broad-spectrum fungicide with activity against many leaf spots, blights, and scabs. It also has suppressant or antifeeding properties on many insects.

Polyram is a foliar and seed protectant fungicide. It controls rusts, downy mildews, leaf spots and blights of vegetables, ornamentals, and certain trees.

Oxyquinoline sulfate (also benzoate and citrate) is used as a soil drench to control damping off and other soil-borne diseases. An oxyquinoline–copper complex has also been used as a seed treatment, as a foliar spray against certain diseases of fruits and vegetables, and as a wood preservative for packing boxes, baskets, crates, etc.

Two cadmium-containing fungicides, Caddy (cadmium chloride) and Cadminate (cadmium succinate), are used for control of turf diseases.

Zinc is sometimes used as zinc naphthenate for disinfestation and preservation of wood.

SOIL FUMIGANTS

They are used primarily for control of plant-parasitic nematodes and are discussed in the chapter on "Plant Diseases Caused by Nematodes."

ANTIBIOTICS

Antibiotics are substances produced by one microorganism and toxic to another microorganism. Most antibiotics known to date are products of Actinomycetes and some fungi, e.g., *Penicillium*, and are toxic mostly against bacteria, including rickettsialike bacteria, mycoplasmas, and also against some fungi. The chemical formulas of most antibiotics are complex and are not, as a rule, related to each other. Antibiotics used for plant disease control are generally absorbed and translocated systemically by the plant. Antibiotics may control plant diseases by acting on the pathogen or on the host directly, or after undergoing transformation within the host.

Among the most important antibiotics in plant disease control are streptomycin, tetracyclines, and cycloheximide.

Streptomycin is produced by the actinomycete *Streptomyces griseus*. Streptomycin or streptomycin sulfate is sold as Agrimycin, Phytomycin, Ortho Streptomycin, Agri-Strep, etc., and as a spray shows activity against a broad range of bacterial plant pathogens causing spots, blights, wilts, rots, etc. Streptomycin is also used as a soil drench, e.g., in the control of geranium foot rot caused by *Xanthomonas* sp., as a dip for potato seed pieces against various bacterial rots of tubers, and as a seed disinfectant against bacterial pathogens of beans, cotton, crucifers, cereals, etc. Moreover, streptomycin is effective against several phycomycetous fungi especially *Pseudoperonospora humuli*, the cause of downy mildew of hops.

Tetracyclines are antibiotics produced by various species of *Streptomyces* and are active against many bacteria, against all mycoplasmas and against some rickettsialike bacteria. Of the tetracyclines, Terramycin (oxytetracycline), Aureomycin (chlortetracycline), and Achromycin (tetracycline) have been used to some extent for plant disease control. Oxytetracycline is often used with streptomycin in the control of fire blight of pome fruits. Tetracyclines injected into trees infected with mycoplasmas or rickettsialike bacteria stop the development of the disease and induce remission of symptoms, i.e., the symptoms disappear and the trees resume growth as long as some tetracycline is present in the trees. Usually one injection at the end of the growing season is sufficient for normal growth of the tree during the following season.

Cycloheximide is produced by *Streptomyces griseus* and is obtained as a by-product in the production of streptomycin. It is sold as Actidione, Actispray, Actidione PM, Actidione RZ, etc., and is effective against many phytopathogenic fungi. Cycloheximide is used for the control of many turf diseases, and of cherry leafspot, caused by *Coccomyces hiemalis*. It is also effective against powdery mildews of many crop and ornamental plants, but its high phytotoxicity limits its usefulness appreciably.

GROWTH REGULATORS

Certain plant hormones have been shown to reduce infection of plants by certain pathogens, e.g., tomato by *Fusarium*, potato by *Phytophthora*, through the increase by these substances of the disease resistance of the host. In tobacco plants treated with maleic hydrazide, a growth retardant, the rootknot nematode, *Meloidogyne*, is unable to induce giant cell formation and is thereby prevented from completing its life cycle and from causing disease. Kinetin treatment of leaves, before or shortly after inoculation with a virus, also reduces virus multiplication, number and size of lesions on local-lesion hosts, and postpones the onset of systemic symptoms and death of the plant. Stunting and axillary bud suppression associated with certain virus and mycoplasma diseases of plants can be overcome with sprays of gibberellic acid. Although treatments with various growth regulators have given encouraging control of some plant diseases in experimental trials, only gibberellic acid sprays are used somewhat for the field control of sour cherry yellows virus on cherries.

MECHANISMS OF ACTION OF CHEMICALS USED TO CONTROL PLANT DISEASES

The complete mechanisms by which the various chemicals applied to plants control plant diseases are yet unknown for most of the chemicals. Some chemicals seem to reduce infection by increasing the resistance of the host to the pathogen. This may be brought about by altering the constitution of the cell walls of the host, by limiting the availability of essential coenzymes in the host, or by altering the rate or the direction of metabolism in the host, which may thus be in a better position to defend itself against the pathogen.

The great majority of chemicals are used for their toxicity directly on the pathogen and are effective only as protectants at the points of entry of the pathogen. Such chemicals act by inhibiting the synthesis by the pathogen of certain of its cell wall substances, by acting as solvents of, or otherwise damaging, the cell membranes of the pathogen, by forming complexes with, and thus inactivating, certain essential coenzymes of the pathogen, or by inactivating enzymes and causing general precipitation of the proteins of the pathogen.

The systemic fungicides and antibiotics are absorbed by the host, are translocated internally through the plant, and are effective against the pathogen at the infection locus. Such chemicals are called chemotherapeutants, and control of plant diseases with such chemicals is called chemotherapy. Once in contact with the pathogen, chemotherapeutants seem to affect pathogens in ways similar to those mentioned above for the nonsystemic chemicals, but systemic fungicides are much more specific in that they apparently affect only one function in the pathogen rather than a variety of them. As a result, new pathogen races resistant to one or another of the systemic fungicides have already appeared.

RESISTANCE
OF PATHOGENS TO
CHEMICALS

Just as human pathogens resistant to antibiotics, and insects and mites resistant to certain insecticides and miticides appeared rather soon after continuous and widespread use of these chemicals, several plant pathogens have also developed strains that are resistant to and therefore unaffected by certain fungicides. For many years, when only protectant fungicides such as thiram, maneb, or captan were used, no such resistant strains were observed, presumably because these fungicides affect several vital processes of the pathogen and it would take too many gene changes to produce a resistant strain. Resistance to some fungicides, all of which contained a benzene ring, began to appear in the 1960s when *Penicillium* strains resistant to diphenyl, *Tilletia* strains resistant to hexachlorobenzene, and *Rhizoctonia* strains resistant to PCNB were found to occur naturally. In some areas these strains became major practical problems. Later, a strain of *Venturia inaequalis* (cause of apple scab) appeared that was resistant to dodine and that excellent chemical became ineffective against the fungus over a large area.

Strains of *Erwinia amylovora*, the fire blight bacterium, that were resistant to the systemic antibiotic streptomycin, had been known for several years. However, it was the introduction and widespread use of the systemic fungicides, especially benomyl, that really triggered the appearance of strains of numerous fungi resistant to one or more of these fungicides. In some cases, strains resistant to the fungicide appeared and became widespread after only two years of use of the chemical, and the chemical had to be abandoned. To date, several of the important fungal pathogens, e.g., *Cercospora*, *Fusarium*, *Sphaerotheca*, *Aspergillus*, *Penicillium*, and *Ustilago*, are known to have produced strains resistant to one or more of the systemic fungicides and it appears that resistant strains of other fungi can be expected to develop in the future. This is apparently because systemic fungicides are specific in their action, i.e., they affect only one or perhaps two steps in a genetically controlled event in the metabolism of the fungus and, as a result, a resistant population can arise quickly either by a single mutation or by selection of resistant individuals in a population.

Good systemic or nonsystemic fungicides that become ineffective because of the appearance of new resistant strains can still be saved, and the resistant strains can still be controlled, either by using mixtures of specific systemic and wide-spectrum protectant fungicides; or by alternating sprays with systemic and protectant fungicides; or by spraying during half of the season with systemic and the other half with protectant fungicides. In each of these schedules, the systemic or specific action chemical carries most of the weight in controlling the disease while the protectant or nonspecific chemical eliminates any strains of the pathogen that may develop resistance to the systemic or specific action chemical.

RESTRICTIONS ON CHEMICAL CONTROL OF PLANT DISEASES

Most chemicals used to control plant diseases are much less toxic than most insecticides but they are, nevertheless, poisonous substances and some of them, especially the nematicides, are extremely poisonous. For this reason, a number of restrictions are imposed in the licensing, registration and use of each chemical.

In the United States, both the Food and Drug Administration (FDA) and the Environmental Protection Agency (EPA) keep a close watch on the registration, production and use of pesticides. It is estimated that only 1 out of 10,000 new compounds synthesized by the pesticide industry turns out to be a successful pesticide and it takes 4 to 8 years and 5 to 6 million dollars from initial laboratory synthesis to government registration. In the meantime, exhaustive biological tests, field testing, crop residue analyses, toxicological tests, environmental impact studies, etc., are carried out. If the compound meets all requirements it is then approved for use on specific food or nonfood crops for which data have been obtained. Clearance must be obtained separately for each crop and each use (e.g., seed treatment, spray, soil drench) for which the chemical is recommended.

Once a chemical is approved for a certain crop, then two important restrictions on the use of the chemical must be observed: (1) the number of days before harvest that use of a particular chemical on the crop must stop; and (2) the amount of the chemical that can be used per acre must not exceed a certain amount. If either of these restrictions are not observed it is likely that, at harvest, the crop, especially vegetables and fruits, carries on it a greater amount than is allowed for the particular chemical, and the crop can be seized. All recommendations contained in bulletins published by the Extension Service are within the tolerances established by FDA and EPA and should be followed carefully.

INTEGRATED CONTROL OF PLANT DISEASES

Control of plant diseases is most successful and economical when all available pertinent information regarding the crop, its pathogens, the environmental conditions expected to prevail, locality, availability of materials, costs, etc., are taken into account in developing the control program. Usually, an integrated control program is aimed against all diseases affecting a crop, e.g., apples, citrus, bananas, potato, beans, etc. Sometimes an integrated control program is aimed against a particularly destructive and common disease, e.g., apple scab, potato late blight, etc.

In an integrated control program of an orchard crop, e.g., apple, peach, or citrus, one must first consider the nursery stock to be used and the location where it will be planted. If the fruit tree is susceptible to certain viruses, mycoplasmas, crown gall bacteria, nematodes, etc., the nursery stock (both the rootstock and the scion) must be free of these pathogens. Stock free of certain viruses and other diseases can usually be bought from selected nurseries whose crops are inspected and certified. If the possibility of nematodes on the roots exists, the stock must be fumigated. The location where the trees will be planted must not be infested with fungi such as *Phytophthora*, *Armillaria*, or serious and numerous nematodes; if it is, it should be treated with fumigants before planting, and varieties grafted on rootstocks resistant to these pathogens should be preferred. The drainage of the location should be checked and improved, if necessary. Finally, the young trees should not be planted between or next to old trees that are heavily infected with canker fungi and bacteria, insect-transmitted viruses and mycoplasmas, pollen-transmitted viruses or with other pathogens.

Once the trees are in place and until they begin to bear fruit they should be fertilized, irrigated, pruned, and sprayed for the most common insects and diseases so that they will grow vigorously and free of infections. Later on, when the trees bear fruit, the care should increase, as should the vigilance to detect and control the diseases that affect any part of the tree. Any trees that develop symptoms of an infectious disease caused by virus, mycoplasma, etc. should be removed as early as possible.

Disease control in an orchard may begin in the winter, when dead twigs, branches or fruit are removed during pruning operations and are buried or burned to reduce the amount of fungal or bacterial primary inoculum that will start infections in the spring. Pruning shears and saws should be disinfected before moving to new trees to avoid spreading any pathogens from tree to tree. Because many fungi and bacteria (as well as insects and mites) are activated in the spring by the same weather conditions that make buds to open, a "dormant" spray, containing a fungicide–bactericide (e.g., Bordeaux mixture), or a plain fungicide plus a miticide–insecticide (e.g., Superior oil), is applied before bud break. After that, as the buds open, the blossoms and leaves that are revealed are usually very susceptible to either fungal or bacterial pathogens, or both, depending on what is present in the particular area. Therefore these organs (blossoms and leaves) must be protected with sprays containing a fungicide and/or a bactericide and, possibly, an insecticide that does not harm bees and/or miticide. It is usually possible to find effective materials compatible with each other and so all of them can be mixed in the same tank and sprayed at once. If one compound, however, must be used to control an existing disease but is incompatible with the other compounds, then a separate spray will be needed. Because flowers appear over a period of several days and the leaves enlarge rapidly at that stage, and because many fungi release their spores and bacteria ooze out most abundantly during and soon after bloom, the blossoms and leaves may have to be sprayed frequently (every 3 to 5 days) so that they will be protected by the fungicide and/or bactericide, especially if it rains often and stays wet for many hours. Insecticides and miticides may still have to be used with the fungicide but these insecticides must not be toxic to bees which must be allowed to pollinate the flowers. The frequent sprays usually continue as long as there are spores being released by fungi, or bacteria oozing out, as long as the weather stays wet, and as long as there are growing plant tissues. Combining weather forecasting with disease control is most helpful.

Once blossoming is over, young fruit appear and these may or may not be affected by the same pathogens and insects as the flowers and leaves. If they are, the same spray schedule with the same materials continues as long as there is inoculum around. But often new pathogens and insects may attack the fruit and the schedule must be adjusted and materials must be included that control the new pathogens.

Usually, fruit becomes susceptible to several fruit-rotting fungi that attack fruit from the early maturity stages through harvest and in storage. Therefore, fruit must be sprayed every 10 to 14 days with materials that will control these fungi until harvest. Most fruit rots start at wounds made by insects and therefore insect control must continue. Also, wounding of fruit during harvesting and handling must be avoided to prevent fungus infections. Fruit-picking baskets and crates must be clean, free of rotten debris which may harbor fruit-rotting fungi, and the packing-house and warehouse must also be clean, free of debris, and preferably fumigated with formaldehyde, sulfur dioxide, etc. Harvested fruit is often washed in a water solution containing a fungicide to further protect the fruit during storage and transportation. During packing, infected fruit is removed and discarded. Storage and transportation, of course, should be refrigerated so that any existing infections will develop slowly and no new infections will get started.

In an integrated control program of an annual crop, e.g., potatoes, one must again start with healthy stock and must plant it in a suitable field. Potato tuber seed may carry several viruses, the late blight fungus, ring-rot bacteria, and several other fungi, bacteria and nematodes. Therefore, starting with clean, disease-free seed is of paramount importance. Certified potato seed is usually free of most such important pathogens and is produced under strict quarantine and inspection rules that guarantee seed free of these pathogens. Healthy seed must then be planted in a field free of old potato tubers that may harbor some of the above pathogens, free of *Verticillium*, *Fusarium*, the root knot nematode, etc. It is best not to follow a potato crop with another and rotation with legumes, corn or other unrelated crops will usually reduce the populations of potato pathogens. Any potato cull piles should be destroyed or sprayed to insure that no *Phytophthora* sporangia will be blown from there to the potato plants in the field later on. Tubers are cut with disinfected knives to reduce spread of ring rot among seed pieces and the seed pieces are usually treated with a fungicide, a bactericide, and an insecticide to protect them from pathogens on their surface or in the soil. The soil may have to be treated with a fumigant if it is known to be infested with the root knot or other nematodes, *Fusarium* or *Verticillium*. The seed pieces are planted at a date when their sprouts are expected to grow quickly since slow growing sprouts in cool weather are particularly susceptible to *Rhizoctonia* attack. The field must, of course, have good drainage to avoid damping off, seed-piece rot, and root rots.

A few weeks after the young plants have emerged they become susceptible to attack by early blight (*Alternaria*) and late blight (*Phytophthora infestans*). If the diseases occur regularly year after year, in addition to using resistant varieties, the grower should start spraying with the appropriate fungicides as soon as the disease appears or even before, and should continue the sprays, especially for late blight, throughout the season whenever the weather is cool and damp. Insecticide sprays control insects and may reduce spread of viruses. Using weather data to forecast disease appearance and development can help in spraying at the right time and in not wasting any sprays. Before harvest, the infected vines must be killed with chemicals to destroy late blight inoculum that could come in contact with the tubers when they are dug up. Tubers must be harvested carefully to avoid wounding that would allow storage-rot fungi such as *Fusarium* and *Pythium* to gain entrance into the tuber. The tubers must then be sorted, and the damaged ones discarded. The healthy tubers are stored at about 15°C for the wounds to heal and then at about 2°C to prevent development of fungus rots in storage. Storage rooms must of course be cleaned and disinfested before the tubers are brought in. Potato cull piles should not be kept near the field but should either be burned or buried as soon as possible.

Thus, in an integrated control program several control methods are employed including regulatory inspections for healthy seed or nursery crop production, cultural practices (crop rotation, sanitation, pruning, etc.), biological control (resistant varieties), physical control (storage temperature), and chemical controls (soil fumigation, seed or nursery stock treatment, sprays, disinfestation of cutting tools, crates, warehouses, washing solution). Each one of these measures must be taken for best results, and the routine use of each of them makes all of them that much more effective.

SELECTED REFERENCES

American Phytopathological Society. 1965–1977. Fungicide–nematicide tests. St. Paul, Minnesota.

Brandes, G. A. 1971. Advances in fungicide utilization. *Ann. Rev. Phytopathol.* **9**:363–386.

Browning, J. A., and K. J. Frey. 1969. Multiline cultivals as a means of disease control. *Ann. Rev. Phytopathol.* **7**:355–382.

Bruehl, G. W. (Ed.). 1975. "Biology and Control of Soil-Borne Plant Pathogens." The American Phytopathological Society. St. Paul, Minnesota. 216 p.

Dekker, J. 1976. Acquired resistance to fungicides. *Ann. Rev. Phytopathol.* **14**:405–428.

Eckert, J. W., and N. F. Sommer. 1967. Control of diseases of fruits and vegetables by postharvest treatment. *Ann. Rev. Phytopathol.* **5**:391–432.

Erwin, D. C. 1973. Systemic fungicides: Disease control, translocation, and mode of action. *Ann. Rev. Phytopathol.* **11**:389–422.

Evans, E. 1968. "Plant Diseases and Their Chemical Control." Blackwell Scientific Publications, Oxford and Edinburgh. 288 p.

Georgopoulos, S. G., and C. Zaracovitis. 1967. Tolerance of fungi to organic fungicides. *Ann. Rev. Phytopathol.* **5**:109–130.

Gram, E. 1960. Quarantines, *in* "Plant Pathology" (J. G. Horsfall and A. E. Dimond, eds.), Vol. 3, pp. 314–356. Academic Press, New York.

Hollings, M. 1965. Disease control through virus-free stock. *Ann. Rev. Phytopathol.* **3**:367–396.

Horsfall, J. G. 1956. "Principles of Fungicidal Action." Chronica Botanica, Waltham, Massachusetts. 279 pp.

Horsfall, J. G., and E. B. Cowling (eds.). 1977. "Plant Disease: An Advanced Treatise. Vol. 1: How Disease Is Managed." Academic Press, New York, 465 p.

Lewis, F. H., and K. D. Hickey. 1972. Fungicide usage on deciduous fruit trees. *Ann. Rev. Phytopathol.* **10**:399–428.

Marx, D. H. 1972. Ectomycorrhizae as biological deterrents to pathogenic root infections. *Ann. Rev. Phytopathol.* **10**:429–454.

National Academy of Sciences. (1). "Plant Disease Development and Control." 1968. (2). "Control of Plant Parasitic Nematodes." 1968. (3). "Effects of Pesticides on Fruit and Vegetable Physiology." 1968. (4). "Genetic Vulnerability of Major Crops." 1972. (5). "Pest Control Strategies for the Future." 1972. Washington, D. C.

Okabe, N., and M. Goto. 1963. Bacteriophages of plant pathogens. *Ann. Rev. Phytopathol.* **1**:397–418.

Sharvelle, E. G. 1969. "Chemical Control of Plant Diseases." University Publishing. College Station, Texas.

Shurtleff, M. C., and D. P. Taylor. 1964. Soil disinfestation. Methods and materials. *Univ. Illinois, Extension Serv. Circ.* **893**:23 pp.

Sijpesteijn, A. K., and G. J. M. van der Kerk. 1965. Fate of fungicides in plants. *Ann. Rev. Phytopathol.* **3**:127–152.

Stevens, R. B. 1960. Cultural practices in disease control, *in* "Plant Pathology" (J. G. Horsfall and A. E. Dimond, eds.), Vol. 3, pp. 357–430. Academic Press, New York.

Torgeson, D. C. (ed.). "Fungicides, An Advanced Treatise." Vol. 1 (1967), Vol. 2 (1969). Academic Press, New York.

Van der Plank, J. E. 1963. "Plant Diseases, Epidemics and Control." Academic Press, New York. 349 pp.

part 2
specific plant diseases

9

environmental factors that cause plant diseases

introduction

Plants grow best within certain ranges of the various factors that make up their environment. Such factors include temperature, soil moisture, soil nutrients, light, air and soil pollutants, air humidity, soil structure, and pH. Although these factors affect all plants growing in nature, their importance is considerably greater for the cultivated plants which are often grown by man in areas barely meeting the requirements for normal growth. Moreover, cultivated plants are frequently grown or kept in completely artificial environments (greenhouses, homes, warehouses, etc.) or are subjected to a number of cultural practices (fertilization, irrigation, spraying with pesticides, etc.) which may affect their growth considerably.

The common characteristic of noninfectious diseases of plants is that they are caused by lack or excess of something that supports life. Noninfectious diseases occur in the absence of pathogens, and cannot, therefore, be transmitted from diseased to healthy plants. Noninfectious diseases may affect plants in all their life stages, such as seed, seedling, mature plant, or fruit, and they may cause damage in the field, storage, or market. The symptoms caused by noninfectious diseases vary in kind and severity with the particular environmental factor involved and with the degree of deviation of this factor from its normal. Symptoms may range from slight to severe, and affected plants may even die.

The diagnosis of noninfectious diseases is sometimes made easy by the presence on the plant of characteristic symptoms known to be caused by the lack or excess of a particular factor (Fig. 23). At other times diagnosis can be arrived at by carefully examining and analyzing the

147

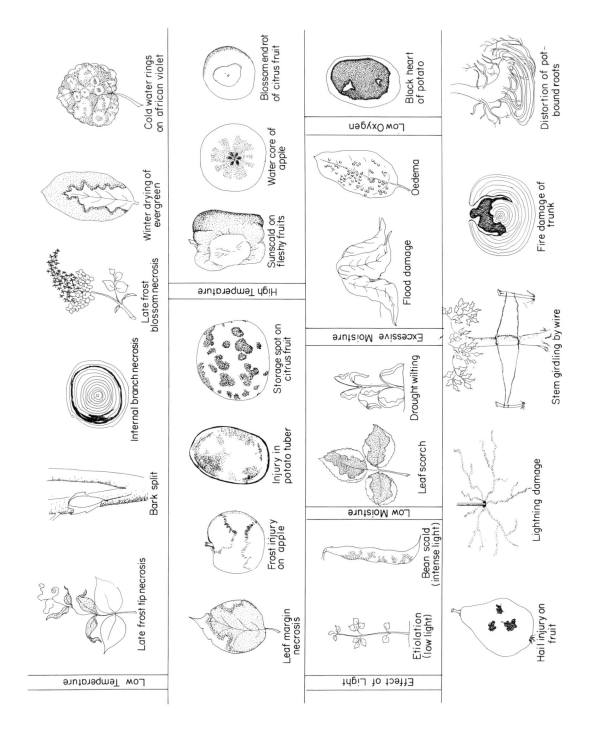

Low Temperature

Late frost tip necrosis

Bark split

Internal branch necrosis

Late frost blossom necrosis

Winter drying of evergreen

Cold water rings on african violet

Leaf margin necrosis

Frost injury on apple

Injury in potato tuber

Storage spot on citrus fruit

High Temperature

Sunscald on fleshy fruits

Water core of apple

Blossom end rot of citrus fruit

Effect of Light

Etiolation (low light)

Bean scald (intense light)

Low Moisture

Leaf scorch

Drought wilting

Excessive Moisture

Flood damage

Oedema

Low Oxygen

Black heart of potato

Hail injury on fruit

Lightning damage

Stem girdling by wire

Fire damage of trunk

Distortion of pot-bound roots

148

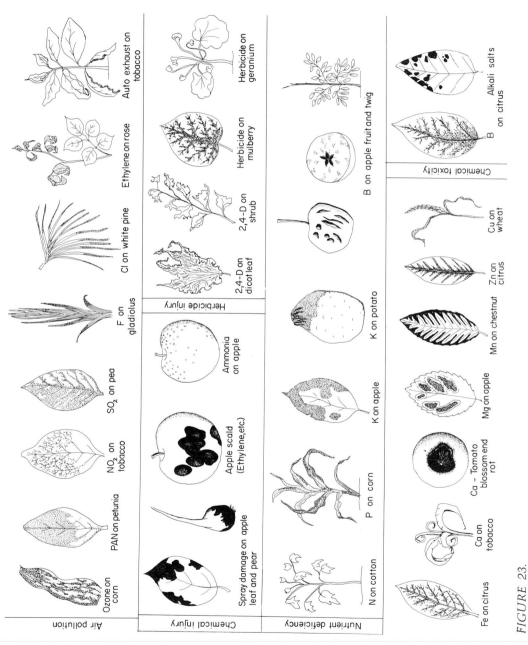

FIGURE 23.
Various types of symptoms caused by different environmental factors.

149

weather conditions prevailing before and during the appearance of the disease, recent changes in the atmospheric and soil contaminants at or near the area where the plants are growing, and the cultural practices, or possible accidents in the course of these practices, preceding the appearance of the disease. Often, however, the symptoms of several noninfectious diseases are too indistinctive and closely resemble those caused by several viruses, mycoplasmas, etc., and by many root pathogens. The diagnosis of such noninfectious diseases then becomes a great deal more complicated. One must then obtain proof of absence from the plant of any of the pathogens that could cause the disease, and must reproduce the disease on healthy plants after subjecting them to conditions similar to those thought of as the cause of the disease. To distinguish further among environmental factors causing similar symptoms, the investigator must cure the diseased plants, if possible, by growing them under conditions in which the degree or the amount of the suspected environmental factor involved has been adjusted to normal.

Noninfectious plant diseases can be controlled by avoiding the extremes of the environmental conditions responsible for such diseases, or by supplying the plants with protection or substances that would bring these conditions to levels favorable for plant growth.

temperature

Plants normally grow at a temperature range from 1 to 40°C, most kinds of plants growing best between 15 and 30°C. Perennial plants and dormant organs, such as seeds and corms, of annual plants may survive temperatures considerably below or above the normal temperature range of 1 to 40°C. The young, growing tissues of most plants, however, and the entire growth of many annual plants, are usually very sensitive to temperatures near or beyond the extremes of this range.

The minimum and maximum temperatures at which plants can still produce normal growth vary greatly with the plant species and with the stage of growth the plant is in during the low or high temperatures. Thus, plants such as tomato, citrus, and other tropical plants grow best at high temperatures and are injured severely when the temperature drops to near, or below, the freezing point. On the other hand, plants such as cabbage, winter wheat, alfalfa, and most perennials of the temperate zone can withstand temperatures considerably below freezing without any apparent ill effects to the plant. Even the latter plants, however, are injured and finally killed if the temperature drops too low.

A plant may also differ in its ability to withstand extremes in temperature at different stages of its growth. Thus, older, hardened plants are more resistant to low temperatures than are young seedlings. Also, different tissues or organs on the same plant may vary greatly in their sensitivity to the same low temperature. Buds are more sensitive than twigs, flowers and newly formed fruit are more sensitive than leaves, and so on.

Plants are generally injured faster and to a greater extent when temperatures become higher than the maximum for plant growth than when they are lower than the minimum. However, too high a temperature rarely occurs in nature. High temperature seems to cause its effects on the plant in conjunction with the effects of other environmental factors, particularly excessive light, drought, lack of oxygen, or high winds accompanied by low relative humidity. High temperatures are usually responsible for sunscald injuries (Fig. 24A) appearing on the sun-exposed sides of fleshy fruits and vegetables, such as peppers, apples, tomatoes, onion bulbs, and potato tubers. On hot, sunny days the temperature of the fruit tissues beneath the surface facing the sun may be much higher than that of those on the shaded side and of the surrounding air. This results in discoloration, water-soaked appearance, blistering, and a desiccation of the tissues beneath the skin which leads to sunken areas on the fruit surface. Succulent leaves of plants may also develop sunscald symptoms, especially when hot sunny days follow periods of cloudy, rainy weather. Irregular areas on the leaves become pale green at first but soon collapse and form brown, dry spots. This is a rather common symptom of fleshy leaved house plants kept next to windows with a southern exposure in early spring and summer when the sun's rays heat the fleshy leaves excessively. Too high a soil temperature at the soil line sometimes kills young seedlings (Fig. 24B) or causes cankers at the crown on the stems of older plants. High temperatures also seem to be involved in the water core disorder of apples (Fig. 24C) and, in combination with reduced oxygen, in the blackheart of potatoes.

Far greater damage to crops is caused by low than by high temperatures. Low temperatures, even if above freezing, may damage warm-weather plants such as corn and beans. They may also cause excessive sweetening and, upon frying, undesirable caramelization of potatoes due to hydrolysis of starch to sugars at the low temperatures.

Temperatures below freezing cause a variety of injuries to plants. Such injuries include the damage caused by late frosts to young meristematic tips (Figs. 25A, C) or entire herbaceous plants, the frost-killing of buds of peach, cherry, and other trees, and the killing of flowers, young fruit, and, sometimes, succulent twigs of most trees. Frost bands, consisting of discolored, corky tissue in a band or large area of the fruit surface, are often produced on apples, pears, etc., following a late frost (Fig. 25D). Low winter temperatures may kill young roots of trees, such as apple, and may also cause bark-splitting and canker development (Figs. 25B and 26) on trunks and large branches, especially on the sun-exposed side, of several kinds of fruit trees. Cross sections of limbs may show a black ring or a "blackheart" condition in the wood. Fleshy tissues, such as potato tubers, may be injured at subfreezing temperatures. The injury varies depending on the degree of temperature drop or the duration of the low temperature. Early injury affects only the main vascular tissues and appears as a ringlike necrosis; injury of the finer vascular elements which are interspersed in the tuber gives the appearance of netlike necrosis. With more general injury, large chunks of the tuber are damaged creating the so-called "blotch-type" necrosis (Fig. 25E).

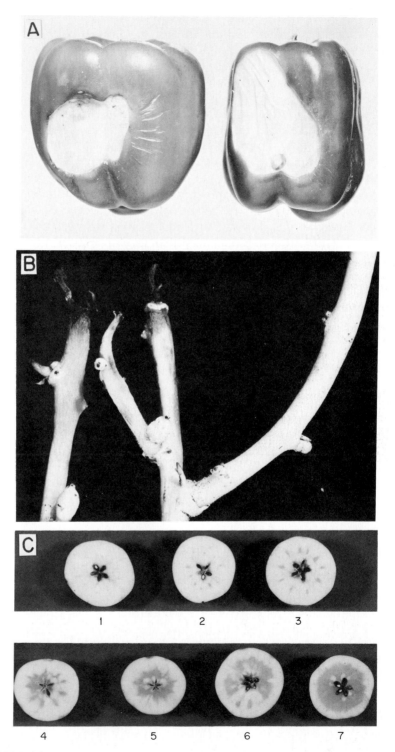

FIGURE 24.
(A) Sunscald injury on pepper fruits. (B) Potato sprouts killed at the soil line by
excessively high temperatures. (C) Stages of watercore development in Delicious
apples. 1 = healthy. (Photos: A—courtesy USDA, B—courtesy Dept. Plant Pathol.,
Cornell Univ., C—courtesy W. J. Lord.)

FIGURE 25.
(A) Chilling injury on leaves and tips of young pea plants due to late frost. (B) Bark split on apple tree trunk due to low winter temperature. (C) Late frost injury on emerging pear leaves. (Left) Discoloration of upper side, (middle) discoloration and necrotic line as seen on upper side of leaf, (right) necrotic line on lower side. (D) Frost injury on apple fruit. (E) Low temperature injury on potato tuber in storage. (Photos A and E courtesy Dept. Plant Pathol., Cornell Univ.)

Indoor plants, whether grown in a home or a greenhouse, are particularly sensitive to low temperatures both where they are growing and during transportation from a greenhouse or florist's shop to a home or from one home to another. Indoor plants are often tropical plants grown far away from their normal climate. Exposure of such plants to low, not necessarily freezing, temperatures may cause stunting, yellowing, leaf or bud drop, etc. Similarly, when grown indoors, even local plants remain in a very succulent vegetative state and are completely unprepared for the stresses of low, particularly subfreezing, temperatures. Plants near windows or doors during cold winter days and, especially, nights are subject to temperatures that are much lower than those away from the window. Also, cracks or breaks in windows, holes of electrical outlets on outside walls, etc., let in cold air that may injure the plants. A drop of night temperatures below 12°C may cause leaves and particularly flower buds

FIGURE 26.
(A) Frost damage on young growth of rhododendron. (B) Cracking of rhododendron stem caused by frost. (Photos courtesy Dept. Plant Pathol., Cornell Univ.)

of many plants to turn yellow and drop. Exposure of indoor plants to subfreezing temperatures for a few minutes or a few hours, e.g., while they are carried or transported in the trunk of a car from the greenhouse to the house, may result in the death of many shoots and flowers or in a sudden shock to the plants from which they may take weeks or months to recover completely. Such a shock is often observed on plants that had been kept indoors and are then transplanted in the field in the spring when temperatures outdoors, although not freezing, are nevertheless much lower than those in the greenhouse. Even without the shock effect, plants growing at temperatures that are generally near the lower—or near the upper—limit of their normal range grow poorly and produce fewer and smaller blossoms and fruits.

The mechanisms by which high and low temperatures injure plants are quite different. High temperatures apparently inactivate certain enzyme systems and accelerate others, thus leading to abnormal biochemical reactions and death of the cell. High temperature may also cause coagulation and denaturation of proteins, disruption of cytoplasmic membranes, suffocation, and possibly release of toxic products into the cell.

Low temperatures, on the other hand, injure plants primarily by inducing ice formation between and/or within the cells. The rather pure water of the intercellular spaces freezes first and at about 0°C, while the water within the cell contains dissolved substances which, depending on their nature and concentration, depress the freezing point of water for several degrees. Furthermore, when the intercellular water becomes ice, more vapor (water) moves out of the cells and into the intercellular spaces, where it also becomes ice. The reduced water content of the cells depresses further the freezing point of the intracellular water and this could continue, up to a point, without damaging the cell. Below that point, however, ice crystals form within the cell, disrupt the plasma membrane, and cause injury and death to the cell. The freezing point of water in cells varies with the tissues and species of the plant; in some tissues of the winter-hardy species of the north, ice probably never forms within the cells regardless of how low the temperatures become. Even when ice forms only in the intercellular spaces, cells and tissues may be damaged either by the inward pressure exerted by the ice crystals, or by loss of water from their protoplasm to the intercellular spaces. This loss causes plasmolysis and dehydration of the protoplasm, which may cause coagulation. The rapidity of the temperature drop in a tissue is also important, since this affects the amount of water remaining in a cell and, therefore, the freezing point of the cell contents. Thus, a rapid drop in temperature may result in intracellular ice formation where a slow drop to the same low temperature would not. The rate of thawing may have similarly variable effects, since rapid thawing may flood the area between cell wall and protoplast and may cause tearing and disruption of the protoplast if the latter is incapable of absorbing the water as fast as it becomes available from the melting of ice in the intercellular spaces.

moisture

Moisture disturbances in the soil are probably responsible for more plants growing poorly and being unproductive annually, over large areas, than any other single environmental factor. Small or large territories may suffer from drought over periods of time. The subnormal amounts of water available to plants in these areas may result in reduced growth, diseased appearance, or even death of the plants. Lack of moisture may also be localized in certain types of soil, slopes, or thin soil layers underlaid by rock or sand and may result in patches of diseased-looking plants, while immediate surrounding areas appear to contain sufficient amounts of moisture and the plants in them grow normally. Plants suffering from lack of sufficient soil moisture usually remain stunted, are pale green to light yellow, have few, small and drooping leaves, flower and fruit sparingly and, if the drought continues, wilt and die (Fig. 27). Although annual plants are considerably more susceptible to short periods of insufficient moisture, even perennial plants and trees are damaged by prolonged periods of drought and produce less growth, small, scorched leaves and short twigs, dieback, defoliation, and finally wilting and death. Plants weakened by drought are also more susceptible to certain pathogens and insects.

Lack of moisture in the atmosphere, i.e., low relative humidity, is usually temporary and seldom causes damage. When combined with high

FIGURE 27.
(A) Healthy fuchsia plant (left), stunted plant due to insufficient water (middle) and plant wilting due to lack of water. (B) Leaf scorch due to insufficient water reaching the leaf. (C) Stunted, wilted and dead corn plants in low part of a field flooded for several days because of heavy rains. (D) Brown, sunken, dry area on orange caused by reduced oxygen during storage.

wind velocity and high temperature, however, it may lead to excessive loss of water from the foliage and may result in leaf scorching or burning, shrivelled fruit, and temporary or permanent wilting of plants.

Conditions of low relative humidity are particularly common and injurious for house plants during the winter. In modern homes and apartments, heating provides comfortable temperatures for plant growth but it often dries the air to relative humidities of 15 to 25 percent which are equivalent to that of desert environments. The air is particularly dry over or near the sources of dry heat, such as radiators. Potted plants kept under these conditions not only use up the water much faster, grow poorly, and may begin to wilt sooner, but the leaves, especially the lower ones, of many kinds of plants become spotted or scorched and fall prematurely, while their flowers suddenly wither and drop off. These effects are particularly noticeable when plants are brought into such a hot, dry house directly from a cool, moist greenhouse or florist's shop. Generally, all house plants prefer high humidity, and certain ones require high humidity if they are to grow properly and produce flowers. Therefore, house plants should never be placed over radiators and humidity should be increased with a commercial humidifier, by occasionally dampening the leaves with water, by placing the pot on a brick, pebbles, etc., in a large pan of water, in a plastic case, etc.

Excessive soil moisture occurs much less often than drought where plants are grown, but poor drainage or flooding of planted fields, gardens, or potted plants may result in more serious and quicker damage, or death, to plants (Fig. 27C) than that from lack of moisture. Poor drainage results in plants that lack vigor, wilt frequently, and have leaves that are pale green or yellowish green. Flooding during the growth season may cause permanent wilting and death of succulent annuals within 2 to 3 days. Trees, too, are killed by waterlogging, but the damage usually appears more slowly and after their roots have been continually flooded for several weeks.

As a result of excessive soil moisture caused by flooding or by poor drainage, the fibrous roots of plants decay, probably due to the reduced supply of oxygen to the roots. Oxygen deprivation causes stress, asphyxiation, and collapse of many root cells. Wet, anaerobic conditions favor the growth of anaerobic microorganisms which, during their life processes, form substances, such as nitrites, that are toxic to plants. Besides, the root cells damaged directly by the lack of oxygen lose their selective permeability and may allow toxic metals, etc. to be taken up by the plant. Also, once parts of roots are killed, more damage is done by facultative parasites which may be greatly favored by the new environment. Thus, the wilting of the plants, which soon follows flooding, is probably the result of lack of water in the aboveground parts of plants caused by the death of the roots, although it appears that translocated toxic substances may also be involved.

In addition to the above, many plants, particularly potted house plants, show several symptoms which are the result of incorrect watering, i.e., either the soil is allowed to dry out too much before it is then repeatedly flooded with water, or the plant is almost constantly overwatered. In either case, overwatered plants may suddenly drop their lower leaves, or their leaves may turn yellow. Sometimes they develop brown or black wet patches on the leaves or stems, or the roots and lower stem may turn black and rot as a result of infection by pathogenic microorganisms encouraged by the excessive watering. Such symptoms can be avoided or corrected by watering only when the topsoil feels dry and then applying enough water to saturate thoroughly the whole mass of soil. Plants should never be watered when the soil is still wet, especially during the winter. When watering, any excess water should be drained through the drainage hole which should always be present in the bottom of the pot. A period of dryness should not be followed with repeated heavy watering but by a gradual return to normal watering. Generally, the supply of water should be maintained as uniform as possible.

Another common symptom of house plants, and sometimes of outdoor plants, that is caused by excessive moisture is the so-called edema (= swelling). Edema appears as numerous small bumps on the lower side of leaves or on stems. The "bumps" are small masses of cells that divide, expand and break out of the normal leaf surface and at first form greenish-white swellings or galls. Later, the exposed surface of the swellings becomes rusty colored and has a corky texture. Edema is caused by overwatering, especially during cloudy, humid weather, and can be avoided by reduced watering and better lighting and air circulation of the plant. Many other disorders are caused by excessive or irregular watering. It is known, for example, that tomatoes grown under rather low moisture conditions at the time they are ripening often crack if they are suddenly supplied with abundant moisture by overwatering or by a heavy rainfall. Also, bitter pit of apples, consisting of small, sunken, black spots on the fruit, is the result of irregular supply of moisture, although excessive nitrogen and low calcium fertilization seem to also be involved in bitter pit development.

inadequate oxygen

Low oxygen conditions in nature are generally associated with high soil moisture and/or high temperatures. Lack of oxyten may cause desiccation of roots of different kinds of plants in waterlogged soils, as was mentioned under moisture effects. A combination of high soil moisture and high soil or air temperature causes root collapse in plants. The first condition, apparently, reduces the amount of oxygen available to the roots while the other increases the amount of oxygen required by the plants. The two effects together result in an extreme lack of oxygen in the roots and cause their collapse and death.

Low oxygen levels may also occur in the centers of fleshy fruit or vegetables in the field, especially during periods of rapid respiration at high temperatures, or in storage of these products in fairly bulky piles (Fig. 27D). The best known such case is the development of the so-called blackheart of potato, in which fairly high temperatures stimulate respiration and abnormal enzymatic reactions in the potato tuber. The oxygen supply of the cells in the interior of the tuber is insufficient to sustain the increased respiration, and the cells die of suboxidation. Enzymatic reactions activated by the high temperature and suboxidation go on before, during, and after the death of the cells. These reactions abnormally oxidize normal plant constituents into dark melanin pigments. The pigments spread into the surrounding tuber tissues and finally make them appear black.

light

Lack of sufficient light retards chlorophyll formation and promotes slender growth with long internodes, thus leading to pale green leaves, spindly growth, and premature drop of leaves and flowers. This condition is known as etiolation. Etiolated plants are found outdoors only when plants are spaced too close together or when they are growing under trees or other objects. Etiolation of various degrees, however, is rather common in house plants, and also in greenhouses, seedbeds, and cold frames, where plants often receive inadequate light. Etiolated plants are usually thin and tall and are susceptible to lodging.

Excess light is rather rare in nature and seldom injures plants. Many injuries attributed to light are probably the result of high temperatures accompanying high light intensities. Excessive light, however, seems to cause sunscald of pods of beans grown at high altitudes where, due to absence of dust, etc., more light of short wavelengths reaches the earth. The pods develop small water-soaked spots which quickly become brown or reddish brown and shrink.

The amount of light is considerably more important in relation to house plants. Some of them prefer shade or semishade during the growth season but full sunlight during the winter. Others prefer shade throughout the year while still others must have sunlight all year long. As a rule, house plants with deep green leaves prefer or tolerate shade much better than do plants with colored leaves, the latter generally doing better when they receive considerable sunlight. Most flowering house plants grow and flower best with full exposure to sunlight at all seasons. Lack of sufficient light for any of these kinds of plants has the same effects as on the outdoor plants, i.e., pale green leaves, spindly growth, leaf drop, few or no flowers, flower drop, etc. On the other hand, excessive sunlight on plants that prefer less light often results in the appearance of yellowish-brown or silvery spots on their leaves. Plants suddenly moved to an area with strikingly different light intensity than the previous one often respond with general defoliation.

air pollution

The air at the earth's surface consists primarily of nitrogen and oxygen (78 and 21 percent, respectively). Much of the remaining 1 percent is water vapor and carbon dioxide. Man's activities in generating energy, manufacturing goods, and disposing of wastes result in the release into the atmosphere of a number of pollutants which may alter plant metabolism and induce disease. Air pollution damage to plants, especially around certain types of factories, has been recognized for about a century. Its extent and importance, however, increased with the industrial revolution and will, apparently, continue to increase with the world's increasing population, industrialization, and urbanization.

Almost all air pollutants causing plant injury are gases, but some particulate matter or dusts may also affect vegetation. Some gas contaminants, such as ethylene, ammonia, chlorine, and sometimes mercury vapors, exert their injurious effects over limited areas only. Most frequently they affect plants or plant products stored in poorly ventilated warehouses in which the pollutants are produced by the plants themselves (ethylene), or from leaks in the cooling system (ammonia).

More serious and widespread damage is caused to plants in the field by chemicals such as hydrogen fluoride, nitrogen dioxide, ozone (Fig. 28),

FIGURE 28.
Flecking on the upper surface of tobacco leaf caused by naturally occurring high concentrations of ozone in the atmosphere.

peroxyacyl nitrates, sulfur dioxide, and particulates. High concentrations of or long exposure to these chemicals cause visible and sometimes characteristic symptoms (e.g., necrosis) on the affected plants. However, when plants are exposed to dosages less than those that cause acute damage, their growth and productivity may still be suppressed due to interference by the pollutants with the metabolism of the plant. The main pollutants, their sources, and their effects on plants are given in Table II.

Exhausts of automobiles and other internal combustion engines are probably the most important sources of ozone and other phytotoxic pollutants. Thousands of tons of incompletely burned hydrocarbons and NO_2 are released into the atmosphere daily by automobile exhausts. In the presence of ultraviolet light from the sun, this nitrogen dioxide reacts with oxygen and forms ozone and nitric oxide. The ozone may react with nitric oxide to form the original compounds:

$$NO_2 + O_2 \xrightleftharpoons{\text{sunlight}} O_3 + NO$$

However, in the presence of unburned hydrocarbon radicals, the nitric oxide reacts with these instead of ozone, and therefore the ozone concentration builds up. Ozone, too, can react with vapors of certain unsaturated hydrocarbons, but the products of such reactions (various organic peroxides) are also toxic to plants. Normally, the noxious fumes produced by automobiles and other engines are swept up by the warm air currents from the earth's surface rising into the cooler air above, where the fumes are dissipated. During periods of calm, stagnant weather, however, an inversion layer of warm air is formed above the cooler air and this prevents the upward dispersion of atmospheric pollutants. The pollutants then are trapped near the ground where, after sufficient buildup, they may seriously damage living organisms.

Peroxyacyl nitrate (PAN) injury has been observed primarily around metropolitan areas where large amounts of hydrocarbons are released into the air from automobiles. The problem is especially serious in areas like Los Angeles and New Jersey, where the atmospheric conditions are conducive to inversion layer formation. Many different kinds of plants are affected by PAN over large geographical areas surrounding the locus of PAN formation, due to diffusion or to dispersal of the pollutant by light air currents.

Sulfur dioxide may injure plants in concentrations as low as 5 to 10 ppm. Since sulfur dioxide is absorbed through the leaf stomata, conditions that favor or inhibit the opening of stomata similarly affect the amount of sulfur dioxide absorbed. After absorption by the leaf, sulfur dioxide reacts with water and forms phytotoxic sulfite ions. The latter, however, are slowly oxidized in the cell to produce harmless sulfate ions. Thus, if the rate of sulfur dioxide absorption is slow enough, the plant may be able to protect itself from the buildup of phytotoxic sulfites.

nutritional deficiencies in plants

Plants require several mineral elements for normal growth. Some elements, such as nitrogen, phosphorus, potassium, calcium, magnesium, and sulfur, needed in relatively large amounts, are called "major" elements, while others, like iron, boron, manganese, zinc, copper, molybdenum, and chlorine, needed in very small amounts, are called "trace" or "minor" elements or "micronutrients." Both major and trace elements are essential to the plant. When they are present in the plant in amounts smaller than the minimum levels required for normal plant growth, the plant becomes diseased and exhibits various external and internal symptoms. The symptoms may appear on any or all organs of the plant, including leaves, stems, roots, flowers, fruits, and seeds.

The kinds of symptoms produced by deficiency of a certain nutrient depend primarily on the functions of that particular element in the plant. These functions presumably are inhibited or interfered with when the element is limiting. Certain symptoms are the same in deficiency of any of several elements, but other diagnostic features usually accompany a deficiency of a particular element. Numerous plant diseases occur annually in most agricultural crops in many locations due to reduced amounts or reduced availability of one or more of the essential elements in the

TABLE II.
AIR POLLUTION INJURY TO PLANTS

Pollutant	Source	Susceptible Plants	Symptoms	Remarks
Ozone (O_3)	Automobile exhausts. Other internal combustion engines. (Released NO_2 combines with O_2 in sunlight $\rightarrow O_3$) From stratosphere. From lightning, from forests.	Expanding leaves of all plants, especially tobacco, bean, cereals, alfalfa, petunia, pine, citrus, corn.	Stippling, mottling, and chlorosis of leaves, primarily on upper leaf surface. Spots are small to large, bleached white to tan, brown, or black (Fig. 28). Premature defoliation and stunting occurs in plants such as citrus, grapes, and vines.	Enters through stomata. It is the most destructive air pollutan to plants. A major component of smog.
Peroxyacyl nitrates (PAN)	Automobile exhausts or other internal combustion engines. (Gasoline vapors and incompletely burned gasoline $\pm O_3$ or $NO_2 \rightarrow$ PAN).	Many kinds of plants, including spinach, petunia, tomato, lettuce, dahlia.	Causes "silver leaf" on plants, i.e., bleached white to bronze spots on lower surface of leaves that may later spread throughout leaf thickness and resemble ozone injury.	Particularly severe near metropolitan areas with smog and inversion layers.
Sulfur dioxide (SO_2)	Stacks of factories. Automobile exhausts, and other internal combustion engines.	Many kinds of plants, including alfalfa, violet, conifers, pea, cotton, bean. Toxic at 0.3 to 0.5 ppm.	Low concentrations cause general chlorosis. Higher concentrations cause bleaching of interveinal tissues of leaves.	It also combines with moisture and forms toxic acid droplets (acid rain).
Nitrogen dioxide (NO_2)	From oxygen and nitrogen in the air by hot combustion sources, e.g., furnaces, internal combustion engines.	Many kinds of plants including beans, tomatoes. Toxic at 2 to 3 ppm.	Causes bleaching and bronzing of plants similar to that caused by SO_2. At low concentration it also suppresses growth of plants.	

Pollutant	Source	Plants affected	Symptoms	Notes
Hydrogen fluoride (HF)	Stacks of factories processing ore or oil.	Many kinds of plants, including corn, peach, tulip. Actively growing, especially wet leaves, are most sensitive. Toxic at 0.1 to 0.2 ppb.	Leaf margins of dicots and leaf tips of monocots turn tan to dark brown, die, and may fall from the leaf. Some plants tolerate HF up to 200 ppm.	HF may evaporate or be washed out of plant and plant recovers slowly.
Chlorine (Cl$_2$) and Hydrogen chloride (HCl)	Refineries, glass factories, incineration of plastics.	Many kinds of plants, usually near the source. Toxic at 0.1 ppm.	Leaves show bleached, necrotic areas between veins. Leaf margins often appear scorched. Leaves may drop prematurely. Damage resembles that caused by SO$_2$.	
Ethylene (CH$_2$CH$_2$)	Automobile exhausts. Burning of gas, fuel oil, and coal. From ripening fruit in storage.	Many kinds of plants. Toxic at 0.05 ppm.	Plants remain stunted, their leaves develop abnormally and senesce prematurely. Plants produce fewer blossoms and fruit. Fruit, e.g., apples, develop depressed, necrotic, dark areas (scald).	Ethylene is a plant hormone with numerous functions.
Particulate matter (dusts)	Dust from roads, cement factories. Burning of coal, etc.	All plants.	Form dust or crusty layers on plant surfaces. Plants become chlorotic, grow poorly, and may die. Some dusts are toxic and burn leaf tissues directly or after dissolving in dew or rainwater.	

TABLE III.
NUTRIENT DEFICIENCIES IN PLANTS

Deficient Nutrient	Functions of Element	Symptoms
Nitrogen N	Present in most substances of cells.	Plants grow poorly and are light green in color. The lower leaves turn yellow or light brown and the stems are short and slender (Fig. 29A).
Phosphorus P	Present in DNA, RNA, phospholipids (membranes), ADP, ATP, etc.	Plants grow poorly and the leaves are bluish-green with purple tints. Lower leaves sometimes turn light bronze with purple or brown spots. Shoots are short and thin, upright, and spindly.
Potassium K	Acts as a catalyst of many reactions.	Plants have thin shoots which in severe cases show dieback. Older leaves show chlorosis with browning of the tips, scorching of the margins, and many brown spots usually near the margins. Fleshy tissues show end necrosis (Fig. 29C and E).
Magnesium Mg	Present in chlorophyll and is part of many enzymes.	First the older, then the younger leaves become mottled or chlorotic, then reddish. Sometimes necrotic spots appear. The tips and margins of leaves may turn upward and the leaves appear cupped. Leaves may drop off (Fig. 29D).
Calcium Ca	Regulates the permeability of membranes. Forms salts with pectins. Affects activity of many enzymes.	Young leaves become distorted, with their tips hooked back and the margins curled. Leaves may be irregular in shape and ragged with brown scorching or spotting. Terminal buds finally die. The plants have poor, bare root systems. Causes blossom end rot of many fruits (Fig. 29F).
Boron B	Not really known. Affects translocation of sugars and utilization of calcium in cell wall formation.	The bases of young leaves of terminal buds become light green and finally break down. Stems and leaves become distorted. Plants are stunted (Fig. 30). Fruit, fleshy roots or stems, etc., may crack on the surface and/or rot in the center. Causes many plant diseases, e.g., heart rot of sugar beets, brown heart of turnips, browning or hollow stem of cauliflower, cracked stem of celery, corky spot, dieback and rosette of apples, hard fruit of citrus, top sickness of tobacco, etc.
Sulfur S	Present in some amino acids and coenzymes.	Young leaves are pale green or light yellow without any spots. The symptoms resemble those of nitrogen deficiency.
Iron Fe	Is a catalyst of chlorophyll synthesis. Part of many enzymes.	Young leaves become severely chlorotic, but their main veins remain characteristically green. Sometimes brown spots develop. Part of or entire leaves may dry. Leaves may be shed (Fig. 29B).

TABLE III. (Continued)

Deficient Nutrient	Functions of Element	Symptoms
Zinc Zn	Is part of enzymes involved in auxin synthesis and in oxidation of sugars.	Leaves show interveinal chlorosis. Later they become necrotic and show purple pigmentation. Leaves are few and small, internodes are short and shoots form rosettes, and fruit production is low. Leaves are shed progressively from base to tip. It causes "little leaf" of apple, stone fruits and grape, "sickle leaf" of cacao, "white tip" of corn, etc.
Copper	Is part of many oxidative enzymes.	Tips of young leaves of cereals wither and their margins become chlorotic. Leaves may fail to unroll and tend to appear wilted. Heading is reduced and the heads are dwarfed and distorted. Citrus, pome, and stone fruits show dieback of twigs in the summer, burning of leaf margins, chlorosis, rosetting, etc. Vegetable crops fail to grow.
Manganese Mn	Is part of many enzymes of respiration, photosynthesis, and nitrogen utilization.	Leaves become chlorotic but their smallest veins remain green and produce a checked effect. Necrotic spots may appear scattered on the leaf. Severely affected leaves turn brown and wither.

soils where the plants are grown. The presence of lower-than-normal amounts of most essential elements usually results in merely a reduction in growth and yield. When the deficiency is greater than a certain critical level, however, the plants develop acute or chronic symptoms and may even die. Some of the general deficiency symptoms caused by each essential element, the possible functions affected, and some examples of common deficiency disorders are given in Table III and shown in Figs. 29 and 30.

soil minerals toxic to plants

Soils often contain excessive amounts of certain essential or nonessential elements, both of which at high concentration may be injurious to the plant. Of the essential elements, those required by plants in large amounts, such as nitrogen and potassium, are usually much less toxic when present in excess than are the elements required only in trace amounts, such as manganese, zinc, and boron. Even among the latter, however, some trace elements such as manganese and magnesium have a much wider range of safety than do others, e.g., boron or zinc. Besides, not only do the elements differ in their ranges of toxicity, but various kinds of plants also differ in their susceptibility to the toxicity to a certain level of a particular element. Concentrations at which nonessential elements are

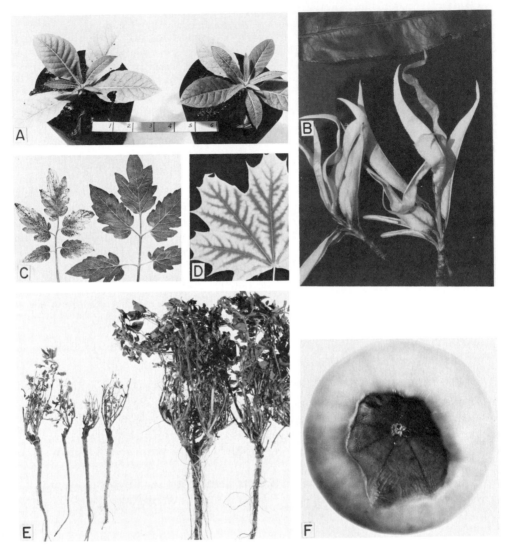

FIGURE 29.
Some examples of nutrient deficiency symptoms in plants. (A) Nitrogen
deficiency on tobacco (left) and one week after fertilization. (B) Iron deficiency on
peach. Note uniform yellowing of affected leaves compared to normal leaf at top
of photo. (C) Potassium deficiency on tomato (left). Healthy leaf at right. (D)
Magnesium deficiency symptoms on maple. (E) Healthy (right) and potassium-
deficient alfalfa plants. (F) Blossom end rot of tomato caused in part by calcium
deficiency.

toxic also vary among elements, and plants in turn vary in their sensitivity to them. For example, some plants are injured by very small amounts
of nickel, but can tolerate considerable concentrations of aluminum.

The injury occurring from excess of an element may be slight or severe
and is usually the result of direct injury by the element to the cell. On the
other hand, the element may interfere with the absorption or function of
another element and thereby lead to the symptoms of a deficiency of the

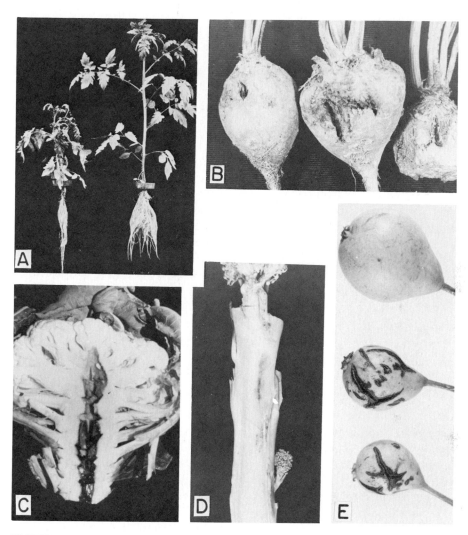

FIGURE 30.
Boron deficiency symptoms on plants. (A) Healthy (right) and stunted tomato plant. (B) Cracking and breakdown of beets. (C) Internal breakdown of cauliflower stem. (D) Corky neck surface and internal breakdown of broccoli stem. (E) Healthy (top) and cracked pears due to boron deficiency aggravated by prolonged drought.

element being interfered with. Thus, excessive sodium induces a deficiency of calcium in the plant, while the toxicity of copper, manganese, or zinc is both direct on the plant and by inducing a deficiency of iron in the plant.

Excessive amounts of sodium salts, especially sodium chloride, sodium sulfate, and sodium carbonate, raise the pH of the soil and cause what is known as alkali injury. This injury varies in the different plants and may range from chlorosis to stunting, leaf burn, wilting, to outright killing of seedlings and young plants. Some plants, e.g., wheat, apple, are very sensitive to alkali injury, while others, e.g., sugar beets, alfalfa, and several grasses, are quite tolerant. On the other hand, when the soil is too acidic, the growth of some kinds of plants is impaired and various symptoms may appear. Plants usually grow well in a soil pH range from 4 to 8, but some plants grow better on the lower pH than others, and vice versa. Thus, blueberries grow well on acid soils, while alfalfa grows best on alkaline soils. The injury caused by low pH is, in most cases, brought about by the greater solubility of mineral salts in acid solutions. These salts then become available in concentrations that, as was pointed out above, either are toxic to the plants or interfere with the absorption of other necessary elements and so cause symptoms of mineral deficiency.

Boron, manganese, and copper have been most frequently implicated in mineral toxicity diseases, although other minerals, e.g., aluminum and iron, also damage plants in acid soils. Excess boron is toxic to many vegetables and trees. Excess manganese is known to cause a crinkle-leaf disease in cotton, and has been implicated in the internal bark-necrosis of Red Delicious apple and in many other diseases of several crop plants. Sodium and chlorine ions also have been shown to cause symptoms of poor growth and decline, like those shown by some of the trees along roads in northern areas where heavy salting is carried out in the winter to remove ice from roads.

herbicide injury

Some of the most frequent plant disorders seem to be the result of the extensive use of herbicides (weed killers). The constantly increasing number of herbicides in use by more and more people for general or specific weed control is creating numerous problems among those who use them, their neighbors, and those who use soil that has been treated with herbicides.

Herbicides are either specific against broadleaved weeds [e.g., 2,4-D, dicamba (Banvel-D)], and these are applied in corn and other small grain fields and on lawns, or specific against grasses and some broadleaved weeds (e.g., Dacthal, Atrazine), and these are applied in orchards, vegetable and truck crop fields; in addition, some herbicides are general weed or shrub killers (e.g., 2,4,5-T, Silvex). Most of the herbicides are safe as long as they are used to control weeds among the right crop plants, at the right time, at the correct dosage and when the correct environmental conditions prevail. When any one of the above conditions are not met, abnormalities develop on the cultivated plants with which the herbicides come in contact. Affected plants show various degrees of distortion or yellowing of leaves (Fig. 31), browning, drying and shedding of leaves, stunting and even death of the plant. Much of this damage is caused by too high doses of herbicides, or when applied too early in the season or on too cold or too hot a day, or when dust or spray droplets of an herbicide are carried by the wind to nearby sensitive plants or to gardens or fields on which plants sensitive to the herbicide are grown. Of course, direct

FIGURE 31.

(A, B, and C) Injury on trees from drift of herbicide applied to lawn or orchard. (A) The leaves are smaller with greatly narrowed interveinal areas. (B) Leaves are rolled and petioles are distorted. Normal leaf at bottom. (C) Yellowing of veins or entire leaves caused by herbicide injury. (D) Leaf distortion of geranium cuttings after transplanting in soil contaminated with herbicide. Note two normal leaves (bottom) developed before transplanting. (E) Frenching of tobacco caused by accumulation of toxic substances in the soil produced by bacteria (*Bacillus cereus*) and fungi (*Aspergillus*) and interfering with amino acid metabolism in plants. (Photo C courtesy W. J. Lord).

application of the wrong pesticide in a field with a particular crop plant will kill the crop just as if it were a weed.

Use of preplant or preemergence herbicides through application to the soil before or at planting time often affects seed germination and growth of the young seedlings if too much or the wrong herbicide has been applied. Most herbicides are used up or are inactivated within a few days

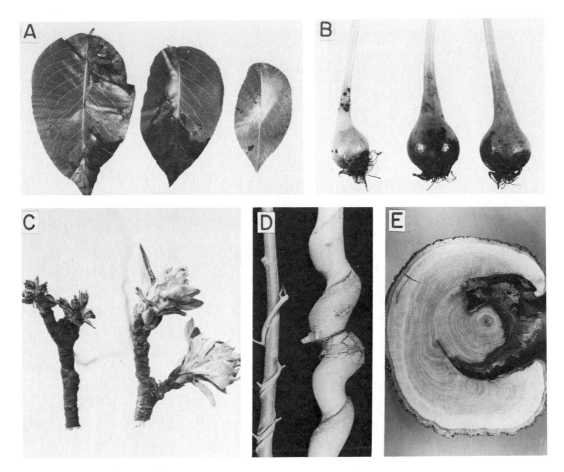

FIGURE 32.
Spray injury on pear leaf (A) and fruit (B), and on apple blossom (C, left). (D)
Distortion of maple stem and twig by the climbing vine of bittersweet, *Celastrus
scandens*. (E) Fire damage on oak trunk.

to a few months from the time of application; some, however, persist in
the soil for more than a year. Sensitive plants planted in fields previously
treated with such a persistent herbicide may grow poorly and may produce various symptoms. Also, home owners, home gardeners, and
greenhouse operators often obtain what looks like good, weed-free soil
from fields that, unbeknown to them, had been treated with herbicides.
Such soil when used to grow potted, bench, or garden plants results in
smaller, distorted, yellowish plants (Fig. 31D) which sometimes shed
some or all of their leaves and either die or finally recover.

other improper agricultural practices

As with herbicides, a variety of other agricultural practices improperly carried out may cause considerable damage to plants and increased financial losses. Almost every agricultural practice can cause damage when done the wrong way, at the wrong time, or with the wrong materials. Most commonly, however, losses result from application of chemicals, such as fungicides, insecticides, nematicides, and fertilizer, at too high concentrations or on plants sensitive to them. Spray injury resulting in leaf burn or spotting or russeting of fruit is common on many crop plants (Fig. 32).

Excessive or too deep cultivation between rows of growing plants may be more harmful than useful because it cuts or pulls many of the plants' roots. Road or other construction often cuts a large portion of the roots of nearby trees and results in their dieback and decline. Inadequate or excessive watering may cause wilting or any of the symptoms described earlier. In the case of African violets, droplets of cold water on the leaves cause the appearance of rings and ringlike patterns reminiscent of virus ringspot diseases. Potatoes stored next to hot water pipes under the kitchen sink often develop black heart. Trees frequently grow poorly and their leaves are chlorotic, curled, or reddened because their trunk is girdled by the fence wire. The roots of plants potted in pots that are too small for their size are often badly distorted and twisted and the whole plant grows poorly (Fig. 23).

SELECTED REFERENCES

Berg, A., Genevieve Clulo (Berg), and C. R. Orton. 1958. Internal bark necrosis of apple resulting from manganese toxicity. *West Va. Agr. Expt. Sta. Bull.* **414** T:22 pp.

Carne, W. M. 1948. The non-parasitic disorders of apple fruits in Australia. *Commonw. Aust., Council Sci. Ind. Res. Bull.* **238**:83 pp., illus.

Daines, R. H., Ida A. Leone, and Eileen Brennan. 1960. Air pollution as it affects agriculture in New Jersey. *New Jersey Agr. Expt. Sta. Bull.* **794**:14 pp.

Darley, E. F., and J. T. Middleton. 1966. Problems of air pollution in plant pathology. *Ann. Rev. Phytopathol.* **4**:103–118.

Jacobson, J. S., and A. C. Hill (eds.). 1970. "Recognition of Air Pollution Injury to Vegetation: A Pictorial Atlas." Air Pollution Control Assoc., Pittsburgh, Penn.

Levitt, J. 1973. "Responses of Plants to Environmental Stresses." Academic Press, New York, 697 p.

McMurtrey, J. E., Jr. 1953. Environmental, nonparasitic injuries. *Yearbook Agr.* (U.S. Dept. Agr.) pp. 94–100.

Wallace, T. 1961. "The Diagnosis of Mineral Deficiencies in Plants by Visual Symptoms," 125 p., illus. Her Majesty's Stationery Office, London.

10 plant diseases caused by fungi

introduction

Fungi are small, generally microscopic, plants lacking chlorophyll and conductive tissues. Most of the 100,000 fungus species known are strictly saprophytic, living on dead organic matter which they help decompose. Some, about 50, species cause diseases in man, and about as many cause diseases in animals, most of them being superficial diseases of the skin or its appendages. More than 8000 species of fungi, however, can cause diseases in plants. All plants are attacked by some kinds of fungi, and each of the parasitic fungi can attack one or many kinds of plants. Some of the fungi can grow and multiply only by remaining in association with their host plants during their entire life (obligate parasites), others require a host plant for part of their life cycles but can complete their cycles on artificial media, and still others can grow and multiply on dead organic matter as well as on living plants (nonobligate parasites).

characteristics of plant pathogenic fungi

MORPHOLOGY

Most fungi have a vegetative body consisting of more or less elongated, continuous filaments which may or may not have cross walls (septa). The body of the fungus is called mycelium, and the individual branches or filaments of the mycelium are called hyphae (Fig. 33). Each hypha or

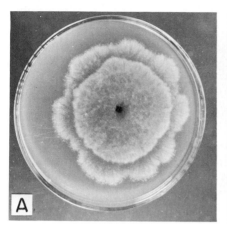
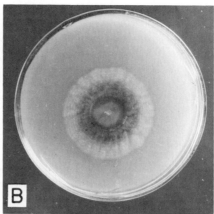

FIGURE 33.
Appearance of the vegetative body (mycelium) of two fungi in culture. (A)
Physalospora. (B) *Phoma.*

mycelium may be uniform in thickness or may taper into thinner or
broader portions. Hyphae of some fungi are only 0.5 μm in diameter,
while in others they may be more than 100 μm thick. The length of the
mycelium may be only a few microns in some fungi, but in others it may
produce mycelial strands several meters long.

In some fungi the mycelium consists of cells containing one or two
nuclei per cell. In others the mycelium is coenocytic, i.e., it contains
many nuclei and either the entire mycelium is one continuous, tubular,
branched or unbranched multinucleate cell, or it is partitioned by several
septa, each segment being a multinucleate hypha. Growth of the
mycelium occurs at the tips of the hyphae.

Some lower fungi lack true mycelium and produce instead a naked,
amoeboid, multinucleate plasmodium (e.g., Myxomycetes) or a system of
strands of grossly dissimilar and continuously varying diameter called a
rhizomycelium (e.g., Chytridiomycetes).

REPRODUCTION

Fungi reproduce chiefly by means of spores (Fig. 34). Spores are
specialized propagative or reproductive bodies consisting of one or a few
cells. Spores may be formed asexually (i.e., through the separation of
minute fragments of the mycelium into spores) or as the result of a sexual
process.

In the lower fungi, asexual spores are produced inside a sac called a
sporangium and are released through an opening of the sporangium or
upon its rupture. Some of these spores are motile by means of flagella and
are, therefore, called zoospores. Other fungi produce asexual spores called
conidia by the cutting off of terminal or lateral cells from special hyphae
called conidiophores. In some fungi terminal or intercalary cells of a

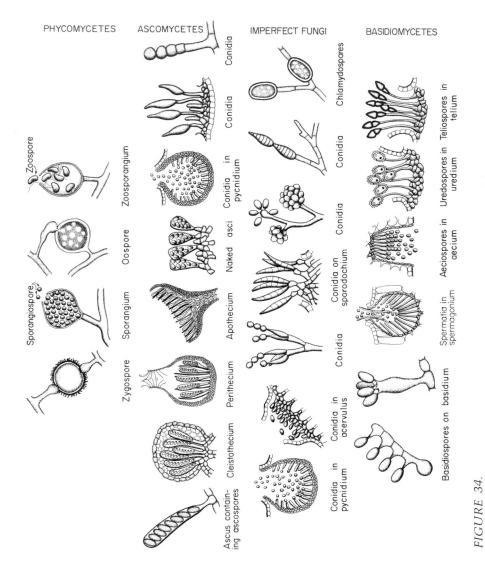

FIGURE 34.
Representative spores and fruiting bodies of the main groups of fungi.

hypha enlarge, round up, form a thick wall and separate to form chlamydospores. In still other fungi, asexual spores (conidia) are produced inside thick-walled structures called pycnidia.

Sexual reproduction, or processes resembling it, occur in most groups of fungi. In some, two cells (gametes) of similar size and appearance unite and produce a zygote, called a zygospore. In others, the gametes are of unequal size and the zygote which they form is called an oospore. In some fungi no definite gametes are produced, but instead one mycelium may unite with another compatible mycelium. In one group of fungi (Ascomycetes) the sexual spores, usually eight in number, are produced within the zygote cell, the ascus, and the spores are called ascospores. In another group of fungi (Basidiomycetes), sexual spores are produced on the outside of the zygote cell called the basidium and the spores are called basidiospores.

For a large group of fungi (Fungi Imperfecti) no sexual reproduction is known either because they do not have one or because it has not yet been discovered. Apparently these fungi reproduce only asexually.

The union of the sexual nuclei in the zygote produces a diploid $(2N)$ nucleus. Usually the first divisions of this nucleus are meiotic so that throughout its life the fungus contains haploid $(1N)$ nuclei, except immediately after the union of the gamete nuclei. In some groups of fungi, however, especially in the Basidiomycetes and to a lesser extent in the Ascomycetes, the cells of the entire mycelium or of parts of the mycelium contain two haploid nuclei which remain separate inside the cell. Such mycelium is called dikaryotic but behaves very much as though it were a diploid mycelium (in which the two nuclei are united).

In most fungi both male and female gametes are produced on the same mycelium (hermaphroditic fungi). When the male gametes can fertilize the female ones of the same mycelium, the fungus is called homothallic. In many cases, however, the male gametes can fertilize only the female gametes of another, sexually compatible mycelium, and the fungus then is called heterothallic.

ECOLOGY AND SPREAD

Almost all plant pathogenic fungi spend part of their lives on their host plants and part in the soil or on plant debris on the soil. Some fungi pass all of their lives on the host and only the spores may land on the soil where they remain inactive until they are again carried to a host on which they grow and multiply. Other fungi (e.g., *Venturia*) must pass part of their lives on the host as parasites and part on dead tissues on the ground as saprophytes in order to complete their life cycle in nature. The latter group of fungi, however, remain continually associated with host tissues, whether living or dead, and, in nature, do not grow on any other kind of organic matter. A third group of fungi grow parasitically on their hosts but they continue to live, grow, and multiply on the dead tissues of the host after its death, and may further move out of the host debris into the soil or other decaying plant material on which they grow and multiply as strict saprophytes. The dead plant material which they colonize need not be related at all to the host they can parasitize. This group of fungi are usually soil pathogens, have a wide host range, and can survive in the soil for many years in the absence of their hosts. They too, however, need to infect a host from time to time in order to increase their populations, since protracted and continuous growth of these fungi as saprophytes in the soil results in more or less rapid reduction in their numbers.

During their parasitic phase fungi assume various positions in relation to the plant cells and tissues. Some fungi (e.g., powdery mildews) grow outside the plant surface but send their feeding organs (haustoria) into the epidermal cells of the plant. Some (e.g., *Venturia*) grow only between the cuticle and the epidermal cells. Others grow between the cells in the intercellular spaces and may or may not send haustoria into the cells. Still others grow between and through the cells indiscriminately. Obligate parasites can grow only in association with living cells, being unable to feed on dead cells. On the other hand, the mycelium of some nonobligate parasites never comes in contact with living plant cells, because their enzymes macerate and kill the plant cells ahead of the mycelium. In most cases, however, regardless of the position of the mycelium in the host, the reproductive bodies (spores) of the fungus are produced at or very near the surface of the host tissues to ensure their prompt and efficient dissemination.

The survival and performance of most plant pathogenic fungi depend greatly on the prevailing conditions of temperature and moisture or the presence of water in their environment. Free mycelium survives only within a certain range of temperatures (-5 to $+45°C$) and in contact with moist surfaces, inside or outside the host. Most kinds of spores, however, can withstand broader ranges of both temperature and moisture and carry the fungus through the low winter temperatures and the dry summer periods. Spores, however, also require favorable temperatures and moisture in order to germinate. Furthermore, lower fungi producing zoospores require free water for the production, movement, and germination of the zoospores.

Zoospores are the only fungus structures that can move by themselves. Zoospores, however, can move for only very short distances (a few millimeters or centimeters, perhaps). Besides, only some myxomycetes and some phycomycetes produce zoospores. The great majority of the plant pathogenic fungi depend for their spread from plant to plant and to different parts of the same plant on chance distribution by agents such as wind, water, birds, insects, other animals, and man. Fungi are disseminated primarily in the form of spores. Fragments of hyphae and hard masses of mycelium known as sclerotia may also be disseminated by the same agents although to a much lesser extent.

Spore dissemination in almost all fungi is passive, although their initial discharge in some fungi is forcible. The distance to which spores may be disseminated varies with the agent of dissemination. Wind is probably the most important disseminating agent of spores of most fungi and may carry spores over great distances. For specific fungi, other agents such as water or insects may play a much more important role than wind in the dissemination of their spores.

classification of plant pathogenic fungi

The fungi that cause diseases on plants are a diverse group, and because of their large numbers and diversity, only a sketchy classification of some of the most important phytopathogenic genera will be presented here.

THE LOWER FUNGI

Class: MYXOMYCETES (The slime molds)—Lack mycelium. Their body is a naked, amorphous plasmodium.

Order: Physarales—Saprophytic plasmodium that gives rise to crusty fructifications containing spores. They produce zoospores.

Genus: *Fuligo, Mucilago,* and *Physarum* cause slime molds on low-lying plants.

Order: Plasmodiophorales—Plasmodia produced within cells of roots and stems of plants. They produce zoospores.

Genus: *Plasmodiophora, P. brassicae* causing clubroot of crucifers.

Polymyxa, P. graminis being parasitic in wheat and other cereals.

Spongospora, S. subterranea causing powdery scab of potato tubers.

Class: PHYCOMYCETES (Algal fungi, the lower true fungi).

Subclass: CHYTRIDIOMYCETES—Have round or elongated mycelium that lacks cross walls.

Order: Chytridiales—Have cell wall but lack true mycelium, at most a rhizomycelium. Zoospores.

Genus: *Olpidium, O. brassicae* being parasitic in roots of cabbage and other plants.

Physoderma, P. maydis causing brown spot of corn.

Synchytrium, S. endobioticum causing potato wart.

Urophlyctis, U. alfalfae causing crown wart of alfalfa.

Subclass: OOMYCETES (The water molds, white rusts, and downy mildews)—Have elongated mycelium. Produce zoospores in zoosporangia. Oospores produced by the union of morphologically different gametes.

Order: Saprolegniales—Have well-developed mycelium. Zoospores produced in long, cylindrical zoosporangia attached to mycelium. Oospores.

Genus: *Aphanomyces,* causing root rot of many vegetables.

Order: Peronosporales—Sporangia, usually zoosporangia produced at tips of hyphae and set free. Oospores.

Family: Pythiaceae

Genus: *Pythium,* causing damping off of seedlings, seed decay, root rots, and cottony blight of turf grasses.

Phytophthora, P. infestans causing late blight of potato, others causing mostly root rots.

Family: Albuginaceae (The white rusts)

Genus: *Albugo, A. candida* causing white rust of crucifers.

Family: Peronosporaceae (The downy mildews)

Genus: *Plasmopara, P. viticola* causing downy mildew of grape.

Peronospora, P. nicotianae causing downy mildew (blue mold) of tobacco.

Bremia, B. lactucae causing downy mildew of lettuce.

Sclerospora, S. graminicola causing downy mildew of grasses.

Pseudoperonospora, P. cubensis causing downy mildew of cucurbits.

Subclass: ZYGOMYCETES (The bread molds)—Terrestrial fungi. Produce nonmotile asexual spores in sporangia. No zoospores. Their resting spore is a zygospore, produced by the fusion of two morphologically similar gametes.

Order: Mucorales—Produce zygospores. Nonmotile asexual spores formed in terminal sporangia.

Genus: *Rhizopus,* causing soft rot of fruits and vegetables.

Choanephora, C. cucurbitarum causing soft rot of squash.

THE HIGHER FUNGI

Class: ASCOMYCETES (The sac fungi)—Produce sexual spores, called ascospores, in groups of eight within an ascus.

Subclass: HEMIASCOMYCETES—Asci naked, not in ascocarps.

Order: Taphrinales—Asci arising from binucleate ascogenous cells.

Genus: *Taphrina*—causing peach leaf curl, plum pocket, oak leaf blister, etc.

Subclass: EUASCOMYCETES—Asci produced in ascocarps.

Series: PYRENOMYCETES (The perithecial fungi)—Asci in fruiting bodies completely closed (cleistothecia) or in fruiting bodies with an opening (perithecia).

Order: Erysiphales (The powdery mildews)—Mycelium and cleistothecia on surface of host plant.

Genus: *Erysiphe,* causing powdery mildew of grasses, cucurbits, etc.

Microsphaera, one species causing powdery mildew of lilac.

Podosphaera, P. leucotricha causing powdery mildew of apple.

Sphaerotheca, S. pannosa causing powdery mildew of roses and peach.

Uncinula, U. necator causing powdery mildew of grape.

Order: Sphaeriales—Perithecia with dark-colored, usually firm walls.

Genus: *Ceratocystis, C. ulmi* causing the Dutch elm disease.

Diaporthe, causing bean pod blight, citrus melanose, and fruit rot of eggplant.

Endothia, E. parasitica causing chestnut blight.

Glomerella, G. cingulata causing many anthracnose diseases and bitter rot of apple.

Gnomonia, causing anthracnose or leaf spot diseases.

Rosellinia, causing root diseases of fruit trees and vines.

Valsa, causing canker diseases of peach and other trees.

Xylaria, causing tree cankers and wood decay.

Order: Hypocreales—Perithecia light-colored, or red or blue.

Genus: *Claviceps, C. purpurea* causing ergot of rye.

Gibberella, causing foot or stalk rot of corn and small grains.

Nectria, causing twig and stem cankers of trees.

Series: PSEUDOSPHAEROMYCETES (The ascostromatic fungi)—Peritheciumlike stromata with asci in separate or single large cavities.

Order: Myriangiales—Cavities arranged at various levels and containing single asci.

Genus: *Elsinoe,* causing anthracnose of grape and raspberry, and scab of citrus.

Order: Dothideales—Cavities arranged in a basal layer and containing many asci. Perithecia lack pseudoparaphyses.

Genus: *Dibotryon, D. morbosum* causing block knot of cherries and plums.

Dothidella, D. ulei causing the leaf spot of rubber trees.

Guignardia, G. bidwellii causing black rot of grapes.

Mycosphaerella, causing leaf spots of many plants.

Order: Pleosporales—Cavities arranged in a basal layer and containing many asci. Perithecia have pseudoparaphyses.

Genus: *Ophiobolus, (Gaeumannomyces)* causing the take-all disease of wheat.

Physalospora, P. obtusa causing black rot of apples.

Venturia, V. inaequalis causing apple scab.

Series: DISCOMYCETES (The cup fungi)—Asci produced at the surface of fleshy cup- or saucer-shaped apothecia.

Order: Helotiales—Asci release spores through an apical, circular perforation.

Genus: *Coccomyces, C. hiemalis* causing cherry leaf spot.

Diplocarpon, D. rosae causing black spot of roses.

Lophodermium, causing pine needle blight.

Monilinia, M. fructicola causing brown rot of stone fruits.

Rhytisma, R. acerinum causing tar spot of maple leaves.

Sclerotinia, S. sclerotiorum causing watery soft rot of vegetables.

Order: Pezizales—Ascospores released through cap- or lidlike structure at tip of ascus.

Genus: *Pseudopeziza, P. medicaginis* causing alfalfa leaf spot.

Class: IMPERFECT FUNGI OR DEUTEROMYCETES (Asexual fungi)—Sexual reproduction and structures lacking or unknown.

Order: Sphaeropsidales—Asexual spores produced in pycnidia.

Genus: *Ascochyta, A. pisi* causing pea blight.

Coniothyrium, causing cane blight on raspberry.

Cytospora, causing canker diseases on peach and other trees. (sex. stage = *Valsa*)

Diplodia, D. zeae causing stalk and ear rot of corn.

Phoma, P. lingam causing black leg of crucifers.

Phomopsis, causing blights and stem cankers of trees.

Phyllosticta, causing leaf spots of many plants.

Septoria, S. apii causing late blight of celery.

Order: Melanconiales—Asexual spores produced in acervulus.

Genus: *Colletotrichum,* causing anthracnose on many field crops.

Coryneum, C. beijerincki causing blight on stone fruits.

Cylindrosporium, causing leaf spots on many kinds of plants.

Gloeosporium, similar if not identical to *Colletotrichum,* causing anthracnose on many plants.

Marssonina, causing leaf and twig blight of poplar, strawberry leaf scorch, and anthracnose of walnuts.

Melanconium, M. fuligenum causing bitter rot of grape.

Sphaceloma, causing anthracnose of grape, raspberry, and scab of citrus and avocado.

Order: Moniliales—Asexual spores produced on or within hyphae freely exposed to the air.

Genus: *Alternaria,* causing leaf spots and blights on many plants.

Asperigillus, causing rots of stored seeds.

Botrytis, *B. cinerea* causing gray mold and blights on many plants.

Cercospora, one species causing early blight of celery.

Cladosporium, *C. fulvum* causing leaf-mold of tomato.

Fusarium, causing wilt and root rot diseases of many annual plants and cankers of forest trees.

Fusicladium, causing apple scab (sex. stage = *Venturia*).

Graphium, *G. ulmi* causing Dutch elm disease (sex. stage = *Ceratocystis*).

Helminthosporium, causing blight of cereals and diseases of turf grasses.

Penicillium, causing blue mold rot of fruits and other fleshy organs.

Phymatotrichum, *P. omnivorum* causing root rot of cotton and other plants.

Pyricularia, causing rice blast and gray leaf-spot of turf grasses.

Strumella, causing cankers on oak.

Thielaviopsis, *T. basicola* causing black root rot of tobacco.

Verticillium, causing wilt of many annuals and perennials.

Order: Mycelia Sterilia—No sexual or asexual spore forms common or known.

Genus: *Rhizoctonia*, causing root rots and crown rots of annals and brown-patch of turf grasses (Perfect stage *Thanatephorus*).

Sclerotium, causing root and stem rots of many plants (Perfect stage *Pellicularia*)

Class: BASIDIOMYCETES (The club fungi)—Sexual spores, called basidiospores or sporidia, are produced externally on a one- or four-celled structure called a basidium.

Subclass: HETEROBASIDIOMYCETES (The rust and smut fungi)—Basidium with cross walls or being the promycelium of a teliospore. Teliospores single or united into crusts or columns, remaining in host tissue or bursting through the epidermis.

Order: Ustilaginales—Fertilization by means of union of compatible spores, hyphae, etc. Only teliospores are produced.

Genus: *Sphacelotheca*, several species causing loose smut of sorghum.

Tilletia, several species causing bunt, or stinking smut, of wheat.

Urocystis, *U. cepulae* causing smut of onion.

Ustilago, causing smut of corn, wheat, barley, etc.

Order: Uredinales—Sperm cells called spermatia or pycniospores fertilize special receptive hyphae in spermagonia (pycnia). Produce aeciospores, uredospores (repeating spores), teliospores, and basidiospores.

Genus: *Cronartium*, *C. ribicola* causing white pine blister rust.

Gymnosporangium, *G. juniperi-virginianae* causing cedar apple rust.

Melampsora, *M. lini* causing rust of flax.

Phragmidium, one species causing rust of roses.

Puccinia, several species causing rust of cereals.

Uromyces, *U. phaseoli* causing rust of beans.

Subclass: HOMOBASIDIOMYCETES (The wood decay and root rot fungi)—Basidium without cross walls. Basidiocarp lacking or present. Include the mushrooms, shelf fungi, puff balls, etc.

Series: HYMENOMYCETAE—Basidia produced in a hymenium becoming exposed to the air before the spores are shot off from the sterigmata.

Order: Exobasidiales—Basidiocarp lacking: basidia produced on surface of parasitized tissue.

Genus: *Corticium*, one species causing the red thread disease of turf grasses.

Exobasidium, causing leaf, flower and stem galls on ornamentals.

Order: Polyporales—Hymenium lining the surfaces of small pores or tubes.

Genus: *Fomes*, causing heart rot of many trees.

Pellicularia (Sclerotium), causing root and stem rots of many plants.

Polyporus, causing root and stem rot of many trees.

Poria, causing wood and root rots of forest trees.

Stereum, causing wood decay and silver leaf disease of trees.

Thanatephorus, *(Rhizoctonia)* causing root and stem rots of many annual plants and brown patch of turf grasses.

Typhula, one species causing snowmold or blight of turf grasses.

Order: Agaricales—Hymenium on radiating gills or lamellae.

Genus: *Armillaria, A. mellea* causing root rots of forest and fruit trees.

Lenzites, causing brown rot of conifers and decay of wood products.

Marasmius, causing the fairy ring disease of turf grasses.

Peniophora, causing decay of conifer logs and pulpwood.

Pholiota, causing brown wood rot in deciduous forest trees.

Pleurotus, causing white rot on many deciduous forest trees.

Schizophyllum, causing white rot in deciduous forest trees.

IDENTIFICATION

Since each fungus disease of plants is usually caused by only one fungus, and since there are more than 100,000 different species of fungi, the identification of the fungus species on a diseased plant specimen or culture of a fungus means that all but one of all the known fungus species must be excluded.

The most significant characteristics of a fungus used for identification are its spores and fructifications, or spore-bearing structures. These are examined under the compound microscope directly after removal from the specimen. The specimen is often kept moist for a few days to promote fructification development, or the fungus may be isolated and grown on artificial media, and identification is made on the basis of the fructifications produced on the media.

The shape, size, color, and manner of arrangement of spores on the sporophores or the fruiting bodies, as well as the shape, color, etc. of the sporophores or fruiting bodies, are sufficient characteristics to suggest, to one somewhat experienced in the taxonomy of fungi, the class, order, family, and genus to which the particular fungus belongs. In any case, these characters can be utilized to trace the fungus through published analytical keys of the fungi to the genus and, finally, to the species to which it belongs. Once the genus of the fungus has been determined, specific descriptions of the species are found in monographs of genera or in specific publications in research journals.

Since there are usually lists of the pathogens affecting a particular host plant, one may use such host indexes as short cuts in quickly finding names of fungus species that might apply to the fungus at hand. Host indexes, however, merely offer suggestions in determining identities, which must ultimately be determined by reference to monographs and other more specific publications.

SYMPTOMS CAUSED BY FUNGI ON PLANTS

Fungi cause local or general symptoms on their hosts and these may occur separately on different hosts, concurrently on the same host, or follow one another on the same host. In general fungi cause local or general necrosis or killing of plant tissues, hypertrophy and hypoplasia or stunting of plant organs or entire plants, and hyperplasia or excessive growth of plant parts or whole plants.

The most common necrotic symptoms are:

Leaf spots—Localized lesions on host leaves consisting of dead and collapsed cells.

Blight—General and extremely rapid browning of leaves, branches, twigs, and floral organs resulting in their death.

Canker—A localized wound or necrotic lesion, often sunken beneath the surface of the stem of a woody plant.

Root rot—Disintegration or decay of part or all of the root system of a plant.

Damping off—The rapid death and collapse of very young seedlings in the seed bed or field.

Basal stem rot—Disintegration of the lower part of the stem.

Soft rots and dry rots—Maceration and disintegration of fruits, roots, bulbs, tubers, and fleshy leaves.

Anthracnose—A necrotic and sunken ulcerlike lesion on the stem, leaf, fruit or flower of the host plant.

Scab—Localized lesions on host fruit, leaves, tubers, etc., usually slightly raised or sunken and cracked, giving a scabby appearance.

Almost all of the above symptoms may also cause pronounced stunting of the infected plants. In addition, certain other symptoms such as leaf rust, mildews, wilts, and even certain diseases causing hyperplasia of some plant organs, such as clubroot, may cause stunting of the plant as a whole.

Symptoms associated with hypertrophy or hyperplasia and distortion of plant parts include:

Clubroot—Enlarged roots appearing like spindles or clubs.

Galls—Enlarged portions of plants usually filled with fungus mycelium.

Warts—Wartlike protuberances on tubers and stems.

Witches'-brooms—Profuse, upward branching of twigs.

Leaf curls—Distortion, thickening and curling of leaves.

In addition to the above, three groups of symptoms may be added:

Wilt—Usually a generalized secondary symptom in which leaves or shoots lose
    their turgidity and droop because of a disturbance in the vascular system of
    the root or of the stem.
Rust—Many, small lesions on leaves or stems, usually of a rusty color.
Mildew—Chlorotic or necrotic areas on leaves, stems, and fruit usually cov-
    ered with mycelium and the fructifications of the fungus.

In many diseases, the pathogen grows or produces various structures on the surface of the host. These structures, which include mycelium, sclerotia, sporophores, fruiting bodies, and spores, are called signs and are distinct from symptoms, which refer only to the appearance of infected plants or plant tissues. Thus, in the mildews, for example, one sees mostly the signs consisting of a whitish, downy growth of fungus mycelium and spores on the plant leaves, fruit, or stem, while the symptoms consist of chlorotic or necrotic lesions on leaves, fruit, and stem, reduced growth of the plant, etc.

isolation of fungi (and bacteria)

Most plant diseases can be diagnosed by observation with the naked eye or with the microscope and, for these, isolation of the pathogen is not necessary. There are many fungal and bacterial diseases, though, in which the pathogen cannot be identified because it is mixed with one or more contaminants, because it has not yet produced its characteristic fruiting structures and spores, because the same disease could be caused by more than one similar-looking pathogen and perhaps by some environmental factor, or because the disease is caused by a new, previously unknown pathogen that must be isolated and studied. Just as often, pathogens of even known diseases must be isolated from diseased plant tissues whenever a study of the characteristics, habits, etc. of these pathogens is to be undertaken.

PREPARING FOR ISOLATION

Even before one attempts to isolate the fungus or bacterium pathogen from a diseased plant tissue, several preliminary operations must be performed. These include:

1. Sterilization of glassware, such as petri dishes, test tubes, pipettes, etc., by
    dry heat (150 to 160°C for 1 hour or more) or autoclaving or by dipping for 1
    minute or more in a potassium dichromate–sulfuric acid solution, or in
    1:1000 mercuric chloride, or 5 percent formalin, or 95 percent ethyl alcohol.
    All chemically treated glassware should be rinsed through at least three
    changes of sterile (boiled or autoclaved) water.

2. Preparation of solutions for treating the surface of the infected or infested tissue so as to eliminate or markedly reduce surface contaminants that could interfere with the isolation of the pathogen. These solutions can be used either as a surface wipe or as a dip. The most commonly used surface sterilants include: 5.75 percent sodium hypochlorite (1 Clorox: 9 water) solution, used both for wiping infected tissues or dipping sections of such tissues in it and for wiping down table or bench surfaces before making isolations; 95 percent ethyl alcohol, which is mild and is used for leaf dips for 3 seconds or more; mercuric chloride 1:1000, for 15 to 45 seconds; or mercuric chloride 1:1000 in 50 percent ethyl alcohol (Rada's solution). The tissues must be blotted dry with a sterile paper towel when treated with the first two solutions but they must be rinsed in 3 changes of sterile water when treated with the last two solutions.

3. Preparation of culture media on which the isolated fungal or bacterial pathogens will grow. An almost infinite number of culture media can be used to grow plant pathogenic fungi and bacteria. Some of them are entirely synthetic—made up of known amounts of certain chemical compounds—and are usually quite specific for certain pathogens. Some are liquid or semiliquid and are used primarily for growth of bacteria but also of fungi in certain cases. Most media contain an extract of a natural source of carbohydrates and other nutrients, such as potato, corn meal, lima bean, or malt extract, to which variable amounts of agar are added to solidify the medium and form a gel on or in which the pathogen can grow and be observed. The most commonly used media are potato dextrose agar (PDA), which is good for most, but not all fungi, water agar or glucose agar (1–3 percent glucose in water agar) for separating some fungi (*Pythium* and *Fusarium*) from bacteria, and nutrient agar, which contains beef extract and peptone and is good for isolating bacterial plant pathogens. Fungi can also be separated in culture from bacteria by adding 1 or 2 drops of a 25 percent solution of lactic acid, which inhibits growth of bacteria, to 10 ml of the medium before pouring it in the plate. Solutions of culture media are prepared in flasks which are then plugged and placed in an autoclave at 120°C and 15 pounds pressure for 20 minutes (Fig. 35). The sterilized media are then allowed to cool somewhat and are subsequently poured from the flask into sterilized petri dishes, test tubes, or other appropriate containers. If agar was added to the medium, the latter will soon solidify and is then ready to be used for growth of the fungus or bacterium. The pouring of the culture medium into petri dishes, tubes, etc. is carried out as aseptically as possible either in a separate culture room or in a clean room free from drafts and dust. In either case, the work table should be wiped with a 10 percent Clorox solution, hands should be clean and tools such as scalpels, forceps or needles should be dipped in alcohol and flamed to prevent introduction of contaminating microorganisms.

It must be kept in mind that, of the different plant pathogens, the bacteria are the only ones whose members can all be grown on culture media. Although most fungi can be cultured on nutrient media with ease, some of them have specific and exacting requirements and will not grow on most commonly used nutrient media. Some groups of fungi, namely the Erysiphales, the causes of the powdery mildew diseases, and the Peronosporales, which cause the downy mildews, are considered strictly obligate parasites and cannot be grown on culture media. Another group of fungi, the Uredinales, which cause the rust diseases of plants, were,

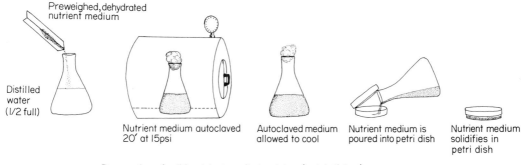

Preparation of solid nutrient media in plates (petri dishes)

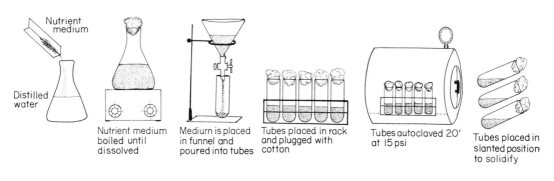

Preparation of solid nutrient media in test tube slants

FIGURE 35.
Preparation of solid nutrient media in plates (petri dishes) and in test tube slants.

until recently, also thought to be strictly obligate parasites. In the last few years, however, it has become possible to grow in culture some stages of some rust fungi by the addition of certain components to the media and so the rust fungi are no longer considered to be obligate parasites. Of the other pathogens, only some spiroplasmas have been grown in culture. None of the other 50 or so mycoplasmalike organisms and none of the rickettsialike bacteria, viruses, nematodes, or protozoa have been grown on nutrient culture media, so far, although it is expected that media for culturing mycoplasmalike organisms and rickettsialike bacteria will soon be discovered.

ISOLATING THE PATHOGEN

FROM LEAVES

If the infection of the leaf is still in progress in the form of a fungal leaf spot or blight and if there are spores present on the surface, a few spores may be shaken loose over a petri plate containing culture medium or picked up at the point of a sterile needle or scalpel and placed on the surface of the culture medium. If the fungus does grow in culture, isolated colonies of mycelium will appear in a few days as a result of germination of the spores. These can be subcultured on separate plates and thus secure some plates that will contain the pathogen free of any contaminants.

Sometimes, isolation of the pathogen from fungal or bacterial leaf spots and blights is made by surface sterilizing the area to be cut with Clorox or Rada's solution, removing a small part of the infected tissue with sterile scalpel, etc., and placing it in a plate containing a nutrient medium.

The most common method, however, for isolating pathogens from infected leaves as well as other plant parts is the one in which several small sections 5- to 10-mm square are cut from the margin of the infected lesion so as to contain both diseased and healthy-looking tissue (Fig. 36). These are placed in one of the surface sterilant solutions, making sure that the surfaces do get wet, and after about 15 to 30 seconds the sections are taken out aseptically one by one and at regular, e.g., 10- to 15-second intervals, so that each of them has been surface-sterilized for different times. The sections are then blotted dry on clean, sterile paper towels or are washed in three changes of sterile water, and are finally placed on the nutrient medium, usually three to five per dish. Those sections surface-sterilized the shortest time usually contain contaminants along with the pathogen, while those surface-sterilized the longest produce no growth at all because all organisms have been killed by the surface sterilant. Some of the sections left in the surface sterilant for intermediate periods of time, however, will allow only the pathogen to grow in culture in pure colonies, since the sterilant was allowed to act long enough to kill all surface contaminants but not too long to kill the pathogen which was advancing alone from the diseased to the healthy tissue. These colonies of

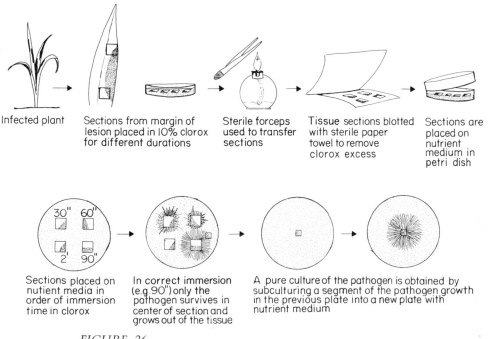

Infected plant

Sections from margin of lesion placed in 10% clorox for different durations

Sterile forceps used to transfer sections

Tissue sections blotted with sterile paper towel to remove clorox excess

Sections are placed on nutrient medium in petri dish

Sections placed on nutient media in order of immersion time in clorox

In correct immersion (e.g. 90") only the pathogen survives in center of section and grows out of the tissue

A pure culture of the pathogen is obtained by subculturing a segment of the pathogen growth in the previous plate into a new plate with nutrient medium

FIGURE 36.
Isolation of fungal pathogens from infected plant tissue.

the pathogen are then subcultured aseptically for further study of the pathogen.

If fruiting structures (pycnidia, perithecia) are present on the leaf, it is sometimes possible to pick them out, drop them in the surface sterilant for a few seconds, and then plate them on the nutrient medium. This procedure, however, requires that most of the work be done under the stereoscopic microscope (binoculars) since the fruiting structures are generally too small to see and to handle with the naked eye. Fruiting structures, after surface sterilization, may also be crushed in a small drop of sterile water and then the spores in the water are diluted serially in small tubes or dishes containing sterile water. Finally, a few drops from each tube of the serial dilution are placed on a nutrient medium and single colonies free of contaminants develop from germinating spores obtained from some of the serial dilution tubes.

The serial dilution method is often used to isolate pathogenic bacteria from diseased tissues contaminated with other bacteria. After surface sterilization of sections of diseased tissues from the margin of the infection, the sections are ground aseptically but quite thoroughly in a small volume of sterile water, and then part of this homogenate is diluted serially in equal volumes or ten times the volume of the initial water. Finally, plates containing nutrient agar are streaked with a needle or loop dipped in each of the different serial dilutions and single colonies of the pathogenic bacterium are obtained from the higher dilutions that still contain bacteria.

FROM STEMS, FRUITS,
AND OTHER AERIAL PLANT PARTS

Almost all the methods described for isolating fungal and bacterial pathogens from leaves can also be used to isolate these pathogens from superficial infections of the above tissues. In addition to these methods, however, pathogens can be often isolated easily from infected stems, fruits, etc., in which the pathogen has penetrated fairly deeply, by splitting the stem or breaking the fruit from the healthy side first and then tearing it apart toward and past the infected margin, thus exposing tissues not previously exposed to contaminants and not touched by hand or knife and therefore not contaminated. Small sections of tissue can be cut from the freshly exposed area of the advancing margin of the infection with a flamed scalpel and can be plated directly on the culture medium.

FROM ROOTS, TUBERS,
FLESHY ROOTS, VEGETABLE FRUITS
IN CONTACT WITH SOIL, ETC.

Isolating pathogens from any diseased plant tissue in contact with soil presents the additional problems of numerous saprophytic organisms invading the plant tissue after it has been killed by the pathogen. For this reason, repeated, thorough washing of such diseased tissues to remove all soil and most of the loose, decayed plant tissue, in which most of the saprophytes are present, is the first step in isolating the pathogen. If the infected root is small, once it is washed thoroughly pathogens can be isolated from it by following one of the methods described for isolating pathogens from leaves. If isolation is attempted from fleshy roots or other fleshy tissues and penetration of the pathogen is slight resulting only in surface lesions, the tissue is washed free from adhering soil, and several bits of tissue from the margin of the lesions are placed in Clorox or Rada's solution. The tissue sections are picked from the solution one by one, blotted or washed in sterile water, and placed on agar in petri plates. If the pathogen has penetrated deeply into the fleshy tissue, the method described above for stems and fruit, i.e., by breaking the specimens from the healthy side first, then tearing toward the infected area and plating bits taken from the previously unexposed margin of the rot, can be used most effectively.

life cycles of fungi

Although the life cycles of the fungi of the different groups vary greatly, the great majority of them go through a series of steps that are quite similar (Fig. 37). Thus almost all fungi have a spore stage with a simple, haploid (possessing one set of chromosomes or $1N$) nucleus. The spore germinates into a hypha which also contains haploid nuclei. The hypha may either produce simple, haploid spores again (as is always the case in the Imperfect Fungi) or it may fuse with another hypha to produce a fertilized hypha in which the nuclei unite to form one diploid nucleus, called zygote (containing two sets of chromosomes, or $2N$). In the Phycomycetes the zygote will divide to produce simple, haploid spores which close the cycle. In a brief phase of most Ascomycetes, and generally in the Basidiomycetes, the two nuclei of the fertilized hypha do not unite, but remain separate within the cell in pairs (dikaryotic or $N + N$) and divide simultaneously to produce more hyphal cells with pairs of nuclei. In the Ascomycetes, the dikaryotic hyphae are found only inside the fruiting body, in which they become the ascogenous hyphae, since the two nuclei of one cell of each hypha unite into a zygote ($2N$) which divides meiotically to produce ascospores that contain haploid nuclei.

In the Basidiomycetes haploid spores produce only short haploid hyphae. Upon fertilization, dikaryotic ($N + N$) mycelium is produced and this develops into the main body of the fungus. Such dikaryotic hyphae may produce, asexually, dikaryotic spores that will grow again into a dikaryotic mycelium. Finally, however, the paired nuclei of the cells unite and form zygotes. The zygotes divide meiotically and produce basidiospores that contain haploid nuclei.

In the Imperfect Fungi, of course, only the asexual cycle (haploid spore → haploid mycelium → haploid spore) is found. Even in the other fungi, however, a similar asexual cycle is the most common one by far, since it can be repeated many times during each growth season. The sexual cycle usually occurs only once a year.

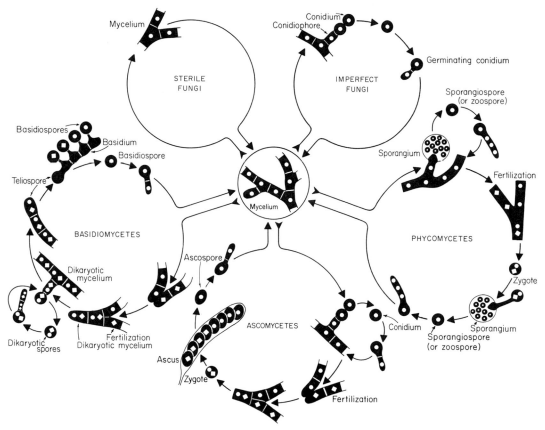

FIGURE 37.
Schematic presentation of the generalized life cycles of the main groups of phytopathogenic fungi.

control of
fungus diseases of plants

The endless variety and the complexity of the many fungus diseases of plants have led to the development of a correspondingly large number of approaches for their control. The particular characteristics in the life cycle of each fungus, its habitat preferences and its performance under certain environmental conditions are some of the most important points to be considered in attempting to control a plant disease caused by a fungus. Although some diseases can be controlled completely by just one type of control measure, a combination of measures is usually necessary for satisfactory control of most diseases.

The use of pathogen-free seed or propagating stock is always recommended and, for control of certain diseases, it is mandatory. Destruction of plant parts or refuse harboring the pathogen, destruction of volunteer plants or alternate hosts of the pathogen, use of clean tools and containers, proper drainage of fields and aeration of plants, are all very important practices in the control of most plant diseases caused by fungi. Crop rotation is helpful in controlling diseases caused by some fungi, but does not satisfactorily control fungi that have wide host ranges, can live saprophytically for a long time, or produce long-lived resting spores.

The use of plant varieties resistant to certain pathogens has found its greatest application in controlling fungus diseases of plants. Some of the most serious fungus diseases (e.g., rusts, *Fusarium* wilts) of the most important crop plants, are successfully controlled today by the use of resistant varieties. Although the degree of control through resistant varieties varies with the crop and the pathogenic fungus involved, its successes as of this time and the very low overall cost of such control make this type of control the most promising for the future.

The most effective method, however, and, sometimes the only one available for controlling most of the fungus diseases of plants, is through application of chemical sprays or dusts on the plants, their seeds, or into the soil where the plants are to grow. Soil-inhabiting fungi may be controlled in small areas by steam or electric heat, and in somewhat larger areas by volatile liquids, such as formaldehyde, chloropicrin, methyl bromide. Some diseases caused by soil-inhabiting fungi can also be controlled, and at a much lower cost, by applying fungicides on the seeds or other propagating materials, such as tubers and corms. Such treatment will also protect the seed from mycelium or spores carried on the seed. Fungicides used for seed treatment include, among others, carboxin, chloroneb, chloranil, dichlone, captan, and thiram.

Most fungicides are used to prevent diseases on the aboveground parts of plants and are applied on the foliage as sprays or dusts. Most of these are protectant, since they can only prevent fungi from causing infection, but cannot stop an infection once it has started. The number of such fungicides is great and includes many inorganic and organic compounds. In recent years several systemic fungicides have been developed, and their use and effectiveness are increasing steadily. In addition to these, certain antibiotics (e.g., cycloheximide) are also effective against certain fungus diseases of plants.

In some diseases (e.g., loose smuts of cereals) the fungus is carried in the seed and control can be obtained only through treatment of the seed with systemic fungicides or hot water. In others, control of the insect vectors may be the only available possibility. In general, great advances have been made toward controlling fungus diseases of plants, especially through resistant varieties and through chemicals, and as a result, these diseases are probably much easier to control than any other group of plant diseases, although the losses caused by fungus diseases of plants are still very great.

SELECTED REFERENCES

Ainsworth, G. C., *et al.* (eds.). 1965–1973. "The Fungi. An Advanced Treatise." Vols. 1–4. Academic Press, New York.

Alexopoulos, C. J. 1962. "Introductory Mycology," Wiley, New York. 613 pp.

Barnett, H. L., and B. B. Hunter. 1972. "Illustrated Genera of Imperfect Fungi," Burgess, Minneapolis, Minnesota. 218 pp.

Barnett, H. L., and F. L. Binder. 1973. The fungal host–parasite relationship. *Ann. Rev. Phytopathol.* **11**:273–292.

Clements, F. E., and C. L. Shear. 1957. "The Genera of Fungi." Hafner, New York. 496 p. 58 pl.

Cummins, G. B. 1959. "Illustrated Genera of Rust Fungi." Burgess, Minneapolis, Minn. 131 p.

Fergus, C. L. 1960. "Illustrated Genera of Wood Destroying Fungi," Burgess, Minneapolis, Minnesota. 132 p.

Griffin, D. M. 1969. Soil water in the ecology of fungi. *Ann. Rev. Phytopathol.* 7:289–310.

Kendrick, B. (Ed.). 1971. "Taxonomy of Fungi Imperfecti." Toronto Univ. Press, Toronto. 309 pp.

Meredith, D. S. 1973. Significance of spore release and dispersal mechanisms in plant disease epidemiology. *Ann. Rev. Phytopathol.* **11**:313–342.

Stevens, F. L. 1913. "The Fungi Which Cause Plant Disease," Macmillan, New York. 754 p.

Tsao, P. H. 1970. Selective media for isolation of pathogenic fungi. *Ann. Rev. Phytopathol.* **8**:157–186.

diseases caused by the lower fungi

DISEASES CAUSED BY MYXOMYCETES

Myxomycetes are fungi whose vegetative body is a plasmodium, i.e., an amoeboid mass of protoplasm that has many nuclei and no definite cell wall. In the true Myxomycetes, also called slime molds, the plasmodium does not invade plant cells and is used up to form superficial fructifications that contain resting spores. In another group of fungi, the Plasmodiophorales, the vegetative body is also a plasmodium but it is produced only in the cells of the host plant and their resting spores are produced in masses but not in distinct fructifications. Both the true slime molds and the Plasmodiophorales produce zoospores that usually have two flagella (Fig. 38).

Two groups of Myxomycetes cause diseases of plants, the one by simply growing externally on the surface of leaves, stems and fruits without parasitizing the plant, and the other by entering and parasitizing the roots and other below-ground parts of plants (Fig. 39). These groups are:

I. Physarales. It includes the true slime mold genera *Fuligo, Mucilago, Physarum,* and others, which cause slime molds on the surface of low-lying plants.

II. Plasmodiophorales. It includes three obligate plant parasitic genera: *Plasmodiophora, Polymyxa,* and *Spongospora.*

THE TRUE SLIME MOLDS

Slime molds appear on plants growing low on the ground, such as turf grasses, strawberries, vegetables, and small ornamentals (Fig. 40). They

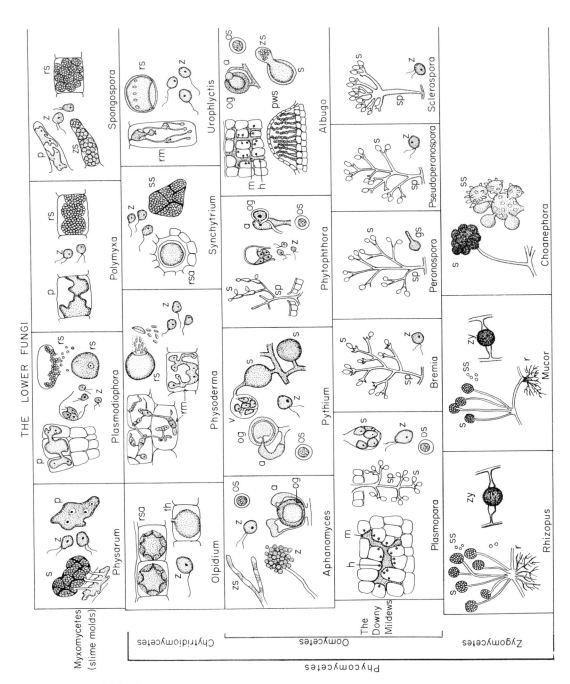

FIGURE 38.
The most common Lower Fungi (Myxomycetes and Phycomycetes) that cause disease in plants: a—antheridium, gs—germinating sporangium, h—haustorium, m—mycelium, og—oogonium, os—oospore, p—plasmodium, pws—pustule with sporangia, rm—rhizomycelium, rs—resting spore, rsa—resting sporangium, s—sporangium, sp—sporangiophore, ss—sporangiospore, th—thallus, z—zoospore, zs—zoosporangium, zy—zygospore.

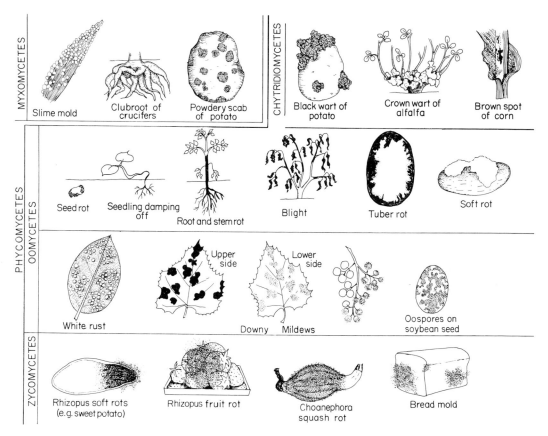

FIGURE 39.

The most common symptoms caused by Myxomycetes and Phycomycetes.

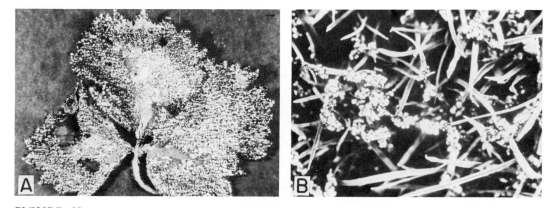

FIGURE 40.

(A) Begonia leaf covered with fructifications (sporangia) of a slime mold. (B) Slime mold fructifications on the blades of a turfgrass.

are most common in warm weather following heavy rains or watering. All aboveground parts of plants in some areas, and even the soil between plants, may be covered by a creamy white or colored slimy growth which later changes to distinct, crusty, ash-gray, or colored fruiting structures. The latter give the affected plants a dull gray appearance.

Slime molds are saprophytic members of the true slime molds (Myxomycetes). Their plasmodium creeps like an amoeba and feeds on decaying organic matter and microorganisms such as bacteria which it simply engulfs and digests. There are many species of slime mold fungi, the most common of which are *Physarum, Fuligo,* and *Mucilago.*

The plasmodium grows mostly in the upper layer of the soil and in the thatch, but during or after warm wet weather it comes to the soil surface and creeps over low-lying vegetation. On the plant surface, which these fungi use merely as a means of support, the plasmodium produces the crusty fruiting structures which vary in size, shape and color depending on the species of slime mold (Figs. 40 and 41). The fruiting structures are sporangia, i.e., containers filled with dark masses of powdery spores, and are easily rubbed off the plant. The spores are spread by wind, water, mowers or other equipment and can survive unfavorable weather. In cool, humid weather, the spores absorb water, their cell wall cracks open and a

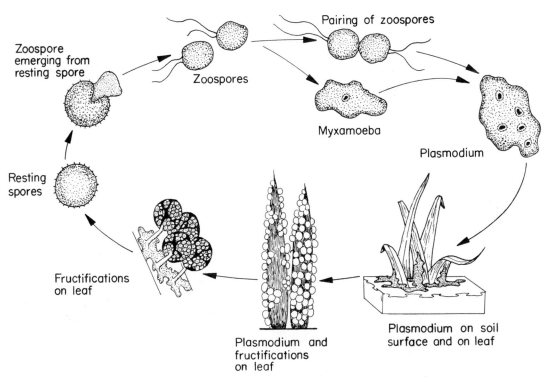

FIGURE 41.
Life cycle of slime mold fungi.

single, naked, motile swarm spore emerges from each. The swarm spores feed like the plasmodium while they undergo several divisions and various changes. Finally they unite in pairs to form amoeboid zygotes which enlarge, become multinucleate and become the plasmodium.

No control is usually considered necessary against slime molds. When they become too numerous and unsightly, breaking up of the spore masses by raking, brushing, or hosing down with water in dry weather, and removal of affected leaves or mowing of grass corrects the problem. Slime mold fungi are generally very sensitive to many fungicides, so if the problem becomes serious, spraying with any fungicide, such as captan, thiram, etc., used to control other diseases of the particular plant should also control the slime molds.

SELECTED REFERENCES

Alexopoulos, C. J. 1962. "Introductory Mycology" (2nd ed.). John Wiley and Sons, Inc., New York. 613 p.

Couch, H. B. 1973. "Diseases of Turfgrasses" (2nd ed.). Krieger Publishing Co., New York. 348 p.

DISEASES OF BELOW-GROUND PLANT PARTS CAUSED BY MYXOMYCETES

Three common and often severe diseases of below-ground parts of plants are caused by Myxomycetes of the order Plasmodiophorales. The fungi involved are:

> *Plasmodiophora*, causing clubroot of crucifers
> *Polymyxa*, causing a root disease of cereals and grasses
> *Spongospora*, causing the powdery scab of potato (Fig. 42)

These fungi are widespread in soils in which they overwinter as resting spores. When the temperature is favorable and moisture is plentiful, the resting spore produces one zoospore which infects a root hair and produces a plasmodium. The latter is transformed into zoosporangia which produce numerous secondary zoospores which, probably after pairing, enter root or tuber tissues, produce plasmodium and cause the typical disease. The plasmodium spreads into the host tissues and is finally transformed into overwintering resting spores.

The pathogens are obligate parasites and, although they can survive in the soil as resting spores for many years, they can only grow and multiply in a rather limited number of hosts. The plasmodium lives off the host cells it invades and it does not kill these cells. On the contrary, in some diseases many invaded and adjacent cells are stimulated by the pathogen to enlarge and divide, thus making available more nutrients for the pathogen. These pathogens spread from plant to plant by means of zoospores and by anything that moves soil or water containing spores, by infected transplants, etc. Control of these pathogens is difficult and depends mostly on avoiding contamination of pathogen-free soils, use of healthy transplants, tubers, etc., crop rotation with nonhost plants, adjustment of soil pH, etc.

FIGURE 42.
Mature stage of powdery scab of potato caused by *Spongospora subterranea*.

Polymyxa and *Spongospora,* in addition to the diseases they cause, can also transmit destructive plant viruses, *Polymyxa* being a vector of soil-borne wheat mosaic virus and *Spongospora* of the potato mop top virus.

- ● *Clubroot of Crucifers*

The clubroot disease of cruciferous plants, such as cabbage and cauliflower, is widely distributed all over the world, found wherever plants of the mustard family grow. It has been observed most frequently in Europe and North America.

Clubroot causes serious losses when susceptible varieties of any cruciferous species are grown in infested fields. Fields once infested with the clubroot pathogen remain so indefinitely and become unfit for cultivation of crucifers.

Symptoms. Infected plants may have pale green to yellowish leaves which may show flagging and wilting in the middle of hot, sunny days but may recover during the night (Fig. 43A). Affected plants show almost normal vigor at first, but then gradually become stunted. Young plants may be killed by the disease within a short time after infection, while older plants may remain alive but fail to produce marketable heads.

The most characteristic symptoms of the disease appear on the roots and sometimes the underground part of the stem (Fig. 43B). The symptoms consist of small or large spindlelike, spherical, knobby, or club-shaped swellings on the roots and rootlets. These malformations may be isolated and cover only part of some roots or they may coalesce and cover the entire root system of the plant. The older and usually the larger clubbed roots disintegrate before the end of the season due to invasion by bacteria and other weakly parasitic soil microorganisms.

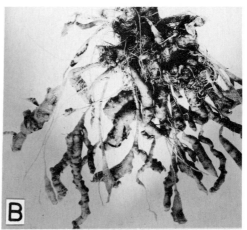

FIGURE 43.
(A) Midday wilting of half-grown cabbage plants that have severely clubbed roots.
(B) Malformed, spindlelike or clubbed roots infected with *Plasmodiophora brassicae.*

The pathogen: Plasmodiophora brassicae. It is a slime mold, the body of which is a plasmodium. The plasmodium gives rise to zoosporangia or to resting spores. Upon germination they produce zoospores. The single zoospore from resting spores penetrates host root hairs and there develops into a plasmodium. After a few days, the plasmodium cleaves into multinucleate portions surrounded by separate membranes; each portion develops into a zoosporangium. The zoosporangia are discharged outside the host through pores dissolved in the host cell wall, and each zoosporangium releases four to eight secondary zoospores. Some of these zoospores fuse in pairs to produce zygotes which can cause new infections and produce new plasmodium. The plasmodium finally turns into resting spores (Fig. 44) which are released into the soil upon disintegration of the host cell walls by secondary microorganisms.

Development of disease. The plasmodium resulting from the germination of the secondary zoospores penetrates young root tissues directly; it can also penetrate secondarily thickened roots and underground stems through wounds. From these points of primary infection the plasmodium spreads to cortical cells and reaches the cambium through direct penetration of host cells (Fig. 45). From the point of infection of the cambium the plasmodium spreads in all directions in the cambium, outward into the cortex and inward toward the xylem and into the medullary rays. Single-point infections result in spindle-shaped clubs, being widest at the point of invasion and tapering off away from it.

As the plasmodia pass through cells, they become established in some of them and stimulate these cells to abnormal enlargement (hypertrophy) and abnormal division (hyperplasia). Infected cells may be five or more

FIGURE 44.
Scanning electron micrograph of resting spores of *Plasmodiophora brassicae* within cells of club roots. (Photo courtesy M. F. Brown and H. G. Brotzman) ×1000.

times larger than adjacent uninfected ones. The infected cells of a club are distributed in small groups throughout the diseased tissue and the groups are usually separated by uninfected cells. The stimulus which is responsible for the abnormal growth of the cells appears to diffuse in advance of the pathogen and acts on the noninvaded cells of diseased tissues as well as on the infected ones. Actively growing and dividing cells, i.e., cambial cells, are more easily invaded by the pathogen and are more responsive to the stimulus than other cells.

In most cases many cells of infected clubs remain free from plasmodia, but in rare instances almost all the cells of a club may be infected. When few cells are infected the plasmodia become large, whereas when many cells are infected, they remain relatively small. Thus, there seems to be a fairly constant ratio between the volume occupied by the plasmodium

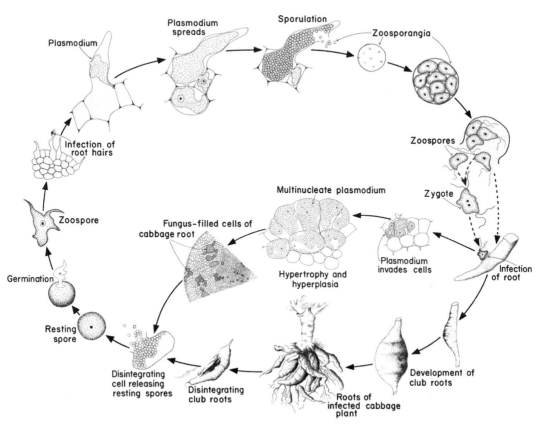

FIGURE 45.
Disease cycle of clubroot of crucifers caused by *Plasmodiophora brassicae.*

and that of the diseased tissue, the former being approximately 30 percent of the latter.

The plasmodium-infected clubs not only utilize much of the food required for the normal growth of the plant, but they also interfere with the absorption and translocation of mineral nutrients and water through the root system resulting in gradual stunting and wilting of the above-ground parts of the plant. Furthermore, the rapidly growing and greatly enlarged cells of the club tissues are unable to form a cork layer at the surface and are easily ruptured and invaded by secondary, weakly parasitic microorganisms. The invasion of clubs by bacteria, and the subsequent disintegration by them, lead to formation of substances toxic to the plant which are partly responsible for the wilting of the tops.

Control. Growing cruciferous crops in fields known to be infested with the clubroot pathogen should be avoided. If that is not possible, cabbage and the other susceptible cruciferous crops should be planted in well-drained fields with a pH slightly above neutral (usually about 7.2) or in fields in which the pH is adjusted to 7.2 by the addition of the proper amount of hydrated lime. The use of soil liming in the control of clubroot is based on the fact that spores of the clubroot organism germinate poorly or not at all in alkaline media.

Although soil fumigants capable of disinfecting fields from the clubroot organism are available, the cost of materials and of application are as yet prohibitive. However, seedbed areas can be kept free of clubroot by treating the soil with chloropicrin, methyl bromide, Mylone, Vapam, or Vorlex approximately 2 weeks before planting. The clean, clubroot-free seedlings should, upon transplanting, be watered with a solution of benomyl (or PCNB = pentachloronitrobenzene) and again with the same solution one month later.

The search for and development of varieties of cruciferous hosts resistant to clubroot has been only partially successful. Such resistance has been most highly developed in varieties of rutabaga and turnip, but the extensive use of resistant varieties in infested soil has resulted in the appearance of highly virulent new races of the pathogen, sometimes within three years from the time of the release of the resistant varieties. Although some varieties of the most popular cruciferous hosts are resistant to certain races of the clubroot organism and can be grown in areas infested with these races, no varieties of cabbage, cauliflower, Brussels sprouts, or broccoli resistant to all the races of *P. brassicae* are presently available.

SELECTED REFERENCES

Colhoun, J. 1958. Clubroot disease of crucifers caused by *Plasmodiophora brassicae. Commonw. Mycol. Inst. Phytopathol. Paper* **3**: 108 p.

Reyes, A. A., *et al.* 1974. Races, pathogenicity and chemical control of *Plasmodiophora brassicae* in Ontario. *Phytopathology* **64**:173–177.

Williams, P. H., N. T. Keen, J. O. Strandberg, and Sharon S. McNabola. 1968. Metabolite synthesis and degradation during clubroot development in cabbage hypocotyls. *Phytopathology* **58**:921–928

Woronin, M. 1878. *Plasmodiophora brassicae*. Urheber der Kohlpflanzen—Hernie. *Jahrb. Wiss. Bot.* **11**:548–574. English Transl. by C. Chupp, in *Phytopathol. Classics* **4**:1934.

DISEASES CAUSED BY PHYCOMYCETES

Phycomycetes contain three classes of plant pathogenic fungi whose body consists of round or elongated mycelium that lacks cross walls. The first two classes, Chytridiomycetes and Oomycetes, produce zoospores, while the third class, Zygomycetes, does not produce zoospores but, instead, produces nonmotile spores in sporangia (Fig. 38). The plant-pathogenic Chytridiomycetes, the Chytridiales, usually known as chytrids, lack true mycelium, having mostly a round or irregularly shaped rhizomycelium that lives entirely within the host cells. Upon maturity, the vegetative body is transformed into one or many thick-walled resting spores or sporangia. The plant-pathogenic Oomycetes, which include the orders Saprolegniales and Peronosporales, have well-developed, elongated mycelium, produce zoospores in zoosporangia; their resting spores are oospores, produced by the union of two morphologically different gametes.

Chytridiomycetes and most Oomycetes are water- or soil-inhabiting fungi. Since they produce zoospores, all require or are favored by free water or a film of water in the soil or on the plant surface. The Peronosporales that attack mostly aboveground parts of plants also require a film of water on the plant surface or very humid atmosphere for infection and further development.

The Zygomycetes have well-developed mycelium, also without cross wall, produce nonmotile spores in sporangia, and their resting spore is a thick-walled zygospore produced by the union of two morphologically similar gametes. The Zygomycetes are strictly terrestrial fungi, their spores often floating around in the air, and are either saprophytes or weak parasites of plants and plant products on which they cause soft rots or molds.

DISEASES CAUSED BY CHYTRIDIOMYCETES

This class contains four plant pathogenic genera: *Olpidium, Physoderma, Synchytrium,* and *Urophlyctis.* Most of them infect belowground parts of plants while one, *Physoderma,* attacks aboveground parts of plants.

DISEASES OF BELOW-GROUND PLANT PARTS CAUSED BY CHYTRIDIOMYCETES Three diseases of moderate importance are caused by Chytridiomycetes that attack roots, tubers, or stems at or below the soil line. The fungi and the diseases they cause are:

Olpidium, causing root diseases of many kinds of plants
Synchytrium, causing black wart of potato
Urophlyctis, causing crown wart of alfalfa

These fungi survive in the soil as resting spores or in host plants as a spherical or irregularly shaped thallus (= rhizomycelium). The resting spores germinate to produce one or many zoospores. These infect plant cells and either produce rhizomycelium directly and cause the typical infection, or they first produce zoosporangia. The latter produce secondary zoospores which then cause the typical infection. Abundant moisture favors the local spread of the pathogens. Over long distances the pathogens are spread in infected plant parts or on contaminated plants and soil. Infected plant cells are not usually killed. Instead, in the diseases caused by *Synchytrium* and *Urophlyctis,* infected tissues are stimulated to divide and enlarge excessively.

Olpidium and *Synchytrium* can also transmit viruses from the hosts in which they are produced to those they infect next. Thus, *Olpidium* is a vector of at least four plant viruses (tobacco necrosis virus, lettuce big vein virus, cucumber necrosis virus, tobacco stunt virus); *Synchytrium* is a vector of potato virus X.

- *Black Wart of Potato*

Black wart is apparently present throughout the world although it seems to be most severe in Europe. It was discovered in a few east-central states of the U.S. in 1918, but it was localized through quarantine measures and has not reappeared since then. Black wart causes losses by reducing the quantity of potato tubers and even more their quality by forming large, unsightly, irregular warts, or galls. The disease also affects tomato and other species of *Solanum*.

Symptoms.   Warts develop on below-ground parts of the stem, on stolons and on tubers but not on roots (Fig. 46). Wart formation begins at the buds of these organs as small swellings. These soon enlarge into roughly spherical or irregular, convoluted, hyperplastic masses of distorted branches and leaves. The warts may be several centimeters in diameter, soft, and at first the color of stolons or tubers. Older warts become darker and are often invaded by secondary organisms that lead to partial disintegration of the warts.

The pathogen: Synchytrium endobioticum.   The vegetative body (= thallus) of the pathogen exists only inside the host cell. Later it produces and is surrounded by a thick wall, thus forming what is called a prosorus. The latter germinates within the host cell, and the protoplast, surrounded by a membrane, comes out of its thick wall, while its nuclei divide repeatedly. The protoplast is divided into 4 to 9 segments each containing 200 to 300 nuclei. Each segment develops into a sporangium, and the mass of sporangia is called a sorus. If moisture and temperature are favorable, the sporangia germinate by means of zoospores which infect

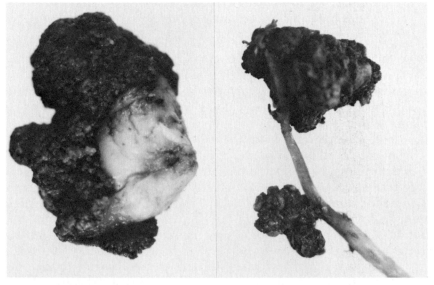

FIGURE 46.
Symptoms of black wart of potato caused by *Synchytrium endobioticum*.

the host and repeat the cycle (Fig. 47). When drought or low temperature sets in, the zoospores from the sporangia fuse in pairs to form zygotes. These infect epidermal cells just like the zoospores. Within the cell they form a thick wall around them converting themselves into resting sporangia which overwinter. The following season the resting sporangia form zoospores which are released and cause new infections. *Synchytrium endobioticum,* in addition to causing the black wart disease, can also act as a vector of the threadlike potato virus X, which the zoospores seem to carry within them from plant to plant.

Development of disease.   The resting sporangia can survive in soil and in old, dry warts for many years. The zoospores produced from resting sporangia or zoosporangia penetrate epidermal cell walls of the bud parts directly. Once inside the cell, they absorb food and secrete substances that stimulate the invaded cell to become greatly enlarged and the surrounding epidermal and cortical cells to divide and enlarge. This results

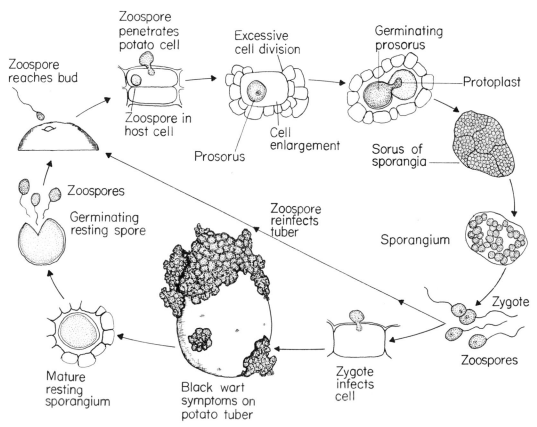

FIGURE 47.
Disease cycle of black wart of potato caused by *Synchytrium endobioticum.*

in the appearance of swellings around the points of infection which soon enlarge further, become intricately branched and produce the warts. Actually, following the initial infection of a bud and the production of several new generations of zoospores from the sporangia, numerous reinfections occur, which stimulate more cells and dormant buds into excessive cell division and enlargement of stem and leaf tissues that result in the soft, convoluted wart. The soft wart tissues lack normal protective epidermal tissues, are easily injured and invaded by secondary microorganisms, their surface disintegrates and the wart appears black.

The pathogen is spread nearby by swimming zoospores, running water or is carried in soil as zoospores or resting sporangia. Over long distances, the pathogen spreads on infected tubers, on latent or indistinct infections of tubers of resistant varieties, and on infested soil or manure carried to other fields or areas.

Control. Quarantine measures have limited the spread of the pathogen from the original areas of infestation. These areas even to date must grow only resistant varieties and export of potatoes from the few *Synchytrium*-infested localities of Pennsylvania, Maryland, and West Virginia is prohibited. Eradication of the pathogen from infested soil, although possible, is too expensive to be practical.

Where the pathogen is already widespread in soils, the best control is through the use of resistant varieties. Fortunately, several potato varieties are resistant to the black wart pathogen, although both the degree and type of resistance seem to vary considerably among varieties and are also affected by the particular race of the pathogen present in the area.

SELECTED REFERENCES

Artschwager, E. F. 1923. Anatomical studies on potato-wart. *J. Agr. Res.* **23**:963–968.

Curtis, K. M. 1921. The life history and cytology of *Synchytrium endobioticum*, the cause of wart disease in potato. *Phil. Trans. Roy. Soc. London, Ser. B* **210**:409–478.

Hampson, M. C. and K. G. Proudfoot. 1974. Potato wart disease, its introduction to North America, distribution and control problems in Newfoundland. *FAO Plant Protection Bull.* **22**:53–64.

Hartman, R. E. 1955. Potato wart eradication program in Pennsylvania. *Am. Pot. J.* **32**:317–326.

Heim, P. 1961. Observations sur l'evolution de l'*Urophlyctis alfalfae. Rev. Mycol.* **26**:3–23.

Kole, A. P. 1965. Resting spore germination in *Synchytrium endobioticum. Neth. J. Plant Pathol.* **71**:72–78.

Teakle, D. S. 1969. Fungi as vectors and hosts of viruses. In "Viruses, Vectors and Vegetation" (K. Maramorosch, ed.). Wiley, New York, pp. 23–54.

Temmink, J. H. M. 1971. An ultrastructural study of *Olpidium brassicae* and its transmission of tobacco necrosis virus. *Meded. Landbouwhogeschool Wageningen* 71–6:135 p.

Diseases of Aboveground Plant Parts Caused by Chytridiomycetes Brown spot of corn, caused by *Physoderma maydis*, is the only disease of some importance in this group.

- *Physoderma Brown Spot of Corn*

The disease occurs mostly in the southeastern U.S. and Mexico, but it has been found occasionally in some midwestern states and in Asia. It causes generally minor losses but localized outbreaks do occur. The losses consist of reduced yields and occasional breakage of the corn stalk at heavily infected nodes.

Symptoms.   Small yellowish spots may appear on all aboveground parts of the plant. The spots later turn brown to reddish-brown, enlarge and often coalesce into irregular blotches. Infected cells break up and expose dark brown sporangia. Heavily infected stalks may break and fall over.

The pathogen: Physoderma maydis.   Its body consists of a multinucleate rhizomycelium consisting of enlarged cells and fine threadlike hyphae (Fig. 48). The enlarged cells form brown sporangia, each of which on germination releases 20 to 50 zoospores that have a single flagellum. The zoospores germinate to produce thin hyphae which finally produce rhizomycelium.

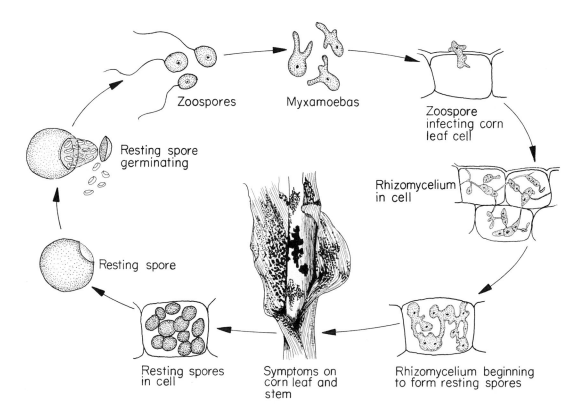

Zoospores   Myxamoebas

Zoospore infecting corn leaf cell

Resting spore germinating

Rhizomycelium in cell

Resting spore

Resting spores in cell

Symptoms on corn leaf and stem

Rhizomycelium beginning to form resting spores

FIGURE 48.
Disease cycle of *Physoderma* brown spot disease of corn.

Development of disease. The pathogen overwinters as resting sporangia in the soil or infected plant debris. In the spring and summer, the sporangia, whether on the soil or on young corn tissues, germinate in the presence of moisture and light and give zoospores. The zoospores, as they move on wet leaf surfaces, form infectious hyphae which penetrate young corn cells, especially within the leaf whorl. The invading mycelium grows within the corn cells in which it becomes rhizo-mycelium and, later, sporangia. This cycle may be completed within 2 to 3 weeks and repeated several times during the growing season, until finally resting sporangia are produced.

Control. No effective, practical control measures are available, but some hybrid varieties are more resistant than others.

SELECTED REFERENCES

Broyles, J. W. 1962. Penetration of meristematic tissue of corn by *Physoderma maydis. Phytopathology* **52**:1013–1016.
Hebert, T. T., and A. Kelman. 1958. Factors influencing the germination of resting sporangia of *Physoderma maydis. Phytopathology* **48**:102–106.
Tisdale, W. H. 1919. Physoderma disease of corn. *J. Agr. Res.* **16**:137–154.

DISEASES CAUSED BY OOMYCETES

The most important plant pathogenic Oomycetes, i.e., fungi that have elongated mycelium without cross walls and produce oospores as their resting spores and zoospores or zoosporangia as their asexual spores, belong to two orders: Saprolegniales and Peronosporales.

Of the Saprolegniales, only the genus *Aphanomyces* is important as a plant pathogen, causing root rot diseases of many annual plants, particularly of pea and sugarbeet.

The Peronosporales include some of the most important plant pathogens known: *Pythium,* the cause of seed rot, seedling damping off, and root rot of most plants, and soft rot of fleshy fruits, etc., in contact with the soil; *Phytophthora,* the cause of late blight of potato and of root rots and blights of many other plants; and several genera comprising the group of fungi causing the very destructive diseases known as downy mildews. Another fungus, *Albugo,* causes the common but usually not serious white rust diseases of cruciferous plants (Fig. 38).

The plant diseases caused by Oomycetes are basically of two types (Fig. 39): (1) Those that affect plant parts present in the soil or in contact with the soil, e.g., roots, lower stems, tubers, seeds, and fleshy fruits lying on the soil. Such diseases are caused by all the species of *Aphanomyces* and *Pythium* and by some species of *Phytophthora.* (2) Those that affect only or primarily aboveground plant parts, particularly the leaves, young stems, and fruits. Diseases of this type are caused by some species of *Phytophthora* and by all the species of *Albugo* and of the downy mildew fungi (*Bremia, Peronospora, Plasmopara, Pseudoperonospora,* and *Sclerospora*).

- *Pythium Seed Rot,*
  Damping off, and Root Rot

Damping-off disease of seedlings is widely distributed all over the world. It occurs in valleys and forest soils, in tropical and temperate climates, and in every greenhouse. The disease affects seeds, seedlings, and older plants of almost all kinds of vegetables, flowers, cereals, and many fruit and forest trees. In all cases, however, the greatest damage is done to the seed and seedling roots during germination either before or after emergence. Losses from this disease vary considerably with soil moisture, temperature, etc. Quite frequently, however, seedlings in seedbeds are completely destroyed by damping off, or they die soon after they are transplanted. In many instances poor germination of seeds or poor emergence of seedlings is the result of damping-off infections in the preemergence stage. Older plants are seldom killed when infected with the damping-off pathogen, but they develop stem lesions or root rots, their growth may be retarded considerably, and their yields may be reduced drastically. Some species of the damping-off fungus also attack the fleshy organs of plants, which rot in the field or in storage.

Symptoms. The symptoms caused by the damping-off fungi vary with the age and stage of development of the plant affected. When seeds of susceptible plants are planted in infested soils and are attacked by the damping-off fungi, they fail to germinate, become soft and mushy, then turn brown, shrink, and finally disintegrate. Seed infections taking place in the soil cannot be observed, and the only manifestations of the disease are poor stands. Poor stands, however, are also the result of infections of the seedling by the damping-off fungus after the seed has germinated but before the seedling has emerged above the soil line. Tissues of such young seedlings can be attacked at any point. The initial infection appears as a slightly darkened, water-soaked spot. The infected area enlarges rapidly, the invaded cells collapse, and the seedling is overrun by the fungus and dies shortly after the beginning of infection. In both cases infection takes place before the seedlings emerge, and this phase of the disease is called preemergence damping off.

Seedlings that have already emerged are usually attacked at the roots and sometimes at or below the soil line (Fig. 49, A–E). The succulent tissues of the seedling are easily penetrated by the fungus, which invades and kills the cells very rapidly. The invaded areas become water soaked and discolored, and the cells soon collapse. At this stage of infection the basal part of the seedling stem is much thinner and softer than the above, yet uninvaded, parts; owing to loss of firmness and supporting power, the invaded portion of the stem cannot support the part of the seedling above it, whereupon the seedling falls over on the soil. The fungus continues to invade the seedling after it has fallen to the ground and the seedling quickly withers and dies. This phase of the disease is called post-emergence damping off.

When older plants are attacked by the damping-off fungus they usually show only small lesions on the stem; these, however, if sufficiently large

or numerous, can girdle the plant and cause stunting or death. More commonly, infections on older plants are limited to rootlets, which are damaged and frequently killed by the fungus; this results in stunting, wilting, and death of the aboveground part of the plant.

Soft fleshy organs of some vegetables, such as cucurbit fruits, green beans, potatoes, and cabbage heads, are sometimes infected by the damping-off fungi during extended wet periods in the field, in storage and transit. Such infections result in a cottony fungus growth on the surface of the fleshy organ, while the interior turns into a soft, watery, rotten mass, called "leak" (Fig. 49F).

The pathogen: Pythium sp.   Several different fungi, e.g., *Rhizoctonia, Fusarium,* can cause symptoms quite similar to one or the other phase of those described above. *Pythium,* however, seems to be the most important cause of the pre- and postemergence phases of damping off. Several species of *Pythium* are involved, but the effect of each one of them on its hosts is usually similar to that of the others.

Pythium produces white, slender, profusely branching, and rapidly growing mycelium. The mycelium gives rise to terminal, or intercalary sporangia which may be spherical, filamentous, or variously shaped. Sporangia germinate directly by producing one to several germ tubes, or by producing a short hypha at the end of which a vesicle is formed (Fig. 50). The protoplasm passes from the sporangium into the vesicle, and there it forms more than 100 zoospores. When the zoospores are released, they swarm about in the water for a few minutes, come to rest, encyst by rounding off, and germinate by producing a germ tube. The germ tube usually penetrates the host tissue and starts new infection, but sometimes it produces another vesicle in which several secondary zoospores are formed, and this may be repeated.

The mycelium also gives rise to spherical oogonia and club-shaped antheridia at the ends of short hyphae. The hypha bearing the antheridium may originate from the hypha bearing the oogonium or from another hypha of the mycelium. Upon contact with the oogonium the antheridium produces a fertilization tube which enters the oogonium. Through this tube the male nuclei of the antheridium move toward the female nuclei of the oogonium, unite with them and form the zygote. A thickening of the wall of the fertilized oogonium takes place, and the thick-walled structure containing the zygote is called an oospore. Oospores are resistant to high or low temperatures and other adverse factors and serve as the overwintering stage of the fungus. Since oospores require a resting period before they germinate, they are also called resting spores. Oospores, too, germinate either by producing germ tubes which develop into mycelium or by producing vesicles in which zoozspores are formed in a way similar to that described for sporangia. The type of germination

---

FIGURE 49.
Damping-off symptoms (A) on tobacco seedlings compared to control (B). (C) Damping-off symptoms on cucumber seedlings. (D) Root and stem rot of Swedish ivy caused by *Pythium*. (E) Damping off on bean. (F) Soft rot of young butternut squash caused by *Pythium*. (Photo E courtesy G. C. Papavizas.)

of both sporangia and oospores is determined primarily by the temperature of the medium, temperatures above 18°C favoring germination by germ tubes, while temperatures between 10 and 18°C induce germination by means of zoospores.

Pythium species are widely distributed in waters and soils throughout the world. They live on dead plant and animal substrata as saprophytes or as low-grade parasites attacking the fibrous roots of plants. When a wet soil is heavily infested with *Pythium,* any seeds, or young plants emerging from seeds, in such a soil may be attacked by the fungus.

Development of disease.   Spore germ tubes or saprophytic mycelium of *Pythium* come in contact with seeds or seedling tissues of host plants either by chance or because the exudates of these plants serve as nutrients and chemotropic stimulants to its zoospores and mycelium which move or grow toward the plants. The fungus enters the seeds by direct penetration of the moistened, swollen seed coats or through cracks, and further penetrates the embryo or emerging seedling tissues through mechanical pressure and dissolution by means of enzymes. Pectinolytic enzymes secreted by the fungus dissolve the middle lamella which holds the cells together, resulting in maceration of the tissues. Further invasion and breakdown of tissues occurs as a result of growth of the fungus between and through the cells. At the points where the hyphae pass through the cell walls they are constricted to approximately half their normal diameter. Proteolytic enzymes break down the protoplasts of invaded cells, while physical forces and, in some cases, fungal cellulolytic enzymes bring about complete collapse and disintegration of the cell walls. The fungus consumes many of the plant cell substances and the products of their breakdown and uses them as building blocks for its own body or as an energy source for its own metabolic activities. Thus infected seeds are killed and turn into a rotten mass consisting primarily of fungus and substances such as suberin and lignin, which this fungus cannot break down.

The infection of roots and stems of young, tender seedlings progresses in essentially the manner described above. The initial infection usually occurs at or slightly below the surface of the soil, depending on moisture level and depth of planting. The mycelium penetrates the epidermal and cortical cells of the stem directly, consumes part or all of their contents and breaks down their cell walls, bringing about the collapse of cells and tissues. In this area vascular tissues may also be invaded, in which case they become discolored even beyond the extent of the cortical lesion. Seedlings so invaded die quickly. When the invasion of the fungus is limited to the cortex of the belowground stem of the seedling, the latter may continue to live and grow for a short time until the lesion extends above the soil line. Then, the invaded, collapsed tissues cannot support the seedling, which falls over and dies (Figs. 49 and 50).

If the initial infection occurs when the seedling is already well developed and has well-thickened and lignified cell walls and active cambium, the advance of the fungus is checked at or near the point of infection, and only relatively small lesions develop. Well-developed, ma-

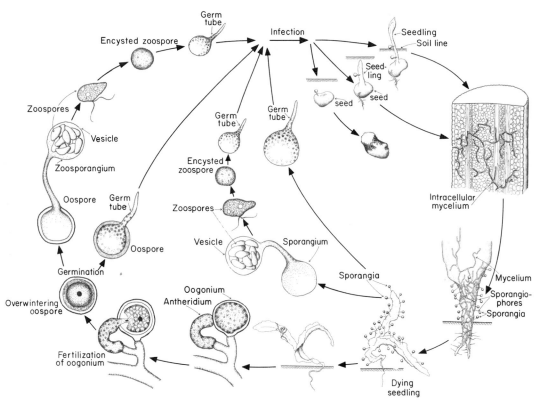

FIGURE 50.
Disease cycle of damping off and seed decay caused by *Pythium* sp.

ture tissues present considerable resistance to the mechanical pressure and to the enzymatic activity of the fungus by means of their increased thickness and modified structure of their cell walls.

Rootlets of most plants can be attacked by *Pythium* at almost any stage of growth. The fungus enters root tips and proliferates in the young cells, causing a rapid collapse and death of the rootlet. Further advance of the fungus into older roots is usually limited to the cortex of the root. Fleshy or relatively young roots are invaded to a much greater extent, the lesions extending to several centimeters in length and through the thickness of the cortex in depth.

Infections by *Pythium* of fleshy vegetable fruits and other organs can take place in the field, in storage, in transit, and in the market. Field infections begin at the point of contact of the fruit with wet soil heavily infested with the fungus. Enzymes secreted by the fungus macerate the tissue, which becomes very soft and watery. An entire cucumber fruit may be invaded by the fungus within 3 days from inoculation. As the mycelium advances through the fleshy tissue it also bursts through the confining epidermis and forms, at first, small patches of aerial mycelium; these soon enlarge, coalesce, and form a continuous, luxuriant, cottony mycelial weft encasing most or all of the fruit. In storage, the aerial mycelium of an infected fruit grows out between adjacent fruits, invests them partially or completely, and soon penetrates, infects, and destroys these fruits.

During the development of infection only mycelium can be found in the infected tissues. As the infection progresses, however, sporangia begin to appear, and they are followed by the production of oospores. Sporangia and oospores can be produced inside or outside the host tissues or both, depending on the particular fungus species involved.

The severity of the diseases and amount of losses caused by *Pythium* infections are greater when the soil is kept wet for prolonged periods, when the temperature is unfavorable for the host plant, i.e., too low for plants requiring high temperatures for optimum growth or too high for plants requiring relatively low temperatures for best growth, when there is an excess of nitrogen in the soil, and when the same crop is planted in the same field for several consecutive years.

Control. Pythium diseases in the greenhouse can be controlled through the use of soil sterilized by steam or dry heat, or with chemicals and through the use of chemically treated seed. In the field, however, where soil sterilization is difficult and expensive, seed treatment with one or more of a number of chemicals is the most important disease preventive measure. The most commonly used materials for seed or bulb treatment include chloranil, thiram, captan, dichlone, ferbam, and diazoben. The systemic fungicide ethazole controls damping off, seedling blights, and root rots caused by *Pythium* and *Phytophthora*. It can be applied as soil or seed treatment. Although more than one of these chemicals may be used on any crop with good results, there are usually some chemical–crop combinations that are much more effective than the rest, while some other combinations may be phytotoxic or ineffective under certain conditions.

Seed treatment is sometimes followed by spraying of seedlings with ziram, chloranil, captan, soluble coppers, etc. This is especially important when the soil is heavily infested with *Pythium* or when the soil stays wet for prolonged periods during the early stages of plant growth.

Certain cultural practices are sometimes helpful in reducing the amount of infection. Good soil drainage is the most important of all. Improvement of heavy soils and improvement of air circulation among plants are advisable; planting should be done when temperatures are favorable for fast plant growth; application of excessive amounts of nitrate forms of nitrogen fertilizers should be avoided. No one crop should be planted in the same field for more than two consecutive years since that would increase the fungus population in the soil unless the soil is sterilized periodically.

SELECTED REFERENCES

Anonymous. 1974. Symposium on the genus *Pythium*. Several papers. *Proc., Am. Phytopath. Soc.* **1**:200–223.
Buchholtz, W. F. 1938. Factors influencing the pathogenicity of *Pythium debaryanum* on sugar beet seedlings. *Phytopathology* **28**:448–475.

Drechsler, C. 1925. The cottony leak of cucumbers caused by *Pythium aphanidermatum. J. Agr. Res.* **30**:1035–1042.

Hendrix, F. F., Jr., and W. A. Campbell. 1973. Pythiums as plant pathogens. *Ann. Rev. Phytopathol.* **11**:77–98.

Labonte, G. A. 1959. Damping-off studies in coniferous seedlings. *Maine Forestry Serv. Bull.* **18**:14 p.

Lumsden, R. D. 1973. Differential isolation of *Pythium* spp. from soil by means of selective media, temperatures and pH. *Phytopathology* **63**:803.

Middleton, J. T. 1943. The taxonomy, host range and geographic distribution of the genus *Pythium. Torrey Bot. Club. Mem.* **20**:1–171.

Mulder, J. 1969. The pathogenicity of several *Pythium* species to rootlets of apple seedlings. *Neth. J. Plant Pathol.* **75**:178–181.

Potter, R. H. 1946. Germinability of treated and untreated lots of vegetable seed in *Pythium*-infested soil and in the field. *Iowa Agr. Expt. Sta. Res. Bull.* **345**:949–983.

PHYTOPHTHORA DISEASES

Species of *Phytophthora* cause a variety of diseases on many different types of plants ranging from seedlings of annual vegetables or ornamentals to fully developed fruit and forest trees. Most species cause root rots, damping off of seedlings, and rots of lower stems, tubers, corms, etc. similar to those caused by *Pythium* sp. Others cause rots of buds or fruits, and some cause foliar blights that attack the foliage, young twigs, and fruit. Some species are host specific, i.e., they attack only one or two species of host plants, but others have a wide host range and may cause similar or different symptoms on many different kinds of host plants. The best-known species is *Phytophthora infestans*, the cause of late blight of potatoes and tomatoes, but several other species cause extremely destructive diseases on their hosts. Some of the other important species and their better-known diseases are listed here.

- *P. cactorum*, causes collar or trunk rot of apple, foot rot of lily and stock, blight of peony, dieback of azalea, stem rot, and wilt of snapdragon, root rot of sweet clover, and blossom blight of tulip.
- *P. capsici*, causes root rot of pepper, carrot and pumpkin, and fruit rot of pepper, cucurbits, eggplant, tomato, etc.
- *P. cambivora*, causes root and crown rot of trees.
- *P. cinnamomi*, causes root rot of avocado, azalea, chestnut, cinnamon, oak, pine, pineapple, and many other trees and shrubs, the "jarrah dieback" of natural forests in Australia.
- *P. citrophthora*, causes foot rot and fruit rot of citrus.
- *P. cryptogea*, causes root rot of tomato and of calla lily, and leaf and stem rot of gloxinia.
- *P. erythroseptica*, causes pink rot of potato, soft rot of sugarcane cuttings, and leaf blight of calla lily.
- *P. fragariae*, causes red stele root rot of strawberry.
- *P. megasperma*, causes root rot of crucifers, carrot, potato, spinach, beet, hollyhock, etc.
- *P. megasperma* var. *sojae*, causes Phytophthora rot of soybeans.
- *P. palmivora*, causes bud rot of coconut and stem rot of dieffenbachia and peperomia.

P. parasitica, causes damping off, leaf blight, stem canker, and buckeye rot of fruit in tomato, crown rot of rhubarb, soft rot of cucurbits, foot rot of citrus.

P. parasitica var. *nicotianae,* causes black shank of tobacco.

P. phaseoli, causes downy mildew of lima bean.

P. syringae, causes lilac blight.

- *Phytophthora Root Rots*

Several species of *Phytophthora,* particularly *P. cinnamomi, P. cryptogea, P. fragariae,* and *P. megasperma* cause root rots on numerous species of plants that include fruit, forest, and ornamental trees and shrubs, annual vegetables and ornamentals, strawberries, etc. The most common *Phytophthora* root rots are littleleaf disease of pine, root rot of avacado and pineapple, root rots of alfalfa, azalea, calla lily, crucifers, tomato, carrot, red stele of strawberries, and many others (Fig. 51). The losses caused by *Phytophthora* root rots are great, especially on trees and shrubs, but the pathogen often goes undetected or unidentified. Plants suffering from such root rots often begin by showing symptoms of drought and starvation, and become weakened and susceptible to attack by other pathogens or various other causes which are then mistakenly taken as the causes of the death of the plants.

Phytophthora root rots cause damage in their hosts in nearly every part of the world where the soil becomes too wet for good growth of susceptible plants and the temperature remains fairly low, i.e., between 15 and 23°C.

Young seedlings of trees and annual plants may be killed by the disease within a few days, weeks, or months, while in older plants the killing of roots may be slow or rapid, depending on the amount of fungus present in the soil and the prevailing environmental conditions. As a result, older trees show sparse foliage, shorter, cupped, and yellow leaves, and dieback of twigs and branches. Such trees increase very little in height and diameter and usually die within 3 to 10 years after infection. Fewer and smaller fruit and seeds are produced each succeeding year.

On all hosts affected by *Phytophthora* root rot, many of the small roots are dead, and necrotic brown lesions are often present on the larger roots. On young plants, or on older succulent plants, the whole root system may decay and this is followed by a more or less rapid death of the plant. In strawberries, as in the other plants, most of the small rootlets rot away, while the larger ones show progressive stages of browning beginning at the tips. In addition, in late spring and before or through harvest, affected larger strawberry roots show a red-colored core or stele, a symptom diagnostic of the strawberry red stele root rot caused by *Phytophthora fragariae* (Fig. 51, A, B).

The behavior of the various *Phytophthora* species that cause root rots is generally similar. The fungus overwinters as oospores, chlamydospores, or mycelium in infected roots or in the soil. In the spring, the oospores and chlamydospores germinate by means of zoospores, while the mycelium grows further and/or produces zoosporangia that release zoospores. The zoospores swim around in the soil water and infect roots of susceptible hosts with which they come in contact. More mycelium and

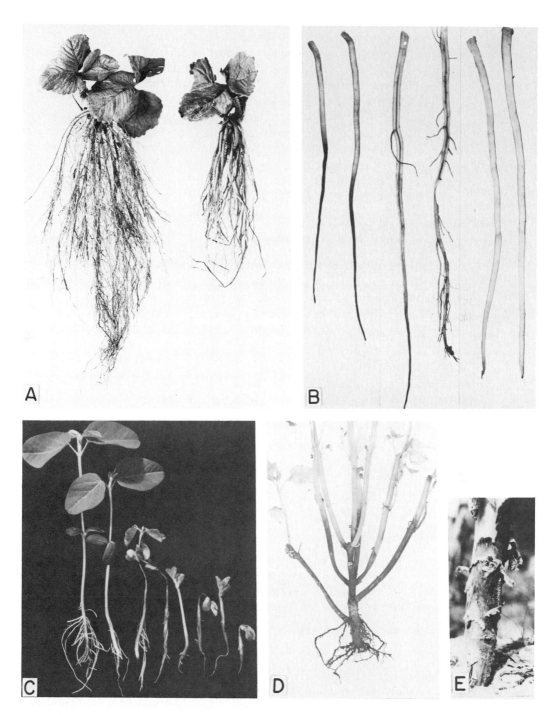

FIGURE 51.

(A) Strawberry plant (right) affected with the red stele disease caused by *Phytophthora fragariae*. Almost all feeder roots have been destroyed. Normal plant at left. (B) Red stele-affected strawberry roots being killed from the tip upwards (left), and roots split lengthwise to show the reddened central portion (middle). Healthy roots split lengthwise at right. (C) Root and stem rot of soybean seedlings caused by *P. megasperma*. Healthy plant at left. (D) Phytophthora root and stem rot of coleus. (E) Tobacco black shank caused by *P. parasitica* var. *nicotiana*. (Photos A, B, and E courtesy U.S.D.A. Photo C courtesy G. C. Papavizas).

zoospores are produced during wet, cool weather and spread the disease to more plants. In dry, hot, or too cold weather the fungus survives as oospores, chlamydospores, or mycelium that can start infections again when the soil is wet and the temperature favorable.

Control of *Phytophthora* root rots depends on planting susceptible crops in soils free of the pathogen or in soils that are light and drain well and quickly. All planting stock should be free of infection and, when available, only resistant varieties should be planted. For plants in pots, greenhouses, or seedbeds, the soil and containers should be sterilized with steam before planting.

- *Phytophthora Foot,*
  Crown, Collar, Stem, or Trunk Rots

Most species of *Phytophthora* listed in the general part cause diseases primarily of the lower stem that are described by one or another of the terms listed here. The most common and most serious of such diseases include the trunk canker or collar rot of apple trees, foot rot of citrus trees (Fig. 52), root and crown rot of cherry trees (Fig. 53) black shank of tobacco

FIGURE 52.
(A) Declining citrus tree as a result of girdling by the foot rot fungus *Phytophthora parasitica* or *P. citrophthora*. (B) Close-up of the rotting of the trunk base caused by the same pathogens. (Photos courtesy Agric. Res. and Educ. Center, Lake Alfred, Fla.)

FIGURE 53.
(A) Sweet cherry trees dying from Phytophthora root and crown rot. (B) Extensive trunk canker of cherry on Mahaleb rootstock caused by *P. cambivora*. (C) Typical crown rot symptoms of cherry on Mazzard rootstock caused by *P. cambivora* and *P. megasperma*. Sporangia (D), oospore with antheridium (E), and oospore (F) of *P. cambivora*. (Photos courtesy S. M. Mircetich, from Mircetich and Matherton, *Phytopathology* **66**:549–558.)

217

(Fig. 51E), foot rot of lily and stock, stem rot and wilt of snapdragon and soybeans, pink rot of potato, soft rot of sugarcane cuttings, stem rot of dieffenbachia and peperomia, and others. In many of these diseases the fungus also attacks the roots, it may attack and kill seedlings before or after emergence above the soil line, thus causing damping-off symptoms, and in some cases the fungus also attacks and causes partial or complete rot of the fruit, as, e.g., in tomato, pepper, cucurbits, citrus, and cacao (Fig. 54).

The general characteristics of *Phytophthora* rots of the lower stem of plants, considering the wide variety of host plants and fungus species involved, appear quite similar. Like the other *Phytophthora* diseases, these are also favored by rather low temperatures and by high soil moisture and atmospheric humidity, and are therefore more common and most severe in low-lying, poorly drained areas. In some host–fungus combinations, the fungus attacks the stem below the soil line or it may first attack the main root and thus cause droughtlike symptoms and general decline of the aboveground parts of the plant before any direct lesions or cankers appear above the soil line. In most cases, however, the fungus attacks the plant at or near the soil line where it causes a water soaking of the bark that appears as a dark area on the trunk. The dark area enlarges in all directions and, if the plant is small and succulent, the darkening may soon encircle the entire stem, after which the lower leaves drop and eventually the whole plant wilts. On larger plants and on trees, the darkening may be on one side of the stem and soon becomes a

FIGURE 54.
Black pod disease of cacao caused by *Phytophthora palmivora*. (Photos courtesy M. O. K. Adegbola and A. Adebayo, Cacao Res. Inst., Nigeria.)

depressed canker below the level of healthy bark. In early stages, the diseased bark is firm and intact while the inner bark is slimy and may produce a moist, gummy exudate. Later, the affected area becomes shrunken and cracked. The collar rot canker may spread through the tissue up into the trunk (Fig. 53) and, sometimes, the branches, or down into the root system. Invasion of the root usually begins at the crown area or at ground level. As the cankers spread and enlarge, they may girdle the trunk, limbs, or roots, at first causing the plant or tree to grow poorly, produce fewer and smaller fruit, show sparse foliage and dieback of twigs, and finally killing the plant parts beyond the infected area.

The *Phytophthora* species that cause these diseases live and reproduce primarily in the soil and usually attack susceptible plants at or below the soil line. Sometimes, however, spores of the fungus may be splashed into injured aboveground bark or low-lying fruit and may cause infections at these points. Fleshy fruit, such as cucurbits, lying on wet soil infested with the fungus are also attacked directly. The fungus overwinters in infected tissues as mycelium, oospores or chlamydospores.

Control of *Phytophthora* diseases of the lower stem requires all precautions described for *Phytophthora* root rots. In addition, it can be improved by applying ethazol to the soil, seeds or transplants. Also on dormant, susceptible plants, primarily trees, and in the soil around them, application of a solution of copper oxychloride or Bordeaux mixture seems to greatly inhibit the growth and activity of the fungus. Resistant varieties should always be preferred, especially for heavy, poorly drained soils. With fruit trees, resistant rootstocks, and sometimes interstocks, offer the most effective means of controlling foot rot or collar rot.

- *Late Blight of Potatoes*

The late blight disease of potatoes is found in nearly all areas of the world where potatoes are grown. It is most destructive, however, in the eastern half of North America and in northwestern Europe, where potatoes are grown in large acreages and where cool, moist weather favors both potato production and the late blight disease. Late blight is also very destructive on tomatoes and on several other species in the family Solanaceae.

Late blight may kill the foliage and stems of potato and tomato plants at any time during the growing season. It also attacks potato tubers and tomato fruits in the field, which rot either in the field or while in storage, transit, and market.

Late blight may cause total destruction of all plants in a field within a week or two when weather conditions are favorable and when no control measures are applied. Losses, however, vary from one area to another and from year to year, depending on the prevailing temperature and moisture at certain periods of the growing season and on the control measures practiced. Even when losses in the field are small, potatoes may become infected during harvest and may rot in storage.

Symptoms. Symptoms appear at first as circular or irregular water-soaked spots, usually at the tips or edges of the lower leaves. In moist weather the spots enlarge rapidly and form brown, blighted areas with indefinite borders. A zone of white, downy fungus growth 3 to 5 mm wide appears at the border of the lesions on the undersides of the leaves (Fig. 55 A, B). Soon the entire leaflet and then all the leaflets on a leaf are infected, die, and become limp. Under continuously wet conditions all tender, aboveground parts of the plants blight and quickly rot away, giving off a characteristic odor. In dry weather the activities of the fungus are checked. Existing lesions stop enlarging, turn black, curl, and wither, and no fungus appears on the underside of the leaves. When the weather becomes moist again the fungus resumes its activities and the disease once again develops rapidly.

Affected tubers at first show more or less irregular, purplish-black or brownish blotches. When cut open, the affected tissue appears water-soaked, dark, somewhat reddish brown, and extendes 5 to 15 mm into the flesh of the tuber (Fig. 55 C, D). Later the affected areas become firm and dry and somewhat sunken. Such lesions may be small or may involve almost the entire surface of the tuber without spreading deeper into the tuber. The rot, however, continues to develop after the tubers are harvested, or infected tubers may be subsequently invaded by secondary fungi and bacteria causing soft rots and giving the rotting potatoes a putrid, offensive odor.

FIGURE 55.
Late blight symptoms on potato (A) and tomato (B) leaves. The whitish zone surrounding the necrotic area of (A) consists of conidiophores and conidia of *Phytophthora infestans*. (C) Exterior and cross-section view of late blight symptoms on potato tubers. (D) Late blight on the stem of a young potato plant originating from mycelium that overwintered in the infected potato seed piece. (E, F) Exterior and cross-section views of late blight symptoms on tomato fruit. (Photos A–D courtesy Dept. Plant Path., Cornell Univ.)

Tomato fruit is attacked and may rot rapidly in the field or in storage (Fig. 55, E, F).

The pathogen: Phytophthora infestans. The mycelium produces branched sporangiophores of unrestricted growth (Fig. 56). Lemon-shaped, papillate sporangia are produced at the tips of the sporangiophore branches, but as the tips of the branches continue to grow the sporangia are pushed aside and later fall off. At the places where sporangia are

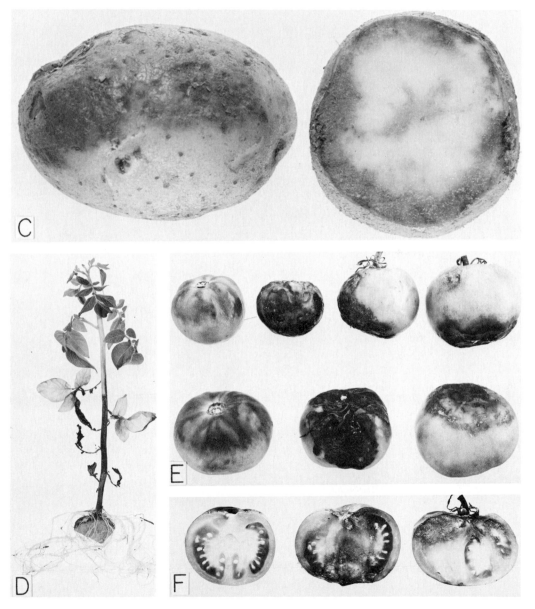

FIGURE 55C–F.

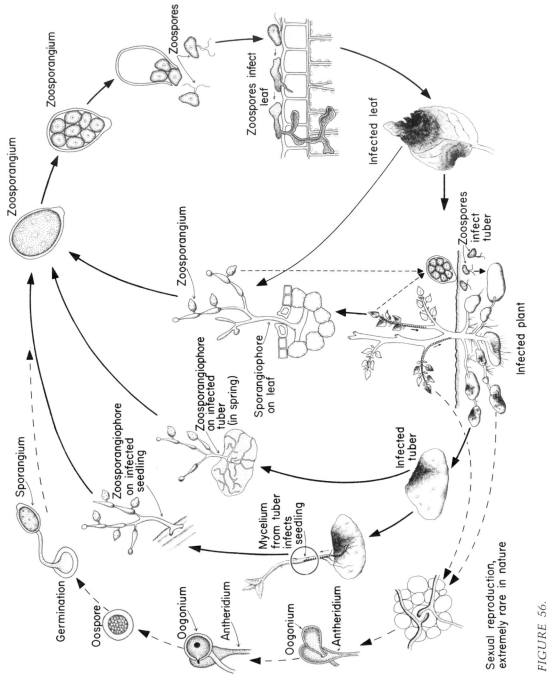

FIGURE 56.
Disease cycle of late blight of potato and tomato caused by *Phytophthora infestans.*

Zoospores

Zoosporangium

Zoosporangium

Zoosporangium

Zoospores infect leaf

Infected leaf

Zoospores infect tuber

Zoosporangium

Zoosporangiophore on infected tuber (in spring)

Sporangiophore on leaf

Infected plant

Zoosporangium

Zoosporangiophore on infected seedling

Infected tuber

Sporangium

Germination

Oospore

Mycelium from tuber infects seedling

Oogonium

Antheridium

Oogonium

Antheridium

Sexual reproduction, extremely rare in nature

produced, the sporangiophores form swellings that are characteristic for this fungus. Sporangia germinate almost entirely by means of zoospores at temperatures up to 12 or 15°C, while above 15°C sporangia may germinate directly by producing a germ tube. Each sporangium produces 3 to 8, sometimes more, zoospores, which are liberated by the bursting of the sporangial wall at the papilla.

This fungus requires two mating types for sexual reproduction, and because only one of them is present in most countries the sexual stage of the fungus has rarely been found. In Mexico, however, both mating types are widely distributed and oospores are common. When the two mating types grow adjacently, the female hypha grows through the young antheridium and develops into globose oogonium above the antheridium. The antheridium then fertilizes the oogonium which develops into a thick-walled and hardy oospore. Oospores germinate by means of a germ tube which produces a sporangium, although at times the germ tube grows directly into mycelium.

Development of disease. The pathogen overwinters as mycelium in infected potato tubers. The mycelium spreads in the tissues of the potato tubers and it finally reaches a few of the shoots produced from infected tubers used as seed, volunteer plants that develop from diseased tubers left in the field or sprouts produced by infected potatoes in cull piles or dumps. The mycelium spreads up the stem most rapidly in the cortical region causing discoloration and collapse of the cells (Fig. 55 D). Later, the mycelium grows between the pith cells of the stem, but it is seldom found in the vascular system. The mycelium grows through the stem and travels up to the surface of the soil. When the mycelium reaches the aerial parts of plants, it produces sporangiophores which emerge through the stomata of the stems and leaves and project into the air (Fig. 56). The sporangia produced on the sporangiophores become detached and drift off when ripe or are dispersed by rain. When the sporangia land on wet potato leaves or stems, they germinate and cause new infections. The germ tube penetrates the leaf cuticle or enters through a stoma and produces mycelium which grows profusely between the cells and sends long, curled haustoria into the cells. The cells on which the mycelium feeds are killed and, as they begin to decay, the mycelium spreads peripherally into fresh tissue. A few days after infection new sporangiophores emerge from the stomata of the leaves and produce numerous sporangia which are spread by the wind and infect new plants. In favorable weather the period from infection to sporangia formation may be as short as 4 days and, therefore, a large number of asexual generations and new infections may be produced in one growing season. With the advance of the disease the established lesions increase in area and new ones develop resulting in premature killing of the foliage and proportional reduction in potato tuber yields.

The second phase of the disease, the infection of tubers, begins in the field when, during wet weather, sporangia are washed down from the leaves and are carried into the soil. The tubers near the surface of the soil are attacked by the emerging zoospores, which germinate and penetrate the tubers through lenticels or through wounds. In the tuber the mycelium grows mostly between the cells and sends its long, sicklelike haustoria into the cells. Tubers are rarely infected by mycelium growing down the stem of a diseased mother plant. If, however, at harvest, the tubers are contaminated with living sporangia still present on the soil or if the tubers are exposed while the fungus is still sporulating on partially diseased foliage, further infection will occur which may not be apparent at the time, but it will develop in storage. Most of the blighted tubers rot in the ground or during storage.

The development of late blight epidemics depends greatly on the effect of humidity and temperature on the different stages of the life cycle of the fungus. The fungus sporulates most abundantly at relative humidity of, or near 100 percent and at temperatures between 16 and 22°C. Sporangia lose their viability in 3 to 6 hours at relative humidities below 80 percent. Germination of sporangia takes place only when free water or dew is present on the leaves and, at 10 to 15°C, it may be completed within 0.5 to 2 hours. After germination a period of 2 to 2.5 hours at 15 to 25°C is required for penetration of the germ tubes into the host tissue. After penetration, the mycelium develops most rapidly at 17 to 21°C, which is also optimal for sporulation. Temperatures above 30°C check the growth of the fungus in the field but do not kill it, and the fungus can start to sporulate again when the temperature becomes favorable, provided of course that the relative humidity is sufficiently high.

Control.   Late blight of potatoes can be successfully controlled by a combination of sanitary measures, resistant varieties, and well-timed chemical sprays. Only disease-free potatoes should be used for seed. Potato dumps or cull piles should be burned before planting time in the spring or sprayed with strong herbicides to kill all sprouts or green growth. All volunteer potato plants in the area, whether in the potato or other fields should be destroyed, since any volunteer potato plant can be a source of late blight infection.

Only the most resistant potato varieties available should be planted. The blight fungus has a number of races or strains differing from each other in the potato varieties that they can attack. Several potato varieties resist one or more races of the late blight fungus. Some of them have resistance to vine infection, but not to tuber infection. New varieties, derived from crosses with *Solanum demissum,* have withstood attack by all known races of the fungus for a while, but were finally attacked by other races not previously distinguished or perhaps not previously exis- tent. Many varieties possess so-called "field resistance," which is only a partial resistance of varying degrees but which is effective against all races of the blight fungus. However, it is not sufficient to rely on varietal resistance to control late blight since, in favorable weather, blight can severely infect these varieties unless they are sprayed with a good protec- tive fungicide. Even resistant varieties should be regularly sprayed with fungicides to eliminate, as much as possible, the possibility of becoming suddenly attacked by races of the fungus to which they are not resistant or by entirely new races.

Chemical sprays with fungicides, if applied properly, generally will keep late blight under control. Spraying should start when potato plants are 15 to 30 cm high or at least 10 days before the date late blight usually appears in the area. Sprays should be applied once every 4 to 5 days when the weather is damp, misty, or rainy and when the nights are cool, and should continue until the foliage dies naturally or is killed artificially by "vine-killers." Proper timing and thorough coverage of old and new foliage are essential if plants are to be protected from the disease. Once late blight becomes established, it is extremely difficult to control unless the weather turns hot (35°C and above) and dry. Materials used for late blight control include several dithiocarbamates, such as mancozeb, captafol, chlorothalonil, polyram, and fentin hydroxide, and several copper materials, such as Kocide, copper oxychloride, and Bordeaux mixture. Protective spraying of foliage usually effects a very considerable reduction in tuber infection. When, however, partially blighted leaves and stems are still surviving at harvest time, it is necessary to remove the aboveground parts of potato plants or destroy them by chemical sprays or mechanical means. Herbicides used for this purpose include copper sulfate, sodium and potassium arsenites, sulfuric acid, and certain dinitro compounds.

SELECTED REFERENCES

Bain, H. F., and J. B. Demaree. 1945. Red stele root disease of the strawberry caused by *Phytophthora fragariae. J. Agr. Res.* **70**:11–30.

Baines, R. C. 1939. *Phytophthora* trunk canker or collar rot of apple trees. *J. Agr. Res.* **59**:159–184.

Berg, A. 1926. Tomato late blight and its relation to late blight of potato. *West Va. Agr. Expt. Sta. Tech. Bull.* **205**:31 p.

Bonde, R., and E. S. Schultz. 1943. Potato cull piles as a source of late-blight infection. *Am. Potato J.* **20**:112–118.

Cox, A. E., and E. C. Large. 1960. Potato blight epidemics throughout the world. *U.S. Dept. Agr., Agr. Handbook* 174, 230 p.

Debruyn, Helena, L. G. 1951. Pathogenic differentiation in *Phytophthora infestans. Phytopathol. Z.* **18**:339–359.

Erwin, D. C., G. A. Zentmyer, J. Galindo, and J. S. Niederhauser. 1963. Variation in the genus *Phytophthora. Ann. Rev. Phytopathol.* **1**:375–396.

Gallegly, M. E. 1968. Genetics of pathogenicity of *Phytophthora infestans. Ann. Rev. Phytopathol.* **6**:375–396.

Gallegly, M. E. 1970. Genetics of *Phytophthora. Phytopathology* **60**:1135–1141.

Hickman, C. J. 1970. Biology of *Phytophthora* zoospores. *Phytopathology* **60**:1128–1135.

Klotz, L. J., and E. C. Calavan. 1969. Gum diseases of citrus in California. *Calif. Agr. Exp. Sta. Ext. Serv. Circ.* 396. 26 p.

Newhook, F. J., and F. D. Podger. 1972. The role of *Phytophthora cinnamomi* in Australian and New Zealand forests. *Ann. Rev. Phytopathol.* **10**:299–326.

Reddick, D., and W. Mills. 1938. Building up virulence in *Phytophthora infestans. Am. Potato J.* **15**:29–34.

Tisdale, W. B., and J. G. Kelley. 1926. A *Phytophthora* disease of tobacco. *Fla. Agr. Exp. Sta. Bull.* 179.

Various authors. 1959. Downy mildew of lima beans. *Plant Dis. Reptr. Suppl.* **257**:177–186.

Waterhouse, G. M. 1956. The genus *Phytophthora*. *Comm. Mycol. Inst. Miscell. publ.* 12.

Waterhouse, G. M. 1970. Taxonomy in *Phytophthora*. *Phytopathology* **60**:1141–1143.

White, R. P. 1937. Rhododendron wilt and root rot. *N. J. Agr. Exp. Sta. Bull.* **615**: 32 p.

Zentmyer, G. A., *et al.* 1971. Avocado root rot. *Calif. Agr. Exper. Sta. Ext. Serv. Circ.* **511**: 16 p.

THE DOWNY MILDEWS

The downy mildews are primarily foliage blights that attack and spread rapidly in young, tender green tissues including leaves, twigs, and fruit. Their development and severity, in areas where susceptible hosts and the respective downy mildew fungus are present, depend greatly on the presence of a film of water on the plant tissues and on high relative humidity in the air during cool or warm, but not hot, periods. The reproduction and spread of the downy mildew fungi are rapid and their diseases can cause heavy losses in short periods of time.

Although even the late blight of potato and tomato looks like and is often called a downy mildew, the true downy mildews are caused by a group of Oomycetous fungi that belong to the family Peronosporaceae. All species of this family are obligate parasites of higher plants and cause downy mildew diseases on a large number of plants including most of the cultivated grain crops and vegetables, and many field crops, ornamentals, shrubs, and vines.

The downy mildew fungi produce sporangia on sporangiophores that are distinct from the mycelium in their way of branching. The sporangia are located at the tips of the branches. Each genus of downy mildew fungi has its own distinctive type of branching of its sporangiophores and this is used for their identification. The sporangiophores are usually long, white at first, emerging in groups from the plant tissues through the stomata. Later on they appear grayish or light brown and form a visible mat of fungus growth on the lower or both sides of leaves and on other affected tissues. Each sporangiophore grows until it reaches maturity and then produces its crop of sporangia, all at about the same time.

In most downy mildews, the sporangia germinate generally by producing zoospores or, at higher temperatures, by producing germ tubes. In the genus *Peronospora*, however, the sporangia germinate only by means of a germ tube. Whenever sporangia germinate by producing a germ tube they are considered spores in themselves rather than sporangia and in that case they are often called conidia, which always germinate by germ tubes.

The oospores of the downy mildews usually germinate by germ tubes, but in a few cases they may produce a sporangium which releases zoospores.

Some of the most common or most serious downy mildew fungi and the diseases they cause are listed below. The structure of their sporangiophores is given in Fig. 38.

Bremia lactucae, causing downy mildew of lettuce.
Peronospora, causing downy mildew of snapdragon (*P. antirrhini*), of onion (*P.

destructor) (Fig. 57, C–D), of soybeans (P. manchurica), mildew (blue mold)
of tobacco (P. nicotianae) (Fig. 57 A–B), mildew of crucifers (P. parasitica),
of alfalfa and clover (P. trifoliorum).

Plasmopara viticola, causing downy mildew of grape (Fig. 58, A–C)

Pseudoperonospora, causing downy mildew of cucurbits (P. cubensis) (Fig. 58,
D–F), of hops (P. humuli).

Sclerospora, causing downy mildew of grasses and millets (S. graminicola), of
cereals and grasses (S. macrospora), of corn (S. maydis), of rice and corn (S.
oryzae), of corn, sorghum, and sugarcane (S. philippinensis), of sugarcane
and corn (S. sacchari), of sorghum and corn (S. sorghi).

- *Downy Mildew of Grape*

Downy mildew of grape occurs in most parts of the world where grapes
are grown under humid conditions. It is most destructive in Europe and in
the eastern half of the United States, where it may cause severe epiphytot-
ics year after year, but it is also known to have caused serious losses in
some years in northern Africa, in South Africa, in parts of Asia, Australia,
and South America. Dry areas are usually free of the disease.

Downy mildew affects the leaves, fruit, and vines of grape plants and
causes losses through killing of leaf tissues and defoliation, through
production of low quality, unsightly, or entirely destroyed grapes, and
through weakening, dwarfing, and killing of young shoots. When weather
is favorable and no protection against the disease is provided, downy
mildew can easily destroy 50 to 75 percent of the crop in one season.

Symptoms.   The disease is usually first observed as small, pale yellow
spots with indefinite borders on the upper surface of the leaves, while on
the under surface of the leaves, and directly under the spots, a downy
growth of the sporophores of the fungus appears (Fig. 58 A–C). Later, the
infected leaf areas are killed and turn brown, while the sporophores of the
fungus on the under surface of the leaves become dark gray. The necrotic
lesions are irregular in outline, and as they enlarge they may coalesce to
form large dead areas on the leaf, frequently resulting in defoliation.

During blossom or early fruting stages, entire clusters or parts of them
may be attacked, are quickly covered with the downy growth, and die. If
infection takes place after the berries are half-grown, the fungus grows
mostly internally, the berries become leathery and somewhat wrinkled
and develop a reddish-marbling to brown coloration.

Infection of green young shoots, tendrils, leaf stems, and fruit stalks
results in stunting, distortion, and thickening (hypertrophy) of the tissues. Entire shoots may be covered with the downy growth of the fungus.
Later the fungus growth breaks down and disappears, and the infected
tissues turn brown and die. In late or localized infections the shoots may
not be killed, but they show various degrees of distortion.

The pathogen: Plasmopara viticola.   The mycelium diameter varies
from 1 to 60 μm because the hyphae take up the shape of the intercellular
spaces of the infected tissues. The mycelium grows between the cells but
sends numerous, globose haustoria into the cells (Fig. 59). In humid
weather the mycelium produces sporangiophores which emerge on the
under side of the leaves and on the stems through the stomata or rarely by

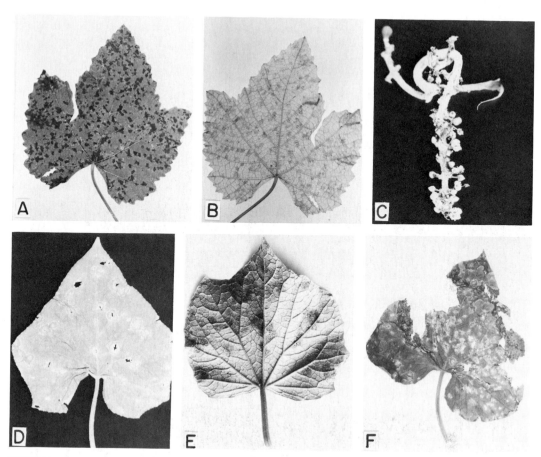

FIGURE 58.

Symptoms of downy mildew of grape (A–C) caused by *Plasmopara viticola*. (A) Upper leaf side. (B) Lower side. (C) Young grape cluster. (D–F) Downy mildew of cucumber caused by *Pseudoperonospora cubensis*. (D, E) Early symptoms as they appear on upper and lower sides of leaves, respectively. (F) Advanced symptoms on cucumber leaf.

pushing directly through the epidermis. In the young fruit, sporangiophores emerge through lenticels. Usually 4 to 6 sporangiophores arise through a single stoma, but sometimes there may be as many as 20. Each produces 4 to 6 branches at nearly right angles to the main stem of the sporangiophore, and each branch produces 2, sometimes 3, secondary branches in a similar manner. At the tips of the secondary branches, single, lemon-shaped sporangia (conidia) are produced. The sporangia are blown by the wind or are transported by water and in the presence of free

FIGURE 57.

(A) Downy mildew on the lower (left) and upper surface of tobacco leaves caused by *Peronospora nicotianae*. (B) Young tobacco plants (left) destroyed by downy mildew. (C) Onion plant affected with downy mildew caused by *Peronospora destructor*. (D) Downy mildew symptoms on onion leaves.

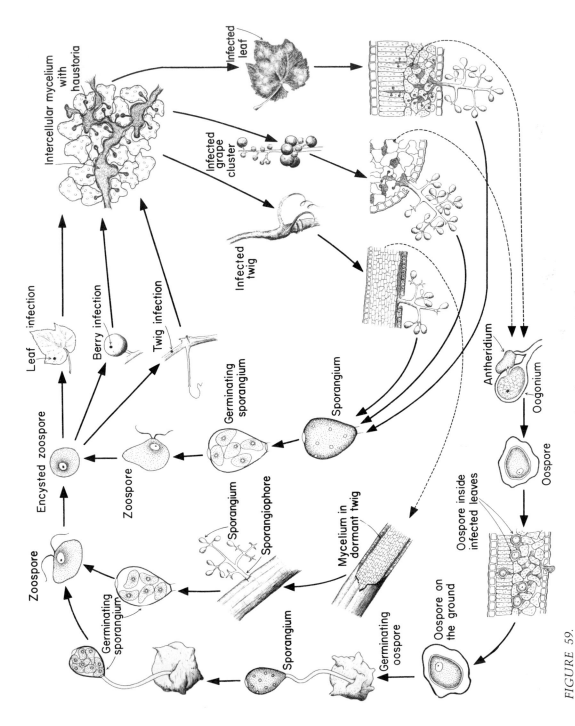

Intercellular mycelium with haustoria

Infected leaf

Infected grape cluster

Infected twig

Antheridium

Oogonium

Oospore

Oospore inside infected leaves

Leaf infection

Berry infection

Twig infection

Germinating sporangium

Sporangium

Encysted zoospore

Zoospore

Sporangium

Sporangiophore

Mycelium in dormant twig

Oospore on the ground

Zoospore

Germinating sporangium

Sporangium

Germinating oospore

FIGURE 59.
Disease cycle of downy mildew of grapes caused by *Plasmopara viticola*.

moisture they germinate. Sporangia generally germinate by means of zoospores, which emerge from the sporangia, swim about for a few minutes, encyst, and then produce a germ tube by which they can infect the plant. In rare instances sporangia may germinate by directly producing a germ tube.

The fungus also produces oospores that germinate by producing a germ tube which terminates in a sporangium. The sporangium then germinates by means of zoospores, as described above.

Development of disease. The pathogen overwinters as oospores embedded in the dead leaves, and occasionally, in dead berries and shoots (Fig. 59). In certain areas, the fungus may also overwinter as mycelium in infected, but not killed, twigs. The dead leaves containing the oospores disintegrate during the winter and liberate the oospores. During rainy periods in the spring the oospores germinate either on the ground or on parts of grape plants on which they are carried by wind or splashing rain drops. The produced sporangium or its zoospores are transported by wind or water to the wet leaves near the ground which they infect. Penetration takes place through stomata of the lower surface of the leaves. The mycelium then spreads into the intercellular spaces of the leaf, obtaining food through globose haustoria which it sends into the cells. The mycelium continues to spread into the tissues and when it reaches the substomatal cavity it forms a cushion of mycelium from which the sporangiophores arise and emerge through the stoma. On these primary lesions, great numbers of sporangia are produced which may be carried, by wind or rain, to nearby healthy plants, germinate quickly, and produce many zoospores. The zoospores then cause secondary infections through stomata or lenticels and thus rapidly spread the disease. The period from infection to new sporangia formation varies from 5 to 18 days, depending on temperature, humidity, and varietal susceptibility.

In the stems the fungus invades the cortex, ray parenchyma, and pith. The distortion and hypertrophy of infected stems is caused by the enlargement of the affected cells and the large volume of mycelium present in the intercellular spaces. Finally the affected stem cells are killed and collapse, producing brown sunken areas in the stem. In the young berries, infection is also intercellular, the chlorophyll breaks down and disappears, the cells collapse and turn brown.

At the end of the growing season the fungus forms oospores in the infected old leaves, and sometimes in the shoots and berries.

Control. Several American grape varieties show considerable resistance to downy mildew, but most European varieties are very susceptible; even the relatively resistant varieties, however, require protection through chemicals.

The most effective fungicides for control of downy mildew are Bordeaux mixture, ferbam, folpet, and captan. The applications begin before bloom and are continued at 7- to 10-day intervals, although the time and number of applications vary with the local conditions, particularly the frequency and duration of rainfall during the growing season.

SELECTED REFERENCES

Anderson, P. J. 1937. Downy mildew of tobacco. *Conn. Agr. Exp. Sta. Bull.* **405.**

Arens, K. 1929. Physiologische Untersuchungen an *Plasmopara viticola*, unter besonderer Berucksichtingung der Infektionsbedingungen. *Jahrb. Wiss. Botan.* **70**:93–157.

Cook, H. T. 1932. Studies on the downy mildew of onions and the causal organism, *Peronospora destructor. N. Y. (Cornell) Agr. Exp. Sta. Mem.* **143.**

DeCastella, F., and C. C. Brittlebank. 1924. Downy mildew of the vine (*Plasmopara viticola* B. and de T.), Dept. Agr., Victoria, Australia, Bull. 49, 45 pp.

Doran, W. L. 1932. Downy mildew of cucumbers. *Mass. Agr. Exp. Sta. Bull.* **283**: 22 p.

Frederiksen, R. A., *et al.* 1973. Sorghum downy mildew, a disease of maize and sorghum. *Tex. Agr. Exp. Sta. Research Monograph* **2**: 32 p.

Gäumann, E. 1923. Beitrage zu einer Monographie der gattung Peronospora. *Beitr. Kryptogamoflora Schweiz* **5**:1–360.

Gregory, C. T. 1915. Studies on *Plasmopara viticola. Int. Congr. Viticult. Rept.* **1915**:126–150.

Klinkowski, M. 1962. Die europäische Pandemic von *Peronospora tabacina*, dem Erreger des Blauschimmels des Tabaks. *Biol. Zentralbl.* **81**:75–89.

Millardet, P. M. A. 1885. (1) Traitement du mildiou et du rot. *J. Agr. Prat.* **2**:513–516. (2) Traitement du mildiou par le mélange de sulphate de cuivre et de chaux. *Ibid.,* pp. 707–719. (3) Sur l'histoire du traitement du mildiou par le sulphate de cuivre. *Ibid.,* pp. 801–805. English transl. by F. L. Schneiderhan, *Phytopathol. Classics* 3:1933.

Royle, D. J., and G. G. Thomas. 1973. Factors affecting zoospore responses towards stomata in hop downy mildew (*Pseudoperonospora humuli*) including some comparisons with grapevine downy mildew (*Plasmopara viticola*). *Physiol. Pl. Pathol.* 3:405–417.

Waterhouse, G. M. 1964. The genus *Sclerospora*. Diagnoses (or descriptions) from the original papers and a key. *Commonw. Mycol. Inst. Misc. Publ.* **17.**

Zachos, D. G. 1959. Recherches sur la biologie et l'epidemiologie du mildiou de la vigne en Grece. Bases de previsions of d'avertissements. *Ann. Inst. Phytopathol. Benaki* 2:196–355.

DISEASES CAUSED BY ZYGOMYCETES

Two genera of Zygomycetes are known to cause disease in living plants or living plant tissue (Figs. 38 and 39). These are: (1) *Choanephora*, which attacks the withering floral parts of many plants after fertilization and from them invades the fruit and causes a soft rot of squash, pumpkin, pepper, okra, etc., but most severely of summer squash (Fig. 60C); and (2) *Rhizopus*, the common bread mold fungus, which in addition causes soft rot of many fleshy fruits, vegetables, flowers, bulbs, corms, seeds, etc. (Figs. 60A, B). Another genus, *Mucor*, also causes molding of bread and other processed plant products, and rarely a rot of sweet potatoes stored at low temperatures. Another genus, *Endogone*, is one of the fungi that become associated with roots of plants and form mycorrhizae that are beneficial to the plant.

The plant-pathogenic Zygomycetes are weak parasites. They grow mostly as saprophytes on dead or processed plant products and, even when they infect living plant tissues, they first attack injured or dead

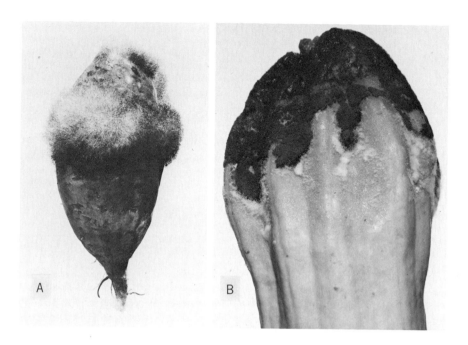

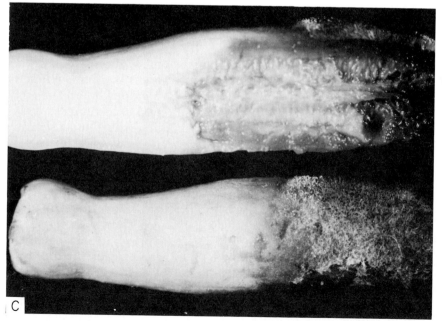

FIGURE 60.
Rhizopus soft rot on sweet potato (A) and on squash (B). *Choanephora* soft rot of young summer squash in the field (C). Sporangiophores and sporangia growing on the surface of sweet potato and squash in the presence of high relative humidity. The maceration and softening of infected tissue can be seen in the longitudinal section of squash (C, upper). (Photo A courtesy U.S.D.A. Photo B courtesy Dept. Plant Path., Cornell Univ.)

plant tissues. In the latter, the fungi build up large masses of mycelium. This secretes enzymes that diffuse into the living tissue, disrupt and kill the cells, and the mycelium then grows into the "living" tissue.

The Zygomycetes have typical elongated mycelium without cross walls. Their asexual spores are produced within sporangia, are nonmotile, and are called sporangiospores, or conidia. They are generally spread by air currents. Their sexual spores are produced through the fusion of two more or less similar-looking sex cells; they are called zygospores. The zygospores are thick walled and can overwinter or withstand other adverse conditions. Under favorable moisture and temperature conditions the zygospores germinate by producing a sporangium containing sporangiospores.

- *Rhizopus Soft Rot of Fruits and Vegetables*

The disease occurs throughout the world on harvested fleshy organs of vegetable, fruit, and flower crops, and is important only, during storage, transit, and marketing of these products. Among the crops affected most by this disease are sweet potatoes, strawberries, all cucurbits, peaches, cherries, peanuts, and several other fruits and vegetables. Corn and some other cereals are affected under fairly high conditions of moisture. Bulbs, corms, and rhizomes of flower crops, e.g., gladiolus and tulips, are also susceptible to this disease.

When conditions are favorable, the disease spreads rapidly throughout the containers, and losses can be very great in a short period of time. (see Fig. 116A, page 365).

Symptoms.   Infected areas of fleshy organs appear water soaked at first, and they are very soft. If the skin of the infected tissues remains intact, the softened fleshy organ gradually loses moisture until it shrivels into a mummy. More frequently, however, the softened skin ruptures during handling or under pressure, e.g., from surrounding fruits, and a whitish-yellow liquid drops out. Soon fungus hyphae grow outward through the wounds and cover the affected portions by producing tufts of whiskerlike gray sporangiophores bearing black sporangia at their tips (Fig. 60). The bushy growth of the fungus extends to the surface of the healthy portions of affected fruit when they become wet with the exuding liquid and even to the surface of the containers. Affected tissues at first give off a mildly pleasant smell, but soon yeasts and bacteria move in and sour odor develops. When loss of moisture is rapid, infected organs finally dry up and mummify; otherwise they break down and disintegrate in a watery rot.

The pathogen: Rhizopus sp.   It is found almost everywhere in nature. It lives usually as a saprophyte and sometimes as a weak parasite on stored organs of plants. The mycelium of the fungus has no cross walls and produces long, aerial sporangiophores at the tips of which black spherical sporangia develop (Figs. 60 and 61). These consist of a thin membrane containing thousands of small spherical sporangiospores. When the membrane of the sporangium is ruptured, the liberated sporangiospores are released and float about in the air or drop to the surface. If

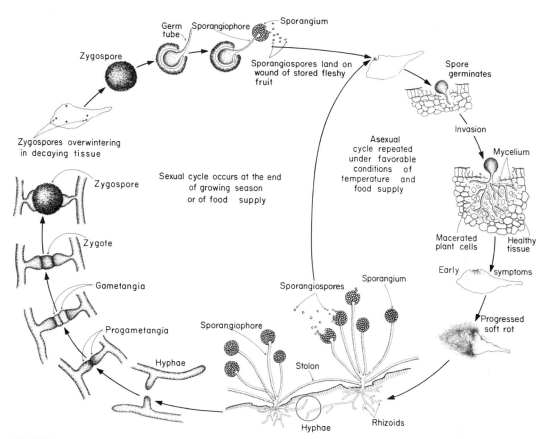

Germ tube — Sporangiophore — Sporangium
Zygospore
Sporangiospores land on wound of stored fleshy fruit
Spore germinates
Zygospores overwintering in decaying tissue
Invasion
Mycelium
Asexual cycle repeated under favorable conditions of temperature and food supply
Zygospore
Sexual cycle occurs at the end of growing season or of food supply
Macerated plant cells
Healthy tissue
Zygote
Early symptoms
Gametangia
Progressed soft rot
Progametangia
Sporangiospores
Sporangium
Sporangiophore
Hyphae
Stolon
Hyphae
Rhizoids

FIGURE 61.
Disease cycle of soft rot of fruits and vegetables caused by *Rhizopus* sp.

they fall on a moist surface or wound of a susceptible plant organ, the sporangiospores germinate and produce mycelium again. When the mycelium grows on a surface, it produces stolons, i.e., hyphae which arch over the surface, and at the next point of contact with the surface produce both rootlike hyphae, called rhizoids, which grow toward the surface, and more aerial sporangiophores bearing sporangia. From each point of contact more stolons are produced in all directions. Adjacent hyphae produce short branches called progametangia, which grow toward each other. When they come in contact, the tip of each hyphae is separated from the progametangium by a cross wall. The terminal cells are the gametangia. These fuse, their protoplasts mix, and their nuclei pair. The cell formed by the fusion enlarges and develops a thick, black and warty cell wall. This sexually produced spore is called a zygospore and is the overwintering or resting stage of the fungus. When it germinates it produces a sporangiophore bearing a sporangium full of sporangiospores.

Development of disease.   Throughout the year, sporangiospores float about and if they land on wounds of fleshy fruits, roots, corms, bulbs, etc., they germinate (Fig. 61). The hyphae thus produced secrete pectinolytic enzymes which break down and dissolve the pectic substances of the middle lamella, i.e., the substances that hold the cells in place in the tissues. This results in loss of cohesion among the cells, since they are now surrounded by liquefied substance, and upon pressure they move freely among each other, resulting in "soft rot."

The pectinolytic enzymes secreted by the fungus advance ahead of the mycelium and separate the cells before the mycelium gets there. Cells in the macerated tissues are then attacked by the cellulolytic enzymes of the fungus, which break down the cellulose of the cell wall, and the cells disintegrate. The mycelium grows intercellularly and does not seem to invade cells until after they are killed and have begun to disintegrate. Thus the mycelium seems never to be in contact with living host cells, but is instead surrounded by dead cells and nonliving organic substances, living more like a saprophyte than a parasite.

The epidermis of infected organs is not usually broken down by the fungus, which continues to grow inside the tissues. The epidermis, however, is softened greatly and breaks easily with the slightest pressure during handling of the product or by the weight of the fruits above it. The fungus then emerges through wounds already existing on the fruit or through subsequent ruptures of the epidermis and produces aerial sporangiophores, sporangia, stolons, and rhizoids, the latter being capable of piercing the softened epidermis. In extremely fleshy fruits, such as peaches and strawberries, the mycelium arising from an infected fruit or from a sporangiospore can penetrate healthy fruit in contact with the infected fruit in the absence of visible wounds. In some cases the epidermis of such fruits has already been softened by the fungal enzymes present in the liquid exuding from the infected fruit, but this is not always necessary.

Sporangiospores are produced on infected tissue within a few days and can cause new infections immediately after they are released. Zygospores, however, are produced later, when the food supply in the infected tissues begins to diminish and, with heterothallic species, only when a compatible strain is also present. Since zygospores germinate after a rest period, most infections in storage are caused by asexual sporangiospores and, within a package of fruit, most infections occur from the hyphae growing from the surface of the previously rotted fruit rather than from spores.

The initiation of infection and the invasion of the tissues by the fungus are influenced greatly by temperature and humidity and by the stage of ripeness of the tissue. Unfavorable combinations of temperature and humidity or insufficient maturity of the fruit slow down the growth and activity of the fungus, and they may allow the host to form layers of cork cells and other histological barriers which retard or completely inhibit further infections by the fungus.

Control. Since the spores of the fungus are present everywhere, wounds of any size can be points of infection. Wounding of fleshy fruits, roots, tubers, bulbs, etc. should, therefore, be avoided as much as possible during harvest, handling, and transportation of these products.

Wounded organs should be discarded or packed and stored separately from healthy ones.

Storage containers and warehouses should be cleaned before use and should be disinfected with a copper sulfate solution, formaldehyde, sulfur fumes, or chloropicrin.

Temperature control of storage rooms and shipping cars is important. With very succulent fruits, such as strawberries, picking of fruit in the morning when it is cool and keeping it at temperatures below 10°C reduces losses considerably. Sweet potatoes and some other not so succulent organs may be protected from the disease by keeping them at 25 to 30°C and 90 percent humidity for 10 to 14 days, during which the cut surfaces cork over and do not allow subsequent penetration by the fungus. After this period is over, temperature is lowered again to about 12°C.

Quite effective control of the disease can be obtained by wrapping susceptible fruit, etc. in paper impregnated with various fungicidal substances such as dichloran.

SELECTED REFERENCES

Adams, J. F., and A. M. Russell. 1920. *Rhizopus* infection of corn on the germinator. *Phytopathology* **10**: 535–543.

Anderson, H. W. 1925. *Rhizopus* rot of peaches. *Phytopathology* **15**: 122–124.

Beraha, L., G. B. Ramsey, M. A. Smith, and W. R. Wright. 1959. Effects of gamma radiation on brown rot and *Rhizopus* rot of peaches and the causal organisms. *Phytopathology* **49**: 354–356.

Luepschen, N. S. 1964. Effectiveness of 2,6-dichloro-4-nitroaniline-impregnated peach wraps in reducing *Rhizopus* decay losses. *Phytopathology* **54**: 1219–1222.

Mirocha, C. J., and E. E. Wilson. 1961. Hull rot disease of almonds. *Phytopathology* **51**: 843–847.

Spaulding, D. H. 1963. Production of pectinolytic and cellulolytic enzymes by *Rhizopus stolonifer*. *Phytopathology* **53**: 929–931.

Srivastava, D. N., and J. C. Walker. 1959. Mechanisms of infection of sweet potato roots by *Rhizopus stolonifer*. *Phytopathology* **49**:400–406.

Stevens, N. E., and R. B. Wilcox. 1917. *Rhizopus* rot of strawberries in transit. *U.S. Dept. Agr. Bull.* **531**: 22 p.

DISEASES CAUSED BY ASCOMYCETES AND IMPERFECT FUNGI

Ascomycetes and Imperfect Fungi are two groups of fungi that closely resemble each other in the structure of the mycelium, the production of asexual spores, and in the kinds of diseases they cause in plants, as well as in the way they induce these diseases. Thus, both produce a haploid mycelium that has cross walls, both produce conidia in identical types of conidiophores or fruiting bodies, and both cause plant diseases that may appear as leaf spots, blights, cankers, fruit spots, fruit rots, anthracnoses, stem rots, root rots, vascular wilts, or soft rots.

The only difference between Ascomycetes and Imperfect Fungi is that the former also produce, regularly or rarely, sexual spores, i.e., ascospores. However, in many Ascomycetes, ascospores are seldom found in nature and seem to play little or no role in the survival of the fungus and in its ability to cause disease in plants. Thus, such Ascomycetes reproduce, spread, cause disease, and overwinter as mycelium and/or conidia, so that they actually behave, and for all practical purposes could be considered, as Imperfect Fungi. On the other hand, for many fungi that were earlier classified as Imperfect Fungi, fruiting bodies containing sexual spores were later shown to be produced by them and since the sexual spores were almost always ascospores, these Imperfect Fungi were then re-classified as Ascomycetes. It would appear, therefore, that Imperfect Fungi are really Ascomycetes that have lost the need for or the ability to produce their sexual stage. Actually, many of the Ascomycetes, because their asexual, imperfect stages are the most common or the only ones found on diseased plants during the growing season, and even during the winter, are usually referred to and are best known by the name of their asexual stage which is generally completely different from the name of the sexual, ascigerous stage. Although usually all species within a genus of an Ascomycete produce the same type of conidia belonging to one genus of an Imperfect Fungus, and various species within a genus of an Imperfect Fungus belong to one genus of an Ascomycete, this is not always so. In many instances, different species within a genus of an Ascomycete have asexual stages that belong to species in other genera of Imperfect Fungi, and vice versa.

Ascomycetes (= the sac fungi) produce mycelium that has cross walls, sexual spores (ascospores) within a sac (ascus), and asexual spores (conidia). The ascus or sexual stage of Ascomycetes is often called the perfect stage, while the conidial or asexual stage is the imperfect stage. In almost all plant pathogenic Ascomycetes during the growing season, the fungus exists as mycelium and reproduces and causes most infections with its asexual, i.e., conidial stage. The sexual or perfect stage is produced on or in infected leaves, fruits, or stems only at the end of the growing season or when the food supply is diminishing. The perfect stage is usually the overwintering stage although in many cases the asci and ascospores do not form and mature until late winter or early spring. In many cases, however, the fungus can overwinter as mycelium and, occasionally, as conidia.

The asci of the Hemiascomycetes. (i.e., the yeasts and leaf curl fungi), arise either directly from zygotes derived from the copulation of two cells or from dikaryotic ascogenous cells formed parthenogenetically. Their ascospores usually multiply by budding. The ascus in the rest of the Ascomycetes, i.e., the Euascomycetes, is generally formed as a result of fertilization of the female sex cell, called an ascogonium, by either an antheridium or a minute male sex spore called a spermatium. In either case, the fertilized ascogonium produces one to many ascogenous hyphae the cells of which are dikaryotic, i.e., contain two nuclei, one male and one female. By a rather complicated process (Fig. 62), the cell at the tip of each ascogenous hypha develops into an ascus, in which the two nuclei

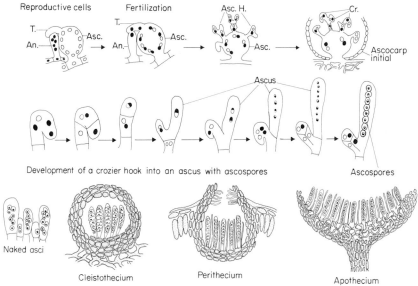

General scheme of sexual reproduction, ascus development and types of ascocarps in the Ascomycetes
An.=Antheridium   Asc.=Ascogonium   T.=Trichogyne   Asc.H.=Ascogenous hyphae   Cr.=Crozier

FIGURE 62.
General scheme of sexual reproduction, ascus development, and types of ascocarps in the Ascomycetes: An.—antheridium; Asc.—ascogonium; T—trichogyne; Asc. H.—ascogenous hyphae; Cr.—crozier.

fuse to produce a zygote which then undergoes meiosis to produce 4 haploid nuclei. The cell, in which these nuclei are, elongates and all 4 nuclei undergo mitosis and produce 8 haploid nuclei. Each nucleus is then surrounded by a portion of the cytoplasm and is enveloped by a wall thus becoming a spore inside an ascus, i.e., an ascospore. There are usually 8 ascospores per ascus (Fig. 62).

The asci in some Ascomycetes, e.g., in the yeasts and leaf curl fungi, are naked, i.e., they are not produced inside fruiting bodies (Fig. 62), but in all other Ascomycetes the asci are produced, singly or in groups, in fruiting bodies called ascocarps. In some, e.g., the powdery mildews, of the Pyrenomycetes (Perithecial Ascomycetes), the ascocarp is a completely closed spherical container and is called a *cleistothecium*. In others, e.g., most of the Pyrenomycetes, the ascocarp is more or less closed but at maturity it has an opening through which the ascospores escape and such ascocarp is called a *perithecium*. In the Pseudosphaeromycetes (Ascostromatic Ascomycetes), the asci are formed directly in cavities within a stroma (= matrix) of mycelium and this ascocarp is called a *perithecium* or an *ascostroma*. Finally, in the Discomycetes (Cup Ascomycetes), the asci are produced in an open, cup- or saucer-shaped ascocarp called an *apothecium* (Fig. 62).

Ascomycetes are characterized and identified by the characteristics of their ascocarps, asci, and ascospores (Fig. 63). However, during the growing season and most of the year, these structures are not produced and cannot be found on diseased plant tissue. What one finds on diseased

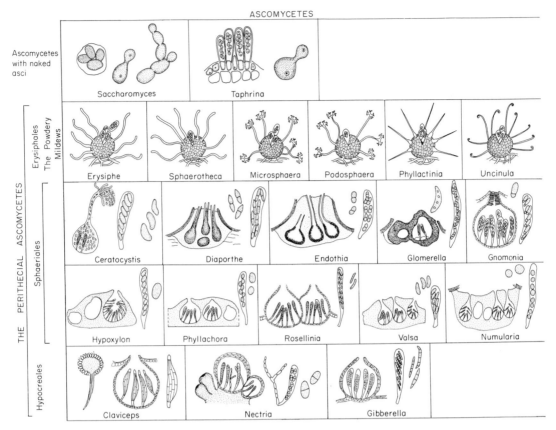

FIGURE 63.
Morphology of fruiting bodies, asci, and ascospores of the main groups and genera of phytopathogenic Ascomycetes.

plants during the growing season are mycelium and asexual spores, i.e., conidia. The conidia themselves often have distinct characteristics which may be sufficient to identify the fungus.

In the Imperfect Fungi and in those of the Ascomycetes that exist primarily as their imperfect stage, the conidial characteristics plus the shape of the conidiophore, i.e., the hypha that produces the conidium, the arrangement of the conidiophores, and the way the conidia are borne on the conidiophore may have to be determined so that the fungus can be identified (Fig. 64). In many cases, the conidia are borne singly or in chains at the tips of conidiophores arising from the mycelium, free from each other. These fungi belong to the order Moniliales of the Imperfect Fungi (Figs. 64 and 65). In the same order, Moniliales, belong the fungi whose conidiophores are produced on a cushion-shaped stroma of mycelium and the whole structure is called a *sporodochium*, or whose conidiophores are cemented together into an elongated spore-bearing structure called *synnema*. In many other cases, conidiophores may be organized into definite fruiting bodies. The most common fruiting body that contains conidiophores and conidia is the *pycnidium*, and fungi that

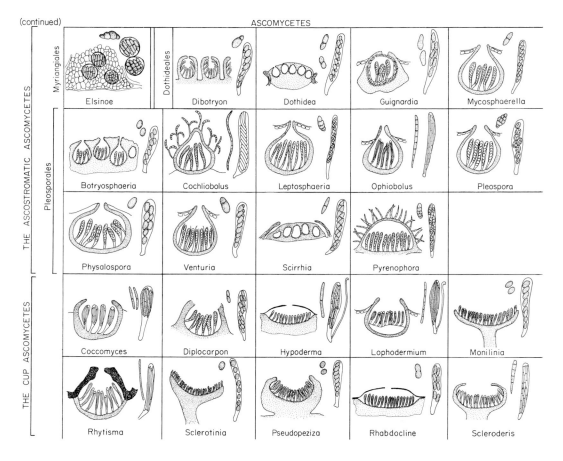

FIGURE 63 (continued).

produce pycnidium belong to the other Sphaeropsidales of the Imperfect Fungi (Figs. 64 and 67). Pycnidium is a hollow, more or less spherical or flask-shaped structure, its walls made of mycelial cells that give rise to the conidiophores. Pycnidia may be almost on the surface of the diseased plant or may be more or less embedded in the diseased tissues. Another type of asexual fruiting body, produced by fungi belonging to the order Melanconiales of the Imperfect Fungi, is the *acervulus*; this is formed by a mat of hyphae below the epidermis or cuticle of the diseased plant giving rise to short conidiophores packed closely together and producing conidia at their tips (Fig. 64 and 66).

As a rule, Ascomycetes overwinter as an ascospore in an ascus. In the spring the ascus releases the ascospores which germinate by one or more germ tubes and cause the primary, i.e., the first infection, of the host plant. The germ tubes grow into hyphae and mycelium that have cross walls. The mycelium soon forms conidiophores that produce conidia. The conidia, produced continually by the mycelium as long as conditions are favorable, spread to other plants, germinate by one or more germ tubes and cause new, secondary infections, which will again produce identical mycelium, conidiophores and conidia. Thus, a large number of asexual

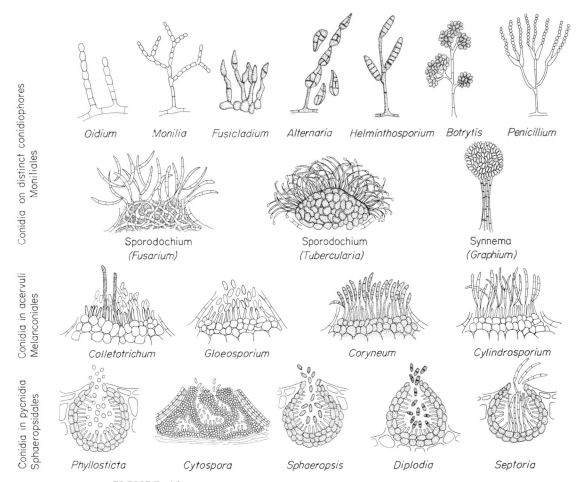

FIGURE 64.
Types of conidia, conidiophores, and fruiting bodies produced by Ascomycetes and Imperfect Fungi and belonging to the three orders of Imperfect Fungi.

generations are produced during the growing season. When conditions become unfavorable, the same mycelium produces the sexual cells that result in the overwintering structure and production of the ascus and ascospores the following spring.

Some Ascomycetes and all Imperfect Fungi overwinter as mycelium or conidia. These then, with or without the presence of ascospores, may start the primary infections. Following infection they proceed to produce mycelium, conidiophores and conidia which then cause the repeated generations of secondary infections.

Many Ascomycetes and Imperfect Fungi cause a variety of diseases in all types of plants (Fig. 68). The most important plant pathogenic Ascomycetes and Imperfect Fungi are briefly discussed below, grouped according to the general symptoms they cause on their hosts.

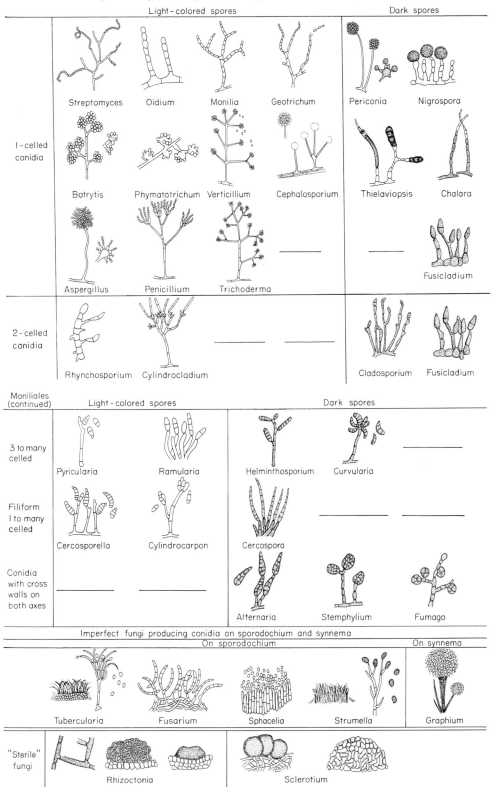

FIGURE 65.
Grouping and morphology of conidiophores and conidia produced by the main genera of phytopathogenic Moniliales of the Imperfect Fungi.

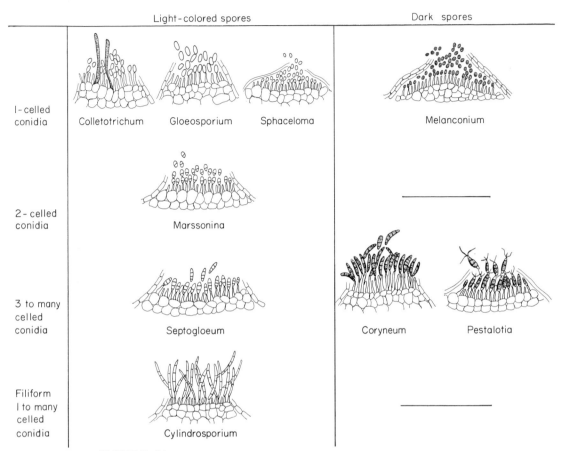

Light-colored spores　　　　　　Dark spores

1-celled conidia — Colletotrichum　Gloeosporium　Sphaceloma　Melanconium

2-celled conidia — Marssonina

3 to many celled conidia — Septogloeum　Coryneum　Pestalotia

Filiform 1 to many celled conidia — Cylindrosporium

FIGURE 66.
Morphology of acervuli and conidia produced by the main genera of Melanconiales of the Imperfect Fungi.

SOOTY MOLDS

Sooty molds appear on the leaves or stems of plants as a superficial, black growth of mycelium forming a film or crust on these plant parts. Sooty molds may be found on all types of plants, including grasses, ornamentals, crop plants, shrubs, and trees. They are most common in warm, humid weather.

Sooty molds are caused by several species of fungi of various types, but primarily dark-colored Ascomycetes of the order Dothideales. These fungi, e.g., *Capnodia (Fumago), Limacinia,* are not parasitic but live off the "honey dew," the sugary deposit forming on plant parts from the droppings of certain insects, particularly aphids and scale insects. The fungal growth is so abundant that it gives the leaf the black, sooty appearance and interferes with the amount of light that reaches the plant. This mycelium sometimes forms a black papery layer that can be peeled

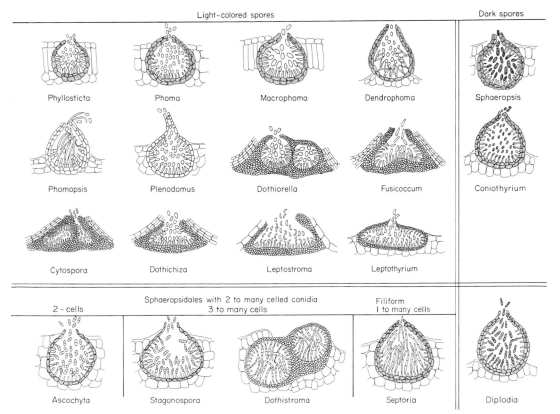

FIGURE 67.
Morphology of pycnidia and conidia of the main genera of Sphaeropsidales of the Imperfect Fungi.

off from the underlying leaf (Fig. 69). The presence of the sooty mold fungi is usually of rather minor importance to the health of the plant but it does indicate the presence of insects and may be a warning of a severe aphid or scale problem.

Sooty molds can be diagnosed easily by the fact that the black sooty mycelial growth can be completely wiped off a leaf or stem with a moistened cloth, paper, or hand, leaving a clean, healthy-looking plant surface underneath.

No control measures are applied against the sooty mold fungi. Since they grow on the excretions of the insects, control of the particular insect involved with the appropriate insecticide or other means also results in the elimination of the sooty mold fungi.

SELECTED REFERENCES

Barr, M. E. 1955. Species of sooty molds from western North America. *Can. J. Bot.* **33**:497–514.

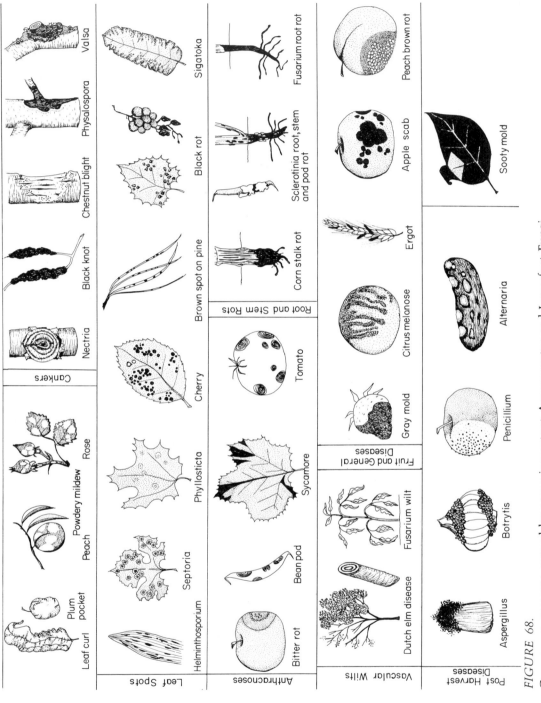

FIGURE 68.
Common symptoms caused by some important Ascomycetes and Imperfect Fungi.

246

FIGURE 69.
Sooty mold on orange leaves. The fungus forms a leathery mat that can be peeled off or washed off. (Photo courtesy Agric. Res. Educ. Center, Lake Alfred, Fla.)

Fraser, Lilian. 1935. An investigation of the "sooty moulds" of New South Wales, I–V. *Proc. Linnean Soc. N. S. Wales* **58**:375–395, 1933; **59**:123–142, 1934; **60**: 97–118, 159–178, 280–290.

Fraser, Lilian. 1937. The distribution of sooty-mould fungi and its relation to certain aspects of their physiology. *Proc. Linnean Soc. N. S. Wales* **62**:25–56.

Webber, H. J. 1897. Sooty mold of the orange and its treatment. *U.S. Dept. Agr., Div. Veg. Phys. Pathol., Bull.* **13**: 34 p.

LEAF CURL DISEASES CAUSED BY TAPHRINA

Several species of *Taphrina* cause leaf, flower, and fruit deformations on several stone fruit and forest trees. Thus *T. deformans* causes peach leaf curl, *T. communis* and *T. pruni* cause plum pocket on American and European plums, respectively, *T. cerasi* causes leaf curl and witches'-broom on cherries, and *T. coerulescens* causes leaf blister of oak. Leaf blister of oak causes local swollen areas on leaves that sometimes result in moderate to heavy (50 to 85 percent) defoliation by midsummer and weakening of the trees. The most important losses, however, are those caused primarily on peach and sometimes plum.

The *Taphrina* diseases are best known in Europe and North America but probably occur all over the world. *Taphrina* causes defoliation of peach trees which may lead to small fruit or fruit drop. In plum, 50 percent or more of the fruit may be affected and lost in years when the disease is severe. In both peach and plum, buds and twigs may also be affected, thus devitalizing the tree.

Symptoms.   In peach, parts of or entire infected leaves are thickened, swollen, distorted and curled downward and inward (Fig. 70). Affected leaves at first appear reddish or purplish but later turn reddish yellow or yellowish gray. At this stage the fungus produces its spores on the swollen areas which appear powdery gray. The leaves later turn yellow to brown and drop. Blossoms, young fruit, and current year's twigs may also be attacked. Infected blossoms and fruit generally fall early in the season. The infected twigs are always stunted but show swelling only in severe infections.

In plum, the disease first appears as small white blisters on the fruit. The blisters enlarge rapidly as the fruit develops and soon involve the entire fruit. The fruit increases abnormally in size and is distorted, the flesh becoming spongy. The seed ceases to develop, turns brown, and withers, leaving a hollow cavity. The fruit appears reddish at first, but later becomes gray and covered with a grayish powder. Leaves and twigs may also be affected, as in peach.

The pathogen: Taphrina sp.   The mycelial cells of *Taphrina* contain two nuclei. These cells may develop into an ascus usually containing 8 uninucleate ascospores. The ascospores multiply by budding inside or outside the ascus producing blastospores (= conidia). The latter may bud again to produce more thin- or thick-walled conidia or may germinate to produce mycelium. Upon germination, the conidial nucleus divides and the two nuclei move into the germ tube. As the mycelium grows, both nuclei divide concurrently, producing the binucleate cells of the

FIGURE 70.
(A) Peach leaf curl caused by *Taphrina deformans*. (B) Oak leaf blister caused by *T. coerulescens*.

mycelium. Mycelial cells near the plant surface separate from each other and produce the asci.

Development of disease. The fungus apparently overwinters as ascospores or thick-walled conidia on the tree, perhaps in the bud scales. In the spring, these spores are blown to young tissues, germinate and penetrate the developing leaves and other organs directly through the cuticle or through stomata (Fig. 71). The binucleate mycelium then grows between cells and invades the tissues extensively inducing excessive cell enlargement and cell division, which result in the enlargement and distortion of the plant organs. Later, numerous hyphae grow outward in the area between the cuticle and epidermis. There they break into their component cells which produce the asci. The asci enlarge, exert pressure on the host cuticle from below and eventually break through to form a compact, feltlike layer of naked asci. The ascospores are released into the air, carried to new tissues, and bud to form conidia. Infection occurs mainly during a short period after the buds open. All organs become

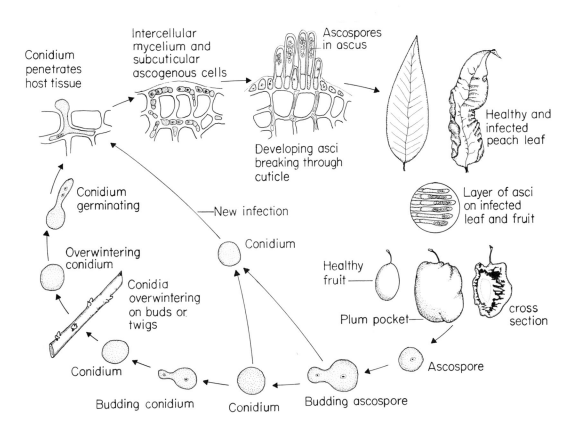

Disease cycle of diseases caused by *Taphrina* sp.

FIGURE 71.
Disease cycle of diseases caused by *Taphrina* sp.

resistant to infection as they grow older. Infection is favored by low temperature and a high humidity at the time the tissues are susceptible.

Control.   *Taphrina* diseases are easily controlled by a single fungicide spray preferably in late fall after the leaves have fallen or in early spring before leaf buds swell. The fungicides most commonly used are ferbam, elgetol, and Bordeaux mixture (8:8:100). Difolatan may also be used as one application only, before leaf drop is complete.

SELECTED REFERENCES

Fitzpatrick, R. E. 1934. The life history and parasitism of *Taphrina deformans. Sci. Agr.* **14**:305–306.

Martin, E. M. 1940. The morphology and cytology of *Taphrina deformans. Am. J. Bot.* **27**:743–751.

Mix, A. J. 1935. The life history of *Taphrina deformans. Phytopathology* **25**:41–46.

Mix, A. J. 1949. A monograph of the genus *Taphrina. Kan. Univ. Sci. Bull.* **33**: 167 p.

Weber, G. F. 1941. Leaf blister of oaks. *Fla. Agr. Exp. Sta. Bull.* **558**: 2 p.

THE POWDERY MILDEWS

Powdery mildews are probably the most common, conspicuous, widespread, and easily recognizable plant diseases. They affect all kinds of plants: cereals and grasses, vegetables, ornamentals, weeds, shrubs, fruit trees, and broad-leaved shade and forest trees.

Powdery mildews are characterized by the appearance of spots or patches of a white to grayish, powdery, mildewy growth on young plant tissues, or of entire leaves and other organs being completely covered by the white powdery mildew. Tiny, pinhead-sized, spherical fruiting bodies, at first white, later yellow-brown, and finally black cleistothecia may be present singly or in groups on the white to grayish mildew in the older areas of infection. Powdery mildew is most commonly observed on the upper side of the leaves but it also affects the under side of leaves, young shoots and stems, buds, flowers, and young fruit.

The fungi causing powdery mildews are obligate parasites—they cannot be cultured on artificial nutrient media. They produce mycelium that grows only on the surface of plant tissues, never invading the tissues themselves. They obtain nutrients from the plant by sending haustoria, i.e., feeding organs, into the epidermal cells of the plant organs. The mycelium produces short conidiophores on the plant surface. Each conidiophore produces chains of rectangular, ovoid, or round conidia that are carried by air currents. When environmental conditions or nutrition become unfavorable, the fungus may produce one or a few asci inside a closed ascocarp, the cleistothecium. The powdery mildew fungi, although they are common and cause serious diseases in cool or warm, humid areas, are even more common and severe in warm, dry climates. This happens because their spores can be released, germinate, and cause infection even when only the relative humidity in the air is fairly high without need for the presence of a film of water on the plant surface. Also because, once infection has begun, the mycelium continues to spread on the plant surface regardless of the moisture conditions in the atmosphere.

The powdery mildew diseases of the various crop or other plants are caused by many species of fungi of the family Erysiphaceae grouped into six main genera. These genera are distinguished from each other by the number (one versus several) of asci per cleistothecium and by the morphology of hyphal appendages growing out of the wall of the cleistothecium. The main genera are illustrated in Fig. 63, and the most important diseases they cause are listed below.

> *Erysiphe cichoracearum*, causes powdery mildew of begonia, chrysanthemum, cosmos, cucurbits, dahlia, flax, lettuce, phlox, zinnia.
> *E. graminis*, causes powdery mildew of cereals and grasses (Fig. 72G).
> *E. polygoni*, causes powdery mildew of beans, soybeans, clovers and other legumes, beets, cabbage and other crucifers, cucumber and cantaloupe, delphinium, hydrangea.
> *Microsphaera alni*, causes powdery mildew of blueberries, catalpa, elm, lilac, linden, oak, rhododendron, and sweet pea.
> *Phyllactinia* sp., causes powdery mildew of catalpa, elm, maple, oak.
> *Podosphaera leucotricha*, causes powdery mildew of apple (Fig. 72F), pear and quince.
> *P. oxyacanthae*, causes powdery mildew of apricot, cherry, peach and plum.
> *Sphaerotheca macularis*, causes powdery mildew of strawberry.
> *S. mors-uvae*, causes powdery mildew of gooseberry and current.
> *S. pannosa*, causes powdery mildew of peach and rose (Fig. 72, A–C).
> *Uncinula necator*, causes powdery mildew of grape, horsechestnut, linden (Fig. 72, D, E).

- *Powdery Mildew of Rose and Peach*

Powdery mildew of roses occurs everywhere in the world where roses are grown. The disease is less common on peaches although it has been found to affect peaches in the U.S., Europe, and Africa.

Powdery mildew is one of the most important diseases of roses, both in the garden and in the greenhouse. The disease appears on roses year after year and causes reduced flower production and weakening of the plants by attacking the buds, young leaves, and growing tips of the plants.

Powdery mildew is usually less severe on peach, but when weather conditions are favorable for infection, peach seedlings are stunted, fruit production is reduced, and the quality of infected fruit is very poor.

Symptoms. On the young leaves the disease appears at first as slightly raised blisterlike areas that soon become covered with a grayish white, powdery fungus growth, and as the leaves expand they become curled and distorted (Fig. 72). On the older leaves, large white patches of fungus growth appear, but there is usually little distortion. Lesions on leaves may appear more or less discolored and may eventually become necrotic.

White patches of fungus growth, similar to those on the leaves, usually appear on young, green shoots, and they may coalesce and cover the entire terminal portions of the growing shoots, which may become

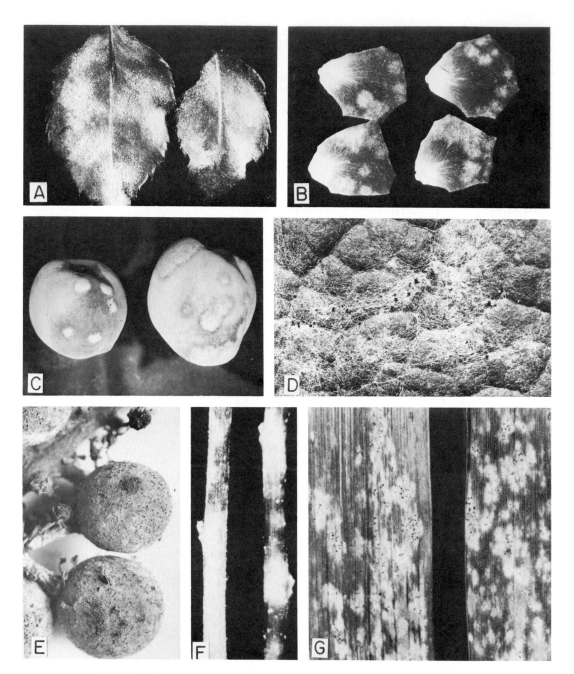

FIGURE 72.
Powdery mildew on rose leaves (A) and petals (B), and on peach fruit (C) caused by *Sphaerotheca pannosa*. (D, E) Mycelium, conidia and dark cleistothecia on grape leaves (D) and fruit cluster (E) caused by the powdery mildew fungus *Uncinula necator*. (F) Powdery mildew on apple twigs caused by *Podosphaera leucotricha*. (G) Powdery mildew on wheat leaves caused by *Erysiphe graminis*. The dark dots are cleistothecia. (Photos D, E courtesy Shade Tree Lab, Univ. of Mass. Photo G courtesy U.S.D.A.)

arched or curved at their tip. Sometimes buds may be attacked and become covered with white mildew before they open, and they either fail to open or open improperly, the infection spreading to the flower parts, which become discolored, dwarfed, and eventually dried.

When peach fruit becomes infected, white circular spots appear which may spread over a large portion or the whole surface of the fruit. The fruit color becomes pinkish at first and later turns to a dark brown, while the fruit surface becomes leathery and hard and sometimes it appears swollen or cracked.

The pathogen: Sphaerotheca pannosa. The pathogen that causes powdery mildew on roses seems to be a distinct variety of *S. pannosa*, since it has been shown in some cases that the fungus from roses does not attack peaches, and vice versa. The life history and the behavior of the fungus, however, are the same in both cases, and therefore they will be described as one.

The mycelium is white and grows on the surface of the plant tissues, sending globose haustoria into the epidermal cells of the plant (Fig. 73).

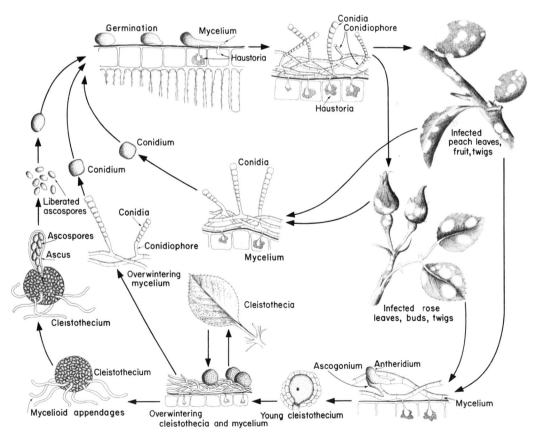

FIGURE 73.
Disease cycle of powdery mildew of roses and peach caused by *Sphaerotheca pannosa*.

The mycelium forms a weft of hyphae on the surface, some of which develop into short, erect, conidiophores. At the tip of each conidiophore egg-shaped conidia are produced which cling together in chains.

With the coming of cool weather late in the season, conidia production ceases and cleistothecia may be formed. The young cleistothecia are globose and at first white, then brown, and finally black when mature. The mature cleistothecia also have several mycelioid appendages, which are flaccid, indefinite hyphae arising from cells of the cleistothecium. The cleistothecia are more or less buried in the mycelial wefts on the plant tissues. The ascospores continue to develop during the fall, and in the spring they are mature and ready for dissemination. In the spring the cleistothecia absorb water and crack open. The single ascus in each cleistothecium protrudes its tip, bursts open and discharges its 8 mature ascospores, which the wind carries away. The ascospores are about the size of conidia and behave exactly like conidia with respect to germination, infection, and formation of subsequent structures.

Development of disease.   On peaches and outdoor roses the fungus apparently overwinters mostly as mycelium in the buds, although on outdoor roses cleistothecia form occasionally on leaves, petals, and stems, especially around the thorns, toward the end of the season. Cleistothecia are much more rare on peach. On greenhouse roses the pathogen overwinters almost exclusively as mycelium and conidia.

When the fungus overwinters as mycelium in dormant buds, shoots arising from such buds become infected and provide inoculum for subsequent secondary mycelial or spore infection and disease development on foliage and fruit. When the fungus overwinters as cleistothecia the discharged mature ascospores also serve as primary inoculum (Fig. 73). Ascospores or conidia are carried by wind to young green tissues, and if temperature and relative humidity are sufficiently high the spores germinate by putting out a germ tube. The germ tube quickly produces a short, fine hypha growing directly through the cuticle and the epidermal cell wall into the epidermal cells. The penetrating hypha enlarges immediately upon entrance into the cell lumen and forms a globose haustorium by which the fungus obtains its nutrients. The germ tube continues to grow and branch on the surface of the plant tissue producing a network of superficial mycelium. As the mycelium spreads on the plant, it continues to send haustoria into the epidermal cells. The absorption of nutrients from the cells depletes their food supply, weakens them, and may sometimes lead to their death. Photosynthesis in the affected areas is greatly reduced and the other functions of the cells are also impaired. Infection of young leaves also causes irritation and uneven growth of the affected and the surrounding cells resulting in slight raised areas on the leaf and distortion of the leaf. The aerial mycelium produces short, erect conidiophores each bearing a chain of 5 to 10 conidia. The conidia are disseminated by air currents and cause new infections on the expanding leaves, shoots, and on peach, on the fruit. Greenhouse roses are susceptible throughout the year. In the field, however, expanding tissues seem to be the most susceptible ones and only under very favorable humidity and temperature do fully developed tissues become infected. Growth of severely infected shoots is inhibited. Infected buds often do not open. If they do open, the flowers become infected and do not develop. Peach fruits are susceptible from the time of their formation until they are approximately 2.5 to 3.0 cm in diameter, beyond which they become resistant to powdery mildew. Once fruit infections are established, however, the lesions continue to enlarge for several weeks after the fruit becomes resistant to new infections. Infected fruits are smaller and unsightly and must be sold at a much lower price.

Control. Powdery mildew can be controlled by application of sulfur, dinocap, benomyl, or certain other fungicides. Sulfur may be used as a spray, as a dust and, in the greenhouse, as a vapor also. Dinocap, benomyl, and cycloheximide are used as sprays. Under most conditions weekly applications give adequate protection, but during rapid development of new growth, temperature fluctuations and frequent rains, more frequent applications may be necessary.

SELECTED REFERENCES

Cherewick, W. J. 1944. Studies on the biology of *Erysiphe graminis. Can. J. Res.* **22**:52–86.

Hills, F. J., *et al.* 1975. Effect of powdery mildew on sugarbeet production. *Plant Disease Reptr.* **59**:513–515.

Keil, H. L., and R. A. Wilson. 1961. Powdery mildew of peach. *Plant Disease Reptr.* **45**:10–11.

Longree, Karla. 1939. The effect of temperature and relative humidity on the powdery mildew of roses. *Cornell Univ. Agr. Expt. Sta. Mem.* **223**:43 p.

Massey, L. M. 1948. Understanding powdery mildew. *Am. Rose Ann.* **33**:136–145.

Moseman, J. G. 1966. Genetics of powdery mildews. *Ann. Rev. Phytopathol.* **4**:269–290.

Salmon, E. S. 1900. A monograph of the Erysiphaceae. *Mem. Torrey Bot. Club Vol.* **9**: 292 p.

Schnathorst, W. C. 1965. Environmental relationships in the powdery mildews. *Ann. Rev. Phytopathol.* **3**:343–366.

Weinhold, A. R. 1961. Temperature and moisture requirements for germination of conidia of *Sphaerotheca pannosa* from peach. *Phytopathology* **51**:699–703.

Yarwood, C. E. 1957. Powdery mildews. *Bot. Rev.* **23**:235–300.

FOLIAR DISEASES
CAUSED BY ASCOMYCETES AND
IMPERFECT FUNGI

Many species of Ascomycetes and of Imperfect Fungi cause primarily foliage diseases but some may also affect blossoms, young stems, and fruit and even roots.

Most of the foliar Ascomycetes reproduce by means of conidia formed on free hyphae or in pycnidia but a few produce conidia in sporodochia, or in acervuli. In many the conidia overwinter; others reproduce by means of conidia during the growing season and by their perfect ascigerous stage at the end of the season and over winter; some produce their ascocarps and ascospores, along with conidia, throughout the growing season. The primary inoculum of these fungi, therefore, may be either ascospores or conidia and usually originates from infected fallen or hanging leaves of the previous year.

Some of the most common Ascomycetes causing primarily foliar diseases are the following:

Coccomyces sp., causing leaf spot or shot-hole of cherries and plums.
Dothidella ulei, causing South American leaf blight of rubber.
Elytroderma deformans, causing a leaf spot and witches'-broom of pines.
Lophodermium pinastri, causing needle blight of pines.
Guignardia, causing leaf spot and black rot of grape, *G. bidwellii,* leaf blotch of horsechestnut (*G. aesculi*), and leaf spots on Boston ivy and Virginia creeper.
Mycosphaerella, causing the extremely destructive Sigatoka disease of banana (*M. musicola*), leaf spots of strawberry (*M. fragariae*), of pear (*M. sentina*), leaf spot and black rot of cucurbits (*M. melonis*), citrus greasy spot (*M. citri*), and other diseases.
Pseudopeziza, causing the common leaf spot of alfalfa (Fig. 74 A) and clovers and the yellow leaf blotch of alfalfa.
Rhabdocline pseudotsugae, causing needle cast of Douglas fir.
Rhytisma, causing tar spot of maple and willow.
Scirrhia, causing brown spot needle blight of pine (*S. acicola,* the conidial stage of which is *Lecanosticta* or *Septoria*), and Dothistroma needle blight of pine (*S. pini,* the conidial stage of which is *Dothistroma pini*).

Some of the most common Imperfect Fungi causing primarily foliar, but also other, symptoms on a large variety of host plants are: *Ascochyta, Cercospora* (Fig. 74B), *Cladosporium* (Fig. 74C), *Helminthosporium, Phyllosticta, Pyricularia* (Fig. 75), and *Septoria.* Many other fungi, e.g., *Alternaria* and *Botrytis,* could be listed here, but they so frequently affect other plant parts that they are discussed elsewhere.

The foliar spots and blights caused by Imperfect Fungi affect numerous hosts and appear in many variations. However, the disease cycles and controls of these diseases are quite similar, although considerable variability may exist between diseases caused by specific fungi on different hosts, especially when the diseases develop under different environmental conditions. Thus, these fungi attack primarily the foliage of annual and/or perennial plants by means of conidia produced on free, single or grouped hyphae (*Cercospora, Cladosporium, Helminthosporium,* and *Pyricularia*) or in pycnidia (*Ascochyta, Phyllosticta,* and *Septoria*). On the infected areas numerous conidia are produced which spread by wind, water, insects, etc. to other plants and cause more infections. In most cases these fungi overwinter primarily as conidia or mycelium in fallen leaves or other plant debris. Some, however, can overwinter as conidia in or on seed of infected plants, or in the soil. When perennial plants are infected, they may overwinter as mycelium in infected tissues of the plant. When these fungi are carried with the seed of annual plants, damping off of seedlings may develop. Control of such diseases is accomplished by using resistant varieties and fungicidal sprays, but disease-free seed and/or re-

FIGURE 74.
(A) Alfalfa leaf spot caused by *Pseudopeziza medicaginis*. (B) Sugar beet leaf spot caused by *Cercospora beticola*. (C) Tomato leaf mold caused by *Cladosporium fulvum*. (Photo C courtesy of U.S.D.A.).

moval and destruction of contaminated debris may be most important in some diseases.

- *Coccomyces Leaf Spot or Shot-hole of Cherries and Plums*

Widespread and serious throughout the world, especially in areas with humid climate, it is most severe on sour cherries and less so on sweet

FIGURE 75.
Foliar symptoms on rice leaves
caused by the rice blast fungus
Pyricularia oryzae. (Photo
courtesy U.S.D.A.)

cherries and plums. In cherries, the symptoms appear as numerous minute, red to purplish-black spots on the upper sides of leaves and, in severe
infections, on the petioles, fruit and pedicels (Fig. 76A). On the under
sides of the spots on the leaves, slightly raised, waxy pustules appear that
after heavy dews or rains produce a white, mildewlike growth. Infected
leaves usually turn yellow and fall, or the spots may drop out and the
leaves have a shot-hole appearance. The shot-hole effect is even more
common and more pronounced on plum and prune leaves which may
appear completely skeletonized. Severe leaf drop and shot-holing early in
the season are common and weaken the trees which may die back during
the same or in subsequent years. The disease is especially common and
serious in nursery plantings.

The fungus, *Coccomyces* sp. (or *Higginsia* sp.), produces ascospores in
apothecia formed in fallen infected leaves and *Cylindrosporium*-type
conidia formed in acervuli on the under side of infected leaves. The
conidia are colorless, threadlike, straight or curved and consisting of one
or several cells. The ascospores are produced in the spring over a period of
6 to 7 weeks and are forcibly ejected when the leaves are thoroughly
soaked. The ascospores are then carried by air currents and cause the
primary infections on the leaves which then produce large numbers of
conidia that are spread by rain from leaf to leaf and cause all the subsequent infections.

Control of the disease is achieved by 4 to 5 sprays starting at petal-fall,
and one postharvest spray with benomyl, captan, or certain other fungicides such as dodine, sulfur, or ferbam.

- *Guignardia Leaf Spot and Black Rot of Grape*

It is present in Europe and in the U.S. and Canada east of the Rocky Mountains. It is probably the most serious disease of grapes where it commonly occurs, particularly in warm, humid regions. In the absence of control measures and in favorable weather, the crop it usually destroyed completely either through direct rotting of the berries or through blasting of the blossom clusters.

The disease causes numerous scattered, circular, red necrotic spots on leaves in late spring (Fig. 76, C–G). The spots form usually between the veins and are most apparent on the upper side of the leaves. Later, when the spots are about 2 to 6 mm or more in diameter, the main area of the spots appears brown to grayish-tan while their margins appear as a black line. Black dotlike, *Phyllosticta*-type pycnidia are formed on the upper side of the spots in a ring near the outer edge of the brown area of the spot. On the shoots, tendrils, the leaf and flower stalks, and on leaf veins, the spots are purple to black, somewhat depressed and elongated, and bear scattered pycnidia. Spots begin to appear on berries when the latter are about half grown. These spots are at first whitish but are soon surrounded by a rapidly widening brown ring with a black margin. The central area of the spot remains flat or becomes depressed and dark pycnidia appear near the center. The whole berry soon becomes rotten and shrinks, and becomes coal black as the surface becomes studded with numerous black pycnidia.

The fungus, *Guignardia bidwellii,* in addition to conidia-bearing pycnidia, also produces ascospores in globose perithecia forming in rotten, mummied fruit. The perithecia supposedly develop from transformed pycnidia. The fungus overwinters mostly as ascospores in perithecia, but conidia can also survive the winter in most locations where grapes grow, so both ascospores and conidia can cause primary infections in the spring (Fig. 77). The release of ascospores and conidia takes place only when the perithecia and pycnidia become thoroughly wet, but while the ascospores are shot out forcibly and may then be carried by air currents, the conidia are exuded in a viscid mass from which they can be washed down or splashed away by rain. Ascospores may be discharged continually through the spring and summer, although most of them are discharged in the spring.

Primary infections, whether from ascospores or conidia, take place on young, rapidly growing leaves and on fruit pedicels. In the ensuing spots, pycnidia are produced rapidly and these provide the conidia for the secondary infections of berries, stems, etc.

Control of the grape leaf spot and black rot depends primarily on timely sprays of grapevines with fungicides. Sprays with ferbam just before bloom, immediately after bloom, and 10 to 14 days later give good control of the disease. Captan, folpet, or copper fungicides are also effective against black rot and are used when downy mildew or powdery mildew are also to be controlled. Where black rot has been severe, another ferbam or captan application should be made during the first part of June.

FIGURE 76.
(A) Cherry leaf spot caused by *Coccomyces hiemalis*. (B) Horsechestnut leaf
blotch caused by *Guignardia aesculi*. (C–G) Black rot of grape caused by
Guignardia bidwellii. (C, D, E) Foliar symptoms of black rot on grape showing
damage to leaves (C, D) and the characteristic circular arrangement of the
Phyllosticta-type pycnidia (E). (F, G) Black rot symptoms on young (F) and
completely rotten berries (G).

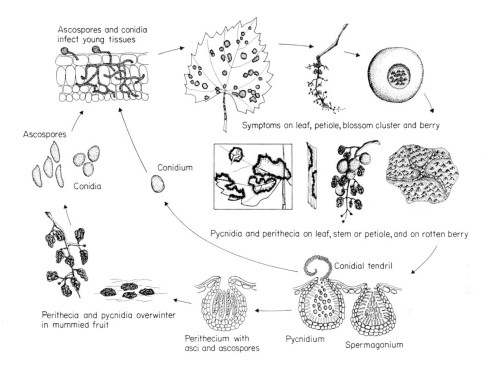

Ascospores and conidia infect young tissues

Symptoms on leaf, petiole, blossom cluster and berry

Ascospores

Conidia

Conidium

Pycnidia and perithecia on leaf, stem or petiole, and on rotten berry

Conidial tendril

Perithecia and pycnidia overwinter in mummied fruit

Perithecium with asci and ascospores

Pycnidium

Spermagonium

FIGURE 77.
Disease cycle of black rot of grapes caused by *Guignardia bidwellii.*

• *Needle Casts and Blights of Conifers*

Several ascomycetous fungi such as *Elytroderma, Hypoderma, Lophodermium,* and *Scirrhia* cause leaf diseases on pine, while *Rhabdocline* infects leaves of Douglas fir. Many other fungi also affect the various conifer hosts and may be more or less damaging depending on the importance of the tree species in the particular area and on the prevailing environmental conditions.

All needle cast and blight diseases have certain common characteristics, although each differs from all others in some respects. Thus, the needlelike leaves of the conifers are infected by the conidia and occasionally by the ascospores of these fungi at some time during the growing season. The type and time of infection may vary with the location where the particular species grows. The fungus enters the needle and usually causes a light green to yellow spot that sooner or later turns brown or red, encircles the needle and kills the part of the needle beyond the spot (Fig. 78). The fungus may spread into the needle, or separate new infections may develop on the needle. As a result, the entire needle is often killed and either clings onto the tree for a while, giving the tree a reddish-brown,

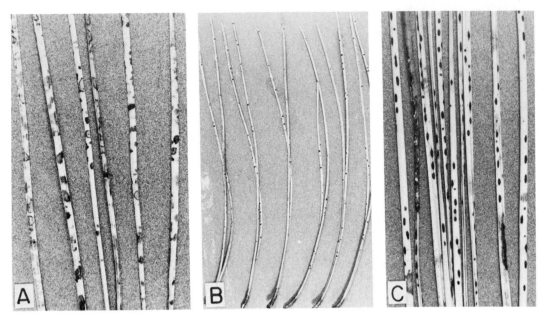

FIGURE 78.
Needle spots and blights of conifers. (A) *Scirrhia* brown spot on long-leaf pine. (B) *Lophodermium* needle cast on red pine. (C) *Hypoderma* needle cast of long-leaf pine.

burned appearance, or it is shed, resulting in partial or total defoliation of the tree. On the infected needles, whether on the tree or on the ground, the fungus produces its conidia and, occasionally, its ascospores in perithecia, which are either released into the air or are exuded during wet weather and are washed down or splashed by the rain onto other needles and trees. In some needle blights, the fungus may overwinter as mycelium in infected but still living needles, while in most cases the fungi overwinter as ascospores or conidia in dead needles on the tree or on the ground.

Needle casts and blights can be destructive on mature trees, especially in plantations of single species, which may be killed following repeated defoliations. Every year, thousands of trees are cut when dead or dying from foliage diseases. These same diseases, however, can be devastating in young or nursery trees which they can kill by the millions in relatively short time if the weather is favorable and no adequate control measures are practiced.

Most, but not all, needle casts and blights can be controlled with fungicidal sprays, especially in the nursery and in young plantation trees. Larger trees are either cut before they die (salvage cutting), or they too may be protected, when possible, with fungicides applied from airplanes. The fungicides most commonly used against leaf cast and blight diseases are Bordeaux mixture (usually a 4:4:100 mixture), other copper fungicides such as tribasic copper sulfate, and maneb, chlorothalonil, and others. In some needle diseases, two sprays either early or late in the season, whenever most of the infections with the particular fungus take place, are sufficient to keep the disease in check, especially in large trees. In most cases, however, nurseries must be sprayed at least every 2 weeks from May through October if the seedlings are to survive the needle attacks by fungi and to grow.

MYCOSPHAERELLA DISEASES

- *Banana Leaf Spot or Sigatoka Disease*

It occurs throughout the world and is one of the most destructive diseases of banana, although its economic importance varies with the region. It causes losses by reducing the functional leaf surface of the plant which results in small, unevenly ripened bananas that may fall and fail to ripen.

The disease first appears as small, indistinct, longitudinal, light yellow spots parallel to the side veins of leaves that unfurled about a month earlier. A few days later, the spots become 1 to 2 cm long, turn brown with light grey centers and become readily visible. Such spots cause little damage, but they soon enlarge further, the tissue around them turns yellow and dies and adjacent spots coalesce forming large dead areas on the leaf (Fig. 79, A, B). In severe infections, entire leaves die within a few weeks. Since it takes at least 12 healthy leaves on mature banana plants to carry fruit to maturity, destruction of most mature leaves by the leaf spot disease may leave only a few functioning leaves that are insufficient to bring the fruit to maturity. As a result, immature fruit bunches on such plants fail to fill out and ripen, and may fall. If the fruit is nearing maturity at the time of heavy infection, the flesh ripens unevenly, individual bananas appear undersized and angular in shape, their flesh develops a buff pinkish color, and they store poorly.

The causal fungus is *Mycosphaerella musicola*. It produces spermatia in spermagonia, ascospores in perithecia, and conidia of the *Cercospora* type in sporodochia. Sporodochia appear while the spots are still light yellow but successive crops of abundant conidia are produced by the same sporodochia only during the brown spot stage of the disease (Fig. 80). Conidia are produced on both sides of the leaf but are usually more abundant on the upper side and are spread by wind and dripping or splashing water. Although conidia are produced throughout the year, their release and germination depends on water or high humidity. Perithecia, formed as a result of fertilization of sexual hyphae by compatible spermatia, are produced during warm humid weather and their ascospores are shot out violently in response to wetting of the perithecia. Ascospores are spread by air currents and are responsible for the long-distance spread of the disease while conidia are generally the most important means of the local spread of the disease. Infection by either ascospores or conidia produces the same type of spot and subsequent development of the disease.

The Sigatoka disease is controlled by a combination of measures including quarantine, sanitation by removal and destruction of badly in-

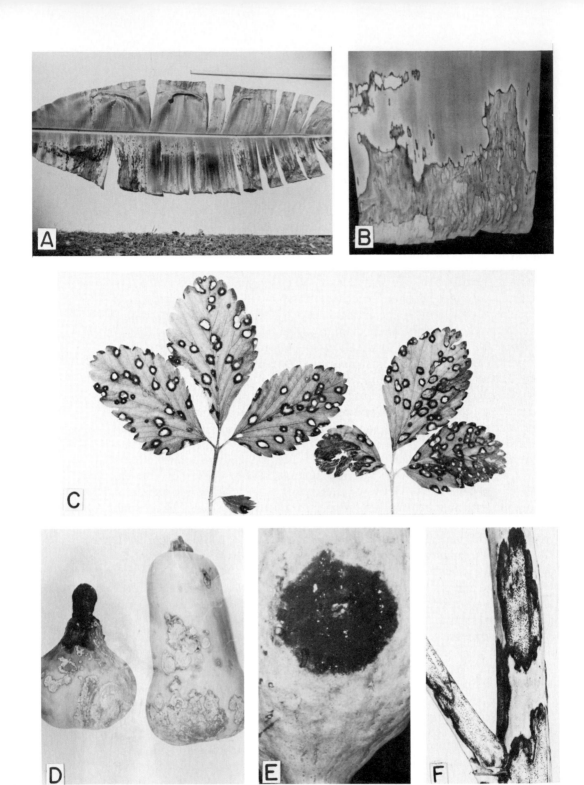

FIGURE 79.

(A, B) Banana leaf spot (Sigatoka disease) caused by *Mycosphaerella musicola*. (C) Strawberry leafspot caused by *M. fragariae*. (D) Superficial and galloping type of black rot on butternut squash caused by *M. melonis*. (E) Black rot lesion on winter squash. (F) Lesions and pycnidia of *Mycosphaerella brassicola* on stem of young cabbage plant. (Photos A and B, courtesy R. H. Stover. Photo C, courtesy U.S.D.A.)

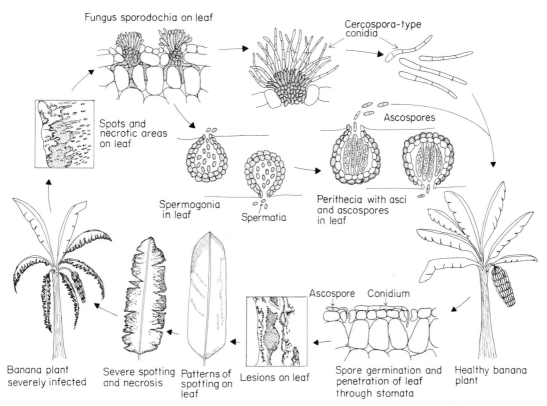

FIGURE 80.
Development of Sigatoka disease of banana caused by *Mycosphaerella musicola*.

fected leaves, and mainly by frequent, year-round application of fungicidal sprays. For many years Bordeaux mixture or copper oxychloride with
or without zineb were the fungicides used. Later it was shown that zineb
or copper oxychloride suspended in mineral oil, or mineral oil alone
(consisting largely of saturated hydrocarbons) gave better and cheaper
control than Bordeaux. The oil sprays, however, are phytotoxic under
certain conditions, may reduce fruit yields by as much as 10 percent, may
cause patchy, retarded ripening when applied directly to fruit and have
resulted in an increase of certain other leaf spot and fruit spot and rot
diseases previously controlled by Bordeaux mixture. To date, maneb or
other fungicides are routinely included in oil–water–fungicide emulsions
for best all-around results. In some areas it is necessary to apply ground or
airplane sprays every 10 to 12 days throughout the year while in other
areas one application every 3 to 4 weeks suffices to maintain control.

• *Strawberry Leaf Spot*

It is worldwide. It appears at first as small, dark purple spots of indefinite
outline. Later, the spots enlarge, a dark purple zone surrounding brown
centers which soon turn light brown in young leaves and white in older

ones (Fig. 79C). Often, spots coalesce and the purple area may extend around several white spots and may even extend to the border of the leaf. Small, black dotlike perithecia appear in the central dead area. The infection may kill parts of leaves or entire leaflets. Although spots may also develop on petioles, fruit, and fruit stems, the frequently severe leaf spot epidemics alone cause a decided reduction in the total yield, reduction in the quality of harvested berries and a weakening of the plants.

The fungus, *Mycosphaerella fragariae,* produces ascospores in perithecia on the upper side of the leaves and great numbers of conidia of the *Ramularia* type are produced on short conidiophores arising over the entire under surface of the diseased area. In most cases the fungus overwinters as mycelium in surviving infected leaves on perennial runners of the strawberry plant. The mycelium produces conidia in the spring which probably cause most of the primary infections and all the secondary ones, although ascospores may also overwinter and cause infection. Both types of spores germinate and penetrate leaves through stomata which are found mostly on the lower side of the leaf. The mycelium then grows between the cells without haustoria, producing conidia on the lower surface throughout the season and perithecia at the upper side during the autumn months.

Control of strawberry leaf spot depends largely on the use of resistant varieties, but when the disease is severe, sprays with benomyl, captan, or thiram at weekly intervals from prebloom to harvest give satisfactory control.

- *Cucurbit Leaf Spot, Gummy Stem Blight,*
  and Black Rot

It is probably worldwide in distribution and in favorable weather the pathogen infects all parts of all cucurbits. Usually, however, it attacks the leaves and stems of watermelons, cucumber, and cantaloupe and the fruit of squash, pumpkin, and gourd. When the fungus is carried in the seed it also causes damping off, killing the seedlings. On the leaves, petioles, and stems, pale brown or gray spots develop. On the stems, the spots usually start at the joints, become elongated and cracked, and exude an amber-colored gummy material. The spots on the leaves enlarge along with the spots on stems and petioles, make the leaves turn yellow and die, and even the whole plant may wilt and die. On the fruit, the spots appear at first as yellowish, irregularly circular areas that later turn gray to brown and may have a droplet of gummy exudate in the center. The spot finally turns black. In some kinds of squash the black rot lesions are superficial and spread over the surface or appear as a dry, brownish, circular mass, the enlargement of which is arrested by a defensive barrier of squash tissue (Fig. 79 D,E). Very often, however, especially in storage, the fungus penetrates through the rind and spreads widely throughout the squash and into the seed cavity. On all the spots, whether on the leaf, stem, or fruit, the fungus produces closely spaced groups of pale-colored pycnidia and dark, globular perithecia that are sometimes arranged in rings and are visible with the naked eye.

The fungus, *Mycosphaerella melonis,* produces ascospores in perithecia and conidia in pycnidia. Both conidia and ascospores are short-lived after they are released and the fungus generally overwinters in diseased plant refuse and in or on the seed. Thus, either type of spore or infected seed can result in primary infections, with the subsequently profusely produced conidia causing the secondary infections.

Control of black rot of cucurbits is difficult, requiring the use of clean or treated seed, long crop rotations, and frequent applications of fungicides such as ziram, captan, and maneb. Good control of the leaf and stem infections reduces fruit infections both in the field and in storage. However, further care is needed to avoid infection in storage. Wounding of stored fruit must be avoided. Curing of squash at 23 to 29°C for two weeks to heal the wounds and subsequent storage at 10 to 12°C are very helpful. If the inoculum present in the field is heavy, dipping of the squash fruit in formaldehyde or Clorox before curing and storage is also helpful.

- *Cercospora Diseases*

They are almost always leaf spots that either stay relatively small and separate or they may enlarge and coalesce resulting in leaf blights. The diseases are generally widespread and among the most common hosts affected are sugar beet, carrot, celery, eggplant, pea, peanut, tomato, rice, corn, sugarcane, and most other cereals and grasses, azalea, Boston ivy, dahlia, geranium, tobacco, and many trees and other crops in temperate zones and in the tropics.

The leaf spots in some plants, e.g., beet and tobacco, are brown, small, about 3 to 5 mm in diameter and roughly circular with reddish-purple borders (Fig. 74B). Later, their centers become ashen-gray, thin, papery and brittle and may drop out leaving a ragged hole, or the spots, if sufficiently numerous, may coalesce causing large necrotic areas. On most other hosts, e.g., celery, carrot, and geranium, the leaf spots are small, reddish, or yellowish at first but they enlarge rapidly, the affected tissue changing to an ashen-gray color and a dry, papery texture. The spots are irregularly circular to angular, with or without a distinct border and often coalesce to form large blighted areas. In monocotyledonous plants the spots are narrow and long, usually 0.5 by 5.0 cm, may coalesce and kill leaves. In humid weather the affected leaf surface on all hosts is covered with an ashen-gray mold barely visible to the naked eye. In severe attacks all the foliage is destroyed and may fall off. On fleshy plants, similar lesions are produced on stems and leaf petioles.

Several species of *Cercospora* are responsible for the diseases on the various hosts. The fungus produces long, slender, colorless to dark, straight to slightly curved, multicellular conidia on long, dark conidiophores. The conidiophores arise from the plant surface in clusters through the stomata and form conidia successively on new growing tips. The conidia are easily detached and often blown long distances by the wind. The fungus is favored by high temperatures and therefore is most destructive in the summer months and in warmer climates. Although *Cercospora* spores need water to germinate and penetrate, heavy dews seem to be sufficient for abundant infection. It overwinters in or on the seed and as minute black stromata in old affected leaves.

Cercospora diseases are controlled by using disease-free seed, or seed at least 3 years old by which time the fungus in the seed has died, crop rotations with hosts not affected by *Cercospora* or by the same *Cercospora* species, and by spraying the plants, both in the seed bed and in the field, with fungicides such as benomyl, dyrene, chlorothalonil, Bordeaux mixture, maneb, dodine, and others.

- *Septoria Diseases*

They occur throughout the world and affect numerous crops on which they cause mostly leaf spots and blights. The most common and serious diseases they cause are leaf blotch and glum blotch of wheat and other cereals and grasses, and leaf spots of celery, beet, carrot, cucurbits, lettuce, tomato, soybean, brambles, aster, azalea, carnation (on which they also cause a corm rot), chrysanthemum, marigolds, and many others.

On cereals and grasses, the leaf spots appear as light-green to yellow or brown spots, first between the veins but soon becoming darker and spreading rapidly to form irregular blotches (Fig. 81C). These may be restricted or may coalesce and cover the entire blade and sheath, depending on the variety and humidity. The blotches often appear speckled due to the more or less abundant, small, submerged brown pycnidia forming on them. In favorable weather, plants become defoliated and the fungus invades the culm causing black necrotic lesions that result in weakened or dead plants and often lodging. Smaller lesions with fewer pycnidia may develop on the floral bracts and on the pericarp of the kernels.

On vegetables, flowers, etc., leaf spots begin as small yellowish specks that later enlarge, turn pale brown or yellowish gray, and finally dark brown, usually surrounded by a narrow yellow zone. The spots, depending on the host and the fungus species, vary in size from those barely visible to those 1 to 2 cm in diameter to occasional individual spots that affect up to one-third of the leaf area. The spots may have distinct margins with a circular outline or may be very irregular with indistinct edges (Fig. 81, A, B, and D). In some hosts, leaves with two or three spots may turn yellow and die, while in others the leaves may have numerous spots before they turn yellow and eventually droop and die. As the spots form, small black pycnidia appear as dots in them. The disease usually starts on the lower foliage and progresses upward.

The fungus, *Septoria* sp., exists in many species that affect the different hosts. It produces long, filiform, colorless, one- to several-celled conidia in dark, globose pycnidia. When the pycnidia become wet, they swell and the conidia are exuded in long tendrils. The conidia are spread by splashing rain, irrigation water, tools, animals, etc. *Septoria* overwinters as mycelium and as conidia within pycnidia on and in infected seed and on diseased plant refuse left in the field (Fig. 82). When the fungus is carried in the seed, it produces seedling infection that may result in damping off or provide inoculum for subsequent infections. Although all *Septoria* species require high moisture for infection and severe disease development, they can cause disease at a wide range of temperatures between 10 and 27°C.

FIGURE 81.
(A, B) Stem lesions and leaf spots caused by the late blight of celery pathogen *Septoria apii*. Black dots are pycnidia. (C) Leaf blotches with pycnidia on wheat and other cereals caused by *Septoria tritici*. (D) Tomato leaf spot caused by *Septoria lycopersici*. (Photos C, D courtesy U.S.D.A.)

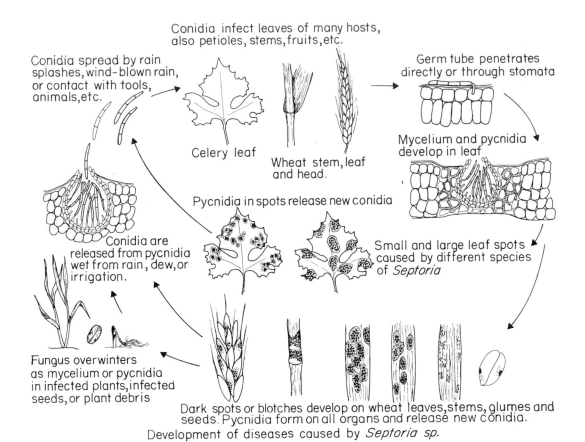

Conidia infect leaves of many hosts, also petioles, stems, fruits, etc.

Conidia spread by rain splashes, wind-blown rain, or contact with tools, animals, etc.

Germ tube penetrates directly or through stomata

Celery leaf

Wheat stem, leaf and head.

Mycelium and pycnidia develop in leaf

Pycnidia in spots release new conidia

Conidia are released from pycnidia wet from rain, dew, or irrigation.

Small and large leaf spots caused by different species of *Septoria*

Fungus overwinters as mycelium or pycnidia in infected plants, infected seeds, or plant debris

Dark spots or blotches develop on wheat leaves, stems, glumes and seeds. Pycnidia form on all organs and release new conidia.

Development of diseases caused by *Septoria sp.*

FIGURE 82.
Development of diseases caused by *Septoria* sp.

Control of *Septoria* diseases depends on the use of disease-free seed in a field free of the pathogen, 2 to 3 year crop rotations, sanitation by deep plowing of plant refuse, use of resistant varieties and chemical sprays of the plants in the seedbed and in the field. The fungicides most commonly recommended for control of *Septoria* diseases include maneb, maneb with zinc, zineb, captan, dichloran, and Bordeaux mixture.

● *Helminthosporium Diseases*

They occur throughout the world and are very common and severe on many important crop plants of the grass family (Gramineae), and, in some areas, they also cause diseases on apple (black pox) and on pear (blister canker). Thus, different species of *Helminthosporium,* which are now known under the names *Bipolaris* or *Drechslera,* cause corn leaf blights, brown spot or blight of rice; crown and root rot of wheat, and the leaf spot of wheat; net blotch of barley, stripe disease of barley, and the spot blotch of barley; the Victoria blight and the leaf blotch of oats; eye spot and the brown stripe of sugarcane; and leaf spots or blights of turf grasses, and the crown and root rots (melting out) of turf grasses.

The leaf spot and blight and also the crown and root rot diseases caused by *Helminthosporium* on the various hosts have many similarities but also some significant differences. Three of them, brown spot or blight of rice, southern corn leaf blight, and *Helminthosporium* blight of oats, caused sudden and catastrophic epidemics that resulted in huge crop losses, human suffering and new approaches to disease control. All *Helminthosporium* diseases destroy various percentages of the leaf area or may attack and destroy part of the stem or roots or may attack the kernels directly, in every case causing considerable yield loss.

Of the diseases on corn, southern corn leaf blight and the leaf spot cause rather small (0.6 × 2.5 cm), tan lesions that may be parallel or elliptical and may be so numerous that they almost cover the entire leaf (Fig. 83A). Some races of the species causing these diseases also attack the stalks, leaf sheaths, ear husks, shanks, ears, and cobs (Figs. 83B and 84). Affected kernels are covered with a black, felty mold and cobs may rot or, if the shank is infected early, the ear may be killed prematurely or drop. Seedlings from infected kernels may wilt and die within a few weeks after planting. A widespread epidemic of race T of the southern corn leaf blight fungus occurred suddenly in 1970 on all corn hybrids containing the Texas cytoplasmic male sterility gene and destroyed about 15 percent of all corn produced in the U.S. that year. The monetary value of the lost crop was estimated at one billion dollars. In the northern corn leaf blight only the leaves are affected; the lesions are grayish-green or tan and range in length

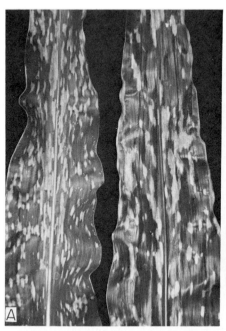

FIGURE 83.
Southern corn leaf blight symptoms on leaves (A) caused by *Helminthosporium maydis*. Symptoms on husks and ears (B) are caused only by race T on corn varieties carrying the gene for Texas cytoplasmic male sterility.

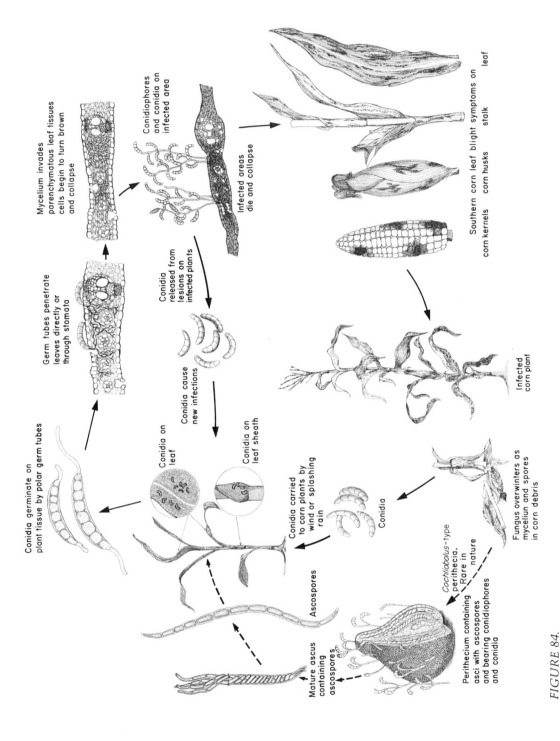

Mycelium invades parenchymatous leaf tissues cells begin to turn brown and collapse

Conidiophores and conidia on infected area

Infected areas die and collapse

Germ tubes penetrate leaves directly or through stomata

Conidia released from lesions on infected plants

Conidia germinate on plant tissue by polar germ tubes

Conidia cause new infections

Conidia on leaf

Conidia on leaf sheath

Southern corn leaf blight symptoms on
corn kernels   corn husks   stalk   leaf

Infected corn plant

Conidia carried to corn plants by wind or splashing rain

Conidia

Fungus overwinters as mycelium and spores in corn debris

Cochliobolus-type perithecia. Rare in nature

Ascospores

Mature ascus containing ascospores

Perithecium containing asci with ascospores and bearing conidiophores and conidia

FIGURE 84.

Disease cycle of southern corn leaf blight caused by *Helminthosporium maydis* race T.

from 2 to 15 cm and are 1 to 3 cm wide. The same fungus causes similar but smaller and darker spots on sorghum.

The brown spot or blight disease of rice appears at first as small, purplish-brown spots on the leaves and glumes (Fig. 85). Later the spots grow larger and have a grey center and brown border. Entire glumes may be covered by several small or one large spot on which a dark brown, velvety felt of conidiophores and conidia is present. The fungus causes damage by infecting the seed directly which then germinates poorly and produces weak seedlings, but primarily by attacking the leaves during the seedling stage either in the seed beds or in the field. In such infections the young foliage is severely damaged, the plants are weakened and the yield drastically reduced. It was such seedling infections that resulted in the Bengal famine in 1942 when approximately two million people died from starvation.

The *Helminthosporium* diseases on barley, wheat, and oats are common, similar, widespread, and cause considerable losses year after year, although the amount of losses varies with the crop, the variety, and the prevailing weather conditions in a particular year. The diseases seem to be more common and more severe on barley than on the other crops. In the net blotch of barley, small, brown, almost square-shaped, netlike blotches appear near the tip of the seedling leaf. The spots enlarge as the leaf grows and spread along the entire leaf blade and, as they become more elongate, the leaf develops a netted appearance (Fig. 86E). In barley stripe, yellow stripes develop along leaf blades and sheaths of the older leaves. Later, these stripes become brown and, near the end of the season, the leaves often split along the stripe lesions and become shredded. Infected plants are stunted and usually do not produce normal heads.

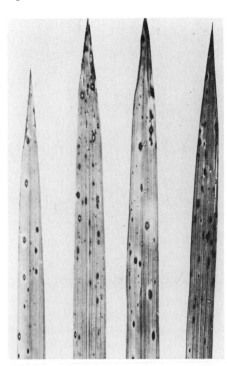

FIGURE 85.
Rice leaves with spots caused by the brown spot fungus *Helminthosporium oryzae*. (Photo courtesy U.S.D.A.)

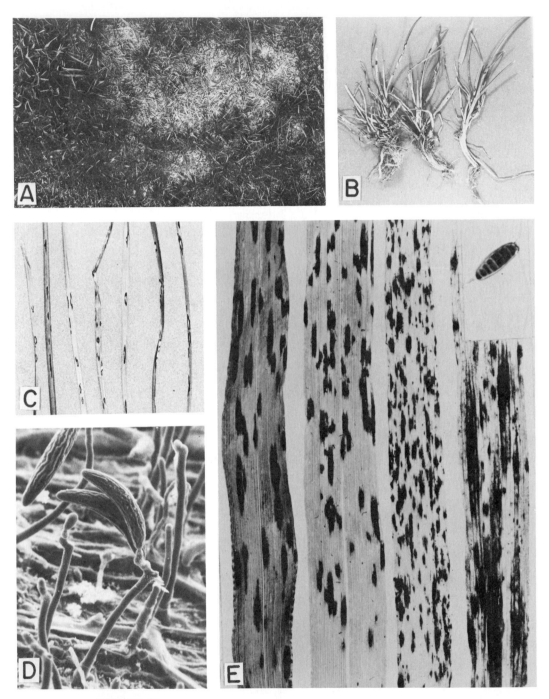

FIGURE 86.
Helminthosporium diseases of turfgrasses and cereals. (A) A small patch of dead turf showing early stages of melting out. (B, C) Plants and leaves from area in (A) showing spotting of leaves and death of sheaths and roots. (D) Scanning electron micrograph of *H. maydis* conidiophores and conidia on corn leaf surface. (E) Spot blotch of cereals and grasses caused by various species of *Helminthosporium* and a spore of one of the species. (Photo D courtesy M. F. Brown and H. G. Brotzman. Photo E courtesy U.S.D.A.)

In the crown rot and root rot of wheat, barley, and other grasses, spots develop on seedlings, plant crowns, stems, leaves, floral parts, and kernels. Dark brown to black lesions develop on seedlings near the soil line and may spread into the leaves. The seedlings may be blighted and killed before or after emergence, or may survive but their growth is retarded. Crowns are infected at or just below the soil line and show a reddish-brown decay which destroys the tiller buds and advances into the root system which it kills. The leaf spots are brown and elongate with definite margins and frequently enlarge and coalesce to form brown irregular stripes or blotches that cover large areas of the leaf blade. Older lesions are covered with a layer of olive-colored conidiophores and conidia of the fungus. Floral parts and kernels also develop lesions or their entire surface may appear dark brown. The embryo end of the kernel is often black and this symptom is characteristic of this disease. Due to crown and root rot and to leaf blotch, surviving plants are shorter; the spikes may be only partially emerged and may be sterile or have poorly filled kernels. Blight of floral parts and kernels also causes sterility or death of individual kernels.

In the Victoria blight of oats, which affects only types of oats bred from the cultivar "Victoria," the fungus causes necrotic lesions at the base of the shoot of seedlings which then develop reddish stripes on the young leaves. Severely affected seedlings die. Those that survive later develop a rot at the base of the stem and in the root and brown lesions on the leaves. The stem is weakened in the infected region and is easily toppled over, while a velvety spore mass develops on its lower blackened nodes. The fungus causing Victoria blight of oats produces a specific toxin that affects and produces the typical stunting and chlorosis symptoms of the disease only on "Victoria" oats.

Lawn and fine golf course grasses are frequently attacked by various *Helminthosporium* species which cause the most common and serious group of diseases known as leaf spots, blights, crown (foot) rots and root rots (melting out) of these grasses (Fig. 86A–C). These diseases resemble in most respects those described above for the small grain crops, although on turf grasses one or more leaf spots often merge and girdle the leaf blade which then turns yellow or reddish-brown and dies back from the tip. In severe infections leaves become completely blighted, wither, die, and drop off. Furthermore, the perennial nature of the turf grasses and the fact that they are mowed and irrigated several times each year create additional opportunities for the diseases to be spread and to become established. The diseases are usually most destructive during wet or humid weather or where the turf is sprinkle-irrigated frequently, especially late in the day. In advanced cases of infection turf areas of various sizes and shapes turn yellow, then brown to straw colored, the plants in these areas are killed and thinned out, and finally all plants die and decay (melting out). The disease, if unchecked and if the weather is favorable, may spread and kill the entire turf in the area.

Quite distinctly from the above diseases, another species of *Helminthosporium* causes a pitted or scaly bark condition and small, black, sunken spots on apple and pear and a blister canker on pear.

The pathogen in the above diseases is one or several of the many species of *Helminthosporium* affecting plants. The fungus produces large, cylindrical, dark, 3- to many (usually 5- to 10)-celled conidia that have thick walls and sometimes are slightly curved (Fig. 86D). The conidia are produced successively on new growing tips of quite dark, septate, irregular conidiophores. Some species of *Helminthosporium* are now known under the name *Drechslera*. Several species have been shown to produce, with more or less regularity, black perithecia containing cylindrical asci within which are formed colorless, threadlike, 4- to 9-celled, coiled ascospores of the genus *Cochliobolus*.

The various *Helminthosporium* species survive the winter as mycelium or spores in or on infected or contaminated seed, in plant debris and in infected crowns or roots of susceptible plants. The fungus is a weak parasite and, when in the soil, it is even weak as a saprophyte probably due to antagonism by soil microorganisms, especially at high nitrogen content. Most *Helminthosporium* species are favored by moderate to warm (18 to 32°C) temperatures and particularly by humid, damp weather. Most *Helminthosporium* diseases, especially leaf spots, are retarded by dry weather, while crown and root affecting fungi may continue their invasion of diseased plants killing out the plants (melting out) in irregular areas. Spread of the fungus is through the seed and infected debris blown or carried away. Over short distances during the growing season the fungus spreads through its numerous conidia which may be carried by air currents, splashing rain, or by clinging to cultivating equipment, feet, animals, etc.

Control of *Helminthosporium* diseases depends on the use of resistant varieties, disease-free seed, seed treatment with fungicides, proper crop rotation and fertilization, plowing under of infected plant debris, and by fungicides. In turf grasses, control of *Helminthosporium* diseases is facilitated by mowing at the recommended maximum height, reducing or removing the accumulated dense thatch, supplying sufficient fertilizer, and by irrigating quickly and sufficiently but in widely spaced (7- to 10-day) intervals. If fungicides are necessary, several of them, including cycloheximide, cycloheximide–thiram, chlorothalonil, Dyrene, maneb, and others can be applied beginning in early spring and continuing at 1- to 2-week intervals for as long as necessary to get the disease under control.

SELECTED REFERENCES

Anonymous. 1965. The rice blast disease. *Proc. Symp. Int. Rice Res. Inst. 1963.* Johns Hopkins Press, Baltimore. 507 p.

Anonymous. 1970. Southern corn leaf blight. *Special Issue, Plant Disease Reptr.* **54**:1099–1136

Baker, K. F., A. W. Dimock, and Lily H. Davis. 1949. Life history and control of the Ascochyta ray blight of chrysanthemum. *Phytopathology* **39**:489–805.

Berger, R. D. 1973. Early blight of celery: Analysis of disease spread in Florida. *Phytopathology* **63**:1161–1165.

Calvert, O. H., and M. S. Zuber. 1973. Ear-rotting potential of *Helminthosporium maydis* race T in corn. *Phytopathology* **63**:769–772.

Chieu, W., and J. C. Walker. 1949. Physiology and pathogenicity of the cucurbit black-rot fungus. *J. Agr. Res.* **78**:589–615.

Childs, T. W. 1968. Elytroderma disease of ponderosa pine in the Pacific Northwest. *USFS Res. Paper* **PNW-69**: 45 p.

Chupp, C. 1953. A monograph of the fungus genus *Cercospora*. Published by the author. Ithaca, N. Y., 186 p.

Chupp, C., and A. F. Sherf. 1960. "Vegetable Diseases and Their Control." The Ronald Press, New York, 693 p.

Cochran, L. C. 1932. A study of two *Septoria* leaf spots of celery. *Phytopathology* **22**:791–812.

Couch, H. B. 1973. "Diseases of Turfgrasses." R. E. Krieger Publishing Co. 348 p.

Eyal, Z., and O. Ziv. 1974. The relationship between epidemics of Septoria leaf blotch and yield losses in spring wheat. *Phytopathology* **64**:1385–1389.

Forsberg, J. L. 1975. "Diseases of Ornamental Plants." Univ. of Ill., Coll. of Agr., Special Publ. No. 3 Rev., 220 p.

Gibson, I. A. S. 1972. Dothistroma blight of *Pinus radiata*. *Ann. Rev. Phytopathol.* **10**:51–72.

Keitt, G. W., E. G. Blodgett, E. E. Wilson, and R. O. Magie. 1937. The epidemiology and control of cherry leaf spot. *Wisc. Agr. Exp. Sta. Res. Bull.* **132**: 117 p.

Klotz, L. J. 1923. A study of the early blight fungus, *Cercospora apii*. *Mich. Agr. Exp. Sta. Tech. Bull.* **63**:1–43.

Meehan, F. L., and H. C. Murphy. 1946. A new *Helminthosporium* blight of oats. *Science (N. S.)* **104**:413–414.

Meredith, D. S. 1970. Banana leaf spot disease (Sigatoka) caused by *Mycosphaerella musicola*. *Commonw. Mycol. Inst. Phytopathol.* Paper No. 11, 147 p.

Nagel, C. M. 1945. Epiphytology and control of sugar beet leaf spot caused by *Cercospora beticola*. *Iowa Agr. Exp. Sta. Res. Bull.* **338**:680–705.

Nicholls, T. H., and H. D. Brown. 1972. How to identify *Lophodermium* and brown spot diseases on pine. U. S. Dept. Agr. Forest Service, North Central For. Exp. Sta. Leaflet, 5 p.

Padnamadhan, S. Y. 1973. The great Bengal famine. *Ann. Rev. Phytopathol.* **11**:11–26.

Peterson, G. W. 1967. Dothistroma needle blight of Austrian and ponderosa pines: epidemiology and control. *Phytopathology* **57**:437–441.

Plakidas, A. G. 1934. The mode of infection of *Diplocarpon earliana* and *Mycosphaerella fragariae*. *Phytopathology* **24**:620–634.

Reddick, D. 1911. The black rot of grapes. *Cornell Univ. Agr. Exp. Sta. Bull.* **293**:289–364.

Rozier, A. 1931. Le black-rot. *Rev. Viticult.* **74**:5–10, 21–25, 37–40, 53–59, 69–71.

Schenck, N. C., and T. J. Stelter. 1974. Southern corn leaf blight development relative to temperature, moisture and fungicide applications. *Phytopathology* **74**:619–624.

Shipton, W. A., W. R. J. Boyd, A. A. Rosielle, and B. L. Shearer. 1971. The common Septoria diseases of wheat. *Bot. Rev.* **37**:231–262.

Shurtleff, M. C., and R. Randell. 1974. "How to Control Lawn Diseases and Pests." Intertec Publishing Corp., Kansas City, Missouri. 97 p.

Shurtleff, M. C., *et al.* 1973. A compendium of corn diseases. The Amer. Phytopathological Society, St. Paul, Minn. 64 p.

Skolnik, M. 1974. Unusual post-infection activity of a piperazine derivative fungicide for the control of cherry leaf spot. *Plant Dis. Reptr.* **58**:326–329.

Sprague, R. 1944. Septoria diseases of gramineae in western United States. *Oreg. State Monog. Stud. Bot.* **6**: 151 p.

Ullstrup, A. J. 1972. The impacts of the southern corn leaf blight epidemics of 1970–1971. *Ann. Rev. Phytopathol.* **10**:37–50.

Western, J. H. (Ed.). 1971. "Diseases of Crop Plants." MacMillan Press Co. 404 p.

STEM AND TWIG
CANKERS CAUSED BY ASCOMYCETES
AND IMPERFECT FUNGI

Cankers are localized wounds or dead areas of the bark which are often sunken beneath the surface of the stem or twigs of woody plants. In some cankers, the healthy tissues immediately next to the canker may increase in thickness and appear higher than the normal surface of the stem.

There are innumerable kinds of pathogens that cause cankers on trees, and cankers can also be caused by factors other than pathogens. Most cankers, however, have many similarities. The most common causes of tree cankers are Ascomycetous fungi, although some other fungi, particularly among the Imperfects, some bacteria, and some viruses also cause cankers.

The basic characteristics of cankers are that they are visible dead areas more or less localized that develop in the bark and, sometimes, in the wood of the tree. Cankers generally begin at a wound or at a dead stub. From that point, they expand in all directions but much faster along the main axis of the stem, branch, or twig. Depending on the host–pathogen combination and the prevailing environnental conditions, the host may survive the disease by producing callus tissue around the dead areas and thus limiting the canker. In infections of large limbs of perennial hosts, concentric layers of raised callus tissue may form. If, however, the fungus grows faster than the host can produce its defensive tissues, either no callus layers form and the canker appears diffuse and spreads rapidly or the fungus invades each new callus layer and the canker grows larger each year. Young twigs are often girdled by the canker and killed soon after infection, but on larger limbs and stems cankers may extend to one or several meters in length although their width extends to only part of the perimeter of the limb. Eventually, however, the limb or entire tree may be killed through girdling either by the original canker or by additional cankers that develop from new infections caused by the spores of the original canker.

Cankers are generally much more serious on fruit trees such as apple, peach, etc., which they debilitate and kill. On forest trees, with the exception of chestnut blight, *Hypoxylon* canker, and *Dothichiza* canker, cankers deform, but do not kill their hosts. They do, however, reduce tree growth and the quality of lumber, result in greater wind breakage, and weaken the trees so that other more destructive wood- or root-rotting fungi can attack the trees.

Although most canker-causing fungi are Ascomycetes, only some of them, e.g., *Dibotryon* and *Nectria,* produce their sexual ascigerous stage regularly. The other canker fungi produce primarily asexual conidia, usually in pycnidia partially or mostly embedded in the bark, and only occasionally do they produce perithecia. For this reason many of these fungi are known by the name that was given them while they were classified as Imperfect fungi, before their sexual stage had been found.

Some of the canker-causing fungi and their most important host plants are listed below.

Botryosphaeria ribis, causes canker on apple, pecan, hickory, sweetgum, redbud, many other trees, and on currant and gooseberry.

Ceratocystis fimbriata, causes canker diseases on cacao, coffee, stone fruits, rubber, poplar, London plane, and sycamore.

Cryptodiaporthe populea (Imperfect stage *Dothichiza populea*) causes the *Dothichiza* canker of poplar.

Dibotryon morbosum, causes black knot of plum and cherry.

Endothia parasitica, causes chestnut blight.

Eutypella parasitica, causes *Eutypella* trunk canker of sugar maple (Fig. 87), red maple and boxelder.

Hypoxylon pruinatum, causes hypoxylon canker of aspen.

Nectria galligena, causes canker of apple and pear, aspen, beech, birch, basswood, black walnut, elm, maple, oak, and other trees.

Urnula craterium (Imperfect stage *Strumella coryneoidea*), causes *Strumella* canker primarily of red and black oak, but also of hickory, beech and maple.

Valsa sp. (Imperfect stage *Cytospora* sp.), causes *Cytospora* canker of peach and many other fruit trees, poplar, willow, and more than 70 species of hardwood trees and shrubs, as well as spruce and some other conifers.

A brief description of the main characteristics of the canker diseases

FIGURE 87.
Eutypella trunk canker of maple. Note size of canker. (Photo courtesy U.S.D.A.)

caused by some of the above fungi, as well as control measures, when possible, are given below.

- *Dothichiza Canker of Poplar*

One of the most important diseases of poplar, especially black, Lombardy and Simon poplar, and of cottonwood, in Europe and North America, it causes cankers on twigs, branches, and stems (Fig. 88, A–C). All the young trees in nurseries and plantations may be killed rapidly by cankers that girdle the stem. On older trees, branch and stem cankers result in a typical dieback. The fungus enters the tree through lenticels, buds, and bark cracks. The developing canker is discolored and sunken. The host checks the advance of the fungus during favorable conditions, but during low temperatures, drought, etc., when the host vigor is reduced, the fungus resumes its activity, the canker enlarges and may girdle and kill the branch or tree. The fungus produces conidia in pycnidia on the diseased parts throughout the growing season. The conidia are spread by rain, water, and wind and can probably cause new infections throughout the season but most infections seem to occur in late spring when conidia are also most abundant. Perithecia and ascospores are seldom produced.

- *Black Knot of Plum and Cherry*

It occurs on cultivated and wild plums and cherries, primarily in the eastern half of the U.S. and New Zealand. It causes conspicuous, 2 to 25 cm or more long, coal-black knotty swellings on one side of, or encircling, twigs and branches (Fig. 88, D, E). They may be several times the diameter of the limbs and make heavily infected trees appear quite grotesque. Infected plants become worthless after a few years as a result of limb death and stunting of the trees. The fungus, *Dibotryon morbosum*, produces *Hormodendrum*-type conidia on free hyphae and ascospores in perithecia formed on the black knots. Both conidia and ascospores are spread by wind and rain, and in early spring they can penetrate healthy and injured woody tissue of the current season's growth. Large limbs are also attacked, especially at points of developing small twigs. The fungus grows into the cambium and xylem parenchyma and along the axis of the twig. After 5 or 6 months, excessive parenchyma cells are produced and pushed outward forming the swelling. The following spring conidia are produced on the knot surface giving it a temporary olive-green velvety appearance. The knots enlarge rapidly during the second summer and, in their surface layer, perithecia are formed that develop during the winter and release ascospores the following spring. The knots continue to expand in following years. The disease can be controlled by pruning and burning of all black knots, and destruction of black knots or of all affected wild plums and cherries near the orchard. Spraying the orchard trees before and during bloom with sulfur or captan, or with fixed copper fungicides to which hydrated lime is added protects trees from infection.

- *Chestnut Blight*

After it was introduced in New York City in 1904, it spread and by 1940 destroyed practically all American chestnut trees throughout their

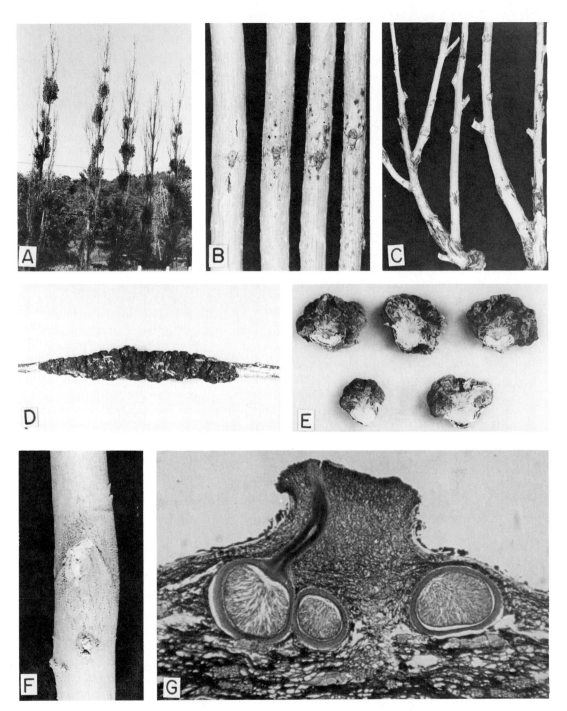

FIGURE 88.

(A–C) *Dothichiza* canker on Lombardy poplar. (A) A row of trees whose twigs
have been killed by cankers like those shown in B and C. Bunches of new twigs
have sprouted in some parts of the trees. (B) Stages in the development of a canker
along the twig. Infection is generally centered around a leaf scar. (C) Advanced
cankers, mostly at the base of twigs. (D, E) External and cross-section views of
black knot canker on cherry caused by *Dibotryon morbosum*. (F, G) Chestnut
blight canker caused by *Endothia parasitica*. (F) Canker on young chestnut stem,
apparently started at broken branch. (G) Perithecia of *Endothia parasitica*
embedded in chestnut bark.

281

natural range in the eastern third of the U.S. from the Canadian border south nearly to the Gulf of Mexico. American chestnuts killed by the blight comprised 50 percent of the overall value of eastern hardwood timber stands. The fungus, *Endothia parasitica*, also attacks oak, red maple, hickory, and, sporadically, other trees, but not nearly as severely as it attacks the American chestnut. It is now present throughout North America, Europe, and Asia. The fungus penetrates the bark of stems through wounds and then grows into the inner bark and cambium. Soon a swollen or sunken canker develops, the bark of which is reddish-orange to yellow-green and covered by pimplelike pycnidia and perithecia (Fig. 88, F, G). Cankers often have long cracks on their surface, may be several inches to many feet long and eventually girdle the stem or branch causing wilting and death of the parts beyond the canker. The pycnidia produce tiny conidia that ooze out as long orange curllike masses during moist weather and are spread by birds, crawling or flying insects, or by splashing rain. The perithecia produce ascospores that are shot forcibly into the air and may be carried by wind over long distances. The fungus survives and continues to invade and produce its spores in trees or parts of trees already killed by the blight. Blighted trees almost always produce sprouts below the basal cankers, but the resulting saplings become blighted in turn by new infections. No control is available against chestnut blight although some new systemic fungicides and some antagonistic or hypo-parasitic strains of the fungus appear promising for isolated trees. Also, no resistant American chestnuts have been found yet.

- *Nectria Canker*

It is one of the most important diseases of apples and pears and of many species of hardwood forest trees in most parts of the world. Losses are greater in young trees because in these the fungus girdles the trunk or scaffold branches, while in older trees only small branches are usually killed directly (Fig. 89, A–D). Cankers on the main stem of older trees, however, reduce the vigor and value or productivity of the tree, and such trees are subject to wind breakage. *Nectria* cankers usually develop around bud scars, wounds, twig stubs, or in the crotches of limbs. Young cankers are small, circular, brown areas. Later, the central area becomes sunken and black while the edges are raised above the surrounding healthy bark. In many hosts and under favorable conditions for the host, the fungus grows slowly, the host produces callus tissue around the canker and the margin of the canker cracks. The tissues under the black bark in the canker are dead, dry and spongy, flake off and fall out revealing the dead wood and the callus ridge around the cavity. In subsequent years the fungus invades more healthy tissue and new, closely packed, roughly concentric ridges of callus tissue are produced every year resulting in the typical open, target-shaped *Nectria* canker. In some hosts, however, and under conditions that favor the fungus, invasion of the host is more rapid, the bark in the cankered area is roughened and cracked but does not fall off and the successive callus ridges are some distance apart.

In hosts such as apple and pear, fruits are also infected and develop a

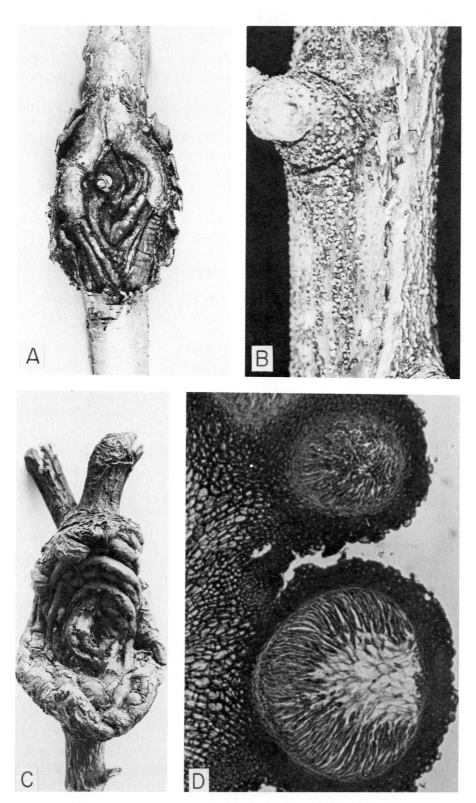

FIGURE 89.
Nectria cankers and fruiting bodies on maple (A) and apple (B–D). (B)
Conidial stage of *Nectria galligena* on apple branch. (D) *Nectria* perithecia.
(Photo A courtesy U.S.D.A.)

circular, sunken, brown rot. White or yellowish pustules producing numerous conidia form on rotted areas. Internally the rotted tissue is soft and has a striated appearance.

The fungi, *Nectria galligena* and some related species, attack the many different tree hosts. All *Nectria* species produce similar, two-celled ascospores in brightly colored perithecia on the surface of a cushion-shaped stroma, but different *Nectria* species produce conidia which in themselves do not appear to be related and are classified as various species of Imperfect fungi. Thus, the conidia of *N. galligena* are either single-celled microconidia or more commonly 2- to 4-celled, cylindrical to crescent-shaped macroconidia of the *Cylindrocarpon* type (Fig. 90). The conidia are produced soon after infection on small, white or creamy-yellow or bright orange-pink sporodochia which appear on the surface of the bark over the infected area or on fruit. The conidia are produced more commonly early in the season but also in the summer and early fall. They are spread by wind and by washing during rainy periods, and perhaps by insects. Perithecia appear in the cankers in late summer and fall and in the same stroma that earlier produced the conidia which they eventually replace. The ascospores are either forcibly discharged and carried by wind or, in moist weather, they ooze from the perithecium and are washed by rain or carried by insects. Ascospores are dispersed more abundantly in late summer and fall but are also released at other times of the year. Sanitation, i.e., removal and burning of cankered limbs or trees where possible,

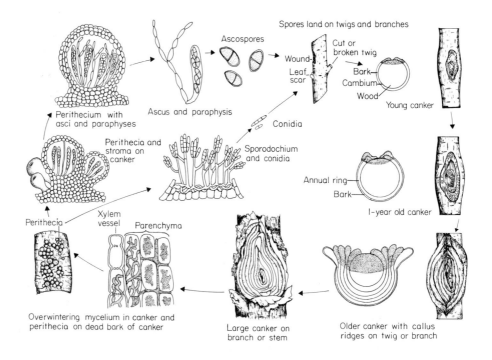

FIGURE 90.
Disease cycle of *Nectria* canker caused by *Nectria galligena*.

is often the only control measure possible. Spraying with a fungicide such as captafol or 8:8:100 Bordeaux mixture immediately after leaf fall helps reduce *Nectria* infections in fruit trees.

- *Valsa or Cytospora Canker*

This worldwide canker disease probably affects more species of trees than any of the previously described. It is estimated that more than 70 species of fruit trees, hardwood forest and shade trees, shrubs, and conifers are attacked by one of several species of the pathogen. The fungus, *Valsa* sp., is most commonly found in its Imperfect stage, *Cytospora* sp., and therefore the disease is usually known as *Cytospora* canker.

Cytospora canker is most serious on peach and the other stone fruits, on apple and pear, and on poplar and willow, but is often serious on many other shade or forest trees (Fig. 91). Actually, few orchards are free from its damaging effects. Many trees are seriously injured by cankers on the trunk, in the main crotch, on the limbs and on the branches. Infected branches of fruit trees often break from the weight of the crop or during storms. *Cytospora* canker is most severe on fruit or shade trees growing under stress, such as those growing on an unfavorable site, or injured by drought, frost, fire, or severe pruning. The fungus is mostly a saprophyte living on dead bark but becomes parasitic when the tree is weakened. However, the presence of *Cytospora* on a dead twig or branch does not mean that this fungus has killed it.

Infection of small twigs and branches results in dieback without definite cankers. The cankers occur and are most pronounced on trunks and large branches. A canker appears first as a gradual circular killing of the bark of a limb or stem. The infected area soon becomes brownish and sunken and the host often produces raised callus tissue around it. In some hosts, particularly the stone fruit trees, the inner diseased bark becomes dark and odorous and a copious flow of gum exudes from the dead tissues. The cankers may be long and slightly sunken or short, deeply sunken, with conspicuously raised callus borders. Later the bark shrivels and separates from the underlying wood and from the surrounding healthy bark. Small, pimplelike pycnidia of the fungus appear on the dead bark and later the shriveled bark may slough off exposing dead wood beneath. The cankers increase in size each year and become unsightly, rough swellings. Many twigs and branches die back as a result of cankers that girdle them completely (Figs. 91 and 92).

Cytospora cankers result from infections by conidia of the fungus *Cytospora*. The fungus produces *Valsa*-type ascospores in perithecia, but the perithecial stage is not common. The conidia are produced inside pycnidia consisting of many connecting cavities and one pore through which the conidia are exuded. The spores are small, hyaline, one-celled, and slightly curved. They are produced in a gelatinous matrix which, during wet weather, absorbs water, swells and oozes out of the pycnidium, carrying with it the masses of spores. The spores may be washed away or splashed by rain or may be spread by insects and man. If it is moist but not rainy, the exuded conidia may form coiled threads of spores

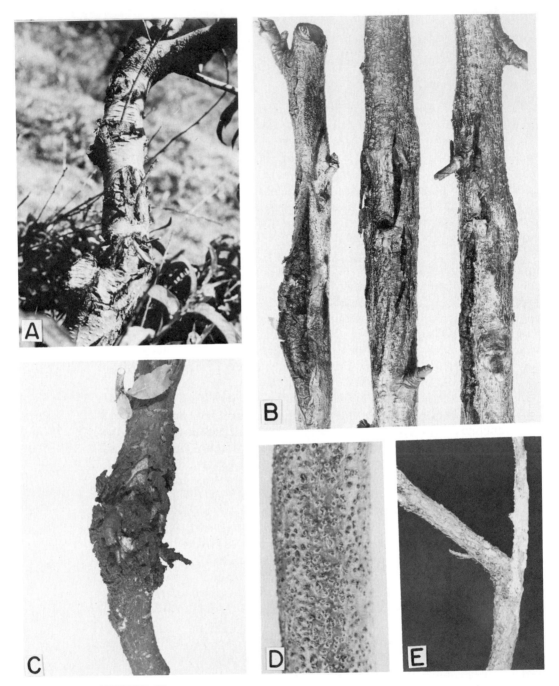

FIGURE 91.
Cytospora canker on peach trunk (A) and twigs (B), and on plum twig (C). (D)
Cytospora pycnidia embedded in the bark of apple twig. (E) *Cytospora* canker
of spruce. Note white pitch flowing out of canker area.

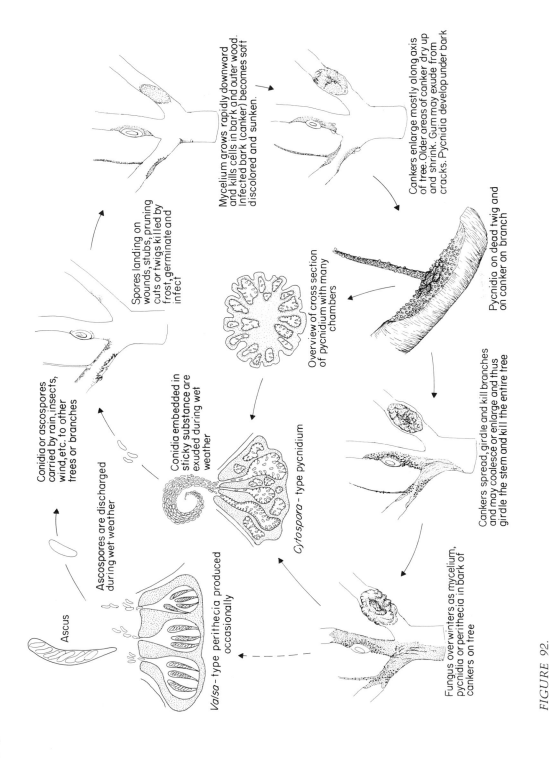

Mycelium grows rapidly downward and kills cells in bark and outer wood. Infected bark (canker) becomes soft discolored and sunken.

Cankers enlarge mostly along axis of tree. Older areas of canker dry up and shrink. Gum may exude from cracks. Pycnidia develop under bark

Spores landing on wounds, stubs, pruning cuts or twigs killed by frost, germinate and infect

Conidia or ascospores carried by rain, insects, wind, etc. to other trees or branches

Conidia embedded in sticky substance are exuded during wet weather

Overview of cross section of pycnidium with many chambers

Pycnidia on dead twig and on canker on branch

Ascospores are discharged during wet weather

Cytospora - type pycnidium

Cankers spread, girdle and kill branches and may coalesce or enlarge and thus girdle the stem and kill the entire tree

Ascus

Valsa - type perithecia produced occasionally

Fungus overwinters as mycelium, pycnidia or perithecia in bark of cankers on tree

FIGURE 92.
Disease cycle of the *Cytospora* (= *Valsa*) canker of peach and most other trees.

287

that dry out and harden and remain on the canker for several days or weeks. Most infections take place in the dormant season, particularly in late fall or early winter and in late winter or early spring. Weakened injured trees, however, may be infected throughout the growing season. Both the mycelium and the conidia of the fungus can live over winter on the infected parts.

Infection of small twigs can occur through injuries or leaf scars. In larger branches, wounds of any kind, broken branches and pruning stubs, winter injuries and sunscalds also make ideal points of entry for the fungus. The fungus becomes established in dead bark and wood and invades the surrounding tissues to form a canker. The fungus grows through the cells in the bark and the outer few rings of the wood.

Control measures for the *Cytospora* canker depend on: good cultural practices such as watering and fertilization to keep the trees in good vigor; avoiding wounding and severe pruning of trees; removing cankers from trunks and large branches during dry weather and treating the wound and all pruning cuts with a disinfectant and a wound dressing; removing and burning cankered and dead branches and twigs; pruning as late in the spring as possible and spraying with Phygon XL immediately after pruning and before it rains. The above practices help but do not completely prevent *Cytospora* canker.

SELECTED REFERENCES

Brandt, R. W. 1964. *Nectria* canker of hardwoods. Forest Pest Leaflet, *U.S. Forest Service* **84**:7 p.

Davidson, A. G., and R. M. Prentice (Eds.). 1967. "Important Forest Insects and Diseases of Mutual Concern to Canada, the United States and Mexico." Dept. For. Urban Development, Canada, 248 p.

Diller, J. D. 1965. Chestnut blight. Forest Pest Leaflet, U.S. Forest Service 94. 7 p.

Dubin, H. J., and H. English. 1974. Factors affecting control of European apple canker by Difolatan and basic copper sulfate. *Phytopathology* **64**:300–306.

Dubin, H. J., and H. English. 1975. Epidemiology of European apple canker in California. *Phytopathology* **65**:542–550.

Fisher, D. F., and E. L. Reeves. 1931. A *Cytospora* canker of apple trees. *J. Agr. Res.* **43**:431–438.

Gross, H. L. 1967. *Cytospora* canker of black cherry. *Plant Dis. Reptr.* **51**:261–263.

Hampson, M. C., and W. A. Sinclair. 1973. Xylem dysfunction in peach caused by *Cytospora leucostoma*. *Phytopathology* **63**:676–680.

Hickey, K. D. 1966. Black knot disease of plum and its control. *N. Y. S. Coll. Agr. (Cornell), Ext. Bull.* **1173**: 7 p.

Hubbes, M. 1959. Untersuchungen über *Dothichiza populea*, den Erreger des Rindenbrandes der Pappel. *Phytopathol. Z.* **35**:58–96.

Jaynes, R. A., and N. K. Van Alfen. 1974. Control of American chestnut blight by trunk injection with methyl-2-benzimidazole carbamate (MBC). *Phytopathology* **64**:1479–1480.

Lortie, M. 1964. Pathogenesis in cankers caused by *Nectria galligena*. *Phytopathology* **54**:261–263.

McCubbin, W. A. 1918. Peach canker. *Can. Dept. Agr. Bull.* **37**: 20 p., plus 30 illustr.

Swinburne, T. R. 1975. European canker of apple (*Nectria galligena*). *Rev. Plant Pathol.* **54**:787–799.

Wainwright, S. H., and F. H. Lewis. 1970. Developmental morphology of the black knot pathogen on plum. *Phytopathology* **60**:1238–1244.

Weaver, D. J. 1974. A gummosis disease of peach trees caused by *Botryosphaeria dothidea*. *Phytopathology* **64**:1429–1432.

Wilson, E. E. 1968. Control of European canker of apple by eradicative and protective fungicides. *Plant Dis. Reptr.* **52**:227–231.

ANTHRACNOSE DISEASES CAUSED BY ASCOMYCETES AND IMPERFECT FUNGI

Anthracnoses are diseases of the foliage, stems, or fruits which typically appear as small or large, dark-colored spots or slightly sunken lesions with a slightly raised rim. In addition to leaf spots, anthracnoses often have a prolonged initial stage in fruit infection and may cause twig or branch dieback. In some anthracnoses, particularly serious on some fruit crops, the symptoms appear as small spots with raised, corky surfaces. Anthracnose diseases of fruit often result in fruit drop and fruit rot. Anthracnoses are caused by fungi that produce their asexual spores, conidia, within small, black acervuli arranged in the lesion concentrically or in a scattered manner.

Four ascomycetous fungi, *Diplocarpon*, *Elsinoe*, *Glomerella*, and *Gnomonia*, belong to this group, although a few of their species produce conidia in pycnidia rather than acervuli. Actually, these fungi are found in nature mostly in their conidial stage and can overwinter as mycelium or conidia, yet, some more frequently than others also produce ascospores in perithecia which may overwinter. The main anthracnose and related diseases caused by these fungi are listed below.

> *Diplocarpon*, causing black spot of rose (*D. rosae*), leaf scorch of strawberry (*D. earliana*), and leaf spot of hawthorn (*D. maculata*).
>
> *Elsinoe* (conidial stage: *Sphaceloma*), causing anthracnose of grape (*E. ampelina*), of dogwood (*E. corni*), of raspberry (*E. veneta*), and scab of citrus (*E. australis* and *E. fawcetti*), of avocado (*E. perseae*), of pansy violet, and of poinsettia.
>
> *Glomerella* (conidial stage: *Colletotrichum* or *Gloeosporium*), causing anthracnose of azalea, cyclamen and sweet pea, bitter rot of apple and cranberry, dieback and canker of camellia and privet, and ripe rot of grape and other fruits (*G. cingulata*), anthracnose of soybean (*G. glycines*), of cotton (*G. gossypii*), and of bean (*G. lindemuthiana*).
>
> *Gnomonia*, causing anthracnose of walnut (*G. leptostyla*), of oak (*G. veneta*) (Fig 93A), of sycamore (*G. platani*) (Fig 93B), of linden (*G. tiliae*), and leaf spot of elm (*G. ulmea*) (Fig 93C) and of hickory (*G. caryae*).

The acervulus-producing Imperfect Fungi make up the order Melanconiales. The most important plant pathogenic fungi that produce acervuli are *Colletotrichum* (=*Gloeosporium*), *Coryneum*, *Cylindrosporium*, *Marssonina*, *Melanconium*, and *Sphaceloma*. Some of these are the conidial stages of some of the Ascomycetes that cause anthracnose diseases. Thus, several *Colletotrichum* or *Gloeosporium* species produce a

FIGURE 93.
(A) Young oak leaves and twigs killed by the anthracnose fungus *Gnomonia veneta*. (B) Defoliated and dead twigs of sycamore tree showing anthracnose caused by *Gnomonia veneta (G. platani)*. (C) Elm leaf spot caused by *Gnomonia ulmea*. (D) Black spot of rose caused by *Diplocarpon rosae*. (E) Scanning electron micrograph of *Marssonina juglandis* acervulus with conidia. (F) Anthracnose of poplar caused by *Marssonina* sp. Note lesions on veins and petiole. (G) Anthracnose symptoms on maple leaf caused by *Gloeosporium apocryptum*. (Photo E courtesy M. F. Brown and H. G. Brotzman.)

Glomerella-type perfect stage and the diseases caused by these species are discussed as Glomerella diseases; some *Cylindrosporium* species have a *Coccomyces* (or *Higginsia*)-type perfect stage and some a *Mycosphaerella*-type perfect stage; some *Marssonina* species have a *Diplocarpon*- or *Gnomonia*-type perfect species; and most *Sphaceloma* species have an *Elsinoe*-type perfect stage. For some species of these same fungi, however, and for all species of *Coryneum* and *Melanconium*, no perfect stage is known yet.

In addition to the diseases described or listed as caused by acervulus-producing fungi that have perfect stages, some of the other important plant diseases caused strictly by the imperfect stages of the fungi are:

> *Colletotrichum* (=*Gloeosporium*), causing anthracnose of cereals and grasses (*C. graminicola*), of cucurbits (*C. lagenarium*), of lima bean (*C. truncatum*), anthracnose of peony, anthracnose or fruit rot of eggplant and of tomato (*C. phomoides*), red rot of sugarcane (*C. falcatum*), anthracnose of pepper, spinach, turnip, cauliflower, etc., onion smudge (*C. circinans*), and black scale of lily bulbs, anthracnose of citrus, fig, olive, avocado, and of many other plants.
>
> *Coryneum*, causing "Coryneum blight," "shot hole" or "fruit spot" of stone fruits, especially peach and apricot.
>
> *Melanconium*, causing bitter rot of grapes.

Anthracnose diseases, particularly those caused by *Colletotrichum* = *Gloeosporium* or *Glomerella* group, are very common and destructive on numerous crop and ornamental plants and their geographic distribution is in most cases worldwide. Although severe everywhere, anthracnose diseases cause their most severe losses in the tropics and subtropics.

- *Diplocarpon Diseases: Black Spot of Rose*

The black spot of rose appears as small to large circular, black lesions on the leaves (Fig. 93D). The spots are consistently black, have fringed margins and may coalesce to produce large, irregular, black lesions. The leaf tissue around the lesions turns yellow and often, when severely infected, entire leaves become yellow and fall off prematurely leaving the canes almost completely defoliated.

The fungus, *Diplocarpon rosae*, produces ascospores in tiny apothecia formed in old lesions, and *Marssonina*-type conidia in acervuli forming between the outer wall and cuticle of the epidermis (Fig. 93E). The conidiophores arise from a thin black stroma, are short, and give rise to successive crops of conidia. The abundant conidia production pushes up and ruptures the cuticle. The fungus overwinters as mycelium or as ascospores and conidia in infected leaves and, in roses, in the canes. Both kinds of spores can cause primary infections of leaves in the spring by direct penetration. The mycelium grows in the mesophyll but within two weeks forms acervuli and conidia at the upper surface. Conidia are produced throughout the growing season and cause repeated infections during warm, wet weather.

Control of *Diplocarpon* diseases is through sanitation, such as removing and burning infected leaves, cutting back the canes of diseased rose plants and spraying with benomyl, zineb, or applying sulfur–copper dust. Applications should begin as soon as new leaves appear in the spring or at the first appearance of black spot on the foliage and then repeated at 7- to 10-day intervals or within 24 hours after each rain.

GLOMERELLA DISEASES

At least four species of *Glomerella* cause serious anthracnose diseases of numerous important annual crop and ornamental plants, while one, *G. cingulata*, also causes cankers and dieback of woody plants such as camelia and privet, bitter rot of apples, and ripe rot of grape, pears, peaches and other fruits.

- *Glomerella Anthracnose Diseases of Annual Plants*

The most important such diseases are those on bean, cotton, soybean, and sweet pea. Although different species of the fungus are responsible for the disease on each of these hosts, the life cycles of the fungi, the symptoms, the disease development, and control are approximately the same in all cases.

The diseases are present wherever their hosts are grown, although they are more severe in warm to cool, humid areas, and generally are not a problem under dry conditions. Plants in all stages of growth are subject to anthracnose. The fungus is often present in or on the seed produced in infected pods or bolls. Infected seed may show yellowish to brown sunken lessions of various sizes. When infected seeds are planted, many of the germinating seeds are killed before emergence. Dark-brown, sunken lesions with a pink mass of spores in the center are often present on the cotyledons of young seedlings. The fungus may destroy one or both of the cotyledons while its spores spread onto the hypocotyl and the mycelium moves into the stem. On the stem the fungus produces numerous small, shallow, reddish-brown specks that subsequently enlarge, become elongated, and finally sunken. The lesions are covered with myriads of pink- to rust-colored spores. If conditions are humid, the lesions may be so numerous that they girdle and weaken the stem to the point where it cannot support the top of the plant. The fugus also attacks the petioles and the veins of the under side of the leaves, on which it causes long, dark, brick-red to purplish colored lesions that later turn dark drown to almost black. Few lesions are produced between the veins in bean, but they are rather common on cotton, and in sweet pea they may involve the entire leaf (Fig. 94, B, C).

The anthracnose fungi also attack and cause their most characteristic symptoms on the bean pods and on the cotton bolls. On the latter, the fungus seems to produce numerous lesions which spread and coalesce and cover most or all the surface of young cotton bolls, with almost continuous masses of spores covering the infected area. On pods, small flesh- to rust-colored elongated lesions appear which later become sunken, circular and about 5 to 8 mm in idameter. Lesions developing on

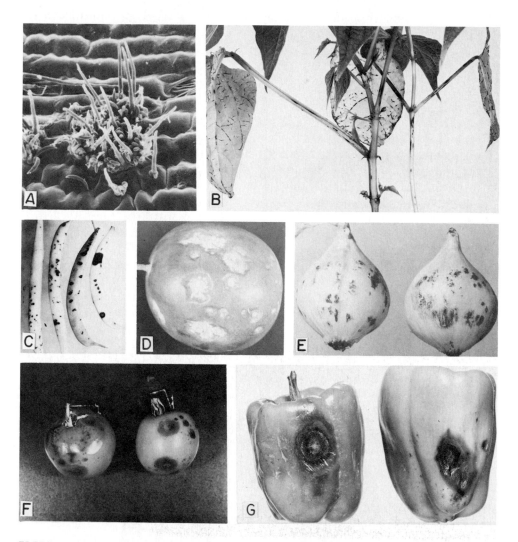

FIGURE 94.
(A) Scanning electron micrograph of acervulus of *Colletotrichum graminicola* showing setae (the long structures) and conidia. (B) Bean anthracnose lesions on stems, petioles and leaves caused by *Glomerella (Colletotrichum) lindemuthiana*. (C) Bean anthracnose on pods. (D) Watermelon anthracnose caused by *Colletotrichum lagenarium*. (E) Onion smudge caused by *Colletotrichum circinans*. (F) Tomato anthracnose caused by *Colletotrichum phomoides*. (G) Anthracnose on peppers caused by *Glomerella cingulata*. (Photo A courtesy M. F. Brown and H. G. Brotzman. Photo B courtesy G. C. Papavizas. Photo E courtesy U.S.D.A.)

young pods may extend through the pod and even to the seed, while in older pods the lesions do not extend beyond the pod. As the pods mature, the margin of the lesions is generally slightly raised, while the pink spore masses of the lesions dry down to gray, brown, or black granulations or to small pimplelike protrusions.

- *Glomerella Fruit Rots*

The most important such diseases are bitter rot of apple and ripe rot of grape.

Bitter rot is worldwide in distribution and in warm, humid weather may cause enormous losses by destroying the entire crop of apples just a few weeks before harvest. Bitter rot symptoms may appear when the fruit is half grown, but more frequently when it approaches its full size. The rot starts as small, light brown areas that enlarge rapidly and become circular and somewhat sunken in the center. The surface of the spots is smooth at first and may be dark brown or black until the spots are 1 to 2 cm in diameter. Then numerous, slightly raised cushions appear mostly near the center of the spots and some extending outward toward the edge of the spots. In humid weather, the cushions produce creamy masses of pink colored spores, sometimes arranged in concentric circles, the rotted area expands rapidly and more rings of spore masses appear (Fig. 95). In older rotted areas the pink masses disappear and the tissue becomes dark brown to black, wrinkled, and sunken. The rot also spreads inward toward the apple core forming a cone of somewhat watery, rotted tissue that may or may not be bitter. When, as it is common, several spots develop on a fruit, they usually enlarge, fuse and rot the entire fruit which

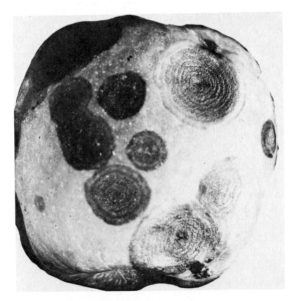

FIGURE 95.
Bitter rot of apple caused by *Glomerella cingulata*. Note the concentric arrangement of *Gloeosporium*-type acervuli. (Photo courtesy U.S.D.A.)

may drop and mummify or mummify and cling on the twig. Bitter rot infections may occur in the fall but fail to develop appreciably during cold storage. When, however, the fruit is marketed and kept at room temperature, bitter rot may develop very rapidly. Bitter rot cankers on the limbs are rare and resemble those produced on branches and stems described above.

Ripe rot of grape and other fruits is also worldwide in distribution but is most serious in areas with warm, humid weather during the ripening of the fruit. Ripe rot appears when the fruit is nearly mature and may continue its destruction of fruit after it has been picked, during shipment, and marketing. On white or light-colored grapes, a small reddish-brown spot appears at first. It soon spreads to over half the berry and becomes darker with a purplish center and a light-brown border. In dark-colored grapes no appreciable change of color is caused by the rot on the berries. As the disease progresses, the whole berry rots, usually in a continuous manner but sometimes marked by a series of concentric zones. By this time, the rotted berry becomes more or less densely covered with numerous, small slightly raised pustules from which, in humid weather, pinkish slimy masses of spores ooze out. Later, the spore masses become darker, almost reddish-brown. The rotted berries then develop a sunken area at the point of infection and gradually become more or less shriveled and mummified while the pustules continue to produce spores. Infected berries often "shell" or drop off before the rot causes them to dry up.

The fungus, *Glomerella* sp., occasionally produces ascospores in perithecia on overwintering fruit and cankers of some hosts, but in other hosts ascospores and perithecia almost never develop. In all hosts, however, the fungus reproduces profusely by forming pink masses of *Colletotrichum*- or *Gloeosporium*-type conidia in acervuli (Fig. 96). The acervuli consist of a cushion stroma of mycelium several to many cells thick and develop just beneath the cuticle which is ruptured by the upward pressure of the developing mass of conidiophores and conidia. The conidia are held together in a sticky mass which is firm and horny in dry weather, but under humid, moist weather the conidia are released in a pinkish mass and may be washed or spattered by raindrops or carried by wind-blown mist and insects.

The fungus usually overwinters in diseased stems, leaves, and fruit as mycelium or spores, in the seed of most affected annual hosts, and in cankers of perennial hosts. In the cases where ascospores are produced in mummified fruit, cankers, etc., they may cause primary infections in the spring, but generally these are not necessary. The surviving mycelium quickly produces conidia in the spring which can themselves cause primary infections and subsequent continuous secondary infections during the entire season as long as temperature and humidity are favorable. Infection is by direct penetration of uninjured tissue and the mycelium grows intercellularly for some time before the cells begin to collapse and rot. The mycelium then produces acervuli and conidia just below the cuticle which upon rupture release conidia again that cause more infections. The fungus generally requires high temperatures and humidity for best activity and although it can attack young cotyledons, stems, and

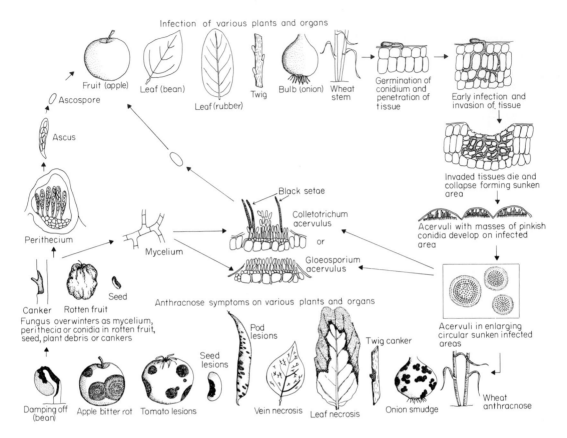

FIGURE 96.
Disease cycle of the anthracnose diseases caused by *Glomerella cingulata* and *Colletotrichum* or *Gloeosporium* sp.

leaves, it cannot attack young fruit until they are past a certain stage of development and maturity.

Control of *Glomerella* diseases depends on the use of disease-free seed grown in the arid west or use of treated seed, and also crop rotation of hosts like bean, soybean, sweet pea, and cotton, the use of resistant varieties when available, removal and burning of dead twigs, branches and fruit infected with the fungus in woody plants, and, finally, on sprays with fungicides. The fungicides most commonly used include Bordeaux mixture, captan, ferbam, Dikar, mancozeb, captafol, and others, the kind of fungicide used and the timing of its application varying considerably with the host treated.

- *Gnomonia Anthracnose and Leaf Spot Diseases*

In Europe, North America and other parts of the world, various species of *Gnomonia* attack mostly forest and shade trees on which they cause symptoms primarily on leaves, e.g., elm and hickory, or more general anthracnose symptoms on the leaves and young shoots, or twigs, e.g., linden, oak, sycamore, and walnut. Gnomonia diseases are favored by wet, humid weather. The various species of the fungus overwinter primarily in fallen leaves or on infected twigs as immature perithecia which mature and produce ascospores in the spring. Ascospores produce most primary infections on young leaves and twigs; then conidia are produced in acervuli, generally of the *Gloeosporium* type, and cause all subsequent infections. Both ascospores and conidia are disseminated only during rainy weather.

In black leaf spot of elm, leaves are infected as they unfold in the spring and produce numerous small, irregular spots on the upper leaf surface. The spots are yellowish to white at first, but later turn grayish and finally black and slightly raised as they enlarge to about 3 mm in diameter (Fig. 93C). Heavily infected leaves turn yellow and during wet seasons may fall off prematurely. Lesions may also form on petioles and shoots. In the spots, shiny black dots, consisting of the acervuli of the fungus, form during the summer while the leaves are on the trees. In moist weather, the tiny acervuli exude a creamy mass of conidia. After leaf fall, ascospores form in black perithecia embedded in the leaf.

In walnut anthracnose, small or large subcircular spots are formed which often merge and produce large necrotic areas on the leaves. These spots or blotches are bordered with yellowish leaf tissue. Current shoots and the hulls of nuts are also attacked. The shoots develop dead sunken areas that are oval to irregularly circular with reddish-brown margins, while the hulls develop sunken, dark, circular spots smaller than those on the leaves. Infected leaves and nuts often drop prematurely, often as early as late July, especially in wet weather. Nutmeats produced by infected trees are dark and shrivelled. The conidia of the pathogen are of the *Marssonina* type and appear as black specks on the under side of the lesions (Fig. 93E).

Anthracnose of oak and sycamore were thought to be caused by the same species of *Gnomonia* (*G. veneta*), but more than one species of the fungus seem to be involved. In some host species only the conidial stage (*Gloeosporium* sp.) has been found. To date, sycamore anthracnose is believed to be caused by *G. platani*.

In oak anthracnose, shoots die back in the spring and irregular, brown diseased areas develop on the leaves along the midrib and the major side veins in the spring and midsummer. If infection occurs early, the diseased areas or blotches may coalesce by late spring or early summer thus forming large, irregular, necrotic areas with a papery texture which are usually confined to the areas bordered by the larger veins (Fig. 93A). Severely infected leaves, found mostly in the lower branches, are frequently killed and fall off. The dark-brown acervuli of the fungus appear as raised pustules late in the season and are usually located on the midrib and the other infected veins of the leaves. The fungus also attacks young twigs on which it produces small, shallow cankers that may result in dieback of the crown.

Sycamore anthracnose is the most important disease of sycamore (*Platanus* sp.). The disease may appear as twig blight before leaf emergence at which time it kills the tips of small, one-year-old twigs. During bud expansion, the disease may appear as bud blight killing the buds before they open. The most frequently observed symptom is shoot blight in which the disease causes the sudden death of expanding shoots and also young leaves. The killing of leaves while they are still very small has often been confused with frost damage. Later the fungus may cause a leaf blight in which irregular brown areas develop adjacent to the midrib, veins, and leaf tips. In moist weather, small, cream-colored acervuli form on the under side of the leaf on dead tissue along the veins. In some years, sycamores may be completely defoliated in the spring by this disease and produce new leaves in the summer. From the buds or leaves, the fungus spreads into the twigs on which it produces cankers, or the fungus may spread through and kill small twigs and form cankers on the branches around the bases of the dead twigs (Fig. 93B). In trees severely affected with anthracnose for several successive years many branches may die. The fungus overwinters as mycelium and as immature perithecia in twig cankers and in fallen leaves. As in the other anthracnose diseases, this too is favored by rainy weather and rather cool temperatures.

Control of *Gnomonia* diseases in trees under forest conditions is not economically feasible. In shade and ornamental trees, however, burning of leaves, pruning of infected twigs, fertilization and watering help reduce the disease. Valuable trees should be sprayed two to four times at 10- to 14-day intervals starting as soon as the buds begin to swell (sycamore) or soon after the buds open. The fungicides most commonly recommended are zineb, Bordeaux mixture, and dodine.

- *Colletotrichum (= Gloeosporium) Diseases*

They are the most common anthracnoses and are very similar, if not identical, to the diseases caused by *Glomerella*. The latter is probably the sexual stage of most or all species of *Colletotrichum* (= *Gloeosporium*).

On cereals and grasses, the fungus, which lives saprophytically on crop residue, may attack young seedlings but, usually, it attacks the tissues at the crown and the bases of stems of more developed plants. Infected areas first appear bleached but later become brown. Brownish blotches often develop at or near diseased joints. Toward maturity of the plant, numerous small black acervuli appear on the stems, lower leaf sheaths, and sometimes on the leaves and on the chaff and spikes of diseased heads (Fig. 94A). Depending on how early the plant is attacked and on the severity of the disease, the plant may show a general reduction in vigor, premature ripening or dying of the head, and shriveled grain. The fungus occasionally causes superficial infection on seeds and it can also overwinter as mycelium on the seed. When seedborne, the fungus may result in root rot and crown rot of the developing plant.

Anthracnose is probably the most destructive disease of cucurbits everywhere, being most severe on watermelon, cantaloupe, and cucumber. All aboveground parts of the plants are affected. Symptoms appear first on the leaves as small, water-soaked, yellowish areas that enlarge from several millimeters to 1 to 2 centimeters and become black in watermelon, while in all other cucurbits they turn brown. Infected tissues dry up and break. Lesions also develop on the petioles, which may result in defoliation of the vine; on the fruit pedicel, which causes the fruit to turn dark, shrivel and die; and on the stem, which weaken or kill whole vines. The fruit becomes susceptible to infection at about the time of ripening. Circular, watery, dark, sunken lesions appear on the surface of the fruit that may be from 5 mm to 10 cm in diameter and up to 8 mm in depth (Fig. 94D). The lesions expand rapidly in the field and during transit or storage, and may coalesce to form larger ones. The sunken lesions have dark centers which in moist weather are filled with pink spore masses exuding from acervuli that break through the cuticle. Severely affected fruits are often tasteless or even bitter and are often invaded by soft-rotting bacteria and fungi that enter through the broken rind. The fungus overwinters in infected debris in the soil and on or in the seed.

In tomato, and other vegetables and fruits, anthracnose or ripe rot results in serious losses of fruit and occasionally in damage to stems and foliage. Canning tomatoes are particularly susceptible to anthracnose before and after harvest but other tomatoes, as well as eggplant, pepper, apple, pear, banana, etc., may be attacked in a similar manner from the time ripening begins through harvest and in storage. In early stages of tomato infection, the symptoms appear as small, circular, sunken, water-soaked spots resembling indentations caused by blunt circular objects. As the fruit softens, the spots enlarge up to 2 to 3 cm in diameter and their central portion becomes dark and slightly roughened due to black acervuli developing just beneath the skin (Fig. 94, F, G). The spots are often numerous and coalesce, leading to the eventual watery softening of the fruit and finally rotting of the fruit, sometimes accelerated by other invading microorganisms. Enormous numbers of conidia are present in acervuli below the skin even in the smallest spots and, under some conditions, pink or salmon-colored masses of spores are also produced on the surface of the spots. The fungus overwinters in infected plant debris and in or on the seed. The fungus may cause light infections of foliage and young stems that may go unnoticed but these enable the fungus to survive and multiply somewhat until the fruit begins to ripen and becomes susceptible to infection. High temperatures and high relative humidity or wet weather at the time of ripening favor infection and spread of the fungus and often lead to destructive epidemics.

In onion smudge, dark-green or black smudges appear on the outer scales or neck of the bulbs primarily of white onions (Fig. 94E), although most colored varieties are also partially susceptible especially in the colorless region of the bulb neck. The smudgy spots first appear as tiny black stromata or visible mycelium beneath the cuticle of the scale and may be scattered over the surface of the bulb; more commonly, they congregate in uniformly black, smudgy, circular areas or are arranged in concentric circles, the outer one being up to two or more centimeters in diameter. In moist weather the stromata produce acervuli filled with cream-colored spore masses and containing numerous black, stiff, bristlelike hairs (= setae) visible with a hand lens. The fungus attacks inner, living scales only under very favorable high moisture and temperature and, usually, underneath spots on outer scales or where the outer scales have cracked and exposed the inner scales to the soil. The fungus overwinters as mycelium or stromata on infected onions, on sets, and in the soil as a saprophyte.

In citrus and especially orange, grapefruit, and lemon, anthracnose affects all mature, weakened, or injured aboveground plant parts, including leaves, twigs and fruit. Anthracnose may occur on trees of any size, in the nursery or in the orchard, but it rarely develops on vigorously growing trees. Anthracnose is common, however, on trees that are weakened or injured from inadequate fertilization, drought, low temperatures, spray injury, insects, other diseases, etc. Anthracnose leaf spots bring about death and drying of infected tissues with minute black acervuli produced in concentric circles in the dead areas. During severe infections, defoliation may occur. Weakened twigs are invaded by the anthracnose fungus and die back slowly or rapidly, resulting in what is called withertip. Leaves in affected twigs turn yellow, wither and drop, or they die quickly and dry before they can fall off. Affected branches also lose fruit. Numerous, small, black, pimplelike acervuli develop on the dead portions of twigs. On dying or dead areas of the surface of citrus fruits, anthracnose fruit spots develop that are circular and sunken and vary from tiny specks to dark-brown or black areas 5 to 10 mm in diameter. The spots become dry and hard and sometimes are dotted with small black acervuli that in humid weather exude masses of pinkish spores. Although anthracnose fruit spots generally affect only the rind, the disease often extends into the pulp and gives it a disagreeable or bitter taste. Some strains of the anthracnose fungus cause a soft decay that results in fruit drop. Also, often other fungi invade the fruit through anthracnose spots and speed up the rate of fruit decay. Overripe fruit is particularly susceptible to anthracnose infection. When spores of the anthracnose fungus are washed over fruits from dead twigs, etc. and then germinate on the fruit, they cause anthracnose russeting of the fruit. The russeting may appear as a large blotch or as a tear stain.

The fungus, *Colletotrichum* or *Gloeosporium* sp., produces colorless, one-celled, ovoid, cylindrical, and sometimes curved or dumbbell-shaped conidia in acervuli. Masses of conidia appear pink or salmon colored. The acervuli are subepidermal and breaking out through the surface of the plant tissue, disk- or cushion-shaped, waxy, with simple short, erect conidiophores. *Colletotrichum* has been distinguished from *Gloeosporium* by the fact that *Colletotrichum* acervuli have dark, long spines, or sterile hairlike hyphae while *Gloeosporium* acervuli do not. However, this is not always so and, therefore, the distinction of the two genera is often impossible and they are often considered as the same fungus. As mentioned earlier, many *Colletotrichum* species also produce an ascigerous perfect stage, usually *Glomerella*, occasionally *Physalospora* or others, while many *Gloeosporium* species have as perfect stage the ascomycetes *Glomerella* or *Gnomonia*, and occasionally others.

The fungus is favored by high temperatures and humid or moist weather. The conidia are released and spread only when the acervuli are wet and are generally spread by splashing and blowing rain or by coming in contact with insects, other animals, tools, etc. The conidia germinate only in the presence of water. Upon germination they produce an appressorium and penetration peg and penetrate the host tissues directly (Fig. 96). In the beginning the hyphae may grow rapidly, both intercellularly and intracellularly, but cause little or no visible discoloration or other symptoms of disturbance. Then more or less suddenly, especially when fruits begin to ripen, the fungus becomes more aggressive and symptoms appear. In many hosts the fungus reaches the seed and is either carried on the seeds or, in some, it may even invade a small number of seeds without causing any apparent injury to them. There is considerable variability in the kinds of host plants each species of *Colletotrichum* or *Cloeosporium* can attack, and there may be several races with varying pathogenicity within each species of the fungus.

Control of *Colletotrichum* or *Gloeosporium* diseases depends on the use of disease-free seed or seed treated with chemicals and hot water, on a two- to three-year crop rotation when possible, on the use of resistant varieties which are available for several annual crops, and the use of fungicides such as benomyl, maneb, zineb, chlorotholonil, captafol, and folpet.

- *Coryneum Blight of Stone Fruits*

It is found throughout the world although its importance varies considerably. It attacks twigs, buds, blossoms, leaves, and fruit, primarily of peach and apricot but also of nectarine and sweet cherry and, less, of other stone fruits. Infected buds die and have a shiny-black appearance due to a glistening layer of gum. On peach, small purplish, raised spots appear that later become black cankers and produce masses of gum in wet weather. Large, gum-exuding cankers may also occur on large limbs and trunks. Severely infected twigs are killed in the late spring and early summer and even branches and sometimes entire trees may die. Blossoms are usually killed as a result of bud infection but sometimes also as a result of cankers developing at the base of the flower pedicel. Leaf symptoms at first appear as scattered brown spots but the affected tissue soon falls out leaving a "shot-hole" effect. Leaf spots sometimes enlarge and involve a considerable area of the leaf surface which also falls out and gives the leaf a ragged appearance. The most distinctive symptoms of *Coryneum* blight occur on the fruit. Small, circular, purple-brown spots with light centers appear on young fruit when it is 1 to 2 cm in diameter. The spots later become slightly raised above the fruit surface and feel rough. The spots are usually more numerous on the upper side of the fruit. On peach fruit the spots are larger than those on apricot, they exude gum freely and result in an unevenly developed and misshapen fruit.

The fungus, *Coryneum beijerinckii*, produces only conidia that are dark olivaceous in color, four- to six-celled, ovoid to spindle-shaped, straight, or slightly curved. The conidia are produced singly on simple, slender conidiophores produced in black, cushion- to disk-shaped acervuli that form below the cuticle or below the outer cortex. The fungus overwinters as mycelium in dormant buds and in cankers on the twigs and branches. *Coryneum* produces conidia throughout the growing season, i.e., from the time the trees bloom until fall. Affected buds produce conidia for two successive years while the cankers produce conidia for three or more years. The conidia are relased from their conidiophores only in the presence of moisture and are splashed or washed by rain over the surface of fruits, leaves, and branches. Conidia germinate on wet plant tissues by producing several germ tubes which penetrate the host directly through the cuticle. The invading hyphae spread within a limited area of the host tissues and then again push to the surface where they produce acervuli and a new crop of conidia.

Control of *Coryneum* blight depends on removal of all dead wood and cankered twigs during pruning and on the application of fungicides such as Bordeaux mixture in the early fall and again at the shuck-fall stage.

SELECTED REFERENCES

Arx, J. A. Von. 1957. Die Arten der Gattung *Colletotrichum* Cda. *Phytopathol. Z.* **29**:413–468.

Baker, K. F. 1948. The history, distribution, and nomenclature of the rose black-spot fungus. *Plant Dis. Reptr.* **32**:260–274, 397.

Berry, F. H. 1960. Etiology and control of walnut anthracnose. *Ind. Agr. Exp. Sta. Bull.* **A 113**: 22 p.

Carter, J. C. 1964. Illinois trees: Their diseases. *Ill. Nat. Hist. Survey Circular* **46**: 96 p.

Drake, C. R. 1968. Grape diseases and their control in Virginia. *Virg. Polytech. Inst. Publication* **32**: 10 p.

Forsberg, J. L. 1975. Diseases of ornamental plants. *Univ. of Ill., Coll. of Agr., Special Publication* **No. 3 Rev.** 220 p.

Gardner, M. W. 1918. Anthracnose of cucurbits. *U.S. Dept. Agr. Bull.* **727**: 68 p.

Hopperstead, S. L., M. W. Goodwin, and K. J. Kadow. 1943. Bitter rot of apples and its control in Delaware. *Del. Agr. Exp. Sta. Bull.* **241**: 23 p.

Howard, C. M. 1972. A strawberry fruit rot caused by *Colletotrichum fragariae*. *Phytopathology* **62**:600–602.

Kendrick, J. B., Jr., and J. C. Walker. 1948. Anthracnose of tomato. *Phytopathology* **32**:247–260.

Layton, D. V. 1937. The parasitism of *Colletotrichum lagenarium*. *Iowa Agr. Exp. Sta. Res. Bull.* **223**:37–67.

Leach, J. G. 1923. The parasitism of *Colletotrichum lindemuthianum*. *Minn. Agr. Exp. Sta. Tech. Bull.* **14**: 41 p., illus.

Ling, L, 1940. Seedling stem blight of soybean caused by *Glomerella glycines*. *Phytopathology* **30**:345–347.

Lyle, E. W. 1938. The black-spot disease of roses and its control under greenhouse conditions. *Cornell Univ. Agr. Exp. Sta. Bull.* **690**: 31 p.

Pomerleau, R. 1938. Recherches sur le *Gnomonia ulmea*. *Contrib. Inst. Bot. Univ. Montreal* **31**: 139 p.

Rhoads, A. S. 1926. Diseases of grapes in Florida. *Univ. Fla. Agr. Exp. Sta. Bull.* **178**:75–156.

Rieman, G. H. 1931. Genetic factors for pigmentation in the onion and their relation to disease resistance. *J. Agr. Res.* **42**:251–278.

Schneider, R. W., *et al.* 1974. *Colletotrichum truncatum* borne within the seedcoat of soybean. *Phytopathology* **64**:154–155.

Schuldt, P. H. 1955. Comparison of anthracnose fungi on oak, sycamore, and other trees. *Contrib. Boyce Thompson Inst.* **18**:85–107.

Shear, C. L. 1929. The life history of *Sphaceloma ampelinum*. *Phytopathology* **19**:673–679.

Shear, C. L., and Anna K. Wood. 1913. Studies of fungus parasites belonging to the genus *Glomerella*. *U.S. Dept. Agr. Bur. Plant Ind. Bull.* **252**:5–110.

Stathis, P. D., and A. G. Plakidas. 1958. Anthracnose of azaleas. *Phytopathology* **48**:256–260.

Sutton, B. C. 1966. Development of fructifications on *Colletotrichum graminicola* and related species. *Can. J. Bot.* **44**:887–897.

Taylor, J. 1971. Epidemiology and symptomatology of apple bitter rot. *Phytopathology* **61**:1028–1029.

Ullstrup, A. J. 1938. Variability of *Glomerella gossypii*. *Phytopathology* **28**:787–798.

Walker, J. C. 1921. Onion smudge. *J. Agr. Res.* **20**:685–722.

Weber, G. F. 1973. "Bacterial and Fungal Diseases of Plants in the Tropics." Univ. of Fla. Press, Gainesville, Fla. 673. p.

Wellman, F. L. 1972. "Tropical American Plant Disease." The Scarecrow Press, Metuchen, N. J., pp. 236–273.

Wheeler, H., D. J. Politis, and C. G. Poneleit. 1974. Pathogenicity, host range and distribution of *Colletotrichum graminicola* in corn. *Phytopathology* **64**:293–295.

Wilson, E. E. 1937. The shot-hole disease of stone fruit trees. *Calif. Agr. Exp. Sta. Bull.* **608**:1–40.

Zaumeyer, W. J., and H. R. Thomas. 1957. A monographic study of bean diseases and methods for their control. *U.S. Dept. Agr. Tech. Bull.* **868**: 255 p.

FRUIT AND GENERAL
DISEASES CAUSED BY ASCOMYCETES
AND IMPERFECT FUNGI

This group of diseases is caused by Ascomycetes and Imperfect Fungi other than those already described and are most commonly found on the fruit or cause most of their damage by their effect on the fruit although they may affect other parts of the plant as well. Most of these fungi differ considerably from each other in fruiting structures, life cycles, and in the diseases they cause although most of these Ascomycetes produce ascospores in perithecia and both, Ascomycetes and Imperfects, produce conidia on free hyphae. The most common ascomycetous fungi and the most important diseases they cause are the following:

Claviceps purpurea, causing ergot of cereals and grasses.

Diaporthe citri, causing pod blight of lima beans, stem canker of soybeans and melanose of citrus fruits.

Physalospora, causing black rot, frogeye leaf spot, and canker of apple (*P. obtusa*), oak canker and dieback (*P. glandicola*), blight and black canker of willow (*P. miyabeana*), canker and dieback of many tropical trees such as citrus, cacao, coconut, rubber, and tropical forest trees (*P. rhodina*).

Venturia inaequalis, causing apple scab.

Monilinia fructicola, causing brown rot of stone fruits.

The most common Imperfect Fungi causing fruit and general diseases on plants are the following:

FIGURE 97 (A,B).
(A) Ergot of rye caused by *Claviceps purpurea*. The long, hornlike structures on the rye head are the sclerotia of the pathogen. (B) Ergot sclerotia of various shapes and sizes.

> *Alternaria*, causing leaf spots, blights, fruits rots, damping off, collar rots, and tuber rots of many vegetables, flowers, fruit trees, etc.
> *Botrytis*, causing blossom blights and fruit rots, but also damping off, stem cankers or rots, leaf spots, and tuber, corm, bulb and root rots of many vegetables, flowers, small fruits, and other fruit trees.
> *Fusicladium*, causing scab of pome fruits, peaches and pecan.
> *Phomopsis*, causing blights of eggplant, carrot, azalea, juniper, stem canker of gardenia, dead arm disease of grape, and stem-end rot of citrus fruits.

- *Ergot of Cereals and Grasses*

Ergot occurs throughout the world, most commonly on rye but also on wheat and less frequently on barley and oats. It is also very common on certain wild and cultivated grasses. An ergot disease affecting corn occurs

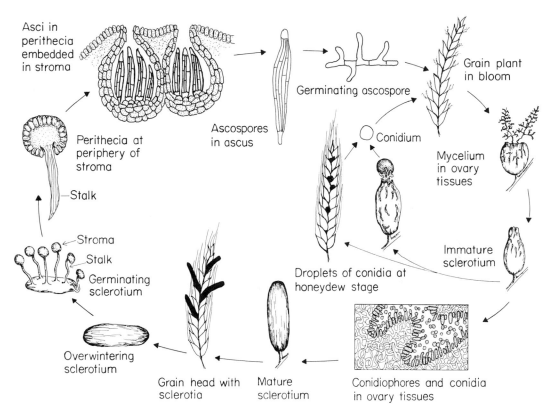

Asci in perithecia embedded in stroma

Perithecia at periphery of stroma

Stalk

Germinating ascospore

Ascospores in ascus

Conidium

Grain plant in bloom

Mycelium in ovary tissues

Stroma

Stalk

Germinating sclerotium

Overwintering sclerotium

Droplets of conidia at honeydew stage

Immature sclerotium

Grain head with sclerotia

Mature sclerotium

Conidiophores and conidia in ovary tissues

FIGURE 97 (C).
Disease cycle of ergot of grains caused by *Claviceps purpurea.*

in Mexico. The disease causes loss of some of the grains in infected heads, occasionally resulting in losses of up to 5 percent in rye and 10 percent in wheat, but its main importance is due to the fact that the characteristic fungal fruiting structures (ergot sclerotia) that replace the grains are poisonous to humans and animals that eat them.

The symptoms of ergot appear at first as creamy to golden-colored droplets of a sticky liquid exuding from young florets of infected heads. These droplets may go unnoticed, but they are each soon replaced by a hard, horn-shaped, purplish-black fungal mass that is usually a few millimeters in diameter but may be from 0.2 to 5.0 cm long. These are the sclerotia of the fungus, sometimes called ergots, that grow in place of the kernel and consist of a hard compact mass of fungal tissue which in cross section appears white or faintly purple (Fig. 97A,B).

Ergot is caused by *Claviceps purpurea* and other *Claviceps* species. The fungus overwinters as sclerotia on or in the ground, or mixed with the seed. In the spring, about the time cereals are in bloom, sclerotia on or near the surface of the soil germinate by forming from one to sixty flesh-colored stalks 0.5 to 2.5 cm tall (Fig. 97C). The tip of each stalk consists of a spherical head at the periphery of which are embedded numerous perithecia each containing many asci. Each ascus contains eight long, multicellular ascospores. The ascospores are discharged forcibly into the air, are picked up by air currents and are carried by wind or insects to young open flowers. There they germinate and infect the ovaries directly or by way of the stigma. Within about a week the fungus in the ovary tissues produces spreading, stromalike sporodochia that produce numerous conidia of the *Sphacelia* type. The conidia are floating in a sticky liquid and exude from the young florets as creamy or golden droplets which are known as the "honey dew" stage. The liquid of the "honey dew" stage is sugary and attractive to insects which visit such infected flowers, become smeared with the conidia of the fungus present in the "honey dew" and carry them to other healthy flowers which the conidia promptly infect. Conidia are probably spread to other flowers by splashing rain resulting in more infections. Gradually, the secretion of the "honey dew" ceases and each infected ovary, instead of producing normal seed, becomes replaced by a hard mass of fungal mycelium which eventually forms the characteristic ergot sclerotium. The sclerotia mature about the same time as the healthy seeds and either fall to the ground where they overwinter, or they are harvested with the grain and may be returned to the land with the seed.

Although ergot does not cause ergotism in humans nearly as often or as severely as it used to, it is probable that it is involved in many otherwise unexplainable poisonings of humans and it certainly continues to be of economic importance as an animal disease. Grain containing more than 0.3 percent by weight of the ergot sclerotia is classed as ergoty and can cause ergotism. Although most of the sclerotia can be removed from ergoty grain with modern cleaning machinery, it is costly and quite often difficult to remove enough sclerotia to meet the legal standards and the remaining traces have often proven toxic to livestock. Feeding livestock with cleanings from contaminated grain or grazing pastures that have infected grass heads can also lead to reproductive failure or gangrene of the peripheral parts of the animals.

Control of ergot depends entirely on cultural and sanitary procedures. Only clean seed or seed that has been freed from ergot should be sown. Sclerotia may be removed from seed by machinery or by soaking contaminated seed for 3 hours in water and then floating off the sclerotia in a solution of about 18 kg salt in 100 liters of water. Ergot sclerotia do not survive for more than a year and do not germinate if buried deep in the ground. Therefore, deep ploughing and/or crop rotation with a noncereal for at least a year helps eliminate the pathogen from a particular field. Wild grasses should be mowed or grazed before flowering to prevent production of ergot sclerotia on them and avoid poisoning of livestock through them and also to prevent the spread of the fungus to cultivated cereals and grasses.

- *Apple Scab*

Apple scab exists in every country where apples are grown. It is most severe in areas with cool, moist springs and summers, however, and may be completely absent in areas with very dry or warm climates. In the United States it is most serious in the north central and northeastern states. The disease affects all apples in the genus *Malus,* but similar scab diseases affect pears and hawthorns.

Scab is the most important disease of apples. Its primary effect is the reduction of quality of infected fruit, but it also reduces fruit size and the length of time infected fruit can be kept in storage. Infection of the stem of young fruit results in premature fruit drop. Severe leaf infections result in reduction of the functioning leaf surface, defoliation, and poor fruit bud development for the next year's crop. Losses from apple scab may be 70 percent or more of the total fruit value.

Symptoms. The first symptoms appear on the under surface of sepals or young leaves of the flower buds as light, somewhat olive-colored, irregular, spots. Soon after, the lesions become olive green with a velvety grayish-dark surface and more circular in outline. Later, the velvety surface disappears, the lesions appear metallic black in color and may be slightly raised. Lesions on leaves that have already unfolded are generally on the upper surface of the leaves (Fig. 98A). The number of lesions per leaf varies with the severity of infection. Lesions may remain distinct or they may coalesce. After severe early leaf infection, leaves may become dwarfed and curled and may later fall off.

Infections of the fruit appear as distinct, almost circular scab lesions which at first are velvety and olive green, but later become darker, scabby, and sometimes cracked (Fig. 98B). The cuticle of the fruit is ruptured at the margin of the lesions. Severe early fruit infections result in misshapen, cracked fruits which frequently drop prematurely. Infections late in the season when the fruit approaches maturity result in small lesions which may even be too small to be visible at harvest, but develop into dark scab spots during storage.

Twig and blossom infections appear as small scab spots, but they are uncommon and of little importance.

The pathogen: Venturia inaequalis. The mycelium is at first light in color, but later turns brownish in the host tissues. In young leaf lesions the mycelium develops radially in branched ribbons of hyphae, but in older leaves and on the fruit the mycelial strands are compact and thick and in several superimposed layers. In living tissues the mycelium is located only between the cuticle and the epidermal cells and produces short, erect, brownish conidiophores which successively give rise to several, one- or two-celled, reddish brown *Fusicladium*-type conidia of variable, but rather characteristic, shape (Fig. 99). In dead leaves the mycelium grows through the leaf tissues. Fertilization takes place by means of ascogonia and antheridia and perithecia form. The latter, when mature, are dark brown to black with a slight beak and a distinct opening (ostiole). Inside the perithecium 50 to 100 asci are formed, each containing eight ascospores. Each ascospore consists of two cells of unequal size, which are hyaline at first, but turn brown when mature.

Development of disease. The pathogen overwinters in dead leaves on the ground as immature perithecia. Perithecia initially develop in the fall and early winter and probably continue to grow during warm periods

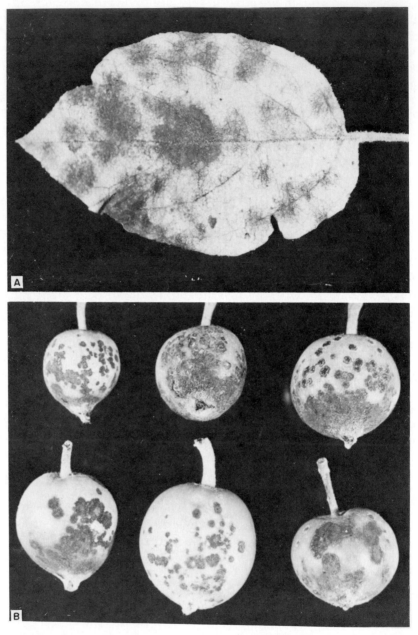

FIGURE 98.
Apple scab symptoms on leaves (A) and young fruit (B) caused by *Venturia inaequalis.*

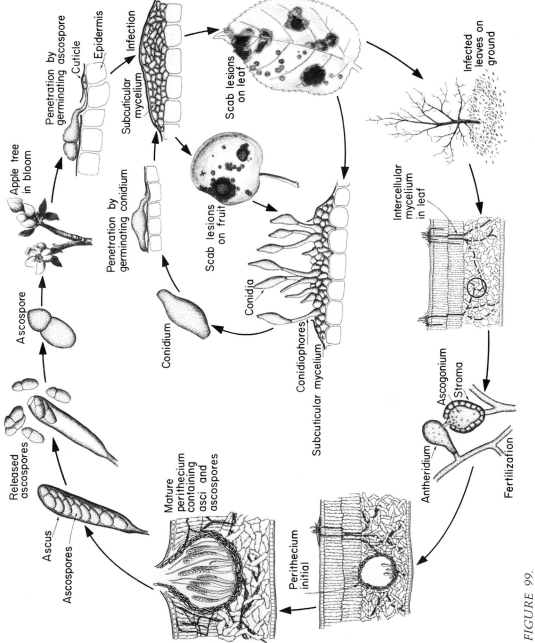

Penetration by germinating ascospore

Epidermis

Cuticle

Infection

Subcuticular mycelium

Scab lesions on leaf

Apple tree in bloom

Infected leaves on ground

Ascospore

Penetration by germinating conidium

Scab lesions on fruit

Intercellular mycelium in leaf

Conidium

Conidia

Released ascospores

Conidiophores

Subcuticular mycelium

Ascus

Ascogonium

Stroma

Ascospores

Mature perithecium containing asci and ascospores

Antheridium

Fertilization

Perithecium initial

FIGURE 99.
Disease cycle of apple scab caused by *Venturia inaequalis*.

309

in the winter and early spring but rapid growth and ascospore maturation occur with the resumption of favorable weather for growth and development of the host (Fig. 99). Neither all the perithecia nor all the asci of a perithecium mature simultaneously, and some may have ascospores before the apple buds start to open in the spring. Most of the ascospores in the perithecia, however, mature in the period during which the fruit buds open.

When dead leaves containing perithecia become thoroughly soaked in the spring, the asci elongate, push through the ostiole, and forcibly discharge the ascospores into the air; air currents may carry them to susceptible green apple tissues. Ascospore discharge may continue for 3 to 5 weeks after petal fall.

The ascospores can germinate and cause infection only when they are kept wet over a certain minimum period of time at temperatures ranging roughly from 6 to 26°C. Thus, for infection to occur, the spores must be continuously wet for 28 hours at 6°C, for 14 hours at 10°C, for 9 hours at 18 to 24°C, and for 12 hours at 26°C.

Upon germination on an apple leaf or fruit, the ascospore produces a disklike appressorium from which a very slender mycelial tube pierces the cuticle, and after developing into a hypha of normal diameter, grows between the cuticle and the outer cell wall of the epidermal cells. For a few days after infection, the epidermal cells show no injury at all, but by the time the lesion appears these cells show a gradual depletion of their contents, and they eventually collapse and die. Soon the palisade, and later the mesophyll, cells exhibit the same reactions, while the fungus still remains largely in the subcuticular position. It is assumed that the fungus obtains its nutrients and causes the death of the cells by secreting enzymes (and possibly toxic substances) that alter the permeability of the cell membranes and break down macromolecular components of the cell into small molecules which then move along osmotic pressure gradients, reach the mycelium and are absorbed by it.

Once the mycelium is established in the host, it produces enormous numbers of conidia which push outward, rupture the cuticle, and within 8 to 15 days from inoculation, form the olive-green, velvety scab lesions. Conidia remain attached to the conidiophores in dry weather, but upon wetting during a rain they are easily detached, and may be washed down or blown away to other leaves or fruit on which they germinate and cause infection in the same way ascospores do. Additional infections by conidia continue throughout the growing season following a rain of sufficient duration. Infections, however, are more abundant in the cool, wet periods of spring and early summer and again in the fall. They are infrequent, if not entirely absent, in the dry, hot summer weather.

After infected leaves fall to the ground, the mycelium penetrates into the interior of the leaf and forms perithecia, which carry the fungus through the winter.

Control.   Apple scab can be thoroughly controlled by timely sprays with the proper fungicides.

For an effective apple scab control program, apple trees must be diligently sprayed or dusted before, during, or immediately after a rain from the time of bud break until all the ascospores are discharged from the perithecia. If these primary infections from ascospores are prevented, there will be less need to spray for scab the remainder of the season. If primary infections do develop, spraying will have to be continued throughout the season. In most areas application of fungicides for scab control is based upon the phenological development of the trees. Sprays begin in the spring when a wetting period (rain) is sufficiently long at the existing temperature to produce an infection, and are repeated every 5 to 7 days, or according to rainfall, until petal fall. One must make sure that the new tissues on rapidly expanding young leaves and fruit are always covered with fungicide during an infection period. After petal fall, and depending on the success of the control program to that point, sprays are usually repeated every 10 to 14 days for several more times.

Several fungicides that give excellent control of apple scab are available. All of them are protectant, since they can protect a plant from becoming infected, but they cannot cure an infection, although some of them have so-called "kickback" action, i.e., ability to stop infections which may have started, and some have an eradicant action, i.e., they can "burn out" young scab spots early in the season. Dodine (= Cyprex) gives excellent scab control and is one of the fungicides exhibiting local systemic action. It reduces spore formation on scab spots and inhibits germination of spores produced on such spots. Excellent scab control is also obtained with captan, ferbam, benomyl, Glyodin, Glyodex (a mixture of dodine and Glyodin), dichlone (= Phygon), thiram, sulfur, captafol, etc. These fungicides may be used alone or in combinations since they differ in their ability to control scab or other diseases, in duration of "kickback" action, in compatibility with other fungicides, and in phytotoxicity. Captafol has recently been used as a single application treatment at the green tip stage of bud development and has been shown to protect new growth from scab up to petal fall.

In some areas, new strains of *Venturia inaequalis* have appeared that are resistant to dodine or to benomyl. These chemicals, therefore, can no longer be relied on to control the disease by themselves.

- *Brown Rot of Stone Fruits*

Brown rot occurs throughout the world where stone fruits are grown and where there is sufficient rainfall during the ripening period of the fruit. It affects peaches, cherries, plums, apricots, and almonds with about equal severity.

Losses from brown rot result primarily by rotting of the fruit in the orchard, although serious losses may also appear during transit and marketing of the fruit. Yields may be reduced also by destruction of the flowers during the blossom blight stage of the disease. In severe infections, and in the absence of good control measures, 50 to 75 percent of the fruit may rot in the orchard and the remainder may become infected before it reaches the market.

Symptoms. The first symptoms of the disease appear on the blossoms (Fig. 100A). Brown spots appear on petals, stamens, or pistils and

FIGURE 100.
(A–E) Brown rot of stone fruits caused by *Monilinia fructicola.* (A) Blossom
blight, twig cankers and killing of the tip. (B) Plums rotting on twig and covered
with tufts of conidia. (C) Mummied peach fruit hanging from the twig and
producing conidia. (D) Pile of peaches discarded because of brown rot infection.
(E) Funnel-shaped apothecia of *M. fructicola* produced on mummied peach fruit
on the ground. (F) Monilia pod rot of cocoa caused by *Monilia roreri.* (Photos A, B
courtesy Dept. Plant Pathol., Cornell Univ. Photos D–F courtesy U.S.D.A.)

they spread rapidly, involving the entire flower and its stem. In humid
weather the infected organs are covered with the grayish-brown conidia
of the fungus and later shrivel and dry up, the rotting mass clinging to the
twigs for some time. Twigs bearing infected flowers develop small, elliptical, sunken, brown cankers around the flower stem which sometimes
may encircle the stem and cause twig blight. In humid weather, gum and
also gray tufts of conidia appear on the bark surface.

Fruit symptoms appear when the fruit approaches maturity. Small, circular, brown spots appear which spread rapidly in all directions, and, depending on the humidity, are sooner or later covered with ash-colored tufts of conidia which break through the skin and are either scattered or arranged in concentric rings on the fruit surface (Fig. 100B). One large or several small rotten areas may be present on the fruit which finally becomes completely rotted and either dries up into a mummy and remains hanging from the tree or falls to the ground, where it also forms a mummy (Figs. 100, C, D).

Sometimes small cankers also develop on twigs or branches bearing infected fruit.

The pathogen: Monilinia fructicola, until recently known as Sclerotinia fructicola.   In addition to *M. fructicola*, two other species, *M. laxa* and *M. fructigena*, cause brown rot of stone fruits. The former occurs occasionally in the west coast and the Wisconsin–Michigan area of the U.S.; the latter is found exclusively in Europe, where it is as serious on apples as it is on stone fruits. With slight differences the development of the disease caused by each species is essentially the same.

The mycelium produces chains of elliptical *Monilia*-type conidia on hyphal branches arranged in groups or tufts. The fungus also produces microconidia (spermatia) in culture and on mummied fruit. The microconidia are borne in chains on bottle-shaped conidiophores, and they do not germinate, but seem to be involved in the fertilization of the fungus. The sexual stage (apothecium) originates at the surface of mummified fruit which is partly or wholly buried in the soil or debris. More than 20 apothecia may form on one mummy (Fig. 100E). Small, bulblike protrusions appear on the mummy and they extend to form the stipe of the fruiting body. As the stipe emerges above ground its upper portion becomes swollen and a depression appears at the tip. Subsequent growth of the sides of the swelling forms the apothecial "cup" which is usually funnel- or, later, disk-shaped. The inside or upper surface of the apothecium is lined with thousands of cylindrical asci interspersed with paraphyses. Each ascus contains eight single-celled ascospores.

Development of disease.   The pathogen overwinters as mycelium or conidia on mummied fruit on the tree and the ground, and on the cankers of infected twigs (Fig. 101). In the spring the mycelium in mummified fruit on the tree and on the ground and in the twig cankers produces new conidia, while the mycelium in mummied fruit buried in the ground produces several apothecia, which will form asci and ascospores.

Both conidia and ascospores can cause blossom infections. The conidia are windblown or may be carried by rain water and splashes or insects to floral parts. The ascospores are forcibly discharged by the ascus, forming a whitish "cloud" over the apothecium. Air currents then carry the ascospores to the flowers. Conidia and ascospores germinate and can cause infection within a few hours.

The mycelium, especially in humid weather, produces short hyphae which group together, push outward through the epidermis and form numerous conidial tufts on the rotten, shriveled floral parts from which new masses of conidia are released. In the meantime, the mycelium

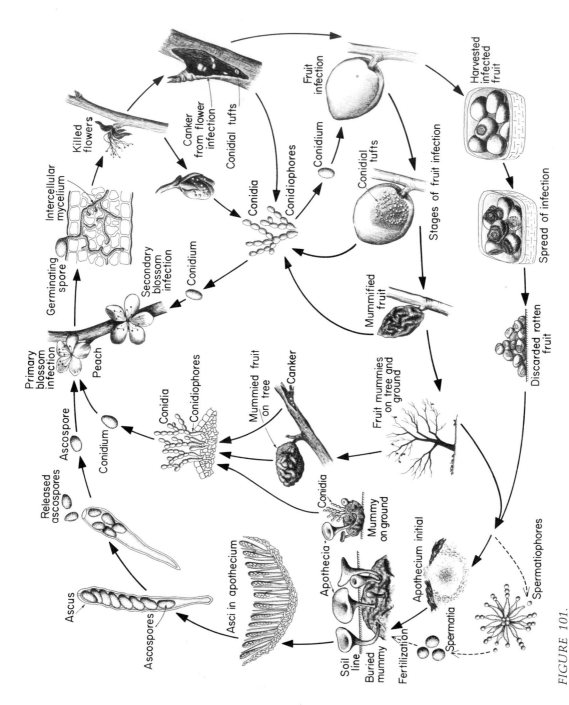

Killed flowers

Canker from flower infection

Conidial tufts

Fruit infection

Conidium

Harvested infected fruit

Intercellular mycelium

Conidia

Conidiophores

Conidial tufts

Stages of fruit infection

Spread of infection

Germinating spore

Secondary blossom infection

Conidium

Conidium

Mummified fruit

Discarded rotten fruit

Primary blossom infection

Peach

Conidia

Conidiophores

Mummied fruit on tree

Canker

Fruit mummies on tree and ground

Ascospore

Conidium

Conidia

Released ascospores

Apothecia

Mummy on ground

Spermatiophores

Ascus

Ascospores

Asci in apothecium

Apothecium initial

Spermatia

Soil line

Buried mummy

Fertilization

FIGURE 101.

Disease cycle of brown rot of stone fruits caused by *Monilinia fructicola*.

advances rapidly down the tissues of the blossom petioles and from there into the fruit spurs and the twigs. In the twig the mycelium causes disintegration and collapse of the cells around the fruit spur, and a depressed, reddish-brown, shield-shaped canker forms. The advance of the mycelium may encircle the twig which then becomes girdled and dies. The surface of the canker is soon covered with conidial tufts, and conidia from these serve as inoculum for fruit infection later in the season.

The brown rot fungus does not infect leaves or bark directly, and the new conidia and ascospores are short lived. The apothecia themselves soon disintegrate and therefore no ascospores are present when the fruit becomes susceptible. The gap between the time the blossoms are infected and when the fruit on the same tree can become infected is bridged by the conidia which are formed on twig cankers during humid weather in the summer. Also, the fruit of some early blooming stone fruits begins to ripen, and therefore becomes susceptible to infection, when some late-blooming ones are still in bloom or have just finished. In the latter case, conidia produced on the flowers of late-blooming trees can be carried to and infect the fruit of early ripening trees.

The susceptibility of the fruit to infection increases with its maturity. Conidia usually penetrate fruit through wounds made by insects, twig punctures, hail, etc., but on some, fruit penetration can take place through stomata or directly through the cuticle. The fungus grows intercellularly at first and through secretion of enzymes causes the maceration and browning of the infected tissues. The invasion of the fruit by the fungus is quite rapid. As the fungus advances into the fruit, it also produces conidial tufts on the already rotted area and the conidia may be carried away and infect more fruit. The entire fruit may become completely rotten within a few days, and it either clings to and hangs from the tree, or falls to the ground. Fruit falling to the ground soon after infection usually disintegrates under the action of saprophytic fungi and bacteria. Fruit left hanging on the tree soon loses moisture and shrivels and by the end of the season it becomes a dry, distorted mummy with a corrugated surface. The skin remains as a covering, and beneath it the remains of the fruit cells are held in place by the mycelial threads which are closely interwoven and form a hard rind. Once the fruit has dried into a mummy, it may fall to the ground, but it is not affected by soil microorganisms and may persist there for two or more years.

Fruit infection also takes place after harvest, in storage, and in transit. Infected fruit will continue to rot after harvest and the mycelium will attack directly healthy fruit in contact with infected ones. Healthy fruit may also be attacked by conidia at any time between harvest and use by the consumer.

Control. Brown rot of stone fruits can be controlled best by completely controlling the blossom blight phase of the disease. This can be done by spraying two to four times with an effective fungicide from the time the blossom buds show pink until the petals fall. Captan, benomyl, dichlone, sulfur, and thiram are the fungicides used for brown rot control.

Twigs bearing infected blossoms or cankers should be removed as early as possible to reduce the inoculum available for fruit infections later on in the season and to eliminate one of the sources of overwintering mycelium and spores that cause blossom infections the following season.

To control brown rot in ripening fruit, captan, thiram, ferban, ziram, wettable sulfur, or sulfur dusts are used in applications on the trees a few weeks before harvest and continued weekly or biweekly until just before harvest. Since most infections of immature fruit and many of mature fruit originate in wounds made by insect punctures, control of insects will also help control the disease.

To prevent infections at harvest and during storage and transit, fruit should be picked and handled with the greatest care to avoid punctures and skin abrasions on the fruit which enable the brown rot fungus to gain entrance more easily. All fruit with brown rot spots should be discarded. Lugs and packing boxes should be dusted both inside and out, particularly in wet weather, each time they are used. Postharvest brown rot can be reduced by dipping or drenching fruit in a benomyl solution before storing and again during the sorting operation; and by hydrocooling or cooling fruit in air before refrigeration at 0 to 3°C.

- *Alternaria Diseases*

They are among the most common diseases of many kinds of plants throughout the world. They affect primarily the leaves, stems, flowers, and fruits of annual plants, especially vegetables and ornamentals, but also may affect parts of trees such as citrus, apple, etc. Alternaria diseases appear usually as leaf spots and blights, but they may also cause damping off of seedlings, collar rots, and tuber and fruit rots. Some of the most common diseases caused by *Alternaria* include early blight of potato and tomato (Fig. 102, A–E), leaf spot of bean, tobacco, geranium, and stock, blight of carrot (Fig. 102F), carnation, chrysanthemum, petunia and zinnia, leaf spot and blight of crucifers (Fig. 102G), onion purple blotch, leaf spot and fruit spot on squash and on apple, core rot of apple, rot of lemons and oranges, and many others.

The leaf spots are generally dark brown to black, often numerous and enlarging, usually developing in concentric rings which give the spots a targetlike appearance (Fig. 102, A–C). Lower, senescent leaves are usually attacked first, but the disease progresses upward and makes affected leaves go into a yellowish, senescent condition and either dry up and droop or fall off. Dark, often targetlike, sunken spots develop on branches and stems of plants such as tomato (Fig. 102E). Stem lesions developing on seedlings may form cankers which may enlarge, girdle the stem and kill the plant, or if present near the soil line they may develop into a collar rot. In belowground parts, e.g., potato tubers, dark, slightly sunken, circular, or irregular lesions develop that may be up to 2 cm in diameter and 5 to 6 mm in depth. Fruits affected by *Alternaria* are usually attacked when they approach maturity, in some hosts at the stem end while in others at the blossom end or at other points through wounds, growth cracks, etc. (Fig. 102D). The spots are brown to black and may be small,

FIGURE 102.
Symptoms caused by species of *Alternaria*. (A–E) Early blight of tomato caused by
A. solani. (A) Typical target-shaped lesions on tomato leaf. (B, C) Leaf spots (B)
and blight (C) on tomato. (D) Typical *Alternaria* lesion at the stem end of
tomato fruit. (E) Lesions on tomato stems. (F) Carrot blight caused by *A. dauci*.
(G) Cabbage leaf spot caused by *A. brassicae*.

sunken, with well-defined margins, or they may enlarge to cover most of
the fruit, have a leathery consistency and a black, velvety surface layer of
fungus growth and spores. In some fruits, such as citrus and tomato, a
small lesion at the surface may indicate an extensive spread of the
infection in the central core and the segments of the fruit.

The pathogen, *Alternaria* sp., has dark-colored mycelium and in older
diseased tissue it produces short, simple, erect conidiophores that bear

single or branched chains of conidia (Fig. 103). The conidia are large, dark, long, or pear shaped and multicellular with both transverse and longitudinal cross walls. The conidia are detached easily and are carried by air currents. *Alternaria* occurs in many species throughout the world. Their spores are present in the air and dust everywhere and are one of the most common fungal causes of hay fever allergies. *Alternaria* spores also land and grow as contaminants in laboratory cultures of other microorganisms and on dead plant tissue killed by other pathogens or other causes. Actually, many species of *Alternaria* are mostly saprophytic, i.e., they cannot infect living plant tissues but they grow only on dead or decaying plant tissues and, at most, on senescent or old tissues such as old petals, old leaves, and ripe fruit. Therefore, it is often difficult to decide whether an *Alternaria* fungus found on a diseased tissue is the cause of the disease or a secondary contaminant.

Plant pathogenic species of *Alternaria* overwinter as mycelium in infected plant debris and as mycelium or spores in or on seeds. If the fungus is carried with the seed, it may attack the seedling, usually after emergence, and cause damping off or stem lesions and collar rot. More

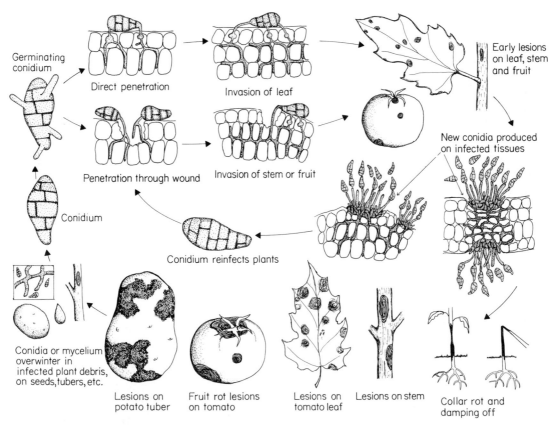

FIGURE 103.
Development and symptoms of diseases caused by *Alternaria*.

frequently, however, spores which are produced abundantly, especially during heavy dews and frequent rains, are blown in from mycelium growing on debris, infected cultivated plants or weeds. The germinating spores penetrate susceptible tissue directly or through wounds and soon produce new conidia that are further spread by wind, splashing rain, tools, etc. With few exceptions, *Alternaria* diseases are more prevalent on older, senescing tissues and particularly on plants of low vigor, poor nutrition or those under some other kind of stress caused by unfavorable environmental conditions, insects, other diseases, etc.

Alternaria diseases are controlled primarily through the use of resistant varieties, of disease-free or treated seed and through chemical sprays with fungicides such as chlorothalonil, maneb, captafol, or maneb–zinc combination. The sprays should begin soon after the seedlings emerge or are transplanted and should be repeated at 1- to 2-week intervals depending on disease prevalence, severity and frequency of rains. Crop rotation, removal and burning of plant debris, if infected, and eradication of weed hosts help reduce the inoculum for subsequent plantings of susceptible crops.

- *Botrytis Diseases*

They are probably the most common and most widely distributed diseases of vegetables, ornamentals, fruits and even field crops throughout the world. They are the most common diseases of greenhouse grown crops. *Botrytis* diseases appear primarily as blossom blights and fruit rots, but also as damping-off, stem cankers or rots, leaf spots and tuber, corm, bulb, and root rots (Figs. 104 and 105). Under humid conditions, the fungus produces a noticeable gray-mold fruiting layer on the affected tissues that is characteristic of the *Botrytis* diseases. Some of the most serious diseases caused by *Botrytis* include gray mold of strawberry, gray mold rot of vegetables such as artichoke, bean, beet, cabbage, carrot, cucumber, eggplant, tip-end rot of bananas, lettuce, pepper, squash, tomato, etc., onion blast and neck rot, calyx end rot of apples, blossom and twig blight of blueberries, blight or gray mold of ornamentals such as African violet, begonia, cyclamen, chrysanthemum, dahlia, geranium, hyacinth, lily, peony, rose, snapdragon, stock and tulip, bulb rot of amaryllis, corm rot and also leaf spot and stem rot of gladiolus, etc. *Botrytis* also causes secondary soft rots of fruits and vegetables in storage, transit, and market.

In the field, blossom blights often precede and lead to fruit rots and stem rots. The fungus becomes established in flower petals, which are particularly susceptible when they begin to age, and there it produces abundant mycelium (Fig. 104B). In cool, humid weather the mycelium produces large numbers of conidia which may cause further infections, but the mycelium itself also grows, penetrates and invades the rest of the inflorescence, which becomes filled and covered with a whitish-gray or light brown cobwebby mold. The fungus then spreads to the pedicel, which rots and lets the buds and flowers lop over. If any fruit has developed, the fungus moves from the petals into the green or ripe fruit and

FIGURE 104.
Scanning electron micrograph of a typical grape clusterlike conidiophore and
conidia of *Botrytis* (A) and various symptoms caused by *Botrytis* (B–I). (B)
Blossom blight of gardenia. (C) Strawberry gray mold. (D) Lettuce gray mold. (E,
F) External (E) and internal (F) symptoms of *Botrytis* blossom end rot of apple. (G)
Stem lesions; note that they originate at a dead leaf. (H): *Botrytis* blast of onion.
(I) *Botrytis* neck rot of onion in storage. The black bodies on the onions are
sclerotia of the fungus. (Photo A courtesy M. F. Brown and H. G. Brotzman.)

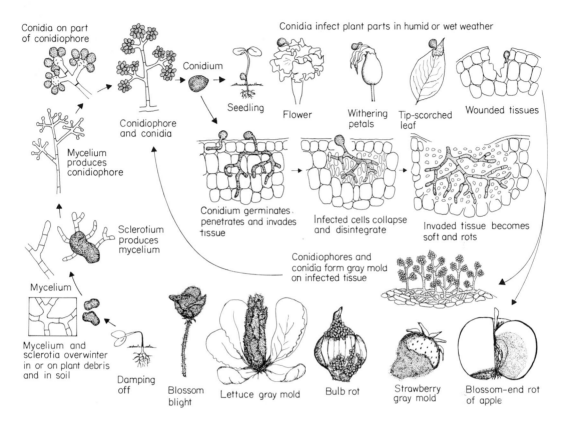

Conidia on part of conidiophore

Conidia infect plant parts in humid or wet weather

Conidium

Conidiophore and conidia

Seedling

Flower

Withering petals

Tip-scorched leaf

Wounded tissues

Mycelium produces conidiophore

Conidium germinates, penetrates and invades tissue

Infected cells collapse and disintegrate

Invaded tissue becomes soft and rots

Sclerotium produces mycelium

Conidiophores and conidia form gray mold on infected tissue

Mycelium

Mycelium and sclerotia overwinter in or on plant debris and in soil

Damping off

Blossom blight

Lettuce gray mold

Bulb rot

Strawberry gray mold

Blossom-end rot of apple

FIGURE 105.
Development of *Botrytis* gray mold diseases.

causes a blossom-end rot of the fruit which advances and may destroy part or all of the fruit and may spread to other fruits touching the diseased one. Infected fruit and succulent stems become soft and watery and, later, invaded tissues appear light brown (Fig. 104, C, E, F). As the tissue rots, the epidermis cracks open and the fungus fruits abundantly. The tissues then become wrinkled and dry, and flat black sclerotia may appear on the surface or are sunken within the tissue.

Damping off of seedlings due to *Botrytis* occurs primarily in cold frames where the humidity is high, but also in the field if the seed is contaminated with sclerotia of the fungus, or fungus mycelium or sclerotia are present in the soil.

Some species of *Botrytis* cause leaf spots on their hosts, e.g., on gladiolus, onions, tulip, the spots being small and yellowish at first but later becoming larger, whitish gray or tan, sunken, coalescing, and frequently involving the entire leaf (Fig. 104, D, H). In many hosts, however, foliage infection occurs only after the fungus has grown on dying parts of the plants or in decaying matter on the soil that comes into contact with healthy leaves.

Stem lesions usually appear on succulent stems or stalks and may be either dark, sunken, elongated lesions with a definite outline or may spread through the stalk and cause it to weaken and break over at the point of infection, as happens with peony, rose, tulip, etc (Fig. 104G). Such stalks are usually susceptible to attack by the fungus along their entire length and in wet weather the diseased parts become covered with the grayish-brown felty coat of fungus spores. Sclerotia may also be produced on infected stems.

Infection of below ground parts, such as bulbs, corms, tubers, roots, etc., may begin while these organs are still in the ground or upon harvest. Lesions may develop at any point of their surface but in most hosts they are more likely to begin at the crown or at the base of these organs. Infected tissues usually appear soft and watery at first but, as the infection progresses, these areas enlarge, turn tan to brown and finally dark brown and become spongy or corky and light in weight. Pockets of mycelium may develop between decayed bulb scales, under the husks of decayed corms, etc., or on the surface of the lesions of such organs. Black sclerotia are often found on the surface or intermingled with the rotted tissues and mycelium (Fig. 104I).

The pathogen, *Botrytis* sp., produces abundant gray mycelium and long, branched conidiophores with rounded apical cells bearing clusters of colorless or gray, one-celled, ovoid conidia (Fig. 104A). The conidiophores and clusters of conidia resemble a grapelike cluster. The conidia are released readily in humid weather and are carried by air currents. The fungus frequently produces black, hard, flat, irregular sclerotia. For some species of *Botrytis* a perfect stage is known, *Botryotinia*, in which ascospores are produced in an apothecium.

Botrytis overwinters in the soil as mycelium growing on decaying plant debris and as sclerotia (Fig. 105). The fungus does not seem to infect seeds but it can be spread with seed contaminated with sclerotia the size of the seed or with bits of plant debris infected with the fungus. The overwintering stages can also be spread by anything that moves soil or plant debris that may carry sclerotia or mycelium. The fungus requires cool (18 to 23°C), damp weather for best growth, sporulation, spore release and germination, and establishment of infection. The pathogen is active at low temperatures and causes considerable losses on crops kept for long periods in storage, even if the temperatures are between 0 and 10°C. Germinating spores seldom penetrate actively growing tissue directly but they can penetrate tissues through wounds or after they have grown for a while and produced mycelium on old flower petals, dying foliage, dead bulb scales, etc. *Botrytis* sclerotia usually germinate by producing mycelial threads that can infect directly, but in a few cases sclerotia germinated by producing apothecia and ascospores.

Control of *Botrytis* diseases is aided by the removal of infected and infested debris from the field and storage rooms, and by providing conditions for proper aeration and quick drying of plants and plant products. In greenhouses, humidity should be reduced by ventilation and heating.

Storage organs such as onion bulbs can be protected by keeping them at 32 to 50°C for 2 to 4 days to remove excess moisture and then keeping them at 3°C in as dry an environment as possible. Control of *Botrytis* in the field through chemical sprays has been only partially successful, especially in cool, damp weather. For *Botrytis* rot of lettuce, sprays with dichloran or zineb are recommended, while difolatan, dyrene, maneb–zinc, maneb or chlorothalonil seem to give better control on crops such as onion and tomatoes. To control fruit rots, such as gray mold of strawberry, sprays or dusts with captan, thiram or benomyl are recommended.

SELECTED REFERENCES

Abdel-Salam, M. M. 1934. *Botrytis* disease of lettuce. *J. Pom. Hort. Sci.* **12**:15–35.

Barger, G. 1931. "Ergot and Ergotism." Gurney and Jackson, London.

Beaumont, A., W. A. R. Dillon Weston, and E. R. Wallace. 1936. Tulip fire. *Ann. Appl. Biol.* **23**:57–88.

Bratley, C. O. 1937. Incidence and development of apple scab on fruit during the late summer and while in storage. *U.S. Dept. Agr. Tech. Bull.* **563**: 45 p.

Dickens, J. S. W., and P. G. Mantle. 1974. Ergot of cereals and grasses. *Min. Agr. Fish Food, Advis. Leaflet* **548.**

Ezekiel, W. N. 1924. Fruit rotting *Sclerotinias*. II. The American brown-rot fungi. *Maryland Agr. Expt. Sta. Bull.* **271**:87–142.

Forsberg, J. L. 1975. Diseases of ornamental plants. Univ. of Ill., Coll. of Agr., Special Publication No. 3 Rev. 220 p.

Foster, H. H. 1937. Studies on the pathogenicity of *Physalospora obtusa*. *Phytopathology* **27**:803–832.

Gould, C. J. 1954. *Botrytis* diseases of gladiolus. *Plant Dis. Reptr. Suppl.* **224**: 1–33.

Guba, E. F. 1950. Spoilage of squash in storage. *Mass. Agr. Exp. Sta. Bull.* **457**: 9–11, 13–15.

Hardison, J. R. 1972. Prevention of ascocarp formation in *Claviceps purpurea* by fungicides applied over sclerotia at the soil surface. *Phytopathology* **62**:609–611.

Hopkins, E. F. 1921. The *Botrytis* blight of tulips. *Cornell Univ. Agr. Exp. Sta. Mem.* **45**:315–361.

Jackson, C. R., and D. K. Bell. 1969. Diseases of peanut (groundnut) caused by fungi. *Georgia Agr. Exp. Sta. Res. Bull.* **56**:60-64, 106–108.

Keitt, G. W., and L. K. Jones. 1926. Studies of the epidemiology and control of apple scab. *Wisc. Agr. Expt. Sta. Res. Bull.* **73**:104 pp.

McClellan, W. D., and W. B. Hewitt. 1973. Early *Botrytis* rot of grapes: time of infection and latency of *Botrytis cinerea* in *Vitis vinifera*. *Phytopathology* **63**:1151–1156.

McColloch, L. P., H. T. Cook, and W. R. Wright. 1968. Market diseases of tomatoes, peppers, and eggplants. *U.S. Dept. Agr., Agr. Handbook 28*: 74 p., 16 plates.

McKeen, W. E. 1974. Mode of penetration of epidermal cell walls of *Vicia faba* by *Botrytis cinerea*. *Phytopathology* **64**:461–467.

Mower, R. L., W. C. Snyder, and J. G. Hancock. 1975. Biological control of ergot by *Fusarium*. *Phytopathology* **65**:5–10.

Nusbaum, C. J., and G. W. Keitt. 1938. A cytological study of host-parasite relations of *Venturia inaequalis* on apple leaves. *J. Agr. Res.* **56**:595–618.

Ogawa, J. M., *et al.* 1975. Monilinia life cycle on sweet cherries and its control by overhead sprinkler fungicide applications. *Plant Dis. Reptr.* **59**:876–880.

Pierson, C. F., M. J. Ceponis, and L. P. McColloch. 1971. Market diseases of apples, pears, and quinces. *U.S. Dept. Agr., Agr. Handbook* **376**: 112 p., 17 pl.

Pound, G. S. 1951. Effect of air temperature on incidence and development of the early blight disease of tomato. *Phytopathology* **41**:127–135.

Powell, D. 1954. The effect of captan on gray mold rot incidence and yield of strawberry. *Plant Dis. Reptr.* **38**:209–211.

Ramsey, G. B., J. S. Wiant, and M. S. Smith. 1949. Market diseases of fruits and vegetables: Potatoes. *U.S. Dept. Agr. Miscell. Publ.* **98**: 60 p.

Rands, R. D. 1917. Early blight of potato and related plants. *Wisc. Agr. Exp. Sta. Res. Bull.* **42**.

Rangel, J. F. 1945. Two Alternaria diseases of cruciferous plants. *Phytopathology* **35**:1002–1007.

Riggs, Rosa K., L. Henson, and R. A. Chapman. 1968. Infectivity of and alkaloid production by some isolates of *Claviceps purpurea. Phytopathology* **58**:54–55.

Roberts, J. W., and J. C. Dunegan. 1932. Peach brown rot. *U.S. Dept. Agr. Tech. Bull.* **328**:1–59.

Rose, D. H., D. F. Fisher, and C. Brooks. 1937. Market diseases of fruits and vegetables: Peaches, plums, cherries and other stone fruits. *U.S. Dept. Agr. Miscell. Publ.* **228**: 26 p.

Rowell, J. B. 1953. Leaf blight of tomato and potato plants; factors affecting the degree of injury incited by *Alternaria dauci* f. *solani. R. I. Agr. Exp. Sta. Bull.* **320**.

Segall, R. H., and A. G. Newhall. 1960. Onion blast or leaf spotting caused by species of *Botrytis. Phytopathology* **50**:76–82.

Smith, M. A., L. P. McColloch, and B. A. Friedman. 1966. Market diseases of asparagus, onions, beans, peas, carrots, celery and related vegetables. *U.S. Dept. Agr., Agr. Handbook* **303**: 65 p., 16 pl.

Smoot, J. J., L. G. Houck, and H. B. Johnson. 1971. Market diseases of citrus and other subtropical fruits. *U.S. Dept. Agr., Agr. Handbook* **398**: 101 p., 13 pl.

Walker, J. C. 1926. *Botrytis* neck rot of onions. *J. Agr. Res.* **33**:893–928.

Wicks, T. 1974. Tolerance of the apple scab fungus to benzimidazole fungicides. *Plant Dis. Reptr.* **58**:886–889.

VASCULAR WILTS
CAUSED BY ASCOMYCETES AND
IMPERFECT FUNGI

Vascular wilts are widespread, very destructive, spectacular, and frightening plant diseases appearing as more or less rapid wilting, browning, and dying of leaves and succulent shoots of plants followed by the final death of the plant. Wilts occur as a result of the presence and activities of the pathogen in the xylem vascular tissues of the plant. Entire plants or plant parts above the point of vascular invasion by the pathogen may die within a matter of weeks in most annuals and some perennials although in certain perennials death may not occur until several years after infection. The pathogen usually continues to spread internally through the xylem vessels as mycelium or conidia until the entire plant is killed. As long as the infected plant is alive, the fungi that cause vascular wilts remain limited to the vascular (xylem) tissues and a few surrounding cells and never come to the surface of the plant—not even to produce spores. Only when the infected plant is killed by the disease do these fungi move into other tissues and sporulate at or near the surface of the dead plant.

There are three genera of fungi that cause vascular wilts: *Ceratocystis, Fusarium,* and *Verticillium.* Each of them causes widespread and severe diseases on several important crop, forest, or ornamental plants.

Ceratocystis causes vascular wilts primarily of trees, e.g., Dutch elm disease (*C. ulmi*) and oak wilt (*C. fagacearum*).

Fusarium causes vascular wilts primarily of annual vegetables and flowers, herbaceous perennial ornamentals, plantation crops, weeds, and of the mimosa tree (silk tree). Most of the vascular wilt-causing *Fusaria* belong to the genus *Fusarium oxysporum.* Different host plants are attacked by special forms or races of the fungus. Thus, the fungus that attacks tomato is designated as *F. oxysporum* f. *lycopersici,* cucurbits *F. o.* f. *niveum,* sweet potato *F. o.* f. *batatas,* onion *F. o.* f. *cepae,* cabbage *F. o.* f. *conglutinans,* banana *F. o.* f. *cubense,* cotton *F. o.* f. *vasinfectum,* carnation *F. o.* f. *dianthii,* chrysanthemum *F. o.* f. *chrysanthemi,* etc.

Verticillium causes vascular wilts of annual vegetables, flowers, crop plants, and weeds, and of perennial ornamentals, fruit trees, forest trees, weeds, etc. It exists as one or two species, *Verticillium albo-atrum* and/or *V. dahliae,* and it attacks hundreds of kinds of plants causing wilts and losses of varying severity.

All vascular wilts, regardless of which pathogen causes them, have certain characteristics in common. The leaves of infected plants or of parts of infected plants lose turgidity, become flaccid, lighter green to greenish yellow, droop and finally wilt, turn yellow, then brown, and die. Wilted leaves may be flat or curled. Young, tender shoots also wilt and die. In cross sections of infected stems and twigs, discolored brown areas appear as a complete or interrupted ring consisting of discolored vascular tissues. In the xylem vessels of infected stems, roots, etc., mycelium and spores of the causal fungus may be present. Some of the vessels may be clogged with mycelium, spores, or polysaccharides produced by the fungus. Clogging is further increased by gels and gums formed by the accumulation and oxidation of breakdown products of plant cells attacked by fungal enzymes. The oxidation and translocation of some such breakdown products seems to also be responsible for the brown discoloration of affected vascular tissues. In newly infected young stems the number of xylem vessels formed is reduced and their cell walls are thinner than normal. Often the parenchyma cells surrounding xylem vessels are stimulated by secretions of the pathogen to divide excessively and this, combined with the thinner and weaker vessel walls, results in reduction of the diameter or complete collapse of the vessels. In some hosts, tyloses are produced by parenchyma cells adjoining some xylem vessels. The balloonlike tyloses protrude into the vessels and contribute to their clogging. Toxins secreted in the vessels by the wilt-causing fungi are carried upward in the water stream and affect living parenchyma cells adjacent to the xylem, thus causing some of the effects described above. Toxins may also be carried to the leaves in which they cause reduced chlorophyll synthesis along the veins (vein clearing) and reduced photosynthesis, disrupt the permeability of the leaf cell membranes and their ability to control water loss through transpiration and thereby result in leaf epinasty, wilting, interveinal necrosis, browning, and death.

Of the wilt fungi, *Fusarium* and *Verticillium* are soil inhabitants and infect plants through the roots which they penetrate directly or through wounds. Large numbers of parasitic nematodes in the soil usually increase the incidence of *Fusarium* and *Verticillium* wilts, presumably by providing more and effective penetration points. Once in the root, the mycelium reaches the xylem vessels and subsequently spreads through them into the stem and the rest of the plant. Both fungi produce only asexual spores and overwinter in the soil or plant debris as mycelium, as thick-walled asexual spores called chlamydospores (*Fusarium*), or as microsclerotia (*Verticillium*). Both are effective saprophytes and, once introduced into a field, they become established there forever although their populations may vary considerably depending on the susceptibility and the duration of cultivation of the host plant grown in the field. *Fusarium* and *Verticillium* spread through the soil to a small extent as mycelium growing through roots or plant debris, but primarily as mycelium, spores, or sclerotia carried in soil water, on farm equipment, transplants, tubers, seeds of some hosts, cuttings of infected plants and in some cases as wind-blown spores or sclerotia.

The other wilt fungus, *Ceratocystis*, lives primarily in the xylem vessels and adjacent parenchyma cells of infected elm and oak trees and in the outer layers of wood and inner layers of bark in elms and oaks killed by it. The fungus produces both ascospores and conidia and spreads from tree to tree either as spores carried by cetain bark-feeding beetles or by natural root grafts. The beetles introduce the fungus into the xylem of young, vigorously growing twigs or larger branches on which they feed and from there, as well as from the natural root grafts, the fungus spreads as mycelium or conidia into the vascular system of the tree and induces wilt.

Vascular wilts are more or less worldwide in distribution causing tremendous losses on most kinds of vegetables and flowers and on field crops such as cotton, alfalfa (*Fusarium* and *Verticillium*), on fruit "trees" such as banana (*Fusarium*) and stone fruits (*Verticillium*), on forest and shade trees (*Verticillium*), and on elm and oak trees, in particular (*Ceratocystis*). *Fusarium* wilts are much more common and destructive in the warmer temperate regions and in the tropics and subtropics, becoming less damaging or rare in colder climates except for greenhouse crops in these areas. *Verticillium* is much more common in the temperate zones and is considerably more cold-resistant than *Fusarium*, especially in its tree hosts and causes diseases in greater latitudes than the latter. *Ceratocystis* wilts are less widespread then the others, their distribution depending on the availability of the hosts (elms and oaks), of the pathogen, and of contaminated insect vectors.

Vascular wilts are among the most difficult to control. The fact that only a single infection of a plant by one spore is sufficient to introduce the pathogen into the plant, in which it then grows and spreads internally, makes prevention of infection and subsequent control with surface fungicides practically impossible. Also the fact that *Fusarium* and *Verticillium* can survive in the soil of a field saprophytically almost forever, makes control through crop rotation or other cultural practices impractical or ineffective. On the other hand, the dissemination of *Ceratocystis* spores over long distances by insect vectors feeding on trees of all sizes makes its control even more problematical.

The most effective means of controlling *Fusarium* and *Verticillium* wilts has been the use of resistant varieties. Due to the relative immobility of the pathogens and therefore slow development and distribution of any new pathogen races, varieties remain resistant for rather long periods of time. Cultural practices such as deep plowing, crop rotation, leaving the soil fallow, or flooding of the field have been helpful in reducing the pathogen populations in the soil but do not eliminate it completely. Soil fumigation has been used with success in some cases but it is too expensive and its effect does not last long enough to make its use profitable. In the greenhouse, soil sterilization gives effective control of both diseases. Control of *Ceratocystis* wilts has been attempted by efforts to control the insect vectors of the pathogen by insecticides sprayed on or injected into the trees, removal and burning of infected trees and logs to eliminate the fungus and the breeding grounds of the insect vectors and through selection of resistant trees. None of these measures has been successful against the *Ceratocystis* wilts and the diseases keep on spreading.

An apparent breakthrough in the control of fungal vascular wilts seems to have been made by the discovery of the systemic fungicides containing thiabendazole or its derivatives, including, and particularly, benomyl in its various formulations. These chemicals, although not yet proven to give complete control of any of these wilts, when injected into elm trees before and sometimes after infection have given promising results with *Ceratocystis ulmi,* the cause of Dutch elm disease.

CERATOCYSTIS WILTS

They include mainly the Dutch elm disease and oak wilt.

● *Dutch Elm Disease*

Dutch elm disease owes its name to the fact that it was first described on elm in Holland in 1921. Since then the disease has spread throughout Europe, parts of Asia, and most of the temperate zones in North America. In the U. S. the disease was first found in Ohio and some states in the east coast in the early 1930s; it has since spread westward to the Pacific coast states.

Dutch elm disease is the most destructive shade tree disease in the U. S. today. It affects all elm species, but most severely the American elm. The disease may kill branches and entire trees within a few weeks or a few years from the time of infection. Hundreds of thousands of elm trees in towns across the country die from Dutch elm disease every year. The cost of cutting down diseased and dead elm trees amounts to many millions of dollars per year. And, of course, no one can estimate the value of the natural beauty of streets and towns destroyed by the disease.

Symptoms.   The first symptoms of the disease appear as sudden or prolonged wilting of the leaves of individual branches or of the entire tree (Fig. 106A). Wilted leaves frequently curl, turn yellow, then brown, and finally fall off the tree earlier than normal (Fig. 106C). Most affected branches die immediately after defoliation. The disease usually appears first on one or several branches and then spreads to other portions of the tree. Thus, many dead branches may appear on a tree or a portion of a tree. Such trees may die gradually, branch by branch, over a period of several years or they may recover. Sometimes, however, entire trees suddenly develop disease symptoms and may die within a few weeks (Fig. 106A, B). Usually trees that become infected in the spring or early summer die quickly, while those infected in late summer are much less seriously affected and may even recover, unless they become reinfected.

When the bark of infected twigs or branches is peeled back, brown streaking or mottling appears on the outer layer of wood. In cross section of the branch, the browning appears as a broken or continuous ring in the outer rings of the wood (Fig. 106D).

The pathogen: Ceratocystis ulmi.   The mycelium is creamy white. While in the vessels, the mycelium produces short hyphal branches on which clusters of *Cephalosporium*-type conidia are formed (Fig. 107). In dying or dead trees the mycelium produces some *Cephalosporium*, but mostly *Graphium*-type spores on coremia developing on bark which is somewhat loose from the wood and in tunnels made in the bark by insects. The coremia consist of hyphae grouped into an erect, dark, solid stalk, and a colorless, flaring head to which the spores adhere, forming a sticky glistening, whitish at first, and later slightly yellowish, droplet.

The fungus is heterothallic and requires contact of two sexually compatible strains for sexual reproduction. Since, frequently, only one of the mating types is found in large areas in nature, sexual reproduction is extremely rare. In the U. S., for example, the fungus rarely reproduces sexually, but it does so rather frequently in Europe. When the two mating types do come in contact, perithecia develop. The perithecia are spherical and black, about 120 μm in diameter, and have a long (about 300 to 400 μm) neck. Perithecia form singly or in groups and in the same areas in the bark as the coremia.

Inside the perithecium many asci develop but as the asci mature, they disintegrate leaving the ascospores free in the perithecial cavity. The ascospores are discharged through the neck canal and accumulate in a sticky droplet.

FIGURE 106.
Dutch elm disease caused by *Ceratocystis ulmi*. (A) Infected elm tree most of which shows thin, yellowish foliage in early summer. (B) The same tree dead and defoliated in late summer. (C) Twig with wilted, rolled, brown leaves. (D) Diagonal and cross sections of elm twig and branch showing a ring of brown discoloration near the surface of the twig and deeper in the wood of the branch. (E) Beetle carriers of the Dutch elm disease. Side and top views of the European (a) and the native (b) elm bark beetles, (c) larva of (a). (F) Galleries beneath the bark of dead elm trees made by the female and the larvae of the European (upper left) and the native (lower right) elm bark beetles. At upper right and lower left of (F) can be seen the bark punctures or wounds on healthy elm made by these beetles. (Photos D–F courtesy Shade Tree Lab., Univ. of Mass.)

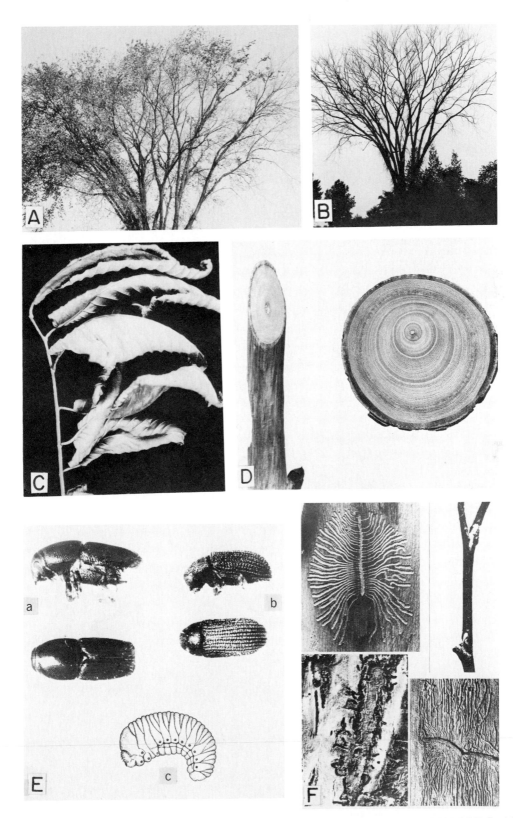

329

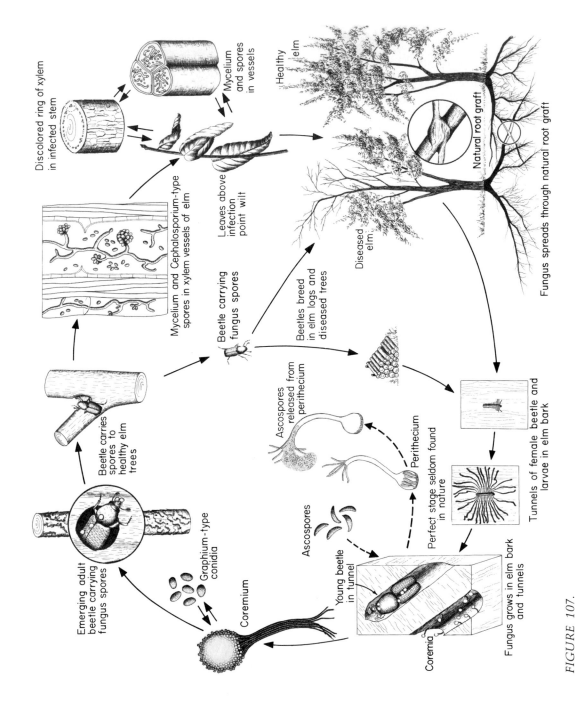

Discolored ring of xylem in infected stem

Mycelium and spores in vessels

Healthy elm

Natural root graft

Fungus spreads through natural root graft

Diseased elm

Mycelium and Cephalosporium-type spores in xylem vessels of elm

Leaves above infection point wilt

Beetle carrying fungus spores

Beetles breed in elm logs and diseased trees

Beetle carries spores to healthy elm trees

Ascospores released from perithecium

Perithecium

Perfect stage seldom found in nature

Tunnels of female beetle and larvae in elm bark

Emerging adult beetle carrying fungus spores

Graphium-type conidia

Coremium

Ascospores

Young beetle in tunnel

Coremia

Fungus grows in elm bark and tunnels

FIGURE 107.
Disease cycle of Dutch elm disease caused by *Ceratocystis ulmi.*

Development of disease.   Dutch elm disease is the result of an un-usual partnership between a fungus and an insect (Fig. 107). Although the fungus alone is responsible for the disease, the insect is the indispensable vector of the fungus, carrying the fungus spores from infected elm wood to healthy elm trees. The insects responsible for the spread of the disease are the European elm bark beetle (*Scolytus multistriatus*) and the native elm bark beetle (*Hylurgopinus rufipes*) (Fig. 106E).

The fungus overwinters in the bark of dying or dead elm trees and logs as mycelium and as spore-bearing coremia. Elm bark beetles prefer to lay their eggs in the intersurface between bark and wood of trees weakened or dying by drought, disease, etc. The adult female beetle tunnels through the bark and opens a gallery parallel with the grain of the wood with *Scolytus,* and at an angle or perpendicular with *Hylurgopinus.* The female lays eggs along the sides of the gallery, the eggs soon hatch, and the larvae open tunnels at right angles to the maternal gallery. If the tree was already infected with the fungus, the latter produces mycelium and sticky, *Graphium*-type spores in the beetle tunnels. When the adult beetles emerge, they carry thousands of fungus spores on and in their bodies. *Scolytus* beetles feed in the crotches of living, vigorous elm twigs. *Hylurgopinus* beetles feed on stems 5 to 30 cm in diameter (Fig. 106F). These beetles burrow into the bark and wood, the spores are deposited in the wounded tissues of the tree, germinate, and grow rapidly into the injured bark and the wood. When the fungus reaches the large xylem vessels of the spring wood, it may produce *Cephalosporium*-type spores which are carried up by the sap stream. These spores reproduce by yeastlike budding, germinate, and start new infections. The extent of symptoms in the crown is correlated with the extent of vascular invasion. In early stages of infection, the mycelium invades primarily the vessels, and only occasionally tracheids, fibers, and the surrounding parenchyma cells. General invasion of tissue begins at the terminal or extensive dieback phase of the disease, at which time there is also considerable intercellular growth of the fungus between the parenchyma cells. Gums and tyloses are produced in the larger vessels and sometimes isolated areas of the sapwood are blocked by a combination of gums, tyloses, and fungal growth. Infection also induces browning of the water-conducting vessels. Infected twigs and branches soon wilt and die.

Infections that take place in the spring or early summer result in invasion of the long vessels of the elm springwood through which the spores can be carried rapidly to all parts of the tree. If vascular invasion becomes general, death of the tree may occur within a few weeks. During later infections, vascular invasion is limited to the outer, shorter vessels of the summerwood in which they move only for short distances. As a result, late infections may produce only localized infections and seldom cause serious immediate damage to the tree, but may kill the tree the next year.

The elm bark beetles feed on living trees for only a few days and then fly back to dying or weakened elm wood in which they construct new galleries and lay eggs. There are usually two generations of beetles per season. In each generation the young adult goes from dead or weakened elm trees to living, vigorous ones on which it feeds, and then returns to the dead or weakened trees to lay its eggs. Therefore, once an insect becomes contaminated with fungus spores, it may carry them either to healthy or diseased wood, in both of which the fungus grows and multiplies and may contaminate all the offspring of the insect as well as any other insects that will visit the infected wood.

Control.   Great efforts have been made and are being continued to find resistant clones within the susceptible American elm species and in other species. Certain Asiatic species such as the Siberian and the Chinese elm are resistant to Dutch elm disease, but produce poor shade trees. Hybrids between various species have shown resistance in varying degrees and some of them look promising, but so far none of them have been planted widely or proven completely resistant.

For a long time, control of Dutch elm disease in the U. S. was attempted primarily through sanitation measures and through chemical control of the insect vectors of the fungus. Sanitation involves the removal and destruction of weakened or dead elm trees and elm logs, thus destroying the larvae contained in them or denying the insect and the fungus their overwintering habitat. Pruning out infected twigs and branches sometimes eliminates the disease. Control of the insect vector by chemicals involves spraying of the healthy elm trees while dormant and in the spring with methoxychlor, but has been only partially effective.

The most promising results for Dutch elm disease control in individual trees have been obtained with trunk or root injections of healthy or diseased elm trees with solubilized benomyl. This systemic fungicide, in various chemical formulations, has arrested the advance of the disease in infected trees and greatly reduced the appearance of new infections on treated healthy trees.

• *Oak Wilt*

It is the most serious disease of oak in the northcentral U. S. and it continues to spread. It is now present in the area from Minnesota to Pennsylvania to South Carolina to Arkansas, Oklahoma, Kansas, Nebraska and Iowa. It affects all oak species and varieties, although red and black oaks are much more susceptible to wilt than are the white oaks. Considering that oaks make up 50 to 80 percent of the woodland trees and lumber production in much of this and other areas and that oaks, especially after the destruction of elm trees by the Dutch elm disease, have become the predominant shade tree species in many residential areas, the potential impact of oak wilt on the economy and the environment of these areas is immeasurable.

The symptoms appear first at or near the top of the trees as wilting yellow, brown, or bronze-colored leaves changing color and becoming dry progressively from edge or tip towards the midrib and base of the leaf. Leaves in all stages of wilting and discoloration are shed off as the disease progresses. The symptoms are most noticeable during late spring or early summer. In red or black oaks the symptoms spread rapidly from the top to the rest of the tree and the trees may be killed in a few weeks. In white oaks the symptoms appear on scattered branches and progress more slowly, the trees surviving for one or more years but eventually being killed by the disease. Brown streaks appear in the outer sapwood of white oaks and less frequently in red oaks. In branches or trees killed by the wilt fungus the bark is often raised and cracked due to the pressure of fungus mats produced between the bark and the wood.

The fungus, *Ceratocystis fagacearum,* produces abundant conidia and less frequently ascospores in perithecia. Both types of spores are produced on mycelial mats formed beneath the bark of infected dying or dead trees. The mats may be up to 10×20 cm in size and their centers form cushion-shaped structures, called pressure pads, which develop sufficient pressure to split open the bark and expose the fruiting surface of the mat. Perithecia are produced when the two compatible strains of the fungus infect the same tree or are brought together by insects.

The fungus spreads locally through natural root grafts and through conidia or ascospores carried by beetles. The same sap feeding (Nitidulidae) and bark feeding (Pseudopityophthorus) beetles also carry both types of spores over long distances and create new infection centers in new areas. The modes of infection and spread of the fungus in the tree are similar to those of *C. ulmi.*

Control of the disease is difficult and still ineffective. Short-distance spread is controlled by cutting all roots of living trees around infected trees with a ditchdigger or killing a portion of the roots halfway between the infected and the surrounding healthy trees with Vapam placed in holes 40 cm deep and 30 cm apart. The infected tree is then cut and burned. Long-distance spread of the fungus by the beetles is difficult to control, the main practice available being the eradication of infected oaks before spores form so that the amount of inoculum can be reduced.

FUSARIUM WILTS

As mentioned above, they affect and cause severe losses on most vegetables and flowers, several field crops such as cotton and tobacco, plantation crops such as banana, plantain, coffee, and sugarcane, and a few shade trees. Fusarial wilts are greatly favored by warm soil conditions and in greenhouses. Since most Fusarial wilts have very similar disease cycles and development, only two such diseases, *Fusarium* wilt of tomato and *Fusarium* wilt of banana, also called Panama disease, will be described in some detail to illustrate this group of vascular wilts.

• *Fusarium Wilt of Tomato*

Fusarium wilt is one of the most prevalent and damaging diseases of tomato wherever tomatoes are grown intensively. The disease is most destructive in warm climates and warm, sandy soils of temperate regions. In the U. S. the disease is most severe in the central states and in the southern regions, whereas in the northern states it can become important only on greenhouse tomatoes.

Great losses may be caused by the disease, especially on susceptible varieties and under favorable weather conditions. *Fusarium* wilt damages plants by causing stunting of plants, which soon wilt and finally die. Occasionally entire fields of tomatoes are killed or severely damaged before a crop can be harvested. Generally, however, the disease does not cause serious losses unless soil and air temperatures are rather high during much of the season.

Symptoms.   The first symptoms appear as slight vein clearing on the outer, younger leaflets, followed by epinasty of the older leaves caused by drooping of the petioles. When plants are infected at the seedling stage, they usually wilt and die soon after appearance of the first symptoms. Older plants in the field may wilt and die suddenly if the infection is severe and if the weather is favorable for the pathogen. More commonly, however, in older plants vein clearing and leaf epinasty are followed by stunting of the plants, yellowing of the lower leaves, occasional forma- tion of adventitious roots, wilting of leaves and young stems, defoliation, marginal necrosis of the remaining leaves, and finally death of the plant (Fig. 108A). Often these symptoms appear on only one side of the stem and progress upward until the foliage is killed and the stem dies. While the plant is still living, no fungus mycelium or fructifications appear on its surface. Fruit may occasionally become infected, and then it rots and drops off without becoming spotted. Roots also become infected, and, after an initial period of stunting, the smaller side roots rot.

In cross sections of the stem near the base of the infected plant a brown ring is evident in the area of the vascular bundles, and the upward extent of the discoloration depends on the severity of the disease.

The pathogen: Fusarium oxysporum f. *lycopersici.*   The mycelium is colorless at first, but becomes cream-colored or pale yellow with age, and under certain conditions it produces a pale pink or somewhat purplish coloration. The fungus produces three kinds of asexual spores (Fig. 109). Microconidia, which are one- or two-celled, are usually the most fre- quently and abundantly produced spores under all conditions. They are the only spores produced by the fungus inside the vessels of infected host plants. Macroconidia are the typical *"Fusarium"* spores, 3- to 5-celled, gradually pointed, and curved toward both ends. They are common on the surface of plants killed by the pathogen and usually appear in sporodochialike groups. Chlamydospores are one- or two-celled, thick- walled, round spores produced terminally or intercalary on older mycelium or in macroconidia. All three types of spores are produced in the soil and in cultures of the fungus.

Development of disease.   The pathogen overwinters in the soil as mycelium and in all its spore forms but most commonly as chlamydo- spores (Fig. 109). It spreads over short distances by means of water and contaminated farm equipment, and over long distances primarily in in- fected transplants or in the soil carried with them. Usually, once an area becomes infested with *Fusarium*, it remains so indefinitely.

When healthy plants grow in contaminated soil, the germ tubes of spores or the mycelium penetrate root tips directly, or enter the roots through wounds or at the point of formation of lateral roots. The

FIGURE 108.
(A) *Fusarium* wilt of tomato. Some shoots are dead but others show only drooping of leaves. (B) *Verticillium* wilt of cotton. Healthy plant at left. (C) *Verticillium* wilt of peach tree. (D) Short brown streaks in branch of tree infected with *Verticillium*. (Photo A courtesy U.S.D.A. Photo B courtesy G. C. Papavizas.)

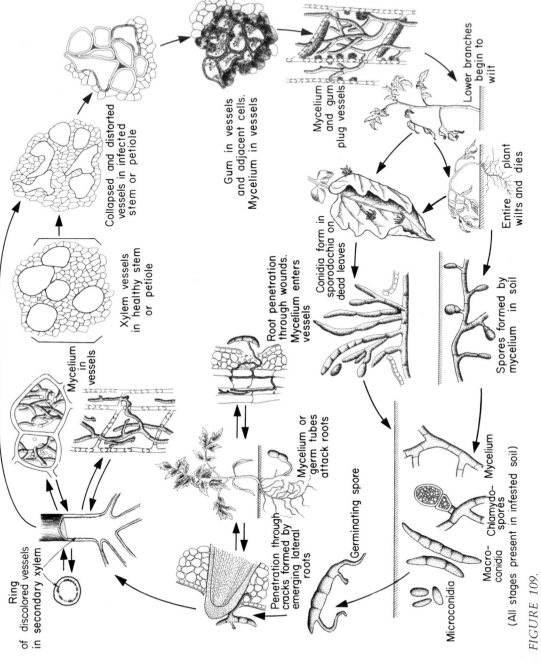

Collapsed and distorted vessels in infected stem or petiole

Gum in vessels and adjacent cells. Mycelium in vessels

Mycelium and gum plug vessels

Lower branches begin to wilt

Entire plant wilts and dies

Xylem vessels in healthy stem or petiole

Mycelium in vessels

Conidia form in sporodochia on dead leaves

Spores formed by mycelium in soil

Root penetration through wounds. Mycelium enters vessels

Ring of discolored vessels in secondary xylem

Mycelium or germ tubes attack roots

Penetration through cracks formed by emerging lateral roots

Germinating spore

Mycelium

Chlamydo-spores

Macro-conidia

Microconidia

(All stages present in infested soil)

FIGURE 109.
Disease cycle of *Fusarium* wilt of tomato caused by *Fusarium oxysporum* f. *lycopersici.*

mycelium advances through the root cortex intercellularly, and, when it reaches the xylem vessels, enters them through the pits. The mycelium then remains exclusively in the vessels and travels through them, mostly upward, toward the stem and crown of the plant. While in the vessels, the mycelium branches and produces microconidia which are detached and carried upward in the sap stream. The microconidia germinate at the point where their upward movement is stopped, the mycelium penetrates the upper wall of the vessel, and more microconidia are produced in the next vessel. The mycelium also advances laterally into the adjacent vessels penetrating them through the pits.

Presumably a combination of the processes discussed earlier—of vessel clogging by mycelium, spores, gells, gums, and tylosis; and crushing of the vessels by proliferating adjacent parenchyma cells—is responsible for the breakdown of the water economy of the infected plant. When the amount of water available to the leaves is below the required minimum for their function, the stomata close, the leaves wilt and finally die, followed in death by the rest of the plant. The fungus then invades the parenchymatous tissues of the plant extensively, reaches the surface of the dead tissues, and there it sporulates profusely. The spores may be disseminated to new plants or areas by wind, water, etc.

Occasionally, the fungus may reach the fruit of infected plants and penetrate or contaminate the seed. This happens primarily when the soil moisture is high and the temperature relatively low, conditions that allow plants to produce good yields although infected with the fungus. Usually, however, infected fruits decay and drop, and, even if harvested, infected seeds are so light that they are eliminated in the procedures of extraction and cleaning of the seed and therefore play little role in the spread of the fungus.

Control.   Use of tomato varieties resistant to the fungus is the only practical measure for controlling the disease in the field. Several such varieties are available today. Most of them are not completely resistant, but under conditions suboptimal for infection will produce good yields even in heavily infested soils. The fungus is so widespread and so persistent in soils that seedbed sterilization and crop rotation, although always sound practices, are of limited value. Soil sterilization is too expensive for field application, but should be always practiced for greenhouse-grown tomato plants. Use of healthy seed and transplants is of course mandatory, and hot-water treatment of seed suspected of being infected should precede planting.

- *Fusarium Wilt of Banana (Panama Disease)*

The disease is very destructive and widespread in most banana-growing countries, including Australia, the Oriental–African tropics, all of Central America, South America, and most tropical islands. The disease is still spreading and, where present, it is an absolute limiting factor in the growing of susceptible bananas.

The symptoms of *Fusarium* wilt of banana appear as a yellowing of the margins of usually the bottom leaves first. The yellowing spreads rapidly toward the midrib and to the upper, younger leaves, and all the leaves wilt, droop, wither and turn brown. In many cases the leaves wilt and die suddenly without prior yellowing. The leaf sheaths at the base of the pseudostem become hardened and develop longitudinal splits. Killed plants remain upright until the wind blows them down. Their rhizomes, which are not always killed throughout immediately, may produce young suckers which later become diseased, remain stunted, and may, for a while, produce abnormally developing and prematurely and irregularly ripening bunches of pithy, acid and yellow bananas, or they, too, wilt and die. Roots of diseased plants usually turn black and rot. When diseased pseudostems are cut open, their vascular bundles appear yellow, reddish or purple, the discoloration appearing first at the outermost sheath of the pseudostem and progressing toward the inner sheaths. The sap of infected plants has a characteristic pungent odor similar to that of a weak alcohol–formalin mixture.

The fungus, *Fusarium oxysporum* f. *cubense*, produces one- or two-celled microconidia in xylem vessels and both microconidia and four- to many-celled, sickle-shaped *Fusarium* macroconidia in sporodochia on leaves and petioles of wilted or dead plants. These spores are disseminated by wind or water, but are rather short lived and they seem to be of limited importance in spreading the disease. The fungus, however, also produces thick-walled chlamydospores that can survive in the soil for several decades and it can live as a saprophyte for considerable periods of time. Although spores may introduce the fungus into new areas, most of the spread of the disease seems to be through infected suckers planted in new areas, and through mycelium or spores carried in infected soil by irrigation or flood water or by mechanical cultivation.

The fungus enters banana plants through the roots and damaged rhizomes, particularly through the slender rootlets that grow on the main roots. Following infection of a rootlet, the fungus reaches the xylem vessels in the stele and grows rapidly backwards toward the rhizome. The fungus grows profusely in the vessels of the rhizome, kills many of its side buds and moves along the vascular bundles to other parts of the rhizome into the leaf sheaths of the pseudostem, then into the aerial, aboveground parts of the banana plant and finally into the leaves. During its growth, the fungus produces microconidia in the vessels that are carried upward and start new infection sites along the vascular bundles. Toxins, enzymes, and other substances secreted by the fungus and by the host in response to infection cause discoloration and clogging of the vessels that eventually contribute to the wilting and death of the plant.

Fungus survival, growth and sporulation are greater in light texture loam and sandy loam soils than in heavy clay soils and also greater in acid (pH 3.8 to 5.0) than in slightly alkaline soils (pH 7.2 to 8.0) that have a high calcium content. The mycelium and conidia survive for only short periods in very dry soil but chlamydospores survive for more than 30 years. As soil moisture increases to saturation, as during flooding, the fungus lives in the surface for at least 2 to 3 months, but it is destroyed at depths much below the surface. Flood-fallowing, i.e., keeping the soil flooded or soaked with water by diking around the field, for up to 18 months seems to kill the fungus in the soil but is not always possible or economically practical and does not always extend the average production life of the field long enough, before the pathogen destroys the crop again, to make it profitable.

No effective control of *Fusarium* wilt of banana is available yet. Some more or less resistant varieties, e.g., Valery and some other selections of the Cavendish group, the Pisang and Lacatan types of banana plants, show various degrees of resistance to Fusarium wilt. On the other hand, the popular Gros Michel bananas are very susceptible and can be grown only in areas still free of the pathogen. Use of disease-free stock in planting new areas is mandatory and immediate eradication of single diseased plants and isolation of the disease focus are absolutely necessary to stop the spread of the disease. Tools used in eradicating diseased plants must be disinfected and the soil of the area must be fumigated or flooded and must remain free of bananas for many years. Liming of the soil also helps keep pathogen populations low. So far, no chemical control of the Fusarium wilt is possible.

VERTICILLIUM WILTS

They are worldwide in distribution but most important in areas of the temperate zones. *Verticillium* attacks more than 200 species of plants, most of them vegetables, e.g., tomato, eggplant, pepper, cantaloupe, and watermelon, flowers, e.g., chrysanthemum, aster, and dahlia, fruit trees, e.g., apricot, cherry, and peach; also strawberries, raspberries and roses, field crops, e.g., cotton, potato, alfalfa, peanuts, and mint, and shade and forest trees, e.g., maple and elm.

The symptoms of *Verticillium* wilts are almost identical to those of *Fusarium* wilts and, on hosts affected by both, the two pathogens cannot be distinguished except by laboratory examinations. In many hosts and most areas, however, *Verticillium* induces wilt at lower temperatures than *Fusarium*, the symptoms develop slower and often appear only on the lower or outer part of the plant or on only a few of its branches. In some hosts, e.g., cotton, *Verticillium* wilt develops primarily in seedlings which usually die shortly after infection while late infections cause epinasty of the upper leaves followed by the appearance in leaves of irregular chlorotic patches which become necrotic. Older plants infected with *Verticillium* are usually stunted in various degrees and their vascular tissues show characteristic discoloration. In many hosts, *Verticillium* infection results in defoliation, gradual wilting and death of successive branches or an abrupt collapse and death of the entire plant (Fig. 108, B–D).

Initial outbreaks of *Verticillium* wilt in a field are typically mild and local. In subsequent years, attacks become severe and widespread until the crop has to be discontinued or is replaced with resistant varieties. The increasing severity of the disease from year to year is brought on by a greater build up in inoculum potential, by the appearance of strains of the fungus more virulent than the original, or by both.

The fungus, *Verticillium albo-atrum* (or *V. dahliae*) produces conidia and microsclerotia either of which can overwinter, as can mycelium, within perennial hosts, propagative organs or plant debris. *Verticillium* penetrates host plants usually through wounds in the roots. The fungus is spread by contaminated seed, by vegetative cuttings and tubers, by scions and buds, by wind and surface ground water and by the soil itself, which may contain up to 100 or more microsclerotia per gram. Many fields have become contaminated with *Verticillium* for the first time by planting infected potato tubers or other crops and it is known that solanaceous crops such as potato, eggplant, and tomato increase the inoculum level in the soil. However, *Verticillium* is often found in uncultivated areas which indicates that the fungus is native to the soils and can attack susceptible crops as soon as they are planted.

Control of *Verticillium* wilts depends on use of disease-free plants in disease-free soil, use of resistant varieties, and by avoiding the planting of susceptible crops where solanaceous crops have been grown repeatedly. Soil fumigation can be profitable when used to protect high value crops but it is too expensive on large areas.

SELECTED REFERENCES

Armstrong, G. M., and J. K. Armstrong. 1975. Reflections on the wilt Fusaria. *Ann. Rev. Phytopathol.* **13**:95–103.

Ashworth, L. J., Jr., *et al.* 1972. Assessment of microsclerotia of *Verticillium albo-atrum* in field soils. *Phytopathology* **62**:715–719; **62**:901–903.

Banfield, W. M. 1941. Distribution by the sap stream of spores of three fungi that induce vascular wilt disease of elm. *J. Agr. Res.* **62**:637–681.

Banfield, W. M. 1968. Dutch elm disease recurrence and recovery in American elm. *Phytopathol. Z.* **62**:21–60.

Beckman, C. H. 1964. Host responses to vascular infection. *Ann. Rev. Phytopathol.* **2**:231–252.

Bohn, G. W., and C. M. Tucker. 1940. Studies on the *Fusarium* wilt of tomato. *Mo. Agr. Exp. Sta. Res. Bull.* **311**: 88 p.

Brandes, E. W. 1919. Banana wilt. *Phytopathology* **9**:339–389.

Brown, M. F., and T. D. Wyllie. 1970. Ultrastructure of microsclerotia of *Verticillium albo-atrum*. *Phytopathology* **60**:538–542.

Buchenauer, H., and D. C. Erwin. 1972. Control of *Verticillium* wilt of cotton by spraying with acidic solutions of benomyl, methyl 2-benzimidazole carbamate, and thiabendazole. *Phytopathol. Z.* **75**:124–139.

Chambers, Henrietta L., and M. E. Corden. 1963. Semeiography of *Fusarium* wilt of tomato. *Phytopathology* **53**:1006–1010.

Dimond, A. E. 1970. Biophysics and biochemistry of the vascular wilt syndrome. *Ann. Rev. Phytopathol.* **8**:301–322.

Engelhard, A.. W. 1957. Host index of *Verticillium albo-atrum* (including *Verticillium dahliae*). *Plant Dis. Reptr. Suppl.* **244**:23–49.

Engelhard, A. W., and S. S. Woltz. 1973. Fusarium wilt of chrysanthemum: complete control of symptoms with an integrated fungicide–lime–nitrate regime. *Phytopathology* **63**:1256–1259.

Fletcher, J. T., and J. A. Martin. 1972. Spread and control of Fusarium wilt of carnations. *Plant Pathol.* **21**:182–187.

Forsberg, J. L. 1975. Diseases of ornamental plants. *Univ. of Ill., Coll. of Agr. Special Publ.* **No. 3 Rev.** 220 p.

Frank, J. A., R. E. Webb, and D. R. Wilson. 1975. The effect of inoculum levels on field evaluations of potatoes for Verticillium wilt resistance. *Phytopathology* **65**:225–228.

Holmes, F. W. 1965. Virulence to *Ceratocystis ulmi*. *Neth. J. Plant Pathol.* **71**:97–112.

Jones, J. P., and P. Crill. 1973. The effect of *Verticillium* wilt on resistant, tolerant and susceptible tomato varieties. *Plant Dis. Reptr.* **57**:122–124.

Jordan, V. W. L. 1974. *Verticillium* wilt of strawberry: cultivar reaction and effect on runner health and production. *Plant Pathol.* **23**:8–13.

Kendrick, J. B. 1944. Fruit invasion and seed carriage of tonato Fusarium wilt. *Phytopathology* **34**:1005–1006.

Kondo, E. S., and G. D. Huntley. 1973. Root-injection field trials of MBC-phosphate in 1972 for Dutch elm disease control. *Can. Forest Service Info. Bull.* **O-X-182**: 17 p.

Krause, C. R., and C. L. Wilson. 1972. Fine structure of *Ceratocystis ulmi* in elm wood. *Phytopathology* **62**:1253–1256.

Magie, R. O., and G. J. Wilfret. 1974. Tolerance of *Fusarium oxysporum* f. sp. *gladioli* to benzimidazole fungicides. *Plant Dis. Reptr.* **58**:256–259.

Merrill, W. 1967. The oak wilt epidemics in Pennsylvania and West Virginia: an analysis. *Phytopathology* **57**:1206–1210.

Nielsen, L. W. 1948. *Verticillium* wilt of potatoes in Idaho. *Id. Agr. Exp. Sta. Res. Bull.* **13**: 23 p.

Papavizas, G. C. 1974. The relation of soil microorganisms to soilborne plant pathogens. *Virg. Polytech. Inst. Southern Coop. Series Bull.* **183**: 98 p.

Parker, K. G. 1959. *Verticillium* hadromycosis of deciduous tree fruits. *Plant Dis. Reptr. Suppl.* **255**:39–61.

Pegg, G. F. 1974. *Verticillium* diseases. *Rev. Plant Pathol.* **53**:157–182.

Pomerleau, R. 1970. Pathological anatomy of the Dutch elm disease. Distribution and development of *Ceratocystis ulmi* in elm tissues. *Can. J. Bot.* **48**:2043–2057.

Rudolph, B. A. 1931. *Verticillium* hadromycosis. *Hilgardia* **5**:201–361.

Stover, R. H. 1962. Fusarial wilt (Panama disease) of bananas and other *Musa* species. *Commonwealth Mycol. Instit. Phytopathol. Paper* **4**: 117 p.

Strong, M. C. 1946. The effect of soil moisture and temperature on Fusarium wilt of tomato. *Phytopathology* **36**:218–225.

Tainter, F. H., and E. D. Gubler. 1973. Natural biological control of oak wilt in Arkansas. *Phytopathology* **63**:1027–1034.

Toussoun, T. A. 1975. *Fursarium*-suppressive soils, *in* "Biology and control of Soil-Borne Plant Pathogens," G. W. Bruehl (Ed.). The Amer. Phytopathol. Soc., St. Paul, Minn, pp. 145–151.

True, R. P., H. L. Barnett, C. K. Dorsey, and J. G. Leach. 1960. Oak wilt in West Virginia. *W. V. Agr. Exp. Sta. Bull* **448T**: 119 p.

Van Alfen, N. K. and G. S. Walton. 1974. Pressure injection of benomyl and methyl-2-benzimidazolecarbamate hydrochloride for control of Dutch elm disease. *Phytopathology* **64**:1231–1234.

Walker, J. C., and R. E. Foster. 1946. Plant nutrition in relation to disease development. III. Fusarium wilt of tomato. *Am. J. Bot.* **33**:259–264.

Wellman, F. L. 1972. "Tropical American plant diseases." The Scarecrow Press, Inc., Metuchen, N. J., pp. 342–393.

Weststeijn, G. 1973. Soil sterilization and glasshouse disinfection to control *Fusarium oxysporum* f. *lycopersici* in tomatoes in the Netherlands. . *Neth. J. Plant Pathol.* **79**:36–40.

Whitten, R. R., and R. U. Swingle. 1964. The Dutch elm disease and its control. *U.S. Dept. Agr. Bull.* **193**: 12 p.

ROOT AND STEM
ROTS CAUSED BY ASCOMYCETES
AND IMPERFECT FUNGI

Several Ascomycetous fungi attack primarily the roots and lower stem of plants, particularly of cereals. The most important of these are:

Cochliobolus, causing root and foot rot and also blight of cereals and grasses
Gibberella, causing seedling blight and foot or stalk rot of corn and small grains
Ophiobolus, (Gaeumannomyces) causing the take-all and whiteheads disease of cereals

Another fungus, *Sclerotinia,* which has recently been renamed *Whetzelinia,* causes crown and root rot of turf grasses, a disease known as dollar spot, and also rots and blights of nearly all kinds of succulent plants, primarily flowers and vegetables.

In few, if any, of the diseases caused by the first two fungi, i.e., *Cochliobolus* and *Gibberella,* does the ascigerous stage play a role of any consequence. In almost all cases, the diseases are caused strictly by the asexual stages of these fungi which are primarily species of *Helminthosporium* for *Cochliobolus* and species of *Fusarium* for *Gibberella.* Therefore the diseases caused by *Cochliobolus* are discussed as diseases caused by *Helminthosporium* (see pages 270–276) and all but one *Gibberella* disease are discussed as diseases caused by *Fusarium.*

Among the Imperfect Fungi causing root and stem rots are several widely distributed and extremely destructive plant pathogens:

Dipodia maydis, causing Diplodia stalk and ear rot of corn, collar rot of peanuts, etc.
Fusarium, causing root rot of bean, asparagus, onion, foot rot of squash, dry rot of potatoes, basal rot of iris and lily, stem rot of carnation and chrysanthemum, corm rot of gladiolus, and seed rot, damping off, and seedling blights of these and numerous other plants
Phoma, causing black leg of crucifers (Fig. 100), root and crown rot of celery and delphinium, heart rot of beet, stem blight and fruit rot of tomato and pepper, etc.
Phymatotrichum omnivorum, causing the Texas root rot of fruit and shade trees, ornamental shrubs, cotton, alfalfa, most flowers and vegetables, and many weeds
Thielaviopsis, causing black root rot and damping off of many vegetables and flowers, particularly bean, beet, carrot, celery, pansy, pea, poinsettia, squash, sweet pea, sweet potato, tomato, and watermelon, and of many field crops including cotton, cowpeas, flax, peanuts, clover, soybean, tobacco, etc.

As a general rule, the root and stem rot diseases caused by the above and by other Ascomycetes and Imperfect Fungi appear on the affected plant organs at first as water-soaked areas that later turn brown to black, although in some diseases they are frequently covered by white fungal mycelium. The roots, stems, etc., are killed more or less rapidly and the entire plant grows poorly or is killed. The fungi that cause these diseases live, grow and multiply in the soil as nonobligate parasites, usually in

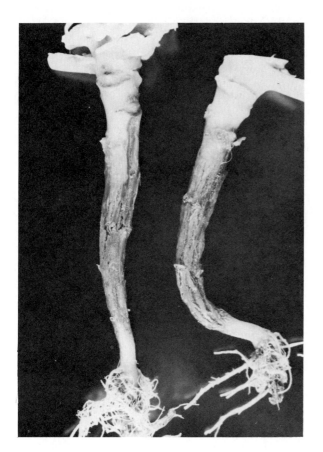

FIGURE 110.
Black-leg disease of cabbage caused by *Phoma lingam*.

association with dead organic matter, and they are favored by high soil moisture and high relative humidity in the air. Most of them produce conidia and some produce ascospores occasionally or regularly. Several produce sclerotia. In all of the above fungi the fungus can overwinter as mycelium in infected plant tissues or debris, as sclerotia, or as any of the kinds of spores the fungus may produce. These same stages also serve as inoculum that can be spread and start new infections.

GIBBERELLA DISEASES

- *Gibberella Stalk Rot, Ear Rot,*
  Root Rot, and Seedling Blight of Corn

The *Gibberella* diseases of corn are worldwide in distribution and cause serious losses. The most important phases of the diseases are stalk rot and ear rot.

In stalk rot, lower internodes become soft and appear tan or brown on the outside while internally they may appear pink or reddish (Fig. 111, A,

FIGURE 111.
(A–C) *Gibberella* stalk and ear rot of corn caused by *Gibberella zeae.* External (A) and internal (B) symptoms. (C) Ear rot. (D) A corn field destroyed by a combination of *Gibberella* and *Diplodia* stalk rots. (E–G) *Diplodia* stalk and ear rot of corn caused by *Diplodia zeae.* (E) External discoloration of stalk (left), stalk with pycnidia appearing as black dots (middle), and internal disorganization of tissues in infected stalk (right). (F) Corn variety resistant (left) and susceptible to *Diplodia* stalk rot, the latter prematurely dead and broken. (G) *Diplodia* ear rot. (Photos courtesy Illinois Agr. Expt. Sta.)

B). The pith disintegrates leaving only the vascular bundles intact. The rot may also affect the roots. Stalk rot leads to a dull gray appearance of the leaves, premature death and stalk breakage (Fig. 111D). Small, black, round perithecia are often produced superficially on rotting stalks.

In ear rot, often called red ear rot, ears develop a reddish mold that often begins at the tip of the ear (Fig. 111C). If infection occurs early, the ears may rot completely and a pinkish to reddish mold grows between the ears and the tightly adhering husks. Perithecia may form on husks and ear shanks. Corn ears infected with *Gibberella* are toxic to humans and certain animals such as hogs.

Gibberella is only one of many fungi causing blight of corn seedlings and may be carried on or in infected seed or it may attack the seed and seedling from the soil. In either case, the germinating seed may be attacked and killed before the seedling emerges from the soil, or after emergence, in which case the seedling may be killed or become dwarfed and chlorotic and later die. Light brown to dark colored lesions are usually evident on the tap and lateral roots and in the lower internode.

Two species of *Gibberella, G. zeae* and *G. fujikuroi* (= *Fusarium moniliforme*) are primarily responsible for the symptoms observed on corn and for those that will be described on small grains below. Both fungi produce ascospores in perithecia and *Fusarium*-type conidia. Perithecia are rather rare in *G. fujikuroi*. The fungi overwinter as perithecia or as mycelium or chlamydospores in infected plant debris, particularly corn stalks. In the spring, during wet, warm conditions, mature ascospores are released and are carried by wind to corn stalks or ears which they penetrate directly or through wounds and cause infections. Conidia may also be produced on infected corn debris, but they are more commonly produced on infected plant parts in moist, warm weather and serve as the secondary inoculum. The diseases are favored by dry weather early in the season and wet weather near or after silking. Also, high plant density, high nitrogen and low potassium in the plant and early maturity of hybrids makes them more susceptible to the diseases.

Control of *Gibberella* diseases of corn depends on the use of resistant varieties, balanced nitrogen and potassium fertilization and lower plant density in the field.

- *Gibberella Scab, Seedling Blight, and Foot Rot of Small Grains*

Also worldwide. Caused by the same fungi causing the diseases in corn and perhaps some additional species of *Gibberella*. Losses may be as high as 50 percent of the yield. In some areas where corn is grown extensively, this disease makes wheat and barley production unfeasible.

Scab or head blight causes severe damage to wheat and the other cereals, especially in areas with high temperature and relative humidity during the heading and blossoming period. Infected spikelets first appear water soaked, then lose their chlorophyll and become straw colored. In warm, humid weather, pinkish-red mycelium and conidia develop abundantly in the infected spikelets and the infection spreads to adjacent spikelets or through the entire head. Purplish perithecia may also develop on the infected floral bracts. Infected kernels become shriveled and discolored with a white, pink, or light brown scaly appearance as a result of the mycelial outgrowths from the pericarp. As with corn, infected kernels of cereals are also toxic to humans, hogs and other animals, containing a substance acting as an emetic.

Seedling blight appears as a light brown to reddish-brown cortical rot and blight either before or after emergence of the seedling above the soil line. In older plants, a foot rot develops appearing as a browning of the basal leaf sheaths or as a pronounced rotting of the basal part of the plant around soil level and for some distance above the soil line.

Control measures against small grain diseases caused by *Gibberella* are the same as those described for the same diseases of corn.

- *Diplodia Stalk and Ear Rot of Corn*

The disease is present throughout the U.S. and until recent years was the most common stalk rot and ear rot in the Corn Belt. Its importance has been reduced in recent years by the use of resistant varieties. It has also been reported from Africa, Australia, the Philippines, and Romania. The disease is most severe on plants affected by foliar pathogens, insects, imbalanced fertilization or dry weather in the early growing season followed by 2 to 3 weeks of wet weather at or after silking.

Symptoms usually appear several weeks after silking. The leaves of infected plants suddenly become grayish green, wilt and become dry as though killed by frost. The lower parts of the stalk become tan or brownish, soft and weakened, and break easily in the wind and rain (Fig. 111F). If the stalk is cut open, the pith appears disclolored and disintegrated and only the vascular bundles are left intact. Minute dark brown to black pycnidia form late in the season just below the epidermis of the lower internode of the stalk (Fig. 111E). White mycelial growth may also be present on the dead stalks.

Ears infected early show bleached or straw-colored husks and the entire ear may become grayish-brown, shrunken, and completely rotted (Fig. 111G). Such ears remain light-weight, upright, and because mycelium grows between the ear and the husks and between the husks, the latter remain stuck tightly together. Black pycnidia may be present at the base of the husks and the sides of the kernels. Ears infected later in the season appear normal on the outside, but when ears are broken, they show a white mold growing between the kernels and the tips of the kernels are discolored. Ear infection, which occurs most commonly in the first 2 to 3 weeks from silking, usually begins at the base of the ear and advances toward the tip. In some cases premature germination of kernels takes place.

The fungus, *Diplodia maydis,* produces only conidia in small, black, flask-shaped pycnidia. The conidia, however, may be of two types. The most common are long, two-celled, olive brown, 6×28 μm spores, while the less common are long, threadlike, narrow, colorless, 1.5×30 μm spores.

Diplodia overwinters as conidia contained in pycnidia produced on infected corn stalk debris, and as mycelium or conidia on seed. When the fungus is carried on the seed, it may cause seedling blight, killing seedlings at any time after germination. Overwintering pycnidia exude sticky masses of conidia when the weather becomes warm and moist. The conidia may be carried to stalks or ears of corn plants by splashing rain or by wind and insects and there they cause infection. The fungus usually attacks corn plants at the crown and spreads for some distance into the stalk and the roots. Although it may attack nodes between the crown and the ear, infection of the ear is almost never by fungus growing through the stalk but rather by spores landing directly on the ear.

Control of *Diplodia* stalk and ear rot depends primarily on the use of resistant hybrids, and also use of disease-free seed, use of balanced nitrogen and potassium fertilization, wider spacing of plants, early harvesting, and proper storage of ears and shelled corn.

SCLEROTINIA DISEASES

Fungi of the genus *Sclerotinia*, which has recently been renamed *Whetzelinia*, especially *S. sclerotiorum*, cause destructive diseases of numerous succulent plants, particularly vegetables and flowers and some shrubs. Another species, *S. homeocarpa*, causes a destructive disease of turf grasses. Sclerotinia diseases are worldwide in distribution and affect plants in all stages of growth including seedlings, mature plants and harvested products in transit and storage.

- *Sclerotinia Diseases of Vegetables and Flowers*

The symptoms caused by *Sclerotinia* vary somewhat with the host or host part affected and with the environmental conditions. The *Sclerotinia* diseases are known under a variety of names and the most common of these, along with some of the host plants most seriously affected, are the following: cottony rot, white mold, or watery soft rot of bean, cabbage, carrot, eggplant, citrus, peanut, potato, stock, tobacco, etc.; stem rot and timber rot of cucumber, squash, bean, artichoke, asparagus, chrysanthemum, dahlia, delphinium, peony, potato, tomato, soybean, sweet potato, etc.; drop of lettuce, broad bean, beet, cabbage, etc.; damping off of celery, lettuce, etc.; crown rot or wilt of columbine, snapdragon, etc.; blossom blight of narcissus, camellia, etc.; pink joint of red pepper, stem canker of hollyhock, root and crown rot of clover, and others.

The most obvious and typical early symptom of Sclerotinia diseases is the appearance on the infected plant of a white fluffy mycelial growth in which soon afterwards develop large, compact resting bodies or sclerotia (Fig. 112, A–D). The sclerotia are white at first but later become black and hard on the outside and may vary is size from 2 to 10 or more mm in diameter, although they are usually more flattened and elongated rather than spherical.

Stems of infected succulent, herbaceous plants at first develop pale or dark brown lesions at their base. The lesions are often quickly covered by white cottony patches of fungal mycelium. In the early stages of lesion development in the stem, the foliage may show little sign of attack and infected plants are easily overlooked until the fungus grows completely through the stem and the stem rots. Then the foliage above the lesion wilts and dies more or less quickly. In some cases the infection may begin on a leaf and then move into the stem through the leaf. The sclerotia of

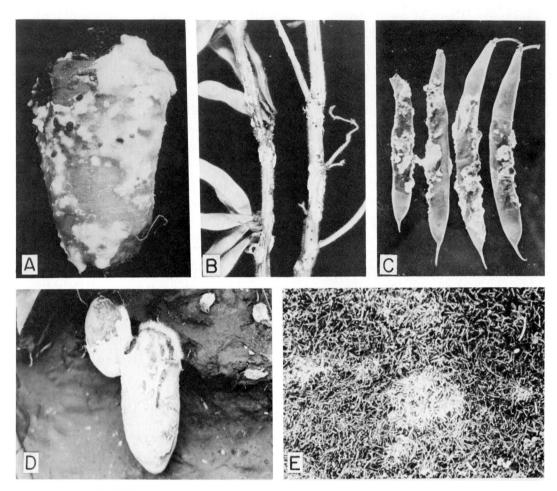

FIGURE 112.
(A–D) *Sclerotinia* diseases of vegetables and field crops caused by *Sclerotinia sclerotiorum*. (A) Cottony rot of carrot. (B) Stem rot of soybean. (C) Watery soft rot of bean. (D) Watery soft rot of cucumber. The black bodies in photos A–C are sclerotia of the fungus. (E) Dollar spot disease of turfgrasses caused by *S. homeocarpa*. (Photos A, C, E courtesy U.S.D.A.)

the fungus may be formed either internally in the pith of the stem, giving no outward signs of their presence there, or they may be formed on the outside of the stem where they are quite apparent.

Leaves and petioles of plants such as lettuce, celery, and beets suddenly collapse and die as the fungus infects the base of the stem and the lower leaves. Rapidly the fungus invades and spreads through the stem, and the entire plant dies and collapses, each leaf dropping downward until it rests on the one below. Mycelium and sclerotia usually appear on the lower surface of the outer leaves, but under moist conditions the fungus invades the plant completely and causes it to rot, producing a white, fluffy, mycelial growth over the entire plant. If dry weather follows infection, the fungus forms cankers in the stem that kill the plant without a soft rot. Attack of celery produces a characteristic pink or reddish-brown, water-soaked area at the base of the affected petioles that is often covered by the white mycelium and the rot may spread through the stalks causing the collapse of the whole plant.

Fleshy storage organs, such as carrots, infected by *Sclerotinia* develop a white, cottony growth on their surface whether they are still in the field or in storage. Black sclerotia are formed externally (Fig. 112A), invaded tissues appear darker than healthy ones and become soft and watery. If the disease develops after harvest in the storage house, the rot spreads to adjacent roots or whatever the storage organs are and produces pockets of rotted organs or all the organs in the crate may become infected and collapse producing a watery soft rot, covered by fungus growth.

Fleshy fruits, such as cucumber, squash, and eggplant, and seed pods of bean, etc. are also attacked by *Sclerotinia* either through their closest point to the ground, at the point of their contact with the ground or through their senescent flower parts. The fungus causes a wet rot that spreads from the tip of the fruit or pod to the rest of the organ which eventually becomes completely rotted and disintegrated (Fig. 112D). The white fungal mycelium and the black sclerotia can usually be seen both externally and within the affected pods and fruits.

Flower infection is important primarily in camellias and narcissus. Few to many small, watery, light-brown spots appear on the petals. The spots may enlarge, coalesce and involve the entire petal, and eventually the entire flower becomes dark brown and drops, but disintegration of the flowers occurs only after they have fallen and in wet weather, when the fungus produces abundant mycelium and sclerotia.

The fungus overwinters as mycelium in dead or living plants and as sclerotia on or within infected tissues or as sclerotia that have fallen on the ground (Fig. 113). In the spring or early summer, the sclerotia germinate and produce one to many slender stalks terminating at a small, 5 to 15 mm in diameter, disk- or cup-shaped apothecium in which asci and ascospores are produced. Large numbers of ascospores are discharged from the apothecia into the air over a period of 2 to 3 weeks. The ascospores are blown away and if they land on susceptible plant parts, they germinate and cause infection. Very frequently the sclerotia cause infection by producing mycelial strands which attack and infect young plant stems directly. Under moist conditions the latter method of infection is probably more common than the one by ascospores.

Control of *Sclerotinia* diseases depends on a number of cultural practices and on chemical sprays. In the greenhouse, soil sterilization with steam eliminates the pathogen. In the field, chemical soil sterilants have been rather ineffective in controlling the disease. Since the disease is favored by high soil moisture and high air humidity and affects many cultivated and wild plants, susceptible crops should be planted only in well-drained soils, the plants should not be spaced too close together for air drainage, and the soil should be kept free of weeds between crops. If the disease has become severe on susceptible crops, infected plants should be pulled and burned to either prevent the fungus from forming sclerotia or to remove from the field as many of the sclerotia as possible. Since sclerotia remain viable in the soil for at least 3 years and since they

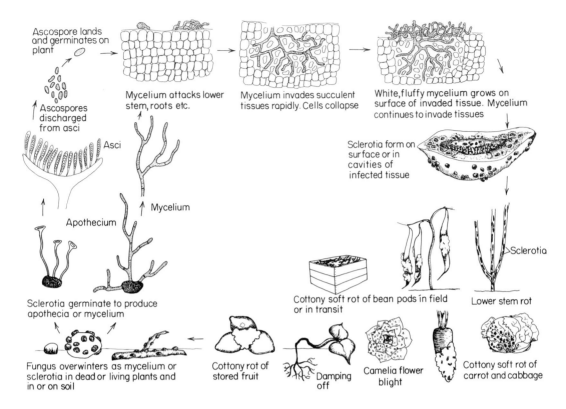

FIGURE 113.
Development and symptoms of diseases of vegetables and flowers caused by *Sclerotinia sclerotiorum.*

do not all germinate or die out at the same time, infested fields should be planted to nonsusceptible crops such as corn and small grains for at least 3 years before susceptible crops are planted again. In several crops, good control of the *Sclerotinia* disease has been obtained by spraying the plants with benomyl before and during their stage of susceptibility to the pathogen.

- *Sclerotinia Dollar Spot of Turf Grasses*

The disease is common and severe on various turf grasses throughout the United States and probably elsewhere. The overall symptoms on closely mowed turf appear as circular, brown to straw-colored, somewhat sunken spots the size of a silver dollar, i.e., about 5 to 7 cm in diameter (Fig. 112 E). In higher mowed turf grasses, the spots are larger, about 10 to 15 cm or more in diameter and are more irregular in outline. In either case, spots may be numerous or they may enlarge and coalesce to form large patches of irregular sunken areas of straw-colored dead turf. When dew is present on the grass in the morning, white, fluffy, mycelial growth can be seen on the affected leaves, but the sun and wind cause most surface mycelium to dry and disappear. When individual grass blades affected by *Sclerotinia* are examined, especially those at the margins of the enlarging spots, they show characteristic light tan lesions with a reddish-brown border that are about 2 to 3 cm long. On fine-leafed turf grasses, the lesions usually extend across the blades and cause girdling of the leaf and death of the distal leaf part resulting in a leaf blight. On coarser grasses, the lesions may occur only along the leaf margins causing only portions of leaves to become necrotic. Although the fungus attacks only the leaves, it secretes in them a toxin that is translocated to the roots and makes them cease elongation and root hair formation and finally makes them degenerate, become necrotic, and wither.

The fungus, *Sclerotinia homeocarpa,* produces small to large paper-thin, flaky, hard, black sclerotia in or on the soil. The fungus overwinters as sclerotia or as mycelium in the crowns and roots of infected plants. In the spring the sclerotia germinate by producing mycelium, which along with the dormant mycelium is responsible for the majority of new infections. The sclerotia may also germinate by producing apothecia containing asci and ascospores, and the mycelium may form conidia but both ascospores and conidia appear to be of minor importance in the spread of the disease. Long-distance spread of the fungus is through transfer of sclerotia or mycelium in infected clippings by mowing or other equipment, while locally the fungus spreads to new plants in a radial fashion by means of mycelial growth. The fungus becomes active when temperature reaches 10 to 15°C but maximum disease development occurs when plant surfaces are wet and the temperature is between 21 and 26°C. The fungus then enters the plant through cut leaf tips and stomata causing the invaded tissue to appear water soaked and dark before it dries up and becomes light tan. If the weather turns dry or fungicides are applied, the advance of the fungus is limited to the leaf blade and the plants may recover quickly, but if the disease is left unchecked, the fungus advances into the crown and kills the infected plants and recovery may be very slow. Plants kept in good vigor by balanced fertilization and proper irrigation are more resistant to dollar spot infection and, if infected, recover more rapidly.

Control of dollar spot depends on keeping the plants in good growth by proper fertilization and irrigation, good soil drainage, pruning of nearby shrubs or trees to increase air circulation and avoiding over watering, especially frequent sprinkling in late afternoon or evening. Kinds of turf grasses that are more resistant to certain strains of *Sclerotinia* should be selected when establishing new lawns. If dollar spot is present, however, the use of fungicides such as benomyl, thiram, cycloheximide, chlorothalonil, dyrene, thiabendazole, cadmium fungicides, or thiophanate methyl is necessary to bring the disease under control. Applications should be made every 3 to 4 days until the plants begin to recover and every 7 to 10 or more days afterwards during moist weather during the growing season when day temperatures average 16 to 26°C. In areas treated with some of the above chemicals, however, the fungus has developed strains resistant to these chemicals, and therefore the use of an integrated control program alternating contact and systemic fungicides rather than using a single fungicide is recommended for the control of dollar spot.

- *Fusarium Root and Stem Rots*

Several *Fusarium* species cause, instead of vascular wilts, rotting of seeds and seedlings (damping off), rotting of roots, lower stems and crowns, and rots of corms, bulbs, tubers, etc. (Fig. 114, A, B). The plants affected belong to widely separated families and may be vegetables, flowers, turf grasses, field crops, and weeds. These diseases are worldwide in distribution and cause severe losses by reducing stands and the growth and yield of infected plants.

In root rots, as those of bean, peanut, soybean, asparagus, and turf grasses, tap roots of young plants at first show a slightly reddish discoloration; this later becomes darker red to brown and larger, more or less covering the tap root and the stem below the soil line without a definite margin, or appearing as streaks extending up to the soil line. Longitudinal fissures appear along the main root while the small lateral roots are killed. Plant growth is generally retarded and in dry weather the leaves may turn yellow and even fall off. Sometimes, infected plants develop secondary roots and a large number of rootlets just below the soil line and these, under favorable moisture conditions, may be sufficient to carry the plant to maturity and to production of a fairly good crop. In many cases, however, infected plants decline and die with or without wilt symptoms.

In stem rots, such as those of carnation, chrysanthemum, etc,. infected plants wilt and die from a stem rot at the base of the plant. Lesions develop on the stem at or below the soil line and their edges often have pink or red discolorations. The lesions develop inward from the outside and usually there is no internal stem discoloration, but in some plants a brown discoloration may extend in the wood of the stem for a considerable distance above the ground. In older plants the roots may have also rotted and may have sloughed off.

Rots of bulbs, corms, tubers, etc. by species of *Fusarium* can occur in the field and in storage. They are common on plants such as onion, iris, lily, and gladiolus. The rot can start at uninjured sides of bulbs, corms, or tubers, but it often starts at wounds, at the base of these organs, at injured or diseased roots or foliage, or through the cuts formed on such tissues during harvest. Invaded bulbs and corms may or may not show outward symptoms although usually the basal plate and fleshy scales, as well as the roots, are brown to black, sunken and decaying and often containing mats of mycelium. The rot is generally dry and firm. The foliage turns yellow, purple or brown and dies prematurely. Tubers usually develop small brown patches that soon enlarge, become sunken, and show concentric wrinkles that contain cavities lined with white mycelium. Eventually large parts of the tuber or entire tubers are destroyed and become hard and mummified, unless it is humid and then they are invaded by soft rotting bacteria .

In the *Fusarium* blight of turf grasses, the fungus attacks the leaves, crown and roots of plants and results at first in small, scattered patches, 5 to 15 cm in diameter, of reddish-brown to tan and finally straw-colored dead plants. The patches enlarge as streaks, crescents, or circular areas up to one meter or more in diameter, and if such patches are numerous they

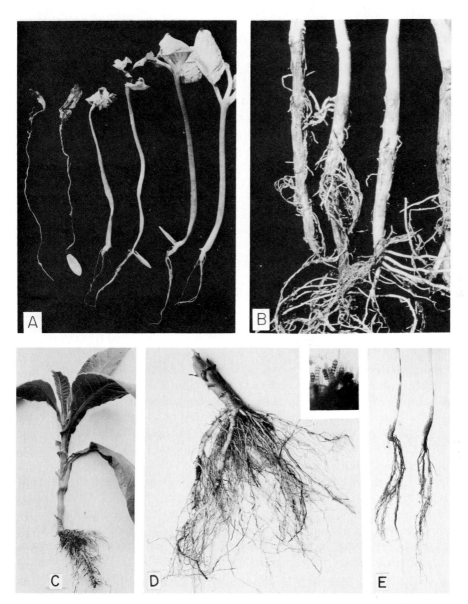

FIGURE 114.
(A) Damping off of cucumber seedlings caused by *Fusarium* sp. Healthy plant at right. (B) *Fusarium* root and stem rot of bean. (C–E) Black root rot of tobacco (C, D) and of bean (E) caused by *Thielaviopsis basicola*. Inset in D shows spores of the fungus. (Photo B courtesy U.S.D.A. Photo E courtesy G. C. Papavizas.)

coalesce and form large areas of dead grass. The disease is most severe at high temperatures and humid weather on sunny exposures of turf where plants die rapidly as the crown and root tissues are destroyed by a brownish dry rot. Pink mycelium of the pathogen is sometimes present on the corms of diseased plants near the soil line.

The fungus, *Fusarium* sp., produces two types of conidia on sporodochia, 1- to 2-celled microconidia, and the typical *Fusarium*-type macroconidia consisting of 3- to 9 (usually 4 to 5)-celled, slightly curved, with more or less pointed ends. *Fusarium* also produces 1- to 2-celled, thick-walled chlamydospores that can withstand drought and low temperatures. The fungus can live on dead plant tissue and can overwinter as mycelium or spores in infected or dead tissues or seed. The spores are easily spread by air, equipment, water, contact, etc., and the fungus is already present in many soils as mycelium or spores.

Control of *Fusarium* rots of root, stem, corms, etc. is possible in the greenhouse through soil sterilization and use of healthy propagative stock, but there are no adequate control measures for these diseases in the field. Rotation with nonsusceptible crops, good soil drainage, use of disease-free or fungicide-treated seed or other propagative stock may help reduce losses. Fertilization with nitrate form of nitrogen also helps reduce disease, as does use of resistant varieties when available. Treatment of propagative stock with benomyl or application of benomyl sprays on the plants in the field or greenhouse has greatly reduced the incidence of Fusarium rots on many kinds of plants, including and especially on turf grasses.

- *Phymatotrichum Root Rot*

Phymatotrichum root rot, usually called Texas root rot or cotton root rot, occurs only in the southwestern U.S. and Mexico. This disease probably affects more kinds of cultivated and wild dicotyledonous plants than any other. Its hosts include many fruit, forest and shade trees, most vegetables and flowers, field crops such as cotton and alfalfa, ornamental shrubs, and many weeds. It causes its greatest losses in cotton in the area from Texas to Arizona and Mexico.

In cotton, for example, infected plants appear in patches in the field and at first show a yellowing and bronzing of the leaves. The leaves then show a slight wilting and turn brown and dry but remain attached to the plant. Below the soil line, and in some plants up to a foot or more above the soil line, the bark and cambium turn brown resulting in a firm brown rot of the root and the lower stem. The surface of rotted roots is usually partly covered by coarse, brown, parallel strands of mycelium and this characteristic helps diagnose the disease.

The fungus, *Phymatotrichum omnivorum*, produces mostly yellowish, thin-celled mycelium but also mycelium made of larger cells. The hyphae grow closely compressed together or interwoven into thick mycelial strands that have characteristic slender, tapering, cross-like side branches. Older strands are dark brown and have few side branches. *Phymatotrichum* produces short, stout, simple, or branched conidiophores that have swollen tips and bear loose heads of dry, colorless, one-celled conidia that apparently do not germinate or cause infection. The fungus also produces numerous small, brown to black sclerotia, singly or in chains, which germinate to produce mycelium. Most of the fungus mycelium and sclerotia are found in soil depths between 30 and 75 cm. Closer to the soil surface the fungus is adversely affected by the normal soil microflora.

The mycelial strands and especially the sclerotia of the fungus can survive in the soil for 5 or more years. The fungus survives best and causes considerably more damage to plants growing in alkaline, black, heavy clay soils that are poorly aerated. The fungus requires high temperature and adequate soil moisture for greatest activity provided the soil pH is near or above neutral.

The fungus enters the plant below the soil line and then grows downward throughout the root and, on some plants, it also invades the lower stem. The fungus spreads from plant to plant through the growth of the mycelial strands and through the spread of such strands or sclerotia by farm equipment, transplants, etc. Once introduced into an area, the fungus can survive on cultivated plants and weeds indefinitely, provided the soil and temperature conditions are favorable. The pathogen cannot stand temperatures below freezing for any appreciable time and its narrow geographical distribution seems to be the result of its high temperature and alkalinity requirements.

Control of *Phymatotrichum* root rot depends on long rotations with grain crops, weed eradication, deep plowing and aeration to keep the soil well aerated, and the use of green manure crops, such as thickly planted corn, sorghum or legumes, which upon decay favor the buildup of large populations of microorganisms that are antagonistic to *Phymatotrichum*. Soil fumigation with ethylene dibromide, vapam, etc., is effective if applied annually and the value of the crop justifies the expense, but it has not generally proven practical because of the rapid spread of the pathogen from deeper in the soil to the root zone once the fumigant has evaporated.

- *Thielaviopsis Black Root Rot*

It is widely distributed in the north temperate zone and affects many vegetables, flowers and field crops, including bean, beet, cotton, peanut, squash, soybean, tobacco, and tomato. *Thielaviopsis* may cause damping off of seedlings of these crops but is much more common as the cause of the black root rot symptoms.

Infected plants make poor growth, have a yellow or chlorotic appearance and may wilt temporarily and then recover when warm weather prevails or they may die. The symptoms of the aboveground parts of the plants are the result of the destruction of the finer roots of the plant by the fungus and the pruning of the larger roots, leaving the black-rotted root stubs that are diagnostic for the disease (Fig. 114, C–E). The pathogen enters the root by direct penetration, through wounds or at the points where branch roots emerge. The plant first produces a gray to reddish lesion, which, as the fungus invades the root cortex, becomes coal black and spreads downward over much of the tap root surface and into the side roots. Small roots and stems are invaded thoroughly by the fungus and rot, while in larger roots the blackening may be limited to the cortex or may also spread into the central stele.

The fungus, *Thielaviopsis basicola*, produces two kinds of conidia: endoconidia or phialospores, which are small, colorless and cylindrical and are produced in great numbers out of a thin, long spore case that looks like regular mycelium with a slightly wider base; and chlamydospores or macroconidia which are large, dark brown, thick-walled and usually 3 to 10 celled (Fig. 114D) before they break apart into smaller cell groups or single, biscuit-shaped chlamydospores. The endoconidia can germinate immediately and cause infection while the chlamydospores require a short resting period before they can germinate.

The fungus overwinters as chlamydospores and as mycelium in infected tissue and in plant debris in the soil. The survival and activity of the fungus are favored by high soil moisture and relatively low temperatures, whether the soil is alkaline or acid but low in humus. *Thielaviopsis* penetrates roots directly and even better through wounds during cool, wet weather. Infection and reproduction of the fungus generally are not rapid, and its parasitic activity is greatly restricted as the soil temperature rises, probably as a result of reduced activity of the pathogen and of increased ability of the plant to defend itself at higher temperatures. The fungus spreads through the soil as mycelium but primarily as spores carried by water, wind, or farm equipment. Long-distance transport is primarily through infected transplants, and in a few crops, such as clover and alfalfa, with the seeds.

Thielaviopsis black root rot is controlled in some crops, such as tobacco, primarily through the use of resistant varieties, while in many other crops no varieties resistant to this disease are known. Good soil drainage is important, while crop rotation is impractical because of the many weeds and crop plants that are susceptible to the disease, although small grains seem to be resistant. Eradication of weeds and late planting when the weather is warm help avoid heavy infections of the crop. Soil sterilization with steam controls the disease in the greenhouse or cold frames. Soil fumigation in the field helps but does not eliminate the disease and is not usually practiced for control of this disease alone.

SELECTED REFERENCES

Abawi, G. S., and R. G. Grogan. 1975. Sources of primary inoculum and effects of temperature and moisture on infection of beans by *Whetzelinia sclerotiorum*. *Phytopathology* **65**:300–309.

Adams, P. B., and C. J. Tate. 1975. Factors affecting lettuce drop caused by *Sclerotinia sclerotiorum*. *Plant Dis. Reptr.* **59**:140–143.

Anderson, A. L. 1948. The development of *Gibberella zeae* head blight of wheat. *Phytopathology* **38**:595–611.

Atanasoff, D. 1920. *Fusarium*-blight scab of wheat and other cereals. *J. Agr. Res.* **20**:1–32.

Baker, K. F., and R. J. Cook. 1974. "Biological Control of Plant Pathogens." W. H. Freeman, San Francisco. 433 p.

Bennett, F. T. 1937. Dollar spot disease of turf and its causal organism, *Sclerotinia homeocarpa* n. sp. *Ann. Appl. Biol.* **34**:236–257.

Booth, C. 1975. The present status of *Fusarium* taxonomy. *Ann. Rev. Phytopathol.* 83–93.

Bruehl, G. W. (Ed.). 1975. "Biology and Control of Soil-Borne Plant Pathogens." The Amer. Phytopathol. Soc., St. Paul, Minn., 216 p.

Burke, D. W., *et al.* 1972. Counteracting bean root rot by loosening the soil. *Phytopathology* **62**:306–309.

Buxton, E. W. 1955. *Fusarium* diseases of gladiolus. *Trans. Brit. Mycol. Soc.* **38**:193–201.

Christensen, J. J., and H. C. H. Kernkamp. 1936. Studies on the toxicity of blighted barley to swine. *Minn. Agr. Exp. Sta. Tech. Bull.* **113.**

Christensen, J. J., and R. D. Wilcoxson. 1966. Stalk rot of corn. Monograph No. 3, Amer. Phytopath. Soc., St. Paul, Minn. 59 p.

Chupp, C., and A. F. Sherf. 1960. "Vegetable Diseases and Their Control." Ronald Press, New York, 693 p.

Clayton, E. E. 1927. *Diplodia* ear rot disease of corn. *J. Agr. Res.* **34**:357–371.

Dickson, J. G. 1956. "Diseases of Field Crops." 2nd ed., McGraw-Hill, New York, 517 p.

Ezekiel, W. M. 1945. Effect of low temperature on survival of *Phymatotrichum omnivorum. Phytopathology* **35**:296–301.

Fajemisin, J. M., and A. L. Hooker. 1974. Predisposition to *Diplodia* stalk rot in corn by three Helminthosporium leaf blights. *Phytopathology* **64**:1496–1499.

Garrett, S. D. 1956. "Biology of Root-Infecting Fungi." Cambridge University Press, 293 p.

Garrett, S. D. 1970. "Pathogenic Root-Infecting Fungi." Cambridge University Press, 294 p.

Johnson, J. 1939. Studies on the nature of brown root rot of tobacco and other plants. *J. Agr. Res.* **58**:843–864.

Koehler, B. 1959. Corn ear rot in Illinois. *Ill. Agr. Exp. Sta. Bull.* **639.**

Koehler, B. 1960. Corn stalk rots in Illinois. *Ill. Agr. Exp. Sta. Bull.* **658.**

Kucharek, T. A., and T. Kommedahl. 1966. Kernel infection and corn stalk rot caused by *Fusarium moniliforme. Phytopathology* **56**:983–984.

Larsen, P. O. 1975. Dollar spot control—Integrating systemics and contacts. *Weeds, Trees and Turf* **13**:46–49.

Lumsden, R. D., and R. L. Dow. 1973. Histopathology of *Sclerotinia sclerotiorum* infections of bean. *Phytopathology* **63**:708–715.

Mitchell, R. B., *et al.* 1941. Soil bacteriological studies on the control of the Phymatotrichum root rot of cotton. *J. Agr. Res.* **63**:535–547.

Moore, W. D., R. A. Conover, and D. L. Stoddard. 1949. The Sclerotiniose disease of vegetable crops in Florida. *Fla. Agr. Exp. Sta. Bull.* **457**: 20 p.

Natti, J. J. 1971. Epidemiology and control of bean white mold. *Phytopathology* **61**:669–674.

Papavizas, G. C. (Ed). 1974. The relation of soil microoganisms to soil borne plant pathogens. Virg. Polyt. Inst. and State Univ., Southern Coop. Series Bull. **183.**

Papavizas, G. C., and J. A. Lewis. 1972. Survival of endoconidia and chlamydospores of *Thielaviopsis basicola* as affected by volatile soil fungicides. *Phytopathology* **62**:417–422.

Papavizas, G. C., and J. A. Lewis. 1975. Effect of seed treatment with fungicides on bean root rots. *Plant Dis. Reptr.* **59**:24–28.

Papendick, R. I., and R. J. Cook. 1974. Plant water stress and development of Fusarium foot rot in wheat subjected to different cultural practices. *Phytopathology* **64**:358–363.

Percich, J. A., and J. L. Lockwood. 1975. Influence of Atrazine on the severity of Fusarium root rot in pea and corn. *Phytopathology* **65**:154–159.

Ramsey, G. B. 1925. *Sclerotinia* species causing decay of vegetables under transit and market conditions. *J. Agr. Res.* **31**:597–632.

Shapavolov, M. 1917. Intoxicating bread. *Phytopathology* **7**:384–386.

Shipton, P. J., R. J. Cook, and J. W. Sitton. 1973. Occurrence and transfer of a biological factor in soil that suppresses take-all of wheat in eastern Washington. *Phytopathology* **63**:511–517.

Stover, R. H. 1950. The black root rot disease of tobacco. *Can. J. Res.* **C 28**:726–738.

Streets, R. B. 1937. *Phymatotrichum* (cotton or Texas) root rot in Arizona. *Ariz. Agr. Exp. Sta. Tech. Bull.* **71.**

Toussoun, T. A., and P. E. Nelson. 1975. Variation and speciation in the Fusaria. *Ann. Rev. Phytopathol.* **13**:71–82.

Western, J. H. (Ed.). 1971. "Diseases of Crop Plants." MacMillan Press, London. 404 p.

Wheeler, J. E., and R. B. Hine. 1972. Influence of soil temperature and moisture on survival and growth of strands of *Phymatotrichum omnivorum. Phytopathology* **62**:828–832.

Williams, G. H., and J. H. Western. 1965. The biology of *Sclerotinia trifoliorum* and other species of sclerotium-forming fungi. I. Apothecium formation from sclerotia. II. The survival of sclerotia in soil. *Ann. Appl. Biol.* **56**:253–268.

POSTHARVEST DISEASES OF PLANT PRODUCTS CAUSED BY ASCOMYCETES AND IMPERFECT FUNGI

Postharvest diseases of plant produce or plant products are those that develop during harvesting and subsequently during grading, packing and transportation of the crop to market; during its storage at shipping points or at the market; and during the various handling operations required to move the crop from the grower to the wholesale dealer, to the retail store and, finally, to the consumer. Postharvest diseases actually continue to develop while the produce is in the possession of the final consumer but is further stored by him at room temperature or under refrigeration until the moment of its actual consumption or use (Figs. 115, 116, 117). During any of these operations the product may show symptoms of diseases that had begun in the field but remained latent till later; the product may be subjected to environmental conditions or treatments that are themselves harmful to the product and impair its appearance and food value; or the product may be subjected to conditions that favor its attack by decay-producing microorganisms which usually cause a portion of it to rot and, in some cases, by secreting toxic substances, make the remainder of the product unfit for consumption or of much lower nutritional and sale value.

All types of plant produce and products are susceptible to postharvest diseases. Generally, the more tender or succulent the exterior of the product, and the greater the water content of the entire product, the more susceptible it is to injury and to infection by fungi and bacteria. Thus, succulent, fleshy fruits and vegetables such as strawberries, peaches, tomatoes, cucurbits, citrus, green vegetables, bananas, onions, and potatoes, as well as cut flowers, bulbs, corms, etc., are affected by postharvest diseases to a greater or lesser extent. The extent of damage or loss depends on the particular product, on the disease organism or organisms involved, and on the storage conditions. Although rotting of fresh fruits and vegetables is much more common and is encountered by everyone at the retail store or at home, rotting of cereal grains and of legumes is also quite common and the losses caused by it quite large. Losses of grains and legumes, however, in spite of their magnitude, occur primarily at the large bins or warehouses of the growers, wholesalers, or manufacturers, and are seldom observed by the general public. In addition, postharvest decays of hay, silage, and other feed stuffs are quite large.

Losses of fresh fruits, vegetables and flowers to postharvest diseases are usually direct, i.e., they result in reduced quality, quantity or both of the affected product. With grains and legumes, however, damage from postharvest diseases, in addition to similar direct losses of quality and quantity, also results from the production by some infecting microorganisms of toxic substances known as mycotoxins. Mycotoxins are poisonous to humans or animals consuming products made from grains or legumes partially or totally infected with such microorganisms. Mycotoxins are also produced by some fungi that infect fresh fruits and vegetables, but in these they are removed during grading, preparation, or before consumption through discarding of the rotten fruits or vegetables or their rotten parts. With increased use by large manufacturers of bulk quantities of fresh fruits and vegetables to make fruit or vegetable juices, purees, cole slaw, baby foods, etc., quality control of individual fruits and vegetables becomes all but economically impractical and, therefore, the importance of postharvest infections and the presence of mycotoxins in bulk-prepared foods is likely to increase in the future.

Postharvest diseases are caused mainly by a relatively small number of Ascomycetes and Imperfect Fungi, by a few Phycomycetes, a few Basidiomycetes, and by a few species of bacteria. The bacteria are primarily of the genera *Erwinia* and *Pseudomonas*. Of the Phycomycetes, *Pythium* and *Phytophthora* cause only soft rots of fleshy fruits and vegetables that are usually in contact with or very near the soil and they may spread to new, healthy fruit during storage. Two other Phycomycetes, *Rhizopus* and, sometimes, *Mucor*, affect fleshy fruits and vegetables after harvest and also stored grains and legumes, as well as prepared foods such as bread, when moisture conditions are favorable (Figs. 115, 116, 117). Of the Basidiomycetes, *Rhizoctonia* and *Sclerotium* cause rotting of fleshy fruits and vegetables, while several fungi, e.g., *Polyporus*, *Poria*, and *Fomes*, cause deterioration of wood and wood products. The Ascomycetes and Imperfect Fungi that cause postharvest diseases are by far the most common and most important causes of postharvest decay and they will be discussed in some detail below.

The fungi and bacteria mentioned above as causing postharvest diseases are usually primary parasites, i.e., they attack healthy, living tissue which they disintegrate and cause to rot. They are often, however, followed into the tissue by other fungi and bacteria that act as secondary parasites, i.e., they live saprophytically on tissues already killed and macerated by the primary parasites. Also, it is not uncommon for more than one of the primary parasites to attack the same tissue concurrently or in sequence. Thus, often some of the primary parasites also act as secondary ones.

Many of the postharvest diseases of fruits, vegetables, grains, and legumes are the results of incipient infections of the plants or their fruits by pathogens in the field while the plants and fruits are still developing, or after the fruits or seeds have matured in the field but before they are

harvested. Symptoms from such "field infections" may be too inconspicuous to be noticed at harvest. In fleshy fruits and vegetables field
infections continue to develop after harvest while in grains and legumes
the development of field infections ceases soon after harvest. In fleshy
fruits and vegetables new infections may be caused in storage by the same
or other pathogens, while in grains and legumes storage infections are
usually caused by pathogens other than those causing field infections.

As with all fungal and bacterial plant diseases, postharvest diseases are
greatly favored by, indeed they depend on, the presence of high moisture
and high temperatures. Fleshy fruits and vegetables contain plentiful
amounts of water and, since they are generally kept at high relative
humidities to avoid shrinkage, they make excellent substrates for attack
by pathogenic microorganisms, provided the latter can penetrate the
outer protective coating of the fruit or vegetable. Wounds, cuts, and
bruises, which are common in fleshy tissues, provide the most common
and effective courts for penetration; but penetration through natural
openings, such as lenticels, also occurs, and direct penetration through
the cuticle and epidermis, especially of fruits and vegetables in contact
with infected ones, is quite common. Once a fresh fruit or vegetable
becomes infected, further development of the infection and spread to
adjacent fruits or vegetables depends mainly on the storage temperature.
Generally, the higher the temperature the faster is the development and
spread of the disease, while at lower temperatures pathogens and the
diseases they cause develop slower or cease to develop at all.

On the other hand, grains and legumes can be, and ordinarily are, kept
for long periods of time because their moisture content is low or can be
reduced to as low as 12 to 14 percent. At such low moisture content
almost none of the fungi that cause field infections can continue to grow
and to cause new infections immediately, or even later when the grains
become remoistened. There are, however, other fungi that can infect
grains and legumes whose moisture content is about or slightly lower
than 14 to 15 percent and the severity and spread of infection increase
drastically with the slightest increase in moisture above that range. High
temperatures favor the infection of grains with high moisture content,
just as they do of fruits and vegetables. Frequently, however, the infection
itself results in a drastic rise of the temperature of the moistened infected
grain due to the heat produced as a result of respiration of the actively
growing fungi and bacteria that cause the infection.

- *Postharvest Decays of Fruits
  and Vegetables*

The most common Ascomycetes or Imperfect Fungi and the main postharvest diseases they cause are the following:

Alternaria.   Its various species probably cause decay on most, if not
all, fresh fruits and vegetables either before or after harvest. The symptoms may appear as brown or black, flat or sunken spots with definite
margins, or they may appear as diffuse, large, decayed areas that are
shallow or extend deep into the flesh of the fruit or vegetable. The fungus

develops well at a wide range of temperatures, even in the refreigerator, although its development is slower at low temperatures (Fig 115F). The fungus may spread into and rot tissues internally with little or no mycelium appearing on the surface, but usually a mat of mycelium that is white at first but later turns brown to black forms on the surface of the rotted area. Some of the most serious diseases caused by *Alternaria* after harvest are *Alternaria* rot of lemons and black rot of oranges, *Alternaria* rot of tomatoes, peppers, eggplant, apples, cucumber, squash, and melons, cabbage, cherries, grapes, strawberries, tuber rot of potatoes, rot of sweet potatoes, purple blotch of onion, etc.

Botrytis.   It causes the "gray molds" or "gray mold rots" of fruits and vegetables, both in the field and in storage. There is practically no fresh fruit, vegetable, bulb, etc. that is not attacked by *Botrytis* in storage. Some of them, e.g., strawberry, lettuce, onion, grape, and apple, are also attacked in the field near maturity or while green. The decay may start at the blossom or stem end of the fruit, or at any wound, crack, or cut of storage tissues. The decay appears as a well-defined water-soaked, then brownish, area that penetrates deeply and advances rapidly into the tissue. In most hosts and under humid conditions a grayish or brownish-gray, granular, velvety mold layer develops on the surface of decaying areas. Gray molds are most severe in cool, humid environments and continue to develop, although slower, even at 0°C. Heavy losses are caused in storage annually by the gray mold fungus on many fleshy fruits and vegetables, particularly pears, apples, strawberries, citrus, tomatoes, onions, and others.

Fusarium.   It causes postharvest "pink or yellow molds" on vegetables and ornamentals, and especially on root crops, tubers, and bulbs, but low-lying crops such as cucurbits and tomatoes are also frequently affected. A brown rot of oranges and lemons held in storage for long periods is also caused by *Fusarium*. With most vegetables, contamination with *Fusarium* takes place in the field before or during harvest but infection may develop in the field or in storage. Losses are particularly heavy with crops, such as potatoes, that are stored for long periods of time. Affected tissues appear fairly moist and light brown at first, but later they become darker brown and somewhat dry. As the decaying areas enlarge, they often become sunken, the skin is wrinkled and small tufts of whitish, pink, or yellow mold appear. Similar mycelial tufts also develop in hollow places formed in decaying tissues. The infection of softer tissues such as tomatoes and cucurbits develops faster and is characterized by pink mycelium and pink, rotten tissues (Fig. 115G).

Geotrichum.   It causes the "sour rots" of citrus fruits, tomatoes, carrots, and other fruits and vegetables. Sour rot is one of the messiest and most unpleasant rots of susceptible fruits and vegetables. Although it may affect tomatoes at the mature green stage, it is the ripe or overripe fruits and vegetables, especially when kept in moisture-holding plastic bags or packages, that are particularly susceptible to sour rot. The fungus is widely distributed in soils and decaying fruits and vegetables and contaminates fruits and vegetables before or during harvest. The fungus

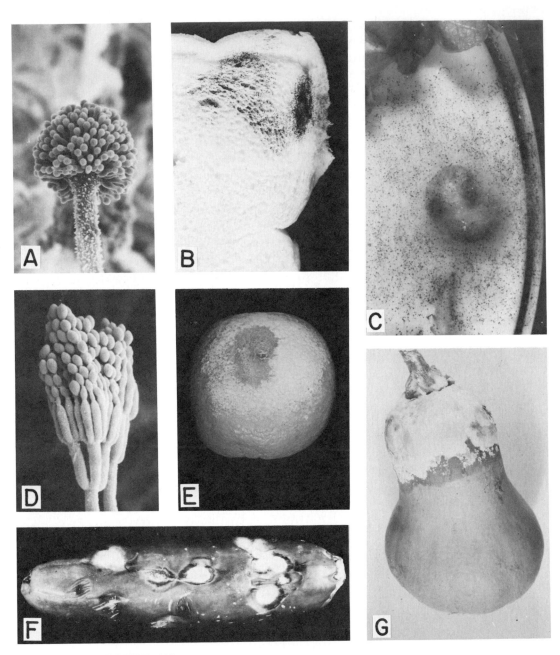

FIGURE 115.
Some fungi causing postharvest diseases, and a few common such diseases. (A) Scanning electron micrograph of conidiophore and conidia of *Aspergillus flavus*. (B) Bread mold caused by *Aspergillus*. (C) *Rhizopus* mycelium and sporangia growing from naturally contaminated seed. (D) Scanning electron micrograph of *Penicillium*. (E) Blue mold of orange caused by *Penicillium italicum*. (F) Cucumber rot during cold storage caused by multiple infections by *Alternaria*. (G) Squash rot in the field or storage caused by *Fusarium*. (Photos A and D courtesy M. F. Brown and H. G. Brotzman.)

penetrates fruits, etc. usually after harvest, at stem scars, skin cracks, cuts, and punctures of various sorts. Infected areas appear water soaked and soft and are easily punctured. The decay spreads rapidly, at first mainly inside the fruit, and eventually involves the whole fruit. Later, the skin frequently cracks over the affected area and is usually filled with a white, cheesy, or scumlike development of the fungus. Also, a thin, water-soaked layer of compact cream-colored fungal growth develops on the surface, while the whole inside becomes a sour-smelling, decayed, watery mass. Fruit flies, which are attracted to tissues affected with sour rot, further spread the pathogen. The fungus prefers high temperatures (24 to 30°C) and humidity but is active at temperatures as low as 2°C.

Penicillium. Its various species cause the "blue mold rots" and the "green mold rots" which are also known as *Penicillium* rots. They are the most common and usually the most destructive of all postharvest diseases affecting all kinds of citrus fruits, apples, pears, and quinces, grapes, onions, melons, figs, sweet potatoes, and many other fruits and vegetables (Fig. 115, D, E). On some fruits, e.g., citrus, some infection may take place in the field, but blue molds or green molds are essentially postharvest diseases and often account for up to 90 percent of decay in transit, in storage, and in the market. *Penicillium* (Fig. 115E) enters tissues through breaks in the skin or rind and even through lenticels. However, it can spread from infected fruit to healthy ones in contact with the infected through the uninjured skin. *Penicillium* rots at first appear as soft, watery, slightly discolored spots of varying size and on any part of the fruit. The spots are rather shallow at first but develop deeper quickly and, at room temperature, most of the fruit or the whole fruit decays in just a few days. Soon after decay develops, a white mold begins to grow on the surface of the skin or rind, near the center of the spot. Later, the fungus growth starts producing spores. The sporulating area has a blue, bluish-green, or olive green color and is usually surrounded by a narrow or wide band of white mycelium with a band of water-soaked tissue ahead of the mycelium. The surface growth of the fungus develops on spots of any size as long as the air is moist and warm. In cool, dry air, surface mold is rare, even when the fruits are totally decayed. Under storage conditions, small, spore-bearing tufts appear on the surface of the spots. Decaying fruit has a musty odor and under dry conditions may shrink and become mummified while under moist conditions, when secondary fungi and yeasts also enter the fruit, it is reduced to a wet, soft mass.

Although most of the damage from blue mold and green mold rots shows up in storage and market, the occurrence of these molds is greater when the fruit is picked and handled during wet, humid weather rather than cool and dry; when fruits are delayed in going into storage; cooled slowly in storage; stored until late in the season, or held at warm temperatures after removal from storage. The most important factor, however, that favors these rots, especially early in the storage season, is mechanical injuries to the fruit surface. Although blue mold and green mold are favored by relatively high storage temperatures, they continue to be slowly active even at temperatures near freezing. Some *Penicillium* species produce ethylene, which diffuses into the container or storage room and increases fruit respiration, affects its coloring and accelerates its maturity and senescence, thus reducing the storage life of the healthy fruit as well.

In addition to the losses caused by the rotting of fruits and vegetables by *Penicillium*, the fungus also produces several mycotoxins, such as patulin, in the affected products, which contaminate juices, sauces, etc. made from healthy and partly rotten fruits. These mycotoxins may cause lesions or degenerations of internal organs such as intestines, kidneys, liver, may affect the nervous system, and some of them also cause cancerous tumours.

Sclerotinia.   It causes the "cottony rot" of citrus fruits, especially lemons, and the "watery soft rot" of bean pods (Fig. 116B), crucifers, cucurbits, strawberries, many other fruits, and practically all vegetables except onions and potatoes. In a moist atmosphere, a characteristic soft, watery decay is produced and the affected tissues are rapidly covered with a white, cottony growth of mycelium that is the outstanding distinguishing characteristic of this decay. The degree of wetness varies with the succulence of the tissues and the humidity of the surrounding air. In moist air, succulent decaying products actually leak and may be completely liquefied leaving a pool of juice. In dry air the water frequently evaporates as fast as it is liberated by the decay and the tissues dry down into a mummy- or parchmentlike remains. Cottony rot is a rapidly spreading, contact decay that attacks both green and mature fruits and vegetables and makes so vigorous and compact a growth as it spreads from one fruit to another that it enmeshes them in its mycelium and creates the so-called "nests." Black, irregularly shaped, sclerotial bodies 2 to 15 mm long later develop in the fungus mat. The fungus is most active and the severity of the rot increases with temperature up to 21 to 25°C but, once started, rotting of tissues continues at temperatures as low as 0°C.

Control of postharvest decays of fresh fruits and vegetables for some diseases depends on effective control of the pathogens that cause the same diseases in the field so that the crop will not be contaminated with the pathogens at harvest and subsequently in storage. The crop should be harvested and handled carefully to avoid wounds, bruises and other injuries that would serve as ports of entry for the pathogen. Harvesting and handling of the crop should be done when the weather is dry and cool to avoid further contamination and infection. The crop should be cooled as quickly as possible to prevent the establishment of new infections and the development of existing ones. All fruits or vegetables showing signs of infection should be removed from the crop that is to be stored or shipped to avoid further spread of the disease. The storage containers, warehouse, shipping cars, etc. should be clean and disinfected with formaldehyde, copper sulfate, or other disinfectant before use. The crop should be stored and shipped at a temperature low enough to slow down development of infections and the physiological breakdown of the tissues, but not too low so as to cause chilling injuries which then serve as ports of entry for fungi. The crop should be free of surface moisture when

FIGURE 116.
Rhizopus soft rot of peaches (A) and *Sclerotinia* cottony soft rot of beans (B) developing during harvest and storage. (Photos courtesy U.S.D.A.)

placed in storage and there should be adequate ventilation in storage to prevent excessively high relative humidity from building up and condensing on the fruit surface. Packaging in plastic bags should be avoided. The crop should be free of insects and other pests when placed in storage and should be kept free of them while in storage to avoid creation of new wounds and development of new infections. Some crops, e.g., sweet potatoes and onions, can be protected from some decay fungi by "curing" at 28 to 32°C for 10 to 14 days, which helps reduce surface moisture and heal any exposed wounds by suberization or wound periderm formation. Hot-air or hot-water treatment is sometimes used to eradicate incipient infections at the surface of some fruits. Gamma rays may be useful in reducing storage rots of some crops. Finally, postharvest decays can be controlled by use of chemical treatments to prevent infection and suppress development of the pathogen on the surface of the diseased host. The chemicals most commonly used for such treatments include sodium-o-phenylphenate, dichloran, 2-aminobutane, thiabendazole, soda ash, borax, etc. These are usually applied as fungicidal wash treatments and are more effective when used "hot" at temperatures between 28° and 50°C, depending on the susceptibility of the crop to injury from heat. Some fungicides, such as dichloran, biphenyl, acetaldehyde vapors, and some ammonia-emitting or nitrogen trichloride-forming chemicals, are used as supplementary, volatile in-package fungistats impregnated in paper sheets during storage and transport.

- *Postharvest Decays of Grains
  and Legumes*

Although several Ascomycetes and Imperfect Fungi such as *Alternaria, Cladosporium, Colletotrichum, Diplodia, Fusarium,* and *Helminthosporium* attack grains and legumes in the field (Fig. 117), they require too high a moisture content in the grain (24 to 25 percent) in order to grow and are, therefore, unable to grow in grains after harvest since grains are usually stored at a moisture content of 12 to 14 percent. Such fungi apparently die out after a few months in storage or are so weakened that they can no longer infect new seeds, but in the meantime they may discolor seeds, kill ovules, weaken or kill the embryos, cause shriveling of seeds, and may produce compounds (mycotoxins) toxic to man and animals.

Most of the decay or deterioration of grains and legumes after harvest, i.e., during storage or transit, is caused by several species of the fungus *Aspergillus* (Fig. 115). Sometimes *Penicillium* infection occurs in grains or legumes stored at low temperatures and with slightly above normal moisture content. Also, *Geotrichum* is often present in seed that has deteriorated in storage due to infection by *Aspergillus,* but it does not itself seem to be important as a cause of seed decay.

Each of the various species or groups of species of *Aspergillus* responsible for seed deterioration has rather definite lower limits of seed moisture content below which it will not grow. They also have less well-defined optimum and upper limits of seed moisture content, these, especially the upper limit, being determined mostly by competition with associated species whose requirement for optimum moisture content coincides with the upper limit at which the former species can survive. Because of competition with field fungi or for other unknown reasons, the storage fungi do not invade grains to any appreciable extent before harvest.

Aspergillus and the few other storage fungi, by invading the embryos of seeds, cause a marked decrease in germination percentage of infected seeds used for planting or in malting barley. Storage fungi also discolor

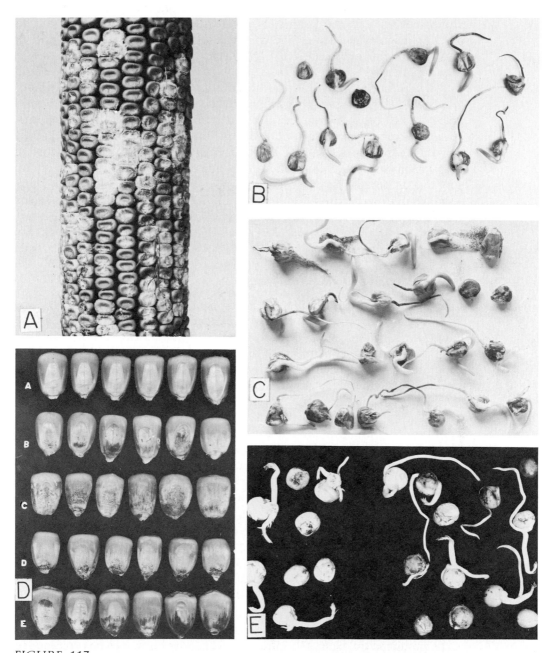

FIGURE 117.
Some seed infections by fungi. (A) Close-up of corn ear infected with *Fusarium moniliforme*. (B) Sweet corn seedlings infected with *Fusarium* carried in corn kernels. (C) Sweet corn seedling infections after 5 days in the laboratory germinator. Top to bottom: *Rhizopus, Diplodia, Penicillium, Fusarium*. (D) Healthy corn kernels (A) and kernels infected with *Diplodia* (B), *Physalospora* (C), *Nigrospora* (D), and *Cladosporium* (E). (E) Pea seeds discolored and infected by different fungi. (Photos A and D courtesy Illinois Agr. Expt. Sta.)

the embryos and the seeds they kill or damage and this reduces the grade and price at which the grain can be sold; flour containing more than 20 percent discolored kernels yields bread of smaller loaf volume and of "off" flavor. In many cases nearly 100 percent of the embryos of wheat may be infected with *Aspergillus* without yet showing discoloration and this wheat is routinely and unknowingly used to make bread, etc., but whether such grain ever poses a health hazard is not known. Infection of grains, hay, feeds, cotton, etc., stored in bulk or during long shipping, results in increased growth and respiration of the fungi and this causes varying degrees of heating of the material; it also produces moisture of respiration which raises moisture in adjacent grain. Although not all spoilage of stored grains results in drastic or even detectable heating, any spoilage in progress produces heat which in some materials may raise the temperature up to 70°C or more. The fungi operate at the lower moisture contents where no free water is available, and bacteria at the higher moisture contents.

- *Mycotoxins and Mycotoxicoses*

One of the more important effects of postharvest decays of fruits and vegetables, and especially of seed and feed deterioration by fungi (Fig. 117), is the induction of mycotoxicoses, i.e., diseases of animals and humans caused by consumption of feeds and foods invaded by fungi that produce toxic substances called mycotoxins. Some mycotoxicoses caused by common and widespread fungi such as *Aspergillus, Penicillium, Fusarium,* and *Stachybotrys* result in severe illness and death. *Aspergillus* and *Penicillium* produce their toxins mostly in stored seeds, hay, or commercially processed foods and feeds, although infection of seeds usually takes place in the field. *Fusarium* produces its toxins primarily on corn and other grains infected in the field or after corn is stored in cribs. *Stachybotrys* produces its toxins while colonizing straw, hay, or other cellulose products used as fodder or bedding for animals.

The mycotoxins produced by each of these fungi may differ from each other in their chemical formula, products in which they are produced, conditions under which they are produced, their effects on various animals and humans, and in degree of toxicity. Several different fungi, however, produce some of the same or closely related toxins. The main mycotoxins produced by the above fungi and some of their properties are listed below.

Aflatoxin. Its name derives from the fact that it was originally found to be produced by *Aspergillus flavus,* but is now known to be produced by other species of *Aspergillus.* Aflatoxin may be produced in infected cereal seeds and most legumes but in these it reaches a rather low (about 50 ppb) and probably nontoxic concentration. However, in peanuts, cottonseed, fishmeal, brazil nuts, copra, and probably other seeds or nuts grown in warm and humid regions, aflatoxin is produced at high concentrations (up to 1000 ppb or more) and causes mostly chronic or occasionally acute mycotoxicoses in humans and domestic animals. Aflatoxin exists in a variety of derivatives with varying effects. Some of these toxins, when ingested with the feed by dairy cattle, are excreted in the milk in still toxic form. The symptoms of mycotoxicoses caused by aflatoxin in animals, and presumably humans, vary widely with the particular toxin and animal, dosage, age of animal, etc. Serious illness and death are not uncommon in young ducklings and turkeys fed high dosages of aflatoxin. Also, weakening, debilitation, abortion, reduced growth, nausea, refusal of feed, and predisposition to other infectious diseases are possible effects of low dosages of aflatoxin fed over long periods to pregnant sows, calves, fattening pigs, mature cattle, and sheep. Moreover, most of the ingested aflatoxin is taken up by the liver and, in some experiments, animals given feed containing even less than the permissible (20 ppb) amount of aflatoxin almost invariably developed liver cancer.

Fusarium toxins.   Two groups of toxins, zearalenone, and its derivatives, and trichothecene and its derivatives, are produced by several species of *Fusarium* primarily in molded corn. Zearalenone, also known as F-2 mycotoxin, is produced by *Fusarium roseum, F. moniliforme, F. tricinctum*, and *F. oxysporum*. It seems to be most toxic to swine in which it causes abnormalities and degeneration of the genital system, the so-called "estrogenic syndrome." Female swine fed zearalenone-containing feed develop swollen vulvas bearing bleeding lesions; atrophying, nonfunctioning ovaries; abortion; and piglets that are born are small and weak. Male swine show signs of feminization—atrophy of the testes and enlargement of the mammary glands. Trichothecins, of which the most common one is known as T-2 mycotoxin, are produced by the same and by several other species of *Fusarium.* Some trichothecenes are also produced in feed infected with *Cephalosporium, Mycothecium, Trichoderma,* and *Stachybotrys.* In nature they are most toxic when fed to swine in which they cause, among other symptoms, listlessness or inactivity, degeneration of the cells of the bone marrow, lymph nodes, and intestines, diarrhea, hemorrhagia, and death. Other animals, however, e.g., cows, chicks, and lambs, are also affected.

Corn infected with *Fusarium* sp. often induces vomiting in swine, or swine refuse to eat it. Although low concentrations of T-2 toxin will induce vomiting in swine, it is likely that other mycotoxins, still unknown, are also involved in inducing vomiting and refusal of corn in swine.

Stachybotrys toxins.   Stachybotryotoxin and several derivatives produced by species of *Stachybotrys* on straw, hay, other fodder, and in animal feeds, bedding, commercial feed, and wheat intended for human consumption, cause a typical chronic form, and a less common acute form of a disease in horses, sheep, swine, poultry, and dogs, and also in man. The symptoms appear as a profuse hemorrhage and necrosis in a variety of body organs such as the stomach, intestines, liver, kidney, and heart. Fumes from burning molded hay may also affect animals and man, and handling of such hay by farm workers causes in them a toxic dermatitis, conjunctivitis, etc.

Other Aspergillus toxins and Penicillium toxins.   In addition to aflatoxins, species of *Aspergillus* also produce other toxins in infected grains, etc. The same or similar toxins are also produced in grains infected by species of *Penicillium*. The most important such toxins are

Ochratoxins, which cause degeneration and necrosis of the liver and kidney, along with several other symptoms, in domestic animals. Some ochratoxins can persist in the meat of animals fed contaminated feed and can be transmitted to the human food chain possibly posing a public health problem.

Yellowed-rice toxins, primarily citreoviridin, citrinin, luteoskyrin, and cyclochlorotine. They are all produced by species of *Penicillium* growing in stored rice, barley, corn, and dried fish, and cause toxicoses associated with various diseases, such as cardiac beri-beri, nervous and circulatory disorders, degeneration of the kidneys and liver, and others.

Tremorgenic toxins, causing marked body tremors and excessive discharge of urine, followed by convulsive seizures that often end in death. They are produced by species of both *Aspergillus* and *Penicillium* infecting foodstuffs in storage and also in refrigerated foods, grains and cereal products. Sheep, horses, and cows seem to be the domestic animals most commonly affected by tremorgenic toxins.

Penicillic acid, which is a carcinogen. It is produced by both *Penicillium* and *Aspergillus* species in molded cereal grains, mixed feeds, and occasionally in other products, e.g., visibly molded tobacco. Not very common in nature and its importance is largely unknown.

Patulin is also a carcinogenic substance produced by *Penicillium* and *Aspergillus*. Patulin is toxic to bacteria, some fungi, to higher plants and animals. It is commonly found to occur naturally in foodstuffs such as fruit or juices made with fruit partly infected with *Penicillium*, in spontaneously molded bread and bakery products and in most commercial apple products. Thus, patulin may constitute a serious health hazard for humans as well as for animals.

- *Control of Grain Decays*

The control of postharvest deterioration and spoilage by fungi of grains, legumes, fodder, commercial feeds, etc., depends on certain precautions and conditions which must be met before and during harvest and then during storage. Provided that the crop was healthy and of high quality when harvested, its subsequent infection and spoilage in storage will be avoided if: (1) The amount of moisture content is kept at levels below the minimum required for the growth of the common storage fungi. Some hardy *Aspergillus* species will grow and cause spoilage of starchy cereal seeds with a moisture content as low as 13.0 to 13.2 percent, and of soybeans with a moisture content of about 11.5 to 11.8 percent. Others require a minimum moisture of 14 percent or more to cause spoilage. (2) The temperature of stored grain is kept as low as possible since most storage fungi grow most rapidly at temperatures between 30 and 55°C, they grow very slowly at 12 to 15°C and their growth almost ceases at 5 to 8°C. Low temperature also slows down respiration of grain and prevents increase of moisture in grain. (3) Infestation of stored products by insects and mites is kept to a minimum through the use of fumigants.

This helps keep the storage fungi from getting started and growing rapidly. (4) The stored grain should not be unripe or too old, should be clean, have good germinability and be free of mechanical damage, broken seeds, etc. Such grain resists infection by storage fungi that could invade otherwise weakened or cracked grain.

In addition to starting with good sound crops free of insects or fumigating to eliminate the insects, the simplest and most common solution to maintaining the grain free of storage fungi is through the use of aeration systems in which air is moved through the grain at relatively low rates of flow. The airflow removes excess moisture and heat. It can be regulated so that it brings the moisture content of the grain mass to the desired level and reduces the temperature to 8 to 10°C at which insects and mites are dormant and storage fungi are almost dormant.

SELECTED REFERENCES

Boyd, A. E. W. 1972. Potato storage diseases. *Rev. Plant Pathol.* **51**:297–321.

Ceponis, M. J., and J. E. Butterfield. 1974. Market losses in Florida cucumbers and bell peppers in metropolitan New York. *Plant Dis. Reptr.* **58**:558–560.

Christensen, C. M. 1975. "Molds, Mushrooms, and Mycotoxins." Univ. of Minnesota Press, Minneapolis. 264 p.

Christensen, C. M., and H. H. Kaufmann. 1965. Deterioration of stored grains by fungi. *Ann. Rev. Phytopathol.* **3**:69–84.

Christensen, C. M., and H. H. Kaufmann. 1969. Grain storage: the role of fungi in quality loss. Univ. Minnesota Press, Minneapolis. 153 p.

Coursey, D. G. and R. H. Booth. 1972. The post-harvest phytopathology of perishable tropical produce. *Rev. Plant Pathol.* **51**:751–765.

Eckert, J. W., and N. F. Sommer. 1967. Control of diseases of fruits and vegetables by postharvest treatment. *Ann. Rev. Phytopathol.* **5**:391–432.

Goldblatt, L. (Ed.). 1969. "Aflatoxin." Academic Press. New York.

Guba, E. F. 1950. Spoilage of squash in storage. *Mass. Agr. Exp. Sta. Bull.* **457**: 52 p.

Harman, G. E., and F. L. Pfleger. 1974. Pathogenicity and infection sites of *Aspergillus* species in stored seeds. *Phytopathology* **64**:1339–1344.

Harvey, J. M., and W. T. Pentzer. 1960. Market diseases of grapes and other small fruits. *U. S. Dept. Agr., Agr. Handbook* **189**: 37 p. 10 pl.

Jackson, C. R., and D. K. Bell. 1969. Diseases of peanut (ground nut) caused by fungi. *Georgia Agr. Exp. Sta. Res. Bull.* **56**:137 p.

Jones, A. L., and C. Burton. 1973. Heat and fungicide treatment to control postharvest brown rot of stone fruits. *Plant Dis. Reptr.* **57**:62–66.

McColloch, L. P., H. T. Cook, and W. R. Wright. 1968. Market diseases of tomatoes, peppers, and eggplants. *U. S. Dept. Agr., Agr. Handbook* **28**: 74 p., 16 pl.

Mirocha, C. J., and C. M. Christensen. 1974. Fungus metabolites toxic to animals. *Ann. Rev. Phytopathol.* **12**:303–330.

Pierson, C. F. 1971. Market diseases of apples, pears and quinces. *U. S. Dept. Agr., Agr. Handbook* **376**: 112 p., 17 pl.

Purchase, I. F. H. (Ed.). 1971. Symposium on mycotoxins in human health. Macmillan, London. 306 p.

Ramsey, G. G., B. A. Friedman, and M. A. Smith. 1967. Market diseases of beets, chicory, endive, escarole, globe artichokes, lettuce, rhubarb, spinach, and sweet potatoes. *U. S. Dept. Agr., Agr. Handbook* **155**: 42 p., 19 pl.

Ramsey, G. B., J. S. Wiant, and G. K. K. Link. 1938. Market diseases of fruits and vegetables: crucifers and cucurbits. *U. S. Dept. Agr., Miscell. Public.* **292**: 74 p., 23 pl.

Ramsey, G. B., J. S. Wiant, and M. A. Smith. 1949. Market diseases of fruits and vegetables: Potatoes. *U.S. Dept. Agr., Miscell. Public.* **98**: 60 p.

Reiss, J. 1972. Nachweis von Patulin in spontan verschimmelten Brot und Geback. *Naturwissenschaften* **1**:1.

Rose, D. H., D. F. Fisher, C. Brooks, and C. O. Bratley. 1937. Market diseases of fruits and vegetables: peaches, plums, cherries and other stone fruits. *U.S. Dept. Agr., Miscell. Public.* **228**: 26 p., 11 pl.

Scott, P. M. 1973. Mycotoxins in stored grains, feeds, and other cereal products, *in* "Grain Storage: Part of a System," R. N. Sinha and W. E. Muir (Eds.). Westport, Conn. 481 p.

Smith, M. A., L. P. McColloch, and B. A. Friedman. 1966. Market diseases of asparagus, onions, beans, peas, carrots, celery, and related vegetables. *U.S. Dept. Agr., Agr. Handbook* **303**: 65 p., 17 pl.

Smoot, J. J., L. G. Houck, and H. B. Johnson. 1971. Market diseases of citrus and other subtropical fruits. *U. S. Dept. Agr., Agr. Handbook* **398**: 115 p.

Spalding, D. H., and W. F. Reeder. 1974. Postharvest control of *Sclerotinia* rot of snap bean pods with heated and unheated chemical dips. *Plant Dis. Reptr.* **58**:59–62.

Wiant, J. S. 1937. Investigations on the market diseases of cantaloupes and honey dew and honey ball melons. *U.S. Dept. Agr. Tech. Bull.* **573**: 47 p., 21 pl.

Wilson, D. M., and G. J. Nuovo. 1973. Patulin production in apples decayed by *Penicillium expansum. Appl. Microbiol.* **26**:124–125.

Zuber, M. S., *et al.* 1976. Preharvest development of aflatoxin B1 in corn in the United States. *Phytopathology* **66**:1120–1121.

DISEASES CAUSED
BY BASIDIOMYCETES

Basidiomycetes are fungi that produce their sexual spores, called basidiospores, on a club-shaped or tubular structure called a basidium (Figs. 118 and 119). Most fleshy fungi, including the common mushrooms, the puffballs, and the shelf fungi or conks, are Basidiomycetes. Their basidia are one-celled, club-shaped structures that bear four external basidiospores on short stalks called sterigmata. These Basidiomycetes belong to the subclass Homobasidiomycetes which includes almost all the wood decaying fungi and a few root-rotting fungi (Figs. 119 and 120). In the other subclass, called Heterobasidiomycetes, the basidium has cross walls that divide it into four cells each of which produces a basidiospore. Such a basidium is often called a promycelium. The Heterobasidiomycetes include two very common and very destructive groups of plant pathogens, the rusts, and the smuts (Figs. 118 and 120).

THE RUSTS

The plant rusts, caused by Basidiomycetes of the order Uredinales, are among the most destructive plant diseases. They have caused famines and ruined economies of large areas and entire countries. They have been most notorious for their destructiveness on grain crops, especially wheat,

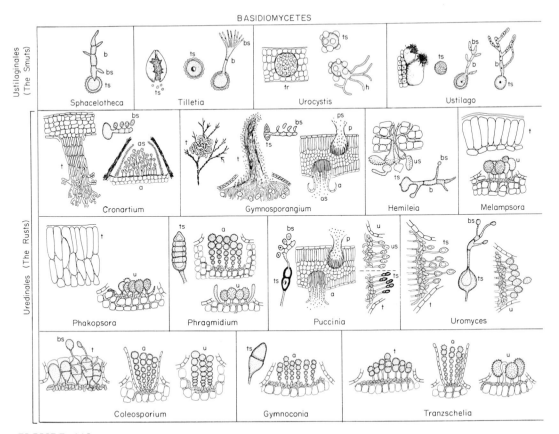

FIGURE 118.
Basidiomycetes: Some common smut and rust fungi. a—aecium, as—aeciospore,
b—basidium, bs—basidiospore, h—hypha, t—telium, tr—teliosorus,
ts—teliospore, u—uredium, us—uredospore.

oats, barley, etc., but they also attack vegetables such as bean and asparagus, field crops such as cotton and soybeans, ornamentals such as
carnation and snapdragon, and have caused tremendous losses on trees
such as pine, apple, coffee, and others.

The rust fungi attack mostly leaves and stems and occasionally floral
parts and fruits. Rust infections usually appear as numerous rusty,
orange, yellow or even white-colored spots that result in rupturing of the
epidermis, in formation of swellings, and even galls. Most rust infections
are strictly local spots but some may spread internally to a more or less
limited extent. The most important rust fungi and the diseases they cause
are the following (Figs. 118 and 120).

> *Puccinia*: causing stem rust of wheat and all other small grains (*P. graminis*);
> yellow or stripe rust of wheat, barley and rye (*P. striiformis*); leaf or brown
> rust of wheat and rye (*P. recondita*) (Fig. 121); leaf or brown dwarf rust of
> barley (*P. hordei*); crown rust of oats (*P. coronata*); corn rust (*P. sorghi*),
> southern or tropical corn rust (*P. polysora*); sorghum rust (*P. purpurea*); and
> sugarcane rusts (*P. sacchari* and *P. kuehnii*).

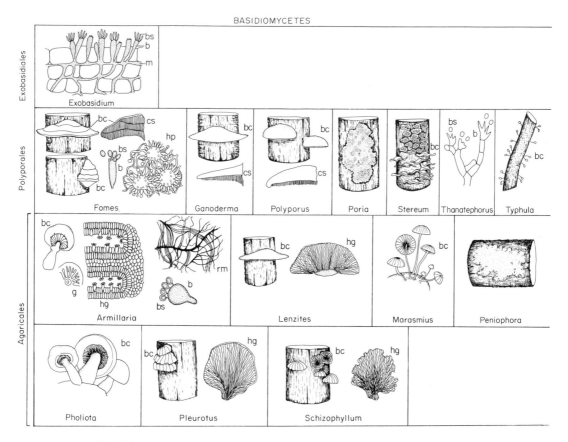

FIGURE 119.
Basidiomycetes: Some of the conk- and mushroom-forming plant pathogens.
b—basidium, bc—basidiocarp, bs—basidiospore, cs—cross section, g—gill,
hg—hymenial gills, hp—hymenial pores, m—mycelium.

Puccinia also causes severe rust diseases on field crops such as cotton (*P. stakmanii*); vegetables such as asparagus (*P. asparagi*); and flowers such as chrysanthemum (*P. chrysanthemi*), hollyhock (*P. malvacearum*), and snapdragon (*P. antirrhini*).

Gymnosporangium, causing the cedar-apple rust (*G. juniperi-virginianae*), and hawthorn-cedar rust (*G. globosum*).

Hemileia, causing coffee leaf rust (*H. vastatrix*).

Phragmidium, causing rust on roses and yellow rust on raspberry.

Uromyces, several species causing the rusts of legumes (bean, broad bean and pea) and one causing rust of carnation (*U. caryophyllinus*).

Cronartium: causing rusts of pines, oaks, and other hosts, such as the white pine blister rust (*C. ribicola*); fusiform rust of pines and oaks (*C. fusiforme*); eastern gall or pine-oak rust (*C. quercuum*); pine-sweet fern blister rust (*C. comptoniae*); pine-Comandra rust (*C. comandrae*); and southern cone rust (*C. strobilinum*).

Melampsora, causing rust of flax (*M. lini*).

Coleosporium, causing blister rust of pine needles (*C. solidaginis*).

Gymnoconia, causing orange rust of blackberry and raspberry.

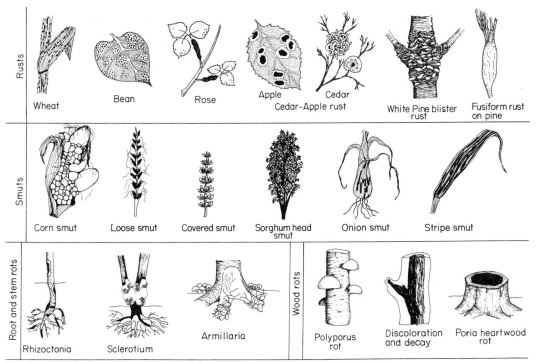

Common symptoms caused by some Basidiomycetes

FIGURE 120.
Common symptoms caused by Basidiomycetes.

Phakopsora, causing soybean rust (*P.pahyrhizi*).
Tranzschelia, causing rust of peach.

Most rust fungi are very specialized parasites and attack only certain host genera or only certain varieties. Rust fungi that are morphologically identical but attack different host genera are regarded as special forms (*formae specialis*), e.g., *Puccinia graminis* f.s. *tritici*. Within each special form of a rust there are many so-called pathogenic (physiologic) races that can attack only certain varieties within the species and can be detected and identified only by the set of differential varieties they can infect. Where sexual reproduction of the rust fungus is rare, the races are more stable and produce large populations over fairly long periods of time, but even so some of these fungi have as many races as those in which sexual reproduction is common.

The rust fungi are obligate parasites, although some of them have now been grown on special culture media in the laboratory. Most rust fungi produce five distinct fruiting structures with five different spore forms which appear in a definite sequence. Some of the spore stages parasitize one host while the others must infect and parasitize a different alternate host. All rust fungi produce teliospores and basidiospores. Rust fungi that produce only teliospores and basidiospores are called microcyclic or short-cycled rusts. Other rust fungi produce, in addition to teliospores

FIGURE 121.
Uredia of the leaf or brown rust on wheat leaves caused by *Puccinia recondita*.

and basidiospores, spermatia (formerly known as pycniospores), aecio-
pores, and uredospores, in that order, and are called macrocyclic or long-
cycled rusts. In some macrocyclic rusts, spermatia and/or uredospores
may be absent. Although basidiospores are produced on basidia, the other
spore forms are produced in specialized fruiting structures called, respectively, spermagonia, aecia, uredia, and telia (Fig. 118).

Basidiospores, aeciospores, and uredospores can attack and infect host
plants. The teliospores serve only as the sexual, overwintering stage
which, upon germination, produce the basidium (promycelium). The
basidium, following meiosis, produces four haploid basidiospores. The

basidiospores, upon infection, produce haploid mycelium that forms spermagonia (formerly known as pycnia) containing haploid spermatia and receptive hyphae. Spermatia are unable to infect plants; their function is the fertilization of receptive hyphae of the compatible mating type and the subsequent production of dikaryotic mycelium and dikaryotic spores. This mycelium forms aecia that produce aeciospores which, upon infection, produce more dikaryotic mycelium that this time forms uredia. The latter produce uredospores which also infect and produce more uredia and uredospores and, near host maturity, telia and teliospores. The cycle is thus completed.

Some macrocyclic rusts, e.g., asparagus rust, complete their life cycles on a single host and are called autoecious. Others, e.g., stem rust of cereals, require two different or alternate hosts for completion of their full life cycle and are called heteroecious.

The rust fungi spread from plant to plant mostly by wind-blown spores, although insects, rain, animals, etc. may play a role. Some of their spores are transported over long distances (several hundred miles) by strong winds and, upon landing (being scrubbed from the air by rain), can start new infections.

Control of rust diseases in some crops, e.g., grains, is achieved by means of resistant varieties. In some vegetable, ornamental and fruit tree rusts, e.g., cedar-apple rust, the disease is controlled with chemical sprays. In others, e.g., white pine blister rust, control has been attempted through removal of the alternate host and avoidance of high rust-hazard zones. With the discovery of several new systemic fungicides effective against rusts, a new impetus has been given toward controlling rust diseases of annual plants as well as trees with these chemicals applied either as sprays, seed dressings, soil drenches, or by injection.

CEREAL RUSTS

Various species or special forms of *Puccinia* attack all cultivated and wild grasses, including all small grains, corn, and sugarcane. They are among the most serious diseases of cultivated plants resulting in losses equivalent to about 10 percent of the world grain crop per year. Rusts may debilitate and kill young plants, but more often they reduce foliage, root growth, and yield by reducing the rate of photosynthesis, increasing the rate of respiration, decreasing translocation of photosynthates from infected tissue, and, instead, diverting materials into the infected tissue. The quantity of grain produced by rusted plants may be reduced greatly and the grain produced may be of extremely poor quality since it may be devoid of starch and may consist mostly of cellulosic materials that are of low or no nutritional value to humans. Some of the most important cereal rusts are as follows.

• *Stem Rust of Wheat and Other Cereals*

Stem rust of wheat is worldwide in distribution and affects wheat wherever it is grown. Similar rusts affect the other cultivated cereals and probably most wild grass genera and species.

The stem rust fungus attacks all the aboveground parts of the wheat plant and causes losses by reducing foliage and root development, and yield and quality of grain. Infected plants usually produce fewer tillers, set fewer seeds per head, and the kernels are smaller in size, generally shriveled, and of poor milling quality and food value. Under extreme situations, heavily infected plants may die. Heavy seedling infection of winter wheat may weaken the plants and make them susceptible to winter injury and to attack by other pathogens. The amount of losses caused by stem rust may vary from slight to complete destruction of wheat fields over large areas, sometimes encompassing several states. Thousands of tons of wheat are lost to stem rust in the U.S. annually, and during years of severe stem rust epidemics the losses are in the millions of tons.

Symptoms. The pathogen causing stem rust of wheat attacks and produces symptoms on two distinctly different kinds of host plants. The most serious, and economically important, symptoms are produced on wheat and certain related cereals (e.g., barley, oats, rye) and other grasses. Symptoms, however, although economically unimportant, are also produced on plants of common barberry (*Berberis vulgaris*) and certain other wild native species of barberry and mahonia.

The symptoms on wheat appear first as long, narrow, elliptical blisters or pustules parallel with the long axis of the stem, leaf, or leaf sheath (Figs. 122A, B and 123A). In later stages of growth, blisters may appear on the neck and glumes of the wheat spike. Within a few days, the epidermis

FIGURE 122.
Stem rust of wheat caused by *Puccinia graminis tritici*. (A) Rust symptoms on wheat stems showing telia. (B) Close-up of infected wheat stem. (C) Barberry leaves with clusters of aecial cups of the stem rust fungus. (Photos courtesy U.S.D.A.)

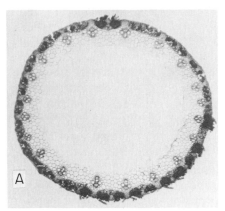
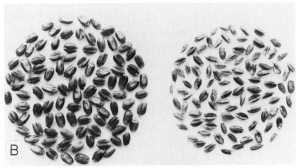

FIGURE 123.
(A) Cross section of wheat stem showing uredia or telia some of which have ruptured the epidermis. (B) Comparison of kernels from healthy (left) and stem rust-infected wheat plant. (Photos courtesy U.S.D.A.)

covering the pustules is ruptured irregularly and pushed back revealing a powdery mass of brick red-colored spores, called uredospores. The pustules, called uredia, vary in size from very small to about 3 mm wide by 10 mm long. Later in the season, as the plant approaches maturity, the rusty color of the pustules turns black as the fungus produces teliospores instead of uredospores and uredia are transformed into black telia. Sometimes telia may develop independently of uredia. Although uredia and telia are rather small, either fruiting structure may exist on wheat plants in such great numbers that large parts of the plant appear to be covered with the ruptured areas that are filled with either the rust-red uredospores or the black teliospores or both.

On barberry, the symptoms appear as yellowish to orange-colored spots on the leaves and sometimes on young twigs and fruits. Within the spots, and in leaves generally on the upper side, appear a few minute dark-colored bodies called spermagonia (or pycnia), usually bearing a small droplet of liquid or nectar. On the lower side of the leaf, beneath the spermagonia, and occasionally on the upper surface, or next to the spermagonia on twigs, fruit, petioles, etc., groups of orange-yellow horn- or cuplike projections, called aecia, appear (Fig. 122C). The host tissue bearing the aecia is frequently hypertrophied. The aecial wall, called a peridium, usually protrudes at the margin of the cups and its light, whitish color is contrasted with the orange-colored aeciospores contained in the aecia.

The pathogen: Puccinia graminis. It is a macrocyclic, heteroecious rust fungus producing spermagonia and aecia on barberry and mahonia, and uredia and telia on wheat and other cereals and grasses.

Development of disease. In cooler, northern regions the fungus overwinters as teliospores on infested wheat debris. Teliospores germinate in the spring only after dormancy is broken by alternate freezing and thawing, occurring naturally in northern areas during the winter and spring. The basidiospores produced by each teliospore are forcefully ejected into the air but, being hyaline and very light sensitive, are carried by air currents for rather short distances, probably no more than a few hundred meters. If barberry plants are growing nearby and the basidiospores land on young barberry leaves, petioles, blossoms, or fruit, the basidiospores germinate and penetrate the epidermal cells directly; after that, the mycelium grows mostly intercellularly with haustoria entering the cells. Within 3 or 4 days the hyphal branches converge toward a point just below the epidermis, where they form a mat of mycelium that develops into a spermagonium (Fig. 124). The outward pressure of the spermagonium ruptures the epidermis, and the ostiole (opening) of the spermagonium emerges on the surface of the plant tissue. Receptive hyphae originating in the spermagonium extend beyond the ostiole, and spermatia embedded in a sticky liquid are exuded through the opening. Insects visiting the infected barberry leaves become smeared with spermatia and carry them to other, possibly sexually compatible spermagonia. Spermatia may also be carried to compatible spermagonia by rainwater or dew running off the plant surface. When a spermatium comes in contact with a receptive hypha of a compatible spermagonium, fertilization takes place. The nucleus of the spermatium passes into the receptive hypha, but it does not fuse with the nucleus already present in the latter. Instead, it migrates passively through the monokaryotic mycelium to the aecial mother cells. There the dikaryotic condition is reestablished and aeciospores formed subsequently will be dikaryotic. This mycelium then grows intercellularly toward the periphery of the spermagonia, present on petioles, fruit, etc., or usually toward the lower side of the leaf bearing the spermagonium, where it forms thick mycelial mats that develop into aecia. In the meantime, the host cells surrounding the mycelium are stimulated to enlarge, and along with the increased volume of the fungus, result in a swelling of the infected area on the lower surface of the leaf.

The aecia form in groups and protrude considerably beyond the hypertrophied leaf or other tissue surface of the barberry plant. The aeciospores are produced in chains on short hyphae inside the aecium, and each spore contains two separate nuclei of opposite mating type. Aeciospores are released in the late spring and are carried by wind to nearby wheat plants on which they germinate. The germ tube penetrates wheat stems, leaves or sheaths through stomata and after the mycelium grows intercellularly for a while, it then grows more profusely toward, but below the surface of, the wheat tissue and forms a mat of mycelium just below the epidermis. Many short hyphae arise from the mycelium, and at the tip of each forms one uredospore. The growth of the sporophores and of the uredospores exert pressure on the epidermis which is pushed outward and forms a pustule manifesting the presence of the uredium. Finally the epidermis is broken irregularly and flaps back revealing several hundred thousand rust-colored uredospores which are easily detached from the sporophores and give a powdery appearance to the uredium.

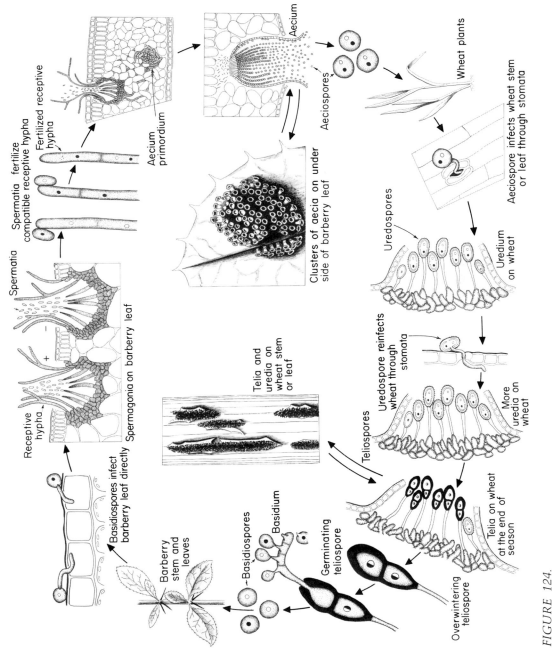

Labels within the figure:

Aecium

Aecium primordium

Fertilized receptive hypha

Spermatia fertilize compatible receptive hypha

Spermatia

Receptive hypha

Spermagonia on barberry leaf

Basidiospores infect barberry leaf directly

Barberry stem and leaves

Basidiospores

Basidium

Germinating teliospore

Clusters of aecia on under side of barberry leaf

Aeciospores

Wheat plants

Aeciospore infects wheat stem or leaf through stomata

Uredospores

Uredium on wheat

Uredospore reinfects wheat through stomata

More uredia on wheat

Teliospores

Telia and uredia on wheat stem or leaf

Telia on wheat at the end of season

Overwintering teliospore

FIGURE 124.
Disease cycle of stem rust of wheat caused by *Puccinia graminis tritici.*

The uredospores are easily blown away by air currents. Stronger winds may carry them many miles, even hundreds of miles, from the point of their origin. The uredospores can reinfect wheat plants. When they land on wheat plants, in the presence of dew, a film of water or relative humidities near the saturation point, they germinate and enter the plant through stomata. The mycelium grows intercellularly again, sends haustoria into the plant cells, and within 8 to 10 days from inoculation it produces a new uredium and more uredospores. Many successive infections of wheat plants by uredospores may take place within one growing season up to the time the plant reaches maturity. Most of the damage caused to wheat growth and yield results from such uredospore infections which may literally cover the stem, leaf, leaf sheaths, glumes, etc. with uredia.

The presence of numerous uredia on wheat plants results in an increased water loss by the plant because of increases in transpiration of water by infected plants and in evaporation of water through the ruptured epidermis. In addition to reduced amounts of water being available to the diseased plant, the fungus itself removes much of the nutrients, and water, that would normally be used by the plant. The respiration of infected plants increases rapidly during the development of the uredia, but a few days after sporulation of the fungus respiration drops to slightly below normal. Photosynthesis of diseased plants is reduced considerably due to destruction of much of the photosynthetic area by the fungus and to interference of the fungal secretions with the photosynthetic activity of the remaining green areas on the plant. The fungus also seems to interfere with normal root development and uptake of nutrients by the roots. All these effects reduce the amount of nutrients available for the production of the normal number and size of seeds on the plant, which are further accentuated by fungus-induced earlier maturity of the plant, resulting in decreased time available for the seed to fill. The total amount of damage depends considerably on the stage of development of the wheat plant at the time rust infection becomes heavy. Thus, heavy rust infections before or at the flower stage are extremely damaging and may cause total yield loss (Fig. 123B), whereas if heavy infections do not occur until late dough stage, the damage to yield is much smaller.

When the wheat plant approaches maturity, or when the plant fails because of overwhelming infection, the uredia produce teliospores instead of uredospores, or new telia may develop from recent uredospore infections. Teliospores do not germinate immediately and do not infect wheat, but are the overwintering stage of the fungus. Teliospores also serve as the stage in which fusion of the two nuclei and meiosis take place and result in the production of new combinations of genetic characters of the fungus through genetic recombination. Several hundred races of the stem rust fungus are known to date and new ones appear every year.

In southern regions the fungus usually overwinters as mycelium on fall-sown wheat which becomes infected by uredospores produced on the previous year's crop. Heavy rust infections in these regions in early spring are important, not only locally, but also for the wheat crop of northern regions, since uredospores produced in the South are carried northward by the warm southern winds of spring and summer and initiate infections of wheat in successively northern regions.

Control. The most effective, and the only practical, means of control of wheat stem rust is through the use of wheat varieties resistant to infection by the pathogen. A tremendous amount of work has been, and is being done for the development of wheat varieties resistant to existing races of the fungus. The best varieties of wheat which combine rust resistance and desirable agronomic characteristics are recommended annually by the state agricultural experiment stations in the U.S. and change periodically in order to meet the existing rust races. Much effort is now directed toward development of varieties with general or nonspecific resistance and toward development of multiline cultivars.

Eradication of barberry, the alternate host of the stem rust fungus, was carried out until recently in most wheat-growing areas of the U.S.; this has reduced losses from stem rust by eliminating the early season infections on wheat in the areas where uredospores cannot overwinter, and by reducing the opportunity for the development of new races of the stem rust fungus through genetic recombination on barberry, thus providing for greater stability in the race population of the pathogen and contributing to the success of breeding of resistant varieties.

Several fungicides, such as sulfur, dichlone, zineb, and mixtures of zinc ion with maneb, can effectively control the stem rust of wheat. In most cases, however, 4 to 10 applications per season are required for complete control of the rust, and because of the low income return per acre of wheat, such a control program is not economically practical. Two applications of zinc ion–maneb mixtures coordinated with forecasts of weather conditions favoring rust epidemics, may reduce damage from stem rust by as much as 75 percent. These chemicals have both protective and eradicative properties and therefore even two sprays, one at trace to 5 percent rust prevalence, and the second 10 to 14 days later, can give economically rewarding control of rust.

Certain systemic fungicides, e.g., oxycarboxin, have also been reported to give experimental control of the stem rust for up to six weeks when applied as seed treatment, or to the soil as granules just before sowing and again in midseason as a spray.

Damage by the stem rust fungus is usually lower in fields in which heavy fertilization with nitrate forms of nitrogen and dense seeding have been avoided.

- *Puccinia Rusts of*
  Vegetables, Field Crops, and Ornamentals

The most common *Puccinia* rusts on plants other than cereals are those on asparagus, peanuts, cotton, chrysanthemum, hollyhock, and snapdragon. The asparagus rust, *Puccinia asparagi*, produces spermagonia, aecia, uredia and telia on asparagus. In the peanut rust (*P. arachidis*), chrysanthemum rust (*P. chrysanthemi*), and snapdragon rust (*P. antirrhini*) (Fig. 125B), spermagonia and aecia are unknown and only uredia and occasion-

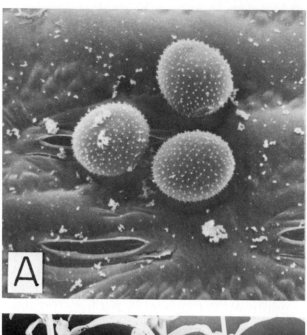

FIGURE 125.
(A) Scanning electron micrograph of uredospores of *Puccinia sorghi* next to a stoma. (B) Snapdragon rust caused by *P. antirrhini*. (Photo A courtesy M. F. Brown and H. G. Brotzman.)

ally telia are produced on the respective hosts. In hollyhock rust (*P. malvacearum*), only telia are produced. In cotton rust (*P. stackmanii*), the disease is caused by the aecial stage, while uredia and telia develop on desert grasses of the genus *Bouteloua* which are common through the southwestern cotton-growing areas of the U.S. and in Mexico.

In most of the above rusts the symptoms appear as rust-colored uredial

spots on leaves and green stems that later in the season may be replaced
or supplemented by black telia. In cotton, the symptoms appear as circular, slightly elevated, orange-yellow aecia mostly on the under surface of
the leaves. Depending on the severity of the infection, plants become
weakened, stunted, and may even be killed.

Control of these rusts, in addition to the use of resistant varieties,
depends primarily on sprays or dusts with fungicides such as polyram,
maneb, zineb, or sulfur. Removal and burning of all infected plant material
helps reduce or eliminate the inoculum in the area and reduces subsequent development of disease.

- *Cedar-Apple Rust*

It is present in North America and in Europe. It causes yellow-to-
orange-colored leaf spots and occasionally fruit spots and premature defoliation on apple (Fig. 126, A, B); and galls, often called cedar apples, that
produce jellylike horns on cedar (Fig. 126, C, D). It can cause considerable
damage to both hosts when they are located near each other. Similar
diseases affect hawthorn and quince.

The fungus, *Gymnosporangium juniperi-virginianae*, overwinters as
dikaryotic mycelium in the galls on cedar trees. Cedar needles or axillary
buds are infected in the summer by wind-borne aeciospores from apple
leaves (Fig. 127). The fungus grows little in the cedar needles during fall
and winter, but the following spring or early summer, galls begin to
appear as small greenish-brown swellings on the upper surface of the
needle. The fungus is present in the galls as mycelium growing between
the cells of cedar. The galls enlarge rapidly and, by fall, they may be 3 to 5
cm in diameter, turn chocolate brown and their surface is covered with
small circular depressions. The cedar-apple rust fungus does not produce
uredia or uredospores. The following spring, however, the small depressions on the galls absorb water during warm, wet weather, swell and
produce orange-brown, jellylike "horns" that are 10 to 20 mm long and
very conspicuous (Fig. 126, C, D). The jellylike horns are columns of
teliospores that germinate in place for several weeks and produce
basidiospores that can infect apple leaves. The galls eventually die but
may remain attached to the tree for a year or more.

Basidiospores are wind borne and may be carried for up to 2 to 3 miles.
Their germ tubes penetrate young apple leaves or fruit directly, and
produce haploid mycelium that spreads through or between the apple
cells. The mycelium forms orange-colored spermagonia on the upper leaf
surface and, presumably following fertilization of receptive hyphae by
compatible spermatia, dikaryotic mycelium ensues which produces
aecial cups in concentric rings on the lower side of leaves and on fruit.
The area of the leaf where spermagonia and aecia are produced is swollen,
especially on the lower side where the clusters of orange-yellow aecial
cups and their white peridia (cup walls) stand out conspicuously (Fig.
126A). In the fruit, spermagonia and aecia are formed in the same areas;
the spermagonia appear first in the center of the spot and the aecia
subsequently in the surrounding area. Infected fruit areas are usually

FIGURE 126.
Cedar-apple rust caused by *Gymnosporangium juniperi-virginianae.* (A)
Clusters of aecial cups on apple leaves. (B) Infected apple fruit with numerous
aecial cups and a few spermagonia in center of spot. (C) Gall on cedar (cedar apple)
in early spring when telial horns are just coming out. (D) Mature telial horns
releasing basidiospores. (Photos A–C courtesy U.S.D.A.)

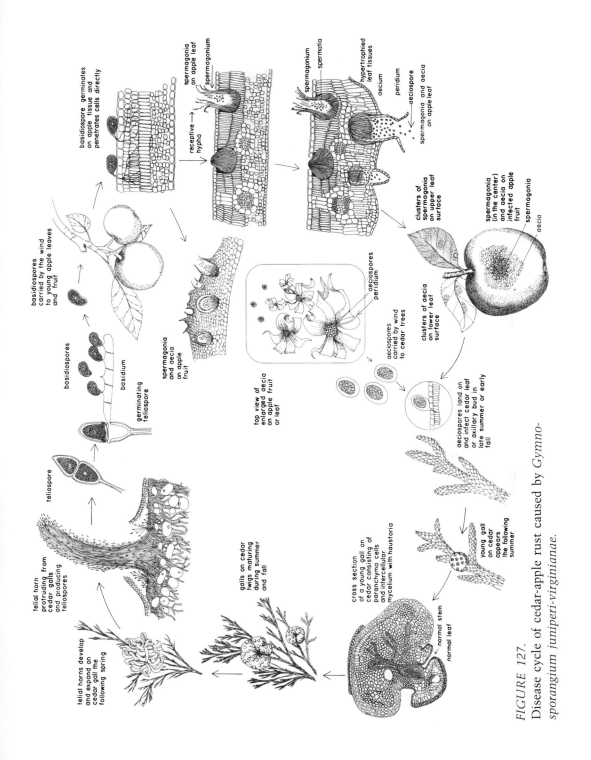

FIGURE 127.
Disease cycle of cedar-apple rust caused by *Gymno-sporangium juniperi-virginianae.*

387

large and flat or depressed rather than swollen (Fig. 126B). The aecio-spores are produced in long chains. They are released in the air during dry weather in late summer and are carried by wind to cedar leaves where they germinate and start new infections.

Control of cedar-apple rust can be effected by keeping apple and cedar trees sufficiently removed from each other so that the fungus cannot complete its life cycle. This, however, is often impossible or impractical and therefore the disease is generally controlled on both hosts with chemical sprays with ferbam, thiram or maneb. Postinfection control of cedar-apple rust has been obtained recently with triforine. Many apple varieties are also quite resistant to rust.

- *Coffee Rust*

Undoubtedly the most destructive disease of coffee, coffee rust damages trees and reduces yields by causing premature drop of infected leaves. Coffee rust has caused devastating losses in all coffee producing countries of Asia and Africa. It attacks all species of coffee but is most severe on *Coffea arabica*. In 1970 the disease appeared for the first time in the Western Hemisphere, in Brazil, and has been steadily spreading into the world's most important coffee producing countries of South America where all commercial coffees are susceptible to the rust.

The symptoms appear as orange-yellow powdery spots on the lower side of the leaves. The spots are circular and small, about 5 mm in diameter, at first, but they often coalesce and form large patches that may be ten times as large. The centers of the spots eventually become dry, turn brownish, and the leaf falls off prematurely. Infected trees produce small yields of poor quality and repeated infections and defoliations result in the death of trees (Fig. 128).

The fungus, *Hemileia vastatrix*, exists primarily as mycelium, uredia, and uredospores which in the tropics, where coffee and the fungus thrive, can perpetuate themselves in infected leaves they continuously and successively infect. The fungus occasionally produces teliospores which upon germination form basidiospores, but the latter do not infect coffee and no alternate host has so far been found. Thus, uredospores are believed to be responsible for all coffee infections. They are easily spread by wind, rain, and perhaps by insects. The spores require high humidity and probably dew for germination and infection. Under favorable conditions they can germinate and enter leaves through the stomata of the lower surface in less than 12 hours. The mycelium grows between the leaf cells and sends haustoria into the cells. Young leaves are generally more susceptible to infection than older ones and new uredia may appear on the lower side of the leaf within 10 to 25 days from infection, depending on the climatic conditions. Once uredia develop, premature falling of infected leaves may occur at any time; sometimes even one uredium is sufficient to cause the leaf to fall. New leaves are affected after the older ones have fallen. The premature shedding of leaves weakens the trees and results in reduced yields, severe dieback of twigs, and death of trees.

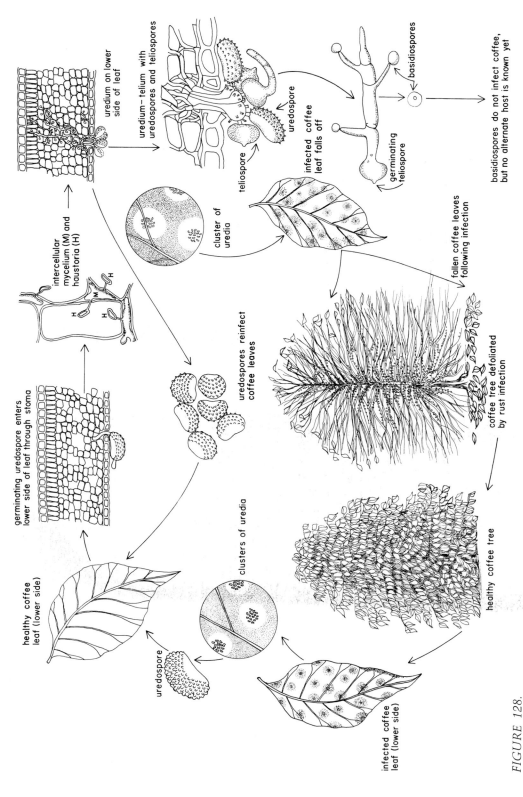

uredium on lower side of leaf

uredium–telium with uredospores and teliospores

uredospore

teliospore

basidiospores

infected coffee leaf falls off

germinating teliospore

basidiospores do not infect coffee, but no alternate host is known yet

intercellular mycelium (M) and haustoria (H)

cluster of uredia

fallen coffee leaves following infection

uredospores reinfect coffee leaves

germinating uredospore enters lower side of leaf through stoma

coffee tree defoliated by rust infection

healthy coffee leaf (lower side)

healthy coffee tree

clusters of uredia

uredospore

infected coffee leaf (lower side)

FIGURE 128.
Disease cycle of coffee rust caused by *Hemileia vastatrix.*

389

Control of coffee rust is difficult, but satisfactory results can be obtained with copper fungicides such as Bordeaux mixture and cuprous oxide, and with zineb. Fungicides must be applied before and during the rainy season at 2- to 3-week intervals or less, depending on weather conditions and the severity of the attack. Sufficient tree pruning, good site selection, and use of the newly found or developed resistant varieties of Arabica coffee in future plantings should help minimize losses from the rust.

RUSTS OF FOREST TREES

Several species of *Cronartium* are responsible for a number of rust diseases that cause major losses in forest trees. Some *Cronartium* species attack the main stem or branches of trees and these are the most destructive; other species attack the needles or leaves and are less serious. All rusts, however, are especially destructive when they attack young trees in the nursery or in recently established plantations. The main economic host of the majority of forest tree rusts and the one to which they cause the most damage is pine. Some of these rusts have oak as their alternate host but the damage to oaks is much less severe. Other pine rusts have as their alternate hosts various wild or cultivated shrubs or weeds.

- *White Pine Blister Rust*

It is native to Asia from where it spread to Europe and about 1900, to North America. It is one of the most important forest diseases in North America where it causes an annual growth loss and mortality of more than 200 million cubic feet, and, if not controlled, it makes white pine growing impossible or unprofitable. White pine blister rust is caused by the fungus *Cronartium ribicola* which produces its spermagonia and aecia on white pine (the 5-needle pines) and its uredia and telia on wild and cultivated currant and gooseberry bushes (*Ribes* sp.). Blister rust kills pines of all ages and sizes. Small pines are killed quickly while larger pines may develop cankers that girdle and either kill the trees or retard their growth and weaken the stems which then break at the canker. Infection of *Ribes* bushes causes relatively little loss through premature, partial defoliation and reduced fruit production.

The symptoms of blister rust on white pine stems or twigs appear first as small, discolored, spindle-shaped swellings (cankers) surrounded by a narrow band of yellow-orange bark. In the canker, small, irregular, dark brown, blisterlike spermagonia appear which rupture, ooze droplets full of spermatia, and then dry. As the canker grows, the margin and the zone of spermagonia expand and the portion formerly occupied by spermagonia is now the area where the aecia are produced. The aecia appear as white sacks or blisters containing orange-yellow aeciospores that push through the diseased bark. The aecial blisters soon rupture (Fig. 129A) and the orange-yellow aeciospores are carried by the wind, sometimes for several hundred miles, some of them landing on and infecting *Ribes* leaves. After the aeciospores have been released, the blisters persist on the bark for a long time although the bark of that area dies. Resin often flows down the

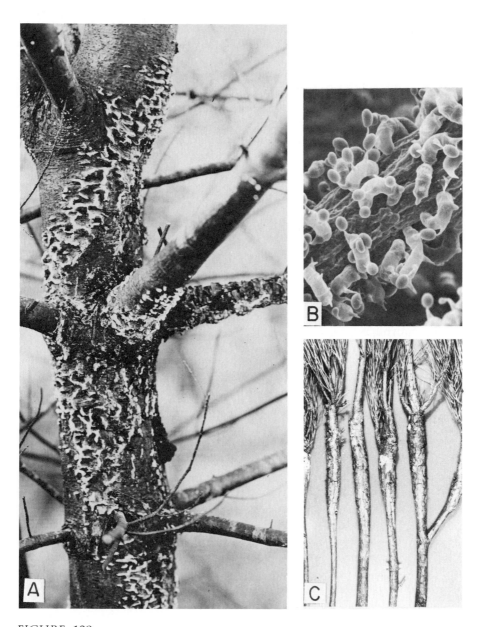

FIGURE 129.
(A) White pine blister rust caused by *Cronartium ribicola*. (B) Scanning electron micrograph of part of telium of *Cronartium fusiforme* showing basidia and basidiospores. (C) Fusiform rust on pine seedlings. Masses of aeciospores can be seen on two of them. (Photo A courtesy U.S.D.A. Photo B courtesy M. F. Brown and H. G. Brotzman.)

stem and hardens in masses characteristic of the disease. The fungus, however, continues to spread into the surrounding healthy bark and the sequence of spore production and bark killing continues in subsequent years until the stem or branch is girdled and killed. The dead branches, called "flags," have dead, brown needles and are visible from a distance.

On currant and gooseberries, the symptoms appear on the undersides of the leaves as slightly raised, yellow-orange uredia grouped in circular or irregular spots. The uredia produce orange masses of uredospores that reinfect *Ribes*. Later, telia develop in the same or new lesions. The telia are slightly darker than the uredia and consist of brownish, hairlike structures up to 2 mm in height that bear the teliospores.

The pathogen, *Cronartium ribicola*, overwinters mostly as mycelium in infected white pines and *Ribes*. Pines are infected only by basidiospores produced by teliospores still in the telia on the undersides of *Ribes* leaves (Fig. 130). The basidiospores are produced only during wet, cool periods, especially during the night, and can be carried by wind and infect pines within a few hundred feet, and generally no more than a mile or so, from the *Ribes* host. The basidiospores infect pine needles through stomata in late summer or early fall. Small, discolored spots may appear on the needles 4 to 10 weeks after infection. The mycelium grows down the conducting tissues of the needle and into the bark of the stem, which it reaches about 12 to 18 months after infection. Spermagonia develop on infected stems or branches in the spring and early summer 2 to 4 years after the needle infection and aecia are produced in the spring, 3 to 6 years from inoculation. Spermatia are short lived and spread over short distances by rain or insects, while aeciospores may live for many months, may overwinter and may be carried by wind over many miles to *Ribes* leaves. On the latter, the aeciospores germinate and infect the leaves which produce uredia and uredospores within 1 to 3 weeks after inoculation. Uredospores can reinfect *Ribes* plants again and again, producing many generations of uredospores in a single growing season. The uredospores can survive for many months and even through winter, and can be spread by wind for a mile or more, but they can infect only *Ribes*. Finally, the same mycelium that produced the uredospores begins to produce telial columns and teliospores. The latter germinate from July to October and produce short-lived basidiospores, which, if blown to nearby white pines, infect the needles and complete the life cycle of the fungus.

Control of white pine blister rust can be obtained by eradication of wild and cultivated *Ribes* bushes mechanically or, better still, with herbicides such as 2,4-D and 2,4,5-T. Pruning infected branches on young trees reduces stem infections and tree mortality. Treatment of cankers or entire trees with antibiotics such as cycloheximide (Actidione) appeared promising for a time but is no longer considered effective. The use of the hyperparasite *Tuberculina maxima*, which parasitizes *Cronartium ribicola* on blister rust cankers, has also been considered as a control measure, but so far its practical value has not been demonstrated. The most promising control for blister rust seems to be the selection and breeding of resistant trees. Seed orchards with trees that have shown resistance to the disease have been established and are expected to produce millions of resistant trees in the near future.

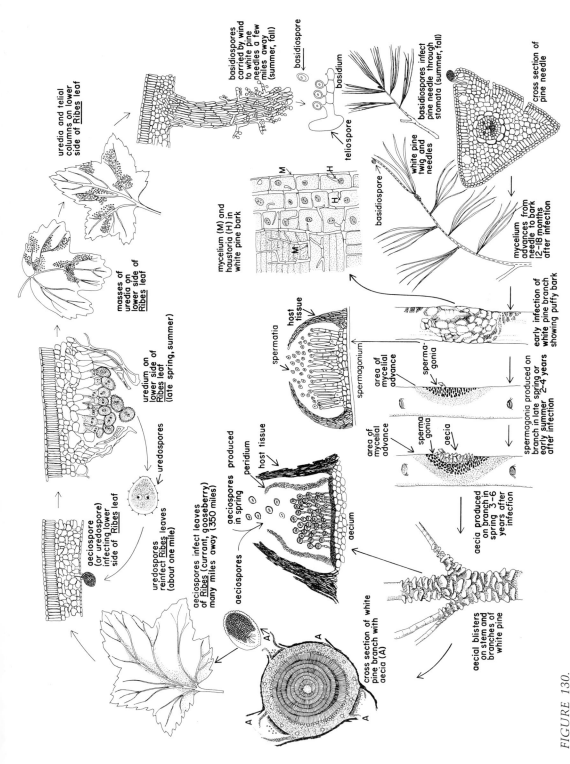

FIGURE 130.
Disease cycle of white pine blister rust caused by *Cronartium ribicola.*

- *Fusiform Rust*

It is one of the most important diseases on southern pines, especially loblolly and slash pines. The disease is present from Maryland to Florida and west to Texas and Arkansas, where it causes tremendous losses in nurseries, young plantations and seed orchards ranging from 20 to 60 percent or even more mortality of young trees. Fusiform rust is caused by *Cronartium fusiforme* which produces spermagonia and aecia on pine stems and branches and uredia and telia on oak leaves. Damage on oak is slight, mostly through occasional partial defoliation.

The symptoms on pine first appear as small, purple spots on needles and succulent shoots. These soon form small galls and later develop into spindleshaped swellings or galls (cankers) on branches and stems of mostly young pines. These galls may elongate from 5 to 15 cm per year and often encircle the stem or branch and cause it to die (Fig. 129C). Infection of young seedlings results in their death within a very few years, while infected young trees may branch excessively for a period and show a bushy growth. On older trees, stem or branch infections lead to weak, distorted boles or, as host tissue is killed, to sunken cankers that break easily during strong winds. Yellowish masses of spermatia and later orange-yellow aeciospores appear on the galls. On oak, the symptoms appear as orange pustules (uredia) and brown, hairlike columns (telia) on the underside of the leaves.

The fusiform rust pathogen, *Cronartium fusiforme*, overwinters as mycelium in the fusiform galls. From February to April, spermagonia and spermatia form and soon aeciospores are produced on the galls. The wind carries the aeciospores to young, expanding oak leaves which they infect. On the oak leaves, orange uredial pustules develop in a few days and produce uredospores from February to May. Uredospores can reinfect more oak leaves and produce more uredospores. The same mycelium also produces brown telia from February to June in place of uredia or in new lesions. The teliospores germinate on the telia (Fig. 129B) and the basidiospores produced are carried by wind to pine needles and shoots which they infect directly. The mycelium grows first in the needles and later spreads into branches or the stem where it induces both hyperplasia and hypertrophy and the formation of the gall.

Fusiform rust infections in the nursery can be prevented by frequent, twice-a-week sprays with ferbam, especially before and during cool wet weather. All infected seedlings should be discarded. In plantations and natural stands, only limited control can be obtained against fusiform rust by either avoiding planting highly susceptible slash and loblolly pines in areas of known high rust incidence or by pruning infected branches before the fungus reaches the trunk. As with white pine blister rust, and perhaps even more so, control of fusiform rust is obtained through selection and breeding of resistant trees, with emphasis on trees possessing general rather than specific resistance.

SELECTED REFERENCES

Allen, Ruth F. 1930. A cytological study of heterothallism in *Puccinia graminis. J. Agr. Res.* **40**:585–614.

Allen, Ruth F. 1933. A cytological study of the teliospores, promycelia and sporidia in *Puccinia malvacearum*. *Phytopathology* **23**:572–586.

Arthur, J. C., and G. B. Cummins. 1962. "Manual of the Rusts in United States and Canada." Hafner Publishing Co., New York, 438 p. plus Suppl. 24 p.

Bliss, D. E. 1933. The pathogenicity and seasonal development of *Gymnosporangium* in Iowa. *Iowa Agr. Exp. Sta. Res. Bull.* **166.**

Browning, J. A. and K. J. Frey. 1969. Multiline cultivars as a means of disease control. *Ann. Rev. Phytopathol.* **7**:355–382.

Burleigh, J. R., A. P. Roelfs, and M. G. Evermeyer. 1972. Estimating damage to wheat caused by *Puccinia recondita tritici*. *Phytopathology* **62**:944–946, 947–953.

Caldwell, R. M., *et al.* 1934. Effect of leaf rust (*Puccinia triticina*) on yield, physical characters, and composition of winter wheats. *J. Agr. Res.* **48**:1049–1071.

Colley, R. H. 1918. Parasitism, morphology, and cytology of *Cronartium ribicola*. *J. Agr. Res.* **15**:619–660.

Crowell, I. H. 1934. The hosts, life history, and control of the cedar-apple rust fungus *Gymnosporangium juniperi-virginianae*. *J. Arnold Arb.* **15**:163–232.

Cummins, G. B. 1959. "Illustrated Genera of Rust Fungi." Burgess Publishing Co., Minneapolis, Minn. 131 p.

Davidson, A. G., and R. M. Prentice (Eds.) 1967. Important forest insects and diseases of mutual concern to Canada, the United States, and Mexico. Department of Forestry and Rural Development, Canada. 248 p.

Dickson, J. G. 1956. "Diseases of Field Crops." McGraw-Hill, New York. 517 p.

Dodge, B. O. 1923. Systemic infections of *Rubus* with the orange rusts. *J. Agr. Res.* **25**:209–242; A new type of orange rust on blackberry, *Ibid.* **25**:491–494.

Flor, H. H. 1971. Current status of the gene-for-gene concept. *Ann. Rev. Phytopathol.* **9**:275–296.

French, D. W., *et al.* 1972. Forest and Shade Tree Pathology. Mimeo. 224 p.

Hamilton, Laura M., and E. C. Stakman. 1967. Time of stem rust appearance on wheat in the western Mississippi basin in relation to the development of epidemics from 1921 to 1962. *Phytopathology* **57**:609–614.

Hare, R. C. 1973. Soil applications of systemics for prevention and eradication of fusiform rust. *Plant Dis. Reptr.* **57**:776–780.

Hart, Helen. 1931. Morphologic and physiologic studies on stem-rust resistance in cereals. *U.S. Dept. Agr. and Minn. Agr. Expt. Sta. Tech. Bull.* **266**: 75 pp.

Harvey, A. E., and J. I. Grasham. 1974. Axenic culture of the mononucleate stage of *Cronartium ribicola*. *Phytopathology* **64**:1028–1035.

Hobbs, C. D., and M. C. Futrell. 1966. Evaluation of nickel and dithiocarbamates for control of wheat stem rust. *Plant Dis. Reptr.* **50**:373–376.

Hogg, W. H., C. E. Hounam, A. K. Mallik, and J. C. Zadoks. 1969. Meteorological factors affecting the epidemiology of wheat rusts. *World Meteorol. Org. Tech. Note* **99**: 143 p.

Hooker, A. L. 1967. The genetics and expression of resistance in plants to rusts of the genus *Puccinia*. *Ann. Rev. Phytopathol.* **5**:163–182.

Johnson, T., G. J. Green, and D. J. Samborski. 1967. The world situation of the cereal rusts. *Ann. Rev. Phytopathol.* **5**:183–200.

Kinloch, B. B., and R. W. Stonecypher. 1969. Genetic variation in susceptibility to fusiform rust in seedlings from a wild population of loblolly pine. *Phytopathology* **59**:1246–1255.

Loegering, W. Z., J. W. Hendrix, and L. E. Browder. 1967. The rust diseases of wheat. *U.S. Dept. Agr. and Wash. State Univ. Agr. Handbook* **334**: 22 pp.

Lopez, A., S. Rajaram, and L. I. DeBauer. 1974. Susceptibility of triticale, rye and wheat to stem rust from these three hosts. *Phytopathology* **64**:266–267.

Mielke, J. L. 1943. White pine blister rust in western North America. *Yale School of Forestry Bull.* 52.

National Academy of Sciences. 1972. "Genetic Vulnerability of Major Crops." Washington, D. C. 307 p.

Palmiter, D. H. 1952. Rust diseases of apples and their control in the Hudson Valley. *N. Y. (State) Agr. Exp. Sta. Bull.* **756.**

Peterson, G. W., and R. S. Smith, Jr. 1975. Forest nursery diseases in the United States. *USDA Forest Service Agr. Handbook* **470**: 125 p.

Peterson, R. S., and F. F. Jewell. 1968. Status of American stem rusts of pine. *Ann. Rev. Phytopathol.* **6**:23–40.

Phelps, W. R., and C. W. Chellman. 1975. Impact of fusiform rust in north Florida slash pine plantations. *Plant Dis. Reptr.* **59**:481–485.

Rijkenberg, F. H. J., and S. J. Truter. 1973. Haustoria and intracellular hyphae in the rusts. *Phytopathology* **63**:281–286.

Roelfs, A. P. 1974. Evidence for two populations of wheat stem and leaf rust in the U.S.A. *Plant Dis. Reptr.* **58**:806–809.

Rowell, J. B. 1971. Chemical control of the cereal rusts. *Ann. Rev. Phytopathol.* **6**:243–262.

Rowell, J. B. 1973. Control of leaf and stem rusts of wheat by seed treatment with oxycarboxin. *Plant Dis. Reptr.* **57**:567–571.

Schieber, E. 1972. Economic impact of coffee rust in Latin America. *Ann. Rev. Phytopathol.* **10**:491–510.

Schieber, E. 1975. Present status of coffee rust in South America. *Ann. Rev. Phytopathol.* **13**:375–382.

Scott, K. J., and D. J. Maclean. 1969. Culturing of rust fungi. *Ann. Rev. Phytopath.* **7**:123–146.

Shaw, M. 1963. The physiology and host-parasite relations of the rusts. *Ann. Rev. Phytopathol.* **1**:259–294.

Simons, M. D. 1975. Heritability of field resistance to the oat crown rust fungus. *Phytopathology* **65**:324–328.

Stakman, E. C. 1914. A study in cereal rusts. Physiological races. *Minn. Agr. Expt. Sta. Bull.* **138**: 56 pp., illus.

Szkolnik, M. 1974. Unique post-infection control of cedar-apple rust on apple with triforine. *Plant Dis. Reptr.* **58**:587–590.

Ward, H. M. 1882. Researches on the life history of *Hemileia vastatrix*, the fungus of the "coffee leaf disease." *Linn. Soc. J. (Bot.)* **19**: 229–335.

Welch, B. L., and N. E. Martin. 1974. Invasion mechanisms of *Cronartium ribicola* in *Pinus monticola* bark. *Phytopathology* **64**:1541–1546.

Western, J. H. 1971. "Diseases of Crop Plants." Macmillan Press, New York, 404 p.

Zadoks, J. C. 1965. Epidemiology of wheat rusts in Europe. *FAO Plant Protect. Bull.* **13**:97–108.

THE SMUTS

The plant smuts caused by Basidiomycetes of the order Ustilaginales occur throughout the world and, until this century, were the causes of serious grain losses that were equal to, or second only to, the losses caused by the rusts. In some respects, the smuts of cereals were dreaded by farmers even more than the rusts because many smuts attack the grain kernels themselves and replace the kernel contents with the black, dusty spore masses that resemble soot or smut. Thus the reduction in yield is conspicuous and direct and the quality of the remaining yield is drastically reduced by the presence of the black smut spores on the surface of the healthy kernels.

In addition to the various cereals, smuts also affect onions and some ornamentals such as carnation.

Most smut fungi attack the ovaries of grains and grasses and develop in them or in the fruit, i.e., the kernels of grain crops, which they destroy completely (Fig. 120). Several smuts, however, attack the leaves, stems or floral parts. Some smuts infect seeds or seedlings before they emerge from the ground and they grow internally in the seedling until they reach the inflorescence; others cause only local infections on leaves, stems, etc. Cells in affected tissues are either destroyed and replaced by black smut spores, or they are first stimulated to divide and enlarge so as to produce a swelling or gall of varying size and are then destroyed and replaced by the black smut spores. The spores are present in masses called sori that may be held together only temporarily by a thin, flimsy membrane or by a more or less durable one. Smut fungi seldom kill their hosts but in some cases infected plants may be severely stunted.

The smut fungi can be grown in culture on artificial media but in nature they exist only as parasites on their hosts. Most smut fungi produce only two kinds of spores: teliospores and basidiospores (Fig. 118). Their teliospores are usually formed from mycelial cells along the length of the mycelium within the smut galls and their basidiospores either bud off laterally from the basidium cells or are produced as a cluster at the tip of a nonseptate basidium. The basidiospores of the smuts are not borne on sterigmata. When basidiospores germinate, they either unite with compatible ones while still on the basidium and then infect, or their germ tubes penetrate tissues directly. Their haploid mycelium, however, cannot invade tissues extensively and does not cause typical infections until two compatible mycelia unite to produce dikaryotic mycelium. The latter, then, invades tissues inter- or intracellularly, generally without haustoria, and produces the typical symptoms and the teliospores. The smut fungi also exist in many races which, however, are not as stable as in the rusts since each generation of smut fungi on the host plant involves meiosis, i.e., genetic recombination, and this results in new races appearing constantly.

The most common smut fungi and the diseases they cause are the following:

> *Ustilago,* causing corn smut (*U. maydis* or *U. zeae*), loose smut of oats (*U. avenae*), of barley (*U. nuda*) and of wheat (*U. tritici*), and covered smut of barley (*U. hordei*), and of oats (*U. kolleri*).
>
> *Tilletia,* causing covered smut or bunt of wheat (*T. caries* and *T. foetida*) and dwarf bunt of wheat (*T. contraversa*).
>
> *Sphacelotheca,* causing the sorghum smuts such as covered kernel smut (*S. sorghi*), loose kernel smut (*S. cruenta*), and head smut (*S. reiliana*).
>
> *Urocystis,* causing onion smut (*U. cepulae*), and leaf or stalk smut of rye (*U. occulta*).
>
> *Neovossia,* causing kernel smut of rice (*N. barclayana*).
>
> *Entyloma,* causing leaf smut of rice (*E. oryzae*) and of some broad-leaved plants such as spinach (*E. ellisii*).

The smuts generally overwinter as teliospores either on contaminated seed, plant debris or in the soil. However, some smuts overwinter as mycelium inside infected kernels or as mycelium in infected plants. The teliospores cannot infect but they produce the basidiospores which upon germination either fuse with compatible ones and then infect, or penetrate the tissue and then fuse to produce dikaryotic mycelium and the typical infection. The smut fungi have only one generation per year, each infection resulting in one crop of teliospores per growing season.

Control of smuts is primarily by resistant varieties and seed treatment. The latter may be either by chemical dusting or dip, if the fungus is present as teliospores on the seed surface or in the soil, or by hot water if the fungus is present as mycelium inside the seed. The discovery of carboxin and other fungicides which are absorbed and translocated systemically by seeds and seedlings allow chemical control by seed treatment of even those smuts present as mycelium inside the seeds. Soil treatments with these and other chemicals are also useful in the control of smut diseases.

GENERAL SMUTS

- *Corn Smut*

Corn smut occurs wherever corn is grown. It is more prevalent, however, in warm and moderately dry areas, where it causes serious damage to susceptible varieties.

Corn smut damages plants and reduces yields by forming galls on any of the aboveground parts of plants, including ears, tassels, stalks, and leaves. The number, size, and location of smut galls on the plant affect the amount of yield loss. Galls on the ear usually destroy it completely, while large galls above the ear cause much greater reduction in yield than do galls below the ear. Losses from corn smut are highly variable from one location to another and may range from a trace up to 10 percent or more in localized areas. Some individual fields of sweet corn may show losses approaching 100 percent from corn smut. Generally, however, over large areas and with the use of resistant varieties, losses in grain yields average about 2 percent.

Symptoms.   When young corn seedlings are infected, minute galls form on the leaves and stem, and the seedling may remain stunted or may be killed. Seedling infection, because of seedling death, is seldom observed in the field.

On older plants, infections occur on the young, actively growing tissues of axillary buds, individual flowers of ear and tassel, leaves, and stalks (Fig. 131).

Infected areas are permeated by the fungus mycelium which stimulates the host cells to divide and enlarge, thus forming overgrowths or galls. Galls are first covered with a greenish white membrane. Later as the galls mature, they reach a size from 1 to 15 cm in diameter, their interior darkens, and they turn into a mass of powdery, dark, olive-brown spores. The silvery gray membrane then ruptures and exposes the millions of the

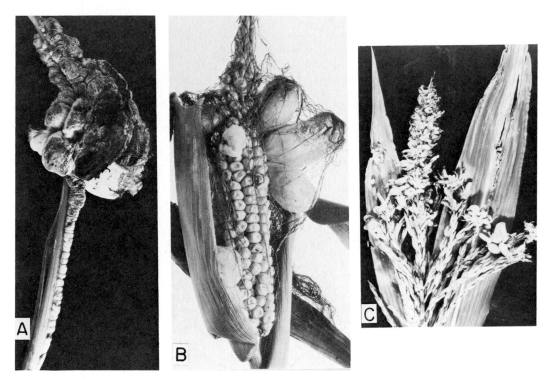

FIGURE 131.
Corn smut symptoms on young stem of corn plant (A), ear of corn (B) and on male inflorescence (C). (Photos A and B courtesy Dept. Plant Path., Cornell Univ.)

sooty teliospores, which are released into the air. Galls on leaves frequently remain very small (about 1 to 2 cm in diameter); they become hard and dry and do not rupture.

The pathogen: Ustilago maydis.   It produces dikaryotic mycelium the cells of which are transformed into black, spherical, or ellipsoidal teliospores that have prominent spinelike protuberances. Teliospores germinate by producing a 4-celled basidium (promycelium), from each cell of which an ovate, hyaline, uninucleate basidiospore (sporidium) develops (Fig. 132).

Development of disease.   The fungus overwinters as teliospores in crop debris and in the soil, where it can remain viable for several years. In the spring and summer teliospores germinate and produce basidiospores which are carried by air currents or are splashed by water to young, developing tissues of corn plants. The basidiospores germinate on the host surface and produce a fine hypha which can enter epidermal cells by direct penetration. After an initial development, however, its growth stops and the hypha usually withers and sometimes dies, unless it contacts and fuses with a haploid hypha derived from a basidiospore of the compatible mating type. If fusion takes place, the resulting hypha enlarges in diameter and becomes dikaryotic. The dikaryotic hypha grows into the plant tissues mostly intercellularly (Fig. 132).

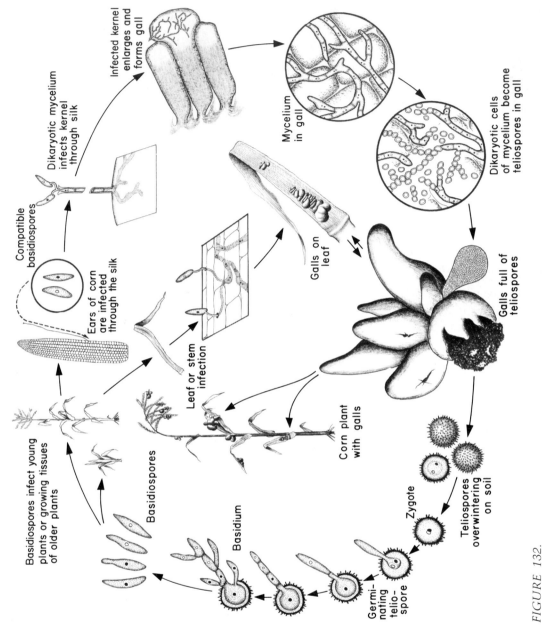

Infected kernel enlarges and forms gall

Dikaryotic mycelium infects kernel through silk

Mycelium in gall

Dikaryotic cells of mycelium become teliospores in gall

Compatible basidiospores

Ears of corn are infected through the silk

Galls on leaf

Galls full of teliospores

Leaf or stem infection

Basidiospores infect young plants or growing tissues of older plants

Corn plant with galls

Basidiospores

Zygote

Teliospores overwintering on soil

Basidium

Germi-nating telio-spore

FIGURE 132.
Disease cycle of corn smut caused by *Ustilago maydis*.

400

The cells surrounding the hypha are stimulated to hypertrophy and hyperplasia, and galls begin to form. Hyperplasia may appear in advance of the actual invasion of the tissues by the fungus, and galls may begin to form even before the fungus actually gets there.

Galls in developed plants seem always to be the result of local infections of plant tissues. Systemic infections seldom occur, and then only in very young seedlings. Frequently, however, only a small number of the actual local infections develop into typical, large galls, the others remaining too small to be visible.

The mycelium in the gall remains intercellular during most of gall formation, but before sporulation, the enlarged corn cells are invaded by the mycelium, collapse, and die. The mycelium utilizes the cell contents for its further growth and the gall then consists primarily of dikaryotic mycelium and cell remnants. Most of the dikaryotic cells subsequently develop into teliospores, and in the process seem to absorb and utilize the protoplasm of the other mycelial cells which remain empty. Only the membrane covering the gall is not affected by the fungus, but finally the membrane breaks and the teliospores are released. Some of the released teliospores, if they land on young, meristematic corn tissues may cause new infections and new galls during the same season, but most of them fall to the ground or remain in the corn debris where they can survive for several years.

Control.   Corn smut may be controlled to a degree through the use of corn hybrids with some resistance to the fungus. No corn varieties or hybrids completely resistant to smut are known. The pathogen, however, shows extreme variability in its pathogenicity and new races appear constantly, making control through resistance difficult. Control through sanitation measures, such as removal of smut galls before they break open, and through crop rotation is possible only where corn is grown in small, rather isolated plots, but is impractical and impossible in large corn-growing areas.

KERNEL SMUTS OF SMALL GRAINS

- *Loose Smut of Cereals*

Loose smut of cereals is worldwide in distribution but is more abundant and serious in humid and subhumid regions.

Loose smut causes damage by destroying the kernels of the infected plants and by smearing and thus reducing the quality of the grain of the noninfected plants upon harvest. Losses from loose smut may be up to 10 or 40 percent in certain localities in a given year, but the overall losses in the U.S. are approximately 2 percent per year.

Symptoms.   Loose smut generally does not produce discernible symptoms until the plant has headed. Smutted plants sometimes head earlier than healthy ones, and smutted heads are often elevated above those of the healthy plants. In an infected plant usually all the heads and all the spikelets and kernels of each head are smutted, although some of them may sometimes escape infection. In infected heads each spikelet is entirely transformed into a smut mass consisting of olive-green spores (Fig. 133). This is at first covered by a delicate grayish membrane which soon bursts and sets the powdery spores free. The spores are then blown off by the wind and leave the rachis a naked stalk.

The pathogens: Ustilago nuda, U. tritici.   The mycelium is hyaline during its growth through the plant, but it changes to brown near maturity. The mycelial cells are transformed into brown, spherical, echinulate teliospores which germinate readily and produce a basidium consisting of one to four cells. The basidium produces no basidiospores, but its cells germinate and produce short, uninucleate hyphae that fuse in pairs and produce dikaryotic mycelium which is capable of infection (Fig. 134).

Development of disease.   The pathogens overwinter as dormant mycelium in the cotyledon (sometimes called the scutellum) of infected kernels. When planted, infected kernels begin to germinate, the mycelium resumes its activity and grows intercellularly through the tissues of the embryo and the young seedling until it reaches the growing point of the plant (Fig. 134). The mycelium then follows closely the growth of the plant and grows best just behind the growing point, while

FIGURE 133.
Loose smut of barley as it appears in the field (A) and on a single head of barley (B). (Photos courtesy Dept. Plant Path., Cornell Univ.)

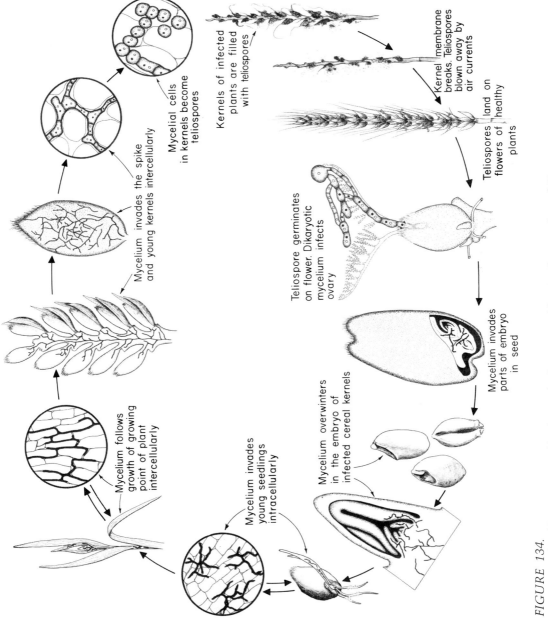

Mycelial cells
in kernels become
teliospores

Mycelium invades the spike
and young kernels intercellularly

Kernels of infected
plants are filled
with teliospores

Kernel membrane
breaks. Teliospores
blown away by
air currents

Teliospores land on
flowers of healthy
plants

Teliospore germinates
on flower. Dikaryotic
mycelium infects
ovary

Mycelium invades
parts of embryo
in seed

Mycelium follows
growth of growing
point of plant
intercellularly

Mycelium invades
young seedlings
intracellularly

Mycelium overwinters
in the embryo of
infected cereal kernels

FIGURE 134.
Disease cycle of loose smuts of barley and wheat caused by *Ustilago nuda* and *U. tritici.*

the hyphae in the tissues of the lower stem atrophy and frequently disappear. When the plant forms the head, and even before it emerges, the mycelium invades all the young spikelets, where it grows intracellularly and destroys most of the tissues of the spike, except the rachis. By this time, most of the infected plants are slightly taller than most healthy plants, probably due to stimulatory action of the pathogen. The mycelium in the infected kernels is soon transformed into teliospores which are contained only by a delicate outer membrane of host tissue. The membranes burst open soon after maturation of the teliospores and the spores are released and blown off by air currents to nearby healthy plants. The spore release coincides with the opening of the flowers of healthy plants. Teliospores landing on flowers germinate through formation of a basidium on which the haploid hyphae are produced. After fusion of the sexually compatible haploid hyphae, the resulting dikaryotic mycelium penetrates the flower through the stigma or through the young ovary walls and becomes established in the pericarp, integuments, etc., and in the tissues of the embryo before the kernels become mature. The mycelium then becomes inactive and remains dormant, primarily in the scutellum, until the infected kernel germinates.

Control. Control of loose smut is now obtained by treating infected seeds with carboxin and other derivatives of 1,4-oxanthiin before planting. These chemicals are absorbed and act systemically in the seed or in the growing plant.

Although some barley and wheat varieties are quite resistant to loose smut, most of the commercial varieties are very susceptible to it.

The best means of controlling loose smut is through the use of certified smut-free seed. Until the discovery of the systemic fungicides, when seed was known to be infected with loose smut mycelium, the best way of disinfecting it was by treating it with hot water. Usually small lots of seed are treated with hot water and planted in isolated fields to produce smut-free seed to be used during the next season. The hot-water treatment consists of soaking the seed, contained in half-filled burlap bags, in 20°C water for 5 hours, draining it for 1 minute, dipping it in 49°C water for about 1 minute, then in 52°C water for exactly 11 minutes, and immediately afterward in cold water for the seed to cool off. The seed is then allowed to dry so that it can be sown. Since some of the seed may be killed by the hot-water treatment, a higher seeding rate may be employed to offset the reduced germinability of the treated seed.

- *Covered Smut, or Bunt, of Wheat*

Covered smut, or bunt, of wheat, sometimes called stinking smut of wheat, is widely distributed in all the wheat-growing areas of the world. In the United States, the disease causes its greatest losses in the Pacific Northwest.

Bunt affects plants by destroying the contents of infected kernels and replacing them with the spores of the fungus. Bunt also causes slight to severe stunting of infected plants, depending on the particular species of the bunt fungus involved. Infected plants are usually more susceptible than healthy plants to certain other diseases and to winter injury. Bunt causes losses in grain yields proportional to the number of plants affected. When bunt is not controlled, it may cause devastating losses, but even with the effective control measures practiced in the U.S. today, the disease continues to cause severe losses. In addition, bunt and the other smuts cause market losses by reducing the quality, and the price, of wheat contaminated with smutted kernels or smut spores because of the discoloration and the foul odor they impart to the whole wheat crop. Bunt, moreover, results in explosions in combines and elevators during threshing or handling of smutted wheat because of the extreme combustibility of the oily smut spores in the presence of sparks from machinery.

Symptoms.   Plants infected with the common bunt fungi are usually a few to several centimeters shorter than healthy plants and may sometimes be only half as tall. Plants infected with the dwarf bunt fungus may be only one-fourth as tall as healthy plants. Infected plants may appear slightly bluish green to grayish green in color, but this is not easily distinguishable. The root system of infected plants is usually poorly developed.

Distinct bunt symptoms, however, appear when the heads of infected plants emerge. Their color is usually bluish green rather than the normal yellowish green, they are slimmer than healthy heads, and the glumes seem to spread apart and form a greater angle with the main axis of the head than they do in healthy plants (Fig. 135). Infected kernels are shorter and thicker than healthy ones and are grayish brown rather than the normal golden yellow or red. When these kernels are broken following their maturation, they are found to be full of a sooty, black, powdery mass of fungus spores which give off a distinctive odor resembling that of decaying fish. During harvest of infected fields, large clouds of spores may be released in the air.

The pathogens: Tilletia caries, T. foetida, and T. contraversa. All three species are similar in their life histories and disease development, but the teliospores of *T. foetida* have smooth walls, while those of *T. caries* have highly reticulate walls. *T. contraversa* is the cause of dwarf bunt, and its spore walls have large, polygonal reticulations.

The mycelium is hyaline and during sporulation most cells are transformed into almost spherical, brownish teliospores. The rest of the mycelial cells remain hyaline, thin walled, and sterile. Upon germination of a teliospore a basidium is produced, at the end of which 8 to 16 basidiospores develop. The basidiospores are usually called primary sporidia. The primary sporidia fuse in pairs through production of lateral branches between compatible mating types and appear as H-shaped structures (Figs. 135 and 136). The nucleus of each primary sporidium divides, and through exchange of one of the nuclei the two fused primary sporidia become dikaryotic. When the primary sporidia germinate, they produce short hyphae on which dikaryotic secondary sporidia are formed. Upon germination the secondary sporidia produce dikaryotic mycelium which can penetrate the plants and cause infection. After systemic development through the plant, the mycelium again forms teliospores.

FIGURE 135.
(A) Covered smut or bunt of wheat caused by *Tilletia* sp. Left two heads show abnormal spread of the glumes due to the bunt balls. Healthy head at right. (B) Germinating teliospore of *T. caries* producing the basidium (promycelium). Primary sporidia are connected by fertilization tubes and form the H-shaped structures shown at right. (C) Covered smut (a) and loose smut (b) of barley. (c) Healthy barley head. (Photos A and C courtesy U.S.D.A. Photo B courtesy M. F. Brown and H. G. Brotzman.)

Development of disease.   The pathogen overwinters as teliospores on contaminated wheat kernels and in the soil. The teliospores of the common bunt fungi are short-lived in wet areas, but those of the dwarf bunt fungus may remain viable in any soil for many years.

When contaminated seed is sown or healthy seed is sown in bunt-infested fields, approximately the same conditions that favor germination of seeds favor germination of teliospores. As the young seedling emerges from the kernel, the teliospore on the kernel or near the seedling also germinates through production of the basidium, primary sporidia, and secondary sporidia (Fig. 136). The secondary sporidia then germinate, and the dikaryotic mycelium they produce penetrates the young seedling directly. After penetration the mycelium grows intercellularly and invades the developing leaves and the meristematic tissue at the growing point of the plant. The mycelium remains dormant in the seedling during the winter, but when the seedling begins to grow again in the spring, the mycelium resumes its growth and grows with the growing point. When the plant forms the head of the grain, the mycelium invades all parts of it

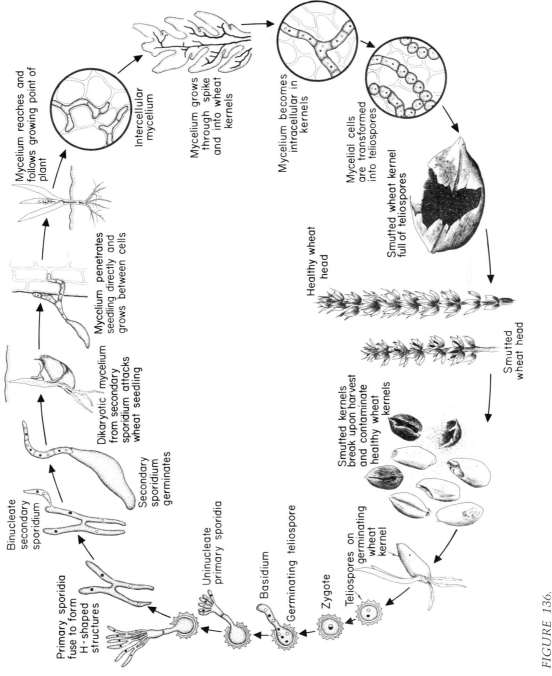

Mycelium reaches and follows growing point of plant

Intercellular mycelium

Mycelium grows through spike and into wheat kernels

Mycelium becomes intracellular in kernels

Mycelial cells are transformed into teliospores

Smutted wheat kernel full of teliospores

Healthy wheat head

Mycelium penetrates seedling directly and grows between cells

Dikaryotic mycelium from secondary sporidium attacks wheat seedling

Secondary sporidium germinates

Binucleate secondary sporidium

Primary sporidia fuse to form H-shaped structures

Uninucleate primary sporidia

Basidium

Germinating teliospore

Zygote

Teliospores on germinating wheat kernel

Smutted kernels break upon harvest and contaminate healthy wheat kernels

Smutted wheat head

FIGURE 136.
Disease cycle of covered smut or bunt of wheat caused by *Tilletia* sp.

407

even before the head emerges out of the "boot." As the head fills and becomes mature, the mycelial threads increase in numbers and soon take over and consume the contents of the kernel cells. The mycelium, however, does not affect the tissues of the pericarp of the kernel which form a rather sturdy covering for the smutted mass they contain. At the same time most hypha cells are transformed into teliospores.

Smutted kernels are usually kept intact while on the plant, but break and release their spores upon harvest or threshing. The liberated spores contaminate the healthy kernels and are also blown away by air currents, thus contaminating the soil.

Control.   Bunt can be controlled by using smut-free seed of a resistant variety treated with an appropriate fungicide. Contaminated seed should be cleaned to remove any unbroken kernels and as many of the smut spores on the seed as possible. The seed is then treated with the fungicide hexachlorobenzene (HCB) or carboxin, although other fungicides, including thiram, chloranil, thiabendazole, and benomyl, give good control of the disease. In dwarf bunt, and in common bunt in drier areas, the spores survive in the soil for long periods and can cause infection of seedlings. In such cases, seed treatment with the systemic fungicides carboxin or thiabendazole are the most effective.

- *Onion Smut*

It is caused by *Urocystis cepulae.* It is probably present throughout the world except the tropics and subtropics. This disease causes important losses in most onion producing areas.

Onion smut appears as dark, elongated, slightly thickened areas beneath the skin of the cotyledon or the leaves and sometimes beneath the scales of the young bulb (Fig. 120). The dark lesions may run almost the entire length of the leaf. Affected leaves may be thicker than healthy ones and tend to twist and bend downwards. The infection spreads from the older outer leaves to the younger inner ones. Most infected plants die within a few weeks after emergence but some may survive for a while longer and develop numerous raised black blisters near the base of the bulb leaves. Some plants may even produce full-sized bulbs which, however, have lesions on the outer fleshy scales and are susceptible to secondary rots in storage.

The dark swollen streaks on the leaves and scales are the sori of the fungus and these rupture to expose and release a black powdery mass of teliospores. The teliospores consist of one thick-walled, dark, fertile cell surrounded by several smaller, almost colorless, thin-walled, sterile cells. The teliospores survive in the soil for up to 20 years and, although they may occasionally be carried on the seed, most infections start from soilborne spores and infected or contaminated onion sets. The teliospores germinate by producing a short, hemisphaerical basidium from which several hyphae develop that later break up into basidiosporelike fragments. Infection of onion takes place only during the first 2 to 3 weeks after sowing, when hyphae enter the cotyledon or first young leaf at or near the soil line and penetrate to the growing point. Plants not infected by the time the cotyledon has reached full size remain healthy. Individual leaves also are susceptible for a short period and, if not infected during that time, they become resistant and cannot be infected later.

Control of onion smut is obtained by seed treatment with thiram, ferbam, captan, hexachlorobenzene, or folpet, each combined with an effective sticker.

SELECTED REFERENCES

Ben-Yephet, Y., Y. Henis, and A. Dinoor. 1975. Inheritance of tolerance to carboxin and benomyl in *Ustilago hordei*. *Phytopathology* 65: 563–567.

Christensen, J. J. 1963. Corn smut caused by *Ustilago maydis*. *Am. Phytopathol. Soc. Monograph* 2: 41 pp.

Churchward, J. G. 1940. The initiation of infection by bunt of wheat (*Tilletia caries*). *Ann. Appl. Biol.* 27:58–64.

Croxall, H. E., and C. J. Hickman. 1953. The control of onion smut by seed treatment. *Ann. Appl. Biol.* 40:176–183.

Davis, G. N. 1936. Some of the factors influencing the infection and pathogenicity of *Ustilago zeae* (Backm.) Unger on *Zea mays* L. *Iowa Agr. Expt. Sta. Res. Bull.* 199:247–278.

Dewey, W. G., and R. S. Albrechtsen. 1974. Effect of thiabendazole seed treatment on the incidence of dwarf bunt and on the yield of winter wheat. *Plant Dis. Reptr.* 58:743–745.

Dickson, J. G. 1956. "Diseases of Field Crops." McGraw-Hill Book Co. 517 p.

Dietrich, S. 1959. Untersuchungen zur Biologie and Bekämpfung von *Ustilago zeae* (Beckm.) Unger. *Phytopathol. Z.* 35:301–322.

Fischer, G. W. 1953. "Manual of the North American Smut Fungi." Ronald Press, New York, 343 pp.

Fischer, G. W., and C. S. Holton. 1957. "Biology and Control of the Smut Fungi." 622 p. Ronald Press, New York.

Halisky, P. M. 1965. Physiologic specialization and genetics of the smut fungi. III. *Bot. Rev.* 31:114–150.

Hoffmann, J. A. 1971. Control of common and dwarf bunt of wheat by seed treatment with thiabendazole. *Phytopathology* 61:1071–1974.

Holton, C. S., J. A. Hoffmann, and R. Duran. 1968. Variation in the smut fungi. *Ann. Rev. Phytopathol.* 6:213–242.

Kendrick, E. L. 1965. The reaction of varieties and hybrid selections of winter wheat to pathogenic races of *Tilletia caries* and *T. foetida. Plant Dis. Reptr.* 49:843–846.

Kuiper, J. 1974. Sodium *p*-(dimethylamino)benzenediazosulfonate, a superior bunticide. *Plant Dis. Reptr.* 58:845–847.

Mosman, J. G. 1968. Fungicidal control of smut diseases of cereals. *U.S D.A. Crops Div., Compiled Crops Rept.* 42:1–36.

Popp, W. 1951. Infection in seeds and seedlings of wheat and barley in relation to development of loose smut. *Phytopathology* 41:261–275.

Spencer, J. L., and H. E. White. 1951. Anther smut of carnation. *Phytopathology* 41:291–299.

Tapke, V. F. 1948. Environment and the cereal smuts. *Bot. Rev.* 14:359–412.

Urech, P. A. 1972. Investigations on the corn smut caused by *Ustilago maydis*. *Phytopathol. Z.* 73:1–26.

ROOT AND STEM ROTS
CAUSED BY BASIDIOMYCETES

Several Basidiomycetes cause serious plant losses by attacking primarily the roots and lower stems of plants (Fig. 120). Some of these fungi, e.g., *Thanatephorus* (= *Rhizoctonia*), *Pellicularia* (= *Sclerotium*), attack primarily herbaceous plants, and *Typhula* attacks only grasses. On the other hand, some other fungi, e.g., *Armillaria*, some species of *Fomes*, particularly *F. annosus*, and of *Poria* and *Polyporus*, attack only roots and lower stems of woody plants, primarily forest trees and certain fruit trees. Another fungus, *Marasmius*, includes species that affect turf grasses, others that cause root or crown rot of banana and sugarcane, and others that cause witches'-broom of cacao and wiry cord blights on the tops of tropical trees.

ROOT AND STEM ROT DISEASES CAUSED BY
THE "STERILE FUNGI" RHIZOCTONIA AND
SCLEROTIUM

The fungi *Rhizoctonia* and *Sclerotium* are soil inhabitants and cause serious diseases on many hosts by affecting the roots, stems, tubers, corms and other plant parts that develop in or on the ground. These two fungi are known as sterile fungi because for many years they were thought to be incapable of producing spores of any kind, either sexual or asexual. It is known now that at least some species within these two genera produce spores, some of them sexual spores and some conidia. Thus, *Rhizoctonia solani* produces basidiospores which thus makes *Rhizoctonia* a basidiomycete and is called *Thanatephorus cucumeris*. On the other hand, the most common species of *Sclerotium*, *S. rolfsii*, also produces basidiospores and has as its perfect stage the basidiomycete *Pellicularia rolfsii*, while two other species, *S. bataticola* and *S. cepivorum*, have been reported to produce conidia belonging to the imperfect fungi *Macrophomina* and *Sphacellia*, respectively. However, the spores of these fungi are either produced only under special conditions in the laboratory or are extremely rare in nature and therefore of little value in diagnosing the fungus. For these reasons, these fungi continue to be considered as sterile mycelia and since, for all practical purposes, they behave as such they continue to be referred to by their names as *Rhizoctonia* and *Sclerotium*.

- *Rhizoctonia Diseases*

They occur throughout the world and cause losses on most annual plants, including weeds, almost all vegetables and flowers, several field crops, and also on perennial plants such as turf grasses, perennial ornamentals, shrubs, and trees. The symptoms of *Rhizoctonia* diseases may vary somewhat on the different crops and even on the same host plant depending on the stage of growth at which the plant becomes infected and the prevailing environmental conditions. The most common symptoms caused by *Rhizoctonia*, primarily *R. solani*, on most plants are damping off of seedlings and root rot, stem rot or stem canker of growing and grown plants. On some hosts, however, *Rhizoctonia* also causes rotting of storage organs and foliage blights or spots, especially of foliage near the ground (Fig. 137).

Damping off is probably the most common symptom caused by *Rhizoctonia* on most plants it affects. It occurs primarily in cold, wet

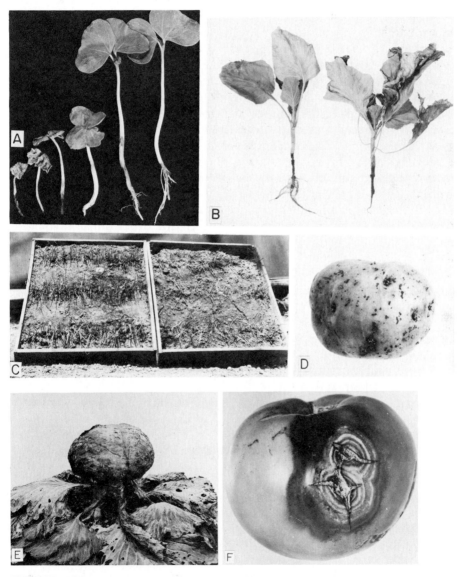

FIGURE 137.
(A) Cotton seedling stem canker ("soreshin") caused by *Rhizoctonia*. (B) "Wire stem" of young cabbage plants caused by *Rhizoctonia*. (C) Onion seedlings damped off by *Rhizoctonia* added to flat at right. (D) Potato showing sclerotia of *Rhizoctonia*. (E) *Rhizoctonia* head rot of cabbage. (F) *Rhizoctonia* soil rot of tomato fruit. (Photos A and B courtesy G. C. Papavizas, E and F courtesy U.S.D.A.)

soils. Very young seedlings may be killed before or soon after they emerge from the soil (Fig. 137C). Before emergence, the fungus attacks and kills the growing tip of the seedling which then soon dies. However, thick, fleshy seedlings such as those of legumes and the sprouts from potato tubers may show noticeable brown, dead tips and lesions before they are killed. After the seedlings have emerged, the fungus attacks their stem and makes it water soaked, soft, and incapable of supporting the seedling which then falls over and dies. Older seedlings may also be attacked but on these the invasion of the fungus is limited to the outer cortical tissues on which the fungus produces elongate, tan to reddish brown lesions. The lesions may increase in length and width until they finally girdle the stem and the plant may die or, as it often happens in crucifers, before the plant dies the stem may turn brownish black and may be bent or twisted without breaking, giving the disease the name wire stem (Fig. 137B).

A seedling stem canker known as "soreshin" is common and destructive in cotton seedlings that have escaped the damping off or seedling blight phase of the disease and develops under conditions that are not especially favorable to the disease. The "soreshin" lesions appear as reddish brown, sunken cankers that range from narrow to completely girdling the stem near the soil line (Fig. 137A). As soil temperature rises later in the season, affected plants may show partial recovery due to new root growth. "Soreshin" also affects tobacco and other crops in the seed bed or in the field. Dark-colored cankers or rotting develops at the base of the stem and may extend into the woody tissues and the pith as well as up the stem and into the lower leaves. The invaded tissues die and collapse, the black rotted area being relatively dry.

Even in the damping off phase of the disease, *Rhizoctonia* frequently attacks the roots at the same time it attacks the stem. On most partly grown or mature plants, the reddish brown lesions usually appear first just below the soil line, but in cool, wet weather the lesions enlarge in all directions and may increase in size and numbers to include the whole base of the plant and most of the roots. This results in weakening, yellowing, and sometimes death of the plant. If infested soil is splashed by rain onto stems and lower branches or their junctions, stem cankers may also develop.

On low-lying plants such as lettuce and cabbage, the lower leaves of which touch the ground or are close to it, *Rhizoctonia* attacks these leaves at the petioles and midribs on which it produces reddish brown, slightly sunken lesions while the entire leaf becomes dark brown and slimy. From the lower leaves the infection spreads upward to the next leaves until most or all leaves and the head may be invaded and rot, with mycelium and sclerotia permeating the tissues or nestled between the leaves (Fig. 137E, F).

On lawn and fine turf grasses, *Rhizoctonia* causes the brown patch disease. Brown patch is particularly severe during periods of hot and humid or wet weather, especially with heavy dew periods. It appears as roughly circular areas, ranging from a few centimeters to one or more meters in diameter, in which the grass blades become water soaked and dark at first but soon become dry, wither and turn light brown. Diseased areas appear slightly sunken, but at the border of the diseased areas, where the fungus is still active and attacks new grass blades making them look water soaked and dark, a characteristic grayish black "smoke" ring 2 to 5 cm wide appears in damp days or in the early morning hours. As the grass dries, the activity of the fungus slows down or stops and the ring disappears. Brown to black, hard, round sclerotia about 2 mm in diameter form in the thatch, diseased plants and soil. In brown patch, *Rhizoctonia* usually kills only the leaf blades, and plants in the affected area begin to recover and grow again from the center outward, resulting in a doughnut-shaped diseased area.

On fleshy, succulent stems and roots, and on tubers, bulbs, corms, etc., *Rhizoctonia* causes brown rotten areas that may be superficial or may extend inward to the middle of the root or stem. The rotting tissues usually decompose and dry, forming a sunken area filled with the dried plant parts mixed with fungus mycelium and sclerotia. The lesions may begin at the top of the fleshy root resulting in a crown rot that, in the field, may cause stunting and yellowing or death of the foliage. Lesions may also develop on the sides of fleshy tissues and may reach various sizes depending on host, weather, presence of cracks, etc. White, cream-colored or brown mycelium may cover the lesions in wet weather, and when the tissues rot and dry, sclerotia also develop.

On potato tubers, *Rhizoctonia* causes characteristic symptoms called "black scurf," in which small, hard, black sclerotia occur on the tuber surface and are not removed by washing (Fig. 137D), or "russeting" or "russet scab" in which the skin becomes roughened in a crisscross pattern resembling the shallow form of common potato scab.

Finally, *Rhizoctonia* causes rots on fruits, pods, etc., lying on or near the soil, such as cucumbers, tomatoes (Fig. 137F), eggplant, beans, etc. These rots develop most frequently in wet, cool weather and appear first in the field, but may continue and spread to other fruits following harvest and during transportation and storage. The lesions appear at first as rather firm, water-soaked areas in which the tissues soon collapse and form a shallow, sunken area. In moist weather, mycelium appears on the spots, white at first but turns brown with age. The affected fruits and pods also turn brown and dry, or they may be invaded by soft-rotting bacteria that cause them to become mushy or watery.

Other *Rhizoctonia* species cause somewhat different symptoms. Thus, *R. crocorum* attacks only underground parts of many vegetables and ornamentals and the diseased plant parts show a violet or red coloration due to the purple color of the superficial growth of the fungus which also contains many closely aggregated and darker sclerotialike bodies.

In the sheath and culm blight of rice and other cereals, different *Rhizoctonia* species cause large, irregular lesions that have a straw-colored center and a wide, reddish-brown margin. Seedling and mature plants may become blighted under favorable conditions for the pathogen.

The pathogen, *Rhizoctonia* sp., and particularly *R. solani*, exists primarily as a sterile mycelium that is colorless when young but turns yellowish or light brown with age. The mycelium consists of long cells and produces branches that grow at approximately right angles to the

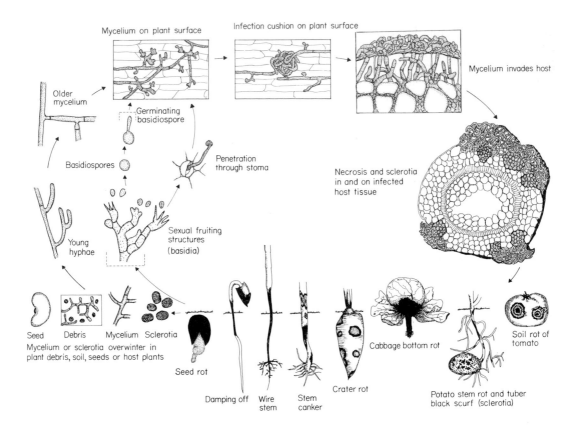

FIGURE 138.
Disease cycle of *Rhizoctonia solani* (*Thanatephorus cucumeris*).

main hypha, are slightly constricted at the junction and have a cross wall near the junction (Fig. 138). The branching characteristics are usually the only ones available for identification of the fungus as *Rhizoctonia*. Under certain conditions the fungus produces sclerotialike tufts of short, broad, ovate- to triangular-shaped cells that function as chlamydospores or, eventually, the tufts develop into rather small, loosely formed brown to black sclerotia which are common on some hosts such as potato. As mentioned earlier, *R. solani* infrequently produces a basidiomycetous perfect stage known as *Pellicularia filamentosa* or *Thanatephorus cucumeris*. The perfect stage forms under high humidity and appears as a thin, mildewlike growth on soil, leaves, and infected stems just above the ground line. The basidia are barrel-shaped, are produced on a membranous layer of mycelium, and have four sterigmata each bearing an ovoid basidiospore.

The pathogen overwinters usually as mycelium or sclerotia in the soil, in or on infected perennial plants or propagative material such as potato tubers. In some hosts, e.g., bean, eggplant, pepper, and tomato, the fungus even invades, and may be carried in the seed. The fungus is present in most soils and, once established in a field, remains there indefinitely. Different races of the fungus with different preferences as to hosts, temperature optimum, etc. exist. The fungus spreads with rain, irrigation, or flood water, with tools and anything else that carries contaminated soil, and with infected or contaminated propagative materials. For most races of the fungus the optimum temperature for infection is about 15 to 18°C, but some races are most active at much higher temperatures—up to 35°C. Disease is more severe in soils that are moderately wet rather than in soils that are waterlogged or dry. Infection of young plants is most severe when plant growth is slow due to adverse environmental conditions for the plant. Rapidly growing plants are likely to escape infection by *Rhizoctonia*, even when moisture and temperature may be favorable for the fungus.

Control of *Rhizoctonia* diseases, when the fungus is carried with the seed, depends on the use of disease-free seed or seed treated with hot water and chemicals. Wet, poorly drained areas should be avoided or drained better and seeds should be planted on raised beds and in soil in the best possible condition to encourage fast growth of the seedling. There should be wide spaces among plants for good aeration of soil surface and of plants. When possible, e.g., in greenhouses and seed beds, the soil should be sterilized with steam or treated with chemicals. Drenching of soil with pentachloronitrobenzene helps reduce damping-off in seed beds and greenhouses. When specific races of the pathogen have built up, a three-year crop rotation with another crop may be valuable. With most vegetables, no effective fungicides are available against Rhizoctonia diseases, although dichloran, ferbam, and some other chemicals are sometimes recommended as sprays on the soil prior to planting and once or twice on the seedlings soon after emergence. On turf grasses, along with proper drainage, removal of thatch, etc., preventive fungicide applications are recommended especially when night temperatures stay above 21°C or day temperatures above 28°C. Several fungicides, including some contact (mancozeb, dyrene, chlorothalonil) and systemic fungicides (carboxin, thiophanate methyl), seem to provide effective control.

- *Sclerotium Diseases*

They appear as damping off of seedlings, stem canker, crown blight, root and crown rot, bulb and tuber rot, and fruit rots. *Sclerotium* frequently causes severe losses of fleshy fruits and vegetables during shipment and storage. Sclerotium diseases are primarily diseases of warm climates affecting plants in countries within 38° latitude on either side of the equator. Because in the United States they are more common and severe in the southern states, they are often called "southern wilts or southern blights." Sclerotium diseases affect a wide variety of plants including vegetables, flowers, cereals, forage plants, and weeds. Some of the most common hosts of *Sclerotium* include legumes, crucifers, cucurbits, carrot, celery, sweet corn, eggplant, lettuce, okra, onion, peppers, potato, sweet potato, tomato, amaryllis, chrysanthemum, delphinium, iris, narcissus, tulip, alfalfa, cereals, cotton, peanuts, tobacco, and many others.

When seedlings are attacked, the fungus invades all the parts of the seedling and the seedlings die quickly. When the fungus attacks plants that have already developed some woody tissue, it does not invade them throughout but it grows into the cortex and slowly or quickly girdles the plants which eventually die. Usually the infection begins on the succulent stem as a dark-brown lesion just below the soil line. The first visible symptoms appear as yellowing or wilting of the lower leaves or dying back of leaves from the tips downward. These symptoms then progress to the upper leaves. In plants with very succulent stems, e.g., celery, the stem may fall over, while in plants with harder stems, e.g., alfalfa, bean, tomato, tobacco, the invaded stem stands upright and begins to lose its leaves or to wilt. In the meantime, the fungus grows upward in the plant covering the stem lesion with a cottony, white mass of mycelium, the upward advance of the fungus depending on the amount of moisture present. The fungus moves even more rapidly downward into the roots and finally destroys the root system. The white mycelium is always present in and on infected tissues and from these it grows over the soil to adjacent plants, starting new infections. Invaded stem, tuber, and fruit tissues are usually pale brown and soft but not watery. The margin between healthy and diseased tissue is often darker than the other tissues. When fleshy roots or bulbs are infected, a watery rot of the outer scales or root tissues may develop or the entire root or bulb may rot and disintegrate and be replaced by debris interwoven with mycelium. If bulbs, roots, fruits, etc., are infected late in their development, symptoms may go unnoticed at harvest but the disease continues as a storage rot.

On all infected tissues, and even on the nearby soil, the fungus produces numerous small sclerotia of uniform size that are roundish or irregular and white when immature, becoming dark brown to black when they mature. The mature sclerotia are not connected with mycelial strands and have the size, shape, and color of mustard seed (Fig. 139).

The fungus, *Sclerotium* sp., produces abundant white, fluffy, branched mycelium that forms numerous sclerotia but is usually sterile, i.e., it does not produce spores. *Sclerotium rolfsii*, which causes the symptoms described above on most of the hosts, occasionally produces basidiospores at the margins of lesions under humid conditions. Its perfect stage is *Pellicularia rolfsii*. Another species, *S. bataticola*, which causes diseases in several different hosts, including ashy stem blight of bean and soybean, charcoal rot of sorghum and corn, stem rot on watermelon, and root and wood rot of citrus, produces numerous, small, black, irregular sclerotia, usually in the pith of stems or stalks of infected plants, and, occasionally, also conidia in pycnidia of the *Macrophomina* type. A third *Sclerotium* species, *S. cepivorum*, which causes the white rot disease of onion and garlic, in addition to sclerotia, also produces occasional conidia on sporodochia; these conidia, however, seem to be sterile.

The fungus seems to overwinter mainly as sclerotia but also as mycelium in infected tissues or plant debris. It is spread by moving water, infested soil, contaminated tools, infected transplant seedlings, infected vegetables and fruits, and, in some hosts, as sclerotia mixed with the seed. Basidiospores and conidia may also participate in the dissemination of

FIGURE 139.
(A) Southern stem blight of tomato caused by *Sclerotium rolfsii*. Note white mycelium and round, uniform sclerotia. (B) Onion white rot caused by *Sclerotium cepivorum*. (C) Stem rot of rice caused by *Sclerotium oryzae*. Numerous sclerotia are showing. (D) Mycelium and sclerotia of *Sclerotium rolfsii* in culture. (E) Cross section of sclerotium showing the compact mass of mycelial cells. (Photos A and C courtesy U.S.D.A. Photo B courtesy P. B. Adams.)

the species that produce them, but their role in this is not clearly established.

The fungus attacks tissues directly. However, it produces a considerable mass of mycelium and kills and disintegrates tissues by secreting oxalic acid, and also pectinolytic, cellulolytic, and other enzymes before it actually penetrates the host. Once the fungus becomes established in the plants, its subsequent advance and production of mycelium and sclerotia are quite rapid, especially during high moisture and high temperature (between 30 and 35°C). The pathogen seems to grow, survive, and attack plants best near the soil line, perhaps because of more favorable temperatures there, more plentiful supply of organic substances the fungus uses for food and, perhaps, less competition or antagonism by other soil organisms. Sclerotia seem capable of surviving for long periods but they may be induced to germinate prematurely by alternate dry and moist periods. The fungus grows at a wide soil pH range (1.4 to 8.8) and is favored by sandy soils low in nitrogen.

Control of *Sclerotium* diseases is difficult and depends partly on crop rotation, partly on cultural practices such as deep plowing to bury surface debris, liming to bring the pH to about 7.0, and, in some cases, by applying fungicides such as pentachloronitrobenzene (PCNB) and dichloran to the soil before planting or in the furrow during planting.

SELECTED REFERENCES

Aycock, R. 1966. Stem rot and other diseases caused by *Sclerotium rolfsii. North Carolina Agr. Exp. Sta. Tech. Bull.* **174**: 202 p.

Baker, K. F., and R. J. Cook. 1974. "Biological Control of Plant Pathogens." W. H. Freeman, San Francisco. 433 p.

Baker, K. F., and W. C. Snyder. 1965. "Ecology of Soil-Borne Plant Pathogens." Univ. Calif. Press, Berkeley and Los Angeles. 571 p.

Benson, D. M., and R. Baker. 1974. Epidemiology of *Rhizoctonia solani* preemergence damping-off of radish: Influence of pentachloronitrobenzene. *Phytopathology* **64**:38–40, 957–962, 1163–1168.

Boosalis, M. G. 1950. Studies on the parasitism of *Rhizoctonia solani* on soybeans. *Phytopathology* **40**:820–831.

Bruehl, G. W. (Ed.) 1975. "Biology and Control of Soil-Borne Plant Pathogens." Amer. Phytopathol. Society, St. Paul, Minn. 216 p.

Christou, T. 1962. Penetration and host-parasite relationships of *Rhizoctonia solani* in the bean plant. *Phytopathology* **52**:381–389.

Chupp, C., and A. F. Sherf. 1960. "Vegetable Diseases and Their Control." The Ronald Press Co., New York, 693 p.

Coley-Smith, J. R. 1959–1960. Studies of the biology of *Sclerotium cepivorum.* III, IV. *Ann. Appl. Biol.* **47**:511–518, **48**:8–18.

Forsberg, J. L. 1975. "Diseases of Ornamental Plants." Univ. of Ill., Col. of Agr., Special Publication No. 3 Rev. 220 p.

Garrett, S. D. 1970. "Pathogenic Root-Infecting Fungi." Cambridge Univ. Press, London. 294 p.

Higgins, B. B. 1927. Physiology and parsitism of *Sclerotium rolfsii. Phytopathology* **17**:417–448.

Keim, R., and R. K. Webster. 1974. Effect of soil moisture and temperature on viability of sclerotia of *Sclerotium oryzae. Phytopathology* **64**:1499–1502.

Neal, D. C. 1942. *Rhizoctonia* infection of cotton and symptoms accompanying the disease in plants beyond the seedling stage. *Phytopathology* **32**:641–642.

Papavizas, G. C. 1974. The relation of soil microorganisms to soilborne plant pathogens. *Southern Coop. Series Bull.* **183**, Virginia Polytech. Inst. 98 p.

Parmeter, J. R., Jr. (Ed.). 1970. "*Rhizoctonia solani*, Biology and Pathology." Univ. of Calif. Press, Berkeley, Los Angeles. 255 p.

Smith, A. M. 1972. Drying and wetting sclerotia promotes biological control of *Sclerotium rolfsii. Soil Biol. Biochem.* **4**:119–123, **4**:125–129, **4**:131–134.

Toussoun, T. A., R. V. Bega, and P. E. Nelson. 1970. "Root Diseases and Soil-Borne Pathogens." Univ. Calif. Press, Berkeley, Los Angeles. 252 p.

Wellman, F. L. 1932. *Rhizoctonia* bottom rot and head rot of cabbage. *J. Agr. Res.* **45**:461–469.

ROOT ROTS OF TREES

- *Armillaria Root Rot of Fruit and Forest Trees*

Armillaria root rot is worldwide in distribution and affects hundreds of species of fruit trees, vines, shrubs, shade and forest trees, as well as other plants such as potatoes and strawberries, both in the temperate and tropical regions. The disease is often known as "shoestring root rot," "mushroom root rot," "crown rot," or "oak root fungus disease." The pathogen, *Armillaria mellea,* is one of the most common fungi in forest soils and the most spectacular losses occur in orchards or vineyards planted in recently cleared forest lands or in forest tree plantations, particularly in stands recently thinned. Most commonly, however, the losses from Armillaria root rot are steady but inconspicuous, appearing as slow decline and death of occasional trees with greater numbers of trees dying from this disease during periods of moisture stress, following defoliation, etc.

The aboveground parts of affected trees show symptoms similar to those caused by other root rot diseases, i.e., reduced growth, smaller, yellowish leaves, dieback of twigs and branches and gradual or sudden death of the tree. Affected trees may be scattered at first, but soon circular areas of diseased trees appear due to the spread of the fungus from its initial infection point. Diagnostic characteristics of Armillaria root rot appear at decayed areas in the bark, at the root collar and on the roots. White mycelial mats, their margins often veined and shaped as fans, form between the bark and wood (Fig. 140B). The mycelium may extend for a few feet upward in the phloem and cambium of the trunk and in some trees, such as oak, sugar maple and hemlock, it may cause a white rot decay. In addition to the mycelial fans, another even more characteristic sign of the disease is the formation of reddish brown to black "rhizomorphs" or "shoestrings," i.e., cordlike threads of mycelium 1 to 3 mm in diameter consisting of a compact outer layer of black mycelium and a core of white or colorless mycelium. These often form a branched network of sorts on the roots, under the bark or in severely decayed wood, with some strands spreading into the soil surrounding the roots (Fig. 140C, D). In areas in which the mycelium has invaded the cambium, cankers form on both hardwoods and conifers, and gum or resin is exuded from the infected area and flows into the soil. As the fungus gradually girdles and kills the tree at the base, infected wood changes from firm and slightly moist to somewhat soft and dry. At the base of dead or dying trees, a few to many honey-colored, speckled mushrooms about 7 or more centimeters tall and with a cap 5 to 15 cm in diameter grow from trunks, stumps or on the ground near infected roots (Fig. 140A). These are the fruiting bodies of *A. mellea* which appear in early fall and within their radial gills produce numerous basidia and basidiospores.

The fungus overwinters as mycelium or rhizomorphs in diseased trees, decaying roots or in the soil. The principal method of tree-to-tree spread of the fungus is through rhizomorphs or direct root contact. Rhizomorphs

FIGURE 140.
(A) *Armillaria mellea* basidiocarps growing at the base of infected tree. (B) Fan-shaped mycelial mats of *Armillaria* advancing on the wood surface of main roots and the trunk. (C) Rhizomorphs of *Armillaria* growing on the wood of tree killed by the fungus. (D) Close-up of rhizomorphs on wood surface of trunk. (Photo A courtesy U.S.D.A. Photo B courtesy U.S. Forest Service.)

grow from roots of infected trees or from decaying roots or stumps through the soil to roots of adjacent healthy trees (Fig. 141). Also, pieces of rhizomorphs may be carried by cultivating equipment into new areas. The fungus can apparently spread by basidiospores, but spores generally colonize dead stumps or woody material first and then rhizomorphs radiating from these attack living roots directly or through wounds. When roots of trees are in contact with infected or decaying roots, mycelium may directly invade healthy roots appressed to diseased roots without forming rhizomorphs. In all cases, trees and roots weakened from other causes are much more easily attacked by *Armillaria* than are vigorous trees.

Control of Armillaria root rot is usually not attempted under forest conditions. Generally, however, losses can be reduced by removing the substrate such as tree stumps and roots and avoiding or delaying planting, for several years, susceptible fruit or forest trees in recently reclaimed forest land originally occupied by oaks or other plants favoring buildup of large amounts of *Armillaria* inoculum. Control of the disease in orchards and occasionally in forest plantations is attempted by digging a trench around an infected tree and its neighbors to prevent growth of rhizomorphs to adjacent trees and local soil fumigation of the infested area to destroy the fungus in the soil before *Armillaria*-killed trees can be replaced.

- *Fomes Annosus Root*
  and Butt Rot of Forest Trees

The disease occurs throughout the north temperate zone of the world and is common and serious on most conifers, particularly pines. Hardwoods may be attacked but they are generally not damaged seriously. Losses of wood from *Fomes annosus* are heavy and result from outright killing of trees, reduced timber due to heart rot, and reduced tree growth.

The fungus, *Fomes annosus*, persists for a long time as a saprophyte on the dead stumps and roots of trees. It produces irregular-shaped conks (sporophores) whose upper surface is gray to brown and the underside white to tan and their size varies from two to many centimeters across. The conks form mostly near the base of the trees and stumps, or in the root crotch, and are often partially buried under litter. Infected trees usually show thin, unhealthy looking foliage at first and are often found in groups or circular pockets in the stand. Sometimes, however, trees with full crowns may have extensive butt and root decay and may die rapidly with a sudden red discoloration of the crown. Internally, the early stages of decay often produce a pink to violet stain in the wood that later gives way to narrow, elongated white pockets and scattered black flecks in the wood. Finally, in the last stages of decay, the wood develops a typical soft yellowish white, stringy rot.

Fomes annosus can spread over long or short distances by airborne basidiospores and can gain entry into plantations by first infecting freshly cut stumps during thinning, and on nursery stock, or on infested material. The fungus grows through the stump and roots and spreads from them to adjacent trees through root grafts and contact points between

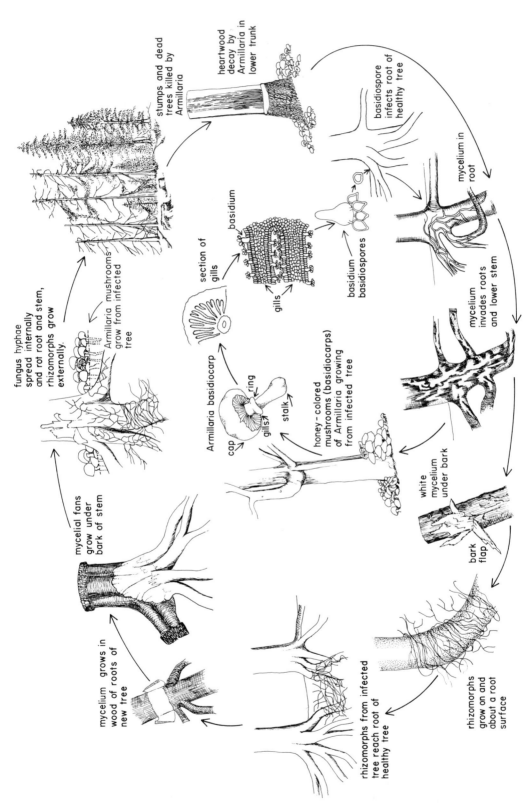

stumps and dead trees killed by Armillaria

heartwood decay by Armillaria in lower trunk

basidiospore infects root of healthy tree

basidium

section of gills

gills

mycelium in root

basidium basidiospores

fungus hyphae spread internally and rot root and stem, rhizomorphs grow externally.

Armillaria mushrooms grow from infected tree

mycelium invades roots and lower stem

cap

ring

stalk

gills

Armillaria basidiocarp

honey-colored mushrooms (basidiocarps) of Armillaria growing from infected tree

white mycelium under bark

mycelial fans grow under bark of stem

bark flap

mycelium grows in wood of roots of new tree

rhizomorphs from infected tree reach root of healthy tree

rhizomorphs grow on and about a root surface

FIGURE 141.
Disease cycle of root rots of trees caused by *Armillaria mellea*.

422

healthy roots and infected wood. It is also possible that basidiospores are washed down into the root zone where they infect dead or weakened roots. The fungus also produces conidia but their role in spreading the disease is not known. Once established in a stand, the fungus persists there indefinitely.

Control of *F. annosus* root and butt rot is difficult. Because the disease is more severe on soils that have low organic matter, new plantations should not be made on such soils. Thinning of trees should be minimized or done during hot summer days in the south, when high temperatures kill the spores and the mycelium of the fungus and thus prevent infection, or during cold weather in the north, where infections are probably limited to the summer months. Stump infection following thinning or harvest may be prevented by treating them with borax powder or urea liquid immediately after cutting or by inoculating them with fungi antagonistic to *F. annosus*, such as *Peniophora gigantea*.

SELECTED REFERENCES

Bliss, D. E. 1951. The destruction of *Armillaria mellea* in citrus soils. *Phytopathology* **41**:665–683.

Driver, C. H., and J. H. Ginns, Jr. 1969. Ecology of slash pine stumps: fungal colonization and infection by *Fomes annosus*. *Forest Sci.* **15**:2–10.

Gremmen, J. 1963. Biological control of the root rot fungus *Fomes annosus* by *Peniophora gigantea*. *Ned. Bosbouw. Tijdschr.* **35**:356–367.

Hodges, C. S. 1969. Modes of infection and spread of *Fomes annosus*. *Ann. Rev. Phytopathol.* **7**:247–266.

Leaphart, C. D. 1963. Armillaria root rot. *U.S.D.A. Forest Service, Forest Pest Leaflet* **78**:8 p.

Mook, P. V., and H. G. Eno. 1961. *Fomes annosus*. What is it and how to recognize it. *Northeast. For. Exp. Sta., Station Paper* **146**: 33 p.

O'Reilly, H. J. 1963. Armillaria root rot of deciduous fruits, nuts and grapevines. *Calif. Agr. Exp. Sta. Ext. Service, Circular* **525**:15 p.

Redfern, D. B. 1973. Growth and behaviour of *Armillaria mellea* rhizomorphs in soil. *Trans. Br. Mycol. Soc.* **61**:569–581.

Redfern, D. B. 1975. The influence of food base on rhizomorph growth and pathogenicity of *Armillaria mellea* isolates, *in* "Biology and Control of Soil-Borne Plant Pathogens," G. W. Bruehl (Ed.). The Amer. Phytopathol. Soc., St. Paul, Minn., pp. 69–73.

Ross, E. W. 1975. *Fomes annosus* in Eastern North America, *in* "Biology and Control of Soil-Borne Plant Pathogens," G. W. Bruehl (Ed.). The Amer. Phytopathol. Soc., St. Paul, Minn., pp. 107–110.

Sinclair, W. A. 1964. Root- and butt-rot of conifers caused by *Fomes annosus*, with special reference to inoculum dispersal and control of the disease in New York. *Cornell Univ. Agr. Exp. Sta. Mem.* **391**: 54 p.

Wargo, P. M., and D. R. Houston. 1974. Infection of defoliated sugar maple trees by *Armillaria mellea*. *Phytopathology* **64**:817–822.

WOOD ROTS AND DECAYS
CAUSED BY BASIDIOMYCETES

Huge losses of timber in the living trees in the forest and in harvested wood or in wood products are caused every year by the wood-rotting

Basidiomycetes (Figs. 119, 142, and 143). In living trees, most of the rotting is confined to the older, central wood of roots, stem or branches sometimes referred to as heartwood (Fig. 142G, H). Once the tree is cut, however, the outer wood, which is sometimes referred to as sapwood, is also attacked by the wood-rotting fungi, as are the wood products made from it—if moisture and certain other conditions remain favorable for the fungi. When large wounds or cuts are present on the tree, discoloration and

FIGURE 142 (A–F).
(A) Large canker and rotten area of trunk originating at decaying smaller branch. (B) Central portion of apple branch rotted by the silverleaf fungus *Stereum purpureum*. (C) Remnants of tree trunk attacked by wood rot fungi. (D) Beech trunk rotted by *Polyporus betulinus* and several basidiocarps of the fungus. (E) The lower side of a basidiocarp of *Polyporus betulinus*. (F) Tangential section of a perennial basidiocarp (conk) of a *Fomes* species.

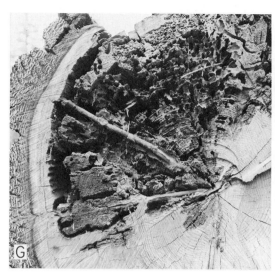 

FIGURE 142 (G, H).
(G) Close-up of wood rot in tree trunk originating at large wound. (H) Most wood of this trunk has been rotted away by wood rot fungi although there are no signs of exterior damage at the level of the cut. (Photos courtesy U.S.D.A.)

decay may also spread into the outer wood, and the entire tree, especially among hardwood trees, becomes of little economic value.

Depending on the host portion attacked, wood rots may be called root rots, root and butt rots, stem rots, or top rots. The fungi that cause these rots or wood product decays grow inside the wood cells and utilize the cell wall components for food and energy. Some of them, the brown-rot fungi, which attack preferably softwoods, can break down and utilize primarily the cell wall polysaccharides (cellulose and hemicellulose), leaving the lignin more or less unaffected. This usually results in rotten wood that is some shade of brown and, in advanced stages, has a cubical pattern of cracking and a crumbly texture (Fig. 143E). Other wood rotters, the white-rot fungi, either decompose lignin and hemicellulose first and cellulose last, or decompose all wood components simultaneously, in either case reducing the wood to a light-colored spongy mass (white rot) with white pockets or streaks separated by thin areas of firm wood (Fig. 143, F, G). The white-rot fungi are able to, or preferably attack hardwoods normally resistant to brown-rot fungi.

It should be noted here that, in addition to the brown rots and white rots caused by Basidiomycetes, wood is also attacked by certain Ascomycetes and Imperfects. Some Ascomycetes, such as *Daldinia, Hypoxylon*, and *Xylaria*, cause a relatively slow white rot with variable black zone lines in and around the rotting wood both in standing hardwood trees and in slash. In standing trees the decay is usually associated with wounds or cankers while in wood pieces the decay is usually at or near the surface of wood with high moisture content. Others, such as species of *Alternaria, Bisporomyces, Diplodia*, and *Paecilomyces*, cause the so-called soft rots of wood that affect the surface layers of wood pieces maintained more or less continuously at a high moisture content. The soft-rot fungi utilize both polysaccharides and lignin. They invade wood preferably through rays or vessels from where they grow into the adjacent tracheids and invade their cell wall. Within the cell wall they produce conical or cylindrical cavities parallel to the orientation of the microfibrils and, with progressing decay, the entire secondary wall is interlaced by confluent cavities. Several types of bacteria also attack wood, primarily in wood parenchyma rays where they break down and utilize the contents and walls of the parenchyma cells and thus increase the porosity and permeability of the wood to liquids, including fungal enzymes. Furthermore, several Ascomycetes and Imperfects result in the appearance of unsightly discolorations in the wood and thus reduce the quality but not the strength of the wood. Some of the wood-staining fungi are simply surface molds that usually grow on freshly cut surfaces of wood and impart to the wood the color of their spores, e.g., *Penicillium* (green or yellow), *Aspergillus* (black, green), *Fusarium* (red), *Rhizopus* (gray), etc. (Fig. 143A, B). Other wood-staining fungi, however, usually called sap-stain or blue-stain fungi, cause discoloration of the sapwood by producing pigmented hyphae that grow mainly in the ray parenchyma but can spread throughout the sapwood and cause lines of discoloration (Fig. 143C, D). Among the blue-stain fungi are species of *Ceratocystis*, *Hypoxylon*, *Xylaria*, *Graphium*, *Diplodia*, and *Cladosporium*.

The bulk of wood rotting, however, is carried out by Basidiomycetes and the most important fungi that rot wood in standing trees or in wood products are the following:

Fomes, causing root and butt rot of conifers (*F. annosus*), mottled white root and butt rot of hardwoods (*F. aplanatus*), white heart rot of deciduous trees (*F. connatus*), spongy white sap and heart rot of birch and aspen (*F. fomentarius*), white heart and sap rot of many living deciduous trees (*F. igniarius*), red heart rot of conifers (*F. pini*), brown cubical sap and heart rot of conifers and hardwoods (*F. pinicola*).

Polyporus, many species causing rot of dead trees or logs (*P. adustus, P. gilvus, P. hirsutus, P. pargamenus, P. picipes* and *P. versicolor*, all of them decaying primarily dead hardwood trees and logs). Many other species attack living as well as dead trees. Some of them attack only or mainly conifers and cause brown cubical root and butt rot (*P. schweinitzii*), red root and butt rot (*P. tomentosus*), white pocket rot of roots and butt (*P. circinatus*), red heart and sap rot of trunks (*P. anceps*). Others attack only or mainly deciduous hardwoods and cause a white heart rot (*P. squamosus* and *P. obtusus*), a brown cubical rot of the trunk (*P. sulphureus*), a spongy white rot (*P. hispidus*), a yellow cubical sap and heart rot of birch (*P. betulinus*), a canker and decay of maple (*P. glomeratus*), a root and butt rot (*P. lucidus*), or a trunk canker and localized decay (*P. hispidus*).

Poria, causing a root rot of most conifers (*P. weirii*), a yellow, feathery root and butt rot of balsam fir (*P. subacida*), and a butt swelling and advanced decay on birch (*P. obliqua*), while another species, *P. incrasata*, causes the common brown cubical rot in buildings and stored lumber, and by means of rhizomorphs can transport water for distances up to 5 meters or more.

Ganoderma, similar to *Fomes*, causing white mottled rot in hardwoods.

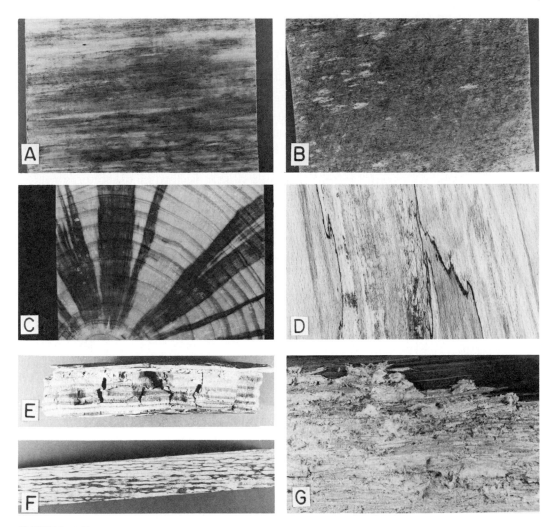

FIGURE 143.
Wood discolorations and decays. (A, B) Surface molds or mildews on wood caused by fungi such as *Penicillium*. (C) Sapstain or blue stain caused by fungi such as *Ceratocystis*. (D) Zone lines of white rot fungus (*Fomes igniarius*) in beech. (E) Brown rot caused by *Poria incrasata*, the water conducting fungus. (F) White pocket rot caused by species of *Polyporus* and *Fomes*. (G) Advanced white rot on wood caused by *Polyporus* and other wood-rotting fungi.

> *Stereum*, causing "silver leaf" of fruit trees as a result of decay of the interior of the tree trunk and branches (*S. purpureum*). Other species cause a white pocket rot of oaks (*S. frustulosum*), a white mottled heart rot of sprout oaks (*S. gausapatum*), and a red top rot of balsam fir (*S. sanguinolentum*).
>
> *Peniophora*, causing decay in coniferous logs and pulpwood.
>
> *Lenzites*, causing a brown cubical rot on coniferous logs, posts, poles, etc. (*L. sepiaria*), and decay of hardwood slash (*L. betulina*).
>
> *Pholiota*, causing brown rot in hardwoods.
>
> *Pleurotus*, causing white rot in hardwoods.
>
> *Schizophyllum*, also causing white rot in hardwoods.

The development of wood rots varies, of course, with the particular fungus involved and the host tree attacked. There are, however, many similarities (Fig. 144). Thus, the wood rot fungi enter trees as germinating basidiospores or mycelium through wounds, dead branches, branch stubs, tree stumps or damaged roots and spread from there to the heartwood and/or sapwood of the tree or tree sprout. Wounds caused by fire and by cutting and thinning operations are the most common points of entry of these fungi. The fungi develop in the wood and spread upward and/or downward in a cylinder much faster than they do radially. In some wood rots, especially those of hardwoods originating from wounds or branch stubs, the rotten cylinder is only a few inches in diameter, forms a column no larger than the diameter of the tree at the time of injury, and may extend to one or a few meters above and below the area where the fungus entered the tree or where its fruiting bodies (conks) appear. In other wood rots, particularly those of conifers, the rotten cylinder enlarges steadily until the tree is killed or blown over by heavy winds and it may extend upward over much of the height of the tree.

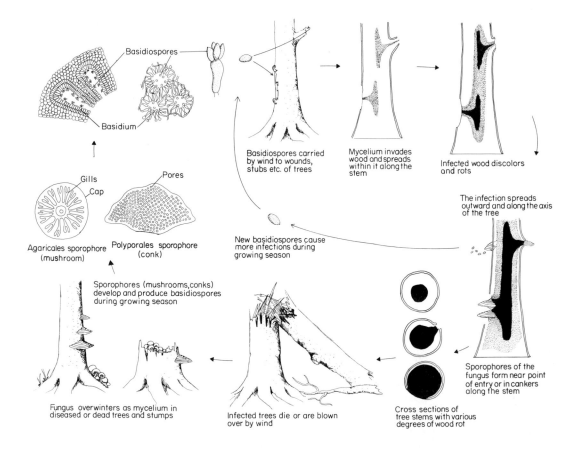

FIGURE 144.
Disease cycle of wood-rotting fungi.

The process of discoloration and decay in the wood of living trees appears to be quite complex, involving a number of successive or overlapping events. First, there must be an injury to the tree that exposes the wood as a result of a dead or broken branch, animal damage, fire burn, or mechanical scraping. The injured cells and those around them undergo chemical changes such as oxidation and become discolored. As long as the wound is open, discoloration advances toward the pith and around the tree, but if the wound is small and occurs early in the season, a new growth ring forms and its cells act as a barrier to the discoloration process. The discoloration moves up and down within the cylinder of barrier cells but not outward into the new and subsequent growth rings. Of course, many microorganisms are likely to land or be brought to the surface of a tree wound and many of them begin to grow on the moist surface. Among these, however, only some bacteria and some Ascomycetes or Imperfect Fungi manage to survive on the discolored wood of the wound. These microorganisms do not cause wood decay but they increase the discoloration and wetness of the wood and erode parts of the cell walls. Such wood is called wetwood, redheart, or blackheart. Finally, however, the wood-rotting Basidiomycetes become active and begin to disintegrate and digest the cell wall components. These wood-rotters attack only the tissues that have already been altered first by the chemical processes and then by the bacteria and the Ascomycetes and Imperfects. So, the wood rotting Basidiomycetes also remain confined to the discolored column within the new growth, being unable to attack the latter. The decay in the discolored column continues until the wood is completely decomposed but the influx of new microorganisms through the wound continues, even after the first decay fungus has caused the tissue to rot, and stops only when all tissues are completely digested. It should be noted that this process of discoloration and decay may take 50 to 100 years to develop. It is most common and rapid in older, larger trees and the older the trees the more likely they are to contain decay columns. The discoloration and decay process starting at a particular wound need not, of course, go through to completion. Quick healing of the wound, antagonisms among the microorganisms involved, natural wood resistance, or other factors may cause the process to stop at any stage. On the other hand, a large tree is likely to be injured many times over its long period of life. The events described above may be repeated many times after each new wound is formed and thus more and more of the wood may be involved in the more or less continuous process of discoloration and decay. The end result is the formation of a single large column or multiple columns of discolored and decayed wood.

The sporophores or conks of the wood-rotting Basidiomycetes appear near the point of entry of the fungus, near the base of the tree, in cankers or swollen knots along the stem of living trees, or along the length of the tree stem following its death. The sporophores of most wood rotting fungi, e.g., *Polyporus* and *Poria*, are formed annually and do not last for more than a year, but those of *Fomes* are perennial, adding a layer of tissue with vertical tubes and pores each year for 50 years or more. The sporophores produce basidiospores during part or most of the growing season and the spores are carried by wind, rain or animals to the nearby trees.

Control of wood rots and decays is impossible in the forest, but losses can be reduced: (1) by management practices that reduce or eliminate the chance of introduction of the fungi into healthy stands, (2) by conducting logging and thinning operations in a way that minimizes breakage of branches or other wounds of remaining trees and conducting them during the dry season to avoid mechanical damage to the root systems of the residual trees, and (3) by harvesting trees prior to the age of extreme susceptibility to wood rot fungi. Damage caused by wood-rotting fungi in shade and fruit trees can be prevented or minimized by avoiding or preventing wounds; by pruning dead and dying branches flush with the main stem; by cleaning wounds through cutting of torn bark and shaping the wound like a vertical ellipse; and by keeping the trees in good vigor through adequate irrigation and proper fertilization. Treating large cuts or wounds with a wound dressing or tree paint has been routinely practiced in the past but its usefulness in preventing wood discoloration and decay is questionable.

Control of discoloration or decays in lumber and wood products is usually accomplished by drying the wood or by treating it with an organic mercuric or a chlorophenate fungicide, or with a mixture of the two. Wood that is likely to be in contact with soil or other moist surfaces should be treated with one of several wood preservatives such as creosote, pentachlorophenol, copper naphthanate, and zinc chromate.

SELECTED REFERENCES

Basham, J. T. 1966. Heart rot of jack pine in Ontario. II. Laboratory studies on the pathogenicity and interrelationships of the principal heartwood-inhabiting fungi. *Can. J. Bot.* **44**:849–860.

Boyce, J. S. 1961. "Forest Pathology." McGraw-Hill, New York. 572 p.

Brooks, F. T., and W. C. Moore. 1926. Silver leaf disease—V. *J. Pomol. Hort. Sci.* **5**:61–97.

Cowling, E. B. 1961. Comparative biochemistry of the decay of sweetgum sapwood by white-rot and brown-rot fungi. *USDA Forest Service Tech. Bull.* **1258**: 79 p.

Eslyn, W. E., T. K. Kirk, and M. J. Effland. 1975. Changes in the chemical composition of wood caused by six soft-rot fungi. *Phytopathology* **65**:473–476.

French, D. W., *et al.* 1972. Forest and Shade Tree Pathology. 224 p. Mimeo.

Greaves, H. 1969. Micromorphology of the bacterial attack of wood. *Wood Sci. Technol.* **3**:150–166.

Hirt, R. R. 1949. Decay of certain northern hardwoods by *Fomes igniarius, Poria obliqua*, and *Polyporus glomeratus. Phytopathology* **39**:475–480.

Kimmey, J. W. 1964. Heart rots of western hemlock. *USDA Forest Service, Forest Pest Leaflet* **90**: 7 p.

Levy, J. F. 1965. The soft rot fungi: their mode of action and significance in the degradation of wood. *Advan. Bot. Res.* **31**:323–357.

Liese, W. 1970. Ultrastructural aspects of woody tissue disintegration. *Ann. Rev. Phytopathol.* **8**:231–258.

Merrill, W. 1970. Spore germination and host penetration by heartrotting Hymenomycetes. *Ann. Rev. Phytopathol.* **8**:281–300.

Peace, T. R. 1962. "Pathology of Trees and Shrubs." Oxford Univ. Press, London and New York. 723 p.

Setliff, E. C., and E. K. Wade. 1973. *Stereum purpureum* associated with sudden decline and death of apple trees in Wisconsin. *Plant Dis. Reptr.* **57**:473–474.

Shigo, A. L. 1967. Successions of organisms in discoloration and decay of wood. Int. Rev. Forest. Res. **2**:237–299.

Shigo, A. L., and P. Berry. 1975. A new tool for detecting decay associated with Fomes annosus in *Pinus resinosa. Plant Dis. Reptr.* **59**:739–742.

Shigo, A. L., and W. E. Hillis. 1973. Heartwood, discolored wood, and microorganisms in living trees. *Ann. Rev. Phytopathol.* **11**:197–222.

Shigo, A. L., and E. vH. Larson. 1969. A photo guide to the patterns of discoloration and decay in living northern hardwood trees. *USDA Forest Service Res. Paper* **NE-127**: 100 p.

Smith, W. H. 1970. "Tree Pathology: A Short Introduction." Academic Press, New York, 309 p.

Tattar, T. A., A. L. Shigo, and T. Chase. 1972. Relationship between the degree of resistance to pulsed electric current and wood in progressive stages of discoloration and decay in living trees. *Can. J. Forest Res.* **2**:236–243.

Verrall, A. F. 1965. Preserving wood by brush, dip, and short-soak methods. *USDA Tech. Bull.* **1334**: 50 p.

Wagener, W. W., and R. W. Davidson. 1954. Heart rots in living trees. *Bot. Rev.* **20**:61–134.

MYCORRHIZAE AND PLANT GROWTH

The feeder roots of most flowering plants growing in nature are generally infected by symbiotic fungi that do not cause root disease but, instead, are beneficial to their plant hosts. The infected feeder roots are transformed into unique morphological structures called mycorrhizae, i.e., "fungus roots." Mycorrizae, known for many years to be common in forest trees, are now considered to be the normal feeder roots for most plants including cereals, vegetables, ornamentals, and, of course, trees.

There are three types of mycorrhizae, distinguished by the way the hyphae of the fungi are arranged within the cortical tissues of the root.

Ectomycorrhizae. These roots are usually swollen and, in some host–fungus combinations, appear considerably more forked than nonmycorrhizal roots. Ectomycorrizae are formed primarily on forest trees by mushroom- and puffball-producing basidiomycetes and by several ascomycetes. Spores of ectomycorrhizal fungi are produced aboveground and are wind disseminated. The hyphae of ectomycorrhizal fungi usually produce a tightly interwoven "fungus mantle" around the outside of the feeder roots, the mantle varying in thickness from one or two hyphal diameters to as many as 30 to 40. These fungi also enter the roots but they only grow around the cortical cells, replacing part of the middle lamella between the cells, and forming the so-called Hartig net. Ectomycorrhizae appear white, brown, yellow, black, etc., depending on the color of the fungus growing on the root.

Endomycorrhizae. These roots externally appear similar to nonmycorrhizal roots in shape and color but, internally, the fungus hyphae grow into the cortical cells of the feeder root either by forming specialized feeding hyphae (haustoria), called arbuscules, or by forming large swollen hyphae, called vesicles. Most endomycorrhizae contain both vesicles and arbuscules and are, therefore, called "vesicular–arbuscular" (VA) mycorrhizae (Fig. 145). Endomycorrhizae are not surrounded by a dense fungal mantle but by a loose mycelial growth on the root surface from which hyphae and large pearl-colored zygospores or chlamydospores are produced underground. Endomycorrhizae are produced on most cultivated plants and on some forest trees by phycomycetes, primarily of the genus *Endogone*, but also other fungi, e.g., *Glomus*.

Ectendomycorrhizae.   They are intermediate between the other two. They are caused by fungi of unknown identity that grow into and also around the cortical cells of the root and may or may not have a fungus mantle on the surface of the feeder roots.

Mycorrhizae apparently improve plant growth by increasing the absorbing surface of the root system; by selectively absorbing and accumulating certain nutrients, especially phosphorus; by solubilizing and making available to the plant some normally nonsoluble minerals; by somehow keeping feeder roots functional longer; and by making feeder roots more resistant to infection by certain soil fungi such as *Phytophthora, Pythium,* and *Fusarium*. It should be kept in mind, however, that there may be many different host–fungus mycorrhizal associations and each combination may have different effects on the growth of the plant. Some mycorrhizal fungi have a broad host range while others are more specific. Also, some mycorrhizal fungi are more beneficial to a certain host than other fungi, and some hosts need and profit from association with a certain mycorrhizal fungus much more than do other hosts. Mycorrhizal fungi also need the host in order to grow and reproduce; in the absence of hosts the fungi remain in a dormant condition as spores or resistant hyphae.

The symbiosis between the host plant and the mycorrhizal fungus is generally viewed as providing equal benefits to both partners. Yet, it is quite probable that under certain nutritional conditions, one of the two partners may dominate and benefit more than the other. It has been suggested that the fungus is most aggressive in its invasion of root tissues when the host is growing at suboptimal nutritional levels (host defenses weak?) and the symbiotic relationship is terminated when nitrogen supply in the host reaches its optimum (host defenses at their best?). If nitrogen supply is again reduced to deficiency levels, the fungus partner begins to dominate and forms in abundance while plant growth is suppressed.

As far as is known, mycorrhizae do not cause disease, but absence of mycorrhizae in certain fields results in plant stunting and poor growth which can be avoided if the appropriate fungi are added to the plants. Also, soil fumigation often results in the eradication of mycorrhizal fungi and this in turn causes plants to remain smaller than plants growing in nonfumigated soil.

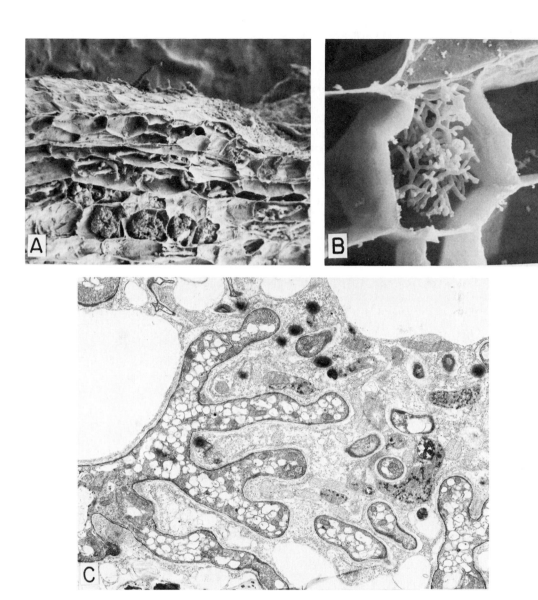

FIGURE 145.
Vesicular-arbuscular mycorrhizae (endomycorrhizae) on yellow poplar
(*Liriodendron tulipifera*) produced by *Glomus mosseae*. (A) Scanning electron
micrograph of interior of mycorrhizal root showing coiled intracellular hyphae in
outer cortical cells and three inner cortical cells which contain arbuscules. Some
external mycelium of the fungus can be seen on the outside of the epidermis (top
center). (B) Scanning electron micrograph of arbuscular morphology in a sample
treated to remove host cytoplasm which surrounds the structure. This is a
mature, viable arbuscule prior to the initiation of degenerative processes which
lead to breakdown of this part of the endophyte. (C) Transmission electron
micrograph of a similar arbuscule in a cortical cell. (Photos courtesy M. F. Brown
and D. A. Kinden.)

SELECTED REFERENCES

Gerdemann, J. W. 1968. Vesicular–arbuscular mycorrhiza and plant growth. *Ann. Rev. Phytopathol.* **6**:397–418.

Hackskaylo, E. 1971. Mycorrhizae. *U.S. Dep. Agric. Misc. Publ.* **1189**: 255 p.

Kleinschmidt, G. D., and J. W. Gerdemann. 1972. Stunting of citrus seedlings in fumigated nursery soils related to the absence of endomycorrhizae. *Phytopathology* **62**:1447–1453.

Marx, D. H. 1972. Mycorrhizae as biological deterrents to pathogenic root infections. *Ann. Rev. Phytopathol.* **10**:429–454.

Mosse, B. 1973. Advances in the study of vesicular–arbuscular mycorrhiza. *Ann. Rev. Phytopathol.* **11**:171–196.

Slankis, V. 1974. Soil factors influencing formation of mycorrhizae. *Ann. Rev. Phytopathol.* **12**:437–457.

plant diseases caused by bacteria

Bacteria are very small, microscopic plants. About 1600 bacterial species are known. The great majority of bacteria are strictly saprophytic and as such are beneficial to man because they help decompose the enormous quantities of organic matter produced yearly by man and his factories as waste products or as a result of the death of plants and animals. Several species cause diseases in man including tuberculosis, pneumonia, and typhoid fever, and a similar number cause diseases in animals, e.g., brucellosis and anthrax. About 200 species of bacteria have been found to cause diseases in plants. All pathogenic bacteria are facultative saprophytes and can be grown artificially on nutrient media.

Bacteria are simple microorganisms usually consisting of single prokaryotic cells, i.e., cells containing a single circular chromosome but no nuclear membranes or internal organelles comparable to mitochrondria or chloroplasts. In fact, bacteria and cellular organelles have much in common and hence antibiotics that affect bacteria often inhibit mitochondria or chloroplasts, but do not interfere with the other functions of the eukaryotic plant cells. Bacteria may be rod shaped, spherical, ellipsoidal, spiral, comma shaped, or filamentous (threadlike). Some bacteria can move through liquid media by means of flagella, while others have no flagella and cannot move themselves. Some can transform themselves into spores and certain filamentous forms can produce spores, called conidia, at the end of the filament. Other bacteria, however, do not produce any spores. The vegetative stages of most types of bacteria reproduce by simple fission. Bacteria multiply with astonishing rapidity and their significance as pathogens stems primarily from the fact that they

435

can produce tremendous numbers of cells in a short period of time. Bacterial diseases of plants occur in every place that is reasonably moist or warm, they affect almost all kinds of plants, and, under favorable environmental conditions, they may be extremely destructive.

characteristics of plant-pathogenic bacteria

MORPHOLOGY

Almost all plant-pathogenic bacteria are rod-shaped (Figs. 146 and 147), the only exceptions being two species of *Streptomyces*, which are filamentous. The rod-shaped bacteria are more or less short and cylindrical, and in young cultures they range from 0.6 to 3.5 μm in length and from 0.5 to 1.0 μm in diameter. In older cultures or at high temperatures,

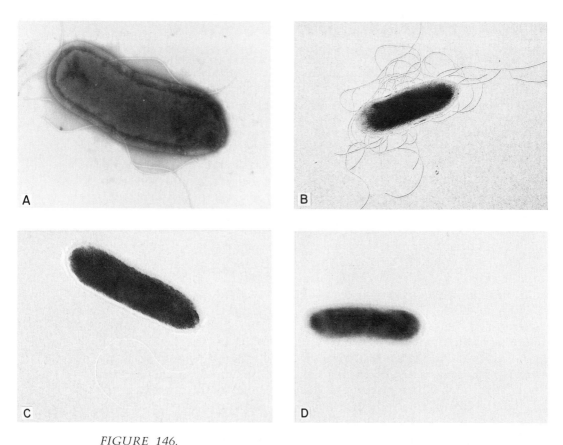

FIGURE 146.
Electron micrographs of some of the most important genera of plant-pathogenic bacteria. (A) *Agrobacterium*. (B) *Erwinia*. (C) *Pseudomonas*. (D) *Xanthomonas*. (Photo A courtesy S. M. Alcorn, B–D courtesy R. N. Goodman and P. Y. Huang.)

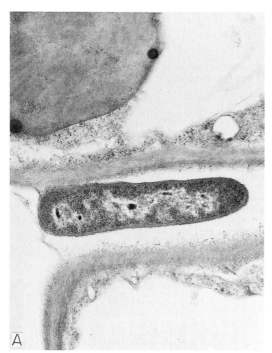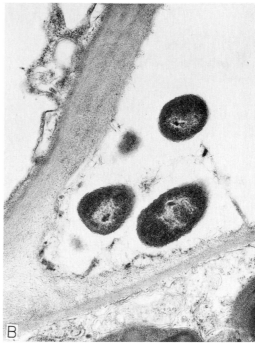

FIGURE 147.
Electron micrographs of longitudinal (A) and cross sections (B) of bacteria
(*Pseudomonas tabaci*) in the intercellular spaces of tobacco leaf mesophyll cells.
(Photos courtesy D. J. Politis and R. N. Goodman.)

the rods of some species are much longer and they may even appear
filamentous. Sometimes deviations from the rod shape in the form of a
club, a Y or V shape, and other branched forms occur, and some bacteria
may occasionally occur in pairs or in short chains.

The cell walls of bacteria of most species are enveloped by a viscous,
gummy material which may be thin (when it is called a "slime layer") or
may be thick, forming a relatively large mass around the cell (when it is
called a "capsule"). Most plant-pathogenic bacteria are equipped with
delicate, threadlike flagella which are usually considerably longer than
the cells by which they are produced. In some bacterial species each
bacterium has only one flagellum, others have a tuft of flagella at one end
of the cell; some have a single flagellum or a tuft of flagella at each end,
and still others have peritrichous flagella, i.e., distributed over the entire
surface of the cell.

In the filamentous *Streptomyces* species, the cells consist of nonseptate branched threads, which usually have a spiral formation and produce
conidia in chains on aerial hyphae (Fig. 148).

Single bacteria appear hyaline or yellowish-white under the compound microscope, and are very difficult to observe in detail. When a
single bacterium is allowed to grow (multiply) on the surface or within a
solid medium, its progeny soon produces a visible mass called a colony.

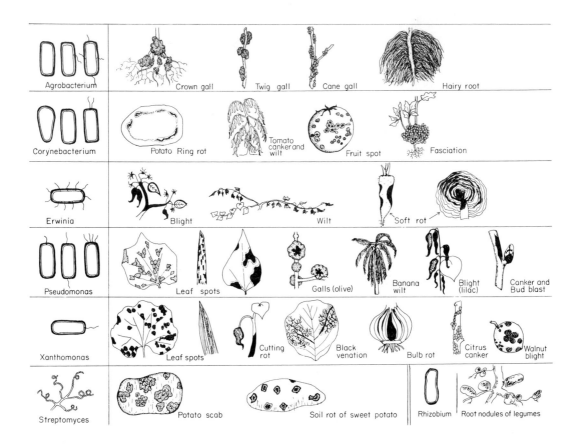

FIGURE 148.
Genera of bacteria and kinds of symptoms they cause.

Colonies of different species may vary in size, shape, form of edges, elevation, color, etc., and are sometimes characteristic of a given species. Colonies may be a fraction of a millimeter to several centimeters in diameter, and they are circular, oval, or irregular. Their edges may be smooth, wavy, angular, etc., and their elevation may be flat, raised, dome shaped, wrinkled, etc. Colonies of most species are whitish or grayish, but some are yellow, red, or other colors. Some produce diffusible pigments into the agar.

Bacterial cells have thin, relatively tough, and somewhat rigid cell walls which seem to be quite distinct from the inner cytoplasmic membrane but which sometimes appear to intergrade and merge with the outer slime layer or capsule. The cell wall contains the cell contents and allows the inward passage of nutrients and the outward passage of waste matter, digestive enzymes, and other products given off by the bacterial cell.

All the material inside the cell wall constitutes the protoplast. The protoplast consists of the cytoplasmic or protoplast membrane, which determines the degree of selective permeability of the various substances into and out of the cell; the cytoplasm, which is the complex mixture of proteins, lipids, carbohydrates, many other organic compounds, and minerals and water; and the nuclear material, which appears as spherical, ellipsoidal, or dumbbell-shaped bodies within the cytoplasm.

REPRODUCTION

Rod-shaped phytopathogenic bacteria reproduce by the asexual process known as "binary fission" or "fission." This occurs by the inward growth of the cytoplasmic membrane toward the center of the cell forming a transverse membranous partition dividing the cytoplasm into two approximately equal parts. Two layers of cell wall material, continuous with the outer cell wall, are then secreted or synthesized between the two layers of membrane. When the formation of these cell walls is completed, the two layers separate, splitting the two cells apart.

While the cell wall and the cytoplasm are undergoing fission, the nuclear material becomes organized in a circular chromosomelike structure which duplicates itself and becomes distributed equally between the two cells formed from the dividing one.

Bacteria reproduce at an astonishingly rapid rate. Under favorable conditions bacteria may divide every 20 minutes, one bacterium becoming two, two becoming four, four becoming eight, and so forth. At this rate one bacterium conceivably could produce one million bacteria in 10 hours. But, because of the diminution of the food supply, the accumulation of metabolic wastes, and other limiting factors, reproduction slows down and may finally come to a stop. Bacteria do reach tremendous numbers in a short time, however, and cause great chemical changes in their environment. It is these changes caused by large populations of bacteria that make them of such a great significance in the world of life in general and in the development of bacterial diseases of plants in particular.

ECOLOGY AND SPREAD

Almost all plant-pathogenic bacteria develop mostly in the host plant as parasites and partly in the soil as saprophytes. There are great differences among species, however, in the degree of their development in the one or the other environment.

Some bacterial pathogens, such as *Erwinia amylovora*, which causes fire blight, produce their populations in the plant host, while in the soil their numbers decline rapidly and usually do not contribute to the propagation of the disease from season to season. These pathogens have developed sustained plant-to-plant infection cycles, often via insect vectors and, either because of the perennial nature of the host or the association of the bacteria with its vegetative propagating organs or seed, they have lost the requirements for survival in the soil.

Some other bacterial pathogens, such as *Agrobacterium tumefaciens*, which causes crown gall, build up their populations within the host plants, but these populations only gradually decline when they are released into the soil. If susceptible hosts are grown in such soil in successive years, sufficiently high numbers of bacteria could be released to cause a net increase of bacterial populations in the soil from season to season. Finally, other bacterial pathogens, such as some of the *Erwinia* and *Pseudomonas* soft rotters, produce some of their populations in the soil.

When in the soil, bacteria live mostly on plant material and less often freely or saprophytically or in their natural bacterial ooze, which protects them from various adverse factors. Bacteria may also survive in or on seeds, other plant parts, insects, etc., found in the soil. On the plants bacteria may survive epiphytically, in buds, on wounds, in their exudate, or inside the various tissues or organs which they infect (Fig. 10).

The dissemination of plant pathogenic bacteria from one plant to another or to other parts of the same plant is carried out primarily by water, insects, other animals, and man (Fig. 9). Even bacteria possessing flagella can move only very short distances on their own power. Rain, by its washing or spattering effect, carries and distributes bacteria from one plant to another, from one plant part to another, and from the soil to the lower parts of plants. Water also separates and carries bacteria on or in the soil to other areas where host plants may be present. Insects not only carry bacteria to plants, but they inoculate the plants with the bacteria by introducing them into the particular sites in plants where they can almost surely develop. In some cases bacterial plant pathogens also persist in the insect and depend on them for their survival and spread. In other cases, insects are important but not essential in the dissemination of certain bacterial plant pathogens. Birds, rabbits, etc. visiting or moving among plants may also carry bacteria on their bodies. Man helps spread bacteria locally by his handling of plants and by his cultural practices, and over long distances by transportation of infected plants or plant parts to new areas or by introduction of such plants from other areas. In cases in which bacteria infect the seeds of their host plants, they can be carried in or on them for short or long distances by any of the agencies of seed dispersal.

CLASSIFICATION AND IDENTIFICATION

In earlier systems of classification of organisms, bacteria comprised the class Schizomycetes and all bacteria causing plant diseases belonged to the orders Pseudomonadales (family Pseudomonadaceae), Eubacteriales (families Rhizobiaceae, Enterobacteriaceae, and Corynebacteriaceae), and Actinomycetales (family Streptomycetaceae).

In the most recent (eighth) edition of *Bergey's Manual of Determinative Bacteriology* (1974), all organisms lacking an organized and bounded nucleus comprise the new kingdom Prokaryotae which is then subdivided as follows:

Division I: Phototrophic prokaryotes ("Photobacteria") which includes the blue green, the red, and the green photobacteria.

Division II: Prokaryotes indifferent to light ("Scotobacteria") which includes:

    Class I: The Bacteria, subdivided in 16 parts

    Class II: The Rickettsias, obligate intracellular scotobacteria in eucaryotic cells

    Class III: Mollicutes, scotobacteria without cell walls

In the new classification the above three orders (Pseudomonadales, Eubacteriales, and Actinomycetales) are eliminated and the plant pathogenic bacteria are classified under three of the sixteen parts.

The families and genera of bacteria that can cause disease in plants are tabulated under each part. The correct total number of plant pathogenic species of bacteria is still not settled among the experts. Bergey's manual lists only a few recognized ("certain") species which are readily determined by physiological tests with the rest being listed as "strains" of others or as "incompletely described." Plant bacteriologists, however, feel that many of the so-called "strains" and the "incompletely described" ones should be considered as separate species because of their unique host ranges, even though the biochemical nature of this adaptation is not yet known.

Part	Family	Genus	Plant-pathogenic species		Gram reaction
			Certain	Possible total	
I. Gram-negative aerobic rods and cocci	(1) Pseudomon-adaceae	*Pseudomonas*	7	80	−
		Xanthomonas	5	75	−
	(2) Rhizobiaceae	*Agrobacterium*	4	6	−
		Rhizobium	—	—	−
II. Gram-negative facultatively anaerobic rods	(1) Enterobacteri-aceae	*Erwinia*	12	16	−
III. Actinomycetes and related organisms	(1) Coryneform group of bacteria	*Corynebacterium*	12	15	+
	(2) Strepto-mycetaceae	*Streptomyces*	1	2	+

The main characteristics of the plant-pathogenic genera of bacteria (Fig. 148) are:

Agrobacterium. The bacteria are rod shaped, 0.8 by 1.5 to 3 μm. They are motile by means of 1 to 4 peritrichous flagella; when only one flagellum is present it is more often lateral than polar. When growing on carbohydrate-containing media the bacteria produce abundant polysaccharide slime. The colonies are nonpigmented and usually smooth. These bacteria are rhizo-sphere and soil inhabitants.

Corynebacterium. Straight to slightly curved rods, 0.5 to 0.9 by 1.5 to 4 μm. Sometimes they have irregularly stained segments or granules and club-shaped swellings. The bacteria are generally nonmotile but some species are motile by means of one or two polar flagella. Gram-positive. Several species of *Corynebacterium* cause diseases in humans and animals.

Erwinia. Straight rods, 0.5 to 1.0 by 1.0 to 3.0 μm. Motile by means of several to many peritrichous flagella. *Erwinias* are the only plant pathogenic bac-teria that are facultative anaerobes. Some authors retain the name *Erwinia* for *Erwinias* that cause necrotic or wilt diseases (e.g., *E. amylovora, E. tracheiphila*), but include the soft-rotting *Erwinias (E. carotovora)* in a new genus, *Pectobacterium*.

Pseudomonas. Straight to curved rods, 0.5 to 1 by 1.5 to 4 μm. Motile by means of one or many polar flagella. Many species are common inhabitants of soil, or of fresh water and marine environments. Most pathogenic *Pseudomonas* species infect plants; few infect animals or humans.

Xanthomonas. Straight rods, 0.4 to 1.0 by 1.2 to 3 μm. Motile by means of a polar flagellum. Growth on agar media usually yellow. Most are slow growing. All species are plant pathogens and are found only in association with plants or plant materials.

Streptomyces. Slender, branched hyphae without cross walls, 0.5 to 2 μm in diameter. At maturity the aerial mycelium forms chains of three to many spores. On nutrient media, colonies are small (1 to 10 mm in diameter) at first with a rather smooth surface but later with a weft of aerial mycelium that may appear granular, powdery, or velvety. The many species and strains of the organism produce a wide variety of pigments that color the mycelium and the substrate; they also produce one or more antibiotics active against bacteria, fungi, algae, viruses, protozoa, or tumor tissues. All species are soil inhabitants. Gram-positive.

Differential media on which the above genera can be separated have been developed.

The genus *Streptomyces* can be easily distinguished from the other bacterial genera because of its much-branched, well-developed mycelium and curled chains of conidia. Identification of bacteria belonging to the rod-shaped genera, however, is a much more complex and difficult pro-cess, and it can be made by taking into consideration not only visible characteristics such as size, shape, structure, and color, but also such obscure properties as chemical composition, antigenic reactivity, nutri-tional versatility, enzymatic action, pathogenicity to plants, susceptibil-ity to certain viruses (bacteriophages), and growth on selective media.

The shape and size of bacteria of a given species in culture can vary with age of the culture, composition, and pH of the medium, temperature, and staining method. Under given conditions, however, the predominat-ing form, size, and arrangement of cells in a pure culture are quite apparent, and they are important and reliable characteristics. The pres-ence, number, and arrangement of flagella on the bacterial cell are also determined, usually after the flagella have been stained with specific stains.

The chemical composition of certain substances in bacterial cells can be detected with specific staining techniques. Information about the presence or absence of such substances is used for identification of bacteria. Gram's staining reaction differentiates bacteria into gram positive and gram negative. In this reaction bacteria are treated with a crystal violet solution for 30 seconds, rinsed gently, treated with iodine solution, and rinsed again with water and then alcohol. Gram-positive bacteria retain the violet-iodine stain combination because it forms a complex with certain components of their cell wall and cytoplasm. Gram-negative bacteria have no affinity for the stain combination, which is therefore removed by the alcohol rinse and the bacteria remain as nearly invisible as before. Unfortunately, of the rod-shaped phytopathogenic bacteria, only the genus *Corynebacterium* is gram positive. *Agrobacterium*, *Erwinia*, *Pseudomonas*, and *Xanthomonas* are gram-negative.

The nutritional spectrum of bacterial cells is studied by recording the substances which the bacteria can or cannot use for food. Extracellular hydrolases, i.e., enzymes produced when the bacteria grow on certain media, are important determinative tools.

Phytopathogenic bacteria are also tested on various species and varieties of host plants for their pathogenicity on them. This test sometimes, and for practical purposes, may be sufficient for tentative identification of the bacterium.

Serological methods have been used for quick and fairly accurate identification of bacteria and have gained popularity in recent years. However, serological methods are not of widespread use in plant pathology because of limited availability of antisera. In a few cases bacterial species and strains can be identified by the bacteriophages (viruses) that infect them.

Recently, a group of compounds called bacteriocins have been used to differentiate or "type" bacterial isolates by their sensitivity patterns to these compounds or by their production of bacteriocins. Bacteriocins are antibacterial substances produced by certain bacteriocinogenic strains of many bacterial species. They are present in cultures of such strains in small amounts, presumably as a result of spontaneous lysis of cells. Bacteriocins are highly specific proteinaceous substances that inhibit and lyse only certain indicator strains of bacteria. Bacteriocins resemble bacteriophage in many respects but differ from them mainly in that they do not reproduce in bacterial host cells. Their production is genetically controlled by extrachromosomal DNA (plasmids) that replicate with the bacterial chromosome and are maintained as long as the bacteriocinogenic strain exists.

An excellent method for isolation and identification of bacteria obtained from plant tissues (Fig. 149) or soil would be through the use of selective nutrient media. Selective media contain nutrients that promote the growth of a particular type of bacterium while at the same time contain substances that inhibit the growth of other types of bacteria. Although some progress towards such selective media has been made, the available selective media for plant pathogenic bacteria are still quite unsatisfactory for routine use in bacterial identification.

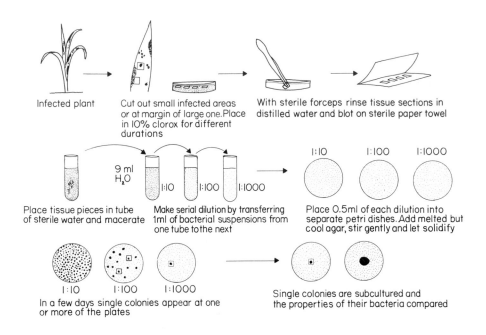

Infected plant

Cut out small infected areas or at margin of large one. Place in 10% clorox for different durations

With sterile forceps rinse tissue sections in distilled water and blot on sterile paper towel

9 ml H₂O

1:10   1:100   1:1000

1:10   1:100   1:1000

Place tissue pieces in tube of sterile water and macerate

Make serial dilution by transferring 1ml of bacterial suspensions from one tube to the next

Place 0.5ml of each dilution into separate petri dishes. Add melted but cool agar, stir gently and let solidify

1:10   1:100   1:1000

In a few days single colonies appear at one or more of the plates

Single colonies are subcultured and the properties of their bacteria compared

FIGURE 149.
Isolation of bacterial pathogens from infected plant tissue.

SYMPTOMS CAUSED BY BACTERIA

Plant-pathogenic bacteria cause the development of almost as many kinds of symptoms on the plants they infect as do fungi. They cause leaf spots and blights, soft rots of fruit, root, and storage organs, wilts, overgrowths, scabs, cankers, etc. (Fig. 148). Any given type of symptom can be caused by bacterial pathogens in several genera, and each genus contains some pathogens capable of causing different types of diseases. Species of *Agrobacterium*, however, can cause only overgrowths or proliferation of organs. On the other hand, overgrowths can also be caused by certain species of *Corynebacterium* and *Pseudomonas*. Also, the two plant pathogenic species of *Streptomyces* cause only scabs or lesions of below-ground crops. *Rhizobium* species induce formation of nodules on the roots of legume plants.

CONTROL OF BACTERIAL DISEASES OF PLANTS

Bacterial diseases of plants are usually very difficult to control. Frequently, a combination of control measures is required to combat a given bacterial disease. Infestation of fields or crops with bacterial pathogens should be avoided by introducing and planting only healthy seeds or plants. Sanitation practices aiming at reducing the inoculum in a field by removing and burning infected plants or branches, and at reducing the spread of bacteria from plant to plant by decontaminating tools and hands after handling diseased plants, are very important. Adjusting certain cultural practices, such as fertilizing and watering, so that the plants will not be extremely succulent during the period of infection may also reduce the incidence of disease. Crop rotation can be very effective with disease-producing bacteria that have a limited host range, but is impractical and ineffective with bacteria that can attack many types of crop plants.

The use of crop varieties resistant to certain bacterial diseases is one of the best ways of avoiding heavy losses. Varying degrees of resistance may be available within the varieties of a plant species, and great efforts are made at crop breeding stations to increase the resistance of, or introduce new types of resistance into, presently popular varieties of plants. Resistant varieties, supplemented with proper cultural practices and chemical applications, are the most effective means of controlling bacterial diseases, especially when environmental conditions favor the development of disease.

The use of chemicals to control bacterial diseases has been, generally, much less successful than the chemical control of fungal diseases.

Soil infested with phytopathogenic bacteria can be sterilized with steam or electric heat and with chemicals such as formaldehyde and chloropicrin, but this is practical only in greenhouses and in small beds or frames.

Seed, when superficially infected, can be disinfected with sodium hypochlorite or HCl solutions or by soaking it for several days in a weak solution of acetic acid. When the pathogen is inside the seed coat and in the embryo, such treatments are ineffective. Treating seed with hot water does not usually control bacterial diseases because of the relatively high thermal death point of the bacteria.

Of the chemicals used as foliar sprays, copper compounds have given the best results. However, even they seldom give satisfactory control of the disease when environmental conditions favor the development and spread of the pathogen. Bordeaux mixture and fixed coppers are most frequently used for the control of bacterial leaf spots and blights. Zineb is also used for the same purpose, especially on young plants which may be injured by the copper compounds.

Antibiotics have been used in recent years against certain bacterial diseases, and the results are encouraging. Some antibiotics are absorbed by the plant and are distributed systemically. They can be applied as sprays or as dips for transplants. The most important antibacterial antibiotics in agriculture are formulations of streptomycin or of streptomycin and oxytetracycline. Several others are presently available, but most of them are still used primarily for experimental purposes.

Since bacteriophages kill their host bacteria and since phages specific against certain phytopathogenic bacteria were found, it was expected that phages would be very valuable in controlling bacterial plant diseases. In some cases, the incidence and severity of some bacterial plant diseases were reduced by spraying the plants with specific bacteriophages or bacteriocins under experimental conditions. To date, however, this means of attack against bacterial diseases has not been developed sufficiently, and it cannot be used against any bacterial disease in the field, although work in this area may prove extremely valuable in the near future.

SELECTED REFERENCES

Buchanan, R. E., and N. E. Gibbons (Eds.). 1974. "Bergey's Manual of Determinative Bacteriology," 8th edition. Williams and Wilkins Co., Baltimore. 1268 p.

Buddenhagen, I. W. 1965. The relation of plant pathogenic bacteria to the soil, in "Ecology of Soil-Borne Plant Pathogens" (K. F. Baker and W. C. Snyder, Eds.), pp. 269–279. Univ. of Calif. Press, Berkeley, California.

Dowson, W. J. 1957. "Plant Diseases Due to Bacteria," 2nd ed. Cambridge Univ. Press, London and New York. 232 pp.

Dye, D. W. 1974. The problem of nomenclature of the plant pathogenic pseudomonads. *Rev. Plant Pathol.* **53**:953–962.

Echandi, E. 1976. Bacteriocin production by *Corynebacterium michiganense*. *Phytopathology* **66**:430–432.

Elliott, Charlotte. 1951. "Manual of Bacterial Plant Pathogens," 2nd ed. Chronica Botanica, Waltham, Massachusetts. 186 p.

Frobisher, M. 1962. "Fundamentals of Microbiology," 7th ed. Saunders, Philadelphia. 610 p.

Gorlenko, M. V. 1961. "Bacterial Diseases of Plants" (translated from Russian, 1963). Jerusalem, Israel. 174 p.

Kado, C. I., and M. G. Heskett. 1970. Selective media for isolation of *Agrobacterium*, *Corynebacterium*, *Erwinia*, *Pseudomonas*, and *Xanthomonas*. *Phytopathology* **60**:969–976.

Moore, L. W., and R. V. Carlson. 1975. Liquid nitrogen storage of phytopathogenic bacteria. *Phytopathology* **65**:246–250.

Okabe, N., and M. Goto. 1963. Bacteriophages of plant pathogens. *Ann. Rev. Phytopathol.* **1**:397–418.

Stapp, C. 1961. "Bacterial Plant Pathogens." Oxford Univ. Press, London and New York. 292 pp.

Vidaver, A. K., *et al.* 1972. Bacteriocins of the phytopathogens *Pseudomonas syringae, P. glycinea* and *P. phaseolicola. Can. J. Microbiol.* **18**:705–713.

Vidaver, Anne, K. 1976. Prospects for control of phytopathogenic bacteria by bacteriophages and bacteriocins. *Ann. Rev. Phytopathol.* **14**:451–465.

bacterial spots and blights

The most common types of bacterial diseases of plants are those that appear as spots of various sizes on leaves, stems, blossoms, and fruits. Some bacterial diseases appear as continuous, rapidly advancing necroses of such organs and are then called blights. It is possible, although not common, that in blights of grown plants, one bacterial infection at one point may spread internally through most or all of the plant and may kill the whole plant. Generally, several infections are involved, even in typical blights such as fire blight, and are responsible for the death of part or all of the plant. Most so-called blights, however, are usually the final expression of severe spot infections on leaves, stems, or blossoms. In severe infections the spots may be so numerous that they destroy most of the plant surface and the plant appears blighted, or the spots may enlarge and coalesce, thus producing large areas of dead plant tissue and blighted plants. The spots are necrotic, usually circular or roughly circular, and in some cases they are surrounded by a yellowish halo. In dicotyledonous plants the development of some bacterial spots is restricted by inter-mediate or large veins and the spots appear typically angular. For the same reason, bacterial spots on leaves and stems of monocotyledonous plants appear as streaks or stripes, the name depending on their length. In humid or wet weather, infected tissue often exudes masses of bacteria that spread to new tissues or plants and start new infections. In such weather, dead leaf tissue often tears up and falls out leaving holes that are round, shot-holelike, or irregular in shape with ragged edges.

Almost all bacterial spots of leaves, stems, fruits, etc. are caused by bacteria of the closely related genera *Pseudomonas* and *Xanthomonas*, while the true blights are caused by species of *Erwinia* and *Pseudomonas*. The most common bacterial spots and blights caused by each of these pathogens are the following:

> *Pseudomonas*, causing wildfire of tobacco (*P. tabaci*), angular leaf spot or blackfire of tobacco (*P. angulata*), angular leaf spot of cucumber (*P. lacry-mans*), halo blight of beans (*P. phaseolicola*), halo blight of oats (*P. coronafaciens*), bacterial blight of peas (*P. pisi*) black spot of delphinium (*P. delphinii*), bacterial leaf spot of carnation (*P. woodsii*) and of gardenia (*P. gardeniae*), bacterial blight of soybeans (*P. glycinea*), fruit spot of apple (*Pseudomonas* sp.), and citrus blast, pear blast, bean leaf spot, and lilac blight (*P. syringae*).

> *Xanthomonas*, causing common blight of beans (*X. phaseoli*), bacterial pustule of soybeans (*X. phaseoli* var. *sojensis*), angular leaf spot of cotton (*X. mal-vacearum*), bacterial leaf blight of rice (*X. oryzae*), bacterial blight or stripe of cereals (*X. translucens*), bacterial leaf streak of rice (*X. translucens* f. sp. *oryzicola*), bacterial spot of stone fruits (*X. pruni*) and of tomato and pepper (*X. vesicatoria*), red stripe and top rot of sugarcane (*X. rubrilineans*), begonia leaf spot (*X. begoniae*), leaf blight of gladiolus (*X. gummisudans*), geranium leaf spot and stem rot (*X. pelargonii*), walnut blight (*X. juglandis*).

> *Erwinia*, causing fire blight of pome fruits (*E. amylovora*), bacterial blight of chrysanthemum (*E. carotovora* var. *chrysanthemi*).

In the bacterial spots and blights routine diagnosis of the disease depends on the morphology of the symptoms, and the absence of pathogenic fungi and presence of bacteria in recently infected tissue. Microscopic distinction among these pathogens is impossible, as it is among all plant-pathogenic bacteria. The bacteria overwinter on infected or healthy parts of perennial plants, on or in seeds, on infected plant debris, on contaminated containers or tools, and on or in the soil. Their spread from the place of overwintering to their hosts and from plant to plant takes place by means of rain, rain splashes, windblown rain, direct contact with the host, insects such as flies, bees, and ants, handling of plants, and with the tools, etc. Penetration takes place through natural openings and wounds and invasion is generally intercellular through parenchymatous tissues. Water soaking of tissues during heavy rains greatly favors penetration and invasion by bacteria. Cells become invaded after part of the cell wall is broken down, presumably by pectinases and cellulases. Control of bacterial spots and blights, in addition to the use of resistant varieties, crop rotation, and sanitation, can be obtained to some extent by spraying several times during the period of plant susceptibility with chemicals such as Bordeaux mixture, other copper compounds, zineb, antibiotics such as streptomycin and tetracyclines, and, in trees, by injecting antibiotics into the trunks.

- *Wildfire of Tobacco*

Wildfire of tobacco occurs in all parts of the world where tobacco is grown. In some regions it occurs year after year and is very destructive whereas in others it appears sporadically and its destructiveness varies. It has been reported to attack other plants; however, it seems to be economically important only on tobacco.

Wildfire causes losses in both seedbed and field. Affected seedlings may be killed. In tobacco plants already in the field, wildfire causes large, irregular, dead areas on the leaves, which may fall off or become commercially worthless.

Symptoms. The first symptoms appear usually on the leaves of young plants in seedbeds, although plants of any age can be attacked. The leaves of poorly growing seedlings show an advancing wet rot at the margins and tips, with a water-soaked zone separating the rotting and the healthy tissues. The whole leaf or only parts of it may rot and fall off. Some seedlings are killed in the seedbed while others may die after they are transplanted.

The most common symptoms appear on leaves of plants in the field and consist of round, yellowish green spots about 0.5 to 1.0 cm in diameter. Within a day or so the centers of the spots turn brown and are surrounded by yellowish green haloes (Fig. 150A). As the disease advances, the brown spots and the chlorotic haloes enlarge. In a few days the brown spots may be 2 to 3 cm in diameter, although they are not always circular. Adjacent spots usually coalesce and form large, irregular, dead areas which may involve a large portion of the leaf (Fig. 150B). In dry weather, these diseased areas dry up and remain intact. But in wet weather they fall off and give a distorted, ragged, and torn appearance to the leaves which thus become worthless. Spots appear less frequently on flowers, seed capsules, petioles, and stems.

The pathogen: Pseudomonas tabaci. This bacterium produces a potent toxin, called tabtoxin or wildfire toxin, in the host plants and on many nutrient media. Only 0.05 μg of this toxin can produce a yellow lesion on a tobacco leaf in the absence of bacteria.

Development of disease. The wildfire bacterium overwinters in plant debris in the soil, in dried or cured diseased tobacco leaves, on seed

FIGURE 150.
Wildfire lesions with chlorotic "haloes" on young tobacco leaf (A) and symptoms
of wildfire on young tobacco plants (B). Healthy plant at right. (Photo B courtesy
G. C. Papavizas.)

from infected seed capsules, on seedbed covers, and in the roots of many
weeds and crop plants. From these sources the bacteria are carried to the
leaves by rain splashes or by wind during wet weather (Fig. 151). They
may also be spread by contaminated tools and hands during handling of
the plants.

Very high humidity or a film of moisture on the plants must be present
for infections to occur and hence for development of epidemics. Water-
soaked areas present in the leaves during long rainy periods or during
rains accompanied by strong winds are excellent infection courts for the
bacterium and result in extensive lesions within 2 to 3 days. The bacteria
enter the leaf through the large stomata and hydathodes and through
wounds caused by insects and other factors. Certain insects such as flea
beetles, aphids, and white flies also act as vectors of this pathogen.

Once inside the leaf tissues the bacteria multiply intercellularly (Fig.
147) at a rapid rate. At the same time they secrete the wildfire toxin
which spreads radially from the point of infection and results in the
formation of the chlorotic halo. This consists of a rather broad zone of
cells which is free of bacteria and surrounds the bacteria-containing spot.

In wet weather the bacteria continue to spread intercellularly and
through the toxin and enxymes they secrete cause the breakdown, collapse, and death of the parenchymatous cells in the leaf tissues they
invade. Collapsed cells are invaded by the wildfire bacteria and also by
saprophytic bacteria and fungi which further disintegrate the tissues.
Dead, disintegrated areas of the leaf are loosely held together and, during
humid weather, they are easily detached from the healthy tissues and fall
to the ground, or are carried by air currents to other plants.

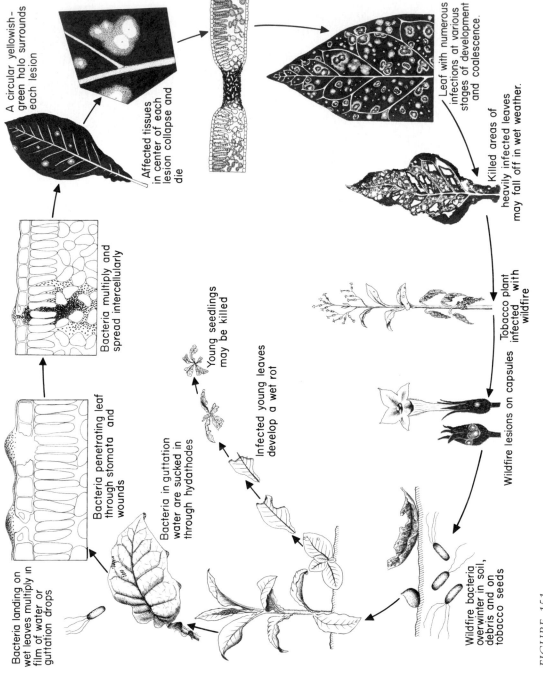

A circular yellowish-green halo surrounds each lesion

Affected tissues in center of each lesion collapse and die

Leaf with numerous infections at various stages of development and coalescence.

Killed areas of heavily infected leaves may fall off in wet weather.

Bacteria multiply and spread intercellularly

Bacteria penetrating leaf through stomata and wounds

Tobacco plant infected with wildfire

Bacteria landing on wet leaves multiply in film of water or guttation drops

Young seedlings may be killed

Bacteria in guttation water are sucked in through hydathodes

Infected young leaves develop a wet rot

Wildfire lesions on capsules

Wildfire bacteria overwinter in soil, debris and on tobacco seeds

FIGURE 151.
Disease cycle of wildfire of tobacco caused by *Pseudomonas tabaci.*

Control. Whenever possible, only resistant varieties should be planted. With susceptible varieties, it is important that control practices begin in the seedbed, since the disease often starts there. Only healthy seed should be used, and if it is suspected of being contaminated with bacteria it should be disinfested by soaking it in a formaldehyde solution for 10 minutes. The seedbed soil should be sterilized, preferably with steam, before planting or with a chemical, such as Vapam, Mylone, or methyl bromide, in the fall. After seedlings emerge, and if wildfire has been present in the area during the previous year, seedbeds should be sprayed with a neutral copper fungicide and streptomycin. The streptomycin sprays should be continued at weekly intervals until plants are transplanted. If isolated spots of wildfire appear, the infected plants plus all healthy plants in a 25-cm band around them should be destroyed by drenching with formaldehyde. Only healthy seedlings should be transplanted into the field and they should be planted only in fields that did not have a diseased crop during the previous year. Overfertilization, especially with nitrogen, should be avoided, since rapidly growing, succulent plants are much more susceptible to the disease than those that have made a slow, normal growth.

- *Bacterial Blights of Bean*

Common blight, caused by *Xanthomonas phaseoli*, and halo blight, caused by *Pseudomonas phaseolicola*. Both diseases occur wherever beans are grown and cause very similar symptoms. The diseases are usually impossible to distinguish from one another in the field. Both affect the leaves, pods, stems, and seeds in a similar way.

The symptoms appear first on the lower sides of the leaves as small, water-soaked spots. The spots enlarge, coalesce, and form larger areas that later become necrotic. The bacteria may also enter the vascular tissues of the leaf and spread into the stem. In halo blight, a halolike zone of greenish yellow tissue 10 mm or more in width forms outside the water-soaked area giving the leaves a yellowish appearance (Fig. 152A). In common blight the infected area, which is surrounded by a much narrower zone of bright, lemon-yellow tissue, turns brown, becomes rapidly necrotic, and, through coalescence of several small spots, may produce large dead areas of various shapes. Both diseases produce identical symptoms on the stems, pods, and seeds but when a bacterial exudate is produced on them, it is yellow in common blight (*Xanthomonas*) and light cream or silver-colored in halo blight (*Pseudomonas*).

The symptoms on the stem appear as water-soaked, sometimes sunken lesions that gradually enlarge longitudinally and turn brown, often splitting at the surface and exuding a bacterial exudate. Such lesions are most common in the vicinity of the first node where they girdle the stem, usually at about the time the pods are half mature. The weighted plant thus breaks at the lesion and this symptom is called girdle stem or joint rot. On the pods, small water-soaked spots also develop that may enlarge, coalesce, and turn brownish or reddish with age (Fig. 152B). Often the vascular systems of the pod sutures become infected causing the adjoin-

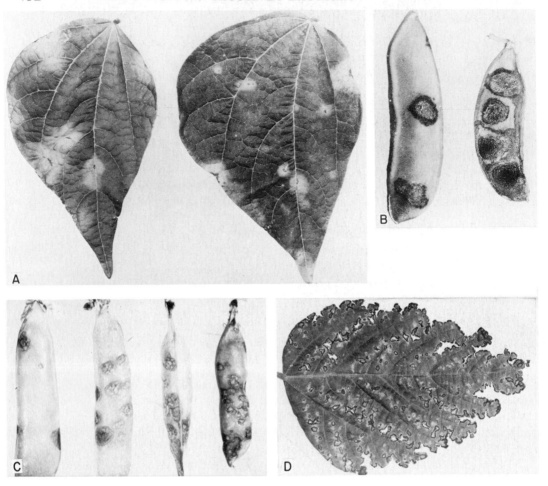

FIGURE 152.
(A) Early foliar symptoms of halo blight of beans caused by *Pseudomonas phaseolicola*. (B) Bacterial blight symptoms on lima bean pods caused by *Xanthomonas phaseoli*. (C) Symptoms on pea pods caused by *P. pisi*. In (D) advanced foliar symptoms of bacterial blight of soybean caused by *P. glycines*. (Photos A–C courtesy U.S.D.A.)

ing tissues to become water soaked and resulting in the infection of the seed through its connection (funiculus) with the pod. Seeds may rot or shrivel if infected quite young or they may show various degrees of shriveling and discoloration depending on the timing and degree of infection. Similar symptoms are caused on pea and soybean by two different species of *Pseudomonas* (Fig. 152C,D).

Both the common and halo blight bacteria overwinter in infected seed and infected bean stems. From the seed, the bacteria infect the cotyledons and from these they either spread to the leaves later on or they enter the vascular system and cause systemic infection producing stem and leaf lesions. Internally, the bacteria move between cells but the latter collapse, are invaded and digested, and cavities form. When in the xylem, the bacteria multiply rapidly and move up or down in the xylem and out into the parenchyma. They may ooze out through stomata or splits in the tissue and may reenter stems or leaves through stomata or wounds.

Control of the bacterial bean blights is through the use of disease-free seed, three year crop rotation, and sprays with copper fungicides.

Bacterial blight of soybeans caused by *Pseudomonas glycinea* and bacterial pustule of soybeans caused by *Xanthomonas phaseoli* var. *sojensis* are similar in almost all aspects to the bacterial blights of bean.

- *Angular Leaf Spot of Cucumber*

It is caused by *Pseudomonas lacrymans.* It affects the leaves, stems and fruit of cucumber, cantaloupe, squash, and some other cucurbits in North America, Europe, and probably elsewhere. At first small and circular spots that soon become large, angular to irregular, water-soaked areas develop on the leaves. In wet weather, droplets of bacterial ooze exude from the spots on the lower leaf sides and later dry into a whitish crust. Later, the infected areas turn gray, die, and shrink, often tearing away from the healthy tissue, falling off, and leaving large, irregular holes in the leaves. Infected fruits show small, almost circular spots that are usually superficial but when the affected tissues die they turn white, crack open, and let soft-rot fungi and bacteria enter and rot the whole fruit.

The bacteria overwinter primarily on contaminated seed and also in infected plant refuse. From the seed or debris the bacteria are splashed to cotyledons and leaves which they penetrate through stomata and wounds. Control is obtained through the use of clean or treated seed, resistant varieties, crop rotation, and somewhat by spraying with fixed copper-containing bactericides.

- *Angular Leaf Spot of Cotton*

It is caused by *Xanthomonas malvacearum*. The disease is present wherever cotton is grown. Small, round, water-soaked spots appear on the undersides of cotyledons and young leaves and on stems of seedlings soon after emergence. Most such leaves and plants are killed. In later stages, the spots on leaves appear as angular, brown to black lesions of varying sizes (Fig. 153A). Infected leaves of some varieties turn yellow, curl, and

FIGURE 153.
Angular leaf spot symptoms on leaves (A) and bolls (B) of cotton caused by *Xanthomonas malvacearum*. (Photo A courtesy G. C. Papavizas. Photo B courtesy U.S.D.A.)

fall. On young stems the lesions become long and black and this has given the name black arm to the disease. Stem lesions sometimes girdle and kill the stems. Angular to irregular black spots also develop on young cotton bolls (Fig. 153B). On these, the spots become sunken and in hot, humid weather the bacteria may invade and rot the bolls, cause them to drop, or to become distorted.

The bacteria overwinter in or on the seed, on the lint, and on undecomposed plant debris. Control is through the use of disease-free or treated seed and of resistant varieties.

- *Bacterial Blight or*
  Stripe of Cereals and Grasses

It is caused by *Xanthomonas translucens*. It is probably worldwide in distribution and affects primarily barley but a similar disease affects the other cereals and many grasses. The diseases occasionally cause reduction in yield, but they are generally of minor importance. The symptoms appear on leaf blades and sheaths as small, linear, water-soaked areas that soon elongate and coalesce into irregular, narrow, yellowish, or brownish glossy stripes having translucent centers (Fig. 154). Droplets of white exudate are common on the stripes. Severe infections cause leaves to turn yellow and die from the tip downward and, along with the lesions on the

FIGURE 154.
Bacterial blight or stripe of barley caused by *Xanthomonas translucens*. (Photo courtesy U.S.D.A.)

leaf sheath and floral bracts, retard spike elongation, and cause blighting. Small lesions also form on the kernels. The disease is favored by and develops mainly in rainy, damp weather. The bacteria overwinter on the seed and in crop residue and spread by rain, direct contact and insects. The main control measures are use of disease-free or treated seed and crop rotation.

- *Geranium Leaf Spot and Stem Rot*

It is caused by *Xanthomonas pelargonii* and is widespread. Symptoms appear as numerous small spots or a few large, angular, dead areas on the leaves, as dead black areas or cankers on the stems and as black rots spreading from the base upward in the cuttings. Infected leaves either die, wilt, and fall off, or hang on the plant for a while before they drop. Stem and cutting rot generally results in death of the plants. The bacteria penetrate leaves and stems through stomata and wounds and grow primarily in parenchyma tissues, but they also invade the xylem vessels and multiply in them. Many apparently healthy cuttings are often taken from symptomless infected plants and may carry the bacteria. Also, the bacteria can survive in the soil, on containers, etc. for several months. The bacteria spread from diseased to adjacent healthy cuttings through the soil, and to the leaves and stems by water splashes, handling, etc. Control of the disease depends on the use of bacteria-free cuttings and sterilized soil.

- *Bacterial Spot of Tomato and Pepper*

It is caused by *Xanthomonas vesicatoria* and is widespread. It causes considerable injury to the leaves and stems, especially of seedlings, but the disease is most noticeable by its effect on the fruit. On the leaves, the symptoms appear as small (about 3 mm), irregular, purplish gray spots with a black center and a narrow yellow halo. Numerous spots may cause defoliation or make the leaves appear ragged. Infection of flower parts usually results in serious blossom drop. On green fruit, small, water-soaked spots appear that are slightly raised, have greenish-white halos, and enlarge to about 3 to 6 mm in diameter (Fig. 155). Soon afterward, the halos disappear and the spots become brown to dark, slightly sunken, with a rough, scabby surface and the fruit epidermis rolled back. The bacteria overwinter on seed contaminated during extraction, in infected plant debris in the soil, and perhaps other hosts. It is spread by rain, wind, or contact and penetrates leaves through stomata and wounds, and fruits through wounds. Control of the disease depends on use of bacteria-free seed and seedlings, and sprays with fixed copper fungicides. The disease, however, after it appears in the field can be controlled with copper fungicides only under reasonably dry weather.

- *Bacterial Spot of Stone Fruits*

It is caused by *Xanthomonas pruni*. It is present in most areas where stone fruits are grown and may cause serious losses by directly reducing

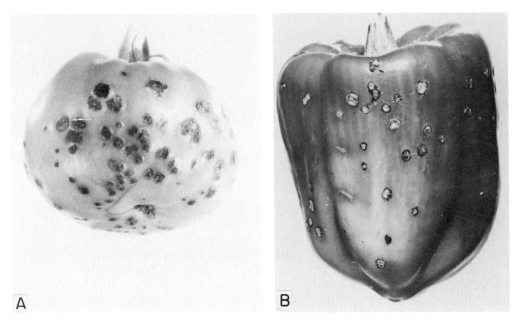

FIGURE 155.
Bacterial spot of tomato (A) and pepper (B) caused by *Xanthomonas vesicatoria*.
(Photos courtesy U.S.D.A.)

the marketability of the fruit and by devitalizing the trees by causing leaf spotting and defoliation, and lesions on twigs. The disease is most severe on peach, plum, and apricot, but it affects all stone fruits.

Symptoms appear on the leaves as small, circular-to-irregular, water-soaked spots that soon enlarge somewhat to about 1 to 5 mm in diameter, become more angular, and turn purple or brown. Often cracks develop around the spots and the affected areas break away from the surrounding healthy tissue, drop out and give a "shot-hole" effect to the leaves (Fig. 156). Several spots may coalesce and may involve large areas of the leaf. Severely affected leaves turn yellow and drop. On the fruit, small, circular, brown, slightly depressed spots appear, usually on a localized area of the fruit. Pitting and cracking occurs in the vicinity of the fruit spots and, following rainy weather, gum may exude from the injured areas. On the twigs, dark purplish to black, slightly sunken, circular-to-elliptical lesions form usually around buds in the spring or on green shoots later in the summer.

The bacteria overwinter in twig lesions and in the buds. In the spring they ooze out and are spread by rain splashes and insects to young leaves, fruits and twigs which they infect through natural openings, leaf scars, and wounds. The disease is more severe on weakened trees than on vigorous ones and, therefore, keeping trees in good vigor helps them resist the disease. Chemical sprays have not been effective so far, but recent application of antibiotics by injection into trees after the fruit has been

FIGURE 156.
Bacterial leaf spot and shot-hole on stone fruits caused by *Xanthomonas pruni.* (A)
On ornamental cherry (*Prunus tomentosa*) leaves where characteristic broad,
light green haloes form around the infected area before all affected tissue falls off.
(B) On peach. (C) On plum. The shot-hole effect is particularly obvious on the
plum leaves.

harvested has given promising control results during the following
season.

- ### Fire Blight of Pear and Apple

Fire blight is the most destructive disease of pear in the eastern half of the
U.S. and also causes damage to pear and apple orchards in other parts of
the U.S., in Canada, New Zealand, Japan and, since 1957, England. It has
been reported from many other parts of the world.

Fire blight is most destructive on pear, making commercial pear growing under certain conditions impossible. Certain apple and quince varieties are very susceptible to the disease and may be damaged as severely
as pear trees. Many other plant species in the rose family (Rosaceae) and
some nonrosaceous hosts are affected by fire blight, including several of
the stone fruits and many cultivated and wild ornamental species. Although most of these other species can serve as hosts for overwintering of
the pathogen and may be affected to varying degrees, only those in the
pome-fruit group are affected seriously.

Fire blight damages susceptible hosts by killing flowers and twigs (Fig.
157), and by girdling of large branches and trunks resulting in the death of
the trees. Young trees in the nursery or in the orchard may be killed to
the ground by a single infection in one season (Fig. 158).

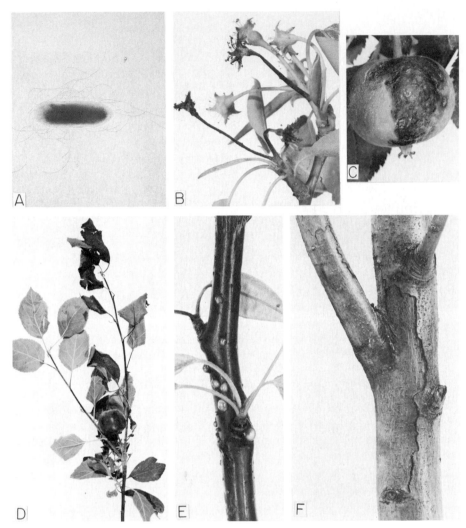

FIGURE 157.
Erwinia amylovora bacterium (A) and fire blight symptoms on pear blossoms (B), fruit (C), and young twig (D). Droplets of bacterial ooze running down the surface of infected pear twig (E). A fire blight canker is shown in (F). (Photo A courtesy R. N. Goodman and P. Y. Huang. Photos B–F courtesy Dept. Plant Path., Cornell Univ.)

Symptoms.   The first symptoms of fire blight appear usually on the flowers, which become water soaked, then shrivel rapidly, turn brownish to black in color and may fall or remain hanging in the tree (Fig. 157). Soon the symptoms spread to the leaves on the same spur or on nearby twigs, starting as brown-black blotches along the midrib and main veins or along the margins and between the veins. As the blackening progresses, the leaves curl and shrivel, hang downward, and usually cling to the curled, blighted twigs.

FIGURE 158.
Young pear tree almost killed by fire blight two months after first appearance of symptoms.

Terminal twigs and watersprouts ("suckers") are usually infected directly and wilt from the tip downward. Their bark turns brownish black and is soft at first but later shrinks and hardens. The tip of the twig is hooked, and the leaves turn black and cling to the twig. From fruit spurs and terminals the symptoms progress down to the supporting branches, where they form cankers. The bark of the branch around the infected twig appears water soaked at first, later becoming darker, sunken, and dry. If the canker enlarges and encircles the branch, the part of the branch above the infection dies. If the infection stops short of girdling the branch, it becomes a dormant or inactive canker, with sunken and sometimes cracked margins (Fig. 157).

Fruit infection usually takes place through the pedicel, but direct infection is not uncommon. Small immature fruit become water soaked, then turn brown, shrivel, mummify, and finally turn black. Dead fruit may also cling to the tree for several months after infection.

Under humid conditions, droplets of a milky colored, sticky ooze may appear on the surface of any recently infected part (Fig. 157E). The ooze usually turns brown soon after exposure to the air. The droplets may coalesce to form large drops which may run off and form a layer on parts of the plant surface.

The pathogen: Erwinia amylovora.   It is a rod-shaped bacterium and has peritrichous flagella (Fig. 157A). Virulent strains of the pathogen growing in host tissue but not on nutrient media result in the production of a toxin called amylovorin which is toxic to susceptible but not to resistant plants.

Development of disease.   The bacteria overwinter at the margins of cankers formed during the previous season, on cankers on other hosts, and possibly in buds and apparently healthy wood tissue. They survive most often in large branches and seldom in twigs less than 1 cm in diameter. In the spring, the bacteria in these "holdover" cankers become active again, multiply, and spread into the adjoining healthy bark. During humid or wet weather, water is absorbed by these bacterial masses, which increase in volume beyond the capacity of the tissues, so that parts of them exude through lenticels and cracks to the surface of the tissue. This gummy exudation, called bacterial ooze or exudate, consists of plant sap, millions of bacteria, and bacterial by-products. The ooze usually appears first about the time when the pear blossoms are opening. Various insects, such as bees, flies, ants, etc., are attracted to the sweet, sticky exudate and become smeared with it. When they visit flowers afterward, they leave some of the bacteria-containing exudate in the nectar of the flower. In some cases bacteria may also be carried from oozing cankers to flowers by splashing rain (Fig. 159). When the ooze dries, it often forms aerial strands which can be spread by wind and serve as inoculum.

The bacteria multiply rapidly in the nectar, reach the nectarthodes, and penetrate into the tissues of the flower. Bees visiting an infected flower carry bacteria from its nectar to all the succeeding blossoms that they visit. Once inside the flower, the bacteria multiply quickly. Through substances they secrete, they break down some of the components of the middle lamella and of the cell walls. The bacteria move quickly, primarily through the intercellular spaces but also through the macerated middle lamella. Sometimes the delicate walls of the flower cells are disrupted, and invasion of the protoplasts follows. Disintegration of several layers of cells can take place in some cases. This results in fairly large-sized cavities filled with bacteria. From the flower the bacteria move down the pedicel into the bark of the fruit spur. Infection of the spur results in the death of all flowers, leaves, and fruit on it (Fig. 159).

Penetration and invasion of leaves, when it happens, is similar to that of flowers. Although stomata and hydathodes may serve as ports of entry for the bacteria, it seems that most leaf infections take place through wounds made by insects, hail storms, etc. The bacteria seem to develop better and faster in the spongy mesophyll than in the palisade parenchyma. From the vein parenchyma the bacteria pass into the petiole and may reach the stem through the petiole.

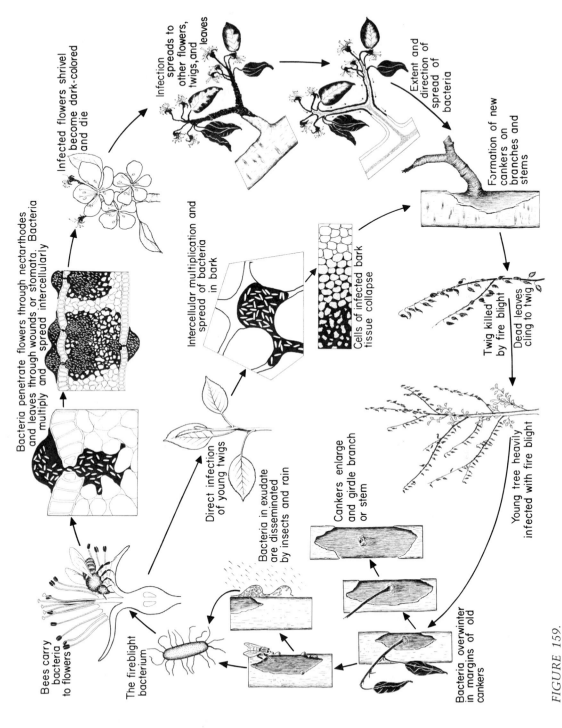

Infected flowers shrivel become dark-colored and die

Infection spreads to other flowers, twigs, and leaves

Extent and direction of spread of bacteria

Formation of new cankers on branches and stems

Bacteria penetrate flowers through nectarthodes and leaves through wounds or stomata. Bacteria multiply and spread intercellularly

Intercellular multiplication and spread of bacteria in bark

Cells of infected bark tissue collapse

Twig killed by fire blight

Dead leaves cling to twig

Direct infection of young twigs

Bacteria in exudate are disseminated by insects and rain

Cankers enlarge and girdle branch or stem

Young tree heavily infected with fire blight

Bees carry bacteria to flowers

The fireblight bacterium

Bacteria overwinter in margins of old cankers

FIGURE 159.
Disease cycle of the fire blight of pear and apple caused by *Erwinia amylovora*.

Young, tender twigs may be infected by bacteria through their lenticels, through wounds made by various agents, and through insects. They may also be infected through flower and leaf infections. In the twig the bacteria travel intercellularly. They soon cause collapse and breakdown of cortical cells, forming large cavities. In young twigs the bacteria may reach the phloem, in which they then are carried upward to the tip of the twig and to the leaves. Invasion of large twigs and branches is restricted primarily to the cortex. Progress of the infection depends on the succulence of the tissues and on the prevailing temperature and humidity. Under conditions adverse for the development of the pathogen, the host may form cork layers around the infected area and may limit the expansion of the canker. In susceptible varieties and during warm, humid weather the bacteria may progress from spurs or shoots into the second-year, third-year, and older growth, killing the bark all along the way.

Control.   Several measures need be taken for a successful fire blight control program.

During the winter all blighted twigs, branches, cankers, and even whole trees, if necessary, should be cut out about 10 cm below the last point of visible infection and burned. Cutting of blighted twigs, suckers, and root sprouts in the summer can reduce the inoculum and prevent the production of large cankers on the branches supporting them. But bacteria are in a very active state in the summer and precautions should be taken not to spread them to new branches or trees. Cutting should be done about 30 cm below the point of visible infection. The tools should be disinfested after each cut by being wiped with a sponge soaked in 10 percent commercial sodium hypochlorite solution. The latter mixture can also be used to disinfect large cuts made by the removal of branches and cankers.

Since fire blight development is greatly favored by the presence of young, succulent tissues, certain cultural practices that favor moderate growth of trees are recommended. These include growing trees in sod, balanced fertilization, especially avoiding the overstimulation of growth by heavy nitrogen applications, and limited pruning. Also a good insect control program should be followed in the post blossom period to reduce or eliminate spread of bacteria by insects to succulent twigs.

No pear or apple varieties are immune to fire blight when conditions are favorable and the pathogen is abundant, but there is a marked difference between the susceptibility of the varieties available. In areas where fire blight is destructive, varieties for new plantings should be chosen from those most resistant to fire blight.

Satisfactory control of fire blight with chemicals can be obtained only in combination with the above-mentioned measures. Dormant sprays with copper sulfate (4 pounds to 100 gallons of water) before bud break, or with Bordeaux mixture (12:12:100) containing 2 percent miscible-type oil in the delayed dormant period offer some, but not much, protection from fire blight to apple trees. Bordeaux (2:6:100) or streptomycin at 100 parts per million (ppm) are the only blossom sprays effective. Bordeaux should be applied during quick drying conditions to avoid possible russeting of fruit. Streptomycin acts systemically to a limited extent and should be applied either when maximum temperatures are above 18°C or during the night, both conditions favoring absorption of streptomycin by the tissues. One to four streptomycin applications may be necessary for satisfactory control of blossom blight. Bordeaux or streptomycin are sometimes used to control twig blight on bearing and nonbearing trees, but none of them gives good control of this phase of the disease. However, in many areas streptomycin-resistant strains of the fire blight bacterium are encountered and that antibiotic is no longer effective in controlling the disease in these areas.

SELECTED REFERENCES

Anderson, P. J. 1924. Overwintering of tobacco wildfire bacteria in New England. *Phytopathology* **14**:132–139.

Baker, K. F. 1971. Fire blight of pome fruits: the genesis of the concept that bacteria can be pathogenic to plants. *Hilgardia* **40**:603–633.

Brinkerhoff, L. A. 1970. Variation in *Xanthomonas malvacearum* and its relation to control. *Ann. Rev. Phytopathol.* **8**:85–110.

Brinkerhoff, L. A., and G. B. Fink. 1964. Survival and infectivity of *Xanthomonas malvacearum* in cotton plant debris and soil. *Phytopathology* **54**:1198–1201.

Burkholder, W. H. 1937. A bacterial leaf spot of geranium. *Phytopathology* **27**:554–560.

Burkholder, W. H., L. A. McFadden, and A. W. Dimock. 1953. A bacterial blight of chrysanthemum. *Phytopathology* **43**:522–526.

Cavadas, D. S. 1924. Le wildfire dans les plantations de tabac de Thrace et de Macedoine. *Rev. Pathol. Veg. Entomol. Agr.* **11**:236–242.

Civerolo, E. L. 1975. Quantitative aspects of pathogenesis of *Xanthomonas pruni* in peach leaves. *Phytopathology* **65**:258–264.

Clayton, E. E. 1936. Water soaking of leaves in relation to development of the wildfire disease of tobacco. *J. Agr. Res.* **52**:239–269.

Coyier, D. L., and R. P. Covey. 1975. Tolerance of *Erwina amylovora* to streptomycin sulfate in Oregon and Washington. *Plant Dis. Reptr.* **59**:849–852.

Daft, G. C., and C. Leben. 1972. Bacterial blight of soybeans: Epidemiology of blight outbreaks. *Also*: Seedling infection after emergence. *Phytopathology* **63**:57–62, 1167–1170.

Dunegan, J. C. 1932. The bacterial spot disease of the peach and other stone fruits. *USDA Tech. Bull.* **273.**

Eden-Green, S. J., and Eve Billing. 1974. Fireblight. *Rev. Plant Pathol.* **53**:353–365.

Feliciano, A., and R. H. Daines. 1970. Factors influencing ingress of *Xanthomonas pruni* through peach leaf scars and subsequent development of spring cankers. *Phytopathology* **60**:1720–1726.

Gardner, M. W., and J. B. Kendrick. 1921. Bacterial spot of tomato. *J. Agr. Res.* **21**:123–156.

Hellmers, E. 1952. Bacterial leaf spot of Pelargonium (*Xanthomonas pelargonii*) in Denmark. *Trans. Danish Acad. Techn. Sci.* **4**: 35 p.

Higgins, B. B. 1922. The bacterial spot of pepper. *Phytopathology* **12**:501–516.

Johnson, J., and S. B. Fracker. 1922. Tobacco wildfire in Wisconsin. *Wisc. Agr. Expt. Sta. Bull.* **348**: 21 pp.

Last, F. T. 1958 and 1959. Stem infection and leaf infection of cotton by *Xanthomonas malvacearum*. *Ann. Appl. Biol.* **46**:321–335, **47**:647–657.

Sands, D. C., and G. S. Walton. 1975. Tetracycline injections for control of eastern X-disease and bacterial spot of peach. *Plant Dis. Reptr.* **59**:573–576.

Schroth, M. N., *et al*. 1974. Epidemiology and control of fire blight. *Ann. Rev. Phytopathol*. **12**:389–412.

Strider, D. L. 1975. Chemical control of bacterial blight of Rieger elatior begonias caused by *Xanthomonas begoniae*. *Plant Dis. Reptr*. **59**:66–70.

Thomson, S. V., *et al*. 1975. Occurrence of fire blight of pears in relation to weather and epiphytic populations of *Erwinia amylovora*. *Phytopathology* **65**:353–358.

Wallin, J. R. 1946. Parasitism of *Xanthomonas translucens* on grasses and cereals. *Iowa State Col. J. Sci*. **20**:171–193.

Wiles, A. B., and J. C. Walker. 1952. Epidemiology and control of angular leaf spot of cucumber. *Phytopathology* **42**:105–108.

Williams, P. H., and N. T. Keen. 1967. Histology of infection by *Pseudomonas lacrymans*. *Phytopathology* **57**:254–256.

Wolf, F. A. 1922. Wildfire of tobacco. *N. C. State Agr. Expt. Sta. Bull*. **246**: 27 pp.

Zaumeyer, W. J., and H. R. Thomas. 1957. A monographic study of bean diseases and methods for their control. *USDA Tech. Bull*. **868**: 255 p.

bacterial vascular wilts

Vascular wilts caused by bacteria affect only herbaceous plants such as several vegetables, field crops, ornamentals and tropical plants.

The bacterial pathogens that cause vascular wilts and the most important diseases they cause are listed below:

Corynebacterium, causing bacterial wilt of alfalfa (*C. insidiosum*) and bean (*C. flacumfaciens*), ring rot of potato (*C. sepedonicum*), and bacterial canker and wilt of tomato (*C. michiganense*).

Erwinia, causing bacterial wilt of cucurbits (*E. tracheiphila*) and Stewart's wilt of corn (*E. stewartii*).

Pseudomonas, causing the southern bacterial wilt of solanaceous crops and the Moko disease of banana (*P. solanacearum*), and bacterial wilt of carnation (*P. caryophylli*).

Xanthomonas, causing black rot or black vein of crucifers (*X. campestris*), and gumming disease of sugarcane (*X. vascularum*).

In vascular wilts, the bacteria enter, multiply in and move through the xylem vessels of the host plants (Fig. 160). In the process, they interfere with the translocation of water and nutrients and this results in the drooping, wilting and death of the aboveground parts of the plants. In these respects bacterial vascular wilts are similar to the fungal vascular wilts caused by *Ceratocystis*, *Fusarium*, and *Verticillium*. However, while in the fungal wilts the fungi remain almost exclusively in the vascular tissues until the death of the plant, in the bacterial wilts the bacteria often destroy (dissolve) parts of cell walls of xylem vessels or cause them to rupture quite early in disease development. Subsequently, they spread and multiply in adjacent parenchyma tissues, at various points along the vessels, kill and dissolve the cells, and cause the formation of pockets or cavities full of bacteria, gums, and cellular debris. In some bacterial vascular wilts, e.g., those of corn and sugarcane, the bacteria, once they reach the leaves, move out of the vascular bundles,

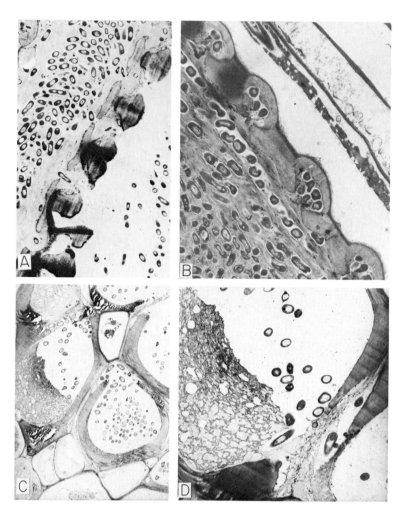

FIGURE 160.
Histopatholoy of cabbage leaf veins infected with black rot caused by
Xanthomonas campestris. (A) Uneven distribution of bacteria in xylem vessels
and passage of bacteria between adjacent vessels. (B) Bacteria in xylem vessel and
in bulges in interspiral regions toward xylem parenchyma cell. (C) Vascular
bundle showing bacteria-containing and apparently bacteria-free vessels
completely or partially occluded with plugging material. (D) A few bacteria and a
mass of plugging material in invaded vessel. (Photos courtesy F. M. Wallis, Univ.
of Natal, S. Africa, from Wallis *et al.*, *Physiol. Plant Pathol.* **3:**371–378.)

spread throughout the intercellular spaces of the leaf and may ooze out
through the stomata or cracks onto the leaf surface. Similarly, in some
cases, as in the bacterial wilt of carnation, the bacteria ooze to the surface
of stems through cracks formed over the bacterial pockets or cavities.
More commonly, however, the wilt bacteria, although they may not be
confined entirely to the vascular elements, do not spread extensively
through the rest of the plant tissues and do not reach the plant surface
until the plant is overcome and killed by the disease.

Bacterial vascular wilts can sometimes be determined by cutting an infected stem with a sharp razor blade and then pulling the two parts apart slowly, in which case a thin bridge of a sticky substance can be seen between the cut surfaces while they are being separated, or better still, by placing small pieces of infected stem, petiole, or leaf in a drop of water and observing it under the microscope, in which case masses of bacteria will be seen flowing out from the cut ends of the vascular bundles.

The mechanisms by which bacteria induce vascular wilt in plants seem to be the same as those operating in the fungal vascular wilts. Thus, the bacterial cells themselves along with their polysaccharides seem to cause occlusion of some vessels. The bacteria also secrete enzymes such as pectinases and cellulases that break down cell wall substances which, when carried in the transpiration stream, collect at vessel ends, form gels and gums that help clog the vessel pores, and thus block movement of water. These enzymes also cause softening and weakening of the cell walls which then collapse and the tissues droop and wilt. Phenoloxidases secreted by the bacteria or released by the disrupted plant cells cause oxidation of phenolics to quinones which then polymerize to form melanoid substances. The latter impart a brown coloration to any cell wall or substance to which they become adsorbed. Growth regulators secreted by bacterial pathogens may cause hyperplasia of xylem parenchyma cells with subsequent crushing of xylem vessels, formation of tyloses, etc. Whether wilt bacteria produce toxins is not known, but many of their secretions certainly have a detrimental effect on plant growth and development.

The wilt bacteria overwinter either in plant debris in the soil, in the seed, in vegetative propagative material or, in some cases, in their insect vectors. They enter the plants through wounds that expose open vascular elements and multiply and spread in the latter. They spread from plant to plant through the soil, through handling and tools, through direct contact of plants, or through insect vectors. Nematode infections, by injuring roots, seem to facilitate infection by wilt bacteria in at least some of the vascular wilts. Control of bacterial vascular wilts is difficult and depends primarily on the use of crop rotation, resistant varieties, the use of bacteria-free seed, or other propagative material, control of the insect vectors of the bacteria when such vectors exist, and through removal of infected plant debris and proper sanitation.

- *Bacterial Wilt of Cucurbits*

Bacterial wilt of cucurbits is found in all the United States, although it is most severe in the eastern half of the country. It also occurs in central and northern Europe, South Africa, and Japan. It affects many cultivated and wild species of plants of the family Cucurbitaceae. Cucumber seems to be the most susceptible host to the disease, followed in susceptibility by muskmelon, squash, and pumpkin. Watermelon is extremely resistant or immune to bacterial wilt.

Bacterial wilt affects plants by causing sudden wilting of foliage and vines and finally death of the plants. It also causes a slime rot of squash fruit in storage. The severity of the disease varies widely in different seasons and localities from an occasional wilted plant up to a destruction of 75 to 95 percent of the crop.

Symptoms. The first symptoms of bacterial wilt appear as drooping of one or more leaves of a vine; this is soon followed by drooping and wilting of all the leaves of that vine and quickly afterward by wilting of all leaves and collapse of all vines of the infected plant (Fig. 161). Wilted leaves shrivel and dry up; affected stems first become soft and pale but later they, too, shrivel and become hard and dry. Symptoms in less susceptible plants or under unfavorable conditions develop slowly and may be accompanied by reduced growth and, occasionally by excessive blossoming and branching of the infected plants. When infected stems are cut and pressed between the fingers, droplets of white bacterial ooze appear on the cut surface. The viscid sap sticks to the finger or to the cut sections and if they are gently pulled apart the ooze forms delicate threads that may be extended for several centimeters. The stickiness and somewhat milky appearance of the sap of infected plants are frequently used as diagnostic characteristics of the disease, although they are not entirely dependable.

The slime of stored squash progresses internally and may cause the

FIGURE 161.
Bacterial wilt of cucumber caused by *Erwinia tracheiphila*. (Photo courtesy Dept. Plant Pathol., Cornell Univ.)

spoilage of every portion of the fruit while the exterior surface of the fruit may appear perfectly sound. Usually, however, as the internal rot progresses there appear on the surface dark spots or blotches which coalesce and enlarge. The disease develops over several months in storage. Infected squash fruits are further invaded by soft-rot microorganisms and are completely destroyed.

The pathogen: Erwinia tracheiphila.   *Erwinia tracheiphila* is very sensitive to drying and does not survive in infected, dried, plant tissue for more than a few weeks. It survives, instead, in the intestines of striped cucumber beetles (*Acalymma vittata*) and spotted cucumber beetles (*Diabrotica undecimpunctata*) on which the cucurbit wilt bacteria are completely dependent for dissemination, inoculation, and overwintering (Fig. 162).

Development of disease.   The cucurbit wilt bacteria hibernate in the digestive tracts of a relatively small number of overwintering striped cucumber beetles and spotted cucumber beetles. In the spring these insects feed on the leaves of cucurbit plants, on which they cause deep wounds. The bacteria are deposited in these wounds with the feces of the insects. Swimming through the droplets of sap present in the wounds, the bacteria enter the xylem vessels where they multiply rapidly and spread to all parts of the plant (Fig. 162). Penetration through stomata does not take place.

As the bacteria multiply in the xylem they cause a mechanical obstruction of the vessels and so reduce the efficiency of the water-conducting system of diseased plants. Furthermore, gum deposits are commonly found in the xylem elements of infected plants and in some wilting plants tyloses are also present. In some instances the presence of gums and/or tyloses appears to be as important in plugging the transpiration stream as the polysaccharides and the bacteria. When wilt symptoms begin to appear, the transpiration rate of infected plants is lower than that of healthy ones and steadily decreases as wilting proceeds. Stems of wilted plants allow less than one-fifth the normal water flow, indicating that an extensive plugging of the vessels is the primary cause of wilting.

Spread of the bacteria from one plant to another is achieved primarily through the striped and the spotted cucumber beetles and to a smaller extent through other insects, such as grasshoppers. When these feed on infected plants, their mouthparts become contaminated with the wilt bacteria. Later the beetles move on to healthy plants and carry with them bacteria which they place in the new wounds they make. Each contaminated beetle can infect at least three or four healthy plants after one feeding on a wilted plant, although some beetles are capable of spreading infection for more than three weeks after one wilt feeding. Only a rather small percentage of beetles, however, become carriers of bacteria. Infections take place only when a film of water is present on the tissues and allows the pathogen to reach the wound and move into the xylem vessels. The first wilt symptoms appear 6 or 7 days after infection and the plant is usually completely wilted by the fifteenth day. The bacteria present in the vessels of infected plants die within one or two months after the dead

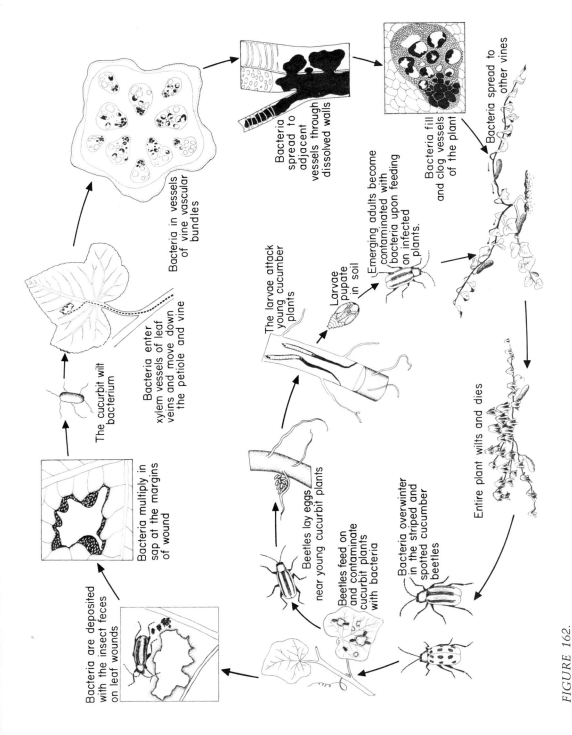

Bacteria in vessels of vine vascular bundles

Bacteria enter xylem vessels of leaf veins and move down the petiole and vine

Bacteria spread to adjacent vessels through dissolved walls

Bacteria fill and clog vessels of the plant

Bacteria spread to other vines

The cucurbit wilt bacterium

Emerging adults become contaminated with bacteria upon feeding on infected plants.

Larvae pupate in soil

The larvae attack young cucumber plants

Bacteria multiply in sap at the margins of wound

Beetles lay eggs near young cucurbit plants

Entire plant wilts and dies

Bacteria are deposited with the insect feces on leaf wounds

Beetles feed on and contaminate cucurbit plants with bacteria

Bacteria overwinter in the striped and spotted cucumber beetles

FIGURE 162.
Disease cycle of bacterial wilt of cucurbits caused by *Erwinia tracheiphila*.

plants dry up. The bacteria are also incapable of surviving the winter in the soil and in or on seeds from infected plants.

Fruit infection of squash plants usually takes place through infected vines, but it is also possible through the blossoms and the rind of the squash by beetles which feed on the blossoms and the fruits during the growing season.

The disease is strongly influenced by certain environmental factors. Thus, the greater the number of cucumber beetles in an area, the younger and more succulent the plants, and the more humid the weather, the greater the number of plants that will become diseased and the more severe the symptoms.

Control.   Bacterial wilt of cucurbits can be controlled best by controlling the cucumber beetles with insecticides, such as carbaryl (Sevin), methoxychlor, and rotenone. Control of the early beetles is most important in limiting or eliminating the primary infections of plants and the multiplication and secondary spread of the pathogen.

To avoid squash rot in storage only fruit from healthy plants should be picked and it should be stored in a clean, fumigated warehouse.

Several varieties within each cucurbit species are resistant to bacterial wilt. These should be preferred to more susceptible ones.

- *Bacterial Wilt*
  or Stewart's Wilt of Corn

It is caused by *Erwinia stewartii*. It is widespread in eastern North America and is present in Central America, Europe and Asia. The disease is much more severe on sweet corn than on field corn.

The symptoms on sweet corn appear as a rapid wilting of the plant and the plants are either killed or remain stunted and produce no ears. Leaves of diseased plants usually show long streaks that have irregular or wavy margins (Fig. 163A) and are pale green to yellowish at first but soon become dry and brown. The length of the streaks varies from small to that of the length of the leaf. The vascular bundles of stalks or leaves of infected plants are filled with bacteria which ooze out on the cut surface as yellowish masses. In severe infections the bacteria invade all vascular tissues, including those of the cob, through which they pass into the kernels, and also spread into parenchyma tissues and produce cavities, particularly in the stalk near the soil line.

In field corn, the disease seldom spreads through the entire plant. Instead, the most common symptoms are the streaks on the leaves which are short or long with irregular, wavy margins but are usually yellow until they die and then they become straw colored. These streaks, which always originate at the points of feeding of the insect vectors, occur rather late, usually after tasseling, and are known as the leaf blight or late-stage phase of the disease. Often large parts of, or entire leaves are killed prematurely thus reducing yields and predisposing the plants to stalk rots.

The bacteria overwinter within the bodies of the corn flea beetle (*Chaetocnema pulicularia*). The bacteria are spread mainly by the corn

FIGURE 163.
(A) Corn leaves showing long yellowish streaks with wavy margins due to infection with the wilt bacterium *Erwinia stewartii*. (B) and (C) Potato tubers showing external and internal symptoms of potato ring rot caused by *Corynebacterium sepedonicum*.

flea beetle but several other beetles and maggots may sometimes play a role in the spread of the disease. The severity of the disease in a given year depends on how mild or severe the previous winter was, severe winters reducing the number of corn flea beetles that survive and are able to cause the first infections of corn seedlings in the spring. Subsequently, the flea beetles spread the disease to more plants by feeding on diseased and then on infected plants. Control of bacterial wilt of corn is obtained through the use of resistant varieties or hybrids, spraying the plants early with insecticides to kill the corn flea beetles, and by providing the plants with high levels of calcium and potassium, the latter somehow helping the plants resist the infection.

- *Ring Rot of Potato*

It is caused by *Corynebacterium sepedonicum*. The disease occurs and causes severe losses in North America and continental Europe. Infected plants usually do not show aboveground symptoms until they are fully grown or the symptoms may occur so late in the season that they are often overlooked or masked by senescence, late blight or other diseases. In years with cool springs and warm summers, however, one or more of the stems in a hill may appear more or less stunted while the rest of the plant appears normal. The interveinal areas of leaflets of affected stems turn yellowish and their margins roll upward and become necrotic. The yellowing of the leaves is accompanied by a progressive wilting which continues until all the leaves of the stem wilt and the stem then dies. Wilted stems do not usually show much internal discoloration but if the stem is cut at the base and is squeezed, a creamy exudate oozes out of the vascular bundles.

The characteristic symptoms of the disease appear in tubers either before or after harvest and may be present in only some of the tubers of a plant. Infection and the symptoms begin to develop at the stem end of the tuber and progress through the vascular tissue. When cut through, infected tubers show at first a ring of light yellow vascular discoloration and some bacterial ooze that may be increased by squeezing the tuber (Fig. 163B, C). As the disease advances, a creamy yellow or light brown crumbly or cheezy rot develops in the region of the vascular ring and, if the tuber is squeezed, a soft, pulpy exudate oozes out from the diseased areas while a more or less continuous ring of cavities is formed by the rotting of tissues in the vascular area. Secondary, soft-rotting bacteria often invade infected tubers and these may cause complete rot of the tuber.

The characteristic morphology of *Corynebacterium* cells and its gram-positive reaction, taken together with the host and the symptoms, are the primary diagnostic tools for this disease.

The ring rot bacteria overwinter mostly in infected tubers and as dried slime on machinery, crates, sacks, etc. They do not overwinter in the soil. The bacteria, however, are easily spread by knives used to cut potato seed pieces and a knife used to cut an infected tuber may infect the next 20 healthy seed pieces cut with it. The bacteria enter plants only through wounds and invade the xylem vessels in which they multiply profusely and may cause plugging. The bacteria also move out of the vessels into the surrounding parenchyma tissues, where they cause cavities, and then again into new vessels. The bacteria also invade the roots and cause the deterioration of the young feeder roots, which contributes to the aboveground symptoms of the plants late in the season.

Control of potato ring rot is through the use of healthy seed tubers wherever available. The bacterium has not been reported to overwinter in soil. If a grower had ring rot in his potato crop the previous year, however, since the bacteria can also overwinter as dried slime on containers or tools, thorough disinfestation of warehouses, crates, equipment, etc. with ethylene oxide, copper sulfate, or formaldehyde must be carried out. Knives used to cut seed tubers should be constantly disinfested by sodium hypochlorite or by boiling water.

• Southern Bacterial Wilt of Solanaceous Plants and Moko Disease of Banana

It is caused by *Pseudomonas solanacearum.* It is present in the tropics and in the warmer climates throughout the world. It causes its most severe losses on banana in the tropics and it is frequently severe on tobacco, tomato, potato, and eggplant in some warm areas outside the tropics. Many other hosts, however, are attacked by the disease, icluding peanuts, soybeans, plantains, and other cultivated and wild herbaceous plants. In the U.S. the disease is most severe in the southeastern states where it is favored by the warm, humid climate and it is known as Granville wilt of tobacco, or as brown rot of potato. At least three races of the pathogen are involved in causing the diseases on the various hosts, one of them attacking all the solanaceous and many nonsolanaceous crops as well as some bananas, another attacking only plants in the banana family, and a third attacking potato and sometimes tobacco.

Symptoms of bacterial wilt on solanaceous crops appear as a rather sudden wilt. Infected young plants die rapidly. Older plants may first show leaf drooping and discoloration, leaf drop, or one-sided wilting and stunting before the plants wilt permanently and die (Fig. 164). In some plants, e.g., tomato, excessive development of adventitious roots may take place. The vascular tissues of stems, roots, tubers, etc. turn brown and in cross sections they ooze a whitish bacterial exudate. Bacterial pockets are commonly present around the vascular bundles in the pith and in the cortex, and roots often rot and disintegrate by the time the plant wilts permanently.

In the Moko disease of banana, young plants wilt rapidly and die, their

FIGURE 164.
Symptoms of southern bacterial wilt on tomato (A) and tobacco (B) caused by *Pseudomonas solanacearum.* (Photo A courtesy U.S.D.A. Photo B courtesy G. C. Papavizas.)

central leaves breaking at a sharp angle without turning yellow. In older plants, first the inner leaf turns a dirty yellow near the petiole, the petiole breaks down and the leaf wilts and dies. In the meantime, more and more of the surrounding leaves droop and die from the center outward until all the leaves bend down and dry out. Fruit growth in infected plants, if it had started, stops. Banana fingers are deformed, turn black, and shrivel. If the fruit was near maturity when infected, it may show no outward symptoms but the pulp of some fingers may be discolored and decaying. In cross section, an infected banana pseudostem shows many discolored, greenish-yellow to reddish-brown or almost black vascular bundles, particularly in the inner leaf sheaths and in the fruit stalk. Pockets of bacteria and decay may be present in the pseudostem, in the rhizome, and most strikingly in individual bananas that become filled with a dark, gummy substance. The pulp of such bananas finally dries out into a gray, crumbly, starchy residue that pours out when the peel splits open.

The *P. solanacearum* bacteria overwinter in diseased plants or plant debris, vegetative propagative organs, e.g., potato tubers, and banana rhizomes, on the seeds of some crops, e.g., soybean and peanut, in wild host plants, and probably in the soil. Injured or decaying infected tissues release bacteria in the soil. The bacteria are spread through the soil water, infected or contaminated seeds, rhizomes, transplants, etc. by contaminated knives used for cutting tubers and rhizomes of for pruning suckers and, in some instances, by insects. The bacteria enter plants through wounds made in roots by cultivating equipment, nematodes, insects, etc., and at cracks where secondary roots emerge. The bacteria reach the large xylem vessels and through them spread into the plant. Along the vessels they escape into the intercellular spaces of the parenchyma cells in the cortex and pith, dissolve the cell walls, and create cavities filled with slimy masses of bacteria and cellular debris.

Control of bacterial wilt of solanaceous plants and banana depends mostly on the use of resistant varieties, when available, and proper crop rotation or fallow. Only bacteria-free rhizomes, transplants, tubers, etc. should be used and tools, such as knives, should be disinfected by dipping for 10 seconds or more in a 10 percent formaldehyde solution when moving from one banana plant to another. Diseased banana plants and rhizomes should be cut up and burned as should plants around them that may be infected but do not yet show symptoms. Infested banana soils can be reclaimed by keeping them fallow for about a year and by frequent disking during the dry season to accelerate desiccation of plant material and apparently the death of the wilt bacteria.

- *Black Rot or Black Vein of Crucifers*

It is caused by *Xanthomonas campestris.* The disease is present throughout the world. It affects all members of the cabbage family and sometimes causes severe losses on these crops. The disease affects plants of any age and primarily the aboveground parts of plants, but in hosts like turnip and radish that have fleshy roots, these organs may also be affected and may develop a dry rot. Infection of young seedlings causes dwarfing, one-sided growth, and drop of the lower leaves. The first symptoms, however, usually appear in the field as large, often V-shaped, chlorotic blotches at the margins of the leaves (Fig. 165A) that progress toward the midrib of the leaf while some of the veins and veinlets within the chlorotic area turn black. The affected area later turns brown and dry. In the meantime, the discoloration of the veins advances to the stem and from there upward and downward to other leaves and roots. When leaves become invaded systemically from bacteria moving upward through the midvein, chlorotic areas may appear anywhere on the leaves. Infected leaves may fall off prematurely one after the other (Fig. 165B). The stem and the stalks of infected leaves appear healthy from the outside but in cross section they show browning or blackening of the vascular tissues and often small yellow slime droplets of bacteria. Sometimes, cavities full of bacteria form in the pith and cortex. Cabbage and cauliflower heads are also invaded and discolored, as are the fleshy roots of turnip, radish, etc. Infected areas are subsequently invaded by soft-rotting bacteria which destroy the tissue and a repulsive odor is given off.

The black rot bacteria overwinter in infected plant debris and on or in the seed. If the bacteria come in contact with or are splashed to cotyledons or young leaves, they infect them through stomata, hydathodes, or wounds and spread through them intercellularly until they reach the open ends of outer vessels which they invade. The bacteria then multiply in the vessels and spread in them throughout the plant (Fig. 160), reaching even the seeds. At the same time, however, disintegration of the xylem occurs in places and the bacteria spread into the intercellular spaces of the surrounding parenchyma. These cells sooner or later are killed and disintegrated, and cavities are formed. In leaf infections, the bacteria reach the surface of the leaves through hydathodes or wounds and are

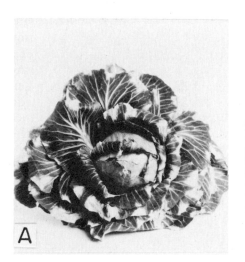

FIGURE 165.
Cabbage black rot caused by *Xanthomonas campestris*. (A) Early stages of infection on margins of leaves. (B) Advanced stages of black rot with many leaves wilting or already fallen off.

subsequently spread by rain splashes and wind, or are carried by equipment to other leaves which they invade through hydathodes, wounds, or insect injuries. In wet, warm weather infection develops rapidly and visible symptoms may appear within hours.

Control of black rot is difficult and depends on the use of bacteria-free seed and transplants planted in soil in which no black rot was present the previous two or three years. So, crop rotation is necessary. Seed treatment with hot water (50°C for 30 minutes) helps insure bacteria-free seed.

SELECTED REFERENCES

Baribeau, B. 1948. Bacterial ring rot of potatoes. *Amer. Pot. J.* **15**:71–82.

Buddenhagen, I., and A. Kelman. 1964. Biological and physiological aspects of bacterial wilt caused by *Pseudomonas solanacearum. Ann. Rev. Phytopathol.* **2**:203–230.

Castor, L. L., *et al.* 1975. Computerized forecasting system for Stewart's bacterial disease on corn. *Plant Dis. Reptr.* **59**:533–536.

Clayton, E. E. 1929. Studies on the black-rot or blight disease of cauliflower. *N. Y. (Geneva) Agr. Exp. Sta. Bull.* **576**: 44 p.

Cook, A. A., J. C. Walker, and R. H. Larson. 1952. Studies on the disease cycle of black rot of crucifers. *Phytopathology* **42**:162–167.

Eddins, A. H. 1936. Brown rot of Irish potatoes and its control. *Florida Agr. Exp. Sta. Bull.* **299.**

Harris, H. A. 1940. Comparative wilt induction by *Erwinia tracheiphila* and *Phytomonas stewartii. Phytopathology* **30**:625–638.

Hunter, J. E., G. S. Abawi, and R. F. Becker. 1975. Observations on the source and spread of *Xanthomonas campestris* in an epidemic of black rot in New York. *Plant Dis. Reptr.* **59**:384–387.

Jenkins, S. F., Jr., D. J. Morton, and P. D. Dukes. 1966. Distinguishing *Pseudomonas solanacearum* infections from other peanut wilt diseases by the use of serological techniques. *Plant Dis. Reptr.* **50**:836–838.

Jones, F. R. 1940. Bacterial wilt of alfalfa and its control. *USDA Circ.* **573.**

Kelman, A. 1953. The bacterial wilt caused by *Pseudomonas solanacearum. N. C. Agr. Exp. Sta. Tech. Bull.* **99**:194 p.

Lelliott, R. A. 1973. A survey in England and Wales for ring rot of potatoes caused by *Corynebacterium sepedonicum. Plant Pathol.* **22**:126–128.

Nelson, P. E., and R. S. Dickey. 1970. Histopathology of plants infected with vascular bacterial pathogens. *Ann. Rev. Phytopathol.* **8**:259–280.

Nuttal, V. W., and J. J. Jasmin. 1958. The inheritance of resistance to bacterial wilt (*Erwinia tracheiphila* E. F. Smith, Holland) in cucumber. *Can. J. Plant Sci.* **38**:401–404.

Pepper, E. H. 1967. "Stewart's Bacterial Wilt of Corn." Monograph No. 4. The American Phytopathological Society: St. Paul, Minn.

Prend, J., and C. A. John. 1961. Method of isolation of *Erwinia tracheiphila* and an improved inoculation technique. *Phytopathology* **51**:255–258.

Rand, F. W., and Ella M. A. Enlows. 1920. Bacterial wilt of cucurbits. *U.S. Dept. Agr. Bull.* **828**:43 pp.

Robert, Alice L. 1955. Bacterial wilt and Stewart's leaf blight of corn. *USDA Farmers' Bull.* No. 2092. 13 p.

Schaad, N. W., and W. C. White. 1974. Survival of *Xanthomonas campestris* in soil. *Phytopathology* **64**:1518–1520.

Sequeira, L. 1958. Bacterial wilt of bananas: dissemination of the pathogen and control of the disease. *Phytopathology* **48**:64–69.

Skaptason, J. B. 1943. Studies on the bacterial ring rot disease of potatoes. *N. Y. (Cornell) Agr. Exp. Sta. Mem.* **250**.

Smith, E. F. 1911. Wilt of cucurbits. *Carnegie Inst. Washington* 2:209–299.

Strandberg, J. 1973. Spatial distribution of cabbage black rot and the estimation of diseased plant populations. *Phytopathology* **63**:998–1003.

Vaughan, E. K. 1944. Bacterial wilt of tomato caused by *Phytomonas solanacearum. Phytopathology* **34**:443–458.

Wallis, F. M., *et al.* 1973. Ultrastructural histopathology of cabbage leaves infected with *Xanthomonas campestris. Physiol. Plant Pathol.* **3**:371–378.

Watterson, J. C., *et al.* 1972. Multiplication and movement of *Erwinia tracheiphila* in resistant and susceptible cucurbits. *Plant Dis. Reptr.* **56**:949–952.

Yu, T. F. 1933. Pathological and physiological effects of *Bacillus tracheiphilus* E. F. Smith on species of cucurbitaceae. *Nanking Univ., Coll. Agr. Forestry Bull.* **5**.

bacterial soft rots

Bacteria are invariably present whenever fleshy plant tissues are rotting in the field or in storage, and the foul smell given off by such rotting tissues is due, usually, to volatile substances released during the disintegration of plant tissues by such bacteria. Rotting tissues become soft and watery, and slimy masses of bacteria and cellular debris frequently ooze out from cracks in the tissues. In many such soft rots, however, the bacteria involved are not plant pathogenic, i.e., they do not attack living cells, but rather they are saprophytic or secondary parasites, i.e., they grow in tissues already killed by other pathogens and environmental causes, or in tissues so weakened or old that they are near their physiological breakdown and are unable to resist attack by any organism.

In addition to these secondary soft rotters, however, there are some bacteria that attack living plant tissues in the field or in storage:

> *Erwinia,* causing soft rot of numerous fleshy fruits, vegetables and ornamentals (*E. carotovora*) and blackleg of potato (*E. carotovora* var. *atroseptica*).
>
> *Pseudomonas,* associated with the pink eye disease of potato and soft rots of other fleshy vegetables (*P. marginalis* and *P. fluorescens*), the slippery skin disease of onion (*P. alliicola*) and the sour skin of onion (*P. cepacia*).

The soft rot bacteria may overwinter in infected tissues, in the soil, and in contaminated equipment and containers. Some of them also overwinter in insects. They are spread by direct contact, hands, tools, soil water, insects, etc. They enter plants or plant tissues primarily through wounds, but in storage uninjured tissues also become infected. Within the tissues they multiply profusely in the intercellular spaces where they produce several kinds of enzymes which, by dissolving the middle lamella and separating the cells from each other, cause maceration and softening of affected tissues. The cells, surrounded as they are by the bacteria and their enzymes, at first lose water and their contents shrivel, but finally parts of their walls are dissolved and the cells are invaded by bacteria. Control of bacterial soft rots is difficult and depends on proper sanitation, avoiding injuries, keeping storage tissues dry and cool, and on crop rotation.

- *Bacterial Soft Rots of Vegetables*

Bacterial soft rots occur most commonly on vegetables (and some annual ornamentals) that have fleshy storage tissues, such as potatoes, carrots, radishes, onions, hyacinths, iris, or fleshy fruit, such as cucumber, squash, eggplant, tomato, or succulent stem, stalk or leaves, such as cabbage, celery, lettuce, or spinach. They are found all over the world and cause serious diseases of crops in the field, in transit, and especially in storage, resulting in greater total loss of produce than any other bacterial disease. Nearly all fresh vegetables are subject to bacterial soft rots and may develop a serious decay within a few hours in storage or during marketing. Bacterial soft rots cause severe economic losses by reducing quantities of produce available for sale, by reducing the quality and thus the market value of the crops, and by greatly increasing expenses for preventive measures against soft rots and for preparation of partially affected produce for use

Symptoms. The soft-rot symptoms produced on vegetable fruits and other fleshy organs in the field or in storage are very similar on all the hosts. At first there appears on the tissue a small water-soaked lesion which enlarges rapidly in diameter and in depth. The affected area becomes soft and mushy (Fig. 166). Its surface may become discolored and somewhat depressed or it may appear wrinkled or blistered. The margins of the lesions usually are well defined at first but later become obscure. The tissues within the affected region become opaque in a short time or appear cream colored and slimy, disintegrating into a mushy mass of disorganized cells. In certain fruits, tubers, etc., the outer surface may remain intact while the entire contents have changed to a turbid liquid. Frequently, however, cracks develop and the slimy mass exudes to the surface where, upon exposure to the air, it turns tan, gray, or dark brown. A whole fruit or tuber may be converted into a soft, watery, colorless, decayed mass within a period of 3 to 5 days. Infected fruits and tubers of many plants are almost odorless until the infected tissues collapse whereupon secondary bacteria, living off the decomposing tissues, produce a foul odor. Cruciferous plants and onions, however, when infected by soft-rot bacteria, almost always give off an offensive sulfurous odor.

When root crops are affected in the field, symptoms may also develop on the lower parts of the stem, which become watery, turn black, and shrivel. This also results in stunting, wilting, and death of the aboveground parts of the plant.

Infections of succulent leaves and stems are seldom important in the field. However, when these parts are infected in storage or in packages, especially in plastic containers, rapid softening and disintegration of the diseased tissues follows and may yield a wet, green, slimy mass within 1 or 2 days.

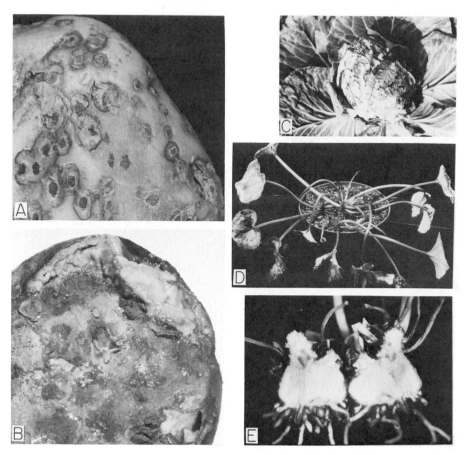

FIGURE 166.
Soft rots caused by *Erwinia carotovora*. (A) Potato tuber with numerous infections. (B) Entire potato tuber rotting. (C) Bacterial soft rot on cabbage. (D, E) Bacterial soft rot on cyclamen. (D) Leaves and stems wilting or rotting. (E) Parts of the corm are rotting and many roots and leaves have rotted and disintegrated. (Photos A and B courtesy Dept. Plant Pathol., Cornell Univ.)

The pathogen: Erwinia carotovora and other species. Although bacteria belonging to other genera (e.g., *Pseudomonas*) can cause soft rots, the most common and the most destructive soft rots are caused by bacteria in the genus *Erwinia*. Of these, *Erwinia carotovora* seems to be the most prevalent. Indeed, it is difficult to differentiate all other specialized soft-rotting *Erwinias* from *Erwinia carotovora*, e.g., *Erwinia carotovora* var. *atroseptica,* the cause of black leg of potato (Fig. 167).

The soft-rot bacteria can grow and are active over a wide range of temperatures. The minumum, optimum, and maximum temperatures for disease development are 5, 22, and 37°C. The bacteria are killed at about 50°C.

Development of disease. The soft-rot bacteria overwinter in infected fleshy organs in storage and in the field, in debris that contains parts

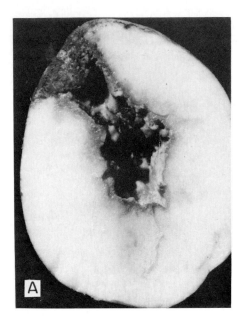

FIGURE 167.
Bacterial soft rot and black leg caused by *Erwinia carotovora* f. *atroseptica*. (A)
Cross section of potato tuber. (B) Bacterial soft rot of potatoes in the field. Note
initiation of infection through the stem end. (Photos courtesy Dept. Plant Pathol.,
Cornell Univ.)

of infected plants, in the soil, in the pupae of the seed-corn maggot
(*Hylemyia cilicrura*), and in the pupae of several other insects (Fig. 168).
The disease may first appear in the field on plants grown from previously
infected seed pieces, as is frequently the case with potato. Some tubers,
rhizomes, bulbs, etc., become infected by the bacterium, which may be
present in the soil, after they are set or formed in the soil. These infections usually take place through wounds. However, under certain conditions the bacteria can invade tubers through lenticels. The inoculation of
bacteria into fleshy organs and their further dissemination are greatly
facilitated by insects which very effectively spread infection both in
storage and in the field. The soft-rot bacteria can live in all stages of the
insect. Moreover, the bodies of the insect larvae (maggots) become contaminated with bacteria when they crawl about in infested soil or on
rotting seed pieces. Therefore, when such insects attack healthy plants
or storage organs by boring holes into them, not only do they carry the
bacteria to the plants, but they put the bacteria into wounds where they
can cause the disease. Even when the plants or storage organs are resistant to soft rot and can stop its advance by formation of wound-cork
layers, if borers are present they destroy the wound cork as fast as it is
formed, so that the wounds never heal and the soft rot continues to
spread.

When the soft-rot bacteria enter wounds, they feed and multiply at first
on the liquids released by the broken cells on the wound surface. Inocula-

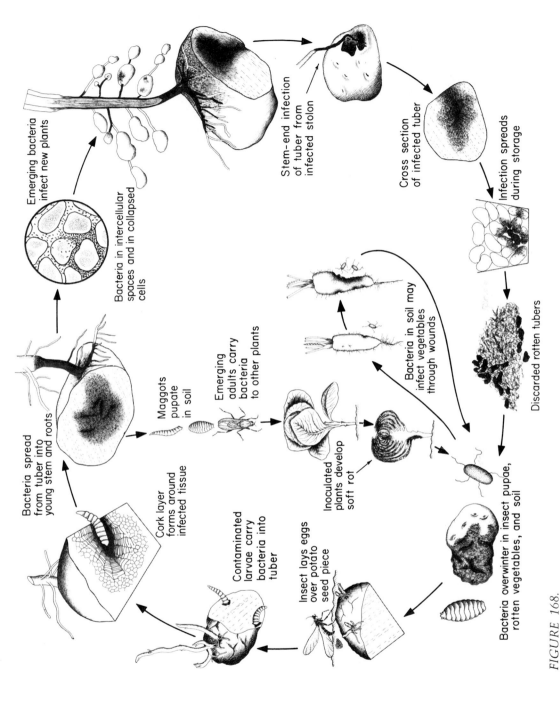

Emerging bacteria infect new plants

Bacteria in intercellular spaces and in collapsed cells

Stem-end infection of tuber from infected stolon

Cross section of infected tuber

Infection spreads during storage

Discarded rotten tubers

Bacteria spread from tuber into young stem and roots

Cork layer forms around infected tissue

Maggots pupate in soil

Emerging adults carry bacteria to other plants

Bacteria in soil may infect vegetables through wounds

Inoculated plants develop soft rot

Contaminated larvae carry bacteria into tuber

Insect lays eggs over potato seed piece

Bacteria overwinter in insect pupae, rotten vegetables, and soil

FIGURE 168.
Disease cycle of bacterial soft rot of vegetables caused by soft-rotting *Erwinia* sp.

481

tion is followed by rapid multiplication of the bacteria, which produce increasing amounts of pectolytic and cellulolytic enzymes. The pectolytic enzymes break down the pectic substances of the middle lamella and of the cell wall and cause maceration of the tissues. The cellulolytic enzymes cause partial breakdown and softening of the cellulose of the cell walls. As a result of the action of these and other enzymes water from the protoplasts diffuses into the intercellular spaces; the cells plasmolyze, collapse, and die. The bacteria continue to move into and to multiply in the intercellular spaces, while their enzymes advance ahead of them and prepare the tissues for invasion.

The liquefaction of pectic substances, and the exosmosis of water from the protoplasts into the intercellular spaces result in the softening of the invaded tissues and their transformation into a slimy mass. This mass consists of innumerable bacteria swimming about in the liquefied substances and among the unbroken walls of collapsed cells or of unaffected lignified tissues. Although the epidermis of most tissues is not attacked by the bacteria, cracks of various sizes are usually present and the slimy mass extrudes through them into the soil or, in storage, it comes into contact with other fleshy organs, which are subsequently infected.

Control.   The control of bacterial soft rots of vegetables is based almost exclusively on sanitary and cultural practices. All debris should be removed from warehouses and the walls should be disinfested with solutions containing formaldehyde, or copper sulfate. Wounding of plants and their storage organs should be avoided as much as possible. Only healthy plants, tubers, fruit, etc. should be stored. When new infections appear in storage, infected organs should be removed quickly and burned. Products to be stored should be dry and the humidity of warehouses should be kept low if infections are to be avoided. Temperatures around 4°C in storage houses inhibit development of new soft-rot infections. Leafy vegetables should be cooled to 4 to 6°C immediately on arrival.

In the field, plants subject to soft rot infection should be kept free of surface moisture by planting in well-drained areas, by allowing sufficient spaces among plants for adequate ventilation, and by avoiding overhead irrigation. Plants very susceptible to soft rot should be rotated with cereals, corn, or other nonsusceptible crops. Although some varieties are less susceptible than others, few are noted for their resistance to soft rot and none is immune.

Chemical sprays are generally not recommended for the control of soft rots, except for soft rot of tomato fruit which can be reduced by repeated applications of Bordeaux mixture. Spraying or dusting with insecticides, however, to control the insects which spread the disease has been effective in reducing infections both in the field and in storage.

SELECTED REFERENCES

Boyd, A. E. W. 1972. Potato storage diseases. *Rev. Plant Pathol.* **51**:297–321.
Burkholder, W. H. 1942. Three bacterial plant pathogens. *Phytopathology* **32**:141–149.
Cuppels, D., and A. Kelman. 1974. Evaluation of selective media for isolation of soft-rot bacteria from soil and plant tissue. *Phytopathology* **64**:468–475.

Dye, D. W. 1969. A taxonomic study of the genus *Erwinia*. II. The "carotovora" group. *New Zeal. J. Sci.* **12**:81–97.

Echandi, E., S. D. Van Gundy, and J. C. Walker. 1957. Pectolytic enzymes secreted by soft-rot bacteria. *Phytopathology* **47**:549–522.

Graham, D. C. 1964. Taxonomy of the soft rot coliform bacteria. *Ann. Rev. Phytopathol.* **2**:13–42.

Harding, H. A., and W. J. Morse. 1910. The bacterial soft rots of certain vegetables. *Vermont Agr. Expt. Sta. Bull.* **147**:243–360.

Howard, C. M., and J. G. Leach. 1963. Relation of the iris borer to bacterial soft rot of iris. *Phytopathology* **53**:1190–1193.

Lauritzen, J. I. 1932. Development of certain storage and transit diseases of carrot. *J. Agr. Res.* **44**:861–912.

Leach, J. G. 1931. Blackleg disease of potatoes in Minnesota. *Minn. Agr. Expt. Sta. Tech. Bull.* **76**:36 p.

Segall, R. H., and A. T. Dow. 1973. Effects of bacterial contamination and refrigerated storage on bacterial soft rot of carrots. *Plant Dis. Reptr.* **57**:896–899.

Smith, M. A. 1944. Bacterial soft-rot of spinach. *Phytopathology* **34**:747–752.

bacterial galls

Galls are produced on the stems and roots of plants infected primarily by bacteria of the genus *Agrobacterium* and by certain species of *Corynebacterium* and *Pseudomonas*. The galls may be amorphous, consisting of overgrowths of more or less anorganized or disorganized plant tissues, as are most *Agrobacterium* and *Pseudomonas* galls, or they may be proliferations of tissues that develop into more or less organized, teratomorphic organs, as are some *Agrobacterium* and *Corynebacterium* galls. The bacterial species that cause galls and the main diseases they cause are the following:

Agrobacterium, causing crown gall of many woody plants primarily stone fruits, pome fruits, willows, brambles, and grapes (*A. tumefaciens*), cane gall of raspberries and blackberries (*A. rubi*), and hairy root of apple (*A. rhizogenes*).

Corynebacterium, causing fasciation or leafy gall on many annual or perennial herbaceous ornamentals (*C. fascians*).

Pseudomonas, causing the olive knot disease (*P. savastanoi*) and the almost identical bacterial gall or canker of oleander (*P. tonelliana*).

The gall-inducing bacteria enter plants through wounds and stimulate cells to divide and enlarge. The *Agrobacterium* and *Corynebacterium* bacteria remain always in the intercellular spaces and never break down cells or result in formation of cavities, while gall-inducing *Pseudomonas* bacteria cause disintegration of cells and formation of cavities in the galls. Crown gall, caused by *A. tumefaciens*, is unique among all other plant galls in that it is malignant, i.e., once the plant cells have been stimulated by the bacteria to divide and enlarge, they continue to divide as long as they can obtain nutrients and do not obey the hormonal controls of the parent plant that regulate growth and differentiation.

Gall bacteria overwinter in galls and in the soil. They are spread by contaminated tools such as budding or pruning knives, by soil water, and rain splashes. Gall bacteria are controlled by avoiding wounding susceptible plants, using only disease-free rootstocks and scions, soil sterilization in greenhouses, crop rotation when possible, and, in olive knot, sprays with Bordeaux mixture.

• *Crown Gall*

Crown gall is worldwide in distribution. It affects many woody and herbaceous plants belonging to 140 genera of more than 60 families. In nature it is found mostly on pome and stone fruit trees, brambles, and grapes.

Crown gall is characterized by the formation of tumors or galls of varying size and form. It is common on the roots and shoots of various nursery plants which are thus unsalable because crown gall is likely to continue on the plants when they are removed to orchards and gardens. Plants with tumors at their crowns or on their main roots grow poorly and produce reduced yields. Severely infected plants or vines may die.

Crown gall tumors have certain similarities to human and animal cancers and, therefore, the cause and mechanism of their formation have been extensively studied. In spite of the apparent similarities to cancer, however, there are many and basic differences between crown gall of plants and malignant tumors of man and animals.

Symptoms.   The disease first appears as small overgrowths on the stem and roots, particularly near the soil line. In early stages of their development the tumors are more or less spherical, white or flesh-colored, and quite soft. Since they originate in a wound, at first they cannot be distinguished from callus. However, they usually develop more rapidly than callus. As the tumors enlarge, their surfaces become more or less convoluted. Later on, the outer tissues become dark brown or black, due to the death and decay of the peripheral cells (Fig. 169). Sometimes there is no distinct line of demarcation between the tumor and the plant proper, the tumor appearing as an irregular swelling of the tissues and surrounding the stem or root. Almost as often, however, the tumor lies outside but close to the outer surface of the host, being connected only by a narrow neck of tissue. Some tumors are spongy throughout and may crumble or become detached from the plant. Others become much more woody and harder, looking knobby or knotty, and reaching sizes up to 30 cm in diameter. Some tumors rot partially or completely from the surface toward the center in the fall and develop again in the same places during the next growing season, or part of the tumor may rot while new tumor centers appear in other parts of the same overgrowth.

Tumors are most common on the roots and stem near the soil line but they can also appear on vines up to 150 cm from the ground, on branches of trees, on petioles, and on leaf veins. Several galls may occur on the same root or stem, continuous or in bunches.

In addition to forming galls, affected plants may become stunted, they produce small, chlorotic leaves and in general are more susceptible to adverse environmental conditions, especially to winter injury.

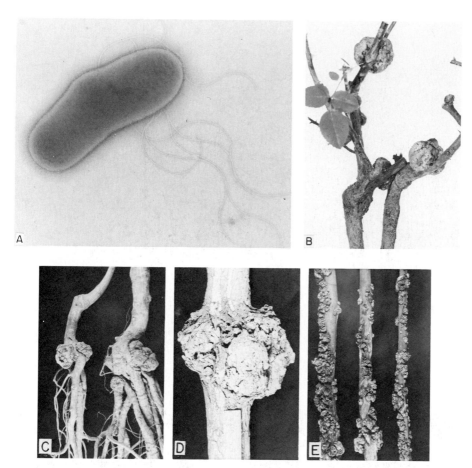

FIGURE 169.
Agrobacterium tumefaciens bacterium (A), and crown gall tumors on rose stems (B), peach root (C) and willow branch (D). E shows cane gall on raspberry caused by *A. rubi.* (Photo A courtesy S. M. Alcorn. Photo B courtesy Dept. Plant Pathol., Cornell Univ.)

The pathogen: Agrobacterium tumefaciens (Fig. 169A). The most characteristic property of this bacterium is its ability to transform normal plant cells to tumor cells in short periods of time. Once the transformation to tumor cells has been completed, these cells become independent of the bacterium and continue to grow and divide abnormally even in the absence of the bacteria.

Development of disease. The bacterium overwinters in infested soils, where it can live as a saprophyte for several years. When host plants are growing in such infested soils, the bacterium enters the roots or stems near the ground through fairly recent wounds made by cultural practices, grafting, insects, etc. Once inside the tissue the bacteria occur primarily intercellularly and stimulate the surrounding cells to divide (Fig. 170). One or more groups or whorls of hyperplastic cells appear in the cortex or

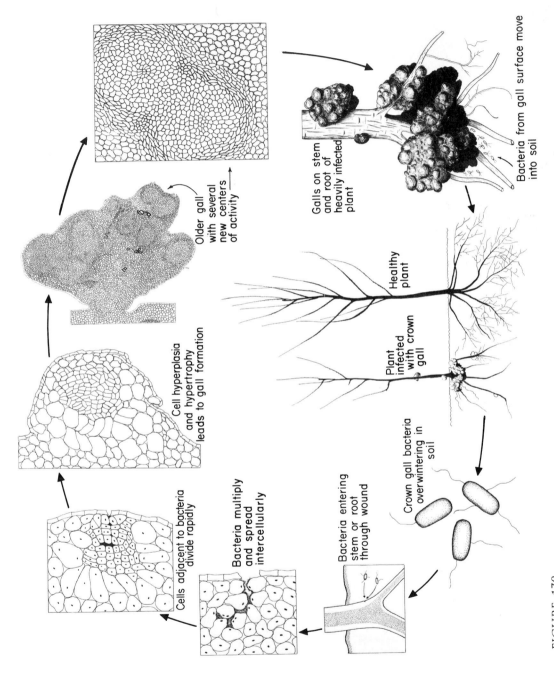

Older gall with several new centers of activity

Cell hyperplasia and hypertrophy leads to gall formation

Cells adjacent to bacteria divide rapidly

Bacteria multiply and spread intercellularly

Bacteria entering stem or root through wound

Crown gall bacteria overwintering in soil

Plant infected with crown gall

Healthy plant

Galls on stem and root of heavily infected plant

Bacteria from gall surface move into soil

FIGURE 170.
Disease cycle of crown gall caused by *Agrobacterium tumefaciens.*

in the cambial layer depending on the depth of the wound. These cells may contain one to several nuclei. They divide at a very fast rate, producing cells that show no differentiation or orientation, and 10 to 14 days after inoculation a small swelling can be seen with the naked eye. As the irregular division and enlargement of the cells continue unchecked, the swelling enlarges, developing into a young tumor. Bacteria are absent from the center of the tumors but can be found intercellularly in their periphery. By this time certain cells have differentiated into vessels or tracheids which, however, are unorganized and with little or no connection with the vascular system of the host plant. As the tumor cells increase in number and size, they exert pressure on the surrounding and underlying normal tissues which may become distorted or crushed. Crushing of xylem vessels by tumors sometimes reduces the amount of water reaching the upper parts of a plant to as little as 20 percent of the normal.

The smooth and soft young tumors are not protected by epidermis and, therefore, are easily injured and attacked by insects and saprophytic microorganisms. These secondary invaders cause decay and discoloration of the peripheral cell layers of the tumors, which turn brown to black. Breakdown of the peripheral tumor tissues releases crown gall bacteria into the soil, where they can be carried in the water and infect new plants.

As the tumors enlarge, they sometimes become woody and hard. The incomplete and disarranged vascular bundles that may be present in the tumor itself are ineffective. When tumors are unable to obtain as much water and nourishment as is required to carry them beyond a certain point in growth, their enlargement stops, decay sets in, and the necrotic tissues are sloughed off. In some cases the tumor regresses and no new one appears. More often, however, some portion of the tumor remains alive and forms additional tumor tissue during the same or the following season.

When very young and expanding tissues are infected, in addition to the primary tumor which develops at the point of infection, secondary tumors appear. These usually form below, but often above, the primary tumor and at varying distances from it. Sometimes the secondary tumors develop at the scars of fallen leaves or at wounds made by various agents. At other times secondary tumors develop on apparently unwounded parts of the stem, on the petiole, and even on leaf midribs or larger veins several internodes above the primary tumor. Their starting point seems to be in the xylem of the vascular bundles. They are free from bacteria, since no bacteria can be isolated when these tumors are plated on nutrient media. When fragments of such bacteria-free tumors are grafted on healthy plants, they develop into large tumors similar in appearance and structure to the primary tumors but remain completely devoid of bacteria. This indicates that the bacteria are important only at the beginning of the disease, presumably by having an irritant effect on the plant cells. Once the cells have been triggered to malignancy, they produce their own irritants and their uncontrolled growth becomes autonomous.

Although the nature of the irritant and the mechanism of transformation of normal plant cells to tumor cells have been the objects of intensive studies, our knowledge on these topics is still rather incomplete. A more detailed discussion of the cause and the physiology of crown gall tumors will be found in the chapter on "growth regulators in plant disease."

Control.   Crown gall control is based primarily on certain cultural and sanitary practices. Susceptible nursery stock should not be planted in fields known to be infested with the pathogen. Instead, infested fields should be planted with corn or other grain crops for several years before they are planted with nursery stock. Since the bacterium enters only through relatively fresh wounds, wounding of the crowns and roots during cultivation should be avoided and root-chewing insects in the nursery should be controlled to reduce crown gall incidence. Nursery stock should be budded rather than grafted because of the much greater incidence of galls on graft than on bud unions. Growers should purchase and plant only crown gall-free trees.

A biological control procedure first developed in Australia offers the most promising control of crown gall so far. Thus, excellent control of crown gall is obtained by soaking germinated seeds or dipping nursery seedlings or rootstocks in a suspension of a particular strain (No. 84) of *Agrobacterium radiobacter* that is antagonistic to most strains of *A. tumefaciens*. Some control is also obtained by treating nongerminated seeds with the antagonist or by drenching the soil with a suspension of the antagonistic bacterium. It is postulated that the antagonist controls crown gall initiation by establishing itself on the surface of the plant tissues where it produces substances, possibly bacteriocins, inhibitory to the virulent *A. tumefaciens* strains.

SELECTED REFERENCES

Banfield, W. M. 1934. Life history of the crown gall organism in relation to its pathogenesis on the red raspberry. *J. Agr. Res.* **48**:761–787.

Dhanvantari, B. N. 1976. Biological control of crown gall of peach in southwestern Ontario. *Plant Dis. Reptr.* **60**:549–551.

Hedgecock, G. G. 1910. Field studies of the crown gall of the grape. *U.S. Dept. Agr. Bull.* **183**:40 p.

Kerr, A. 1972. Biological control of crown gall: seed inoculation. *J. Appl. Bacteriol.* **35**:493–497, **35**:279–287.

Lelliott, R. A. 1971. A survey of crown gall in rootstock beds of apple, cherry, plum and quince in England. *Plant Pathol.* **20**:59–63.

Lippincott, J. A., and Barbara B. Lippincott. 1975. The genus *Agrobacterium* and plant tumorigenesis. *Ann. Rev. Microbiol.* **29**:377–405.

Miller, H. N. 1975. Leaf, stem, crown and root galls induced in chrysanthemum by *Agrobacterium tumefaciens*. *Phytopathology* **65**:805–811.

Muncie, J. H. 1926. A study of crown gall caused by *Pseudomonas tumefaciens* on rosaceous hosts. *Iowa State Coll. J. Sci.* **1**:67–117.

Riker, A. J., and G. W. Keitt. 1926. Studies on crown gall and wound overgrowth on apple nursery stock. *Phytopathology* **16**:765–808.

Riker, A. J., *et al.* 1946. Some comparisons of bacterial plant galls and of their causal agents. *Bot. Rev.* **12**:57–82.

Schroth, M. N., and W. J. Moller. 1976. Crown gall controlled in the field with a nonpathogenic bacterium. *Plant Dis. Reptr.* **60**:275–278.

Smith, E. F., Nellie A. Brown, and C. O. Townsend. 1911. Crown gall of plants: Its cause and remedy. *U.S. Dept. Agr. Bull.* **213**:215 p.

bacterial cankers

Relatively few canker diseases of plants are caused by bacteria, but some of them are widespread and devastating so that great losses are caused by them or great efforts are required to protect the plants from them. The bacteria and the most important cankers they cause are the following:

Corynebacterium, causing the bacterial canker and wilt of tomato (*C. michiganense*).

Pseudomonas, causing the bacterial canker of stone fruit and pome fruit trees (*P. morsprunorum* and *P. syringae*).

Xanthomonas, causing the bacterial canker of citrus (*X. citri*).

In all bacterial cankers, the canker symptoms on stems, branches, or twigs are only part of the disease syndrome, and direct symptoms on fruits, leaves, buds, or blossoms may be at least as important in the overall effect of the disease on the plant as are cankers. Also, bacterial cankers are not always sunken and soft, as is the case with fungal cankers, but they may also appear as splits in the stem, as necrotic areas within the woody cylinder, or as scabby excrescences on the surface of the tissue. In some bacterial cankers, soft decayed tissue and bacterial cavities that ooze either a slimy exudate or a dark gummy substance may be present in the stem, but during much of the year populations of bacteria in woody cankers are low and their isolation from them is erratic.

The canker bacteria overwinter in perennial cankers, in buds, in plant refuse and, in tomatoes, in or on the seed. They are spread by rain splashes or runoff water, wind-blown rain, handling of plants, on contaminated tools, and on infected plant material. The bacteria enter tissues primarily through wounds, but in young plants they may also enter through natural openings. Control of bacterial cankers is through proper sanitation and eradication practices, through use of bacteria-free seeds or budwood, and somewhat through several sprays with Bordeaux mixture, other copper formulations, or antibiotics.

- *Bacterial Canker and Gummosis of Stone Fruit Trees*

This disease apparently occurs in all major fruit growing areas of the world. It affects primarily stone fruit and citrus trees. The same pathogen also affects pear, lilac, rose, and many other annuals and perennial ornamentals, some vegetables, and some small grains. The pathogen causing bacterial canker and gummosis occurs in numerous strains, each of which attack some but not all the hosts that develop the disease, although the host range of an isolate usually overlaps somewhat with the host ranges of other isolates. The disease is also known as bud blast, blossom blast, dieback, spur blight, and twig blight.

Bacterial canker and gummosis is one of the most important diseases of stone fruit trees in many fruit-growing areas. Exact losses are difficult to assess because of serious damage to trees as well as reduction of yields. The disease causes cankers on branches and main trunks, kills young trees, and reduces the yield of or kills older ones. Tree losses from 10 to

75 percent have been observed in young orchards. Bacterial canker and gummosis also kills buds and flowers of trees, usually resulting in yield losses of 10 to 20 percent but sometimes up to 80 percent. Leaves and fruits are also attacked, resulting in weaker plants and in low quality or unsalable fruit.

Symptoms. The most characteristic symptom of the disease, although not always the most common or the most destructive on all hosts, is the formation of cankers accompanied by gum exudation (Figs. 171 and 172). Cankers usually develop at the base of an infected spur. They then spread mostly upward and to a lesser extent down and to the sides. Infected areas are slightly sunken and darker brown in color than the surrounding healthy bark. The color of the cortical tissues of the cankered area varies from bright orange to brown. Narrow brown streaks extend into the healthy tissue above and below the canker. Cankers are first noticed in late winter or early spring. As the trees break dormancy in the spring, gum is produced by the tissues surrounding most cankers, breaks through the bark, and runs down on the surface of the limbs. Cankers in which gum is not produced are similar, but usually are softer,

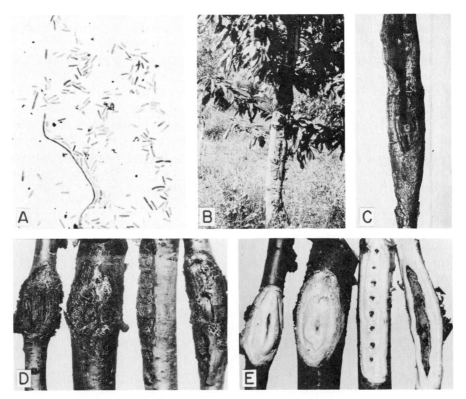

FIGURE 171.
Pseudomonas syringae bacteria (A) and cankers on cherry trunk (B), branch (C) and twigs (D, E). E shows the same twigs as in D but with the bark removed. (Photo A courtesy H. R. Cameron.)

FIGURE 172.
Young cherry tree killed by girdling of the trunk by a *Pseudomonas syringae*
canker (A), side view of exposed canker (B), and brown streaks of bacteria
extending into healthy tissue above rapidly advancing canker (C). (D) Leaf
symptoms of *P. syringae* on cherry. From left to right: Small water-soaked spots,
large angular necrotic spots, and shot-hole or tattered appearance. (Photos
courtesy H. R. Cameron.)

more moist, sunken, and may have a sour smell. When the trunk or
branch of a tree is girdled by a canker, the leaves above the girdle show an
inward curling and drooping, then a light green color, and then yellow.
Within a few weeks the branch or entire tree above the canker is dead
(Figs. 171 and 172).

Dormant bud blast is especially serious on cherry, apricot, and pear. In
some areas great numbers of buds are killed. Isolated buds are often killed
or fail to develop on year-old twigs. When sectioned, infected buds show
brown areas at the bud scales extending across the base of the bud. The

entire bud eventually dies (Fig. 173). Both flower and leaf buds are equally affected. The damage to buds becomes most obvious in the orchard during full bloom when the light bloom of infected trees is most conspicuous (Fig. 173E).

Infection of flowers occurs under favorable weather conditions and it can be very severe. Infected flowers appear water soaked, turn brown, wilt, and hang on the twig. From the flower, infections may spread into the twig and cause twig blight or they may spread into the spur and cause canker formation.

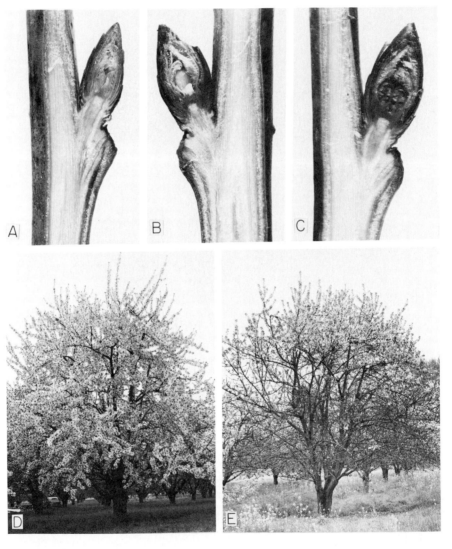

FIGURE 173.
Healthy bud (A), base and scales of bud infected (B), and entire bud killed by *Pseudomonas syringae* (C). (D) Healthy cherry tree. (E) Cherry tree in same orchard with most of the lower buds killed by *P. syringae*. (Photos courtesy H. R. Cameron.)

Leaf infections appear as water-soaked spots about 1 to 3 mm in diameter. As the leaves mature the spots become brown, dry, and brittle. Eventually infected areas fall out and the leaves have a shot-hole or tattered appearance (Fig. 172D).

When the fruit is infected, flat, superficial, dark brown spots develop. The spots are 2 to 3 mm deep, depressed, and may have underlying gum pockets on cherry, while on peach they may be 2 to 10 mm in both diameter and depth. The underlying tissue is dark brown to black and sometimes spongy.

The pathogen: Pseudomonas syringae (Fig. 171A). Many strains of *Pseudomonas syringae* are known. A related species, *Pseudomonas morsprunorum*, causes canker and gummosis symptoms on stone fruits identical to those caused by *P. syringae* and is considered by some to be the main causal agent of the disease. These two species, and the diseases they cause, however, are so similar that *P. morsprunorum* can be considered for all practical purposes to be a variant of *P. syringae.*

Development of disease. The bacteria overwinter in active cankers, in infected buds and leaves, on limbs of infected or healthy trees, and possibly on weeds and on nonsusceptible hosts (Fig. 174).

Infection of limbs usually takes place during the fall and winter months. Bacteria enter the limbs through the bases of infected buds or spurs and also through pruning cuts, leaf scars, and injuries caused by various agents. The bacteria move intercellularly and advance into the bark and into the medullary rays of the phloem and xylem. In advanced stages of infection, the bacteria invade and break down parenchyma cells. Lysogenic cavities full of bacteria develop. Xylem vessels are sometimes invaded by bacteria but the bacteria do not seem to move far through the vessels.

Cankers develop rather rapidly in the fall after the trees have gone into dormancy but before the onset of low winter temperatures. During the cold winter periods canker development is slow. Cankers develop most rapidly in the period between the end of the cold weather and the beginning of rapid tree growth in the spring. Cankers on the south side of trees are usually larger, due to warming by the sun during the dormant season. Cankers appear either brown, with well defined margins, or watery and gum soaked, with brownish strands. The advance of the canker is checked by the advent of higher temperatures and the beginning of active growth in the spring, when the host usually forms callus tissue around the canker and the canker becomes inactive. Some cankers are permanently inactivated but others, in which the encirclement of the canker by the callus tissue is incomplete, become active again the following year and continue to spread in succeeding years. Infections during the active growing season are seldom of any consequence and apparently are isolated very quickly by callus tissue. The ability to wall infection seems to be correlated with varietal resistance, but is also affected by the age and succulence of the plant, the temperature and rainfall during a season, and the type of rootstock on which the tree is growing. Infections of buds seem to originate at the base of the outside bud scales and then spread throughout the base of the bud, killing the tissues across the base and

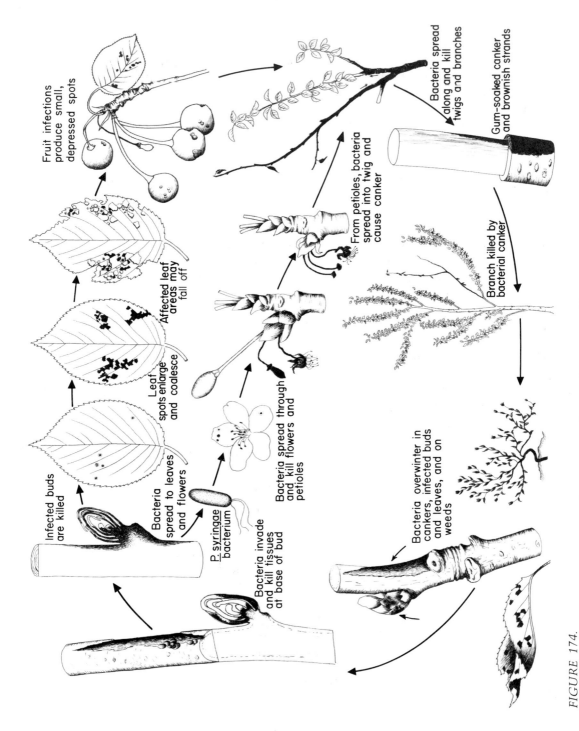

Fruit infections produce small, depressed spots

Bacteria spread along and kill twigs and branches

Gum-soaked canker and brownish strands

Affected leaf areas may fall off

Leaf spots enlarge and coalesce

From petioles, bacteria spread into twig and cause canker

Branch killed by bacterial canker

Infected buds are killed

Bacteria spread to leaves and flowers

P. syringae bacterium

Bacteria invade and kill tissues at base of bud

Bacteria spread through and kill flowers and petioles

Bacteria overwinter in cankers, infected buds and leaves, and on weeds

FIGURE 174.
Disease cycle of bacterial canker and gummosis of stone fruits caused by *Pseudomonas syringae*.

separating the growing point from the rest of the plant. Most bud infections are initiated between November and February but symptoms are first observed in mid-February and continue to develop at least through March. The bacteria sometimes spread downward and kill stem tissues around the base of the bud.

Flower infection is rare, but whenever it occurs it seems to take place through natural openings and through wounds made by insects or wind-blown rain. Under very humid conditions the bacteria spread through the floral parts quickly and may advance into the spur and twig, where they initiate canker formation.

Leaf infections appear on young, succulent leaves. They occur most frequently in areas with cool, wet springs and during periods of high winds and continued moisture. Infection takes place through stomata. The bacteria spread intercellularly and cause collapse and death of the cells, resulting in small angular leaf spots. During wet weather bacteria ooze out of the spots and are spread to other leaves by direct contact, by visiting insects, by rain, etc. As leaves mature, however, they become less susceptible and leaf infections late in the season are rare. While most of the bacteria inside or on the surface of infected leaves are dead in the fall, sufficient numbers survive to initiate new bud and stem infections.

Control. Although no complete control of bacterial canker and gummosis of fruit trees can be obtained as yet by any single method, certain cultural practices and control measures help keep down the number and severity of infections.

Only healthy budwood should be used for propagation. Susceptible varieties should be propagated on rootstocks resistant to the disease and should be grafted as high as possible. Only healthy nursery trees should be planted in the orchard.

Chemical control of the canker phase of the disease both in the nursery and in the orchard is based on sprays with fixed copper or Bordeaux mixture (10:15:100) in the fall and in the spring (at 6:9:100 strength) before blossoming. Streptomycin applied in the spring is more effective in reducing leafspot than is Bordeaux mixture, but it does not seem to control canker initiation and development.

Cankers on trunks and large branches can be controlled by cauterization with a hand-held propane burner. The flame is aimed at the canker and especially its margins for 5 to 20 seconds until the underlying tissue begins to crackle and char. The treatment is carried out in early to midspring and, if necessary, should be repeated 2 to 3 weeks later.

- *Bacterial Canker and Wilt of Tomato*

It is caused by *Corynebacterium michiganense.* It has been reported from many parts of the world and causes considerable losses particularly in field-grown tomatoes. The disease appears as spotting of leaves, stem, and fruit and as wilting of the leaves and shoots. Cankers are usually very small and indiscernible but may occur on stems and leaf veins.

The first noticeable symptoms are spotting or wilting of leaflets at the outer and lower parts of the plant. Leaf spotting occurs during wet weather and appears initially as white blisterlike spots which become brown with age and may coalesce. Wilting leaves curl upward and inward and later turn brown and wither but do not fall off. Often only the leaflets on one side of the leaf are affected or only on one side of the plant. The wilt may develop gradually from one leaflet to the next or it may become general and destroy much of the foliage. In the meantime, light-colored streaks appear on the stems, shoots, and leaf stalks, usually at the joints of petioles and stems. Later, cracks may appear in the streaks and these form the cankers (Fig. 175A). Through them, in humid or wet weather, slimy masses of bacteria ooze to the surface of the stem from which they are spread to leaves and fruits and cause secondary infections. The symptoms on the fruit appear as small, shallow, water-soaked, white spots

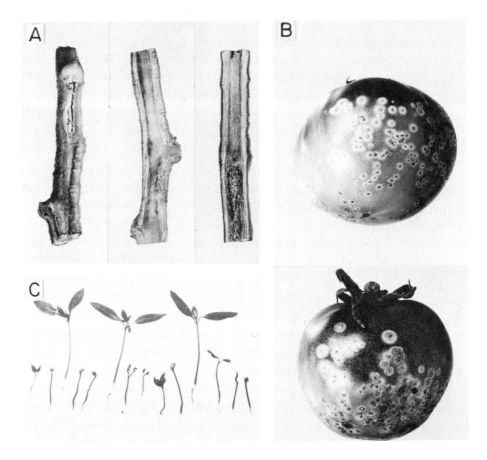

FIGURE 175.
Tomato stems, fruit and seedlings infected with bacterial canker caused by *Corynebacterium michiganense.* (A) Stems showing open canker (left) and split lengthwise to show discoloration and decay of vascular tissues and pith. (B) Tomatoes showing bird's-eye-like spots with dark rough centers and white haloes at the margins. (C) Three healthy and several diseased tomato seedlings infected through the seed. (Photos A and B courtesy U.S.D.A.)

whose centers later become slightly raised, tan colored, and rough. The final, bird's-eye-like appearance of the spots, which have brownish centers about 3 mm in diameter and white halos around them, is quite characteristic of the disease (Fig. 175B).

In longitudinal sections of infected stems a creamy white, yellow, or reddish-brown line can be seen just inside the woody tissue and along the phloem. The vascular tissues show a brown discoloration and large cavities are present in the pith and in the cortex and extend to the outer surface of the stem where they form the cankers. The discoloration of the vascular tissues extends all the way to the fruits both outward toward the surface and inward toward the seeds, and small dark cavities may develop in the centers of such fruits.

The bacteria overwinter in or on seeds and, in some areas, in plant refuse in the soil. Primary infections may result from spread of the bacteria from the seed to cotyledons or leaves (Fig. 175C), but most infections result from penetration of bacteria through wounds of roots, stems, leaves, and fruits. The bacteria are spread to them through handling during transplanting, by soil water, or by wind-blown rain. Once inside the plant, these bacteria enter the vascular system and move and multiply primarily in the phloem tissues. The bacteria also invade the spiral xylem vessels and move through them and out of them into the phloem, pith, and cortex where they form the large cavities that result in the cankers.

Control of the disease is through use of bacteria-free seed, protective application of copper or streptomycin in the seed bed, and soil sterilization of the seedbeds.

SELECTED REFERENCES

Agrios, G. H. 1972. A severe new canker disease of peach in Greece. *Phytopathol. Mediterr.* **11**:91–96.

Bryan, Mary K. 1930. Studies on bacterial canker of tomato. *J. Agr. Res.* **41**:825–851.

Cameron, H. R. 1962. Diseases of deciduous fruit trees incited by *Pseudomonas syringae* van Hall. *Oregon Agr. Expt. Sta. Tech. Bull.* **66**:64 p.

Crosse, J. E. 1954. Bacterial canker, leaf spot, and shoot wilt of cherry and plum. *Ann. Rept. East Malling Res. Sta.,* 1953, Sect. IV, pp. 202–207.

Crosse, J. E. 1966. Epidemiological relations of the Pseudomonad pathogens of deciduous fruit trees. *Ann. Rev. Phytopathol.* **4**:291–310.

Dowler, W. M., and D. H. Petersen. 1967. Transmission of *Pseudomonas syringae* in peach trees by bud propagation. *Plant Dis. Reptr.* **51**:666–668.

Forster, R. L., and E. Echandi. 1973. Relation of age of plants, temperature, and inoculum concentration to bacterial canker development in resistant and susceptible *Lycopersicon* spp. *Phytopathology* **63**:773–777.

Grogan, R. G., and J. B. Kendrick. 1953. Seed transmission, mode of overwintering and spread of bacterial canker of tomato caused by *Corynebacterium michiganense. Phytopathology* **43**:473.

Hawkins, J. E. 1976. A cauterization method for the control of cankers caused by *Pseudomonas syringae* in stone fruit trees. *Plant Dis. Reptr.* **60**:60–61.

Jones, A. L. 1971. Bacterial canker of sweet cherry in Michigan. *Plant Dis. Reptr.* **55**:961–965.

Knorr, L. C., R. F. Suit, and E. P. DuCharme. 1957. Handbook of citrus diseases in Florida. *Fla. Agr. Exp. Sta. Bull.* **587**:157 p.

Pine, T. S., R. G. Grogan, and W. B. Hewitt. 1955. Pathological anatomy of bacterial canker of young tomato plants. *Phytopathology* **45**:268–271.

Strider, D. L. 1969. Bacterial canker of tomato caused by *Corynebacterium michiganense*. A literature review and bibliography. *N. C. Agr. Exp. Sta. Bull.* **193**:110 p.

Wilson, E. E. 1933. Bacterial canker of stone fruit trees in California. *Hilgardia* **8**:83–123.

Wormald, H. 1931. Bacterial diseases of stone fruit trees in Britain. III. The symptoms of bacterial canker in plum trees. *J. Pomol.* **9**:239–256.

bacterial scabs

This group of diseases includes mainly diseases that affect below-ground parts of plants and whose symptoms consist of more or less localized scabby lesions affecting primarily the outer tissues of these parts. The scab bacteria and the diseases they cause are:

> *Streptomyces*, causing the common scab of potato and of other below-ground crops (*S. scabies*), and the soil rot or pox of sweet potato (*S. ipomoeae*). *Pseudomonas*, causing scab of gladiolus (*P. marginata*).

The scab bacteria survive in infected plant debris and in the soil and penetrate tissues through natural openings or wounds. In the tissues, the bacteria grow mostly in the intercellular spaces of parenchyma cells but these cells are sooner or later invaded by the bacteria and break down. In typical scabs, healthy cells below and around the lesion divide and form layers of corky cells. These push the infected tissues outward and give them the scabby appearance. Scab lesions often serve as points of entry for secondary parasitic or saprophytic organisms that may cause the tissues to rot.

- *Common Scab of Potato*

It is caused by *Streptomyces scabies* and occurs throughout the world. It is most prevalent and important in neutral or slightly alkaline and light sandy soils, especially during relatively dry years. The same pathogen also affects garden beets, sugar beets, radish, and other crops. The disease, by its usually superficial blemishes on tubers, roots, etc., reduces the value rather than the yield of the crop, although severe root infection may reduce yields, and deep scabs increase the waste in peeling.

The symptoms of common scab of potato are observed mostly on tubers and at first they consist of small, brownish, and slightly raised spots, but later they may enlarge, coalesce, and become very corky. Frequently, the lesions extend below the tuber surface and, when the corky tissue is removed, 3 to 4 mm deep pits are present in the tuber. Sometimes the lesions appear as small russeted areas and are so numerous that they almost cover the tuber surface, or they may appear as slight protuberances with depressed centers covered with a small amount of corky tissue (Fig. 176).

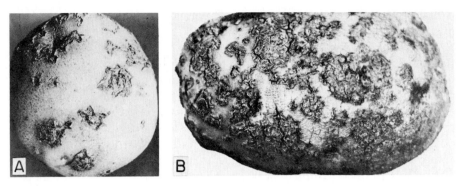

FIGURE 176.
Early (A) and advanced (B) symptoms of common scab of potato caused by
Streptomyces scabies. (Photos courtesy U.S.D.A.)

The pathogen, *S scabies,* is a hardy saprophyte that can survive indefinitely either in its vegetative mycelioid form or as spores in most soils
except the most acidic ones. The vegetative form consists of slender
(about 1 μm thick), branched mycelium with few or no cross walls. The
spores are cylindrical or ellipsoid, about 0.6 by 1.5 μm, and are produced
on specialized spiral hyphae that develop cross walls from the tip towards
their base and, as the cross walls constrict, spores are pinched off at the
tip and eventually break away from the hypha. The spores germinate by
means of one or two germ tubes which develop into the mycelioid form
(Fig. 177).

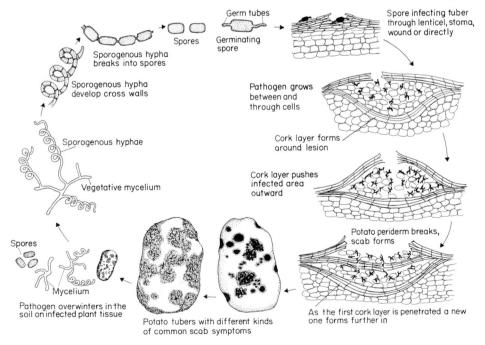

FIGURE 177.
Disease cycle of the common scab of potato caused by *Streptomyces scabies.*

The pathogen is spread through soil water, wind-blown soil, and on infected potato seed tubers. It penetrates tissues through lenticels, wounds, stomata and, in young tubers, directly. Young tubers are more susceptible to infection than older ones. Following penetration the pathogen apparently grows between or through a few layers of cells, the cells die and the pathogen then lives off them as a saprophyte. In the meantime, however, the pathogen apparently secretes a substance that stimulates the living cells surrounding the lesion to divide rapidly and to produce several layers of cork cells that isolate the pathogen and several plant cells. As the cells that are cut off by the cork layer die, the pathogen subsists on them. Usually, several such groups of cork cell layers are produced, and as they are pushed outward and sloughed off, the pathogen grows and multiplies in the additional dead cells and thereby large scab lesions develop. The depth of the lesion seems to depend on the variety, soil conditions, and on the invasion of scab lesions by other organisms, including insects. The latter apparently break down the cork layers and allow the pathogen to invade the tuber in greater depth.

The severity of common scab of potato increases as the pH of the soil increases from 5.2 to 8.0 and decreases beyond these limits. The disease develops most rapidly at soil temperatures of about 20 to 22°C but it can occur between 11 and 30°C. Potato scab incidence is greatly reduced by high soil moisture during the period of tuber initiation and for several weeks afterwards. Potato scab is also lower in fields following certain crop rotations and the plowing under of certain green manure crops, probably as a result of inhibition of the pathogen by antagonistic microorganisms.

Control of the common scab of potato is through the use of certified scab-free seed potatoes or through seed treatment with pentachloronitrobenzene (PCNB) or with maneb–zinc dust. If the field is already infested with the pathogen, a fair degree of disease control may be obtained by using certain crop rotations, bringing and holding the soil pH to about 5.3 with sulfur, irrigating for about 6 weeks during the early stages of tuber development, and using resistant or tolerant potato varieties.

SELECTED REFERENCES

Jones, A. P. 1931. The histogeny of potato scab. *Ann. Appl. Biol.* **18**:313–333.

KenKnight, G. 1941. Studies on soil *Actinomyces* in relation to potato scab and its control. *Mich. Agr. Exp. Sta. Tech. Bull.* **178**:48 p.

Labruyere, R. E. 1971. Common scab and its control in seed-potato crops. *Versl. Landbouwk. Agr. Res. Rept.* **767**:71 p.

Lapwood, D. H. 1973. Irrigation as a practical means to control potato common scab (*Streptomyces scabies*): Final experiment and conclusions. *Plant Pathol.* **22**:35–41.

Lapwood, D. H., and M. J. Adams. 1975. Mechanisms of control of common scab by irrigation, *in* "Biology and Control of Soil-Borne Plant Pathogens," G. W. Bruehl (Ed.). The Amer. Phytopathol. Soc., St. Paul, Minn., pp. 123–129.

Nelson, R. 1948. Diseases of gladiolus. *Mich. Agr. Exp. Sta. Special Bull.* **350**:63 p.

Person, L. H. 1946. The soil rot of sweet potatoes and its control with sulfur. *Phytopathology* **36**:869–875.

root nodules of legumes

Root nodules are well-organized structures produced on the roots of most legume plants following inoculation with certain species of bacteria of the genus *Rhizobium*. Root nodules, although they are the result of infection of legumes by bacteria, are considered a condition of symbiosis rather than of disease. The infecting bacteria fix (trap) atmospheric nitrogen and make it available to the plant in a utilizable organic form and the plant profits from this nitrogen more than it loses as sugars and other nutrients to the bacteria. Unfortunately, not all root nodule bacteria are beneficial to the legume host. Some nodule bacteria are apparently strictly parasites since they form nodules on the roots but fail to fix nitrogen. Therefore, the number of root nodules does not always indicate their value to the plant unless the strain of bacteria is known to be effective in fixing nitrogen. As a result legume seeds are routinely inoculated commercially with appropriate strains of root nodule bacteria to improve plant growth and yields.

Nodules are produced on taproots as well as lateral roots of legumes and may vary in size from 1 mm to 2 to 3 cm (Fig. 178A). Nodules may be round or cylindrical and as large or larger than the root diameter on which they form. Their number and size varies with the plant, bacterial strain, age of infection, etc. On herbaceous plants nodules are fragile and short-lived while on woody plants they may persist for several years. Each nodule consists of an epidermal layer, a cortical layer and the bacteria-containing central tissue, each of them consisting of several layers of cells (Fig. 178B). Vascular bundles are present in the cortical layer just outside the central tissue. In elongated nodules, the tip of the nodule farthest away from the root consists of a zone of meristematic cells through which the nodule grows. In rounded nodules, the meristematic region is laid around the nodule except at the neck.

The organism. The root nodule bacteria, *Rhizobium* sp., vary in size and shape with age, the typical bacteria being rod-shaped (1.2 to 3.0 by 0.5 to 0.9 μm) or irregular, club-shaped forms. They have no flagella, and they are gram-negative. The bacteria survive in roots of susceptible legumes and, for varying periods of time, in the soil. Continued growth of the same legume in the soil tends to build up the population of nodule bacteria affecting that legume. Not all nodule bacteria affect all legumes. For example, the bacteria that grow on alfalfa and sweet clover do not grow on clovers, beans, peas, soybeans, etc., and vice versa. Strains of nodule bacteria often show definite varietal preferences, e.g., some soybean bacteria work better on one or two soybean varieties than on others.

Development of nodules. The bacteria penetrate root hairs or young epidermal cells directly. Within the cell the bacteria become embedded in a double-walled, tubular, mucoid sheath called an infection thread. The infection thread, which contains the bacteria, penetrates into the cortical parenchyma cells and branches along the way, with terminal and lateral vesicles forming on the strands. These vesicles soon break and

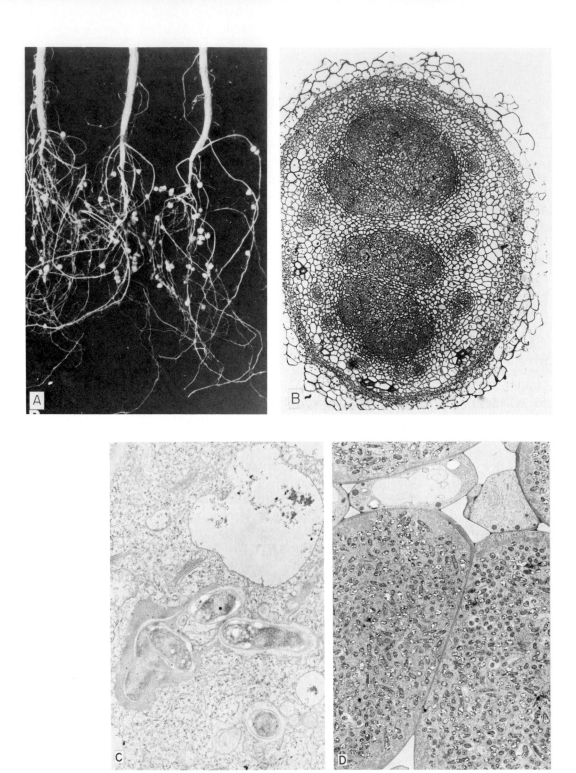

FIGURE 178.

(A) "Healthy" soybean roots bearing numerous bacterial nodules. (B) Cross section of a developing soybean nodule, 12 days after inoculation. There are at least three central areas containing bacteroids apparently as a result of several closely adjacent infections. (C, D) Electron micrographs of sections of a soybean root nodule. (C) Area of an infection thread where bacteria are apparently being release the bacteria mostly within the cells (Fig. 178C). The released bacteria then enlarge and become enclosed in a membrane envelope (Fig. 178D). These membrane-enclosed bacteria are called bacteroids. In the meantime, the cortical parenchyma cells along the path of bacterial invasion begin to divide and the invaded cells increase in size as the bacteroids appear. The increased meristematic activity and cell enlargement of cortical cells result in the formation of the nodule which grows outward from the root cortex. At the same time differentiation of vascular tissues, both xylem and phloem, takes place in the nodule. The vascular tissues of the nodule are not connected directly with those of the root.

While the outermost tip or layer of the nodule remains meristematic and continues to grow and thus to increase the size of the nodule up to a certain point, many of the cortical cells behind the meristematic zone and in all the central tissue of the nodule are uniformly enlarged and infected with several bacteroids. In the most recently infected cells, each bacteroid is enclosed in a membrane envelope, while in earlier infected cells several bacteroids may be enclosed in a membrane envelope. In cells that have been infected even longer than the latter, the bacteroids lack a membrane envelope and the host cellular membrane system also has deteriorated. It appears that the membraneless bacteroids which occur in the advanced stages of infection, and which increase in numbers while the nodule is still growing, lack the ability to fix nitrogen. Therefore, the efficiency of root nodules in nitrogen fixation is proportional to the number of enveloped bacteroids they contain and not necessarily to the size of the nodules. As the nodules age, first cortical cells in the earliest-infected areas and then in the entire central area of the nodule disintegrate and collapse. The bacteroids, which have by now lost their membrane envelope, either disintegrate or become intercellular bacteria and are finally released into the soil as the nodule cortex and epidermis disintegrate.

SELECTED REFERENCES

Bieberdorf, F. W. 1938. The cytology and histology of the root nodules of some Leguminosae. *J. Agron.* **30**:375–389.

Erdman, L. W. 1967. Legume inoculation: What it is—what it does. *USDA Farmers' Bull.* **2003**:10 p.

Jordan, D. C., I. Cinyer, and W. H. Coulter. 1963. Electron microscopy of infection threads and bacteria in young root nodules of *Medicago sativa. J. Bacteriol.* **86**:125–137.

Mosse, B. 1964. Electron microscope studies of nodule development in some clover species. *J. Gen. Microbiol.* **36**:49–66.

Tu, J. C. 1975. Rhizobial root nodules of soybeans as revealed by scanning and transmission electron microscopy. *Phytopathology* **65**:447–454.

released. (D) Infected and uninfected cells in a young nitrogen-fixing nodule. Membrane envelopes are visible around some bacteria. The electron-lucent granules in the bacteria consist of poly-beta-hydroxybutyrate. (Photo A courtesy U.S.D.A. Photos B–D courtesy B. K. Bassett and R. N. Goodman.)

plant diseases caused
by rickettsialike bacteria

INTRODUCTION

The rickettsialike organisms (RLO) that cause plant diseases are still of uncertain identity. Most researchers studing these organisms in plants consider them to belong to some group of cultivable bacteria, while others believe them to be "obligate parasites" more closely related to rickettsiae and therefore call them rickettsiae or rickettsialike organisms. Even the recognized rickettsiae of arthropods, however, are considered by many researchers studying them to be parasitic bacteria which simply have not yet been grown on culture media in the absence of host cells.

Organisms formally classified as Rickettsiae are one of the groups of prokaryotic organisms, i.e., organisms without an organized and bounded nucleus. The other such prokaryotic groups are the bacteria and the mycoplasmas. Rickettsiae differ from bacteria in that rickettsiae are obligate intracellular prokaryotic parasites in cells of higher organisms containing typical nuclei, and differ from mycoplasmas in their possession of a cell wall, a structure absent in mycoplasmas.

Rickettsiae make up the class of Rickettsias which has two orders: Rickettsiales and Chlamydiales. The Rickettsiales are mostly rod shaped, coccoid, or pleomorphic microorganisms that have cell walls but no flagella, are gram-negative and multiply only inside host cells. They multiply by binary fission. The Chlamydiales are intracellular coccoid microorganisms, gram-negative, and nonmotile. They multiply only within the cytoplasm of host cells through a unique growth cycle that includes a small (0.2 to 0.4 μm) "elementary body" that is infectious and develops intracellularly into the "initial body." The latter may be 0.8 to 1.5 μm in diameter, has a thin reticulated wall, and divides intracellularly by fission and produces an "intermediate body" which finally produces "elementary bodies" again. The Chlamydiales are sensitive to tetracyclines, penicillin, and 5-fluorouracil.

The Rickettsiales are similar to bacteria in most respects: the majority of them are rod shaped, spheroidal, or pleomorphic microorganisms that have typical bacterial cell walls, have no flagella, divide by binary fission, and are gram-negative. Rickettsiae generally multiply only inside living host cells. However, several exceptions are known and some microorganisms are considered as rickettsiae although they have a ring shape, or possess a flagellum, or are gram-positive, or multiply on bacteriological media of varying complexity. Rickettsiae are parasitic or symbiotic. The parasitic ones affect reticuloendothelial and vascular endothelial cells or erythrocytes of vertebrates and various organs of arthropods; many cause disease in man and other vertebrate and invertebrate animals. The symbiotic forms affect insects and may be essential for development and reproduction of certain insect hosts. Most rickettsiae are transmitted by arthropods such as lice, fleas, ticks, and mites. Some rickettsiae grow only in the cytoplasm, some in nuclei, and others in the vacuoles of host cells, while a minority may grow in an extracellular environment within the host. Although most rickettsiae parasitize arthropods, some also attack man in whom they cause diseases such as the epidemic typhus, scrub typhus, rocky mountain spotted fever, trench fever, and Q fever.

The Rickettsiales are 0.2 to 0.8 by 0.8 to 2.0 μm in size, with some of them up to 4 μm long prior to fission. They seem to have a cytoplasmic membrane and a well-defined cell wall, which is often, but not always, rippled. Internally they consist of electron dense polar or peripheral zones corresponding to ribosome-filled cytoplasm, and a light central zone corresponding to DNA strands. Rickettsiae grow best at 32 to 35°C. Their growth is inhibited by p-aminobenzoic acid and several antibiotics but not by sulfonamides. The antibiotics most effective in inhibiting growth of rickettsiae are oxytetracycline, chlortetracycline, erythromycin, and chloramphenicol. Penicillin and streptomycin also inhibit growth of rickettsiae but to a lesser extent than the other antibiotics.

THE PLANT-PATHOGENIC RICKETTSIALIKE ORGANISMS

Rickettsialike organisms were first observed in plants in 1972, in the phloem of clover and periwinkle plants infected with the clover club leaf disease. The following year rickettsialike organisms were observed in plants affected by Pierce's disease of grape and alfalfa dwarf, ratoon stunting of sugarcane, phony disease of peach, and in a disease of cereals (Fig. 179). Subsequently, similar organisms were observed in plum leaf scald, in almond leaf scorch, and in certain other diseases.

The rickettsialike organisms of plants resemble the typical rickettsias in shape, size, possession of generally rippled cell wall, lack of flagella, intracellular growth, and inability to grow on artificial nutrient media (Fig. 180). In the cases in which the vector of the plant pathogen is known, it is a leafhopper. The pathogens are obtained by the insect vector following feeding of the latter on diseased plants for 12 hours or more. Vectors require an incubation period of several days to several weeks before they can transmit the pathogen to healthy plants. The rickettsialike organisms infect and multiply inside the various organs of the vector which can transmit the pathogen throughout its life. In at least one disease, the club leaf of clover, the pathogen is even passed from the mother to the progeny insects through the eggs (transovarial transmission).

In some of the diseases, the rickettsialike organisms are confined to the xylem vessels (e.g., in Pierce's disease of grape, almond leaf scorch, plum leaf scald, phony peach, and ratoon stunting), while in clover club leaf and most of the others the rickettsialike organisms seem to be confined to the phloem sieve elements. In some diseases, e.g., club leaf of clover, the symptoms receded during treatment of the plants with penicillin or tetracycline hydrochloride, but reappeared when treatments were halted. In others, e.g., Pierce's disease, the appearance of symptoms was delayed for more than two years when the inoculated plants were

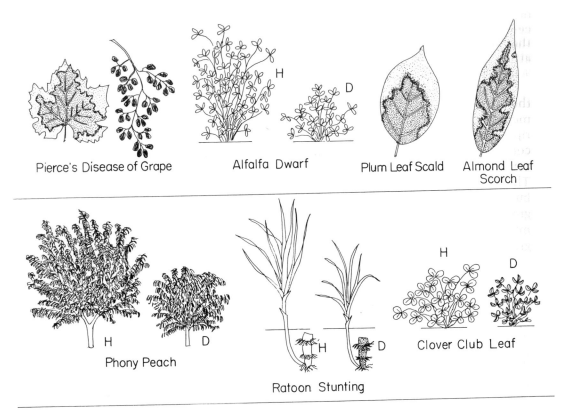

Pierce's Disease of Grape   Alfalfa Dwarf   Plum Leaf Scald   Almond Leaf Scorch

Phony Peach   Ratoon Stunting   Clover Club Leaf

FIGURE 179.
Symptoms caused by rickettsialike bacteria.

receiving weekly and biweekly drenches of oxytetracycline or tetracycline hydrochloride solutions around their roots. Plants infected with rickettsialike diseases can be freed of these organisms by immersion of the entire plant in water at 45 to 49°C for 2 to 3 hours, or treating the plants with hot air at 50 to 55°C for several hours. In some hosts infected with rickettsialike organisms, purplish red streaks develop in the woody cylinder of sections immersed for 10 to 30 minutes in acidified methyl alcohol.

To date, there are many uncertainties as to the nature of the pathogens described here as rickettsialike and as to their taxonomic position in regard to the bacteria or within the bacteria. Also, with the exception of the ratoon stunting organism, no distinction can be made among such organisms from their various hosts and, therefore, it is not known whether some of the pathogens already detected on different hosts belong to the same or different species.

Among the most important plant diseases caused by rickettsialike organisms are Pierce's disease of grape (= alfalfa dwarf), phony peach disease, sugarcane ratoon stunting, citrus greening, and almond leaf

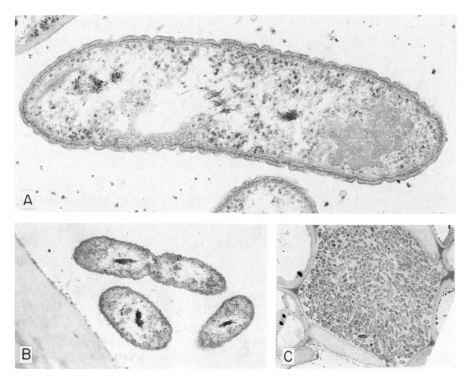

FIGURE 180.
Morphology, multiplication, and distribution of plant pathogenic rickettsialike bacteria, the causal agent of Pierce's disease of grapes. (A) Typical cell showing rippled cell wall. (B) Rickettsialike bacteria in xylem vessel, one of them undergoing binary fission. (C) Rickettsialike bacteria in a tracheary element of a leaf vein. (Photos courtesy H. H. Mollenhauer, from Mollenhauer and Hopkins, *J. Bacteriol.* **119**:612–618.)

scorch. Some of these, e.g., Pierce's disease of grape, almond leaf scorch and phony peach, may be caused by the same or closely related pathogens since they have a common leafhopper vector and react identically to the phony peach chemical test. On the other hand, in ratoon stunting of sugarcane, gram-positive, smooth-walled coryneform bacteria have been found in the xylem of affected palnts. It is expected that many more plant diseases will be shown to be caused by rickettsialike organisms or by other walled microorganisms with heretofore unknown properties and taxonomic relationships.

- *Pierce's Disease of Grape
  and Alfalfa Dwarf*

The disease is present in the southern U.S. from California to Florida where it kills grapevines and makes large areas unfit for grape culture. In some areas the disease is endemic and no grapes can be grown while in others it breaks out as infrequent epidemics. Alfalfa plants remain dwarfed and die 6 to 8 months after infection. Many other annual and perennial kinds of plants of some 28 families, including grasses, herbs, shrubs, and trees are affected by the disease.

Infected grapevines may die within a few months or may live for several years after the onset of infection. Some varieties survive infections longer than others, older plants survive infections longer than young vigorous ones, and vines in colder regions survive infections longer than vines in hot areas.

In grapes the symptoms appear first as a sudden drying, i.e., scalding, of part or most of the margin area of the leaf while the leaf is still green (Fig. 181). Scalded areas advance toward the central area of the leaf and later turn brown. In late season affected leaves usually drop, leaving the petioles attached to the canes. Grape clusters on vines with leaf symptoms stop growth, wilt, and dry up. Infected canes mature irregularly, forming patches of brown bark among areas of cortex that remain immature and green. During the following season(s), infected plants show delayed spring growth, dwarfed vines, and greenish vein banding in the first few leaves, but later in the season leaves and fruit show the same symptoms as during the first season. Decline of the top is followed by dieback of the root system. Internally, the current-season wood of all parts of infected vines shows yellow to brown streaks that are readily seen in longitudinal and cross sections. In the same wood, gum forms in vessels and other types of cells and tyloses develop in vessels of all sizes. Both gum and tyloses cause plugging of vessels sufficient to account for many of the external symptoms of diseased plants.

FIGURE 181.
Leaf scorch symptom of Pierce's disease in grape. (Photo courtesy
A. C. Goheen.)

In alfalfa, infected plants decline in vigor gradually. After each cutting they show very slow recovery, produce excessive numbers of small spindly stems and small dark green leaves. Six to eight months after infection the plants die. Internally, the woody portion of the crown and roots shows a brown to yellow discoloration due to deposits of gumlike materials in the vessels.

The pathogen has been described as a ricketssialike organism measuring 0.4 to 0.5 by 1.0 to 3.2 μm and having a typically rickettsialike rippled cell wall (Fig. 180). This pathogen is found in xylem vessels of grapevines in large numbers, and was recently cultured on nutrient media.

The pathogen is transmitted by grafting and by many species of leafhoppers, most commonly by species of *Carneocephala* and *Draeculacephala*. The leafhopper vectors acquire the pathogen after feeding on infected hosts for about 12 hours and may continue to transmit it to healthy hosts for the rest of their lives. The presence of the pathogen in a plant is determined by the symptoms of the disease, if present, or by transmission to indicator plants either by grafting or by insect vectors. The grape varieties most commonly used as indicators are Carignane, Emperor, and Palomino.

There is no practical control of Pierce's disease of grape and alfalfa dwarf in the field. Control of the insect vector and roguing of infected plants have not been effective. All commercial grape varieties are susceptible to the disease but some grape and alfalfa varieties carry some resistance. Drench treatments with 4 liters of 50 and 100 ppm tetracycline solutions applied around the base of infected plants twice weekly, weekly, or biweekly, inhibited symptom development on most plants for one or two years but such treatments are not commercially feasible. Individual plants can be freed of the pathogen by immersing the entire plant in water at 45°C for 3 hours, 50°C for 20 minutes, or 55°C for 10 minutes. Such treatments are also of little help to the grower. Best is to plant in areas remote from natural reservoirs of the pathogen.

SELECTED REFERENCES

Davis, R. E., and R. F. Whitcomb. 1971. Mycoplasmas, Rickettsiae, and Chlamydiae: Possible relation to yellows diseases and other disorders of plants and insects. *Ann. Rev. Phytopathol.* **9**:119–154.

Davis, J. M., A. H. Purcell, and S. V. Thomson. 1978. Pierce's disease of grapevines: Isolation of the bacterium. *Science* **199**:75–77.

Esau, K. 1948. Anatomic effects of the viruses of Pierce's disease and phony peach. *Hilgardia* **18**:423–482.

Giannotti, J., *et al.* 1970. Infection de plante par un type inhabituel de microorganisme intracellulaire. *C. R. Acad. Sci. Ser. D.* **271**:2118–2119.

Goheen, A. C., G. Nyland, and S. K. Lowe. 1973. Association of a rickettsialike organism with Pierce's disease of grapevines and alfalfa dwarf and heat therapy of the disease in grapevines. *Phytopathology* **63**:341–345.

Hewitt, W. B. 1970. Pierce's disease of *Vitis* species, *in* "Virus Diseases of Small Fruits and Grapevines," N. W. Frazier, *et al.* (Eds.). Univ. of Calif., Berkely, pp. 196–200.

Hopkins, D. L., and H. H. Mollenhauer. 1973. Rickettsia-like bacterium associated with Pierce's disease of grapes. *Science* **179**:298–300.

Hopkins, D. L., H. H. Mollenhauer, and W. J. French. 1973. Occurrence of a rickettsia-like bacterium in the xylem of peach trees with phony disease. *Phytopathology* **63**:1422–1423.

Hopkins, D. L., J. A. Mortensen, and W. C. Adlerz. 1973. Protection of grapevines from Pierce's disease with tetracycline antibiotics. *Phytopathology* **63**:443.

Hutchins, L. M., L. C. Cochran, and W. F. Turner. 1951. Phony, *in* "Virus Diseases and Other Disorders with Viruslike Symptoms of Stone Fruits in North America." *USDA Agr. Handbook*, Vol. 10, pp. 17–25.

Kitajima, E. W., M. Bakercic, and M. V. Fernandez-Valiela. 1975. Association of rickettsia-like bacteria with plum leaf scald disease. *Phytopathology* **65**:476–479.

Laflèche, D., and J. M. Bové. 1970. Structures de type mycoplasme dans les feuilles d' orangers atteints de la maladie du "greening." *C. R. Acad. Sci. Ser. D.* **270**:1915–1917.

Latorre, B. A., *et al.* 1975. Characterization of the bacterium associated with Pierce's disease. *Proc. Amer. Phytopath. Soc.* **2**:67.

Lowe, S. K., G. Nyland, and S. M. Mircetich. 1976. The ultrastructure of the almond leaf scorch bacterium with special reference to topography of the cell wall. *Phytopathology* **66**:147–151.

Maramorosch, K. 1974. Mycoplasmas and rickettsiae in relation to plant diseases. *Ann. Rev. Microbiol.* **28**:301–324.

Mircetich, S. M., *et al.* 1976. Etiology of almond leaf scorch disease and transmission of the causal agent. *Phytopathology* **66**:17–24.

Nyland, G., *et al.* 1973. The ultrastructure of a rickettsialike organism from a peach tree affected with phony disease. *Phytopathology* **63**:1275–1278.

Ormsbee, R. A. 1969. Ricketssiae (as organisms). *Ann. Rev. Microbiol.* **23**:275–292.

Pierce, N. B. 1892. The California vine disease. *USDA, Div. Veg. Pathol. Bull.* **2**: 222 p.

Saglio, P. D., *et al.* 1971. Isolement, culture et observation au microscope électronique des structures de type mycoplasme associées a la maladie du stubborn des agrumes et leur comparison avec les structures observées dans le cas de la maladie du greening des agrumes. *Physiol. Vég.* **9**:569–582.

Teakle, D. S., P. M. Smith, and D. R. L. Steindl. 1973. Association of a small coryneform bacterium with the ratoon stunting disease of sugarcane. *Aust. J. Agric. Res.* **24**:869–874.

Windsor, I. M., and L. M. Black. 1972. Clover club leaf: a possible rickettsial disease of plants. *Phytopathology* **62**:1112.

Windsor, I. M., and L. M. Black. 1973. Remission of symptoms of clover club leaf following treatment with penicillin. *Phytopathology* **63**:44–46.

Windsor, I. M., and L. M. Black. 1973. Evidence that clover club leaf is caused by a rickettsia-like organism. *Phytopathology* **63**:1139–1148.

Worley, J. F., and A. G. Gillaspie, Jr. 1975. Electron microscopy in situ of the bacterium associated with ratoon stunting disease in sudangrass. *Phytopathology* **65**:287–295.

12
plant diseases caused by mycoplasmalike organisms

In 1967, wall-less microorganisms resembling mycoplasmas were seen with the electron microscope in the phloem of plants infected with one of several yellows-type diseases. Such diseases, up to that moment, were thought to be caused by viruses. That same year, similar microorganisms were seen in the insect vectors of these diseases. Furthermore, it was shown that these microorganisms were susceptible to tetracycline but not to penicillin antibiotics and that the symptoms of infected plants could be suppressed, at least temporarily, by treatment with antibiotics.

Since then, more than 75 distinct plant diseases affecting several hundred genera of plants have been shown to be caused by mycoplasmalike organisms. Among them are some very destructive diseases, especially of trees, e.g., pear decline, coconut lethal yellowing, X-disease of peach, apple proliferation, etc., but also of herbaceous annual and perennial plants such as aster yellows of vegetables and ornamentals, and stolbur. Furthermore, several diseases, e.g., citrus stubborn and corn stunt, were shown to be caused by spiroplasmas. The main characteristics of yellows-type diseases are a more or less gradual, uniform yellowing or reddening of the leaves, smaller leaves, shortening of the internodes and stunting of the plant, excessive proliferation of shoots and formation of witches'-brooms, greening or sterility of flowers and reduced yields and, finally, a more or less rapid dieback, decline, and death of the plant (Fig. 182).

Although mycoplasmalike organisms have been observed in the phloem of diseased plants, in sap extracted from such plants, and in the insect vectors of some of them, the true nature of mycoplasmalike organisms and their taxonomic position among the lower organisms is still uncertain. Morphologically, the organisms observed in plants resemble the typical mycoplasmas found in animals and humans and those living saprophytically, but the mycoplasmalike organisms of plants cannot be grown on artificial nutrient media. Also, so far, no plant disease has been reproduced on healthy plants inoculated directly with mycoplasmalike organisms obtained from diseased plants. The pathogens of at least two diseases, citrus stubborn and corn stunt, have been grown on artificial nutrient media and have even reproduced the disease in plants when inoculated by insects injected with the organism from culture; however, these pathogens differ from all other plant mycoplasmalike organisms and from the true mycoplasmas in that they have a helical structure, they are motile, and in some other characteristics. At present it is believed that most of the plant mycoplasmalike organisms will be proved to be similar to the true mycoplasmas, belonging to a new taxon rather than to either of the two mycoplasma genera, i.e., *Mycoplasma* and *Acholeplasma*, while the citrus stubborn and the corn stunt organisms and any others like them are placed in a newly created genus of mycoplasmas called *Spiroplasma*.

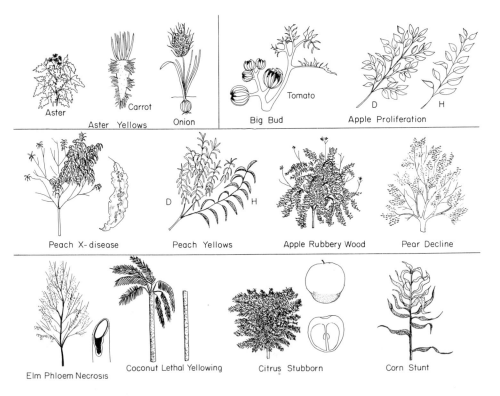

FIGURE 182.
Symptoms caused by mycoplasmalike organisms.

properties
of mycoplasmas

Since, until recently, only mycoplasmas that infect animals and man or are saprophytic were known, all information regarding mycoplasmas has been obtained from the study of these types only.

TRUE MYCOPLASMAS

Mycoplasmas are the third group of prokaryotic organisms, i.e., organisms without an organized and bounded nucleus. The other such prokaryotic groups are the bacteria and the rickettsiae.

The mycoplasmas make up the class of Mollicutes which has one order, Mycoplasmatales. The order has three families: Mycoplasmataceae, composed of one genus, *Mycoplasma*, Acholeplasmataceae, also composed of one genus, *Acholeplasma*, and Spiroplasmataceae, also composed of one genus, *Spiroplasma*. *Mycoplasma* differs from *Acholeplasma* in that its species require sterol for growth and are sensitive to digitonin, while species of *Acholeplasma* do not require sterol for growth and are resistant to digitonin. Also, species of *Mycoplasma* have only half as much DNA (5×10^8 daltons) as do species of *Acholeplasma* (10^9 daltons), the amount of DNA in *Acholeplasma* being about half, or at most equal to that of the smallest bacteria (1.5×10^9 daltons). The genome of *Spiroplasma* is 10^9 daltons.

Mycoplasmas lack a true cell wall and the ability to synthesize the substances required to form a cell wall. Mycoplasmas, therefore, are bounded only by a single triple-layered "unit" membrane. They are small, sometimes ultramicroscopic cells containing cytoplasm, randomly distributed ribosomes and strands of nuclear material. They measure from 175 to 250 nm in diameter during reproduction but grow into various sizes and shapes later on. The shapes range from coccoid or slightly ovoid to filamentous. Sometimes they produce branched mycelioid structures. The size of fully developed coccoid mycoplasmas may vary from one to a few microns, while slender branched filamentous forms may range in length from a few to 150 μm. Mycoplasmas seem capable of reproducing by budding and by binary transverse fission of coccoid and filamentous cells. Mycoplasmas have no flagella, produce no spores, and are gram-negative. Nearly all mycoplasmas parasitic to man and animals and all saprophytic ones can be grown on more or less complex artificial nutrient media in which they produce minute colonies that usually have a characteristic "fried egg" appearance. Mycoplasmas have been isolated mostly from healthy and or diseased animals and humans suffering from diseases of the respiratory and urinogenital tracts; they have been associated with some arthritic and nervous disorders of animals; and some have been found to exist as saprophytes. Most mycoplasmas are completely resistant to penicillin but they are sensitive to tetracycline, chloramphenicol, some to erythromycin and to certain other antibiotics.

MYCOPLASMALIKE ORGANISMS OF PLANTS

The organisms observed in plants and insect vectors, with the exception of spiroplasmas, resemble the mycoplasmas of the genera *Mycoplasma* or *Acholeplasma* in all morphological aspects. They lack cell wall, they are bounded by a triple-layered "unit" membrane, and have cytoplasm, ribosomes and strands of nuclear material. Their shape is usually spheroidal to ovoid or irregularly tubular to filamentous and their sizes comparable to those of the typical mycoplasmas (Fig. 183).

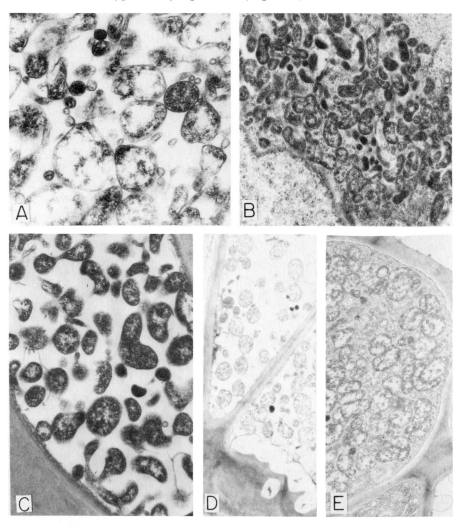

FIGURE 183.
Aster yellows mycoplasma. (A) Typical large mycoplasmalike bodies bound by a unit membrane and containing strands resembling DNA. The smaller particles contain ribosomes. (B) Mycoplasmalike bodies in cytoplasm of infected phloem parenchyma cell. (C, D) Several polymorphic mycoplasmas (C) and some apparently undergoing binary fission or budding (C, D). (E) Invagination of some mycoplasmalike bodies by others indicating the extreme pliability of the organisms. (Photos courtesy J. F. Worley.)

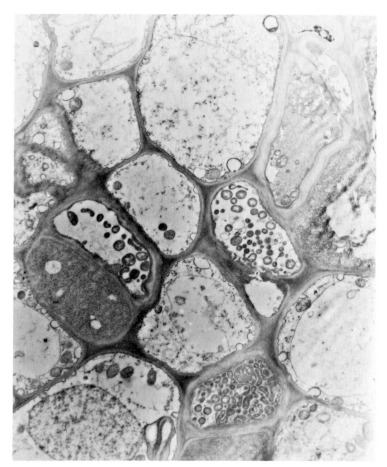

FIGURE 184.
Distribution of the peach yellows mycoplasma within vascular tissues of diseased cherry showing symptoms. (Photo courtesy J. F. Worley.)

Plant mycoplasmalike organisms are generally present in the sap of a small number of phloem sieve tubes (Fig. 184). However, unlike the typical mycoplasmas, which grow only on, not in, cells, the plant mycoplasmalike bodies also grow in the cytoplasm of phloem parenchyma cells and of their insect vectors. In such cases, the mycoplasmalike organisms often appear in tightly packed colonies and consist of a single type of spherical or ovoid cell.

Most plant mycoplasmalike organisms are transmitted from plant to plant by leafhoppers (Fig. 185), but some are transmitted by psyllids, treehoppers, planthoppers (see Fig. 215), and some possibly by aphids and mites. Some of the pathogens are known to infect various organs of their leafhopper or psyllid vectors and to multiply in their cells.

The insect vectors can acquire the pathogen after feeding on infected plants for several hours or days. The insect may also become an infective vector if it is injected with extracts from infected plants or vectors. More insects become infective vectors when feeding on young leaves and stems of infected plants than on older ones. The vector cannot transmit the mycoplasma immediately after feeding on the infected plant but it begins to transmit it after an incubation period of 10 to 45 days, depending on the temperature; the shortest incubation period occurs at about 30°C, the longest at about 10°C. The incubation period in insects, however, can be shortened by injecting them with high doses of extracts from infective insects.

The incubation period is required for the multiplication and distribution of the mycoplasma within the insect (Fig. 185). If the mycoplasma is acquired from the plant, it multiplies first in the intestinal cells of the vector; it then passes into the hemolymph and internal organs are infected. Eventually the brain and the salivary glands are invaded. When the concentration of the mycoplasma in the salivary glands reaches a certain level, the insect begins to transmit the pathogen to new plants and continues to do so more or less efficiently for the rest of its life. Insect vectors usually are not affected adversely by the mycoplasmas but in some cases they show severe pathological effects. Mycoplasmas can usually be acquired as readily or better by nymphs as by adult leafhoppers and survive through subsequent molts, but are not passed from the adults to the eggs and to the next generation, which, therefore, must feed on infected plants in order to become infective vectors.

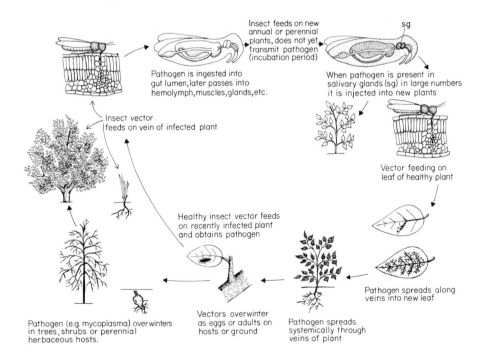

FIGURE 185.
Sequence of events in the overwintering, acquisition, and transmission of viruses, mycoplasmas, and rickettsialike bacteria by leafhoppers.

In spite of countless attempts by numerous investigators to culture plant mycoplasmalike organisms on artificial nutrient media, including the media on which all typical mycoplasmas grow, this has not yet been possible. Although some investigators have recently reported successful culture of mycoplasmalike organisms, these reports either could not or have not yet been verified and it is possible that they describe maintenance rather than growth of these organisms in culture.

Plant mycoplasmalike organisms, like the typical mycoplasmas, are attacked by viruses. Thus, in plants affected by one of several diseases, the mycoplasmalike organisms present in the plants are infected with rod-shaped or bacilliform viruslike particles. Since treatment of the plants with antibiotics results in remission of disease and disappearance of the mycoplasmalike organisms but not of the viruslike particles, it is assumed that the viruslike particles in mycoplasmas are not the causal agents of the disease of the mycoplasma-infected plant.

In many cases, plants are infected with both a mycoplasmalike organism and a virus but the two pathogens are usually present in different phloem sieve elements and only rarely in the same cells. They apparently act independently of each other. In some diseases both pathogens are transmitted by the same leafhopper vector. Some dual infections, however, seem to cause external symptoms that do not occur in plants infected by either pathogen singly.

Mycoplasmalike organisms are sensitive to antibiotics, particularly those of the tetracycline group. When infected plants are immersed periodically into tetracycline solutions, the symptoms, if already present, recede or disappear and, if not yet present, their appearance is delayed. Foliar application of tetraylcines on infected plants is ineffective as are soil drench applications. In trees, application of antibiotics is most successful by direct injection into the trunk by pressure or by gravity flow and results in the alleviation or disappearance of symptoms for many months. None of these treatments has, so far, cured any plants from the disease: the symptoms reappear soon after treatment stops. Generally, treatment of plants during the early phases of the disease is much more effective than treatment of plants in advanced stages of the disease. Infected growing plants or dormant propagative organs can be totally freed of mycoplasmalike organisms by heat treatment. This can be applied as hot air in growth chambers at 30° to 37°C for several days, weeks or months, or as hot water in which dormant organs are immersed at 30 to 50°C for as short as 10 minutes at the higher temperatures and as long as 72 hours at the lower temperatures.

SPIROPLASMAS

Spiroplasmas have so far been found in citrus plants infected with the stubborn disease (*Spiroplasma citri*), in corn plants infected with the stunt disease, in Bermuda grass showing witches'-broom symptoms, and in the ornamental cactus *Opuntia tuna monstrosa*. Recently, a spiroplasma was reportedly obtained and grown in culture from aster plants infected with what was assumed to be the aster yellows mycoplasma. Two spiroplasmas have also been found in animals: one, the suckling mouse cataract agent, isolated from ticks and infected mice; the other, the sex ratio organism (SRO), found in several species of *Drosophila*, is inherited maternally and kills the males in the progeny of infected females. Spiroplasmas are pleomorphic cells that vary in shape from spherical or slightly ovoid, 100 to 250 nm or larger in diameter, to helical and branched nonhelical filaments that are about 120 nm in diameter and 2 to 4 μm during active growth and considerably longer in later stages of their growth. Unlike the mycoplasmalike organisms described above, spiroplasmas can be obtained from their host plants or their insect vectors and cultured on nutrient media (Figs. 186 and 187). They produce mostly helical forms in liquid media. It is not yet known with certainty how they multiply, but it is likely that they multiply by binary fission. They lack a true cell wall and are bounded by a single triple-layered "unit" membrane

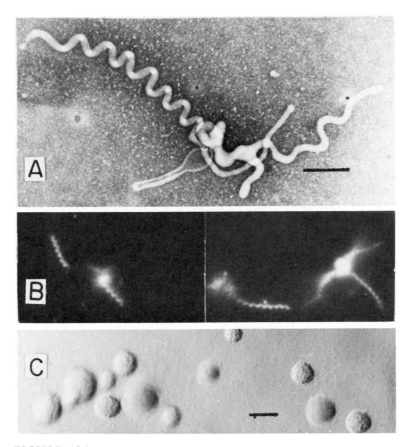

FIGURE 186.
Corn stunt spiroplasmas isolated from infected corn plants and grown on nutrient media. (A) Electron micrograph of a spiroplasma showing typical helical morphology (scale bar, 0.5 μm). (B) Living spiroplasmas from liquid cultures observed by dark field microscopy. (C) Colonies of corn stunt spiroplasma on agar plates 14 days after inoculation (scale bar, 0.05 mm). (Photos courtesy T. A. Chen, from Chen and Liao, *Science* **188**:1015–1017. Copyright © 1975 by the American Association for the Advancement of Science.)

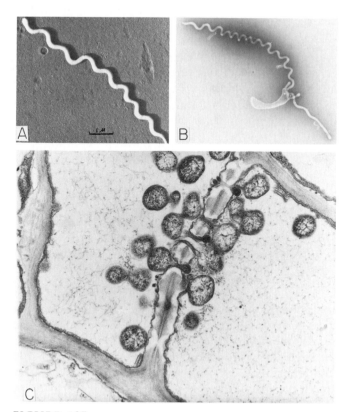

FIGURE 187.

(A) A replicative form of *Spiroplasma citri* isolated from stubborn-infected citrus. Platinum shadowing from a log-phase culture fixed with glutaraldehyde. (B) *Spiroplasma citri* obtained from its leafhopper vector, *Circulifer tenellus*, and grown in broth culture. Note presence of bleb. (C) *Spiroplasma citri* in sieve plate in midvein of sweet orange leaf. (Photos courtesy E. C. Calavan.)

but on the surface of the membrane they also have an additional outer layer of periodically repeated short projections. Spiroplasmas, unlike the true mycoplasmas and the mycoplasmalike organisms, are gram-positive or gram-variable. The helical filaments are motile, moving by a slow undulation of the filament and probably by a rapid rotary or "screw" motion of the helix. There are no flagella. Colonies of spiroplasmas on agar have a diameter of about 0.2 mm; some have a typical "fried egg" appearance, but others are granular (Fig. 186C). Spiroplasmas require sterol for growth. They are resistant to penicillin but inhibited by erythromycin, tetracycline, neomycin, and amphotericin.

The amount of DNA in spiroplasmas is equal to that of *Acholeplasma* and of the smallest bacteria. The citrus spiroplasma has been shown to be attacked by at least three different kinds of viruses.

Most known plant spiroplasmas, e.g., *Spiroplasma citri* and the corn stunt spiroplasma, have been obtained from their respective hosts and vectors, have been grown as pure cultures on nutrient media, and have been injected into or fed to their insect vectors which then, upon feeding on the host plants, transmitted the organisms to the plants. The hosts thus inoculated developed typical symptoms of the disease. The pathogens could then be recovered from such plants again and grown and observed in culture. Thus, the spiroplasmas are definitively the causes of their respective diseases. Although some of these organisms are serologically related they are by no means identical; also they seem to be distinct from each other in the hosts they infect and in certain nutrients required for growth of each in culture.

other organisms that resemble mycoplasmas: L-forms of bacteria

In addition to the three types of organisms described above, the bacteria often produce variants that fail to produce cell walls. The progeny of such variants comprise populations of wall-less bacteria, called L-form or L-phase bacteria, that are morphologically indistinguishable from mycoplasmas and the mycoplasmalike organisms observed in plants. L-form, i.e., wall-less, bacteria are usually produced under laboratory conditions when penicillin or other substances that inhibit cell wall production are added to the culture medium. They can apparently also develop in living organisms during treatment with certain antibiotics.

L-form bacteria are either unstable and revert to the original bacterial form when the substance inhibiting bacterial cell wall formation is removed from the medium, or they are stable, i.e., they are unable to revert to the original bacterium. L-form bacteria can be cultured on the same simple nutrient media as the original bacteria, but they usually lose any pathogenicity the original bacteria may have had. It is still uncertain, however, whether the L-form bacteria might not play a role in the persistence of disease agents during antibiotic treatments, in the recurrence of disease, and in latency by exhibiting a high degree of resistance to antibiotics acting on cell wall synthesis and by reverting to the original pathogenic bacteria at the termination of the antibiotic treatment. It is also conceivable that, in vivo, the L-form bacteria may themselves induce disease without reversion to the bacterial parents. Usually, however, L-form bacteria become more permeable, and thereby more sensitive, to antibiotics that affect other cell functions besides cell wall synthesis.

Although the ability to form L-forms is now accepted as a general property of bacteria, the only plant pathogenic bacterium reported to produce L-forms is *Agrobacterium tumefaciens*, the cause of crown gall disease. Moreover, the L-forms of this bacterium retained the pathogenicity of the parent bacteria, produced tumors identical to those produced by the bacteria and could be reisolated and cultured from such tumors.

Since L-form bacteria are morphologically indistinguishable from mycoplasmas and plant mycoplasmalike organisms, their diagnosis depends on their ability to grow on simple nutrient media and to revert to the original bacteria in culture. In some cases, more complex tests involving a comparison of DNA composition and homologies may be necessary to provide a definitive diagnosis.

examples of plant diseases caused by mycoplasmalike organisms

Among the most important plant diseases caused by mycoplasmalike organisms are aster yellows, apple proliferation, coconut lethal yellowing, elm phloem necrosis, grapevine flavescence dorèe, papaya bunchy top, peach X-disease, peach yellows, pear decline, stolbur diseases of solanaceous plants, tomato big bud, and many more. In addition, two diseases caused by spiroplasmas, citrus stubborn and corn stunt, are also of great economic importance.

- *Aster Yellows*

Aster yellows occurs in North America, Europe, and Japan. The pathogen attacks many vegetables, ornamentals, and weeds belonging to more than forty families. Some of the most severely affected hosts are carrot, lettuce, onion, spinach, potato, barley, flax, aster, gladiolus, tomato, celery, and phlox.

Aster yellows affects plants by causing a general yellowing (chlorosis) and dwarfing of the plant, abnormal production of shoots, sterility of flowers, malformation of organs, and a general reduction in the quantity and quality of yield (Fig. 188). Losses from aster yellows vary considerably among the different host crops, the greatest losses being suffered by carrot, in which losses between 10 and 25 percent are rather common and occasional losses of 80 to 90 percent have been reported. The products of infected plants, in addition to being reduced in size or quantity, also have an unpleasant flavor.

Symptoms. Although the general effects of aster yellows on most kinds of host plants are similar, each kind of host plant also produces characteristic symptoms which may differ appreciably from those produced on other hosts.

On carrot, symptoms appear first as a veinclearing and then yellowing of the younger leaves at the center of the crown. Soon after, infected plants produce many adventitious shoots that give the tops the appearance of a witches'-broom (Fig. 188A). The internodes of such shoots are short, as are the petioles of the leaves. The young leaves are generally smaller and more narrow and often become dry. The petioles of the older leaves become twisted and eventually break off. In the later part of the season, the remaining older leaves usually become bronzed and reddened.

FIGURE 188.
Symptoms of aster yellows disease on various hosts. (A) On carrots. Note the
bushy tops and the stunted, hairy, tapered roots of the infected carrots. Healthy
plant at right. (B) On onion. Diseased flower cluster showing varying degrees of
distortion and sterility. (C) Distortion, yellowing, and stunting of tomato plant
and fruits. (D) Yellowing and stunting of lettuce. (Photo A courtesy N. Y. Agric.
Expt. Sta., Geneva. Photos B–D, Dept. Plant Pathol., Cornell Univ.)

Plants infected while very young may die before reaching maturity. In
later infections, the resulting bushy tops not only make the plants unsightly and reduce their market value, but they also make mechanical
harvesting of carrots difficult or impossible and, furthermore, predispose
the roots to various soft rots in the field and in storage. The floral parts of
infected plants are deformed.

Aster yellows also affects the carrot roots, and the earlier the infection
takes place the more severe the damage produced. Infected carrots are
generally small, tapered, abnormally shaped, and have a variable number
of woolly secondary roots on which the soil clings tenaciously when the

plant is pulled from the ground. The crown of infected roots, instead of being flat or hollowed as in healthy plants, bulges upward, forming a conelike neck. In longitudinal section, the xylem or core of infected carrots appears enlarged while the cortex zone is much narrower than in healthy carrots. The core of infected carrots appears translucent and also somewhat lighter in color. Infected carrots have an unpleasant flavor, the degree of which is proportional to the severity of the disease. In processed carrots (canned or frozen purees), the presence of even 15 percent of yellows-infected carrots imparts an objectionable off flavor to the entire processed product.

The pathogen: aster yellows mycoplasmalike organism.   The aster yellows pathogen has the morphology, size, and properties of the mycoplasmalike organisms of plants described earlier in this chapter (Fig. 183). However, it was reported recently that the aster yellows organism was grown in culture in which it produced typical spiroplasmas.

Aster yellows has been transmitted by budding or grafting, by the leafhopper *Macrosteles fascifrons,* and by several other leafhoppers.

Development of disease.   The mycoplasma overwinters in perennial or biennial, ornamental, vegetable, and weed plants. A few of the most important weeds that serve as a reservoir for the pathogen are thistle, wild chicory, wild carrot, dandelion, field daisy, black-eyed Susan, and wide-leafed plantain.

The vector leafhopper acquires the mycoplasma while feeding by inserting its stylet into the phloem of infected plants and sucking the mycoplasma with the plant sap. At the end of the incubation period in the insect, during the feeding of the insect on healthy plants, the mycoplasma is injected through the stylet into the phloem of the healthy plants, where it establishes infection and multiplies. When inoculation takes place in the leaf the mycoplasma moves out of the leaf and into the rest of the plant occasionally within 8 hours but generally within 24 hours after inoculation. Infected plants usually do not show symptoms until after an incubation period of at least 8 to 9 days at 25°C, and 18 days at 20°C, while no symptoms develop at 10°C. The length of incubation period in plants is independent of the number of feeding vector insects and of the length of their feeding.

Aster yellows mycoplasma seems to be limited primarily to the phloem of infected plants. Some cells adjacent to the young sieve elements first become altered physiologically, become hypertrophied, and then die. Surviving cells become hyperplastic, but these too soon die. Cells surrounding the necrotic areas then begin to divide and enlarge excessively, producing abnormal sieve elements, while the phloem elements within the necrotic areas become degenerate and collapse.

Control.   Several measures help reduce losses from aster yellows, although none of them will control the disease completely. Eradication of perennial or biennial weed hosts from the field, roadways, and fences, and avoidance of planting a susceptible crop next to a crop harboring the pathogen, help eliminate a large source of mycoplasma inoculum. Control of the leafhopper vector in the crop and on nearby weeds with insecticides as early in the season as possible helps reduce transmission of the mycoplasma to the crop plants and thus reduces incidence of the disease. Certain varieties of plants are more resistant to the disease than others, but none are immune; during severe outbreaks of the disease, they, too, suffer serious losses.

Experimentally, aster yellows can be controlled by immersing the roots of newly infected plants in solutions of tetracyclines. Weekly immersions suppressed symptom development in infected plants, but symptoms usually reappeared within 2 to 4 weeks after treatment was discontinued. Foliar sprays or drenching of the soil around the plant with tetracyclines had little or no effect in controlling the disease.

• *Lethal Yellowing of Coconut Palms*

Lethal yellowing appears as a blight that kills palm trees within 3 to 6 months after first appearance of symptoms. The disease is present in Florida and most Caribbean islands, and probably in Panama, Venezuela, West Africa, and elsewhere. The disease was first identified in Key West in 1955 and in the next five years killed about three-fourths of the coconut palms in Key West. Lethal yellowing appeared in the Miami area of the Florida mainland in the fall of 1971 and it had killed an estimated 15,000 trees by October 1973 and 40,000 coconut palms by August 1974. By August 1975, 75 percent of the coconut palms in Dade County (Miami area) were reported to have been killed by, or be dying of, the lethal yellowing disease. In addition to coconut palm (*Cocos nucifera*), the disease apparently affects several other kinds of palms growing in south Florida, including *Veitchia, Pritchardia, Phoenix, Corypha*, and others. All the diseased palms appear to be infected with mycoplasmalike organisms, and decline and die with symptoms similar to lethal yellowing.

The symptoms of lethal yellowing appear at first as a premature drop of coconuts of any size. Then, the next inflorescence that appears has blackened tips, almost all its male flowers are dead and black (Fig. 189B), and sets no fruit. Soon the lower leaves turn yellow and the yellowing progresses upward from the older to the younger leaves. The older leaves then die prematurely, turn brown, and cling to the tree while the younger leaves are turning yellow (Fig. 189A). Before long, however, all the leaves die, as does the vegetative bud. Finally, the entire top of the palm falls away and leaves nothing but the tall trunk of the palm tree which by now looks like a telephone pole.

The pathogen is a mycoplasmalike organism morphologically similar to all other such organisms observed in plants. The pathogen occurs mainly in young phloem cells (Fig. 190). Although the disease is obviously spreading rapidly in nature, the vector is not yet known.

Control of lethal yellowing is so far being attempted primarily by sanitation measures, i.e., removal and burning of diseased palms as soon as symptoms appear, to reduce the source of inoculum from which the vector(s) can transmit the pathogen to healthy trees. Among the various coconut palms, only certain Malayan dwarf varieties appear to be resistant or immune to lethal yellowing and thousands of such trees are now

FIGURE 189.
(A) Coconut palm in late stage of decline as a result of infection with lethal yellowing. (B) Necrosis of inflorescence of coconut palm, an early diagnostic symptom of lethal yellowing. (Photos courtesy R. E. McCoy.)

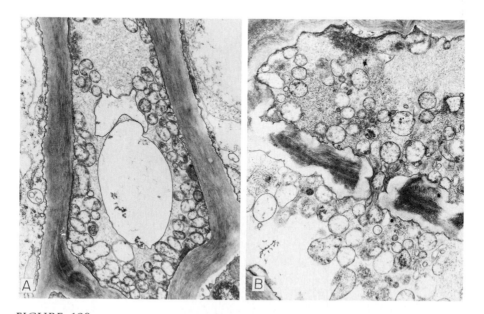

FIGURE 190.
(A) Mycoplasmalike organism in sieve element of young infloresence of coconut palm infected with lethal yellowing. (B) Lethal yellowing mycoplasmas passing through a sieve-plate pore lined with callose. (Photos courtesy M. V. Parthasarathy, from *Phytopathology* **64:**667–674.)

planted to replace the other coconut palms wherever lethal yellowing exists.

Very encouraging results of lethal yellowing control have been obtained by treating infected trees with solutions of tetracycline antibiotics. When 0.5 to 20 g of oxytetracycline hydrochloride is injected into trunks of diseased palms through either gravity flow or pressurized trunk injection, symptom expression is arrested or slowed down for several months and healthy new inflorescences and leaves grow in treated palms 3 to 4 months after initial treatment. Palms respond much better when treated with tetracycline in the early or preyellowing stages of disease development than in more advanced stages of the disease. The higher dosages of antibiotic (6 to 20 g per tree) are more effective in inducing remission in intermediate to advanced stages of the disease and their effect is longer lasting than that of lower dosages.

- *Elm Phloem Necrosis*

The disease occurs in about 15 central and southern states and was recently found in Pennsylvania and New York. Elm phloem necrosis epidemics have killed thousands of trees in each of numerous communities.

The symptoms consist of a general decline of the tree in which the leaves droop and curl, turn bright yellow, then brown, and finally fall (Fig. 191A, B). Some trees are killed within a few weeks and most trees that

FIGURE 191 (A, B).
Elm phloem necrosis. (A) Diseased tree. (B) Discolored inner bark of diseased tree. (Photos courtesy W. A. Sinclair).

show symptoms in June or July die in a single growing season. If the tree was infected late, it may live through the winter but then in the spring it produces a thin crop of small leaves and dies soon after. In later stages of the disease the inner layers of peeled bark (phloem) at the base of the stem show a butterscotch-brown color and have a faint odor of wintergreen. The latter characteristics are often used for a quick diagnosis of the disease. The discoloration of the phloem is apparently the result of rapid deposition of callose within the sieve tubes and then collapse of sieve elements and companion cells. The cambium produces replacement phloem but its cells become quickly necrotic also.

The pathogen is a mycoplasmalike organism present in the phloem of infected trees. It is transmitted from diseased to healthy trees by the leafhopper *Scaphoideus luteolus.*

Injection of tetracyclines into recently infected trees causes remission of symptoms for several months and up to three years. Severely diseased or dead trees should be removed and burned.

- *Peach X-disease*

X-disease, including western X-disease, occurs in the northwestern and northeastern parts of the U.S. and the adjacent parts of Canada. It also occurs in Michigan and several other states. Where present, X-disease is one of the most important diseases of peach. Affected trees become commercially worthless in 2 to 4 years. Young peach trees are rendered useless within one year of inoculation. X-disease of peach also attacks sweet and sour cherries, nectarines and chokecherries.

FIGURE 191 (C).
Peach tree showing symptoms of X-disease. (Photo courtesy D. Sands).

The symptoms of X-disease of peach appear as a slight mottle and reddish purple spots on the leaves of some or all branches. The spots die and fall out, giving a "shot-hole" appearance to the leaf. The leaves take on a reddish coloration and roll upward. Later, most leaves on affected branches drop except the ones at the tips (Fig. 191C). The fruits on affected branches usually shrivel and drop soon after the symptoms appear on the leaves. Any fruits remaining on the trees ripen prematurely, have an unpleasant taste and are unsalable. No seeds develop in the pits of affected fruit. Fruits on healthy looking parts of infected trees show no signs of the disease.

The pathogen is a mycoplasmalike organism found in the phloem sieve tube elements of diseased trees. The pathogen is transmitted by several species of leafhoppers of the genera *Colladonus*, *Scaphytopius*, etc., and, of course, by budding and grafting. Peach trees inoculated early in the season develop symptoms in less than two months while later inoculations may not produce symptoms until the following season. The insect vector can transmit the pathogen within trees of the same species and between trees of different species, e.g., from chokecherry to peach. It appears, however, that in the northeastern U. S. either the vector strain or the pathogen strain is different in that the pathogen is transmitted from chokecherry to peach or to chokecherry but not from peach to peach. This property has allowed successful control of X-disease on peach by eradicating chokecherry from the vicinity of peach orchards in a zone within about 200 meters from the orchard. Additional controls include the use of disease-free scion wood and rootstocks, and removal of diseased trees. Injections of tetracyclines into diseased trees results in temporary remission of X-disease symptoms and in reduced transmission of the disease by leafhoppers that obtain the inoculum from treated trees.

• *Pear Decline*

The pear decline occurs in the Pacific coast and some east coast states, in Europe, and probably in other continents where similar declinelike disorders of pear have been observed but whose relationship to pear decline has not been established. Pear decline causes either a slow, progressive weakening and final death of trees or a quick, sudden wilting and death of trees. The disease can be extremely catastrophic, having killed more than 1,100,000 trees in California between 1959 and 1962. Pear decline affects all pear varieties when they are grafted on certain rootstocks. Although oriental rootstocks such as *Pyrus serotina* and *P. ussuriensis* are affected the most, pear decline has also been observed on pear varieties grafted on *P. communis*, *P. betulaefolia*, and on quince.

The symptoms of pear decline in the "slow decline" syndrome appear as a progressive weakening of the trees which, however, may continue to live for many years, but eventually, despite occasional apparent improvement, are killed by the disease. During this period there is little twig growth, the leaves are few, small, pale green and leathery and roll slightly upward (Fig. 192). Such leaves often turn reddish in late summer and drop off prematurely in the fall. In the early stages of the disease the trees

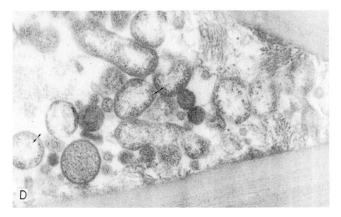

FIGURE 192.
(A–C) Pear decline. (A) "Bartlett" pear tree on *Pyrus serotina* rootstock showing symptoms of chronic pear decline. (B) As in (A) but chronic decline is more advanced—poor tree growth; few, small fruits. (C) "Bartlett" tree recovering after injection with 1 g oxytetracycline hydrochloride the previous September. Note new elongated shoots and denser leaf growth. (D) Mycoplasmalike organisms in phloem sieve tube of leaf of pear tree affected by pear decline. Arrows point to "unit" membrane. (Photos A–C courtesy G. Nyland, D courtesy Hibino and Schneider, from *Phytopathology* **60:**499–501.)

produce abundant blossoms but as the disease progresses the trees produce fewer blossoms, set fewer fruit, and the fruits are small. By this time, although starch accumulates above the graft union, it is almost absent below the union and most of the feeder roots of the tree are dead.

In the "quick decline" syndrome the trees wilt suddenly and die within a few weeks. Quick decline is more common in trees grafted on oriental rootstocks while trees grafted on other rootstocks usually develop the slow decline syndrome.

Both types of pear decline can be detected and verified by a microscopic observation of the phloem below the graft union: In diseased trees the current season's ring of phloem immediately below the graft union degenerates and the degeneration becomes more pronounced as the season progresses. Also, at the graft union of diseased trees, narrow, small sieve tube elements (replacement phloem) are produced rather than normal ones. Pear decline can be transmitted by budding or grafting, although only about one-third of the buds seems to transmit the disease. Pear decline is also transmitted naturally by pear psylla (*Psylla pyricola*), large numbers of which are responsible for the outbreaks of the disease.

The pathogen is a mycoplasmalike organism present in the phloem sieve elements. The bodies are mostly spherical to oblong particles of about 50 to 800 nm; relatively few are elongated (Fig. 192D). In pear psylla, the pathogens are present in various organs of the insect but seem to be almost five times more common in the cephalic part of the foregut than they are in the salivary glands.

A certain degree of control of pear decline is obtained by growing pear varieties on resistant rootstocks such as *Pyrus communis* and by avoiding the highly sensitive oriental rootstocks. Control of the pear psylla vector has not been successful. However, injection of 6 to 8 liters of a tetracycline solution (100 mg/liter) in the trunk of infected trees soon after harvest prevents leaf curl symptoms in the fall of the current season and greatly stimulates shoot and spur growth of treated trees the following season. When two or three such treatments are made in the fall, previously severely diseased trees are restored to normal or near normal condition (Fig. 192C). Antibiotic treatments must continue annually, however, or the disease will reappear.

- *Citrus Stubborn Disease*

Citrus stubborn is present in all the Mediterranean countries, southwestern U.S., Brazil, Australia, and possibly in South Africa. In some Mediterranean countries and in California, stubborn is regarded as the greatest threat to production of sweet oranges and grapefruit. Because of the slow development of symptoms and the long survival of affected trees, spread of stubborn is insidious and its detection difficult. However, yields are reduced drastically; the trees produce fewer fruits and many of those are too small to be marketable. In California, approximately two million orange, grapefruit, and tangelo trees are so severely affected that they are practically worthless and many more trees are infected in one or several branches but are not yet severely damaged.

The symptoms of stubborn disease appear on leaves, fruit, and stems of all commercial varieties regardless of rootstock (Fig. 193). Symptoms, however, vary a great deal and frequently only a few are expressed at one time on an entire tree or parts of a tree. In general, affected trees show a bunchy upright growth of twigs and branches, with short internodes and an excessive number of shoots; multiple buds and sprouts are common. Some of the affected twigs die back. The bark is thickened and sometimes pinholed. The trees show slight to severe stunting and often appear flat topped. The leaves are small or misshapen or both, often mottled or chlorotic. Excessive winter defoliation is common in infected trees. Affected trees bloom at all seasons, especially in the winter, but produce

FIGURE 193.
(A) Six-year-old sweet orange trees. Healthy at left, stubborn diseased at right showing extreme dwarfing. (B) Tangerine showing stylar-end greening and acorn shape (left) and healthy fruit. (C) Leaf symptoms of stubborn on sweet orange. (D) Four stubborn and one normal fruit of sweet orange. (Photos courtesy E. C. Calavan.)

fewer fruits. Some of the fruits are very small, lopsided, or otherwise deformed, frequently resembling acorns. Such fruits have normally thick rind from the stem end to the fruit equator, and abnormally thin rind from there to the stylar end. The rind is often dense or cheesy. Some fruits show greening of the stylar end or inverted development of ripe coloration in which, normally, color appears first at the stylar end. Affected fruit tends to drop prematurely and an excessive number of them become mummified. Fruits are usually sour or bitter and have an unpleasant odor and flavor. Also, fruits from affected trees or parts of trees tend to have many poorly developed, discolored, and aborted seeds.

The pathogen is *Spiroplasma citri* (Fig. 187). It is found in the sieve tubes of stubborn-diseased citrus phloem from which it can be obtained and cultured readily on artificial media. *Spiroplasma citri* was the first mycoplasmalike pathogen of a plant disease to be cultured. Within phloem sieve tubes the pathogen is present mostly as what appear to be spherical, ovoid, or elongated forms and occasionally as helical filaments. In liquid cultures, the pathogen appears primarily as motile helical filaments that are sometimes connected to irregularly shaped main bodies. The pathogen may lose its helical structure and motility in older cultures or on solid agar media, and then appears mostly as irregular filaments and blebs. The pathogen is gram-positive, has a layer of surface projections on the cytoplasmic membrane, and is usually found infected with one of three distinct kinds of virus. The pathogen has a sharp optimum temperature for growth at about 30 to 32°C, while little growth occurs at 20°C and none at 37°C. The pathogen is insensitive to penicillin but is highly sensitive to tetracycline and less so to amphotericin, neomycin, and digitonin.

Citrus stubborn disease is transmitted with moderate frequency by budding and grafting. It is also known to spread naturally in citrus orchards and a vector has been suspected and searched for intensively. Recently, *Spiroplasma citri* was isolated and cultured from *Circulifer tenellus* leafhoppers found feeding on sweet orange seedlings and nearby weeds. In other transmission experiments, the leafhopper *Scaphytopius nitridus*, which reproduces on citrus trees, acquired the pathogen from stubborn-diseased citrus and transmitted it to healthy sweet orange seedlings. Also, when the leafhoppers *S. nitridus* and *Euscelis plebejus* were injected with a pure culture of the pathogen and were then allowed to feed on sweet orange seedlings, they transmitted the pathogen to the seedlings and the latter developed stubborn disease symptoms. Moreover, the pathogen can survive from about 2 weeks to as long as 3 months *in vivo* in several species of leafhoppers, planthoppers, psyllids, plant bugs, and even *Drosophila*, following injection of these insects with a pure culture of *Spiroplasma citri*. However, none of these experimental hosts is known to be a vector of the disease in nature.

Stubborn disease can be detected and diagnosed by the symptoms it causes on trees in the field; by indexing on seedlings of several varieties of sweet orange, tangelo, grapefruit and other citrus which usually develop symptoms within 2 to 8 months in the greenhouse and within 15 to 24 months in the field; and by remission of stubborn disease symptoms of diseased trees following injection of erythromycin, tylosin, or tetracycline into the trunks of the trees.

Control of citrus stubborn depends on the use of spiroplasma-free scionwood and rootstocks, detection through indexing and removal of infected trees, and through antibiotic treatments of affected trees after the fruit has been picked.

- *Corn Stunt Disease*

Corn stunt occurs in the southern U.S., Central America, and northern South America. The disease causes severe losses in most areas where it occurs, although disease severity varies with the variety and the stage of host development at the time of infection.

The symptoms consist at first of faint yellowish streaks in the youngest leaves. As the plant matures the yellowing becomes more apparent and more general over the leaves; soon much of the leaf area turns red to purple, especially on the upper leaves. Infected plants remain stunted due to shorter stem internodes, particularly in the part of the plant produced after infection, which gives the plants a somewhat bunchy appearance at the top (Fig. 194). Infected plants often have more ears than do healthy plants but the ears are smaller and bear little or no seed. Tassels of infected plants are usually sterile. There is also a proliferation of sucker shoots and, in severe infections, of roots.

The corn stunt pathogen was the first spiroplasma discovered. Its morphology is very similar to the one causing the stubborn disease of citrus (Fig. 194B and Fig. 186). The corn stunt spiroplasma has been grown on artificial nutrient media and its pathogenicity proven either by injecting leafhopper vectors with, or allowing them to feed on, pure cultures of the spiroplasma and then allowing them to feed on healthy corn seedlings. The inoculated plants developed typical corn stunt symptoms and the spiroplasma was reisolated and cultured from such plants.

Corn stunt is transmitted in nature by the leafhoppers *Dalbulus elimatus, D. maidis, Graminella nigrifrons*, and others. The leafhoppers

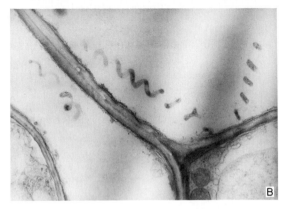

FIGURE 194.
(A) Corn stunt disease in corn. Leaves show chlorotic streaks, plant is stunted, proliferation begins at nodes, and tassel is sterile. (B) Portions of corn stunt spiroplasmas as seen in a section of phloem tissue from a corn stunt-infected plant. (Photos courtesy R. E. Davis, B from *Phytopathology* **63**:403–408.)

must feed on diseased plants for several days before they can acquire the spiroplasma and an incubation period of two to three weeks from the start of the feeding must elapse before the insects can transmit the spiroplasma to healthy plants. A feeding period of a few minutes to a few days may be required for the insects to inoculate the healthy plants with the spiroplasma. Plants show corn stunt symptoms 4 to 6 weeks after inoculation.

Where the corn stunt spiroplasma overwinters is not known with certainty although it was previously believed to overwinter in Johsongrass and possibly other perennial plants. In the tropics, it perpetuates itself in continuous croppings of corn.

Control of corn stunt depends on the planting of corn hybrids resistant to corn stunt.

SELECTED REFERENCES

Chen, T. A., and C. H. Liao. 1975. Corn stunt spiroplasma: Isolation, cultivation, and proof of pathogenicity. *Science* **188**:1015–1017.

Cole, R. M., *et al.* 1973. Morphology, ultrastructure and bacteriophage infection of the helical mycoplasma-like organism (*Sprioplasma citri* gen. and sp. nov.) cultured from "stubborn" disease of citrus. *J. Bacteriol.* **115**:367–386.

Cousin, Marie-Thérèse. 1972. Les mycoplasmes vegetaux. "Le selectionneur Français." *Versailles* No. 15:26 p.

Davis, R. E., *et al.* 1972. Helical filaments produced by a mycoplasma-like organism associated with corn stunt disease. *Science* **176**:521–523.

Davis, R. E., and R. F. Whitcomb. 1971. Mycoplasmas, rickettsiae, and chlamydiae: Possible relation to yellows diseases and other disorders of plants and insects. *Ann. Rev. Phytopathol.* **9**:119–154.

Davis, R. E., and J. F. Worley. 1973. Spiroplasma: Motile, helical microorganism associated with corn stunt disease. *Phytopathology* **63**:403–408.

Doi, Y., *et al.* 1967. Mycoplasma- or PLT group-like microorganisms found in the phloem elements of plants infected with mulberry dwarf, potato witches'-broom, aster yellows, or paulownia witches'-broom. *Ann. Phytopathol. Soc. Japan* **33**:259–266.

Filer, T. H., Jr. 1973. Suppression of elm phloem necrosis symptoms with tetracycline antibiotics. *Plant Dis. Reptr.* **57**:341–343.

Fisher, J. B. 1973. Report of the lethal yellowing symposium at Fairchild Tropical Garden, Miami. *Principes—J. Palm Soc.* **17**:151–159.

Fudl-Allah, A. A., and E. C. Calavan. 1974. Cellular morphology and reproduction of the mycoplasma-like organism associated with citrus stubborn disease. *Phytopathology* **64**:1309–1313.

Granados, R. R., K. Maramorosch, and E. Shikata. 1968. Mycoplasma: Suspected etiologic agent of corn stunt. *Proc. Nat. Acad. Sci.* **60**:841–844.

Hayflick, L. (Ed.). 1969. "The Mycoplasmatales and the L-Phase of Bacteria." North-Holland Publishing Co., Amsterdam.

Hervey, G. E. R., and W. T. Schroeder. 1949. The yellows disease of carrot. *N.Y. State Agr. Expt. Sta. (Geneva) Bull.* **737**: 29 p.

Hibino, H., and H. Schneider. 1970. Mycoplasmalike bodies in sieve tubes of pear trees affected with pear decline. *Phytopathology* **60**:449–501.

Hibino, H., G. H. Kaloostian, and H. Schneider. 1971. Mycoplasma-like bodies in the pear psylla vector of pear decline. *Virology* **43**:34–40.

Hirumi, H., and K. Maramorosch. 1973. Intracytoplasmic mycoplasmalike bodies in phloem parenchyma cells of aster yellows-infected *Nicotiana rustica*. *Phytopathol. Z.* **77**:71–83.

Hull, R. 1971. Mycoplasma-like organisms in plants. *Rev. Plant Pathol.* **50**:121–130.

Ishiie, T., *et al.* 1967. Suppressive effects of antibiotics of tetracycline group on symptom development of mulberry dwarf disease. *Ann. Phytopathol. Soc. Japan* **33**:267–275.

Klotz, L. J., E. C. Calavan, and L. G. Weathers. 1972. Virus and viruslike diseases of citrus. *Calif. Agr. Expt. Sta. Ext. Service Circulat* **559**:42 p.

Knorr, L. C. 1965. Serious diseases of citrus foreign to Florida. *Fla. Dept. Agr., Div. Pl. Industry, Bull.* **5**: 59 p.

Kuhn, C. W., and M. D. Jellum. 1970. Evaluations for resistance to corn stunt and maize dwarf mosaic diseases in corn. *Univ. Ga., Coll. Agr. Exp. Sta., Research Bull.* **82**: 37 p.

Kunkel, L. O. 1926. Studies on aster yellows. *Am. J. Bot.* **13**:646–705.

Lovisolo, O. 1973. I micoplasmi, microrganismi associati a malattie delle piante. *Informat. Bot. Ital.* **5**:59–70.

Maramorosch, K. 1953. Incubation period of aster yellows virus. *Am. J. Bot.* **40**:797–809.

Maramorosch, K. (Ed.). 1973. Mycoplasma and mycoplasma-like agents of human, animal, and plant diseases. *Ann. N. Y. Acad. Sci.* **225**: 532 p.

Maramorosch, K. 1974. Mycoplasmas and rickettsiae in relation to plant diseases. *Ann. Rev. Microbiol.* **28**:301–324.

Maramorosch, K., R. R. Granados, and H. Hirumi. 1970. Mycoplasma diseases of plants and insects. *Advan. Virus Res.* **16**:135–193.

Marchoux, G., F. Leclant, and P. J. Mathai. 1970. Maladies de type jaunisse et maladies voisines affectant principalement les solanacées et transmises par des insectes. *Ann. Phytopathol.* **2**:735–773.

McCoy, R. E. 1975. Effect of oxytetracycline dose and stage of disease development on remission of lethal yellowing in coconut palm. *Plant Dis. Reptr.* **59**:717–720.

Mullin, R. S., and D. A. Roberts. 1972. Lethal yellowing of coconut palms. *Fla. Coop. Ext. Serv., Inst. Food and Agr. Sci., Circular* **358**: 4 p.

Nasu, S., D. D. Jensen, and J. Richardson. 1970. Electronmicroscopy of mycoplasmalike bodies associated with insect and plant hosts of peach western X-disease. *Virology* **41**:583–595.

Nichols, C. W., *et al.* 1960. Pear decline in California. *Calif. Dept. Agr. "The Bulletin"* **49**:186–192.

Nyland, G., and W. J. Moller. 1973. Control of pear decline with a tetracycline. *Plant Dis. Reptr.* **57**:634–637.

Posnette, A. F. (Ed.). 1963 and Supplements 1969. Virus diseases of apples and pears. *Commonw. Bur. Hort. and Plantation Crops. Tech. Commun.* **30**:141 p.

Saglio, P., *et al.* 1971. Isolement et culture *in vitro* des mycoplasmes associés au "stubborn" des agrumes et leur observation au microscope electronique. *C. R. Acad. Sci (Paris) Ser. D* **272**:1387–1390.

Sinclair, W. A. 1972. Phloem necrosis of American and slippery elms in New York. *Plant Dis. Reptr.* **56**:159–161.

Sinha, R. C., and L. N. Chiykowski. 1967. Initial and subsequent sites of aster yellows virus infection in a leafhopper vector. *Virology* **33**:702–708.

Smith, P. F. 1971. "The Biology of Mycoplasmas." Academic Press, New York, 258 p.

Stoner, W. N. 1964. Corn stunt disease. *Miss. Agr. Expt. Sta. Info. Sheet* **244**: 4 p.

Thomas, D. L. 1974. Mycoplasmalike bodies associated with declining palms in south Florida. *Proc. Am. Phytopathol. Soc.* **1**:97.

Tully, J. G., R. F. Whitcomb, D. L. Williamson, and H. F. Clark. 1976. Suckling mouse cataract agent is a helical wall-free prokaryote (*Spiroplasma*) pathogenic for vertebrates. *Nature (London)* **259**:117–120.

U. S. Dept. Agr. 1951. Virus Diseases and Other Disorders with Virus-like Symptoms of Stone Fruits in North America. *Agr. Handbook* **10**: 276 p.

Valenta, V., M. Musil, and S. Misica. 1961. Investigations on European yellows-type viruses. I. The stolbur virus. *Phytopathol. Z.* **42**:1–38.

Whitcomb, R. F., and R. E. Davis. 1970. Mycoplasma and phytarboviruses as plant pathogens persistently transmitted by insects. *Ann. Rev. Ent.* **15**:405–464.

Williamson, D. L., and R. F. Whitcomb. 1975. Plant mycoplasmas: A cultivable spiroplasma causes corn stunt disease. *Science* **188**:1018–1020.

Wilson, C. L., and C. E. Seliskar. 1976. Mycoplasma-associated diseases of trees. *J. Arbor.* **2**:6–12.

13
plant diseases caused by parasitic higher plants

introduction

More than 2500 species of higher plants are known to live parasitically on other plants. These parasitic plants produce flowers and seeds similar to those produced by the plants they parasitize. They belong to several widely separated botanical families and vary greatly in their dependence on their host plants. Some, e.g., orchids, are epiphytes rather than parasites since they have chlorophyll and roots and can, therefore, manufacture their own food from carbon dioxide and water but depend on their hosts for certain dissolved minerals and probably some organic substances. Others (e.g., mistletoes) have chlorophyll but no roots and depend on their hosts for water and all minerals although they can produce all the carbohydrates in their green leaves and stems. Some other parasitic higher plants, however, having little or no chlorophyll nor true roots, depend entirely on their hosts for their existence (e.g., dodder).

Relatively few of the known parasitic higher plants cause important diseases on agricultural crops or forest trees. The most common and serious parasites belong to the following botanical families and genera:

Convolvulaceae
    Genus: *Cuscuta*, the dodders
Loranthaceae
    Genus: *Arceuthobium*, the dwarf mistletoes of conifers
                Phoradendron, the American true mistletoes of broadleaved trees
                Viscum, the European true mistletoes
Orobanchaceae
    Genus: *Orobanche*, the broomrapes of tobacco
Scrophulariaceae
    Genus: *Striga*, the witchweeds of many monocotyledonous plants

- *Dodder*

Dodder is widely distributed in Europe and in North America. In the U.S., it is most serious in the southern half of the country and in the north central states, but crops like alfalfa and clover raised for seed may be destroyed by dodder wherever they are grown. Other crops which suffer losses from dodder include lespedeza, onions, flax, sugar beets, several ornamentals, and potatoes.

Dodder affects the growth and yield of infected plants and causes losses which range from slight to complete destruction of the crop in the infested areas. Names, such as strangleweed, pull-down, hellbind, devil's hair, and hailweed, by which dodder is referred to in different areas, are descriptive of the ways in which dodder affects its host plants.

Dodder may also serve as a bridge for transmission of viruses from virus-infected to virus-free plants as long as both plants are infected by dodder.

Symptoms. Orange or yellow vine strands grow and entwine around the stems and the other aboveground parts of the plants. Dodder forms dense tangles of leafless strands on and through the crowns of the host plants (Fig. 195A). The growing tips reach out and attack adjacent plants, until a gradually enlarging circle of infestation, up to 10 feet in diameter, is formed by a single dodder plant. Dodder-infested areas appear as patches in the field (Fig. 195B) which continue to enlarge during the growth season and, in perennial plants such as alfalfa, become larger every year. During late spring and in the summer, dodder produces massed clusters of white, pink, or yellowish flowers which soon form seed. The infected host plants become weakened by the parasite, their vigor declines, and they produce poor yields. Many are pulled to the ground and may be killed to the roots by the parasite. As the infection spreads, several patches coalesce and large areas may be formed which are easily seen by the yellowish color of the parasitic vine that covers them.

FIGURE 195.
(A) Dodder on alfalfa. (B) Patches of dodder in a heavy infestation of an alfalfa field. (Photos courtesy U.S.D.A.)

The pathogen: Cuscuta sp.   Three species of dodder, largeseed (*Cuscuta indicora*), smallseed (*C. planiflora*), and field dodder (*C. campestris*), are important in the U.S. The first two show preference for legumes, but the third attacks many other broadleaf plants as well as legumes.

Dodder is a slender, twining plant (Fig. 196). The stem is tough, curling, threadlike, and leafless, bearing only minute scales in place of leaves. The stem is usually yellowish or orange in color, sometimes tinged with red or purple; sometimes it is almost white. Tiny flowers massed in clusters occur on the stem from early June until frost. Gray to reddish-brown seeds are produced in abundance by the flowers and mature within a few weeks from bloom.

Development of disease.   Dodder seed overwinters in infested fields or mixed with the seed of crop plants. Sometimes dodder stems may also overwinter on debris on the ground. During the growing season the seed germinates and produces a slender yellowish shoot but no roots (Fig. 196). This leafless shoot rotates as though in search of a host. If no contact with a susceptible plant is made, the stem falls to the ground, where it lies dormant for a few weeks and then dies.

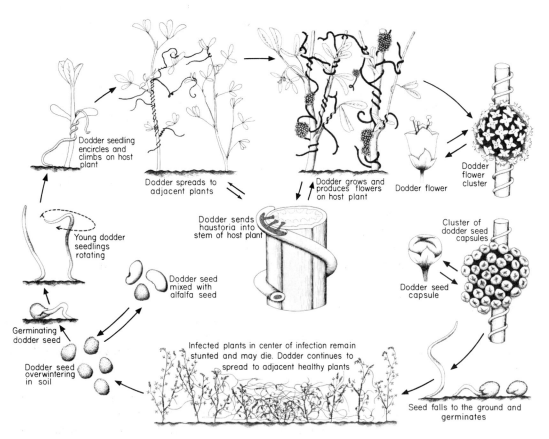

FIGURE 196.
Disease cycle of dodder (*Cuscuta* sp.) on alfalfa.

If the dodder stem comes in contact with a susceptible host, however, the stem immediately encircles the host plant, sends haustoria into it, and begins to climb the plant. The haustoria penetrate the stem or leaf and reach into the fibrovascular tissues. Foodstuffs and water are absorbed by the haustoria and are transported to the dodder stem where they are utilized for further growth and reproduction.

Soon after contact with the host is established, the base of the dodder shrivels and dries, so that it loses all connection with the ground and becomes completely dependent on the host for nutrients and water. The dodder continues to grow and expand at the expense of the host and the twisting tips of dodder reach out and attack adjacent plants. Thus the infection spreads from plant to plant and patches of infected plants are formed. The growth of infected plants is suppressed and they may finally die.

In the meantime, the dodder plant has developed flowers and produced seeds. These fall to the ground where they either germinate immediately or remain dormant until next season. The seed may be spread to new nearby areas by animals, water, equipment, etc., and over long distances it is distributed mixed with the crop seed.

Control. Dodder is best controlled by preventing its introduction into a field by the use of dodder-free seed, by cleaning equipment thoroughly before moving it from dodder-infested fields to new areas, and by limiting the movement of domestic animals from infested to dodder-free fields.

If dodder is already present in the field, scattered patches may be sprayed early in the season with contact herbicides such as diesel oil fortified with DNBP (4,6-dinitro-*o-sec*-butylphenol), PCP (pentachlorophenol), or 2,4-D. Such treatment, or cutting or burning of patches, kill both the dodder and the host plants, but prevent dodder from spreading and from producing seed. When dodder infestations are already widespread in a field, dodder can be controlled by frequent tillage, flaming, and use of soil herbicides such as CIPC or Cholor IPC (isopropyl *N*-(3-chlorophenyl) carbamate), DCPA or Dacthal (dimethyl 2,3,5,6-tetrachloroterephthalate), dichlobenil, or Casoron (2,6-dichlorobenzonitrile). These chemicals kill the dodder plant upon its germination from the seed but before it becomes attached to the host.

• Witchweed

Witchweed was known as a serious parasitic weed in Africa, Asia, and Australia before 1900. In 1956 the weed was discovered for the first time in America, in North and South Carolina. Due to effective federal and state quarantines the spread of the parasite has been largely limited to the area of the original infestations.

Witchweed parasitizes important economic plants such as corn, sugarcane, rice, tobacco, and some small grains. Witchweed causes its host plants to become stunted and chlorotic. Heavily infected plants usually wilt and die. Losses vary with the degree of infestation in a field and may range from slight to 100 percent.

Symptoms.   Affected plants develop symptoms resembling those produced by acute drought. The plants remain stunted, wilt, and turn yellowish. Death may follow these symptoms if the plants are heavily parasitized. Infected roots of host plants bear a large number of witchweed tentacles or haustoria which are attached to the root and feed upon it. One to several witchweed plants may be growing above ground next to the infected plants, although roots of many more witchweed plants, which do not survive to reach the surface, may parasitize the roots of the same host (Fig. 197).

The pathogen: Striga asiatica.   It is a small, pretty plant with bright green, slightly hairy stem and leaves. The weed grows 15 to 30 cm high. It produces multiple branches both near the ground and higher on the plant. The leaves are rather long and narrow in opposite pairs (Fig. 198).

The flowers are small and usually brick red or scarlet, although some may be yellowish-red, yellowish, or almost white, always having yellow centers. Flowers appear just above the leaf attachment to the stem and are produced throughout the season. After pollination seed pods or capsules develop, each containing more than a thousand tiny brown seeds. A single plant may produce from 50,000 to 500,000 seeds.

The root of witchweed is watery white in color and round in cross section. It has no root hairs, for it obtains all nutrients from the host plant through haustoria.

The life cycle of the parasite, from the time a seed germinates until the developing plant releases its first seeds, takes 90 to 120 days. Although after emergence the plant turns green and can probably manufacture some of its own food, it appears that it still continues to depend upon the host not only for all its water and minerals, but for organic substances as well.

FIGURE 197.
Witchweeds parasitizing roots of
corn plant. (Photo courtesy U.S.D.A.)

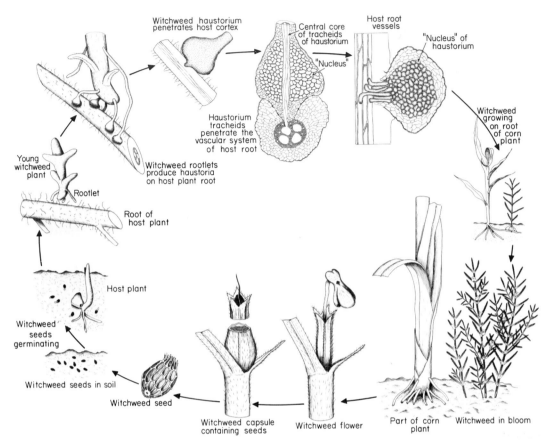

FIGURE 198.
Disease cycle of witchweed (*Striga asiatica*) on corn.

Development of disease.   The parasite overwinters as seeds, most of which generally require a rest period of 15 to 18 months before germination, although some can germinate without any dormancy. Seeds within a few millimeters from host roots germinate and grow toward these roots, probably in response to stimulants contained in the exudates of the host roots. As soon as the witchweed rootlet comes in contact with the host root, its tip swells into a conical or bulb-shaped haustorium which presses against the host root. The haustorium dissolves host cells through enzymic secretions and penetrates the host roots within 8 to 24 hours. The haustorium advances into the roots through dissolution of host cell walls. Finally its leading cells, usually tracheids, reach the vessels of the host roots (Fig. 198). The tracheids eventually dissolve the vessel walls or force their way into the vessel, from which they absorb water and nutrients. Although xylem vessels are present in the haustorium, no typical phloem cells develop, but cells in the "nucleus" of the haustorium seem to connect the phloem of host and parasite. It has been shown that the chlorophyll of witchweed plants is functional, but still manufactured foodstuffs move from the host plant into the parasite even when the latter is fully developed.

From the initial rootlet the weed produces more roots which move parallel to the roots of the host plant and send more haustoria into them. Furthermore, several hundred separate witchweed plants may parasitize the roots of a single host plant at once, although relatively few of these survive to reach the surface because the host plant cannot support so many.

The disease spreads in the field in a circular pattern. The circle of infected plants, however, increases year after year as the witchweed seeds spread in increasingly larger areas. The seeds are spread by wind, by water, by contaminated tools and equipment, or by contaminated soil carried on farm machinery.

Control.   Witchweed can be controlled by preventing its movement from infested areas into uninfested ones on transplants, agricultural products and machinery, or in any other way. Catch crops, consisting of host plants, may be planted to force the germination of witchweed seed, and the witchweed plants then can be destroyed by plowing under or by the use of weed killers such as 2,4-D. Trap crops, consisting mostly of nonhost legumes, may be used to stimulate germination of witchweed seeds, which, however, cannot infect the trap plants and therefore starve to death. Usually, a combination of the above methods is required to prevent witchweed plants from flowering and seeding, but with appropriate timing and perseverance an area can be completely freed from the parasite.

• Broomrapes

They are widely distributed in Europe, the U.S., Africa and Asia. They attack several hundred species of herbaceous crop plants including tobacco, potato, tomato, hemp, clover, and alfalfa. In some areas of the world, broomrapes cause losses varying from 10 to 70 percent of the crop.

Plants affected by broomrapes usually occur in small patches and may be stunted in various degrees depending on how early in their lives and by how many broomrapes the host plants were infected.

The broomrape pathogen, *Orobanche* sp., is a whitish to yellowish annual plant that may be 15 to 50 cm tall. It has a fleshy stem and scalelike leaves, and produces numerous pretty, white, yellow-white, or slightly purple, snapdragonlike flowers arising singly along the stem. The broomrapes produce ovoid seed pods about 5 mm long that contain several hundred minute seeds.

Broomrapes overwinter as seeds which may survive in the soil for more than ten years. Seeds germinate only when roots of certain plants grow near them although not all these plants are susceptible to the pathogen. Upon germination the seed produces a radicle which grows toward the root of the host plant, becomes attached to it and produces a shallow disk- or cuplike appressorium that surrounds the root, penetrates and absorbs nutrients and water from it. Soon the parasite begins to develop a stem which appears above the soil line and looks like an asparagus shoot. Meanwhile, the original root produces secondary roots that grow outward until they come in contact with other host roots to which they become attached and subsequently infect. From these points of contact, new roots and stems of the parasite are produced and result in the appearance of the typical clusters of broomrape plants arising from the soil around infected host plants. Several such broomrapes may be growing concurrently on the roots of the same host plant. The broomrape stems continue to grow and produce flowers and seeds which mature and are scattered over the ground in less than two months from the emergence of the stems.

Control of broomrapes depends on prevention of introduction of its seeds in new areas, planting of nonsusceptible crops in infested fields, frequent weeding and removal of broomrapes before they produce new seed, and, where feasible, soil fumigation with methyl bromide.

- *Dwarfmistletoes of Conifers*

Dwarfmistletoes occur in all parts of the world where conifer trees grow. In the U.S. they are more prevalent and most serious in the western half of the country, especially in the states along the Pacific coast, but they also cause appreciable losses in the northeastern and southeastern states.

The damage caused by dwarfmistletoes in coniferous forests is extensive although not always spectacular. Trees of any age may be retarded, deformed, or killed. Height growth of trees may be reduced by 50 to 80 percent. Timber quality is reduced by numerous large knots and by abnormally grained, spongy wood. Seedlings and saplings, as well as trees of certain species, are frequently killed by dwarfmistletoe infections.

Symptoms. Simple or branched shoots of dwarfmistletoe plants occur in tufts or scattered along the twigs of the host (Fig. 199). If the shoots have dropped off, small basal "cups" appear on the bark. Infected twigs and branches develop swellings and cankers on the infected areas. Cross sections at the swellings of infected branches reveal yellowish, wedge-shaped haustoria of the parasite which grow into the bark, cambium, and xylem of the branch. Large swellings or flattened cankers may

FIGURE 199.
Dwarfmistletoe on ponderosa pine. (A) Female plant (left) and a male plant on side limb. (B) Swellings and distortion of pine branches parasitized by dwarfmistletoe. (Photos courtesy U.S. Forest Service.)

also develop on the trunks of some infected trees. Infected branches often produce witches'-brooms. Tree stands with light or moderate infections are difficult to distinguish from healthy stands except for the presence of cankers, swellings, and brooms. Heavily infected stands, however, contain deformed, stunted, dying, and dead trees, or trees broken off at trunk cankers.

The pathogen: Arceuthobium sp. Some species produce shoots up to 10 cm long, while others are usually no more than 1.5 cm.

The mistletoe stems are yellowish to brownish-green or olive green. The shoots may be simple or branched and they are jointed. The leaves are inconspicuous, scalelike, in opposite pairs, and of the same color as the stem. Mistletoe plants also produce a complex, ramifying system of haustoria which consists of a longitudinal system of strands, external to and more or less parallel to the host cambium, and a radial system of "sinkers" produced by the former and oriented radially into the phloem and xylem.

The plants are either male or female, and produce flowers when they are 4 to 6 years old. After flowering the male shoots die; the female shoots die after the seeds are discharged. Fruits mature 5 to 16 months after pollination of the flowers. The fruit at maturity is a turgid, elliptical berry. On ripening, the fruit develops considerable internal pressure and, when disturbed, expels the seed upward or obliquely at lateral distances up to 15 meters. The seed is covered with a sticky substance and adheres to whatever it comes in contact with. This is the main means of spread of the parasite.

Development of disease. When a mistletoe seed lands on and becomes attached to the bark of a twig or a young branch of a susceptible host, it germinates and produces a germ tube or radicle. This grows along the bark surface until it meets a bud or a leaf base, at which point the radicle becomes broad and flattened on the side of the bark. A rootlike haustorium is then produced from the center of the flattened area of the radicle, which penetrates the bark directly and reaches the phloem and the cambium. From this haustorium develops the system of longitudinal strands and radial sinkers, all of which absorb from the host the nutrients needed for the development of the parasite (Fig. 200). The strands that reach the cambium of the host become permanently embedded in the wood as the latter is laid down each year, but they always retain their connections with the strands in the phloem. After the haustorial system is well established and developed in the host, it produces buds from which shoots develop the following year or several years later. The shoots first appear near the original point of infection, but later more shoots emerge in concentric zones of increasing diameter. The center of the infection usually deteriorates and becomes easily attacked by various decay-producing fungi. If witches'-brooms are produced on the affected area, the haustoria pervade all branches and produce mistletoe shoots along the proliferating host branches.

The parasite removes nutrients from the host and so starves and kills the portion of the branch lying beyond the point of infection. It also saps

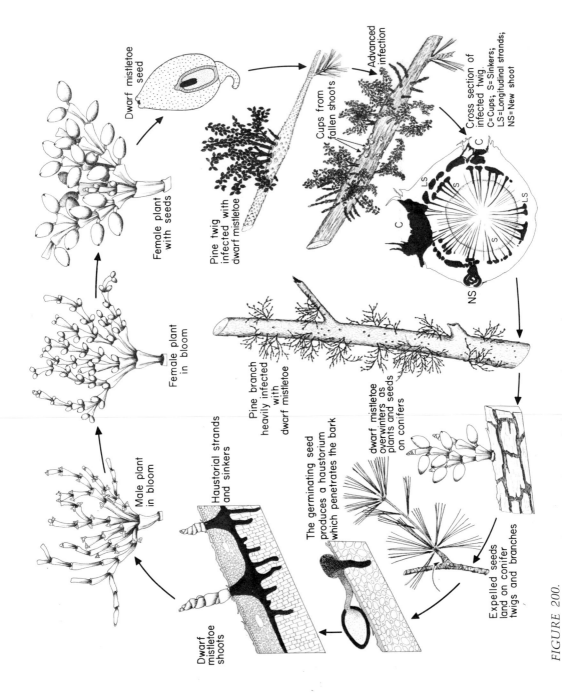

Dwarf mistletoe seed

Female plant with seeds

Female plant in bloom

Male plant in bloom

Pine twig infected with dwarf mistletoe

Cups from fallen shoots

Advanced infection

Cross section of infected twig.
C=Cups; S=Sinkers;
LS=Longitudinal strands;
NS=New shoot

C

LS

S

LS

C

NS

Pine branch heavily infected with dwarf mistletoe

Haustorial strands and sinkers

Dwarf mistletoe shoots

The germinating seed produces a haustorium which penetrates the bark

dwarf mistletoe overwinters as plants and seeds on conifers

Expelled seeds land on conifer twigs and branches

FIGURE 200.
Disease cycle of dwarfmistletoe (*Arceuthobium sp.*) on conifers.

the vitality of the branch and, when sufficiently abundant, of the whole tree. Furthermore, it upsets the balance of hormonal substances of the host in the infected area and causes hypertrophy and hyperplasia of the cells with resulting swellings and deformities of various shapes on the branches. This hormonal imbalance also stimulates the normally dormant lateral buds to excessive formation of shoots, forming a dense growth of abnormal appearance. Heavy dwarfmistletoe infections weaken trees and predispose them to wood-decaying and root pathogens, to beetles, and to windthrow and breakage.

Control. The only means of controlling dwarfmistletoes is by physical removal of the parasite. This is done either by pruning infected branches or by cutting and removing entire infected trees. Uninfected stands can be protected from the dwarfmistletoe infections by maintaining a protective zone free of the parasite between the diseased stand and the stand to be protected.

- *True or Leafy Mistletoes*

They occur throughout the world, particularly in warmer climates. They attack primarily hardwood forest and shade trees, but also many of the common fruit and plantation trees such as apple, cherry, citrus, rubber, cacao, and coffee, and even some gymnosperms such as juniper and cypress. They cause serious economic losses in some areas, although not nearly as severe as those caused by the dwarfmistletoes.

The symptoms are quite similar to those caused by dwarfmistletoes. Infected areas become swollen and produce witches'-brooms. The mistletoe plants sometimes are so numerous that they make up almost half of the green foliage of the tree and, in the winter, they make deciduous trees appear like evergreens with the normal tree branches appearing as though they have died back. Infected trees may survive for many years but they show reduced growth and portions of the tree beyond the mistletoe infection often become deformed and die.

The pathogens are *Phoradendron flavescens* in North America and *Viscum album* in Europe and the other continents. These mistletoes are parasitic evergreens that have well-developed leaves and stems less than 1 or 2 cm in diameter (Fig. 201). In some species of true mistletoes, however, the stems may be up to 30 cm or more in diameter. The height of plants varies from a few centimeters to a meter or more. The true mistletoes produce typical green leaves that can carry on photosynthesis, usually small, dioecious flowers, and berrylike fruits containing a single seed. They produce haustorial sinkers, rather than roots, however, which grow in branches and stems of trees and absorb water and nutrients from them.

True mistletoes are spread by birds that eat the seed-containing berries and excrete the sticky seeds in the tops of taller trees on which they like to perch. From that point on, infection, disease development, and control of true mistletoes are almost identical to those of dwarfmistletoes. Control in isolated shade or fruit trees can be obtained by pruning of infected branches or periodic removal of mistletoe stems from the branches or trunks.

FIGURE 201.
True or leafy mistletoe growing on branch of a hardwood tree.

SELECTED REFERENCES

Anonymous. 1957. Witchweed. *U.S. Dept. Agr., Agr. Res. Serv. Spec. Rept.* **22–41**: 17 p.

Dawson, J. H., W. O. Lee, and F. L. Timmons. 1969. Controlling dodder in alfalfa. *U.S. Dept. Agr. Farmer's Bull.* **2211**: 16 p.

Garman, H. 1903. Broom-rapes. *Ky. Agr. Exp. Sta. Bull.* **105**: 32 p.

Gill, L. S. 1935. *Arceuthobium* in the United States. *Conn. Acad. Arts Sci. Trans.* **32**:111–245.

Gill, S. L. 1953. Broomrapes, dodders and mistletoes, *in* "Plant Diseases." U.S.D.A. Yearbook, pp. 73–77.

Hansen, A. A. 1921. Dodder. *U.S. Dept. Agr. Farmer's Bull.* **1161**.

Hawksworth, F. G., and D. Wiens. 1970. Biology and taxonomy of the dwarf mistletoes. *Ann. Rev. Phytopathol.* **8**:187–208.

Kuijt, J. 1960. Morphological aspects of parasitism in the dwarf mistletoes (*Arceuthobium*). Univ. Calif. Publ. *Botany* **30**:337–436.

Kuijt, J. 1969. "The Biology of Parasitic Flowering Plants." Univ. of Calif. Press, Berkeley. 246 p.

Leonard, O. A. 1965. Translocation relationships in and between mistletoes and their hosts. *Hilgardia* **37**:115–153.

Rogers, W. E., and R. R. Nelson. 1962. Penetration and nutrition of *Striga asiatica*. *Phytopathology* **52**:1064–1070.

Saunders, A. R. 1933. Studies in phanerogamic parasitism with particular reference to *Striga lutea*. *Union South Africa Dept. Agr. Sci. Bull.* **128**.

Scharpf, R. F., and J. R. Parmeter, Jr. 1967. The biology and pathology of dwarfmistletoe *Arceuthobium campylopodium* f. *abietinum* parasitizing true firs (*Abbies* spp.) in California. *U.S. Dept. Agr., Forest Service Tech. Bull.* **1362**:42 p.

Scharpf, R. F., and F. G. Hawksworth. 1974. Mistletoes on hardwoods in the United States. *U.S.D.A. Forest Service, Forest Pest Leaflet* **147**: 7 p.

Thoday, Mary G. 1911. On the histological relations between *Cuscuta* and its host. *Ann. Bot.* **25**:655–682.

Wilhelm, S., *et al.* 1959. Large-scale fumigation against broomrape. *Phytopathology* **49**:530.

14
plant diseases caused by viruses

introduction

Viruses are entities that are too small to be seen with a light microscope, multiply only in living cells, and have the ability to cause disease. All viruses are parasitic in cells and cause a multitude of diseases to all forms of living organisms, from single-celled plants or animals to large trees and mammals. Some viruses attack man and/or animals and cause diseases such as influenza, polio, rabies, small pox, and warts; others attack plants; and still others attack microorganisms, e.g., bacteria and mycoplasmas. The total number of viruses known to date is well over a thousand, and new viruses are described almost every month. More than half of all known viruses attack and cause diseases of plants. One virus may infect one or dozens of different species of plants, and one plant may be attacked by one or many different viruses. A plant may also commonly be infected by more than one virus at the same time.

Although viruses are agents of disease and share with other living organisms genetic functions and the ability to reproduce, they also behave as chemical molecules. At their simplest, viruses consist of nucleic acid and protein, with the protein wrapped around the nucleic acid. Although viruses can take any of several forms, they are mostly either rod shaped or polyhedral, or variants of these two basic structures. There is always only RNA or only DNA in each virus and, in most plant viruses, only one kind of protein. Some of the larger viruses, however, may have several different proteins, each probably having a different function.

Viruses do not divide and do not produce any kind of specialized reproductive structures such as spores, but they multiply by inducing host cells to form more virus. Viruses cause disease not by consuming cells or killing them with toxins, but by upsetting the metabolism of the cells which, in turn, leads to the development by the cell of abnormal substances and conditions injurious to the functions and the life of the cell or the organism.

characteristics of plant viruses

Plant viruses differ greatly from all other plant pathogens not only in size and shape, but also in the simplicity of their chemical constitution and physical structure, methods of infection, multiplication, translocation within the host, dissemination, and the symptoms they produce on the host. Because of their small size and transparency of their bodies, viruses cannot even be viewed and detected by the methods used for other pathogens. Viruses are not cells, nor do they consist of cells.

DETECTION

When a plant disease is caused by a virus, individual virus particles cannot be seen with the light microscope, although some virus-containing inclusions or crystals may be seen in virus-infected cells. Examination of sections of cells or of crude sap from virus-infected plants under the electron microscope may or may not reveal viruslike particles. Virus particles are not always easy to find under the electron microscope, and even in the rare cases in which such particles are revealed, proof that the particles are a virus, and that this virus causes the particular disease, requires much additional work and time.

A few plant symptoms, such as oak-leaf patterns on leaves and chlorotic or necrotic ring spots, can be attributed to viruses with some degree of certainty. Most other symptoms caused by viruses resemble those caused by mutations, nutrient deficiencies or toxicities, insect secretions, by other pathogens, and other factors. The determination, therefore, that certain plant symptoms are caused by viruses involves the elimination of every other possible cause of the disease, and the transmission of the virus from diseased to healthy plants in a way that would exclude transmission of any of the other causal agents.

The present methods for detection of plant viruses involve primarily the transmission of the virus from a diseased to a healthy plant by budding, grafting, or by rubbing with plant sap. Certain other methods of transmission, such as by dodder or insect vectors, are also used to demonstrate the presence of a virus. Most of these methods, however, cannot distinguish whether the pathogen is a virus, mycoplasma or rickettsialike bacterium; only transmission through plant sap is presently considered as proof of the viral nature of the pathogen. The most definitive proof of the presence of a virus in a plant is provided by its purification, electron microscopy, and/or serology.

MORPHOLOGY

Plant viruses come in different shapes and sizes, but they are usually described as elongate (rigid rods or flexuous threads), as rhabdoviruses (bacilluslike), and as spherical (isometric or polyhedral) (Figs. 202, 203, and 204).

Some elongated viruses like tobacco mosaic virus and barley stripe mosaic virus, have the shape of rigid rods with measurements about 15 × 300 nm and 20 × 130 nm, respectively. Most of the elongated viruses appear as long, thin, flexible threads that are usually 10 to 13 nm wide and range in length from 480 nm (potato virus X) to 2000 nm (tristeza virus). Many of the elongated viruses seem to occur in particles of differing lengths, and the number given usually represents the length that is more common than any other.

The rhabdoviruses are short, bacilluslike rods, approximately three to five times as long as they are wide, as in the cases of potato yellow dwarf

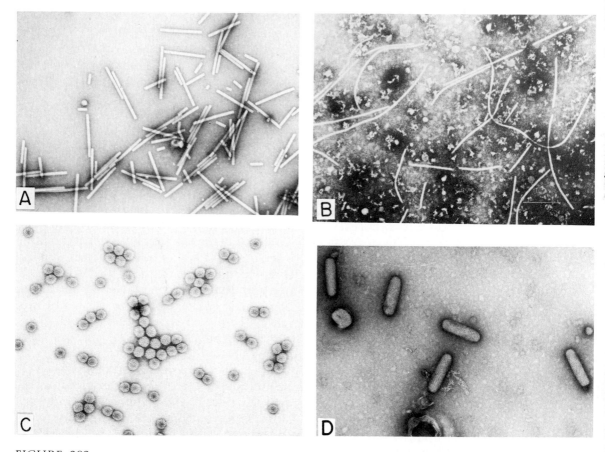

FIGURE 202.
Electron micrographs of the various shapes of plant viruses. (A) Rod shaped (tobacco mosaic). (B) Flexuous thread (maize dwarf mosaic). (C) Isometric (cowpea chlorotic mottle). (D) Rhabdovirus (broccoli necrotic yellows). (Photo D from Lin and Campbell, *Virology* **48**:30–40, 1972.)

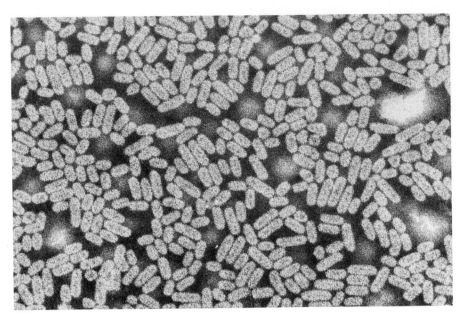

FIGURE 203.
Electron micrograph of alfalfa mosaic virus showing the various sizes of the five components of this virus. ×168,000. (Photo courtesy E. M. J. Jaspars, Univ. of Leiden, The Netherlands.)

virus which measures 75 × 380 nm, wheat striate mosaic virus (65 × 270 nm), and the lettuce necrotic yellows virus (52 × 300).

Most, and probably all, spherical viruses are actually polyhedral, ranging in diameter from about 17 nm (tobacco necrosis satellite virus) to 60 nm (wound tumor virus). Tomato spotted wilt virus seems to have a flexible, spherical shape 70 to 80 nm in diameter.

Many plant viruses consist of more than one component. Thus, tobacco rattle virus consists of two rods, a long one measuring 195 × 25 nm and a shorter one varying in length from 43 to 110 × 25 nm; alfalfa mosaic virus consists of five components measuring 58 × 18, 54 × 18, 42 × 18, 30 × 18, and 18 × 18 nm (Fig. 203). Also, many isometric viruses have two or three different components of, usually, the same size but different weights as they contain different amounts of nucleic acid. In all the above cases more than one of the components must be present in the plant for the virus to multiply and perform in its usual manner.

The surface of both the elongated and the spherical viruses consists of a definite number of protein subunits, which are spirally arranged in the elongated viruses, and packed on the sides of the polyhedral particles of the spherical viruses (Fig. 204). In cross sections, the elongated viruses appear as hollow tubes with the protein subunits forming the outer coat and the nucleic acid, also spirally arranged, embedded between the inner ends of two successive spirals of the protein subunits. The spherical viruses may or may not be hollow, the visible shell consisting of the protein subunits, with the nucleic acid inside the shell and arranged in a yet unknown manner.

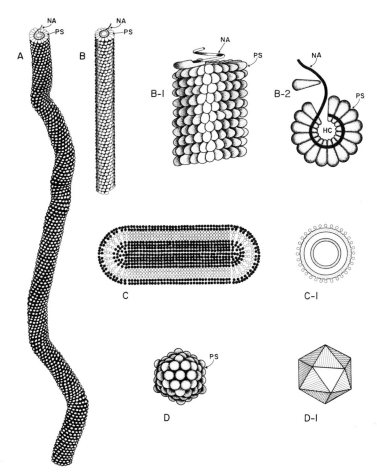

FIGURE 204.
Relative shapes, sizes, and structures of some representative plant viruses. (A) An elongate virus appearing as a flexuous thread. (B) A rigid rod-shaped virus. (B-1) Side arrangement of protein subunits (*PS*) and nucleic acid (*NA*) in viruses A and B. (B-2) Cross section view of the same viruses. *HC* = hollow core. (C) A short, bacillus-like virus. (C-1) Cross-section view of such a virus. (D) A polyhedral virus. (D-1) An icosahedron, representing the 20-sided symmetry of the protein subunits of the polyhedral virus.

The rhabdoviruses, potato yellow dwarf virus, lettuce necrotic yellows, etc. are provided with an outer envelope or membrane bearing surface projections. Inside the membrane is the nucleocapsid, consisting of helically arranged nucleic acid and associated protein subunits.

COMPOSITION AND STRUCTURE

Each plant virus consists of at least a nucleic acid and a protein. Some viruses consist of more than one size of nucleic acid and proteins, and some of them contain additional chemical compounds, such as polyamines, lipids, or specific enzymes.

The proportions of nucleic acid and protein vary with each virus, nucleic acid making up 5 to 40 percent of the virus and protein making up the remaining 60 to 95 percent. The lower nucleic acid and the higher protein percentages are found in the elongated viruses, while the spherical viruses contain higher percentages of nucleic acid and lower percentages of proteins. The total weight of the nucleoprotein of different virus particles varies from 4.6 million molecular weight units (bromegrass mosaic virus) to 39 million (tobacco mosaic virus) to 73 million (tobacco rattle virus). The weight of the nucleic acid alone, however, ranges only between 1 and 3 million ($1-3 \times 10^6$) molecular weight units per virus particle, compared to 0.5×10^9 for mycoplasmas, 1×10^9 for spiroplasmas, and more than 1.5×10^9 for bacteria.

COMPOSITION AND STRUCTURE OF VIRAL PROTEIN

Viral proteins, like all proteins, consist of amino acids. The sequence of amino acids within a protein is dictated by the genetic material, which in viruses is either deoxyribonucleic acid (DNA) or ribonucleic acid (RNA), and determines the nature of the protein.

The protein components of plant viruses are composed of repeating subunits. The amino acid content and sequence is constant for the identical protein subunits of a virus, but may vary for different viruses, different strains of the same virus, and even for different proteins of the same virus particle. The content and partial sequences of amino acids are known for the proteins of several viruses, but only for the protein of tobacco mosaic virus (TMV) and of turnip yellow mosaic virus (TYMV) is the complete sequence of amino acids known. Thus, the protein subunit of TMV consists of 158 amino acids in a constant sequence. Similarly, the protein subunit of TYMV has 189 amino acids.

In TMV the protein subunits are arranged in a helix containing 16⅓ subunits per turn (or 49 subunits per three turns). The central hole of the virus particle down the axis has a diameter of 40 Å, while the maximum diameter of the particle is 180 Å. Each TMV particle consists of approximately 130 helix turns of protein subunits. The nucleic acid is packed tightly between the helices of protein subunits. In the rhabdoviruses the helical nucleoproteins are enveloped in a membrane.

In the polyhedral plant viruses the protein subunits are tightly packed in arrangements that produce 20, or some multiple of 20, facets and form a shell. Within this shell the nucleic acid is folded or otherwise organized.

COMPOSITION AND STRUCTURE OF VIRAL NUCLEIC ACID

The nucleic acid of most plant viruses consists of RNA. To date only three plant viruses (cauliflower mosaic, dahlia mosaic, and carnation etched ring) have been shown to contain DNA. Both RNA and DNA are long, chainlike molecules consisting of hundreds or, more often, thousands of units called nucleotides. Each nucleotide consists of a ring compound called the base attached to a 5-carbon sugar [ribose (I) in RNA, deoxyribose (II) in DNA], which in turn is attached to phosphoric acid. The sugar of one nucleotide reacts with the phosphate of another nucleotide, and this is repeated many times, thus forming the RNA or DNA strand. In viral RNA, one of only four bases can be attached to each ribose molecule. These bases are adenine, guanine, cytosine, and uracil. The first two, adenine and guanine, are purines, while cytosine and uracil are pyrimidines. The chemical formulas of the bases and one of their possible relative positions in the RNA chain, are shown in structure (III). DNA is similar to RNA with two small, but very important differences: the oxygen of the sugar hydroxyl is missing; and the base uracil is replaced by the base methyluracil, better known as thymine (IV).

The sequence and the frequency of the bases on the RNA strand vary from one RNA to another, but they are fixed within a given RNA and determine its properties. RNA, whether in RNA viruses or in healthy cells, is usually found as single strands, although in several viruses it exists as double-stranded RNA. Of course, DNA exists always as a double-stranded helix, except in single-stranded DNA viruses.

the biological
function of viral components
—coding

Although apparently each virus produces its own distinct protein coat, the only known function of the protein is to provide a protective sheathing for the nucleic acid of the virus. Protein itself has no infectivity, although its presence generally increases the infectivity of the nucleic acid. In inoculations with intact virus particles (virions) the protein does not seem to assist or to affect the nucleic acid either in its functions or its composition, since inoculations with nucleic acid alone can cause infection and lead to synthesis of new nucleic acid and also of new protein, both being identical with those of the original virus. The synthesis, composition, and structure of the protein, on the other hand, depend entirely on the nucleic acid component which alone is responsible for the synthesis and assembly of both the RNA and the protein.

Infectivity of viruses in most cases is strictly the property of their nucleic acid, which in most plant viruses is RNA. Some viruses require and carry within them an RNA transcriptase enzyme in order to multiply and infect. The capability, however, of the viral RNA to reproduce both itself and its specific protein, indicates that the RNA carries the genetic determinants of the viral characteristics. The expression of each inherited characteristic depends on the sequence of nucleotides within a certain area (cistron) of the viral RNA which determines the sequence of amino acids in a particular protein, either structural or enzyme. This is called coding and seems to be identical in all living organisms and the viruses.

The code consists of coding units called codons. Each codon consists of three adjacent nucleotides and determines the position of a given amino acid.

The amount of RNA, then, contained in each virus indicates the approximate length of, and the number of nucleotides in, the viral RNA. This in turn determines the number of codons in each RNA and, therefore, the number of amino acids that can be coded for. Since the protein subunit of viruses contains relatively few amino acids (158 in TMV), the number of codons utilized for its synthesis is only a fraction of the total number of codons available (158 out of 2130 in TMV). The remaining codons are presumably involved in the synthesis of several other proteins, either structural or enzymes, and it is these proteins that are apparently responsible to a large extent for the diseased conditions produced in many virus infections of plants.

virus infection and virus synthesis

Plant viruses enter cells only through wounds made mechanically or by vectors, or by deposition into an ovule by an infected pollen grain.

The nucleic acid (RNA) of the virus is first freed from the protein coat. It then induces formation by the cell of enzymes called RNA-polymerases (= RNA-synthetases = RNA-replicases). These enzymes, in the presence of the viral RNA acting as a template and of the nucleotides that compose RNA, produce additional, RNA. The first new RNA produced is not the viral RNA but a strand that is a mirror image of the virus, and which, as it is formed, is temporarily connected to the viral strand (Fig. 205). Thus, the two form a double-stranded RNA that soon separates to produce the original virus RNA and the mirror image (−) strand, the latter then serving as a template for more virus (+ strand) RNA synthesis.

The replication of some single-stranded RNA viruses that have parts of their RNA in two or more virus particles, of some rhabdoviruses, and of some double-stranded RNA viruses differs considerably from the above. In viruses in which the different RNA segments are present within two or more virus particles, all or most of the particles must be present in the same cell for the virus to replicate and for infection to develop. In the single-stranded RNA rhabdoviruses the RNA is not infectious because it is the (−) strand. This RNA must be transcribed by the virus-carried enzyme called transciptase into a (+) strand RNA in the host and the latter RNA then replicates as above. In the double-stranded RNA isometric viruses, the RNA is segmented within the same virus, is noninfectious and depends for its replication in the host on a transciptase enzyme also carried within the virus.

As soon as new viral nucleic acid is produced it induces the host cell to produce the protein molecules that will be the protein subunits and that will form the protein coat of the virus. Apparently, only a part of the viral RNA strand is needed to participate in the formation of the viral protein. Since each amino acid on the protein subunit molecule is "coded" by three nucleotides of the viral RNA, for TMV, whose RNA consists of 6400 nucleotides and its protein of 158 amino acids, only 474 nucleotides are required to code the arrangement of the amino acids in the protein subunit.

Protein synthesis in healthy cells depends on the presence of amino acids and the cooperation of ribosomes, messenger RNA, and transfer RNAs. Each transfer RNA is specific for one amino acid which it carries toward and along the messenger RNA. Messenger RNA, which is produced in the nucleus and reflects part of the DNA code, determines the kind of protein that will be produced by coding the sequence in which the amino acids will be arranged. The ribosomes seem to travel along the messenger RNA and to provide the energy for the bonding of the prearranged amino acids to form the protein (Fig. 206).

For virus protein synthesis, the part of the viral RNA coding for the viral protein plays the role of messenger RNA. The virus utilizes the

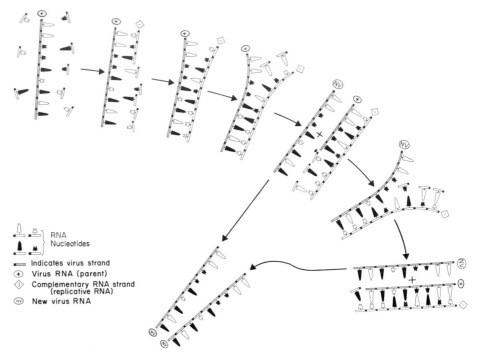

RNA
Nucleotides

▨ Indicates virus strand
⊕ Virus RNA (parent)
◇ Complementary RNA strand
   (replicative RNA)
Ⓝⓥ New virus RNA

FIGURE 205.
Hypothetical schematic representation of viral RNA replication.

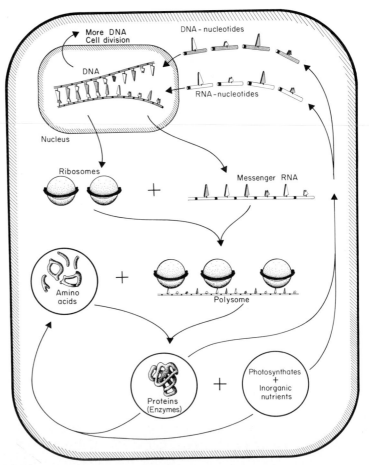

More DNA
Cell division

DNA - nucleotides

DNA

RNA - nucleotides

Nucleus

Ribosomes   +   Messenger RNA

Amino   +
acids   Polysome

Proteins   +   Photosynthates
(Enzymes)   +
                        Inorganic
                        nutrients

FIGURE 206.
Schematic representation of the basic functions in a living cell.

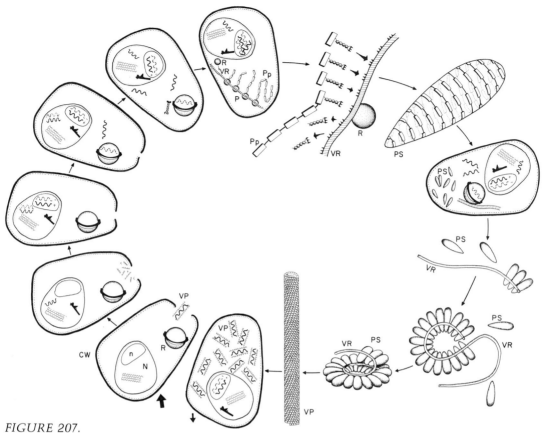

FIGURE 207.

Sequence of events in virus infection and biosynthesis. CW = cell wall, R = ribosome, N = nucleus, n = nucleolus, P = polyribosome (polysome), Pp = protein subunit, VP = viral particle. ▭ Amino acid, ◣▬ Viral RNA replicase, ∞∞∞E Transfer RNA, ∿∿ or VR Viral RNA.

amino acids, ribosomes, and transfer RNAs of the host, but it becomes its own blueprint (messenger RNA), and the protein formed is for exclusive use by the virus as a coat (Fig. 207) or other functions.

During virus synthesis, parts of its nucleic acid also become involved with synthesis of proteins other than the viral coat protein. Some of these proteins are enzymes, either of the kinds already present in the host cell or entirely new, which may activate or initiate in the cell chemical reactions that, in turn, may affect the physiological functions of the cell.

When new virus nucleic acid and virus protein subunits have been produced, the nucleic acid seems to organize the protein subunits around it, and the two are assembled together to form the complete virus particle, the virion.

The site or sites of the cell in which virus RNA and protein are synthesized and in which these two components are assembled to produce the virions have not yet been determined with absolute certainty. Studies with TMV suggest that the virus RNA, after it is freed from the protein coat, moves into the nucleus and possibly the nucleolus, where it replicates itself. The new virus RNA is then released into the cytoplasm, where it serves as a messenger RNA and, in cooperation with the ribosomes and transfer RNAs, produces the virus protein subunits. The assembly of virions follows, also in the cytoplasm. In other viruses, the synthesis of viral nucleic acid and protein, as well as their assembly into virions, seems to take place in the nucleus, from which the virus particles are then released into the cytoplasm.

The first intact virions appear in plant cells approximately 10 hours after inoculation. The virus particles may exist singly or in groups and may form amorphous or crystalline inclusion bodies within the cell areas (cytoplasm, nucleus, nucleolus) in which they happen to be.

translocation and distribution of viruses in plants

For infection of a plant by a virus to take place, the virus must move from one cell to another and must multiply in most, if not all, cells into which it moves. In their movement from cell to cell, viruses follow the pathways through the plasmodesmata connecting adjacent cells (Fig. 208). Viruses, however, do not seem to move through parenchyma cells unless they infect the cells and multiply in them, thus resulting in continuous

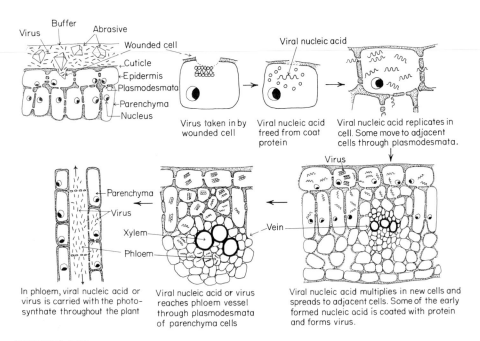

FIGURE 208.
Mechanical inoculation and early stages in the systemic distribution of viruses in plants.

and direct cell-to-cell invasion. In leaf parenchyma cells the virus moves approximately 1 mm, or 8 to 10 cells, per day.

Although some viruses appear to be more or less restricted to cell-to-cell movement through parenchyma cells, a large number of viruses are known to be rapidly transported over long distances through the phloem. Transport of viruses in the phloem apparently occurs in the sieve tubes, in which they can move as rapidly as 15 cm in the first 6 minutes. However, most viruses require 2 to 5 or more days to move out of an inoculated leaf. Once the virus has entered the phloem, it moves rapidly in the phloem toward growing regions (apical meristems) or other regions of food utilization in the plant, such as tubers and rhizomes (Fig. 209). For example, when potato virus is introduced into the basal leaves of young potato plants, it moves rapidly up the stem, but when plants already forming tubers are similarly inoculated, the virus does not move upward for more than 30 days while it moves downward into the tubers. Once in the phloem, the virus spreads systemically throughout the plant and reenters the parenchyma cells adjacent to the phloem through plasmodesmata.

The distribution of viruses within plants varies with the virus and the plant. The development of local lesion symptoms has been considered as an indication of the localization of the virus within the lesion area (Fig. 210); although this is probably true in some cases, in several diseases the lesions continue to enlarge and, sometimes, development of systemic symptoms follows, indicating that the virus continued to spread beyond the borders of the lesions.

In systemic virus infections, some phloem-translocated viruses seem to be limited to this tissue and to a few adjacent parenchyma cells. These include such diseases as potato leaf roll, cereal yellow dwarf, etc. Viruses causing mosaic-type diseases are not generally tissue-restricted, although there may be different patterns of localization. Mosaic virus-infected plant cells have been estimated to contain between 100,000 and 10,000,000 virus particles per cell. Systemic distribution of some viruses is quite thorough and may involve all living cells of a plant. Other viruses, however, seem to leave segments or gaps of tissues that are virus free. Some viruses invade newly produced apical meristematic tissues almost immediately, while in other cases growing points of stems or roots of affected plants apparently remain free of virus.

symptoms caused
by plant viruses

The most common and sometimes the only kind of symptoms produced is reduced growth rate of the plant, resulting in various degrees of dwarfing or stunting of the entire plant. Almost all viral diseases seem to cause some degree of reduction in total yield, and the length of life of virus-infected plants is usually shortened. These effects may be severe and easily noticeable, or they may be very slight and easily overlooked.

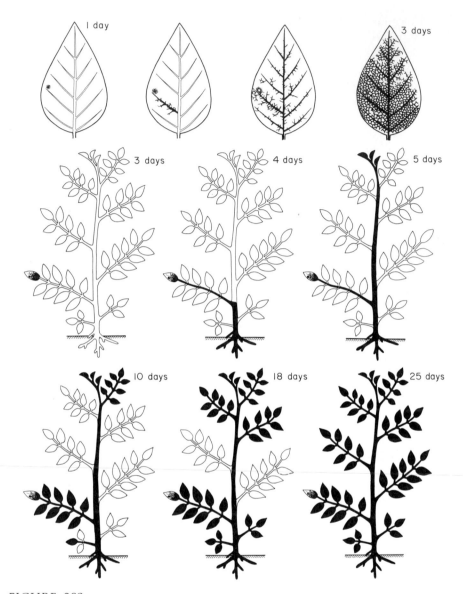

FIGURE 209.
Schematic representation of the direction and rate of translocation of a virus in a plant. (Adapted from G. Samuel (1934) *Ann. Appl. Biol.* **21**:90–111.)

The most obvious symptoms of virus-infected plants are usually those appearing on the foliage, but some viruses may cause striking symptoms on the stem, fruit, and roots, with or without symptom development on the leaves (Fig. 211A, B). In almost all virus diseases of plants occurring in the field, the virus is present throughout the plant (systemic infection) and the symptoms produced are called systemic symptoms. In many plants inoculated artificially with certain viruses, and probably in some natural infections, the virus causes the formation of small, usually necro-

FIGURE 210.
Local lesions caused by two strains of a virus (tobacco ringspot) on mechanically inoculated leaves (cowpea).

tic lesions only at the points of entry (local infections), and the symptoms are called local lesions. Many viruses may infect certain hosts without ever causing development of visible symptoms on them. Such viruses are usually called latent viruses, and the hosts are called symptomless carriers. In other cases, however, plants that usually develop symptoms upon infection with a certain virus may remain temporarily symptomless under certain environmental conditions (e.g., high or low temperature), and such symptoms are called masked. Finally, plants may show acute or severe symptoms soon after inoculation that may lead to death of the host; if the host survives the initial shock phase, the symptoms tend to become milder (chronic symptoms) in the subsequently developing parts of the plant, leading to partial or even total recovery. On the other hand, symptoms may progressively increase in severity and may result in gradual (slow) or quick decline of the plant.

The most common types of plant symptoms produced by systemic virus infections are mosaics and ringspots.

Mosaics, characterized by light-green, yellow, or white areas intermingled with the normal green of the leaves or fruit, or of whitish areas intermingled with areas of the normal color of flowers or fruit. Depending on the intensity or pattern of discolorations, mosaic-type symptoms may be described as mottling, streak, ring pattern, line pattern, veinclearing, veinbanding, chlorotic spotting, etc.

Ringspots, characterized by the appearance of chlorotic or necrotic rings on the leaves and sometimes also on the fruit and stem. In many ringspot diseases the symptoms, but not the virus, tend to disappear after onset and to reappear under certain environmental conditions.

A large number of other less common virus symptoms have been described (Fig. 211) and include stunt (e.g., tomato bushy stunt), dwarf

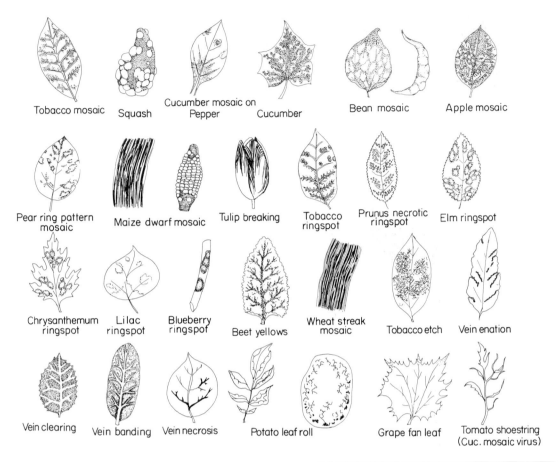

FIGURE 211.
Kinds of symptoms caused by viruses in plants.

(e.g., barley yellow dwarf), leaf roll (e.g., potato leaf roll), yellows (e.g., beet yellows), streak (e.g., tobacco streak), pox (e.g., plum pox), enation (e.g., pea enation mosaic), tumors (e.g., wound tumor), pitting of stem (e.g., apple stem pitting), pitting of fruit (e.g., pear stony pit), and flattening and distortion of stem (e.g., apple flat limb). These symptoms may be accompanied by other symptoms on other parts of the same plant.

physiology of virus-infected plants

Plant viruses do not contain any enzymes, toxins, or other substances considered to be involved in the pathogenicity of other types of pathogens, and yet cause a variety of symptoms on the host. The viral nucleic

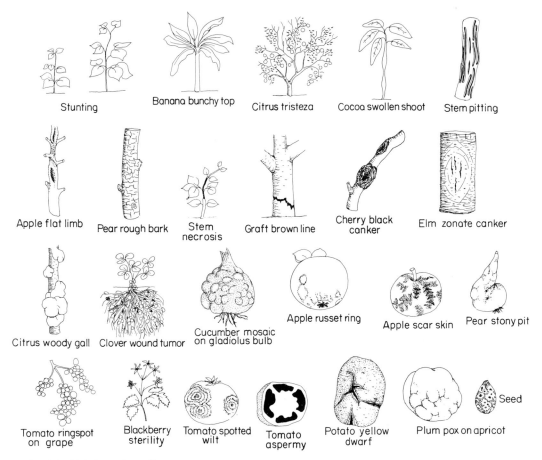

Stunting Banana bunchy top Citrus tristeza Cocoa swollen shoot Stem pitting

Apple flat limb Pear rough bark Stem necrosis Graft brown line Cherry black canker Elm zonate canker

Citrus woody gall Clover wound tumor Cucumber mosaic on gladiolus bulb Apple russet ring Apple scar skin Pear stony pit

Tomato ringspot on grape Blackberry sterility Tomato spotted wilt Tomato aspermy Potato yellow dwarf Plum pox on apricot Seed

FIGURE 211 (continued)

acid (RNA) seems to be the only determinant of disease, but the mere presence of RNA or complete virus in a plant, even in large quantities, does not seem to be sufficient reason for the disease syndrome, since some plants containing much higher concentrations of virus than others may show milder symptoms than the latter or they may even be symptomless carriers. This indicates that viral diseases of plants are not due primarily to depletion of nutrients that have been diverted toward synthesis of the virus itself, but to other more indirect effects of the virus on the metabolism of the host. These effects are brought about probably through the virus-induced synthesis of new proteins by the host, some of which are biologically active substances (enzymes, toxins, etc.) and can interfere with the normal metabolism of the host.

Viruses generally cause a decrease in photosynthesis through a decrease in chlorophyll per leaf, in chlorophyll efficiency, and in leaf area per plant. Viruses usually cause a decrease in the amount of growth-regulating substances (hormones) in the plant, frequently by inducing an increase in growth-inhibiting substances. A decrease in soluble nitrogen during rapid virus synthesis is rather common in virus diseases of plants, and in the mosaic diseases there is a chronic decrease in the levels of carbohydrates in the plant tissues.

Respiration of plants is generally increased immediately after infection with a virus, but after the initial increase the respiration of plants infected with some viruses remains higher, while with other viruses it becomes lower than that of healthy plants, and with still other viruses it may return to normal.

The amounts of nonvirus nitrogenous compounds in diseased plants seem to be generally lower than those found in healthy plants, probably because the virus, which in some virus–host systems may account for 33 to 65 percent of the total nitrogen in the plant, is formed at the expense of the normal levels of nitrogenous compounds in the plant. When the plant, however, is provided with high nitrogen nutrition, the amount of total nitrogen in diseased plants may be higher than that in healthy plants, especially after completion of the phase of rapid virus synthesis.

It appears, therefore, that many of the functional systems of the plant are directly or indirectly affected by virus infection. Certain degrees or types of such metabolic derangements can probably be tolerated by the plant and do not cause any symptoms, while others probably have a deleterious effect on the host and contribute to symptom development. The effects of virus on nitrogenous compounds, on growth regulators, and on phenolics, have often been considered to be the immediate causes of various types of symptoms, since the first two are so profoundly involved in anything concerned with plant growth and differentiation, and since the oxidized products of phenolics may themselves, because of their toxicity, be responsible for the development of certain kinds of necrotic symptoms.

transmission
of plant viruses

Plant viruses rarely, if ever, come out of the plant spontaneously. For this reason, viruses are not disseminated as such by wind or water, and even when they are carried in plant sap or debris they generally do not cause infections unless they come in contact with the contents of a wounded living cell. Viruses, however, are transmitted from plant to plant in a number of ways such as vegetative propagation; mechanically through sap; and by seed, pollen, insects, mites, nematodes, dodder, and fungi.

TRANSMISSION OF
VIRUSES BY VEGETATIVE
PROPAGATION

Whenever plants are propagated vegetatively by budding or grafting, by cuttings, or by the use of tubers, corms, bulbs, or rhizomes, any viruses present in the mother plant from which these organs are taken will almost always be transmitted to the progeny (Fig. 212). Considering that almost all fruit and many ornamental trees and shrubs are propagated by budding, grafting, or cuttings, and that many field crops, e.g., potatoes, and most florist's crops are propagated by tubers, corms, or cuttings, this means of transmission of viruses is the most important for all these types of crop plants. Transmission of viruses by vegetative propagation not only makes the new plants diseased, but in the cases of propagation by budding or grafting, the presence of a virus in the bud or graft may result in appreciable reduction of successful bud or graft unions with the rootstock and, therefore, in poor stands.

Transmission of viruses may also occur through natural root grafts of adjacent plants, particularly trees, the roots of which are often intermingled and in contact with each other. For several tree viruses, natural root grafts are the only known means of tree-to-tree spread of the virus within established orchards.

MECHANICAL TRANSMISSION OF VIRUSES THROUGH SAP

Mechanical transmission of plant viruses in nature by direct transfer of sap through contact of one plant with another is uncommon and relatively unimportant. Such transmission may take place between closely

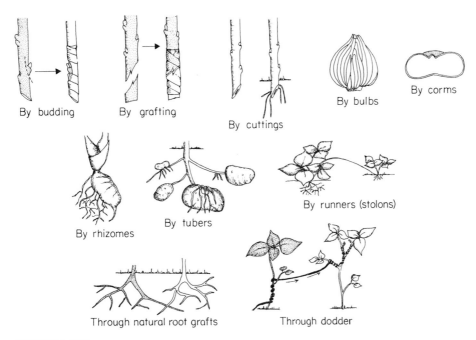

FIGURE 212.
Transmission of viruses, mycoplasmalike organisms, and other pathogens by vegetative propagation, natural root grafts, and through dodder.

spaced plants following a strong wind that could cause the leaves of adjacent plants to rub together and, if wounded, to exchange some of their sap, and thus transmit any virus present in the sap (Figs. 213 and 214). Potato virus X (PVX) seems to be one of the viruses most easily transmitted that way. When plants are wounded by man during cultural practices in the field or greenhouse and some of the virus-infected sap adhering to the tools, hands, or clothes is accidentally transferred to subsequently wounded plants, virus transmission through sap may be rapid and widespread and, as in the case of TMV on tobacco and tomato, may result in very serious losses. Virus-infected sap transferred from plant to plant on the mouthparts or body of animals feeding on and moving among the plants may on rare occasions lead to virus transmission.

The greatest importance of mechanical transmission of plant viruses stems from its indispensability in studying almost every facet of the viruses that cause plant diseases, since all investigations of virus outside the host are dependent on the ability to demonstrate and measure the infectiousness of the material.

For mechanical transmission of a virus from one plant to another, tissues of the infected plant believed to contain a high concentration of the virus, i.e., young leaves and flower petals, are ground with a mortar and pestle or with some other grinder (Fig. 213). Breakage of the cells

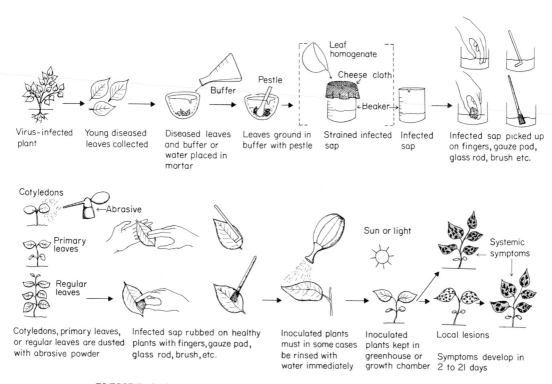

FIGURE 213.
Typical mechanical or sap transmission of plant viruses.

results in release of the virus in the sap. Sometimes a buffer solution, usually phosphate buffer, is added for stabilization of the virus. The expressed sap is then strained through cheesecloth and is centrifuged at low speeds to remove tissue fragments, or at alternate low and high speeds if concentration or purification of the virus is desired. The crude or partially purified sap is then applied to the surface of leaves of young plants which have been previously dusted with an abrasive such as 600-mesh Carborundum added to aid in wounding of the cells. Application of the sap is usually made by gently rubbing the leaves with a cheesecloth or gauze pad dipped in the sap, with the finger, a glass spatula, a painter's brush, or with a small sprayer. In successful inoculations, the virus enters the leaf cells through the wounds made by the abrasive or through broken leaf hairs and initiates new infections. In local-lesion hosts, symptoms usually appear within 3 to 7 or more days, and the number of local lesions is proportional to the concentration of the virus in the sap. In systemically infected hosts, symptoms usually take 10 to 14 or more days to develop. Sometimes the same plants may first develop local lesions and then systemic symptoms. In mechanical transmission of viruses, the taxonomic relationship of the donor and receiving (indicator) plants is unimportant, since virus from one kind of plant, whether herbaceous or a tree, may be transmitted to dozens of unrelated herbaceous plants (vegetables, flowers, or weeds).

Although viruses are almost always transmitted by budding or grafting, several viruses, especially of woody plants, have not yet been transmitted mechnically. The possible reasons for this failure seem to be that some viruses are not present in high enough concentration in the donor plant, they are unstable in sap or are quickly inactivated by inhibitory substances released or formed upon grinding of the cells, and also because some viruses, e.g., those causing yellows-type diseases, apparently require that they be introduced into specific tissues (phloem) if they are to cause infection.

SEED TRANSMISSION

About one hundred viruses have been reported to be transmitted by seed. As a rule, however, only a small portion (1 to 30 percent) of the seeds derived from virus-infected plants transmit the virus, and the frequency varies with the host–virus combination (Fig. 214). In a few cases, e.g., tobacco ringspot virus in soybean, the virus may be transmitted by almost 100 percent of the seeds of infected plants, and in others, seed transmission may be quite high, e.g., 28 to 94 percent in squash mosaic virus in muskmelon, 50 to 100 percent in barley stripe mosaic virus in barley. Even within a species, however, different varieties or plants inoculated at different stages of their growth may vary in the percentages of their seeds that transmit the virus.

In most seed-transmitted viruses, the virus seems to come primarily from the ovule of infected plants, but several cases are known in which the virus in the seed seems to be just as often derived from the pollen that fertilized the flower.

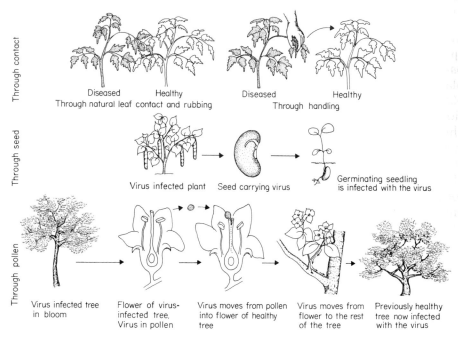

FIGURE 214.
Virus transmission through direct contact, handling, seed, and pollen.

POLLEN TRANSMISSION

Virus transmitted by pollen may infect not only the seed and the seedling that will grow from it, but more important, it can spread through the fertilized flower and down into the mother plant, which thus becomes infected with the virus (Fig. 214). Such plant-to-plant transmission of virus through pollen is known to occur, for example, in stone fruit ringspot virus in sour cherry.

Although pollination of flowers with virus-infected pollen may result in considerably lower fruit set than is produced with virus-free pollen, transmission of pollen-carried virus from plant-to-plant is apparently quite rare or it occurs with only a few of the viruses.

INSECT TRANSMISSION

Undoubtedly the most common and economically most important means of transmission of viruses in the field is by insect vectors. Members of relatively few groups of insects, however, can transmit plant viruses (Fig. 215). The order Homoptera, which includes both aphids

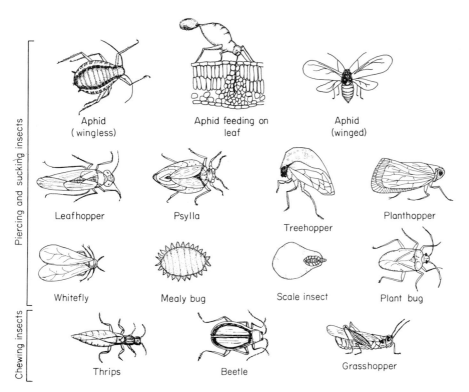

FIGURE 215.
Insect vectors of plant viruses. Insects in second row also transmit mycoplasmas and rickettsialike bacteria.

(Aphidae) and leafhoppers (Cicadellidae or Jassidae), contains by far the largest number and the most important insect vectors of plant viruses. Certain species of several other families of the same order also transmit plant viruses, but neither their numbers nor their importance compare with the Aphidae and Cicadellidae. Among these families are the white flies (Aleurodidae), the mealy bugs and scale insects (Coccoidae), and the treehoppers (Membracidae). A few insect vectors of plant viruses belong to other orders, such as the true bugs (Hemiptera), the thrips (Thysanoptera), the beetles (Coleoptera), and the grasshoppers (Orthoptera). The most important virus vectors, i.e., aphids, leafhoppers, and the other groups of Homoptera, as well as the true bugs, have piercing and sucking mouthparts; all the other groups of insect vectors have chewing mouthparts and virus transmission by the latter is much less common.

Insects with sucking mouthparts carry plant viruses on their stylets (style-borne or nonpersistent viruses) or they accumulate the virus internally and, after passage of the virus through the insect tissues, they introduce the virus into plants again through their mouthparts (circulative or persistent viruses). Some circulative viruses may multiply in their respective vectors and are then called propagative viruses. Viruses transmitted by insects with chewing mouthparts may also be circulative or they may be carried on the mouthparts.

Aphids are the most important insect vectors of plant viruses and transmit the great majority of all stylet-borne viruses (Fig. 216). As a rule several aphid species can transmit the same stylet-borne virus and the same aphid species can transmit several viruses, but in many cases the vector–virus relationship is quite specific. Aphids generally acquire the stylet-borne virus after feeding on a diseased plant for only a few seconds (30 seconds or less) and can transmit the virus after transfer to and feeding on a healthy plant for a similarly short time of a few seconds. The length of time aphids remain viruliferous after acquisition of a stylet-borne virus varies from a few minutes to several hours, after which they can no longer transmit the virus. In the few cases of aphid transmission of circulative viruses, aphids cannot transmit the virus immediately but must wait several hours after the acquisition feeding, but once they start to transmit the virus, they continue to do so for many days following the removal of the insects from the virus source. In aphids transmitting stylet-borne viruses, the virus seems to be borne on the tips of the stylets, it is easily lost through the scouring that occurs during probing of host cells, and it does not persist through the molt or egg.

At least 10 plant viruses are transmitted by leafhoppers, including viruses with double-stranded RNA, bacilliform viruses, and small isometric viruses.

Leafhopper-transmitted viruses cause disturbances in plants that arise primarily in the region of the phloem. All leafhopper-transmitted viruses are circulatory, several are known to multiply in the vector (propagative), and some persist through the molt and are transmitted to a greater or

FIGURE 216.
A winged aphid vector of a plant virus (barley yellow dwarf) sucking up juices, and possibly virus, from an oat stem. (Photo courtesy Dept. Plant Pathol., Cornell Univ.)

lesser degree through the egg stage of the vector. Most leafhopper vectors require a feeding period of one to several days before they become viruliferous, but once they have acquired the virus they may remain viruliferous for the rest of their lives. There is usually an incubation period of 1 to 2 weeks between the time a leafhopper acquires a virus and the time it can transmit it for the first time.

MITE TRANSMISSION

Mites of the family Eriophyidae have been shown to transmit nine viruses, including agropyron mosaic, currant reversion, wheat streak mosaic, peach mosaic, and fig mosaic viruses. These mites have piercing and sucking mouthparts (Fig. 217). Virus transmission by eriophyid mites seems to be quite specific, since each of these mites has a restricted host range and is the only known vector for the virus or viruses it transmits. Some of the mite-transmitted viruses are stylet borne, while others are circulatory and, of the latter, at least one persists through the molts.

In addition to the eriophyid mites, one mite of the family Tetranychidae (spider mites) has also been known to transmit a plant virus, potato virus Y.

NEMATODE TRANSMISSION

Approximately one dozen plant viruses have been shown to be transmitted by one or more species of three genera of soil-inhabiting, ectoparasitic nematodes (Fig. 217). Nematodes of the genera *Longidorus* and

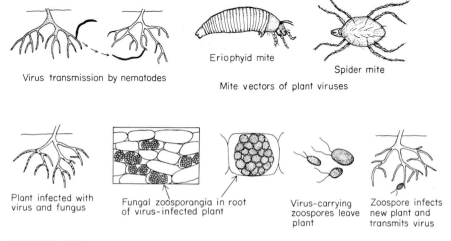

Virus transmission by nematodes

Eriophyid mite

Spider mite

Mite vectors of plant viruses

Plant infected with virus and fungus

Fungal zoosporangia in root of virus-infected plant

Virus-carrying zoospores leave plant

Zoospore infects new plant and transmits virus

FIGURE 217.
Transmission of plant viruses by nematodes, mites, and fungi.

Xiphinema are vectors of polyhedral-shaped viruses such as tobacco ringspot, tomato ringspot, raspberry ringspot, tomato black ring, cherry leaf roll, brome mosaic, grape fanleaf, and other viruses, while nematodes of the genus *Trichodorus* transmit two rod-shaped viruses, tobacco rattle, and pea early browning viruses. Nematode vectors transmit viruses by feeding on roots of infected plants and then moving on to roots of healthy plants. Larvae as well as adult nematodes can acquire and transmit viruses, but the virus is not carried through the larval molts or through the eggs, and after molting, the larvae or the resulting adults must feed on a virus source before they can transmit again.

FUNGUS TRANSMISSION

The root-infecting fungus *Olpidium* transmits at least four plant viruses, tobacco necrosis, cucumber necrosis, lettuce big vein, and tobacco stunt viruses. Four other fungi, *Synchytrium*, *Polymyxa*, *Spongospora*, and *Pythium*, transmit respectively, potato virus X, wheat mosaic virus, potato mop top virus, and beet necrotic yellow vein virus, and pea false leaf roll. Some of these viruses apparently are borne internally in, and others on, the resting spores and the zoospores, which upon infection of new host plants introduce the virus and cause symptoms characteristic of the virus they transmit (Fig. 217).

DODDER TRANSMISSION

Several plant viruses can be transmitted from one plant to another through the bridge formed between the two plants by the twining stems of the parasitic plant dodder (*Cuscuta* sp.)(Fig. 212). A large number of viruses have been transmitted in this way, frequently between plants belonging to families widely separated taxonomically. The virus is usually transmitted passively in the food stream of the dodder plant, being acquired from the vascular bundles of the infected plant by the haustoria of dodder and, after translocation through the dodder phloem, it is introduced in the next plant by the new dodder haustoria produced in contact with the vascular bundles of the inoculated plant.

purification
of plant viruses

Isolation or, as it is usually called, purification of viruses is most commonly obtained by ultracentrifugation of the plant sap. This involves 3 to 5 cycles of alternate high (40,000 to 100,000 g or more) and low (3000 to 10,000 g) speeds. Ultracentrifugation concentrates the virus and separates it from host cell contaminants. Several modifications of the ultracentrifugation technique, particularly density-gradient centrifugation, are

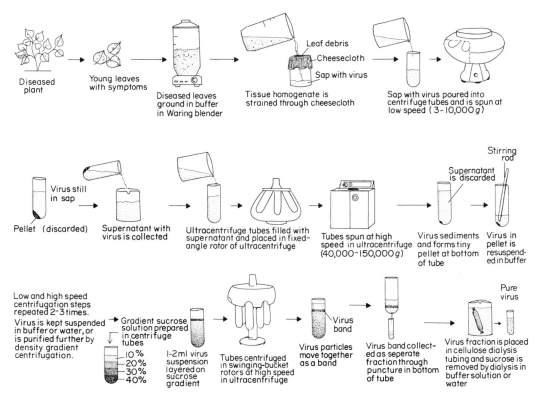

FIGURE 218.
Steps in the purification of plant viruses.

presently employed in virus purification with excellent results (Fig. 218). In all these methods, the virus is finally obtained as a colorless pellet in a test tube and may be used for infections, electron microscopy and serology.

serology of plant viruses

When a virus protein or any other foreign protein (antigen) is injected into a mammal (rabbit, mouse, horse), or bird (chicken, turkey), it results in the appearance of substances (antibodies) in the blood serum which react specifically with the antigen injected. The virus and its antibody are brought together in several ways, the most common being the precipitin reaction. In this, the antibodies and antigens are mixed in solution (precipitin test), or they meet at the interface between two solutions containing each separately (ring interface test), or they diffuse toward each other through an agar gel and meet in a zone in suitable concentrations (Ouchterlony test). Sometimes the antigen is absorbed on the surface of a large particle such as a cell or plastid and these are precipitated by addition of antibodies (agglutination reaction). In all these tests the reac-

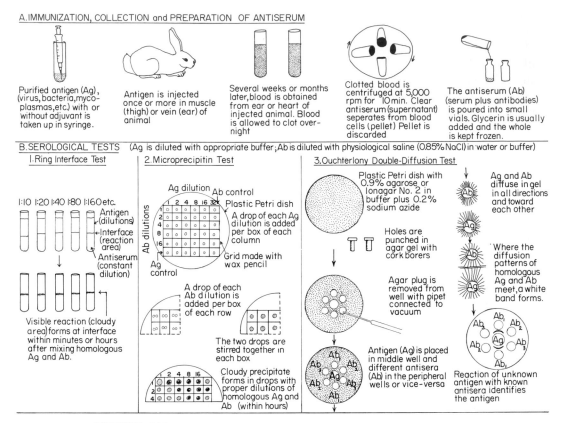

FIGURE 219.
Production of antisera and serological tests for identification of unknown pathogens.

tion of antigen and antibody becomes visible either by precipitation of the two on the bottom of the test tube or by formation of a band at the interface where the two meet (Fig. 219).

The uses of plant virus serology are numerous. Thus, it is used to determine relationships between viruses, to identify a virus causing a plant disease, to detect virus in foundation stocks of plants, and to detect symptomless virus infections. It can also be used to measure virus quantitatively, to locate the virus within a cell or tissue, to detect plant viruses in insects, and to purify a virus.

nomenclature and classification of plant viruses

Naming of plant viruses usually has been based on the most conspicuous symptom they cause on the first host they have been studied in. Thus, a virus causing a mosaic on tobacco is called tobacco mosaic virus, while the disease itself is called tobacco mosaic; another virus causing ringspot symptoms on tomato is called tomato ringspot virus, and the disease is called tomato ringspot, and so forth. Considering, however, the variability of symptoms caused by the same virus on the same host plant under different environmental conditions, by different strains of a virus on the same host, or by the same virus on different hosts, it becomes apparent that this system of nomenclature leaves much to be desired.

All viruses belong to the kingdom VIRA. Within the kingdom there are two virus divisions, DNA viruses and RNA viruses, depending on whether the nucleic acid of the virus is DNA or RNA. Viruses within each division are either helical or cubical (polyhedral). Within each subdivision there may be viruses possessing one or two strands of RNA or DNA, possessing or lacking a membrane around the protein coat, containing or lacking certain substances, having certain symmetry of helix in the helical viruses or number of subunits in the cubical (polyhedral) viruses, size of the virus, and, finally, any other physical, chemical, or biological properties.

In many plant diseases assumed to be caused by viruses, no virus has yet been observed and it is possible that some of these diseases will be proven later to be caused by pathogens other than viruses or by as yet uncharacterized viruses. For those plant diseases, however, proven to be caused by viruses, a system of nomenclature and classification has been proposed (Fig. 220), in which the viruses are grouped according to the above-listed criteria, according to size and several additional criteria unique to plant viruses. The groups are named after a typical virus in the group and are accompanied by a cryptogram indicating whether the virus contains RNA (R) or DNA(D), single (1) or double stranded (2); the percent RNA or DNA and the molecular weight of the virus; whether the virus and its nucleocapsid are elongated (E) or spherical (S); and the kind of host(s) (S = plants) and of vector, if any.

identification of plant viruses

Once the cause of a disease has been established as a virus, a series of tests may be necessary to determine its identity. The host range of the virus, i.e., the hosts on which the virus induces symptoms and the kinds of symptoms produced, may help to differentiate this virus from several others. Transmission studies should indicate whether the virus is transmitted mechanically and to what hosts, or by insects and which insects, and so on; each new property discovered helping to further characterize the virus. If the virus is transmitted mechnically, certain properties of the virus such as its thermal inactivation point, i.e., the temperature required for complete inactivation of the virus in untreated crude juice during a 10-minute exposure, its longevity *in vitro,* its dilution end point, i.e., the highest dilution of the juice at which the virus can still cause infection, may be used to narrow the possibilities to just a few viruses. If, at this stage, the identity of the virus is suspected, serological tests may be used and if they are positive, a tentative identification may be made. Examination of the virus in the electron microscope, and inoculation of certain

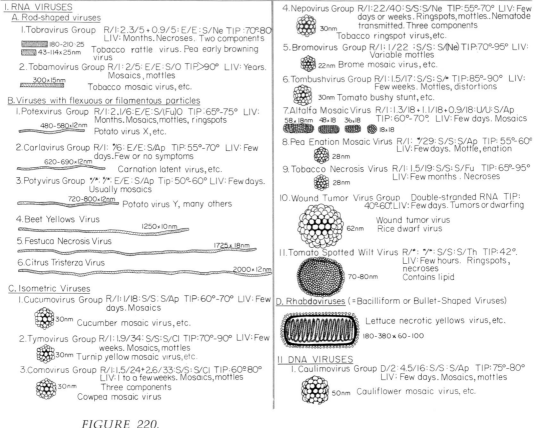

FIGURE 220.
Groups of plant viruses.

plant species, is also usually sufficient for a tentative identification of the virus.

In the viruslike diseases of woody (and other) plants in which no pathogens have been observed so far, identification of the pathogens, which are at present presumed to be viruses, is made strictly by indexing. The latter involves inoculation by grafting, etc. (Fig. 221) of certain plant species or varieties called indicators; these are sensitive to specific viruses and upon inoculation with these viruses develop characteristic symptoms and vice versa, i.e., development of the characteristic symptoms by an indicator identifies the virus with which the indicator was inoculated.

economic importance of plant viruses

Viruses attack all forms of plant life, from mycoplasmas, bacteria, fungi, and algae to herbaceous plants and trees.

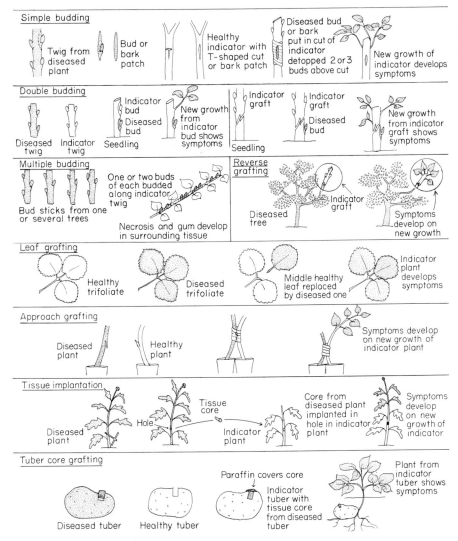

FIGURE 221.
Indexing for viral, mycoplasmal, and rickettsialike diseases.

Plant virus diseases may damage leaves, stems, roots, fruits, seed, or flowers and may cause economic losses by reduction in yield and quality of plant products. Losses may be catastrophic or they may be very mild and insignificant. On a nationwide basis, viruses account for a considerable portion of the losses suffered annually from diseases by the various crops.

The severity of individual virus diseases may vary with the locality, the crop variety, and from one season to the next. Some virus diseases have destroyed entire plantings of certain crops in some areas, e.g., plum pox, hoja blanca or rice, sugar beet yellows, and citrus tristeza. Most virus diseases, however, occur year after year on crops on which they cause small and unspectacular losses, sometimes without even inducing any visible symptoms. For example, potato virus X, which used to be present in all potato plants grown in the U.S., reduces yields by about 10 percent although the potato plants show no obvious symptoms in the field.

control of plant viruses

The best way to control a virus disease is to keep it out of an area through quarantine, inspection, and certification systems. The existence of symptomless hosts, the incubation period following inoculation, and the absence of obvious symptoms in seeds, tubers, bulbs, and nursery stock make quarantine sometimes ineffective. Eradication of diseased plants to eliminate inoculum from the field may, in some cases, help to control the disease. Plants may be protected against certain viruses by protecting them against the virus vectors. Controlling the insect vectors and removing weeds which serve as hosts may help in controlling the disease.

The use of virus-free seed, tubers, budwood, etc. is the single most important measure for avoiding virus diseases of many crops, especially those lacking insect vectors. Periodic indexing of the mother plants producing such propagative organs is necessary to ascertain their continuous freedom from viruses. Several types of inspection and certification programs are now in effect in various states producing seeds, tubers, and nursery stock used for propagation.

Although health or vigor of host plants confers no resistance or immunity to virus disease, breeding plants for hereditary resistance to virus is of great importance, and many plant varieties resistant to certain virus diseases have already been produced.

Once inside a plant some viruses can be inactivated by heat. Dormant, propagative organs are usually dipped in hot water (35 to 54°C) for a few minutes or hours, while actively growing plants are usually kept in greenhouses or growth chambers at 35 to 40°C for several days, weeks, or months, after which the virus in some of them is inactivated and the plants are completely healthy. Plants free of virus may also be produced from virus-infected ones by culture of short (0.1 mm to 1 cm or more) tips of apical and root meristems.

No chemical substances (viricides) are yet available for controlling virus diseases of plants. Foliar application of certain growth-regulating substances, such as gibberellic acid, has been effective in stimulating growth of the virus-suppressed axillary buds in sour cherry yellows, resulting in increased fruit production. Similarly, sprays with gibberellic acid can overcome the stunting induced by severe etch virus on tobacco.

SELECTED REFERENCES

Anonymous. 1966. Index of plant virus diseases. *U.S. Dept. Agr. Handbook* **307**:446 p.

Atabekov, J. G. 1975. Host specificity of plant viruses. *Ann. Rev. Phytopathol.* **13**:127–145.

Bald, J. G. 1966. Cytology of plant virus infections. *Advan. Virus Res.* **12**:103–126.

Bos, L. 1963. "Symptoms of Virus Diseases in Plants." Inst. Phytopathol. Res., Centre Agr. Publications and Documents, Wageningen.

Esau, Katherine. 1967. Anatomy of plant virus infections. *Ann. Rev. Phytopathol.* **5**:45–76.

Fraenkel-Conrat, H. (Ed.). 1968. "The Molecular Basis of Virology." Reinhold, New York. 656 p.

Hamilton, R. I. 1974. Replication of plant viruses. *Ann. Rev. Phytopathol.* **12**:223–245.

Hollings, M. 1965. Disease control through virus-free stock. *Ann. Rev. Phytopathol.* **3**:367–396.

Kado, C. I., and H. O. Agrawa (Eds.). 1972. "Principles and Techniques in Plant Virology." Van Nostrand Reinhold Co., New York. 688 p.

Knudson, D. L. 1973. Rhabdoviruses. *J. Gen. Virol.* **20**: Suppl. 1–130.

Maramorosch, K. (Ed.). 1969. "Viruses, Vectors and Vegetation." Interscience, New York. 666 p.

Maramorosch, K., and H. Koprowski (Eds.). 1967–1970. "Methods in Virology." Academic Press, New York. 5 vols.

Matthews, R. E. F. 1970. "Plant Virology." Academic Press, New York. 778 p.

Matthews, R. E. F. 1973. Induction of disease by viruses, with special reference to turnip yellow mosaic virus. *Ann. Rev. Phytopathol.* **11**:147–170.

McWhorter, F. P. 1965. Plant virus inclusions. *Ann. Rev. Phytopathol.* **3**:287–312.

Mundry, K. W. 1963. Plant virus-host cell relations. *Ann. Rev. Phytopathol.* **1**:173–196.

Schlegel, D. E., S. H. Smith, and G. A. deZoeten. 1967. Sites of virus synthesis within cells. *Ann. Rev. Phytopathol.* **4**:223–246.

Schneider, I. R. 1965. Introduction, translocation and distribution of viruses in plants. *Advan. Virus Res.* **11**:163–222.

Smith, K. M. 1965. Plant virus–vector relationships. *Advan. Virus Res.* **11**:61–96.

Smith, K. M. 1972. "A Textbook of Plant Virus Diseases." Academic Press, New York. 684 p.

Van Kammen, A. 1972. Plant viruses with a divided genome. *Ann. Rev. Phytopathol.* **10**:125–150.

Wetter, C. 1965. Serology in virus-disease diagnosis. *Ann. Rev. Phytopathol.* **3**:19–42.

• *Tobacco Mosaic*

Tobacco mosaic is worldwide in distribution. It is known to affect more than 150 genera of primarily herbaceous, dicotyledonous plants including many vegetables, flowers, and weeds. It causes serious losses on tobacco, tomato, and some other crop plants, but is almost symptomless on crops like grape and apple.

Tobacco mosaic affects plants by damaging the leaves, flowers, and fruit and by causing stunting of the plant. It almost never kills plants. In tobacco, the disease lowers the quantity and especially the quality of the crop, particularly when plants are infected while young. Thus, plants inoculated at transplanting time, a month later or at topping time produced yields that were lower than those of healthy plants by 33, 20, and 5 percent, while the quality of the crop, as measured by its market value, was reduced by 50, 42, and 23 percent, respectively. In tomato, also, yield reductions may vary from 5 to more than 25 percent and the fruit quality is proportionally lower, depending on the age of plants at the time of infection and on environmental conditions.

Symptoms. The symptoms of tobacco mosaic virus-infected plants consist of various degrees of chlorosis, curling, mottling, dwarfing, distortion, and blistering of the leaves, dwarfing of the entire plant, dwarfing, distortion, and discoloration of flowers, and, in some plants, development of necrotic areas on the leaf.

The most common symptom on tobacco is the appearance of mottled dark-green and light-green areas on leaves developing after inoculation (Fig. 222A). The dark green areas are thicker and appear somewhat elevated in a blisterlike fashion over the thinner, chlorotic, light-green areas. Stunting of young plants is common, and is accompanied by a slight downward curling and distortion of the leaves, which may also become narrow and elongated rather than the normal oval shape. Inoculation of plants approaching maturity usually causes no symptoms on the older leaves, but it does affect any new ones that may be produced.

On tomato, mottling of the older leaves and mottling with or without malformation of the leaflets are also produced. Leaflets become long and pointed and, sometimes, shoestringlike. Infections of young plants reduce fruit set and may occasionally cause blemishes and internal browning on the fruit that does form.

The pathogen: Tobacco mosaic virus (TMV). Tobacco mosaic virus (TMV) is rod shaped, 300 nm long by 15 nm in diameter (Fig. 222B, C). Its protein consists of approximately 2130 protein subunits, and each subunit consists of 158 amino acids. The protein subunits are arranged in a helix. The TMV nucleic acid is single-stranded ribonucleic acid (RNA) and consists of approximately 6400 nucleotides. The RNA strand also forms a helix which is parallel with that of the protein and is located on the protein subunits and approximately 20 Å out from the inner end of the protein subunits. The weight of each virus particle is between 39 and 40 million molecular weight units.

TMV is one of the most thermostable viruses known, the thermal inactivation point of the virus in undiluted plant juice being 93°C. In dried, mosaic-infected leaves, however, the virus retains its infectiousness even when heated at 120°C for 30 minutes. TMV-infected tobacco plants may contain up to 4 g of virus per liter of plant juice, and the virus retains its infectivity even at dilutions of 1:1,000,000. The virus is inactivated in 4 to 6 weeks in ordinary plant sap, but in sterile, bacteria-free sap the virus may survive for five years, and in TMV-infected leaves kept dry in the laboratory the virus remains infectious for more than 50 years. Tobacco mosaic virus is transmitted readily through sap, grafting, and dodder, and, in some hosts such as apple, pear, and grape, through seed. Tobacco mosaic virus is not transmitted by insects, except occasionally by contaminated jaws and feet of insects feeding on TMV-infected and healthy plants. The most common means of transmission of TMV in the field and in the greenhouse is through the hands of workers handling infected and healthy plants indiscriminately.

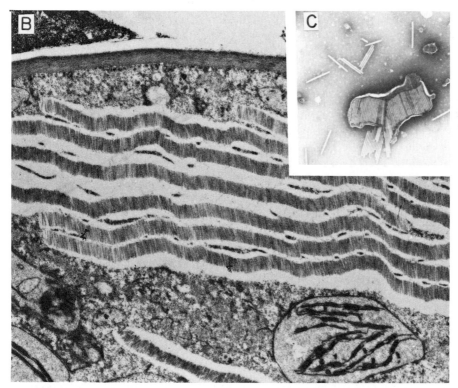

FIGURE 222.
(A) Symptoms of tobacco mosaic (upper) and healthy tobacco leaf. (B) Layers of tobacco mosaic virus particles in a tobacco epidermal leaf cell. (C) TMV particles in sap from an infected tobacco leaf. The virus was negatively stained with phosphotungstate. (Photos B and C courtesy of H. E. Warmke.)

Tobacco mosaic virus exists in numerous strains which differ from each other in one or more characteristics.

Development of disease.   Tobacco mosaic virus overwinters in infected tobacco stalks and leaves in the soil, on the surface of contaminated tobacco seeds, on contaminated seedbed cloth, and in natural leaf and manufactured tobacco, including cigarettes, cigars, and snuff. Contact of the virus with wounded tissues of tobacco seedlings in the seedbed or of transplants in the fields results in initial infections of a few plants. These subsequently serve as a source of inoculum for further spread of the virus to more plants through contaminated hands, tools, or equipment during handling of tobacco plants in the routine cultural practices with that crop. The virus can, of course, be introduced into a field on transplants already infected in the seedbed. The spread of the virus in the field continues throughout the season, the number of plants infected increasing progressively during the season and doubling approximately with each handling or cultivating of the crop.

In almost all host plants, TMV produces systemic infections, invading all parenchymatous cells of the plant. The virus moves from cell to cell and through the phloem.

Within the cell, TMV seems to occur primarily in the cytoplasm as individual particles, as crystalline aggregates (Fig. 222B), and as amorphous bodies (X-bodies), ranging in size from submicroscopic to those seen with the light microscope.

The TMV-infected leaves show thin, light-green or yellowish areas intermixed with thicker, dark green areas. In the light-green areas both palisade and spongy parenchyma cells are round rather than their normal elongated shape and, due to reduction of the intercellular spaces among such cells, they are arranged much more compactly than cells in the dark-green areas or in healthy leaves. In the light areas, the number of chloroplasts is reduced appreciably; they seem to contain much less chlorophyll than those of healthy or dark-green areas. Thus, chlorophyll synthesis is impaired while some of the chlorophyll produced is destroyed or its activity is impaired as a consequence of virus infection. This leads to reduced photosynthesis and, therefore, reduced levels of carbohydrates in TMV-infected plants. Tobacco mosaic virus also induces alterations in a number of other physiological processes in infected plants.

Control.   Sanitation and use of resistant varieties are the two main means of control of TMV in tobacco and tomato fields or greenhouses.

Tobacco should not be grown for at least 2 years in seedbeds or fields where a diseased crop was grown. Removal of infected plants and of certain solanaceous weeds that harbor the virus early in the season helps reduce or eliminate the subsequent spread of the virus to other plants during the various cultural practices. The chewing and smoking of tobacco during cultural practices requiring handling of tobacco and other susceptible plants should be avoided. Workers using tobacco products or involved in removing TMV-infected plants should wash their hands with soap and water before handling healthy tobacco or tomato plants.

Several TMV-resistant varieties of tobacco have been developed but are generally of low quality. Tomato varieties resistant to TMV are also available. Tomatoes in greenhouses, which in most cases become infected with TMV, are in some countries protected against virulent strains of TMV by infecting young plants with a mild strain. This practice results in an increase in yield of up to 15 percent.

Because infection by TMV is inhibited by milk, some states are now recommending spraying the plants with milk before transplanting or otherwise handling them, or dipping the hands in milk during transplanting and handling, since these practices greatly reduce the spread of TMV from plant to plant.

SELECTED REFERENCES

Allard, H. A. 1914. The mosaic disease of tobacco. *U.S. Dept. Agr., Agr. Bull.* **40**:33 p.

Esau, Katherine, and J. Cronshaw. 1967. Relation of tobacco mosaic virus to the host cells. *J. Cell Biol.* **33**:665–678.

McMurtrey, J. E., Jr. 1929. Effect of mosaic diseases on yield and quality of tobacco. *J. Agr. Res.* **38**:257–267.

Markham, R., J. Hitchborn, G. Hills, and S. Frey. 1964. The anatomy of the tobacco mosaic virus. *Virology* **22**:342–359.

Zaitlin, M., and H. W. Israel. 1975. Tobacco mosaic virus (type strain). *C.M.I/ A.A.B. Descriptions of Plant Viruses*, No. 151, 5 p.

- *Cucumber Mosaic*

Cucumber mosaic is worldwide in distribution. The virus causing cucumber mosaic has, perhaps, a wider range of hosts and attacks a greater variety of vegetables, ornamentals, and other plants than any other virus. Among the most important vegetables and ornamentals affected by cucumber mosaic are cucumbers, melons, squash, peppers, spinach, tomatoes, celery, beets, bean, banana, crucifers, delphinium, gladiolus, lilies, petunias, zinnias, and many weeds.

Cucumber mosaic affects plants by causing mottling or discoloration and distortion of leaves, flowers, and fruit. Infected plants may be greatly reduced in size or they may be killed. Crop yields are reduced in quantity and are often lower in quality. Plants are seriously affected in the field as well as the greenhouse. In some localities one-third to one-half of the plants may be destroyed by the disease, and susceptible crops, such as cucumbers, may have to be replaced by other crops.

Symptoms. Young cucumber seedlings are seldom attacked in the field during the first few weeks. Most general field infections of cucumber occur when the plants are about 6 weeks old and growing vigorously. Four or five days after inoculation, the young developing leaves become mottled, distorted, and wrinkled, and their edges begin to curl downward (Fig. 223). All subsequent growth is reduced drastically and the plants appear dwarfed as a result of shorter stem internodes and petioles and of leaves developing to only half their normal size. Such plants produce few runners and also few flowers and fruit. Instead, they have a bunched or bushy appearance, with the leaves forming a rosettelike clump near the ground. The older leaves of infected plants develop at first chlorotic and then necrotic areas along the margins which later spread over the entire leaf. The killed leaves hang down on the petiole or fall off, leaving part or most of the older vine bare.

Fruit produced on the plant after the infection shows pale green or white areas intermingled with dark green, raised areas; the latter often form rough, wartlike projections and cause distortion of the fruit. Cucumbers produced by the plants in the later stages of the disease are somewhat misshapen but have smooth gray-white color with some ir- regular green areas and are often called "white pickle." Cucumbers in- fected with cucumber mosaic often have a bitter taste and upon pickling become soft and soggy.

The pathogen: Cucumber mosaic virus (CMV). Cucumber mosaic virus is polyhedral, with a diameter of approximately 30 nm (Fig. 223A). The virus consists of 180 protein subunits, single-stranded RNA and a

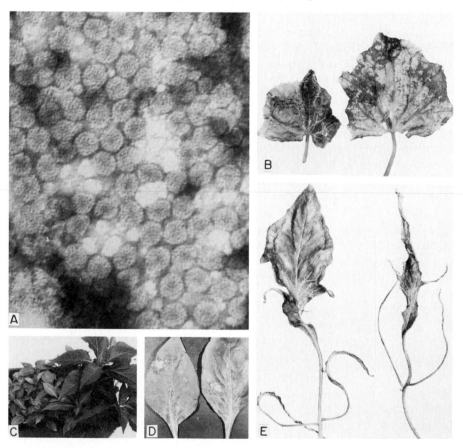

FIGURE 223.
Cucumber mosaic virus (A) and some of the symptoms it causes. Cucumber mosaic on cucumber leaves (B). Stunting of infected pepper plants is shown at C (left) compared to two healthy plants, and leaf symptoms on pepper (D). CMV-infected tomato leaves often become filiform or shoestring-like (E.)

hollow core. The molecular weight of CMV falls in the range of 5.8 to 6.7 million, of which 18 percent is RNA and the remaining 82 percent protein. The thermal inactivation point of the virus is about 70°C, while its dilution end point is about 1:10,000. Cucumber mosaic virus exists in numerous strains that differ somewhat in their hosts, in the symptoms they produce, in the ways they are transmitted, and in other properties and characteristics.

The virus is readily transmitted by sap and also by many aphids such as the common green peach aphid.

Development of disease. Cucumber mosaic virus overwinters in many perennial weeds, flowers, and crop plants. Perennial weeds such as white cockle, wild ground cherry, horse nettle, milkweed, ragweed, pokeweed, nightshade, and the various mints harbor the virus in their roots during the winter and carry it to their top growth in the spring from which aphids transmit it to susceptible crop plants. Once a few cucumber plants have become infected with CMV, insect vectors, and man during his cultivating and handling of the plants, especially at picking time, spread the virus to many more healthy plants. Entire fields of cucurbits sometimes begin to turn yellow with mosaic immediately after the first pick has been made, indicating the ease and efficiency of transmission of CMV mechanically through sap carried on the hands and clothes of the workers.

Whether the virus is transmitted by insects or through sap, it produces a systemic infection of curcurbit and most other host plants. Older tissues and organs developed prior to infection are not, as a rule, affected by the virus, but young active cells and tissues developing after infection may be affected with varying severity. The virus concentration in CMV-infected plants continues to increase for several days following inoculation and then it decreases until it levels off or until the plant dies.

Control. Cucumber mosaic in vegetables and flowers can be controlled primarily through the use of resistant varieties, elimination of weed hosts, and control of the insect vectors.

Varieties resistant to CMV have been developed for several host crops, including cucumber and spinach.

Transplant crops kept in greenhouses should be isolated from other plants such as geraniums, lilies, and cucumbers that may harbor the virus, and when transplanted they should not be planted near early, susceptible crops or near woods in which there may be weeds harboring the virus. Perennial weeds should be eradicated from around greenhouses, cold frames, gardens, and fields to eliminate the source of CMV likely to be carried to crop plants by insects or sap. Since most of the early virus infections are initiated by insects, early sprays with insecticides to control the aphid vectors before they carry the virus into the young, rapidly growing plants have been very helpful.

SELECTED REFERENCES

Doolittle, S. P. 1920. The mosaic disease of cucurbits. *U.S. Dept. Agr. Bull.* **879**: 69 p.

Doolittle, S. P., and M. N. Walker. 1925. Further studies on the overwintering and dissemination of cucurbit mosaic. *J. Agr. Res.* **31**:1–58.

Gibbs, A. J., and B. D. Harrison. 1970. Cucumber mosaic virus. *C.M.I./A.A.B. Descriptions of Plant Viruses*, No. 1. 4 p.

Porter, C. A. 1954. Histological and cytological changes induced in plants by cucumber mosaic virus. *Contrib. Boyce Thomp. Inst.* **17**:453–471.

Sherf, A. F. 1965. Cucumber mosaic virus in New York vegetables. *N.Y. State Agr. Expt. Sta. Cornell Extension Bull.* **1144**:8 p.

- *Bean Common Mosaic and*
  Bean Yellow Mosaic Diseases

Both diseases occur wherever beans are grown. Bean common mosaic affects primarily the French or snap beans (*Phaseolus vulgaris*) but also other *Phaseolus* sp., while bean yellow mosaic affects the above and, in addition, peas, clovers, vetch, black locust, gladiolus, and yellow summer squash. Both diseases are widespread in bean fields with common mosaic being more widespread than yellow mosaic. Often, 80 to 100 percent of the plants in some fields are infected. Depending on the growth stage of the plant at the time of infection, the plants may be stunted to a smaller or greater extent and losses may vary from slight up to 35 percent for common mosaic and up to 100 percent for yellow mosaic. Usually both diseases occur in the same fields and often on the same plants.

Symptoms.   Bean common mosaic causes stunting of the plants and mottling and malformation of the leaves (Fig. 224A). The leaves show mild mottling or they have rather large, irregularly shaped light-yellow and light-green areas. Often, leaves are narrower and longer than normal and show considerable puckering, consisting of raised dark-green areas along the main veins, while the leaf margins curl downward. The younger the bean plants at the time of infection the more dwarfed and spindling they remain and the smaller the crop they produce. Pods may be mottled or malformed and the seeds are shriveled and undersized. In some varieties the roots turn dark to almost black and become necrotic. External discoloration may also develop on young stems and petioles while vascular necrosis is evident in the root, stem, leaves, and pods.

Bean yellow mosaic produces symptoms similar to the above and in the field it is usually impossible to distinguish it from bean common mosaic. Generally, however, bean yellow mosaic produces a much more yellow mottling of the leaves with an intense contrast between the yellow and the green areas. Also, bean plants infected with yellow mosaic are much more dwarfed and bunchy than those infected with common mosaic. Yellow mosaic also produces greater leaf malformation and pod distortion than common mosaic (Fig. 224B, C). In both diseases, however, symptoms vary greatly with the variety and with the virus strain prevailing in the area.

The pathogens: Bean common mosaic virus and bean yellow mosaic virus.   Both viruses are filamentous and measure 750×15 nm. Their dilution end point is 10^{-3} to 10^{-4} and their thermal inactivation point is about 60°C although it may range from 50 to 70°C. Their longevity *in vitro* is usually 1 to 4 days.

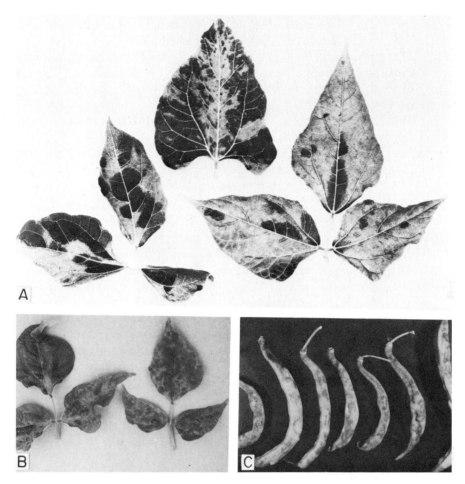

FIGURE 224.
Symptoms of bean common mosaic on bean leaves (A), and of bean yellow mosaic on bean leaves (B) and pods (C). (Photo A courtesy U.S.D.A.)

Both viruses are transmitted by several species of aphids in the nonpersistent manner, most of the vectors being common to both viruses. Both viruses are also readily transmitted by sap inoculation. Bean common mosaic is, moreover, transmitted readily through bean seeds, especially when the mother plants are infected while young. As much as 83 percent of the seed of diseased plants may produce virus-infected plants. Seed transmission is the most important source of initial crop infection in bean fields. Bean common mosaic virus is also transmitted to new plants through pollen. Bean yellow mosaic virus is not transmitted through the seed in beans but is transmitted in about 3 to 6 percent of the seeds of several other legumes.

Although bean common mosaic and bean yellow mosaic viruses differ in the kinds of hosts they attack and in seed transmissibility, they resemble each other in numerous characteristics, they are serologically related, and in some hosts they cross-protect against each other, indicating that the two viruses may be distant strains of the same virus.

Development of disease. Bean common mosaic virus overwinters in infected bean seed. When such seed is planted, the virus multiplies in the cells of the growing plant and the produced plant is infected with the virus. Subsequently, aphid vectors feeding on such plants acquire the virus within a few seconds and transmit it to healthy bean plants on which they move to feed.

Bean yellow mosaic virus overwinters primarily in perennial hosts such as clovers and gladiolus from which it is spread to beans and other annual hosts by its aphid vectors. The same vectors, of course, transmit both viruses from bean to bean. Bean plants inoculated by insects with either the common or the yellow mosaic viruses usually develop mosaic symptoms within 10 days from infection. Symptoms, however, may be mild and almost unnoticeable if the weather remains cool.

Control. The best control of bean common mosaic is obtained through the use of virus-free seed and when only varieties resistant to bean common mosaic virus are planted. Several varieties resistant to common mosaic are available. Control of bean yellow mosaic is more difficult because few bean varieties show only some resistance to some, but not all, of the strains of the virus and because of the overwintering of the virus in perennial hosts such as clovers and gladiolus. Planting beans in fields removed from gladiolus fields and destruction of clovers from around bean fields are recommended, but such control measures are difficult to carry out in practice and, besides, their effectiveness is not always apparent.

SELECTED REFERENCES

Bos, L. 1970. Bean yellow mosaic virus. *C.M.I./A.A.B. Descriptions of Plant Viruses* **No. 40**, 4 p.
Bos, L. 1971. Bean common mosaic virus. *Ibid*. No. 73, 4 p.
Pierce, W. H. 1934. Viruses of the bean. *Phytopathology* **24**:87–115.

- *Curly Top of Sugar Beets*

Curly top occurs primarily and is most destructive in the western half of the U.S., but it has been found in some eastern states, in Turkey and South America. Curly top is most destructive on sugar beet, bean, tomato, flax, melons, and spinach. The virus infects more than 150 species of herbaceous plants belonging to more than 50 families.

Curly top damages plants by killing young plants and causing stunting, malformations, reduced yields, and lower quality in older plants. Losses from curly top of sugar beets and tomato have sometimes been so severe that vast areas previously planted to these crops had to be completely abandoned after years of destructive outbreaks of curly top.

Symptoms. The first symptoms appear as a clearing and swelling of the veins of the younger leaves, the edges of which begin to roll and curl inward. If the plant is infected at the seedling stage, while the root is still about 1 cm or less in diameter, it makes little further growth; it either dies shortly or remains as a tight ball of stunted leaves for several weeks or months, and finally dies. Infected larger plants remain stunted and produce more but smaller leaves than healthy plants. Most of these leaves become curled and their veins become swollen and give rise to little nipplelike swellings on the lower side of the leaves (Fig. 225). At times, a sticky brownish fluid may be exuded from the veins, and this collects in droplets along the leaf stalks. Affected leaves generally remain dark green for a time, but they eventually become yellow, then brown, and usually die prematurely. Leaves that were mature at the time of infection develop no curling or swelling of the veins, but they soon turn yellow and die. Infection of fully matured plants late in the season usually has little or no effect on the appearance or yield of the plant.

The roots of curly-top-infected sugar beets are affected almost proportionally to the damage caused by the disease on the tops of the plants. The younger the plant at the time of infection, the smaller the root. In many instances the roots of the diseased beets are exceedingly hairy, and the root tissue is woody and tough. In cross sections, infected roots show brownish rings indicating degenerative changes in the vascular tissues. In longitudinal sections, the same tissues appear as a discolored line (Fig. 225C).

FIGURE 225.
(A) Sugar beet plant infected with curly top virus. (B) Reaction of sugar beet plants of the same variety to three strains of the virus of different degrees of virulence: mild, moderately severe and severe. (C) Longitudinal sections of CTV-infected (left) and healthy sugar beet roots. The dark lines in the diseased root represent necrotic vascular bundles. (D) Bean plant infected with CTV compared to healthy plants on both sides. (Photos A, B, C courtesy C. W. Bennett, D courtesy U.S.D.A.)

The pathogen: Curly top virus (CTV).   Curly top virus is a polyhedral particle 20 to 28 nm in diameter. The virus is not sap transmitted. In nature it is transmitted by the leafhopper *Circulifer tenellus*. The thermal inactivation point of curly top virus is known to be between 75 and 80°C. The dilution end point of curly top virus in beet extracts is about 1:1000, while in leafhopper extracts it is 1:24,000. In beet leaf juice the virus retains its infectivity for about 7 days, but it can remain infectious for 4 months in dried beet leaves, and 6 months in dried insect vectors.

The leafhopper, *Circulifer tenellus*, transmits the virus after feeding for short periods or after incubation periods varying from 4 hours to more than 5 days depending on the concentration of the virus in the vector's food. Viruliferous leafhoppers transmit the virus efficiently for 2 or 3 days after the incubation period, but subsequently their efficiency declines steadily. Thus, although the virus appears to be both stylet borne and circulative, it does not multiply in the vector.

In the plant the virus seems to be limited almost entirely to the phloem and adjacent parenchyma cells.

Development of disease.   The virus overwinters primarily in infected perennial and biennial weeds such as plantago, pepper grass, Russian thistle, and filaree. It also overwinters in perennial ornamental hosts, in annuals in the greenhouse, and occasionally in the overwintering adults of the insect vector. The insects feed on the infected wild plants in the winter and spring, become viruliferous, and carry the virus to cultivated crops in late spring or summer. The insect feeds by inserting its stylet into the phloem of infected or healthy plants and transmits the virus in the process. Once inside the phloem, the virus moves rather rapidly through it, i.e., about 2 to 3 cm per minute, while at the same time it causes destructive changes in the sieve elements of the phloem.

After inoculation, the first symptoms on the plant may appear within 24 hours when the temperature is high, but usually there is an incubation period of 7 to 14 days under normal temperatures, and even longer during cool weather. The virus, however, spreads throughout the plant quickly, so that a plant may become a source of virus for new leafhoppers within 5 hours from its inoculation.

Curly top virus-infected plants exhibit hypertrophy, hyperplasia, and necrosis of the phloem elements. The hyperplastic sieve elements apparently are not functional and sometimes spread beyond the limits of the phloem, into the cortex and the xylem. Hypertrophy and hyperplasia also occur in parenchyma cells adjacent to the phloem. These cells become closely packed, leave no intercellular spaces, and their chloroplasts are few, small and pale and result in the appearance of vein clearing. Further hypertrophy and hyperplasia of these cells produce thickening and distortion of the veins, result in the formation of protuberances, and, since they occur primarily on the underside of the leaves, cause upward rolling of the leaves. Degeneration and necrosis of phloem also occurs in the stem and root. The latter is retarded in growth and produces numerous laterals. Although hypertrophy and hyperplasia occur in the phloem and adjacent parenchyma cells, most other cells remain hypoplastic and result in dwarfing and stunting of the whole plant.

Control. Insecticide sprays carried out systematically and over a large area simultaneously have been effective in controlling the vector. Statewide programs to eradicate the leafhopper by mapping and spraying the breeding ground of the leafhopper with insecticides have markedly reduced the disease in some areas.

The most effective and most widespread means of curly top control today is through the use of resistant varieties. Several sugar beet varieties resistant to curly top are available. Resistant varieties to curly top have also been developed for tomato, for bean, and for other crops.

SELECTED REFERENCES

Bennett, C. W. 1971. "The Curly Top Disease of Sugarbeet and Other Plants." Monograph No. 7. The American Phytopathological Society, St. Paul, Minn. 81 p.

Carsner, E., and C. F. Stahl. 1924. Studies on curly-top disease of the sugar beet. *J. Agr. Res.* **28**:297–319.

Hoefert, L. L., and Katherine Esau. 1967. Degeneration of sieve element plastids in sugar beet infected with curly top virus. *Virology* **31**:422–426.

Mumford, D. L., and also Mink, G. I., and P. E. Thomas. 1974. Purification of curly top virus. *Phytopathology* **64**:136–139, 140–142, respectively.

• *Barley Yellow Dwarf*

Barley yellow dwarf occurs throughout the world. The barley yellow dwarf virus attacks a wide variety of gramineous hosts, including barley, oats, wheat, rye, and many lawn, weed, pasture, and range grasses.

Barley yellow dwarf affects plants by causing stunting, reduced tillering, suppressed heading, sterility, and failure to fill the kernels. In some localities, plant damage may be so severe that entire fields are destroyed and the crops are not worth harvesting. Of the three main crops, barley, oats, and wheat, oats is the most severely affected and suffers serious losses annually. In years of barley yellow dwarf outbreaks, some states reported yield losses ranging from 30 to 50 percent of their entire oat crop, while barley and wheat losses ranged between 5 and 30 percent. To the losses in yield of these cereals should be added losses in quality of the grain and losses in forage crops from the resulting failure or reduced productivity of pasture, range, and meadow grasses.

Symptoms. The first symptoms on barley yellow dwarf-infected plants appear as yellowish, reddish, or purple areas along the margins, tips, or lamina of the older leaves. The discolored areas soon enlarge and frequently surround still unaffected green areas. The tissues along the midrib usually remain green longer than the rest, but finally they, too, become discolored. In late infections the flag leaf may be the only one that develops the characteristic discoloration. In seedling infections, leaves may emerge distorted, curled and with serrations. Leaves developing after the infection are progressively shortened, narrower and stiffer than normal, and grow more upright than normal.

The stem internodes of infected plants are shorter, and sometimes the head fails to emerge (Fig. 226). Tillering is reduced or completely suppressed in oat and wheat plants, but severely stunted barley plants may

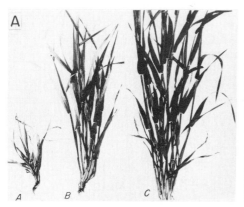
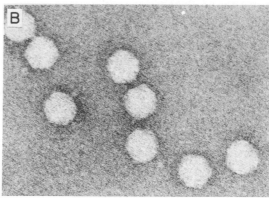

FIGURE 226.
(A) Symptoms of barley yellow dwarf virus on barley plants infected in early
tillering stage (left), and in jointing stage (middle). Healthy plant at right. (B)
Particles of barley yellow dwarf virus. (Photo A courtesy U.S.D.A. Photo B taken by
H. W. Israel and supplied courtesy W. F. Rochow.)

show excessive tillering. Inflorescences of diseased plants emerge later
and are smaller. Many of the flowers are also sterile and the number and
weight of kernels are reduced. The root systems of diseased plants are
drastically reduced in weight but show no characteristic symptoms.

The pathogen: Barley yellow dwarf virus (BYDV).   The virus is a
polyhedral particle about 22 nm in diameter (Fig. 226B). The virus concentration in infected plants is very low; 1 liter of plant juice contains
only about 25 to 50 micrograms (μg) of virus. The thermal inactivation
point of BYDV appears to be between 65 and 70°C. Barley yellow dwarf
virus is not mechanically transmissible to plants and, therefore, its chemical and most of its physical properties are unknown. The virus, however,
is transmitted by dodder, and in nature it is readily transmitted by
insects. At least 14 species of aphids (Fig. 216) serve as vectors of BYDV,
including the apple green aphid (*Rhopalosiphum fitchii*), the English grain
aphid (*R. maydis*), and the green bug (*Schizaphis graminum*). Most aphids
require an acquisition feeding period of about 24 hours and an inoculation
feeding period of 4 to 8 hours or more. Some of the aphid vectors seem to
be much more efficient vectors of BYDV than others. BYDV is
circulative in its vectors, all viruliferous aphids remaining so for 2 to 3
weeks. The virus is not passed from adults to their progeny, but it does
persist through molting. Barley yellow dwarf virus can be transmitted
mechanically from aphid to aphid by injection of infectious extracts but
not from aphid to plant.

Barley yellow dwarf virus consists of numerous strains which differ in
their relative virulence on different host varieties, in the symptoms they
produce, and in their transmission by different aphid vectors.

Development of disease.   Barley yellow dwarf virus in the northern
areas overwinters in perennial grass hosts while in the south it may
overwinter in annual grasses and fall-sown cereals and in viruliferous
adult aphids. The spread of the virus depends on the spread of the aphid

vectors. In a few areas, such as parts of Oregon and Washington, the climate is favorable or tolerable for some of the aphid vectors throughout the year and they annually carry the virus from wild grasses to cultivated cereals causing frequent outbreaks of the disease and subsequent serious losses. In most of the other main cereal-producing areas of the U.S. and Canada, the winter temperatures are too low to allow overwintering of the adult aphids, but they can overwinter in the egg stage. It appears, however, that the aphid populations resulting from eggs is rather small, and since they must first feed on virus-infected perennial grasses in order to become viruliferous, the rate of BYDV spread by them is rather limited and does not result in severe outbreaks of the disease unless weather conditions become extremely favorable for aphid multiplication and a virus reservoir is plentiful and readily accessible to the aphids. The worst epidemics, however, develop from virus brought into cereal fields in the spring by viruliferous aphids migrating northward from the south. Winter survival of large populations of aphids in north Texas and Oklahoma, followed by properly timed south winds, could move the aphids northward in stages so as to capitalize on the northward progression of spring and the successive appearance of wheat, oat, and other susceptible grass seedlings on which these aphids thrive. Barley yellow dwarf epidemics generally occur when the spring and early summer weather is cool and moist.

Following inoculation of a grass plant with BYDV by a viruliferous aphid, the first symptoms appear within 7 to 28 days, the shorter incubation periods occurring at lower temperatures (about 16°C), and the longer periods at about 27°C. The severity of symptoms is also greater at 16°C, while at 27°C they are mild and at 32°C they are masked.

The stage of host development at the time of infection is a crucial factor in disease development. The most severe symptoms result only from infection of the annual cereals in the seedling stage. Infected seedlings may die as in a seedling "blight" or they may survive for a time with the third or fourth leaf emerging distorted. Such plants usually fail to head, and if they do, the inflorescence and entire plant are extremely small. In later stages of infection, in which the virus has progressively less time in which to affect the host, the disease severity is reduced proportionately, and only the last formed leaf may show mild symptoms. In fall-sown cereals, BYDV infections increase winter killing of plants as well as reduce yields, and the effects are much more pronounced in young seedling infections than in infections of more developed plants.

Control.   Control of barley yellow dwarf through control of the aphid vectors with insecticides has been attempted repeatedly, but the results have been disappointing.

The main hope for control of BYDV is the use of resistant varieties. Most of the commercial varieties of oats, barley, and wheat commonly grown in the United States are susceptible to BYDV, but some are less susceptible than others. A number of varieties have been found or developed that show some resistance to BYDV. An extensive breeding program to develop varieties of the three main cereals that can withstand heavy barley yellow dwarf epidemics is presently being carried out.

SELECTED REFERENCES

Endo, R. M., and C. M. Brown. 1963. Effects of barley yellow dwarf virus on yield of oats as influenced by variety, virus strain, and developmental stage of plants at inoculation. *Phytopathology* **53**:965–968.

Esau, Katherine. 1957. Anatomic effects of barley yellow dwarf virus and maleic hydrazide on certain Gramineae. *Hilgardia* **27**:15–69.

Orlob, C. B., and D. C. Arny. 1961. Influence of some environmental factors and growth substances on the development of barley yellow dwarf. *Plant Dis. Reptr.* **45**:192–195.

Rochow, W. F. 1970. Barley yellow dwarf virus. *C.M.I./A.A.B. Descriptions of Plant Viruses*, No. 32. 4 p.

- *Necrotic Ring Spot of Stone Fruits*

Necrotic ring spot occurs worldwide in temperate regions. The disease affects most cultivated stone fruits, including sour cherry, cherry, almond, peach, apricot, and plum, many wild and flowering cherries, peaches, and plums, and also some ornamental species such as rose. Necrotic ring spot virus is present in all trees infected with cherry yellows (prune dwarf virus) and it is frequently associated with other viruses.

Necrotic ring spot is the most widespread virus disease of stone fruit trees. In the fruit producing areas almost all orchard trees in production are infected. The losses caused by necrotic ring spot vary with the *Prunus* species or variety affected and with the time from inoculation with the virus. Successful commercial budding is lower in combinations in which the bud or the rootstock carry the virus than when both are virus free or virus infected. The growth of virus-infected trees may be reduced by 10 to 30 percent or more, while the yield of virus-infected trees may be 20 to 56 percent or more lower than that of healthy trees. Trees affected with necrotic ring spot also show increased susceptibility to winter injury.

Symptoms.   The first symptoms appear as a pronounced delayed foliation of individual branches or entire trees. Leaves on affected branches are small and show light green spots and dark rings 1 to 5 mm in diameter. In later stages of the disease affected areas may become necrotic, fall out, and give a "shredded leaf" or "tatter-leaf" effect (Fig. 227 A). Such symptoms, called shock or acute symptoms, are usually limited to the first leaves that unfold, while leaves formed later generally do not show marked symptoms. Affected trees, however, usually have fewer leaves and therefore have a thin appearance.

Blossoms of affected trees have short or no pedicels, the calyx and corolla may be twisted and distorted, and the sepals may develop chlorotic or necrotic rings or arcs. Such severely affected blossoms ordinarily do not set fruit and occasional fruits also develop small rings similar to those on the leaves.

As a rule, trees severely affected one year show few or no symptoms in subsequent years except for the thinness of foliage. If severe symptoms are present only on a few branches the first year, other branches may show striking symptoms the following year. In many areas, however, trees may continue to show striking ring symptoms and wavy leaf margins for 4 to 6 years or more.

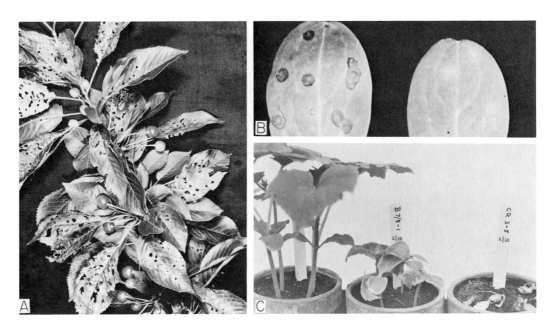

FIGURE 227.
(A) Necrotic ringspot symptoms ("shock" phase) in sweet cherry. (B) Local lesion reactions of two PNRV strains in watermelon cotyledons about 10 days after inoculation. (C) Cucumber plants inoculated with sap from cherry infected with a mild (middle) and a severe (right) strain of PNRV. Healthy plant at left. (Photos courtesy R. M. Gilmer.)

The pathogen: Prunus necrotic ringspot virus (PNRV). Necrotic ringspot virus is a small polyhedral particle about 23 nm in diameter. The virus is very unstable in undiluted plant extracts, beginning to lose infectivity within a few minutes and being completely inactivated in a few hours. The thermal inactivation points for various isolates of PNRV range between 55 and 62°C.

PNRV can be transmitted by budding and grafting and by sap from virus-infected tree leaves or petals to leaves of cucumber and of several other herbaceous plants (Fig. 227B,C). PNRV is transmitted through seed, the percentage of transmission varying among various species from 5 to 70 percent. The virus is also transmitted through pollen to seeds and to pollinated plants. No insect vector of PNRV is known, but a mite, *Vasates fockeui*, and a nematode, *Longidorus macrosoma*, have been reported to transmit this virus.

Development of disease. The virus overwinters in infected stone fruit trees from which it spreads to healthy trees in the spring primarily through infected pollen. PNRV spreads very slowly in orchards under four years old, but can spread very rapidly in older orchards, probably because older trees have more bloom and therefore are much more subject to infection through pollen than young ones. PNRV can spread over a distance of at least 800 meters, but most infections occur within 15 meters of a known infected tree. Symptoms on trees infected by virus-infected pollen usually develop in the spring one year after inoculation.

Although PNRV becomes systemic and spreads throughout the tree in one or, at most, two seasons, it is not known whether the virus moves from cell to cell, is spread through the phloem, or both. The virus seems to move in the tree at first upward from the point of inoculation, then on parts below this point along a direct path to the base of the tree without affecting side branches. Leaf buds that are just opening at the time the virus reaches them react with development of acute symptoms. Leaves that are fully opened at the time of invasion by the virus develop no symptoms during that growing season. Invasion of the new buds formed at the base of these leaves results in symptoms the following spring when the buds will produce new leaves or flowers. The virus, however, even in systemic infections, does not invade all parts of the tree, but it may leave short gaps along twigs or branches and therefore some buds may be virus free although the entire tree seems to be virus infected.

Control. The control of necrotic ringspot of stone fruits is based almost exclusively on starting with virus-free nursery stock and on eliminating PNRV-infected *Prunus* trees from the area where the virus-free trees are grown.

FIGURE 228.
Indexing of PNRV on Shirofugen. (A) Buds grafted on twig on left were virus free and are growing. Buds grafted on other twigs were infected with PNRV, caused local necrosis on Shirofugen, and failed to grow. (B) Necrosis of bark around the areas on which PNRV-infected buds had been grafted (right), compared to healthy (left). (C) Localized necrosis and gumming on Shirofugen inoculated with PNRV. (Photos A and B courtesy Dept. Plant Pathol., Cornell Univ. Photo C courtesy R. M. Gilmer.)

The production of virus-free nursery stock depends on the use of rootstock seedlings derived from virus-free seed and of scion buds derived from virus-free trees of the desired variety. Since no PNRV symptoms are usually present on trees several years after infection, the mother trees providing the seed or the buds are indexed on PNRV-sensitive indicator hosts which reveal the presence or absence of PNRV in the mother trees. The indicator hosts most commonly used for detection of PNRV are seedlings of peach, *Prunus tomentosa*, or cucumber, and limbs of Shirofugen flowering cherry (*Prunus serrulata*). Peach, *P. tomentosa*, and Shirofugen are inoculated with buds taken from the trees being indexed and inserted into the stem of the seedlings or along the limbs of Shirofugen. Inoculated peach and *P. tomentosa* seedlings produce shock symptoms on the foliage and dieback of terminals, while inoculated limbs of Shirofugen produce a local necrotic reaction around the inserted buds (Fig. 228). Cucumber seedlings are inoculated at the cotyledon stage with sap obtained from immature leaves of the suspected trees and react by producing local lesions, systemic mottling, or death of the cucumber seedlings (Fig. 227 B,C). Seed or scion mother trees indexing positive for PNRV are destroyed; those indexing negative for PNRV are used for propagation, but they must be reindexed annually or biannually since they may become infected with the virus through virus-infected pollen.

After a new orchard has been established with virus-free trees, it is necessary to remove all wild *Prunus* trees from a radius of about 200 meters around the periphery of the orchard to avoid spread of the virus into the orchard. A new orchard should not be planted next to an older one containing infected trees, and any infected trees appearing in the new orchard should be removed immediately to prevent further spread of the virus in the orchard.

SELECTED REFERENCES

Agrios, G. N., and W. F. Buchholtz. 1967. Virus effect on union and growth of peach scions on *Prunus besseyi* and *P. tomentosa* understocks. *Iowa State J. Sci.* **41**:385–391.

Berkeley, G. H., D. Cation, E. M. Hildebrand, G. W. Keitt, and J. D. Moore. 1951. Necrotic ring spot. *U.S. Dept. Agr. Agr. Handbook* **10**:164–170.

Davidson, T. R., and J. A. George. 1964. Spread of necrotic ring spot and sour cherry yellows viruses in Niagara Peninsula orchards. *Can. J. Plant Sci.* **44**:471–484.

Davidson, T. R., and J. A. George. 1965. Effects of necrotic ring spot and sour cherry yellows on the growth and yield of young sour cherry trees. *Can. J. Plant Sci.* **45**:525–535.

Fulton, R. W. 1970. Prunus necrotic ring spot virus. *C.M.I./A.A.B. Descriptions of Plant Viruses*, No. 5, 4 p.

George, J. A., and T. R. Davidson. 1963. Pollen transmission of necrotic ring spot and sour cherry yellows viruses from tree to tree. *Can. J. Plant Sci.* **43**:276–288.

Gilmer, R. M., K. D. Brase, and K. G. Parker. 1957. Control of virus diseases of stone fruit nursery trees in New York. *N. Y. State Agr. Expt. Sta. (Geneva) Bull.* **799**:53 p.

- *Tristeza Disease of Citrus*

Tristeza occurs in all citrus-growing areas of the world. Tristeza affects practically all kinds of citrus plants but mostly orange, grapefruit, and lime.

Tristeza causes collapse and decline of trees through a more or less sudden wilting and drying of the leaves followed by death of the tree or by dieback of twigs and partial recovery. Many host plants of tristeza develop stem pitting, followed by poor growth and decline of the tree. Although certain citrus trees can be affected by tristeza even when they are grown on their own roots (seedlings), tristeza causes its most severe damage on trees budded or grafted to rootstocks of certain other species, e.g., sweet orange trees growing on sour orange rootstocks. Losses from tristeza vary greatly with the particular scion–rootstock combination grown in a particular area, the strain of the tristeza virus, and the abundance and efficiency of the vectors. In the first 15 years from the discovery of the disease in California, almost 400,000 trees were destroyed or made worthless by tristeza even before the disease had invaded some of the important citrus-growing areas. In the state of Sao Paolo, Brazil, 9,000,000 trees or about 75 percent of the orange trees of the state were destroyed by tristeza within 12 years from the appearance of the disease in that state.

Symptoms. Tristeza symptoms vary on different hosts and even on the same hosts if they are grown on different rootstocks. The typical tristeza symptoms appear as a quick or chronic decline of trees budded on susceptible rootstocks (Fig. 229, A–C), but seedling or budded trees may also develop stem pitting (Fig. 229E), and seedling trees of various varieties may develop yellows symptoms (see Fig. 229D).

The typical tristeza symptoms in older orange trees (Fig. 229, A–C) appear as suppression of new growth and bronzed to yellow leaves that tend to stand upright. As the disease progresses, the older leaves begin to fall, abscission often taking place between the petiole and the leaf blade, leaving the twigs defoliated or with a few younger leaves. Twigs begin to die back from the tip and later smaller limbs die and only a few weak shoots on the main limbs still have leaves. Twig growth becomes weaker each season until the tree dies, but some trees seem to linger on for many years. In some cases, affected trees collapse quickly following a sudden wilting and drying of the leaves. Tristeza-affected trees also show root symptoms consisting of a marked depletion of starch (Fig. 229F), death and decay of the feeder rootlets, and the injury later extends to the larger roots.

Seedlings and budded trees of many different varieties of citrus, especially lime and grapefruit, develop stem pitting as a result of infection with tristeza. Stem pitting consists most commonly of longitudinal grooves or depressions in the stem paralleling the grain of the wood (Fig. 229E). Stem pitting symptoms are always associated with decreased vigor of the trees, poor bushy growth, and small and distorted fruit.

Seedlings of certain citrus species, e.g., sour orange, lemon, grapefruit, and citron, when inoculated with tristeza develop the so-called seedling

FIGURE 229.
Tristeza disease of citrus. (A) Healthy and diseased sweet orange trees, the latter
showing quick decline. (B) Chronic decline of sweet orange tree. (C) Vein clearing
of Mexican lime trees caused by tristeza. (D) Effects of three different isolates
(A–C) of tristeza on Eureka lemon seedlings. Healthy seedling at right. (E)
Stem pitting symptoms on Mexican lime caused by tristeza. (F) Split trunks and
roots of sweet orange on sour orange rootstock treated with potassium iodide and
showing absence of starch below bud union of tristeza-infected tree and normal
starch throughout in healthy tree. (Photos courtesy J. M. Wallace. Photo B taken
by L. J. Klotz.)

yellows symptoms. The leaves produced subsequent to inoculation are
small and yellow (Fig. 229D). Growth ceases after the first few leaves are
formed, though a certain amount of restricted growth may occur at a later
stage.

The pathogen: Citrus tristeza virus. The citrus tristeza virus is one
of the longest plant viruses, appearing as a threadlike particle, approximately 2000 nm long by 12 nm in diameter, and is present only in phloem
cells. The virus is not transmitted mechanically but is transmitted by
grafting and, in nature, by insects. The insect vectors of tristeza are all
aphids, the most important vector being the tropical citrus aphid, *Tox-*

optera citricidus. The tristeza virus is stylet borne; its vectors become viruliferous after feeding for a few seconds and transmit the virus after equally short feedings.

Tristeza virus exists in nature in numerous strains. At least three distinct strains inducing typical tristeza symptoms are known, and many are known to induce the stem-pitting or yellows syndrome or both. Each strain may infect trees alone or in combination with one or more of the others.

Development of disease.   The citrus tristeza virus or virus complex is widely distributed among citrus trees in the citrus-growing areas. In some localities 100 percent of the trees carry the virus. The spread of the virus is accomplished readily through the use of tristeza-infected propagative material, both scion and rootstocks, and through the insect vectors. The tristeza virus is not transmitted through seed.

Infection of citrus plants with tristeza virus apparently occurs only when the virus is introduced into phloem sieve tubes. The virus seems to be limited to a few of the phloem cells in each bundle and this may account for the ability of a second strain of the virus to infect the same plant by multiplying in some of the remaining phloem cells.

Following infection of citrus plants with the tristeza virus, cells adjacent to a sieve tube begin to degenerate and become necrotic. In hosts that develop stem pitting, the degeneration spreads first into the cambium and inhibits the formation of normal xylem and phloem cells. The tissue produced in the lesion is soft and disorganized and usually remains attached to the bark so that when the bark is removed it leaves a pit in the wood. In some feeder roots the cambium is affected in its entirety rather than in localized areas, and therefore no normal xylem or phloem is produced after the infection.

In hosts that develop seedling yellows, the degeneration appears in cells adjacent to phloem sieve tubes which in leaves and stems are only midly affected, but most of those in feeder roots become extensively necrotic. Abnormal cambium activity and phloem formation are followed by the eventual deterioration of entire clusters of feeder roots.

In trees budded to rootstocks on which typical tristeza symptoms develop, sieve-tube necrosis appears below the bud union about 7 or 8 months after inoculation with the virus and top symptoms about 10 to 23 months later. During this time the root tissues utilize and finally exhaust the reserve starch previously stored in them (Fig. 229F). Also, during the growing season, new phloem is produced intensively and functions for a time before becoming necrotic. When the reserve starch is finally depleted, the roots rot and decline, or collapse follows.

Tristeza is primarily a disease of citrus trees on sour orange stocks caused by a virus that is transmitted by grafting and by some kinds of vectors. The tristeza symptoms shown by scion trees of certain combinations of scion and rootstock are due not to the susceptibility of the scions but to harmful effects produced by the virus on the phloem cells of the rootstock just below the union.

Control.   Control of tristeza disease on existing plantings of susceptible scion–stock combinations is very difficult or impossible. In tristeza-affected areas, some success has resulted from top working the existing, still healthy sweet orange trees on sour orange rootstocks to tristeza-resistant tops such as lemon, and also from changing existing trees over to resistant rootstocks by inarching. Both practices, however, are expensive and time consuming, and their success is influenced by several factors, particularly environment. For these reasons it is generally more satisfactory and economical to remove susceptible trees after they become infected and to replant with a resistant combination.

Avoiding losses in new citrus plantings depends mainly on the use of tolerant scion–stock combinations. The rootstocks most generally recommended are sweet orange, Rough lemon, Cleopatra mandarin, and Troyer citrange. Rootstocks of several other species or varieties are also tolerant to tristeza, but they are objectionable because of their susceptibility to other viruses or to root diseases caused by other pathogens.

SELECTED REFERENCES

Bennett, C. W., and A. S. Costa. 1949. Tristeza disease of citrus. *J. Agr. Res.* **78**:207–237.

McClean, A. P. D. 1963. The tristeza virus complex: Its variability in field-grown citrus in South Africa. *S. African J. Agr. Sci.* **6**:303–332.

Price, W. C. 1970. Citrus tristeza virus. *C.M.I./A.A.B. Descriptions of Plant Viruses* No. 33, 4 p.

Price, W. C. (Ed.). 1972. "Proceedings of the 5th Conference of Citrus Virologists." Univ. of Florida Press, Gainesville, Florida. 301 p.

Schneider, H. 1954. Anatomy of bark of bud union, trunk, and roots of quick decline affected sweet orange trees on sour orange rootstocks. *Hilgardia* **22**:567–581.

Wallace, J. M. (Ed.). 1957. "Citrus Virus Diseases." Univ. of Calif. Division Agr. Sci. 243 p.

plant diseases caused by viroids

To date, five plant diseases, potato spindle tuber, citrus exocortis, chrysanthemum stunt, chrysanthemum chlorotic mottle, and cucumber pale fruit, have been shown to be caused by viroids. A sixth plant disease, the cadang-cadang disease of coconut palms, is presently suspected to be caused by a viroid. So far, among the animal and human diseases, only the scrapie disease of sheep is suspected to be caused by a viroid. It is likely, however, that viroids will be soon implicated as the causes of several "unexplained" diseases in plants, animals, and humans. Almost all the information on viroids up to now has been obtained from studies with the potato spindle tuber viroid and the citrus exocortis viroid.

Viroids are small, low-molecular-weight ribonucleic acids (RNA) that can infect plant cells, replicate themselves and cause disease (Fig. 230).

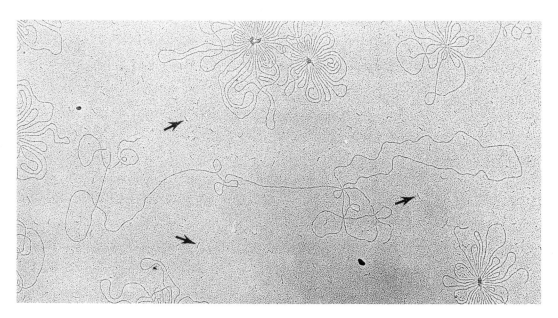

FIGURE 230.
Electron micrograph of potato spindle tuber viroids (arrows) mixed with a double-stranded DNA of a bacterial virus (T₇) for comparison. (Photo taken by T. Koller and J. M. Sogo, and supplied by courtesy of T. O. Diener.)

Viroids differ from viruses in at least two main characteristics: (1) the size of RNA, which has a molecular weight of 75,000 to 120,000 in viroids compared to 1,000,000–10,000,000 for self-replicating viruses, (2) the fact that virus RNA is enclosed in a protein coat while the viroids lack a protein coat and apparently exist as free RNA.

The small size of RNA of viroids indicates that they consist of about 250 to 350 nucleotides and therefore lack sufficient information to code for even one enzyme (replicase) that may be required to replicate the viroid. The existence of viroids as free RNAs rather than as nucleoproteins necessitates the use of quite different methods of extraction, isolation, and purification than those used for viruses, and makes their visualization with the electron microscope extremely difficult even in purified preparations, while in plant tissues or plant sap their detection with the electron microscope is currently impossible.

Viroids appear to be short, single-stranded RNA molecules with extensive base-pairing in parts of the RNA strand. The base-pairing results in some sort of hairpin structure with single-stranded and double-stranded regions on the same viroid. Although viroids have many of the properties of single-stranded RNAs, at least one of them, the potato spindle tuber viroid, which has been seen with the electron microscope, appears about 50 nm in length and has the thickness of double-stranded DNA rather than single-stranded RNA (Fig. 230).

Viroids seem to be associated with the cell nuclei, particularly the chromatin, and, possibly, with the endomembrane system of the cell.

How viroids replicate themselves is still not known. Their small size is barely sufficient to code for a very small protein and such a protein would be considerably smaller than known RNA polymerase (replicase) subunits and would therefore be unable to carry out the replication of the viroid. Besides, viroids have been shown to be inactive as a messenger RNA in several *in vitro* protein-synthesizing systems and no new proteins could be detected in viroid-infected plants. It is known, however, that at least 60 percent, and possibly all, of the potato spindle tuber viroid hybridizes with the host plant DNA, i.e., 60 percent or more of the sequence of the bases of the viroid correspond to a complementary sequence of bases on a segment of the DNA of the host. This suggests that the capacity to reproduce viroids is encoded in part of the DNA of the host plant. Presumably, the genetic information in that segment of the host DNA is completely repressed in uninoculated organisms and is triggered to action by the introduced viroid RNA or by the protein coded by it. Other possible mechanisms of viroid replication may involve: (1) the production of a new DNA as a consequence of infection, the viroid RNA then being synthesized from this DNA in the usual way; (2) the viroid RNA replicates itself by having the small protein for which it can code combine with one or more proteins of the cell and produce a large enough specific enzyme (replicase) that can carry out the replication of the viroid.

How viroids cause disease is also not known. Viroid diseases show a variety of symptoms (Fig. 231) that resemble those caused by virus infections. The amount of viroids formed in cells seems to be extremely small and it is therefore unlikely that they cause a shortage of RNA nucleotides in cells. Besides, as with viruses, many infected hosts show no obvious damage although viroids seem to be replicated in them as much as in the sensitive hosts. So, viroids apparently interfere with the host metabolism in ways resembling those of viruses but which ways are also unknown.

Viroids are spread from diseased to healthy plants primarily by mechanical means, i.e., through sap carried on the hands or tools during propagation or cultural practices and, of course, by vegetative propagation. Some viroids, e.g., potato spindle tuber, chrysanthemum stunt, and chrysanthemum chlorotic mottle viroids, are transmitted through sap quite readily while others, e.g., citrus exocortis, are transmitted through sap with some difficulty. Some viroids, e.g., potato spindle tuber, are transmitted through the pollen and seed in rates ranging from 0 to 100 percent. No specific insect or other vectors of viroids are known although viroids seem to be transmitted on the mouthparts or feet of some insects.

Viroids apparently survive in nature outside the host or in dead plant matter for periods of time varying from a few minutes to a few months. Generally, they seem to overwinter and oversummer in perennial hosts, which include the main hosts of almost all known viroids. Viroids are usually very resistant to high temperatures and cannot be inactivated in infected plants by heat treatment.

Control of diseases caused by viroids is based on the use of viroid-free propagating stock, removal and destruction of viroid-infected plants, and

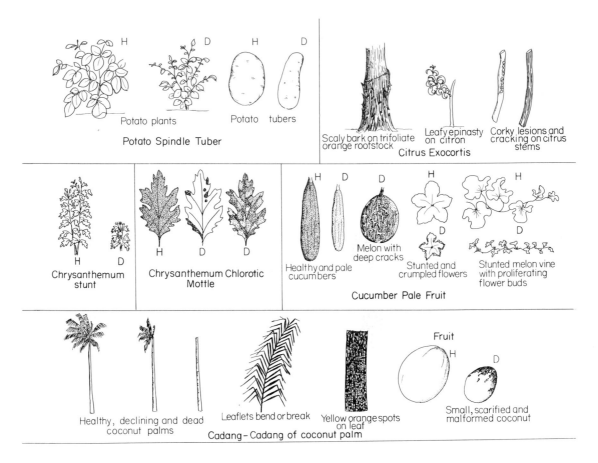

FIGURE 231.
Kinds of symptoms caused by viroids.

washing of hands or sterilizing of tools after handling viroid-infected plants, before moving on to healthy plants.

- *Potato Spindle Tuber*

The potato spindle tuber disease occurs in the U.S. and Canada, Russia, and South Africa. It causes quite severe losses and, in some regions, it is one of the most destructive diseases of potatoes. It attacks all varieties, spreads rapidly, and often occurs in combination with virus diseases. It also attacks tomato but seems to be of little economic importance in that crop.

Symptoms.   Infected potato plants appear erect, spindly and dwarfed (Fig. 232). The leaves are small and erect and the leaflets are darker green and sometimes show rolling and twisting. The tubers are elongated, with a cylindrical middle and tapering ends. Tubers are smoother, with a more tender skin and flesh, but tuber eyes are more numerous, more conspicuous and shallower. Yields are reduced considerably, often by 25 percent or more. Susceptible tomato plants are stunted and have smaller rugose

FIGURE 232.
Symptoms caused by potato spindle tuber viroid. (A) Diseased potato plant (left) showing stunted and upright growth. (B) Diseased tubers (below) are spindle shaped and smaller, compared to healthy tubers (above). (C) Tomato plants cultivar Rutgers. L: Healthy. R: Twenty days after inoculation with potato spindle tuber viroid. (Photo C courtesy T. O. Diener.)

leaves with necrosis of petioles and veins. Diseased tomatoes have a bunchy-top appearance.

The pathogen: Potato spindle tuber viroid (PSTV). It is the first recognized viroid and many of its properties are still being elucidated. PSTV is an infectious RNA of low molecular weight, approximately 80,000 daltons. The RNA is a single-stranded molecule with extensive regions of base pairing. Under the electron microscope, purified but apparently denatured PSTV appears as short strands about 50 nm long and has the thickness of a double-stranded DNA (Fig. 230). Sap from infected plants is still infective after dilution of 1:1000 to 1:10,000, and after heating for 10 minutes at 75 to 80°C. PSTV is quickly inactivated in expressed sap of infected plants but the infectivity can be preserved by treatment of the sap with phenol. Phenol inhibits the activity of the enzyme ribonuclease that breaks down the viroid RNA.

PSTV is mechanically transmissible and is spread primarily by knives used to cut healthy and infected potato "seed" tubers and during handling and planting of the crop. PSTV seems to also be transmitted by pollen and seed and by several insects including some aphids, grasshoppers, flea beetles, and bugs. Insect transmission is apparently nonspecific and incidental, i.e., on contaminated mouthparts and feet of insects visiting the plants.

Development of disease. Following inoculation of a tuber with PSTV by means of a contaminated knife, or of a growing plant with sap from an infected plant, the viroid replicates itself and spreads systemically throughout the plant. There is no information on the mechanism of replication or spread of the viroid within the plant nor on the mechanism(s) by which the viroid brings about development of symptoms in infected plants.

Control. Potato spindle tuber can be controlled effectively by planting only PSTV-free potato tubers in fields free of diseased tubers that may have survived from the previous year's crop.

• *Citrus Exocortis*

Exocortis is worldwide in distribution and affects trifoliate oranges, citranges, Rangpur and other mandarin and sweet limes, some lemons, and citrons. Orange, lemon, grapefruit, and other citrus trees grafted on exocortis-sensitive rootstocks show slight to great reductions in growth, and yields are reduced by as much as 40 percent.

Symptoms. Infected susceptible plants show vertical splits in the bark and narrow, vertical, thin strips of partially loosened outer bark that give the bark a cracked and scaly appearance (Fig. 233). Since many of the exocortis-susceptible plants, such as trifoliate orange, are used primarily as rootstocks for other citrus trees, and because the scions make poor growth on such rootstocks, the enlarged, scaly rootstocks have given the disease the name "scaly butt." Infected exocortis-susceptible plants may also show yellow blotches on young infected stems, and some citrons show leaf and stem epinasty, and cracking and darkening of leaf veins and petioles. All infected plants usually appear stunted to a smaller or greater extent and have lower yields.

The pathogen: Citrus exocortis viroid (CEV). It is apparently similar to, but not identical with the potato spindle tuber viroid. CEV is readily transmitted from diseased to healthy trees by budding knives, pruning shears, or other cutting tools, by hand, and possibly by scratching and gnawing of animals; CEV is also transmitted by dodder and by sap to *Gynura, Petunia,* and other herbaceous plants. On contaminated knife blades CEV retains its infectivity for at least 8 days and, when partially purified, CEV remains infective at room temperature for several months. The thermal inactivation point of extracted sap is about 80°C for 10 minutes, but partially purified CEV remains infectious even after boiling for 20 minutes. The viroid also survives brief heating of contaminated blades in the flame of a propane torch (blade temperature about 260°C!) and flaming of blades dipped in alcohol. The viroid also survives on contaminated blades treated with almost all common chemical sterilants except sodium hypochlorite solution.

Development of disease. CEV survives in most citrus and many

FIGURE 233.
Exocortis symptoms on the trifoliate rootstock portion of an orange tree. (Photo courtesy L. C. Knorr, Agric. Res. Educ. Center, Lake Alfred, Fla.)

herbaceous hosts and is spread to healthy citrus plants by budding or grafting and by contaminated cutting tools or other cultivating equipment. The viroid apparently enters the phloem elements and spreads in them throughout the plant. The viroid seems to be associated with the nuclei and internal membranes of host cells and results in aberrations of the plasma membranes. Although the viroid apparently lacks the ability to serve as a messenger molecule or as an amino acid acceptor, it brings about several metabolic changes in infected plants. These changes include an increase in oxygen uptake and respiration, and also in sugars and certain enzymes. Marked changes also occur in several amino acids.

Control. Exocortis can be controlled only by propagating exocortis-free nursery trees from certified healthy foundation stock and use of sanitary budding, nursery, and field practices. Tools should be disinfected between cuts into different plants by dipping in a 10 to 20 percent solution of household bleach (sodium hypochlorite).

- *Chrysanthemum Stunt*

It occurs in the U.S., Canada, England, and the Netherlands. It causes mild to severe losses in florists' and garden-type chrysanthemums and, unless closely watched, it can reach epidemic proportions.

Chrysanthemum plants and their flowers are smaller, paler, and of inferior quality compared to normal ones. Some flowers may appear bleached. Diseased flowers open 7 to 10 days earlier than normal. Axillary buds often grow prematurely and produce an excessive number of branches and stolons. Some varieties show white flecks or yellow blotches on the leaves. Cuttings from infected plants root poorly.

The pathogen: Chrysanthemum stunt viroid (ChSV). It is transmitted through sap. It has a dilution end point of 10^{-4}, a thermal inactivation point of 96 to 100°C for 10 minutes and retains its infectivity for 2 months in sap and for 2 years in dried leaves. ChSV is spread readily in sap carried on the fingers or on knives or tools used during cultural practices such as pruning or pinching plants, taking cuttings, cutting flowers, etc. ChSV is not transmitted by insect or other vectors. ChSV moves slowly through a plant, often taking 5 to 6 weeks to move out of an inoculated leaf into the stem, and new symptoms develop 3 to 4 months from inoculation. ChSV survives mainly in infected plants which, being perennial, carry it over to the next season. Plants may also be contaminated with ChSV surviving in dead plant parts in the soil.

Control of ChSV is obtained only through use of certified viroid-free propagating stock. Plants infected with ChSV should be removed and destroyed.

SELECTED REFERENCES

Dickson, E., W. Prensky, and H. D. Robertson. 1975. Comparative studies of two viroids: Analysis of potato spindle tuber and citrus exocortis viroids by RNA fingerprinting and polyacrylamide-gel electrophoresis. *Virology* **68**:309–316.

Diener, T. O. 1971. Potato spindle tuber "virus." IV. A replicating, low molecular weight RNA. *Virology* **45**:411–428.

Diener, T. O. 1974. Viroids: The smallest known agents of infectious disease. *Ann. Rev. Microbiol.* **28**:23–39.

Diener, T. O., and R. H. Lawson. 1973. Chrysanthemum stunt: A viroid disease. *Virology* **51**:94–101.

Folsom, D. 1923. Potato spindle tuber. *Maine Agr. Expt. Sta. Bull.* **312**:4 p., illus.

Hall, T. C., *et al.* 1974. Functional distinctions between the ribonucleic acids from citrus exocortis viroid and plant viruses: cell-free translation and aminoacylation reactions. *Virology* **61**:486–492.

Horst, R. K., and P. E. Nelson. 1975. Diseases of chrysanthemum. *Cornell Univ. Information Bull.* **85**: 36 p.

Price, W. C. 1971. Cadang-cadang of coconut—a review. *Plant Sci.* **3**:1–13.

Price, W. C. 1972. "Proceedings, 5th Conf. Int. Organ. Citrus Virologists." Univ. Florida Press, Gainesville. 301 p.

Randles, J. W. 1975. Association of two ribonucleic acid species with cadang-cadang disease of coconut palms. *Phytopathology* **65**:163–167.

Romaine, C. P., and R. K. Horst. 1975. Suggested viroid etiology for chyrsanthemum chlorotic mottle disease. *Virology* **64**:86–95.

Semanchik, J. S., *et al.* 1976. Exocortis disease: subcellular distribution of pathogenic (viroid) RNA. *Virology* **69**:669–676.

Takahashi, T., and T. O. Diener. 1975. Potato spindle tuber viroid XIV. Replication in nuclei isolated from infected leaves. *Virology* **64**:106–114.

Van Dorst, H. J. M., and D. Peters. 1974. Some biological observations on pale fruit, a viroid-incited disease of cucumber. *Neth. J. Pl. Pathol.* **80**:85–96.

15
plant
diseases
caused
by
nematodes

introduction

Nematodes are one of the plant parasites belonging to the animal kingdom that are studied in plant pathology. Nematodes, sometimes called eelworms, are wormlike in appearance but quite distinct taxonomically from the true worms. Most of the several thousand species of nematodes live in great numbers freely in fresh or salt waters or in the soil feeding on microscopic plants and animals. Numerous species of nematodes attack and parasitize man and animals, on which they cause various diseases. Several hundred species, however, are known to feed on living plants as parasites and to cause a variety of plant diseases.

characteristics of plant-pathogenic nematodes

MORPHOLOGY Plant-parasitic nematodes are small, 300 to 1000 μm with some up to 4 mm long by 15 to 35 μm wide (Fig. 234). Their small diameter makes them invisible to the naked eye, but they can be easily observed under the microscope. Nematodes are, in general, eel-shaped, and round in cross section, with smooth, unsegmented bodies, without legs or other appendages. The females of some species, however, become swollen at maturity and have pear-shaped or spheroid bodies (Fig. 235).

ANATOMY The nematode body is more or less transparent (Fig. 234). It is covered by a colorless cuticle which is usually marked by striations or other markings. The cuticle molts when nematodes go through their

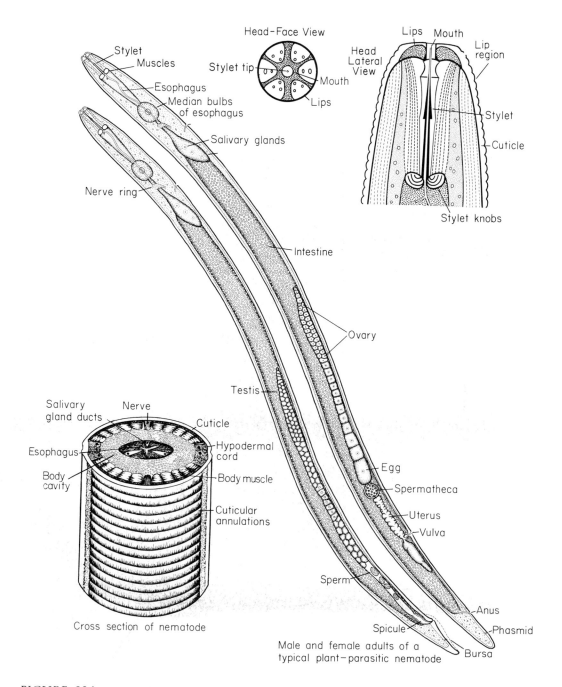

FIGURE 234.
Morphology and main characteristics of typical male and female plant parasitic nematodes.

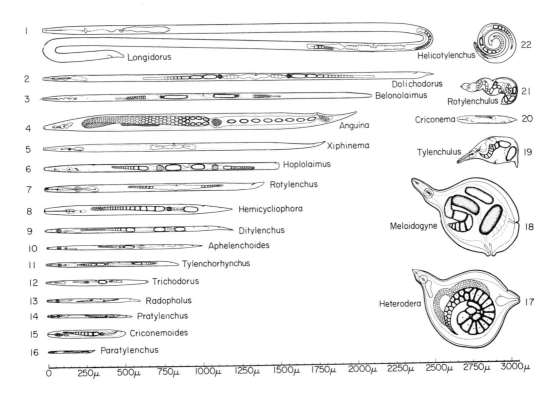

FIGURE 235.
Morphology and relative size of the most important plant-parasitic nematodes.

successive larval stages. The cuticle is produced by the hypodermis, which consists of living cells and extends into the body cavity as four chords separating four bands of longitudinal muscles. These muscles enable the nematode to move. Additional specialized muscles exist at the mouth and along the digestive tract and the reproductive structures.

The body cavity contains a fluid through which circulation and respiration take place. The digestive system is a hollow tube extending from the mouth through the esophagus, intestine, rectum, and anus. Lips, usually six in number, surround the mouth. All plant-parasitic nematodes have a hollow stylet or spear which is used to puncture plant cells.

The reproductive systems are well developed. Female nematodes have one or two ovaries followed by an oviduct and uterus terminating in a vulva. The male reproductive structure is similar to the female but there is a testis, seminal vesicle, and a terminus in a common opening with the intestine. A pair of protrusible, copulatory spicules are also present in the male. Reproduction in nematodes is through eggs and may be sexual, hermaphroditic, or parthenogenetic. Many species lack males.

LIFE CYCLES   The life histories of most plant parasitic nematodes are, in general, quite similar. Eggs hatch into larvae, whose appearance and structure are usually similar to those of the adult nematodes. Larvae grow in size and each larval stage is terminated by a molt. All nematodes have four larval stages, with the first molt usually occurring in the egg. After the final molt the nematodes differentiate into adult males and females. The female can then produce fertile eggs either after mating with a male or, in the absence of males, parthenogenetically, or can produce sperm herself.

A life cycle from egg to egg may be completed within 3 or 4 weeks under optimum environmental, especially temperature, conditions, but will take longer in cooler temperatures. In some species of nematodes the first or second larval stages cannot infect plants and depend for their metabolic functions on the energy stored in the egg. When the infective stages are produced, however, they must feed on a susceptible host or starve to death. Absence of suitable hosts may result in the death of all individuals of certain nematode species within a few months, but in other species the larval stages may dry up and remain quiescent, or the eggs may remain dormant in the soil for years.

ECOLOGY AND SPREAD   Almost all plant-pathogenic nematodes live part of their lives in the soil. Many of these live freely in the soil, feeding superficially on roots and underground stems, but even in the specialized sedentary parasites, the eggs, the preparasitic larval stages, and the males are found in the soil for all or part of their lives. Soil temperature, moisture, and aeration affect survival and movement of nematodes in the soil. Nematodes occur in greatest abundance in a layer of soil from 0 to 15 cm deep, although distribution of nematodes in cultivated soil is irregular and is greatest in or around roots of susceptible plants, which they follow sometimes to considerable depths (30 to 150 cm or more). The greater concentration of nematodes in the region of host plant roots is due primarily to their more rapid reproduction on the food supply available and also to attraction of nematodes by substances released into the rhizosphere. To these must be added the so-called hatching factor effect of substances originating from the root which diffuse into the surrounding soil and markedly stimulate the hatching of eggs of certain species. Most nematode eggs, however, hatch freely in water in the absence of any special stimulus.

Nematodes spread through the soil very slowly under their own power. The overall distance traveled by a nematode probably does not exceed a meter per season. Nematodes move faster in the soil when the pores are lined with a thin (a few micrometers) film of water than when the soil is waterlogged. In addition to their own movement, however, nematodes can be easily spread by anything that moves and can carry particles of soil. Farm equipment, irrigation, flood or drainage water, animal feet, and dust storms spread nematodes in local areas, while over long distances nematodes are spread primarily with farm produce and nursery plants. A few nematodes that attack aboveground parts of plants not only spread through the soil as described above, but they are also splashed to the plants by falling rain or overhead watering, or they ascend wet plant stem or leaf surfaces on their own power. Further spread takes place upon contact of infected plant parts with adjacent healthy plants.

CLASSIFICATION   All plant-parasitic nematodes (Fig. 235) belong to the phylum Nemathelminthes, class Nematoda. Most of the important parasitic genera belong to the subclass Secernentea, order Tylenchida:

Superfamily: Tylenchoidea
  Family:
    Tylenchidae
      Genus: *Anguina*, wheat or seed-gall nematode
            Ditylenchus, stem or bulb nematode of alfalfa, onion, narcissus, etc.
            Tylenchorhynchus, stunt nematode of tobacco, corn, cotton, etc.
    Heteroderidae
      Genus: *Heterodera*, cyst nematode of potato, tobacco, soybean, sugar beets, cereals, etc.
            Meloidogyne, root-knot nematode of almost all crop plants
    Hoplolaimidae
      Genus: *Helicotylenchus* and *Rotylenchus*, spiral nematodes of various plants
            Hoplolaimus, lance nematode of corn, sugarcane, cotton, alfalfa, etc.
            Pratylenchus, lesion nematode of almost all crop plants and trees
            Radopholus, burrowing nematode of banana, citrus, coffee, sugarcane, etc.
            Rotylenchulus, reniform nematode of cotton, papaya, tea, tomato, etc.
            Belonolaimus, sting nematode of cereals, legumes, cucurbits, etc.
            Dolichodorus, awl nematode of celery, corn, bean, etc.
    Tylenchulidae
      Genus: *Tylenchulus*, citrus nematode of citrus, grapes, olive, lilac, etc.
    Criconematidae
      Genus: *Criconema* and *Criconemoides*, ring nematodes of woody perennials, turf, peanuts, etc.
            Paratylenchus, pin nematode of various plants
            Hemicycliophora, sheath nematode of various plants
Superfamily: Aphelenchoidea
  Family:
    Aphelenchoididae
      Genus: *Aphelenchoides*, foliar nematode of chrysanthemum, strawberry, begonia, rice, coconut, etc.

Three important genera of nematodes belong to the subclass Adenophorea, order Dorylaimida:

  Family:
    Tylencholaimidae
      Genus: *Longidorus*, needle nematode of some plants
            Xiphinema, dagger nematode of trees, woody vines, and of many annuals
    Trichodoridae
      Genus: *Trichodorus*, stubby root nematode of vegetables and field crops

In terms of habitat, pathogenic nematodes are either *ectoparasites*, i.e., species that do not normally enter root tissue but feed only on the cells near the root surfaces, or *endoparasites*, i.e., species that enter the host and feed from within. Both of these can be either *migratory*, i.e., they live freely in the soil and feed on plants without becoming attached, or move around inside the plant, or *sedentary*, i.e., species that, once within a root, do not move about. The ectoparasitic nematodes include the ring nematodes (sedentary) and the dagger, stubby root, and sting nematodes (all migratory). The endoparasitic nematodes include the root knot, cyst, and citrus nematodes (all sedentary), and the lesion, stem and bulb, burrowing, leaf, stunt, lance, and spiral nematodes (all somewhat migratory). Of these, the cyst, lance, and spiral nematodes may be somewhat ectoparasitic, at least during part of their lives.

isolation of nematodes

Plant-parasitic nematodes are generally isolated from the roots of plants they infect or from the soil surrounding the roots on which they feed (Fig. 236). A few kinds of nematodes, however, attack aboveground plant parts, e.g., chrysanthemum foliar nematode, grass and gall nematode, and the stem, leaf, and bulb nematode, and these can be isolated primarily from the plant parts they infect.

FROM SOIL

Using a freshly collected soil sample of about 100 to 300 cc, the nematodes in it can be isolated by either the Baermann funnel method or by sieving.

A Baermann funnel consists of a fairly large (12- to 15-cm diameter) glass funnel to which a piece of rubber tubing is attached, with a pinch-cock placed on the tubing. The funnel is placed on a stand and filled with water. The soil sample is placed in the funnel on porous, wet-strength paper, sometimes supported by a 5- to 6-cm circular piece of screen, or in a beaker over which a piece of cloth is fastened with a rubber band. The beaker is then inverted in the funnel with the cloth and all the soil below the surface of the water and allowed to stand overnight or for several hours. The live nematodes move actively and migrate through the cloth or porous paper into the water and sink to the bottom of the rubber tubing just above the pinchcock. Over 90 percent of the live nematodes are recovered in the first 5 to 8 ml of water drawn from the rubber tubing and this sample is placed in a shallow dish for examination and, if desired, single nematode isolation.

The sieving method is based on the fact that when a small soil sample, e.g., 300 cc, is mixed with considerably more water, e.g., 2 liters, the nematodes float in the water and can be collected on sieves with pores of certain sizes. Thus, the soil–water mixture is stirred and then allowed to

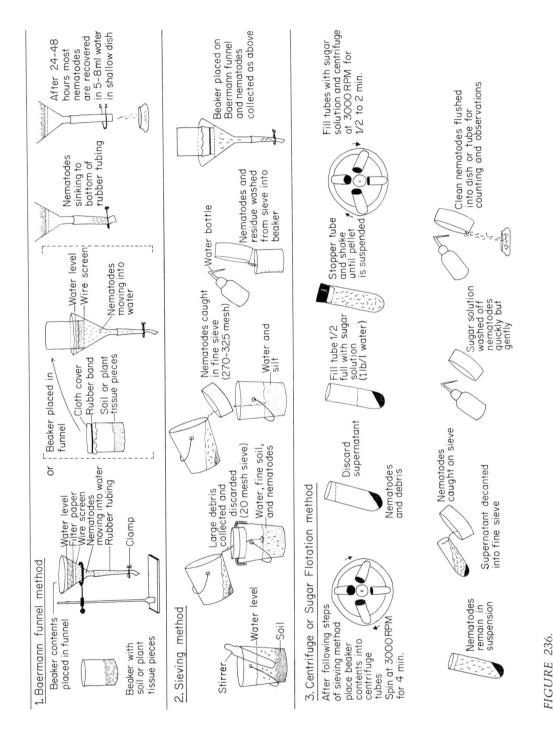

FIGURE 236.
Methods of isolation of nematodes from soil or plant tissues.

stand for 30 seconds. The supernatant is poured through a 20-mesh sieve (20 holes per sq. inch) which holds large debris but allows the nematodes to pass into a bucket. The liquid containing the nematodes is then poured through a 60-mesh sieve which holds the larger nematodes and some debris but lets the smaller ones pass through into another bucket. The latter is then passed through a 200-mesh sieve which holds the small nematodes and some debris. Both the 60- and the 200-mesh sieves are washed 2 or 3 times to remove as much of the debris as possible and the nematodes are then washed into shallow dishes for direct examination and further isolation.

FROM PLANT MATERIAL

Regardless of the type of plant material containing the nematodes, it is cut into very small pieces by hand or by use of a blender for a few seconds, and is then placed in the Baermann funnel as described above. The nematodes leave the tissue and move into the water in the tubing from where they are collected in a shallow dish.

symptoms
caused by nematodes

Nematode infections of plants result in the appearance of symptoms on roots as well as on aboveground parts of plants (Fig. 237). Root symptoms may appear as root knots or root galls, root lesions, excessive root branching, injured root tips, and root rots when nematode infections are accompanied by plant-pathogenic or saprophytic bacteria and fungi.

These root symptoms are usually accompanied by noncharacteristic symptoms in the aboveground parts of plants appearing primarily as reduced growth, symptoms of nutrient deficiencies such as yellowing of foliage, excessive wilting in hot or dry weather, reduced yields, and poor quality of products.

Certain species of nematodes invade the aboveground portions of plants rather than the roots, and on these they cause galls, necrotic lesions and rots, twisting or distortion of leaves and stems, and abnormal development of the floral parts. Certain nematodes attack grains or grasses forming galls full of nematodes in place of seed.

how nematodes affect plants

Nematodes damage plants only slightly by direct mechanical injury inflicted upon the plants during feeding. Most of the damage seems to be caused by a secretion of saliva injected into the plants while the

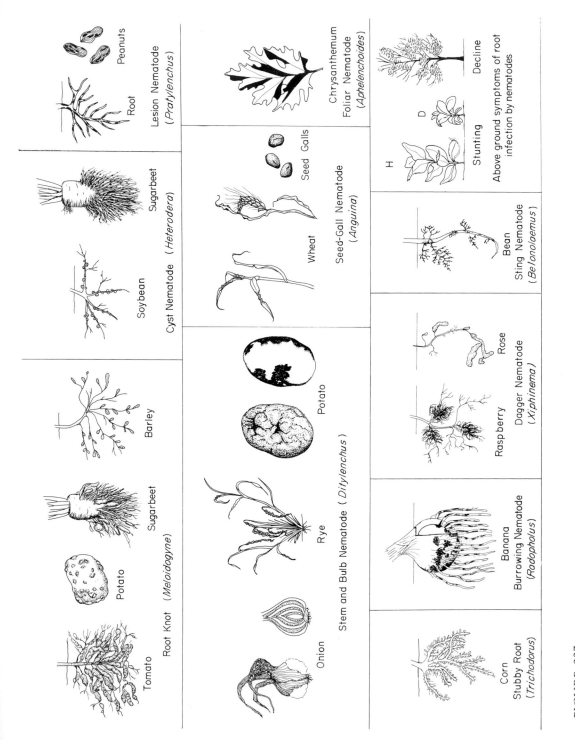

FIGURE 237.
Types of symptoms caused by the most important plant parasitic nematodes.

nematodes are feeding. Some nematode species are rapid feeders. They puncture a cell wall, inject saliva into the cell, suck part of the cell contents, and move on within a few seconds. Others feed much more slowly and may remain at the same puncture for several hours or days. These, as well as the females of species which become permanently established in or on roots, inject saliva intermittently as long as they are feeding.

The feeding process causes the affected plant cells to react resulting in dead or devitalized root tips and buds, lesion formation and tissue breakdown, swellings and galls of various kinds, and crinkled and distorted stems and foliage. Some of these manifestations are caused by dissolution of infected tissues by nematode enzymes, which, with or without the help of toxic metabolites, cause tissue disintegration and death of cells. Others are caused by abnormal cell enlargement (hypertrophy), by suppression of cell divisions, or by stimulation of cell division proceeding in a controlled manner and resulting in the formation of galls or of large numbers of lateral roots at or near the points of infection.

Plant disease syndromes caused by nematodes are complex. Root-feeding species probably decrease the ability of plants to take up water and nutrients from soil and thus cause symptoms of water and nutrient deficiencies in the aboveground parts of plants. However, it is the plant–nematode biochemical interactions, which impair the overall physiology of plants, and the role nematodes play in providing courts for entry of other pathogens that are primarily responsible for plant injury; the mechanical damage or withdrawal of food from plants by nematodes is generally less significant.

interrelationships between nematodes and other plant pathogens

Although nematodes can cause diseases to plants by themselves, most of them live and operate in the soil where they are constantly surrounded by fungi and bacteria, many of which can also cause plant diseases. In many cases an association develops between nematodes and certain of the other pathogens. Nematodes then become a part of an etiological complex resulting in a combined pathogenic potential far greater than the sum of the damage either of the pathogens can produce individually.

Several nematode–fungus disease complexes are known. *Fusarium* wilt of several plants increases in incidence and severity when the plants are also infected by the root knot, lesion, sting, reniform, burrowing, or stunt nematodes. Similar effects have also been noted in disease complexes involving nematodes and *Verticillium* wilt, *Pythium* damping off, *Rhizoctonia* and *Phytophthora* root rots, and in some other instances. In none of these cases is the fungus transmitted by the nematode. However, plant varieties susceptible to the respective fungi are damaged even more when the plants are infected with the nematodes, the combined damage being considerably greater than the sum of the damage caused by each pathogen acting alone. Also, varieties ordinarily resistant to the fungi apparently become infected by them after previous infection by nematodes. The importance of nematodes in these complexes is indicated by the fact that soil fumigation aimed at eliminating the nematode, but not the fungus, greatly reduces the incidence and the damage caused by the fungus-induced disease.

Although it seems quite probable that the mechanical wounding caused to plants by nematodes is an important factor in providing avenues of entry for the fungus, the continuation of the effect that nematodes have on host susceptibility in later stages of plant development suggests that the nematodes may also cause some host response that lowers natural resistance to the fungus. It should also be noted that, in at least some such complexes, there is a greater mass of mycelium present in nematode-infected than in nematode-free tissues of the same plant and also that higher populations of nematodes are present in fungus-infected than in fungus-free tissues of a diseased plant.

Relatively few cases of nematode–bacterial disease complexes are known. Thus, the root-knot nematode increases the frequency and severity of the bacterial wilt of tobacco caused by *Pseudomonas solanacearum*, of the bacterial wilt of alfalfa caused by *Corynebacterium insidiosum*, and of the bacterial scab of gladiolus caused by *Pseudomonas marginata*. In most of these the nematode role seems to be that of providing the bacteria with an infection court and to assist bacterial infection by wounding the host. On the other hand, root infection of plum trees with the ring nematode *Criconemoides xenoplax* changed the physiology of the trees and resulted in the development of more extensive cankers by the bacterium *Pseudomonas syringae* on branches of nematode-infected trees than on nematode-free trees.

Much better known are the interrelationships between nematodes and viruses. Several plant viruses such as grapevine fanleaf, arabis mosaic, tobacco ringspot, tomato ringspot, tomato black ring, raspberry ringspot, tobacco rattle, and pea early browning virus arc transmitted through the soil by means of nematode vectors. All these viruses, however, are transmitted by only one or more of the three genera of dagger, needle, and stubby root nematodes: *Xiphinema*, *Longidorus*, and *Trichodorus*. *Xiphinema* and *Longidorus* transmit only round, i.e., polyhedral viruses, which include most of the nematode-transmitted viruses, while *Trichodorus* transmits two rod- or tubular-shaped viruses, tobacco rattle and pea early browning viruses. These nematodes can transmit some of the viruses after feeding on infected plants for as short a time as one hour, but the percentage of transmission increases with longer feedings up to four days. Once they have acquired virus from an infected plant the nematodes remain infective for periods of 2 to 4 months and sometimes even longer. All stages, larval and adult nematodes, can transmit viruses, but the virus is not carried from one larval stage to another and to adults through molts nor does the virus pass from adults, through eggs, to larvae. Although nematodes can ingest and carry within them several plant viruses, they can only transmit certain of them to healthy plants, which suggests that there is a close biological association between the nematode vectors and the viruses they can transmit.

control of nematodes

Several methods of effectively controlling nematodes are available although certain factors, such as expense and types of crops, limit their applicability in some cases. Four general types of control methods are employed. Control through cultural practices, biological control through resistant varieties and certain other means, control by means of physical agents, e.g., heat, and control through chemicals. Usually a combination of several of these methods is employed for controlling nematode diseases of plants in practice.

CULTURAL PRACTICES

They result in partial or complete control of nematodes and include:

CROP ROTATION TO NONHOST PLANTS   Several nematode species can infect only a few crops. Since plant-pathogenic nematodes are obligate parasites, the absence of susceptible hosts from the soil for 2 to 3 years results in elimination of the nematodes from that area through starvation and inability to reproduce. This method, of course, requires knowledge of the kinds of nematodes present in the soil and of what plants are resistant or susceptible to them. It is limited by the fact that many other nematodes can attack a wide variety of plants, thus limiting the choice of nonhost plants; it is impractical for permanent plants, such as orchards, and when the nonhost crops are drastically different from the one previously grown and require specialized production know-how and equipment.

SANITARY PRACTICES   They include cleaning of all machinery thoroughly before moving into an uncontaminated area; taking care not to bring nematodes into a field by means of contaminated nursery stock, seed, containers, etc.; maintaining soil free from host plants which deprive nematodes of roots on which to feed.

FLOODING   Some species of nematodes have become adapted to living in the normal soil under ordinary moisture content and depend on a certain amount of aeration present in it. Flooding of the land for a period of several months results in the death of these nematodes and thus frees the land from these pathogens. However, in only very few fields is it possible or practical to flood the land for long periods and, therefore, the applicability of this control method is quite limited.

RESISTANT VARIETIES

Several crop varieties resistant to nematodes are available and others are under development. When resistant varieties with the desired horticultural qualities are available, their cultivation, instead of the susceptible ones, provides the most convenient and least expensive way of combatting nematodes. Experimental control of nematodes has also been obtained by interplanting with plants such as marigolds that are toxic to nematodes or with nematode-trapping plants such as *Hesperis* for the sugarbeet nematode and *Crotalaria* for the root-knot nematode. Trap crops attract the nematodes away from the susceptible crop plants and although the nematodes enter the trap plants they fail to develop in them. Addition of organic matter or mulching has also controlled nematodes experimentally, presumably by increasing the populations of certain fungi and predatory nematodes which feed on plant parasitic nematodes.

HEAT TREATMENT

Two types of heat treatment are effective in controlling nematodes. Raising the temperature of the soil to about 50°C for 30 minutes by means of steam or hot water is sufficient to kill most nematodes and nematode eggs. The commonly practiced soil "sterilization" at 82°C for 30 minutes eradicates all nematodes along with practically all other soil organisms. Heat treatment is the most effective and most commonly used method for treating soil to be used for a container or greenhouse bench crop. It is sometimes used for treatment of ground beds and small outdoor areas but is too expensive and impractical for use in large acreages. Its use, even in the greenhouse, is limited to preplanting applications.

Hot-water dip treatments are used to a limited extent to eradicate nematodes from within roots, bulbs, etc., and also those clinging to the surfaces of roots or other propagative materials of vegetatively propagated nursery stock before they are planted in nematode-free soil. Temperatures varying from 43 to 53°C for periods of time varying from a few minutes to 30 minutes and up to 4 hours for the lower temperatures are usually employed for this purpose. The ability to withstand such temperatures varies from plant to plant and with the stage of growth of the plant, dormant plants or plant organs being more tolerant to high temperatures than actively growing ones. Frequently the margin of safety between the temperature lethal to nematodes and injurious to plants is very narrow, and extreme precautions must be taken to avoid damaging the plants treated.

CHEMICAL CONTROL

The most promising method of controlling nematodes in the field has been through the use of chemicals called nematicides. Some of these, including chloropicrin, methyl bromide, Mylone, Vapam, and Vorlex, give off

gases after application to the soil and are general-purpose preplant fumigants; they are effective against a wide range of soil microorganisms
including, in addition to all nematodes, many fungi, insects, and weeds.
Other nematicides, e.g., Nemagon, Zinophos, Dasanit, Furadan, Mocap,
and Temik, are of low volatility, are effective against nematodes and
insects, and can be applied before and after planting of many, particularly
non-food, crops which are tolerant to these chemicals.

Nematicides used as soil fumigants are available as liquids, emulsi-
fiable concentrates, or granules. Application of nematicides in the soil is
made either by applying the chemical evenly over the entire field (broad-
cast) or by applying it only to the rows to be planted with the crop (row
treatment). In both cases the fumigant is applied through delivery tubes
attached at the back of tractor-mounted chisel-tooth injection shanks or
disks spaced at variable widths and usually reaching 6 inches below the
soil surface. The nematicide is covered instantly by a smoothing and
firming drag or can be mixed into the soil with disk harrows or rototillers.
Highly volatile nematicides should be immediately covered with
polyethylene sheeting (Fig. 238), and this should be left in place for at
least 48 hours. When small areas are to be fumigated, the most convenient method is through injection of the chemical with a hand applicator
or by placement of small amounts of granules in holes 6 inches deep, 6 to
12 inches apart, and immediately covering the holes with soil. In all cases

FIGURE 238.
Soil fumigation for the control of nematodes. Plastic sheet covers soil to keep
volatile nematicides from escaping too soon. (Photo courtesy U.S.D.A.)

of preplant soil fumigation with phytotoxic nematicides, at least 2 weeks must elapse from the time of treatment before seeding or planting in the field to avoid plant injury.

In the above types of nematicide application, only a small portion of the soil and its microorganisms come in contact with the chemical immediately. The effectiveness of the fumigants, however, is based on the diffusion of the nematicides in a gaseous state through the pores of the soil throughout the area in which nematode control is desired. The distance of movement of the vapors is influenced by the size and continuity of soil pores, soil temperature (best range between 10 and 20°C), soil moisture (best at about 80 percent of field capacity), and by the type of soil (more material is required for soils rich in colloidal or organic matter). Nematicides with low volatility, such as Nemagon, Furadan, and Zinophos, do not diffuse through the soil to any great extent and must be mixed with the soil mechanically or by irrigation water or rainfall. Most nematicides, with the exception of the highly volatile ones, can be applied in irrigation water as soaks or drenches, but only low-volatility nematicides can be applied through overhead sprinkler systems.

The most common nematicides and some of their properties are listed in Table IV. In practice, nematode control in the field is generally obtained by preplant soil fumigation with one of the nematicides listed in Table IV as "applied only before planting." These chemicals are nonspecific, i.e., they control all types of nematodes, although some nematodes are harder to control than others no matter what the nematicide. Chloropicrin, methyl bromide, Mylone, and Vapam are expensive, broad spectrum nematicides that must be covered upon application and are therefore used for treatment of seedbeds and small areas. On the other hand, DD and EDB are cheaper, need not be covered upon application and are used for treatment of large fields. DD and EDB, however, control only nematodes and, therefore, mixtures of these with Vapam, Chloropicrin, or methyl bromide are often used to increase their fungicidal action. The chemicals listed as "contact nematicides applied before or after planting" can be used as preplant soil treatment of all types or crops, but their use after planting is limited to nonfood crops such as turf, ornamentals, nurseries, and young nonbearing orchard trees, and to a few, specific food crops for which each chemical has received clearance by the Food and Drug Administration. It should be noted that most of the contact nematicides were developed as insecticides, and nematicidal dosages are much higher than insecticidal ones. All nematicides are extremely toxic to humans and animals and should be handled with great caution.

SELECTED REFERENCES

Anonymous. 1972 and annually afterwards. "Commonw. Inst. of Helminthology Descriptions of Plant-Parasitic Nematodes." Commonw. Agric. Bureaux, England.

Christie, J. R. 1959. "Plant Nematodes, Their Bionomics and Control." Fla. Univ. Agr. Expt. Sta. (Gainesville). 256 p.

Endo, B. Y. 1975. Pathogenesis of nematode-infected plants. *Ann. Rev. Phytopathol.* **12**:213–238.

TABLE IV.
NAMES AND PROPERTIES OF THE MOST COMMON NEMATICIDES

Trade Name	Chemical	Control	Volatility	Form[a]
I. *Soil fumigants applied only before planting*				
Chloropicrin, Larvacide 100, Picfume, Chlor-O-Pic	Chloropicrin	Nematodes, soil fungi, soil insects, weed seeds	Very high	L or G
DD or Telone, Vidden-D, Telone-II	Dichloropropene–Dichloropropane	Nematodes, soil insects	Moderate	L
EDB, Dowfume 85, Soilbrom 85	Ethylene dibromide	Nematodes, soil insects	Moderate	L
Methyl-bromide, MC-2 or Brom-O-Gas, MC-33, Brozone, Terrogas 70	Methyl bromide (usually with a small amount of chloropicrin added)	Nematodes, soil insects, soil fungi, weed seeds	Very high	L or G
Mylone	Dimethyltetrahydro-thiadiazinethione, or DMTT, dazomet	Nematodes, soil fungi, weed seeds, soil insects	High	WP or GR
Vapam	Sodium methyl-dithio-carbamate, or SMDC	Nematodes, some soil fungi, germinating weed seeds, soil insects	High	L
Vorlex	Methyl isothiocyanate–dichloropropene mixture, or MITC	Nematodes, soil fungi, weed seeds, soil insects	High	L
II. *Contact nematicides applied before or after planting*				
DBCP, Nemagon, Fumazone	Dibromochloro-propane	Nematodes, damping-off fungi (Pythium)	Low to moderate	L or GR
VC-13, Mobilawn	Dichlofenthion	Nematodes, soil insects	Low	L
Zinophos, Cynem, Nemaphos	Thionazin	Nematodes, soil insects	Low	L or GR
Dasanit, Terracur	Fensulfothion	Nematodes, soil insects	Low	L or GR
Systox, Demox	Demeton	Nematodes, soil insects	Low	L
Di-syston	Disulfoton	Nematodes, soil insects	Low	L or GR
Mocap	Ethoprop	Nematodes, soil insects	Low	L or GR
Furadan, Curaterr	Carbofuran	Nematodes, soil insects	Low	GR
Vydate	Oxamyl	Nematodes, soil insects	Low	L or GR
Temik	Aldicarb	Nematodes, soil insects	Low	GR
Nemacur	Phenamophos	Nematodes, soil insects	Low	L or GR

[a] L = liquid; G = gas; WP = wettable powder; GR = granules.

NOTE: Some of the above chemicals are presently being reviewed by FDA and EPA for possible side effects and may be discontinued.

Jenkins, W. R., and D. P. Taylor. 1967. "Plant Nematology." Reinhold, New York, 270 p.

Krusberg, L. R. 1963. Host response to nematode infection. *Ann. Rev. Phytopathol.* **1**:219–240.

Nusbaum, C. J., and H. Ferris. 1973. The role of cropping systems in nematode population management. *Ann. Rev. Phytopathol.* **11**:423–440.

Peachy, J. E. (Ed.). 1969. "Nematodes of Tropical Crops." *Tech. Commun. Commonw. Bur. Helminth.* No. 40. 355 p.

Powell, N. T. 1971. Interactions between nematodes and fungi in disease complexes. *Ann. Rev. Phytopathol.* **9**:253–294.

Rohde, R. A. 1972. Expression of resistance in plants to nematodes. *Ann. Rev. Phytopathol.* **10**:233–252.

Southey, J. F. (Ed.). 1959. "Plant Nematology." *Ministry of Agr., Fisheries Food. Tech. Bull.* **7**.

Smart, G. C., Jr., and V. G. Perry (Eds.). 1968. "Tropical Nematology." Univ. Fla. Press, Gainesville. 153 p.

Thorne, G. 1961. "Principles of Nematology." McGraw-Hill, New York, 553 p.

Webster, J. M. 1969. The host–parasite relationships of plant-parasitic nematodes. *Advan. Parasitol.* **7**:1–40.

Webster, J. M. (Ed.). 1972. "Economic Nematology." Academic Press, New York, 563 p.

Zuckerman, B. M., W. F. Mai, and R. A. Rohde (Eds.). 1971. "Plant Parasitic Nematodes." Academic Press, New York, 2 volumes.

• Root-Knot Nematodes: Meloidogyne

Root-knot nematodes occur throughout the world, but are found more frequently and in greater numbers in areas with warm or hot climates and short or mild winters. Root-knot nematodes are also found in greenhouses everywhere when nonsterilized soil is used. They attack more than 2000 species of plants including almost all cultivated plants.

Root-knot nematodes damage plants by devitalizing root tips and either stopping their growth or causing excessive root production, but primarily by causing formation of swellings of the roots which not only deprive plants of nutrients but also disfigure and reduce the market value of many root crops. When susceptible plants are infected at the seedling stage, losses are heavy and may result in complete destruction of the crop. Infections of older plants may have only slight effects on yield or they may reduce yields considerably.

Symptoms.   The aboveground symptoms are similar to those caused by many other root diseases or environmental factors that result in reduced amounts of water available to the plant. Infected plants show reduced growth and fewer, small, pale green, or yellowish leaves that tend to wilt in warm weather. Blossoms and fruits are either lacking or are dwarfed and of poor quality. Affected plants usually linger through the growing season and are seldom killed prematurely.

The most characteristic symptoms of the disease are those appearing on the underground parts of the plants. Infected roots swell at the point of invasion and develop into the typical root-knot galls which are two or three times as large in diameter as the healthy root (Fig. 239A). Several infections take place along the same root and the developing galls give the

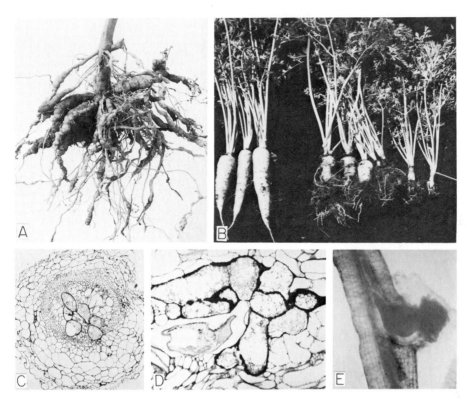

FIGURE 239.
(A) Galls on the roots of tomato plant caused by the root-knot nematode
Meloidogyne sp. (B) Healthy and root-knot nematode-infected carrots. (C) Cross
section of young tomato root showing part of root-knot nematode (arrow) and
giant cells in the stele. (D) Section of tomato root showing a root-knot nematode
feeding on the giant cells surrounding its head. (E) Female root-knot nematode
feeding on young root and laying its egg mass in a matrix outside the root. (Photos
A and B courtesy U.S.D.A. Photos C, D, and E courtesy R. A. Rohde.)

root a rough, clubbed appearance. Roots infected by certain species of this
nematode develop, in addition to galls, several short root branches which
rise from the upper part of the gall and result in a dense, bushy root
system (Fig. 239B). Usually, however, infected roots remain smaller and
show various stages of necrosis. Rotting of the roots frequently develops,
particularly late in the season. When tubers or other fleshy underground
organs are attacked, they produce small swellings over their surface which
become quite prominent at times and may cause distortion of the organs
or cracking of their skin.

The pathogen: Meloidogyne sp.   The adult male and female root-knot
nematodes are easily distinguishable morphologically (Figs. 239 and 240).
The males are wormlike and about 1.2 to 1.5 mm long by 30 to 36 μm in
diameter. The females are pear shaped and about 0.40 to 1.30 mm long by
0.27 to 0.75 mm wide. Each female lays approximately 500 eggs in a
gelatinous substance produced by the nematode. The first-stage larva

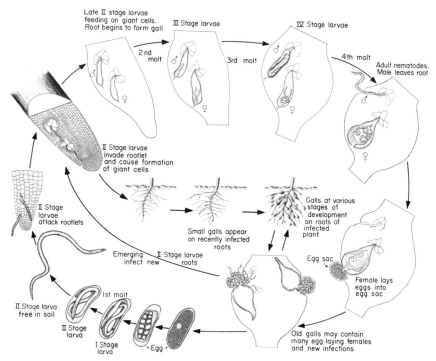

FIGURE 240.
Disease cycle of root knot caused by nematodes of the genus *Meloidogyne*.

develops inside each egg and after undergoing the first molt within the egg it becomes second-stage larva. The latter emerges from the egg into the soil, where it moves until it finds a susceptible root. The second-stage larva is wormlike and is the only infective stage of this nematode. If a susceptible host is present in its vicinity, the larva enters the root, becomes sedentary, and grows in thickness, assuming a sausage-shaped form. The nematode feeds on the cells around its head by inserting its stylet and secreting saliva into these cells. The saliva stimulates cell enlargement and also liquefies part of the contents of the cells, which are then sucked by the nematode through its stylet. The nematode undergoes a second molt and gives rise to the third-stage larva, which is similar to, but lacks a stylet and is stouter than, the second-stage larva. The third-stage larva goes through the third molt and gives rise to the fourth-stage larva, which can be distinguished as either male or female. A male fourth-stage larva becomes wormlike and is coiled within the third cuticle. It undergoes the fourth and final molt and emerges from the root as the wormlike adult male which becomes free living in the soil. The fourth-stage female larva continues to grow in thickness and somewhat in length, undergoes the fourth and final molt, and becomes an adult female which appears pear shaped. The adult female continues to swell and, with or without fertilization by a male, produces eggs which are laid in a gelatinous protective coat. The eggs may be laid inside or outside the root tissues depending on the position of the female. Eggs may hatch immediately or they may overwinter and hatch in the spring. A life cycle is completed in 25 days at 27°C, but it takes longer at lower or higher temperatures. When the eggs hatch, the infective second-stage larvae may migrate from within galls to adjacent parts of the root and cause new infections in the same root, or they may emerge from the root and infect other roots of the same plants or roots of other plants. The greatest numbers of root-knot nematodes are usually in the root zone from 5 to 25 cm below the surface, but galls have been found on peach and other roots 2 to 2.5 m deep. The ability of root-knot nematodes to move on their own power is limited, but they can be spread by water or by soil clinging to farm equipment or otherwise transported into uninfested areas.

Development of disease.   Infective second-stage larvae usually enter roots behind the root tip, and push their way between or through cells until they reach positions behind the growing point. There they become permanently established with their head in the plerome (Fig. 240). In older roots the head is usually in the pericycle. Some cell damage occurs along the path of the larva and, if several larvae have entered, the cells near the root tip cease to divide and growth of the root stops. On the other hand, cortical cells near the point of entry begin to enlarge as sometimes do cells of the pericycle and endodermis near the path of the larvae. Two or three days after the larva has become established, some of the cells around its head begin to enlarge. Their nuclei divide but no cell walls are laid down. The existing walls between some of the cells break down and disappear and the protoplasmic contents of several coalesce, giving rise to giant cells (Figs. 239, C, D, and 240). Enlargement and coalescing of cells continues for 2 to 3 weeks, and the giant cells invade the surrounding tissues irregularly. Each gall usually contains 3 to 6 giant cells, which may form in the cortex as well as in the stele. The enlargement of the cells seems to be brought about by the substances contained in the saliva secreted by the nematode in the giant cells during feeding. The giant cells degenerate when nematodes cease to feed or die. When giant cells form in the stele, irregular xylem elements develop or their development may be interrupted. Xylem elements already present may be crushed by the mechanical pressure exerted by the enlarging cells. In the early stages of gall development the cortical cells enlarge in size but, during the later stages, they also divide rapidly. Swelling of the root results also from hypertrophy and hyperplasia of the vascular parenchyma, pericycle, and endodermis cells surrounding the giant cells and from enlargement of the nematode. As the females enlarge and egg sacs are formed, they push outward, split the cortex, and may become exposed on the surface of the root or may remain completely covered, depending on the position of the nematode in relation to the root surface.

In addition to the disturbance caused to plants by the nematode galls themselves, frequently damage to infected plants is increased by certain parasitic fungi, which can easily attack the weakened root tissues and the hypertrophied, undifferentiated cells of the galls. Moreover some fungi, e.g., *Pythium, Fusarium*, and *Rhizoctonia*, grow and reproduce much faster in the galls than in other areas of the root, thus inducing an earlier breakdown of the root tissues.

Control.   Root knot can be effectively controlled in the greenhouse with steam sterilization of the soil or soil fumigation with nematicides. In the field the best control of root knot is obtained by fumigating the soil with chemicals such as DD, DBCP, or EDB. Several newer nematicides such as aldicarb, oxamyl, and phenamiphos are being used effectively. Each treatment usually gives satisfactory control of root knot for one season. Varieties resistant to root-knot nematodes are also available in several crops.

SELECTED REFERENCES

Bird, A. F. 1974. Plant response to root-knot nematode. *Ann. Rev. Phytopathol.* **12**:69–85.

Carter, W. W., and S. Nieto, Jr. 1975. Population development of *Meloidogyne incognita* as influenced by crop rotation and fallow. *Plant Dis. Reptr.* **59**:402–403.

Christie, J. R. 1936. The development of root-knot nematode galls. *Phytopathology* **26**:1–22.

Dropkin, V. H., and P. E. Nelson. 1960. The histopathology of root-knot nematode infections in soybeans. *Phytopathology* **50**:442–447.

Sasser, J. N. 1954. Identification and host-parasite relationships of certain root-knot nematodes (*Meloidogyne* sp.). *Maryland Agr. Expt. Sta. Bull.* **A-77**:30 p.

- *Cyst Nematodes: Heterodera*

Cyst nematodes cause a variety of plant diseases mostly in temperate regions of the world. Some species of cyst nematodes attack only a few plant species and are present over limited geographic areas while others attack a large number of plant species and are widely distributed. The most common cyst nematodes and their most important hosts are *Heterodera avenae* on cereals, *H. glycines* on soybeans, *H. rostochiensis* on potato (Fig. 241A), tomato, and eggplant, *H. schachtii* on sugar beets (Fig. 241B), crucifers, and spinach, *H. tabacum* on tobacco, and *H. trifolii* on clover. The diagnostic feature of cyst nematode infections is the presence of cysts on the roots and usually the proliferation of roots and production of shallow, bushy root systems.

- *Soybean Cyst Nematode: Heterodera glycines*

The soybean cyst nematode has been found in northeastern Asia, Japan, and in the U.S. in an area from Virginia to Florida to Arkansas to Missouri and Illinois. It continues to spread slowly to new areas in spite of the strict quarantine measures imposed on the presently infested areas. The most severely affected host is soybean, but several other legumes, such as common bean, vetch, lespedeza, lupine, and a few nonleguminous plants are also attacked by this nematode. Depending on the degree of infestation, it can cause losses varying from slight to complete destruction of the crop. Usually, however, in heavily infested fields yield is reduced from 30 to 75 percent.

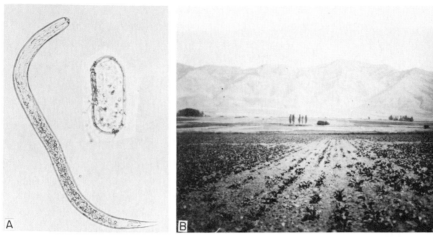

FIGURE 241.
(A) Larva and egg of the golden nematode of potato (*Heterodera rostochiensis*). (B) Bare spots in sugar beet field caused by injury by the sugar beet nematode *Heterodera schachtii*. (Photos courtesy U.S.D.A.)

Symptoms. Infected soybean plants appear stunted and have an unthrifty appearance. The foliage turns yellow prematurely and falls off early. The plants bear only a few flowers and a few small seeds. Infected plants growing on sandy soil usually die. Infected plants growing on fertile soils with plenty of moisture may show only slight chlorosis of the older leaves, little or no stunting, and may produce a nearly normal yield for a year or two. In subsequent years, however, due to the tremendous buildup of nematodes in the soil, plants in these areas also become severely chlorotic and dwarfed.

The root system of infected plants appears smaller than that of healthy plants, but no macroscopic lesions, galls, or other type of abnormalities are evident on infected roots. Roots of infected plants usually have considerably fewer bacterial nodules than those of healthy plants. The most characteristic symptom of this disease is the presence of female nematodes in varying stages of development and of cysts attached on the soybean roots (Fig. 242). Young females are small, white, and partly buried in the root with only part of them protruding on the surface. Older females are larger, almost completely on the surface of the root, and appear yellowish or brown depending on maturity. Dead, brown cysts are also present on the roots.

The pathogen: Heterodera glycines. The soybean cyst nematode overwinters as a brown cyst in the upper 90 to 100 cm of soil. The cysts are the leathery skins of the females and are filled with eggs. The eggs contain fully developed second-stage larvae (Fig. 243). When temperature and moisture become favorable in the spring, the larvae emerge from the cysts and infect roots of host plants.

At 4 to 6 days after penetrating the roots, the larvae molt and produce the third-stage larvae. The third-stage larvae are much stouter than the second-stage larvae and 5 to 6 days later fourth-stage larvae begin to appear. The female fourth-stage larva loses its somewhat slender appear-

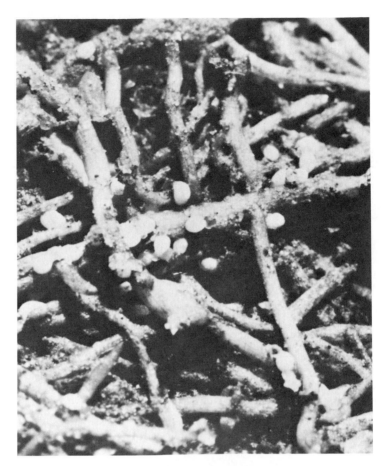

FIGURE 242.
Lemon-shaped encysted female nematodes attached to soybean roots. (Photo courtesy U.S.D.A.)

ance and develops the typical flask shape, measuring approximately 0.40 mm in length by 0.12 to 0.17 mm in width. By day 12 to 15, adult males and females appear.

The adult male is wormlike, about 1.3 mm long by 30 to 40 μm in diameter. The males remain in the root for a few days, during which they may or may not fertilize the females, then move into the soil and soon die.

The adult females when fully developed are lemon shaped, measuring 0.6 to 0.8 mm in length and 0.3 to 0.5 mm in diameter. They are white to pale yellow at first, becoming yellowish-brown as they mature. The body cavity of the female is almost completely filled by the ovaries, and as the ova gradually develop into fully formed eggs, the body cavity of the female becomes completely filled with eggs. As the female body distends during egg production, it crushes cortical cells, splits the root surface, and protrudes until it is almost entirely exposed through the root surface. A gelatinous mass, usually mixed with dirt and debris, surrounds the poste-

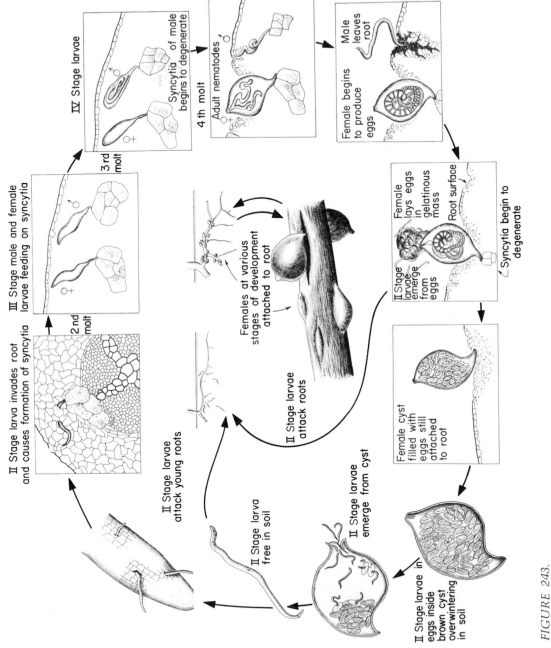

IV Stage larvae

Syncytia of male begins to degenerate

3 rd molt

4 th molt

Adult nematodes

Male leaves root

Female begins to produce eggs

III Stage male and female larvae feeding on syncytia

2 nd molt

II Stage larva invades root and causes formation of syncytia

Females at various stages of development attached to root

Female lays eggs in gelatinous mass

Root surface

II Stage larvae emerge from eggs

Syncytia begin to degenerate

II Stage larvae attack roots

II Stage larvae attack young roots

Female cyst filled with eggs still attached to root

II Stage larva free in soil

II Stage larvae emerge from cyst

II Stage larvae in eggs inside brown cyst overwintering in soil

FIGURE 243.
Disease cycle of the soybean cyst nematode *Heterodera glycines.*

635

rior end of the females and the nematodes deposit some of their eggs in it. Each female produces 300 to 600 eggs, most of which remain inside her body when the female dies. Eggs in the gelatinous matrix may hatch immediately and the emerging second-stage larvae may cause new infections. Finally, the old body wall, darkening to brown, becomes the cyst. Approximately 21 to 24 days are required for the completion of a life cycle of this nematode. The cyst consists of the female cuticle transformed through the secretions of the nematode into a tough, brown sac that persists in the soil for many years and protects the eggs which have been formed within the body.

Development of disease.   The infective second-stage larvae penetrate young primary roots or apical meristems of secondary roots directly (Fig. 243). The advance into the cortex is mostly intracellular and results in distortion and death of invaded cells. The larvae often pass through the cortex and pierce their stylets into cells of the endodermis or the pericycle. Within 2 days from penetration larvae come to rest and feed on cells of the cortex and stele tissues causing the enlargement of these cells. Such groups of enlarged cells, called syncytia, are surrounded by a single layer of small, hyperplastic cells the walls of which undergo further dissolution and allow enlargement of the syncytia. During the development of the third larval stage cortical cells surrounding the nematode are crushed by the expanding nematode body, especially by developing females. Syncytial development is either restricted largely to pericyclic tissue or occurs in tissues of the phloem and secondary cambium. Syncytia in contact with developing third- or fourth-stage males begin to show signs of degeneration indicating cessation of feeding. Syncytia in contact with females remain active up to and beyond the stage of egg deposition. Degeneration of syncytia is accompanied by reduction of syncytial volume and results in the receding and collapse of the syncytial wall. The resulting space is only partly occupied by surrounding parenchymatous tissue.

When soybean varieties resistant to the soybean cyst nematode are attacked, there is no apparent inhibition of penetration of the organism into the host tissues. Syncytia are formed within 2 to 3 days from inoculation, but by day 5 many of them degenerate and most second-stage larvae associated with them are dead. A few nematodes advance to the third stage, but no adult males or females are produced. Development of syncytia and subsequent degeneration and necrosis is restricted to the periphery of the stele and to regions in the cortex that are invaded and stimulated by infective larvae. The root regions vacated by degenerate syncytia are quickly filled by adjacent rejuvenated parenchyma cells.

Syncytial development into the secondary cambial region of susceptible varieties results in inhibition of secondary growth of both phloem and xylem. Since a short portion of a root may be attacked by many larvae, the large number of syncytia that develop may cause widespread reduction of the conductive elements, resulting in the restricted growth and yield of soybean plants, especially under stresses of moisture.

Control.   Soil fumigation of soybean cyst nematode-infested fields with a variety of nematocides temporarily increases plant growth and soybean yield. Nematode cysts and larvae, however, are almost never eradicated from a field completely by fumigation and a small nematode population left over after fumigation can build up rapidly on the vigorous soybean grown in newly fumigated soil. In addition, the cost of fumigation per acre makes its use impractical.

The most practical method of control of the soybean cyst nematode is through a 2- to 3-year crop rotation, since some legumes are the only other cultivated crops that are hosts of this nematode. The effectiveness of crop rotation is increased by planting the more resistant soybean varieties which do not allow a quick and excessive buildup of nematode populations.

Quarantine regulations are presently enforced to prevent the parasite from spreading into nematode-free areas by means of contaminated soil, products, machinery, or other articles.

- *Sugar Beet Nematode: Heterodera schachtii*

It occurs wherever sugarbeets are grown in North America, Europe, Middle East, and Australia and is the most important nematode pest of sugar beet production. It also affects spinach and crucifers. The sugar beet nematode causes yield losses of 25 to 50 percent or more, especially in warmer climates or late planted crops. The losses on sugar beet are mostly the result of reduced root weight but in warm climates the sugar content is also reduced and, generally, the nematode agravates losses caused by other pathogens such as *Cercospora, Rhizoctonia*, and beet viruses. In fields infested with the sugar beet nematode, small to large patches of wilting or dead young plants or stunted older sugar beets appear (Fig. 241B). The latter have an excessive number of hairlike roots. Small white or brownish cysts of female nematodes can be seen clinging to the roots. The morphology, biology, and spread of the sugar beet nematode is similar to that of the soybean cyst nematode. Control of the sugar beet nematode is based on early sowing so that plants can grow as much as possible at temperatures at which the nematodes are more or less inactive, on crop rotations with alfalfa, cereals, or potatoes which are not hosts of this nematode, and soil fumigation with DCP, EDB, or DBCP. No sugar beet varieties resistant to this nematode are commercially available yet.

SELECTED REFERENCES

Anonymous. 1961. Soybean cyst nematode. *U.S. Dept. Agr., Agr. Res. Serv. Spec. Reptr.* **22–72**:20 p.

Endo, B. V. 1964. Penetration and development of *Heterodera glycines* in soybean roots and related anatomical changes. *Phytopathology* 54:79–88.

Endo, B. V. 1965. Histological responses of resistant and susceptible soybean varieties and backcross progeny to entry and development of *Heterodera glycines*. *Phytopathology* **55**:375–381.

Franklin, Mary T. 1972. *Heterodera schachtii. C.I.H. Descriptions of Plant-Parasitic Nematodes*. Set 1, No. 1. 4 p.

Gipson, Ilene, K. S. Kim, and R. D. Riggs. 1971. An ultrastructural study of syncytium development in soybean roots infected with *Heterodera glycines*. *Phytopathology* **61**:347–353.

Raski, D. J. 1950. The life history and morphology of the sugar beet nematode *Heterodera schachtii*. *Phytopathology* **40**:135–152.

Stone, A. R. 1973. *Heterodera rostochiensis. C.I.H. Descriptions of Plant-Parasitic Nematodes.* Set 2, No. 16, 4 p.

- ## *The Citrus Nematode: Tylenchulus semipenetrans*

It is present wherever citrus trees are grown. In some regions in addition to citrus the nematode also attacks grapevines, olive, lilac, and other plants. Infected trees show a slow decline, i.e., they grow poorly, their leaves turn yellowish and drop early, their twigs die back, and fruit production is gradually reduced to unprofitable levels.

The pathogen, *Tylenchulus semipenetrans*, is a semiendoparasitic sedentary nematode. The larvae and males are wormlike but the female body is swollen irregularly behind the neck. The nematodes measure about 0.4 mm long by 18 to 80 μm in diameter, the larger diameters found only in the maturing and mature females. The females, whose front end of the body is buried in the root tissue and the rear end remains outside (Fig. 244), lay eggs in a gelatinous substance. The life cycle of *T. semipenetrans* is completed within 6 to 14 weeks at 24°C. The eggs hatch and second-stage larvae emerge. The male larvae and adult do not feed and apparently do not play a role either in the disease or the reproduction of the nematode. The second-stage female larva is the only infective stage of the nematode and cannot develop without feeding, but it can survive for several years. In the soil, the citrus nematode occurs as deep as 4 meters.

FIGURE 244.
Tylenchulus semipenetrans females feeding on citrus roots with their heads embedded in individual cells. (Photo courtesy U.S.D.A.)

The female second-stage larvae usually attack the 4- to 5-week-old feeder roots and feed on the surface cells of the roots. There they undergo the three additional molts and produce adult females. The young females then penetrate deeper into the cortex and may reach as deep as the pericycle. The head of the nematode develops a tiny cavity around it and feeds on the surrounding 3 to 4 layers of parenchyma cells known as "nurse cells." Later on, the cells around the feeding site become disorganized and break down. Following invasion by secondary fungi and bacteria, the affected areas turn into dark necrotic lesions which may be so numerous that they give the root a darkened appearance. In severe infections, one hundred or more females may be feeding per centimeter of root. The females, along with the soil particles that cling to the gelatinous substance of the egg mass, result in dark, bumpy, and often decayed young roots.

The spread of the nematode through the soil is slow, the rate being approximately 1.5 cm per month when the roots of adjacent citrus plants are in contact. The nematode, however, is spread over long distances by movement of nematode-infested soil on equipment, animals, by irrigation water, etc., and to even longer distances by transfer of infested citrus nursery plants. The nematodes reach high populations in infected trees which begin to show decline 3 to 5 years after the initial infection. When the trees show advanced stages of decline, the nematode populations also decline in numbers.

Control of the citrus nematode is based on preventing its introduction into new areas by growing nursery stock in nematode-free fields and by treating nursery stock with hot water at 45°C for 25 minutes or with DBCP, fensulfothion, or thionazin. Due to the great depth at which the citrus nematode can survive, soil fumigation is not always effective. Satisfactory control has been obtained by preplant fumigation with DD, methyl bromide and DBCP; also by postplant treatment with DBCP applied as drench treatments, by chisel injection or through sprinkler irrigation systems. Some citrus clones are resistant to the nematode populations of some regions but not to those of others.

SELECTED REFERENCES

Cohn, E. 1972. Nematode diseases of citrus, *in* "Economic Nematology." J. M. Webster (Ed.). Academic Press, New York, pp. 215–244.

Siddiqi, M. R. 1974. *Tylenchulus semipenetrans. C.I.H. Descriptions of Plant-Parasitic Nematodes.* Set 3, No. 34. 4 p.

Van Gundy, S. D. 1958. The life history of the citrus nematode *Tylenchulus semipenetrans. Nematologia* 3:283–294.

- *Lesion Nematodes: Pratylenchus*

Lesion or meadow nematodes occur in all parts of the world, where they attack the roots of all kinds of plants, e.g., field crops, such as tobacco, alfalfa, cotton; cereal crops, such as wheat, corn, oats; vegetable crops, such as tomato, potato, carrot; fruit trees, such as apple, peach, cherry; and many ornamentals, both herbaceous and shrubs.

The severity of damage caused by the lesion nematodes is difficult to estimate. It varies with the crop attacked and is greater in subtropical than in temperate regions. The damage to plants consists in root reduction or inhibition by formation of local lesions on young roots which may be followed by root rotting due to secondary fungi, bacteria, etc. As a result of the root damage affected plants grow poorly, produce low yields, and may finally die.

Symptoms. Susceptible herbaceous host plants affected by lesion nematodes appear stunted and chlorotic as though they are suffering from mineral deficiencies or lack of water. Usually several plants are affected in one area, producing patches of plants with reduced growth and yellowish-green color which can be seen from a distance. As the season progresses, stunting becomes more pronounced, the foliage wilts during hot summer days, and the color of the leaves becomes yellowish brown. Such plants can be easily pulled from the soil because of the extensive destruction of the root system. Yields of affected plants are reduced in varying degrees, and in severe infections the plants are killed.

When shrubs or trees are attacked by lesion nematodes, damage is usually slow to appear; it is less obvious than that on herbaceous hosts, and it rarely kills the plants. The symptoms usually consist of isolated trees or patches of trees gradually becoming unthrifty and producing poor crops. The leaves are smaller in size, their color being a dull green or yellow. Terminal branches may lose their leaves prematurely and die back. The whole appearance of affected trees indicates that the trees are weakened and are in a condition of decline. The patches of affected trees may slowly increase in size, although this happens over a rather long period.

The root symptoms of affected plants consist of lesions which first appear as tiny, elongate, water-soaked, or cloudy yellow spots, but which soon turn brown to almost black. The lesions appear mainly on the young feeder roots and they are most concentrated in the area of the root hairs, but they may appear anywhere along the roots. The lesions enlarge mostly longitudinally following the root axis and they may coalesce with other lesions, but at the same time they slowly expand laterally until they finally girdle the entire root, which they kill. As the lesions enlarge, the affected cells in the cortex collapse and the discolored area appears constricted. Secondary fungi and bacteria usually accompany nematode infections in the soil and contribute to further discoloration and rotting of the affected root areas, which may slough off. Moderately affected plants exhibit varying degrees of root survival, and in some hosts production of adventitious roots may be stimulated by the infection; but generally the individual roots are discolored and stubby, and the whole root system is severely reduced by the root pruning that results from the formation of lesions (Fig. 245A).

The pathogen: Pratylenchus sp. The nematodes are approximately 0.4 to 0.7 mm long and 20 to 25 μm in diameter. They appear as stout, cylindroid nematodes with a blunt head, strong, stout spear, and bluntly rounded tail (Fig. 245B). They are migratory, endoparasitic nematodes

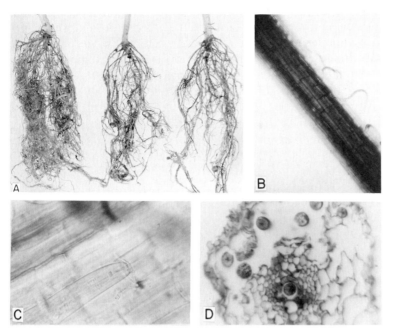

FIGURE 245.
(A) Roots of bean plants showing symptoms caused by lesion nematodes. Healthy
plant on left. (B) Lesion nematodes (*Pratylenchus* sp.) penetrating young carrot
root. (C) Lesion nematode feeding within a young carrot root. (D) Cross section of
young carrot showing six lesion nematodes in the destroyed cortex and one in the
stele. (Photo A courtesy Dept. Plant Pathol., Cornell Univ. Photos B, C, and D
courtesy R. A. Rohde.)

affecting the roots of many kinds of plants. Development and reproduction of *Pratylenchus* nematodes is rather slow, the life cycle of the
various species being completed within 45 to 65 days. These nematodes
overwinter in infected roots or in soil as eggs, larvae, or adults, except for
the egg-producing females which seem to be unable to survive the winter.
Adults and larvae of various ages can enter and leave roots of susceptible
hosts. The females, with or without fertilization, lay their eggs singly or
in small groups inside infected roots. The eggs remain in the roots and
hatch there, or, when the root tissues break down, they are released into
the soil. The first larval stage and the first molt occur in the egg. The
emerging second-stage larva moves about in the soil or enters the root, in
either case developing into the subsequent larval stages and the adults.
When in the soil the nematodes are susceptible to drying and during
periods of drought they lie quiescent until the moisture increases and the
plants resume growth.

Development of disease. Larvae and adult *Pratylenchus* nematodes
enter roots usually in a radial direction anywhere along the roots (Fig.
245B). Intracellular penetration is accomplished by a persistent thrusting
of the stylet and head which seems to soften and break the cell wall. The

cell walls and the adherent cytoplasm usually turn light brown in color and appear as small, discolored spots within a few hours after inoculation. The nematodes move into the cortex where they feed and reproduce. (Fig. 245C, D). The endodermis is not attacked even when the nematodes completely fill the area between the endodermis and the epidermis (Fig. 246). Necrosis of cortical cells follows the path of nematodes, but discoloration of the affected and adjoining cells varies with the host plant. Sometimes only 1 or 2 cells on each side of the nematode tunnels are affected, but at other times the lesion involves over half the circumference of the root. The part of the endodermal layer adjacent to the nematode also takes on a deep brown color which extends into rather large groups of cells. As the feeding of the nematode on cortical cells continues, cell walls break down, and cavities appear in the cortex with their walls sometimes lined with brown deposits.

Each lesion is usually inhabited by more than one nematode, and sometimes single host cells are simultaneously transversed by four or more nematodes. The females lay their eggs in the cortex, and frequently eggs, larvae, and a few adults form "nests"; these occur in great numbers in the cortex. Upon hatching of the eggs, the nematodes feed on the parenchyma cells and move mostly lengthwise within the cortex, thus

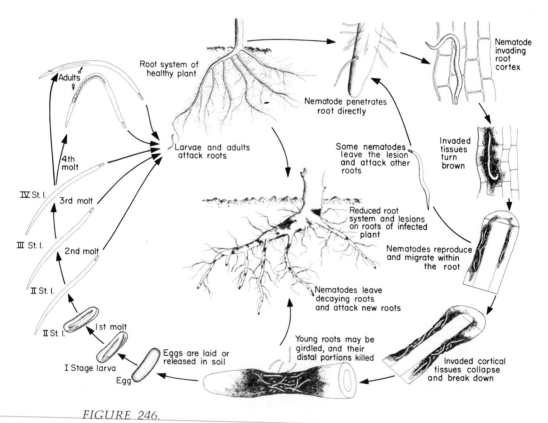

FIGURE 246.
Disease cycle of the lesion nematode *Pratylenchus* sp.

enlarging the lesion (Fig. 246). Some of the nematodes leave the lesion, emerge from the root, and travel to other points of the root or other roots where they cause new infections. Necrotic cortical tissues of large lesions are sloughed off or are invaded by secondary fungi and bacteria with resulting rotting and breakdown of the root tissues around the point of infection and subsequent death of the distal part of the root beyond the point of infection. Thus, the number of functioning roots of the plant is drastically reduced, absorption of water and nutrients becomes insufficient, and the aboveground parts of the plant become stunted and chlorotic, showing symptoms of water and nutrient deficiencies.

Control. Root-lesion nematodes can best be controlled by overall or row treatment of the soil with nematicides before the crop is planted. DD, EDB, and Brozone give good control of these nematodes, but they usually fail to eradicate them completely.

In hot and dry climates fairly good control of lesion nematodes can be achieved by summer fallow, which reduces nematode populations by exposing them to heat and drying and by eliminating host plants. Control through crop rotation is at present rather unsuccessful due to the wide host ranges of the lesion nematodes and to the lack of information on their precise host preferences.

SELECTED REFERENCES

Dickerson, O. J., H. M. Darling, and G. D. Griffin. 1964. Pathogenicity and population trends of *Pratylenchus penetrans* on potato and corn. *Phytopathology* **54**:317–322.

Good, J. M., L. W. Boyle, and R. O. Hammons. 1958. Studies of *Pratylenchus brachyurus* on peanuts. *Phytopathology* **48**:530–535.

Mountain, W. B., and Z. A. Patrick. 1959. The peach replant problem. VII. The pathogenicity of *Pratylenchus penetrans. Can. J. Bot.* **37**:459–470.

Parker, K. G., and W. F. Mai. 1974. Root diseases of fruit trees in New York State. VI. Damage caused by *Pratylenchus penetrans* to apple trees in the orchard growing on different rootstocks. *Plant Dis. Reptr.* **58**:1007–1011.

• *The Burrowing Nematode: Radopholus similis*

It occurs widely in tropical and subtropical regions of the world and in greenhouses in Europe. It is the most important banana root pathogen in most banana-growing areas, where it causes the so-called root rot, black-head, toppling disease or decline of banana. It also causes the spreading decline disease of citrus in Florida, a decline of avocados in Florida and of tea in Ceylon, and the yellows disease of black pepper in Indonesia. Furthermore, it attacks coffee and other fruit, ornamental and forest trees, sugarcane, corn, vegetables, grasses, and weeds.

Infected banana plants appear to be growing poorly, have fewer and smaller leaves, show premature defoliation and reduced weight of fruits. Often entire banana plants topple over. Banana roots at first show browning and cavities in the cortex and these are followed by deep cracks with raised margins on the root surface (Fig. 247D). The nematodes, along with fungi and bacteria that invade the cracked roots, cause the roots to rot and

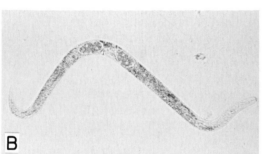

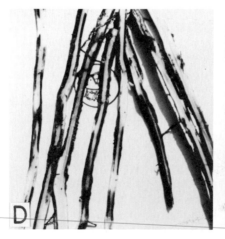

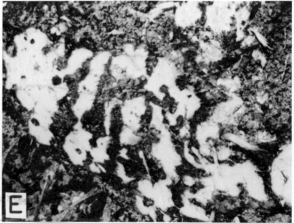

eventually only a few short root stubs remain which cannot anchor the plant sufficiently and the latter topples over. Diffuse black, rotten areas also develop in the rhizome cortex surrounding infected roots (Fig. 247E). As a result of this disease the profitable life of a banana plantation in many areas is decreased from indefinite to as little as one year and the costs of annual replanting and losses in production are tremendous.

In the spreading decline of citrus, blocks of affected trees also have fewer and smaller leaves and fruits, and many of the twigs and branches die back (Fig. 247A). Yields of infected trees are reduced by 40 to 70 percent. Even during periods of mild moisture stress, infected trees wilt readily but they generally do not die and often recover temporarily after rainy periods. The symptoms of decline spread steadily to more trees each year, the diameter of the decline area increasing approximately 10 to 20 meters per year. The symptoms on the aboveground parts follow infection of the roots by about a year. Infected feeder roots have numerous lesions that appear puffy and cracked and are often invaded by primary and secondary fungal parasites that result in the rotting and destruction of the feeder roots. Feeder roots seem to be attacked and destroyed most at depths of 50 cm or more leaving less than half of the feeder roots functional.

The pathogen, *Radopholus similis*, usually known as the burrowing nematode, is wormlike, measuring about 0.65 mm long by 25 μm wide (Fig. 247B, C). It spends its life and reproduces inside cavities in the root cortex where it completes a life cycle in about 20 days. All larvae and the adults can infect roots and, although they can emerge from the roots and spread through the soil, most of the spread of the nematode from plant to plant is through root contact or near contact. Long-distance spread of the nematode is primarily with infected plant material, such as infected banana sets. Although the nematodes infecting banana and citrus are morphologically identical, the "banana race" can attack banana but not citrus while the "citrus race" can attack citrus as well as banana and several other hosts. The citrus race, however, is so far known to occur only in Florida. Other races probably exist in other parts of the world.

The burrowing nematode enters feeder roots and moves intercellularly in the cortical parenchyma feeding on nearby cells, destroying them, and causing the formation of cavities (Fig. 248). As the nematodes continue to feed, the cavities enlarge and coalesce with others forming long and lateral tunnels. In banana, the tunnels are limited to the cortex between the epidermis and the endodermis. From the feeding roots the nematodes also move into the rhizome. In citrus, however, the nematodes not only form cavities in the cortex but they also enter the stele through endoder-

FIGURE 247.
(A) Healthy and infected orange trees at the margin of a spreading decline area caused by *Radopholus similis*. (B) *R. similis* female containing two eggs. (C) *R. similis* and egg in banana cells. (D) Lesions on banana roots caused by *Radopholus similis*. (E) Lesions on banana rhizome caused by *R. similis*. (Photo A courtesy Agric. Res. Educ. Center, Lake Alfred, Fla. Photos B–E courtesy R. H. Stover.)

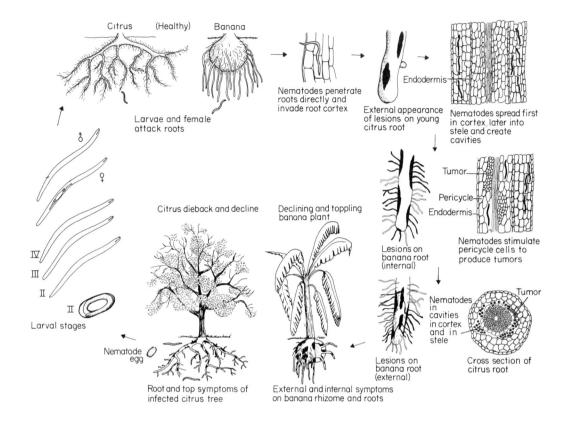

FIGURE 248.
Disease cycle of the burrowing nematode *Radopholus similis* in citrus, banana, etc.

mal passage cells. There they accumulate in the phloem and cambium which they destroy in time and form nematode-filled cavities. At the same time, gum is deposited in the cortex and cells of the pericycle divide excessively and produce groups of tumorlike cells. The cavities within the roots appear externally as brown-reddish lesions throughout the cortex and three to four weeks from infection the lesions develop one or more deep cracks. Each female lays one or a few eggs per day for many days and as they hatch, develop, and reproduce, nematode populations increase rapidly. As many as 800 nematodes may be present in a single lesion and a single tree may be supporting hundreds of thousands of burrowing nematodes. Fungi such as *Fusarium*, *Sclerotium*, and others invade nematode-infected roots much more readily and further increase their rotting and destruction.

Control of the burrowing nematode in banana can be obtained by removing discolored tissues from banana sets by paring and then dipping the sets in DBCP solution; by hot water treatment of sets at 55°C for 20 minutes; by flooding the field for 5 to 6 months where possible; and by soil fumigation with DBCP, DD, or EDB.

Control of spreading decline of citrus is much more difficult and depends primarily on: (1) preventive regulatory measures that inhibit the spread and establishment of the nematode in new areas by treating nursery trees with hot water at 50°C for 10 minutes or dipping them in nematicides such as Cynem, Dasanit, and Mocap; disinfesting equipment; and fumigating every 6 months with EDB 5-meter-wide strips of land around the area of decline and keeping it free of weeds with cultivation or herbicides. (2) Fumigation of decline areas with heavy doses of DD following removal of all declining trees and at least two rows around them. (3) Use of tolerant or resistant rootstocks.

SELECTED REFERENCES

Blake, C. D. 1972. Nematode diseases in banana plantations, in "Economic Nematology," J. M. Webster (Ed.). Academic Press, New York, pp. 245–267.

DuCharme, E. P. 1959. Morphogenesis and histopathology of lesions induced on citrus roots by *Radopholus similis. Phytopathology* **49**:388–395.

Poucher, C., *et al.* 1967. Burrowing nematode in citrus. *Fla. Dept. Agric. Bull. No.* 7: 63 p.

Williams, K. J. O., and M. R. Siddiqi. 1973. *Radopholus similis. C.I.H. Descriptions of Plant-Parasitic Nematodes.* Set 2, No. 27. 4 p.

- *Stem and Bulb Nematode: Ditylenchus*

Of the several species of *Ditylenchus* that cause diseases in plants, *D. dipsaci* is the most common and most serious of all and is the one generally referred to as the stem and bulb nematode while *D. destructor* is known as the potato rot nematode.

The stem and bulb nematode is worldwide in distribution but is particularly prevalent and destructive in areas with temperate climate. It is one of the most destructive plant-parasitic nematodes. It attacks a large number of host plants including onions, narcissus, hyacinth, tulip, oats, rye, alfalfa (Fig. 249A), red clover, strawberry, and phlox. Although all these hosts are attacked by nematodes belonging to one species, different populations or biotypes of the stem and bulb nematode have certain host preferences and, when transferred to them, will attack one or a few of these hosts but not the others. On most crops it causes heavy losses by killing seedlings, dwarfing plants, destroying bulbs or making them unfit for propagation or consumption, by causing the development of distorted, swollen and twisted stems and foliage, and, generally, reducing yields greatly.

Symptoms. In fields infested with stem and bulb nematodes, emergence of seedlings such as onion is retarded and stands are reduced considerably. Half or more of the emerging seedlings may be diseased, appearing pale, twisted, arched, and with enlarged areas along the cotyledon. Cotyledons are usually puffy with the epidermis cracked in a lacelike fashion. Most infected seedlings die within 3 weeks from planting and the remainder usually die later.

When bulbs are planted in infested soil, the developing plants within about 3 weeks show stunting, light yellow spots, swellings ("spikkles"),

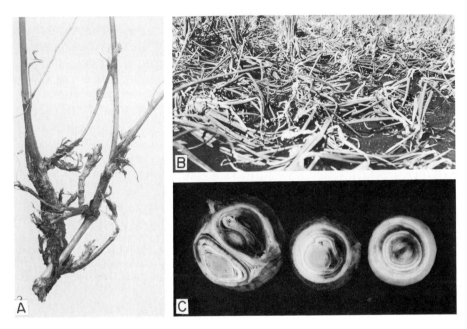

FIGURE 249.
(A) Alfalfa plant with shortened and thickened stems and small crowded leaves as a result of infection with the stem nematode *Ditylenchus dipsaci*. (B) Infected onion plants showing the prostrate condition, thin stand, stunting, and many dead outer leaves. (C) Infected narcissus bulbs showing complete or incomplete brown rings. (Photo A courtesy U.S.D.A. Photos B and C courtesy Dept. Plant Pathol., Cornell Univ.)

and open lesions on the foliage. Young plants and sprouts develop swellings on the stem and foreshortening and curling of the leaves. Many outer leaves often become flaccid, their leaf tips die back, and the leaves are so weakened that they cannot maintain their erect growth and fall to the ground (Fig. 249B). The stem and the neck of the bulb become softened, and the softening gradually proceeds downward into individual scales, which become soft, loose, and pale gray in color. Affected scales appear as discolored rings in cross sections of infected bulbs (Fig. 249C) and as discolored, unequal lines in longitudinal sections. In more advanced cases large areas or the whole bulb may be affected. Infected bulbs may also split and become malformed or may produce sprouts and double bulbs. The outer scales may be loosened and detached by applying a little oblique pressure with the thumb on the upper half of the bulb and reveal the soft mealy, frosty-looking tissue beneath. In dry weather the bulbs become desiccated, odorless, and very light in weight. In wet seasons a soft rot due to secondary invaders sets in, destroying the bulb and giving off a foul odor. Infected bulbs are sometimes superficially healthy, but they continue to decay in storage, where the outer scale often sloughs off exposing the lower puffy, soft scales with the characteristic frosty mealiness.

The pathogen: Ditylenchus dipsaci. The nematode is 1.0 to 1.3 mm long and about 30 μm in diameter (Fig. 250). Its cylindroid body is marked by faint transverse lines, about 1 μm apart. The tail of both males and females is acute. Each female lays 200 to 500 eggs. The first molt occurs in the egg. Second-stage larvae emerge from the egg and quickly undergo the second and third molt and produce the preadult or infective larva. The latter can withstand adverse conditions of freezing and of extreme drying for long periods in fragments of plant tissue, stems, leaves, bulbs, seeds, or in the soil. Under favorable moisture and temperature conditions the preadult larvae become active, enter the host, pass through the fourth molt, and become males and females. The females then lay eggs, mostly after fertilization by the males. A complete cycle from egg to egg usually lasts about 19 to 25 days. Reproduction takes place in succulent, rapidly growing tissues or in storage organs and continues throughout the year, although it is retarded or stopped by low temperatures. *Ditylenchus dipsaci* is an internal parasite of bulbs, stems, and leaves and passes generation after generation in these tissues escaping to the soil only when living conditions in the plant tissues become unfavorable. When heavily infected bulbs decay, preadult larvae pass out and sometimes accumulate about the basal plates of dried bulbs as grayish-white, cottony masses, called nematode "wool," where they can remain alive for years.

Development of disease. When nematodes attack a germinating seed or young seedlings, they enter near the root cap or at points still within the seed. The nematodes remain mostly intercellular, feeding on the parenchymatous cells of the cortex. Cells near the heads of the nematodes lose all or a portion of their contents, while cells surrounding these divide and enlarge, resulting in development of swellings on the seedlings. The seedling may become malformed. Splitting of the epidermis often follows the enlargement and opens the way to secondary invaders, such as bacteria and fungi.

In older seedlings or young plants, nematodes enter the leaves through stomata or penetrate directly through the epidermis in leaf bases (Fig. 250). Cell enlargement, disappearance of chloroplasts, and an increase of intercellular spaces in parenchyma tissue follow penetration. The nematodes usually remain and reproduce in the intercellular spaces feeding on the nearby parenchyma cells whose contents they consume without causing appreciable discoloration on the cell remains. As the bulbs enlarge, the nematodes migrate down from the leaves either intercellularly or by traveling on the surface of the leaves and entering again at the outer sheaths of the stem or neck, through which they infect the outer scales of the bulbs. Heavily infected stems become soft and puffy due to the formation of large cavities through the breakdown of the middle lamella and of the cells the nematodes feed on. Such stems can no longer remain rigid under the weight of the foliage and they frequently collapse. The nematodes continue their progress intercellularly through the outer scales of the bulbs by breaking down the parenchymatous tissue. Parenchyma cells are separated from each other and from the vessels, the latter giving a lacy appearance to the scale. The macerated parenchyma cells

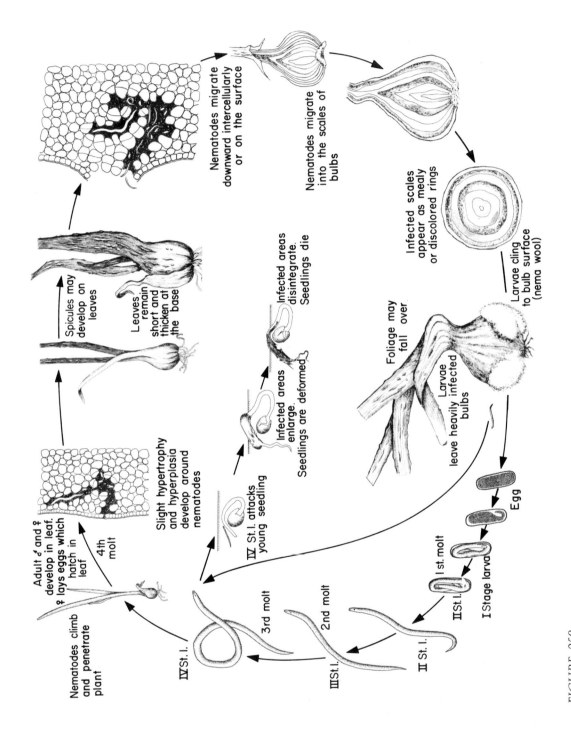

Nematodes migrate downward intercellularly or on the surface

Nematodes migrate into the scales of bulbs

Infected scales appear as mealy or discolored rings

Larvae cling to bulb surface (nema wool)

Spicules may develop on leaves

Leaves remain short and thicken at the base

Infected areas disintegrate. Seedlings die

Infected areas enlarge. Seedlings are deformed

Foliage may fall over.

Larvae leave heavily infected bulbs

Adult ♂ and ♀ develop in leaf. ♀ lays eggs which hatch in leaf

4th molt

Slight hypertrophy and hyperplasia develop around nematodes

IV St. l. attacks young seedling

Egg

I st. molt

II St. l.

I Stage larva

Nematodes climb and penetrate plant

IV St. l.

3rd molt

III St. l.

2nd molt

II St. l.

FIGURE 250.
Disease cycle of the stem and bulb nematode *Ditylenchus dipsaci*.

have a white mealy texture at first, but secondary invaders usually set in and cause them to turn brown. In early stages of infection the nematodes remain within individual scales, and in sections the infection appears as complete or incomplete rings of frosty white or brownish tissue. In later stages of infection the nematodes pass from one scale to the next, and thus more scales may be involved in one ring. The spread of the infection within a bulb continues in the field and in storage until, usually, the entire bulb becomes affected.

Control. Populations of *Ditylenchus dipsaci* parasitic on certain crops can be reduced by long (2 to 3 years at least) rotations with resistant crops, such as spinach, carrots, potatoes, and lettuce. Since this nematode also overwinters in infected bulbs and seeds, the need for use of nematode-free sets or seeds is obvious. Infested seeds or bulbs can be disinfested by treating them in hot water for an hour at 46°C. Onion seed can also be freed from nematodes by enclosing it with methyl bromide gas for 24 hours at 24°C in a gas-tight container, and flower bulbs can be successfully disinfested by placing them in 0.5 percent formaldehyde solution at 43°C for 4 hours.

Control of this nematode in large fields, although often too expensive, can be achieved by fall fumigation of the soil with the nematicides DD, Vidden D, or Telone at 50 gallons per acre, by preplant row treatment with Nemafos 10G, and by treatment at or soon after planting with granules of dazomet, aldicarb, or thionazin.

SELECTED REFERENCES

Chitwood, B. G., A. G. Newhall, and R. L. Clement. 1940. Onion bloat or eelworm rot, a disease caused by the bulb or stem nematode, *Ditylenchus dipsaci* (Kuhn) Filipjev. *Proc. Helminthol. Soc. Wash.* **7**:44–51.

Krusberg, L. R. 1961. Studies on the culturing and parasitism of plant-parasitic nematodes, in particular *Ditylenchus dipsaci* and *Aphelenchus ritzemabosi* on alfalfa tissues. *Nematologica* **6**:181–200.

Newhall, A. G. 1943. Pathogenesis of *Ditylenchus dipsaci* in seedlings of *Allium cepa. Phytopathology* **33**:61–69.

Sayre, R. M., and W. B. Mountain. 1962. The bulb and stem nematode (*Ditylenchus dipsaci*) on onion in Southwestern Ontario. *Phytopathology* **52**:510–516.

Seinhorst, J. W., and J. L. Koert. 1971. Stem nematodes in onion seed. *Gewasbescherming* **2**:25–31.

• *Seed-Gall Nematodes: Anguina*

They were the first recorded plant-parasitic nematodes, discovered in 1743 when an infected wheat seed (seed gall) was crushed in a drop of water under a microscope. Several species of *Anguina* are known and all of them cause formations of galls on seeds, leaves, and other aboveground parts of plants. Of the seed-gall nematodes, *A. agrostis* is probably the most widespread in Europe and North America and causes severe injury to bentgrasses (*Agrostis* spp.), but *A. tritici* has been important on wheat and rye in the past and in some areas still is. The wheat seed gall, which is described below, is present wherever wheat is grown but in most countries it is quite rare due to the use of fresh and cleaned seed. The wheat seed-gall nematode is still common, however, in some Mediterranean countries, eastern Europe, and Asia.

The symptoms appear on plants in all growth stages. Infected seedlings are more or less severely stunted and show characteristic rolling, twisting, curling, or wrinkling of the leaves (Fig. 251). A rolled leaf often traps the next emerging leaf or the inflorescence within it and causes it to become looped or bent and badly distorted. Stems are often enlarged near the base, frequently bent, and generally stunted. Diseased heads are shorter and thicker than healthy ones and the glumes are spread further apart by the nematode-filled seed galls. A diseased head may have one, a few, or all of its kernels turned into nematode galls. The galls are shiny green at first, but turn brown or black as the head matures. Diseased heads remain green longer than healthy ones and galls are shed off the heads more readily than kernels. Mature galls are hard, dark, rounded, and shorter than normal wheat kernels and often resemble cockle seeds, smutted grains or ergot sclerotia.

The pathogen, *Auguina tritici*, is a large nematode about 3.2 mm long by 120 μm in diameter. The nematode lays its eggs and produces all its larval stages and the adults in seed galls.

The seed-gall nematode overwinters as second-stage larvae in seed galls or in plants infected in the fall. Galls fallen to the ground or sown

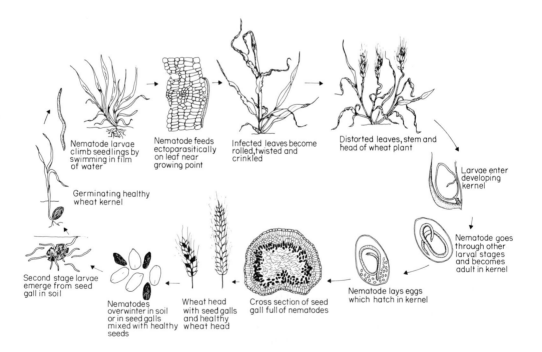

FIGURE 251.
Disease cycle of wheat seed gall caused by *Anguina tritici.*

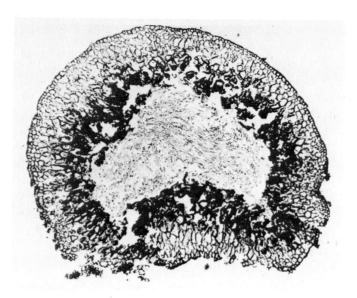

FIGURE 252.
Cross section of a wheat seed gall showing the white mass of dormant nematodes in the center. (Photo courtesy U.S.D.A.)

with the seed soften during warm moist weather and release infective second-stage larvae. When a film of water is present on the surface of the plants, the larvae swim upward and feed ectoparasitically on the tightly compacted leaves near the growing point causing the leaves and stem to become malformed. When the inflorescence begins to form, the larvae enter the floral primordia and produce the third- and fourth-stage larvae and the adults. Each infected floral primordium becomes a seed gall and may contain 80 or more adults of both sexes. Each of the females then lays up to 2000 eggs over several weeks within the freshly formed gall so that each gall contains 10,000 to 30,000 eggs. The adults die soon after the eggs are laid. The eggs then hatch and the first-stage larvae emerge but these soon molt and by harvest produce the second-stage larvae which are very resistant to desiccation and can survive in the galls for up to 30 years (Fig. 252). The seed-gall nematode produces only one generation per year. The nematode is spread in infected seed.

Control of the seed gall nematode depends on the use of clean seed free of nematode-containing galls. Fields infested with seed-gall nematodes should not be planted to wheat or rye for at least a year. In moist weather the seed galls release the second-stage larvae and, if no susceptible hosts are present, they die before they can infect and reproduce. In dry weather, however, nematodes can survive in the seed galls for many years.

SELECTED REFERENCES

Leukel, R. W. 1929. The nematode disease of wheat and rye. *U.S. Dept. Agr. Farmers' Bull.* **1607**:12 p.

Southey, J. F. 1972. *Anguina tritici. C.I.H. Descriptions of Plant-Parasitic Nematodes.* Set 1, No. 13, 4 p.

• *Foliar Nematodes: Aphelenchoides*

Several species of *Aphelenchoides* feed ectoparasitically and endoparasitically on aboveground plant parts. Some of the most important species are: *A. ritzemabosi*, the chrysanthemum foliar nematode; *A. fragariae*, the spring crimp or spring dwarf nematode of strawberry and which also attacks many ornamentals; and *A. besseyi*, the nematode causing summer dwarf or crimp of strawberry and white tip of rice.

The foliar nematode of chrysanthemums, also known as chrysanthemum eelworm, is known to be widespread in the U.S. and in Europe. It results in fairly severe losses. In addition to chrysanthemum, the foliar nematode or closely related species attack several other plants including aster, dahlia, delphinium, phlox, zinnia, and sometimes tobacco and strawberry.

Symptoms.   Affected buds or growing points of stems produce short and often abnormally bushy-looking plants with short internodes. The growing point may be so damaged that the shoot does not grow and turns brown. The leaves produced from infested buds are small and distorted. The stem and petioles show brown scars caused by nematodes feeding externally on these tissues while still in the bud. Shoots so infested seldom develop into normal plants. Uninfested shoots may exist on the same stool with infested ones and they develop into normal new plants. As the season progresses, however, nematodes climb up the stem and attack first the lower and then the upper leaves on which they cause small, yellowish spots later turning brownish black. These spots soon coalesce and form large blotches which at first are contained between the larger leaf veins (Fig. 253). Eventually the entire leaf is covered with spots or blotches, and it soon shrinks, becomes brittle, and falls to the ground. Defoliation, like infection, progresses from the lower to the upper leaves. The nematodes also infest the ray flowers and prevent their development. Severely infected plants die without producing much normal foliage or marketable flowers.

The pathogen: Aphelenchoides ritzemabosi.   It is a long, slender nematode measuring about 1 mm long by 20 μm in diameter.

Aphelenchoides ritzemabosi may live its entire life inside leaves or at the surface of other plant organs. The female adult lays its eggs in the intercellular spaces of leaves. The eggs hatch and produce the 4 larval stages, and finally adults, all inside the leaf. The life cycle is completed in about 2 weeks. These nematodes do not have to spend any part of their lives in the soil but are frequently found there carried by infected, dead leaves fallen to the ground, or washed down by rain or irrigation water when they happen to be on the surface of plant tissues. The foliar nematodes overwinter as adults in dead leaves or between the scales of buds of infected tissues.

FIGURE 253.
Discolored areas on chrysanthemum leaves caused by the foliar nematode
Aphelenchoides ritzemabosi. (Photo courtesy Dept. Plant Pathol., Cornell Univ.)

Development of disease.   Nematodes overwintering between the bud
scales or the growing point of shoots become activated in the spring and
feed ectoparasitically, by inserting their stylets into the epidermal cells of
the organs in their vicinity. Thus, stem areas near infested buds and the
petioles and leaves derived from such buds show brown scars consisting
of groups of cells killed by the nematodes. In addition to direct killing of
cells the nematodes, through their secretions, cause shortening of the
internodes, which results in a bushy appearance of the plant; also browning and failure of the shoot to grow (blindness); production of low, premature side-shoots; and development of distorted leaves.

Nematodes infest new, healthy plants by swimming up the stem when
it is covered with a film of water during rainy or humid weather. When
they reach the leaves, the nematodes enter through the stomata (Fig. 254).
The presence of nematodes between the leaf cells causes browning in
cells. Cells in the mesophyll begin to break down creating large cavities
in the mesophyll. In the early stages of infection the cells of the vein

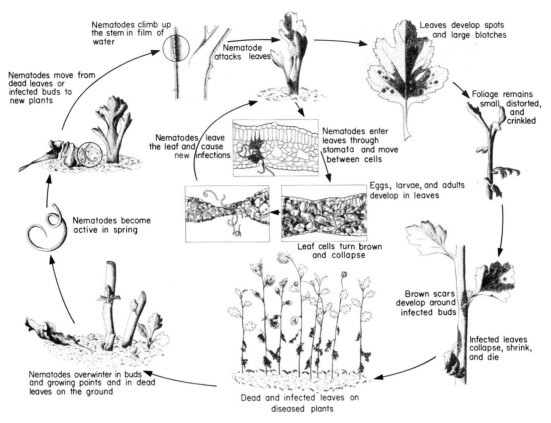

FIGURE 254.
Disease cycle of the foliar (chrysanthemum) nematode *Aphelenchoides ritzemabosi.*

sheath, which in large veins extend from the upper to the lower surface, do not allow penetration of the nematodes through their intercellular spaces and thus block extension of leaf necrosis across the veins. In advanced stages of infection even these cells break down and the nematodes and leaf necrosis spread across the veins over the entire leaf. In heavily infected leaves the disintegrated cells have a thick layer of brown substance on their walls, the epidermis is broken down in places, and the leaves shrink and after clinging to the stem for some time, fall to the ground.

Control. Several sanitary practices are very helpful and very important in controlling the foliar nematode. The leaves and stem should be kept dry, especially indoors, to prevent movement and spreading of the nematodes. Cuttings should be taken only from the tops of long, vigorous branches, not from shoots near the base of the plant. Early in the spring the soil surface around chrysanthemums should be mulched to cover old infested leaves and prevent nematodes overwintering in them from reaching the lower leaves. Suspected dormant cuttings or stools may be disinfested by dipping in hot water (50°C) for 5 minutes or at 44°C for 30 minutes. Excellent control of this nematode can be obtained by spraying plants with parathion or malathion from July to early September, and by applying thionazin as a drench twice with a 2-week interval.

SELECTED REFERENCES

Bryden, J. W., and W. E. H. Hodson. 1957. Control of chrysanthemum eelworm by parathion. *Plant Pathol.* **6**:20–24.

French, N., and Ruth M. Barraglough. 1964. Observations on eelworm on chrysanthemum stools. *Plant Pathol.* **13**:32–37.

Hesling, J. J., and H. R. Wallace. 1961. Observations on the biology of chrysanthemum eelworm, *Aphelenchoides ritzemabosi* (Schwartz) Steiner in florist's chrysanthemum. I. Spread of eelworm infestation. *Ann. Appl. Biol.* **49**:195–203, 204–209.

Voss, W. 1930. Beitrage zur Kenntnis der Aelchenkrankheit der Chrysanthemum. *Z. Parisitenk.* **2**:310–356.

• Stubby-Root Nematodes: Trichodorus

Stubby-root nematodes occur all over the world. They attack a wide variety of plants including oats, cabbage, tomato, clover, corn, bean, grape, peach, and many others. They affect plants by devitalizing root tips and stopping their growth, resulting in reduction of the root system of plants. This results in severe stunting and chlorosis of the whole plant, reduced yields, and poor quality of produce. Infected plants, however, are seldom, if ever, killed by these nematodes.

Symptoms. Infected plants appear stunted within 2 to 3 weeks from inoculation. They have fewer and smaller leaves and branches than healthy plants, although at first their color appears normal. As the growing season progresses, the difference in size between healthy and stubby-root infected plants increases, healthy plants being 3 to 4 times as large as infected ones. The latter also begin to show changes in color, appearing chlorotic instead of normal green.

The root symptoms of infected plants appear as an abnormal growth of lateral roots and proliferation of branch roots. Parasitized root tips show no necrosis or other injury although they are usually darker than normal in color. In such root tips meristematic activity and root growth stops, but cells already formed may enlarge abnormally and cause swelling of the root tip (Fig. 255). Frequently, affected roots produce numerous lateral roots, which are in turn attacked by nematodes. Repeated infections of lateral roots and their branches produce a smaller root system devoid of feeder roots and characterized instead by short, stubby, swollen root branches, the growth of which was stopped by the nematode infections (Figs. 255 and 256).

The pathogen: Trichodorus christiei. It is a small nematode about 0.65 mm long by 40 μm wide. It lives in the upper 30 cm of the soil. It is an ectoparasite, feeding on the epidermal cells at or near the root-tip region, never entering the root tissue (Fig. 255C). It lays eggs in the soil which hatch to produce larvae and then adults. The life cycle of this nematode is completed within about 20 days (Fig. 256). Populations of *T.*

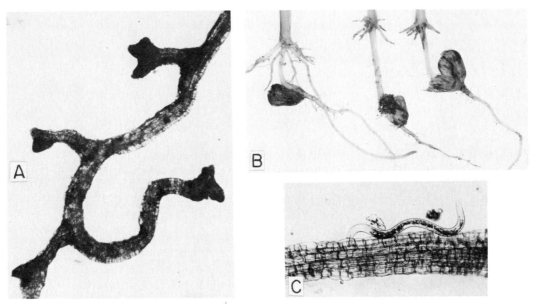

FIGURE 255.
(A) Blackberry root showing stubby-root symptoms caused by *Trichodorus christiei*. (B) Roots of young corn plants injured by the stubby-root nematode. (C) *Trichodorus christii* nematode feeding externally on blueberry root. (Photos A and C courtesy B. M. Zuckerman. Photo B courtesy U.S.D.A.)

christiei build up quickly around susceptible hosts but decline when host plants become old and do not produce new root tips or when good host plants are absent. Eggs, larvae, and adults are usually found in the soil throughout the year, although preadults and eggs seem to be the stages found mostly during winter.

Several species of *Trichodorus*, *T. christiei* included, are capable of transmitting plant viruses from one plant to another. Tobacco-rattle virus and pea early browning virus, both rod-shaped viruses, are the only ones known to be transmitted by *Trichodorus* nematodes. Several other plant viruses, all polyhedral shaped, are transmitted by the only other nematode vectors: *Longidorus* and *Xiphinema*.

Development of disease. When susceptible host plants are growing in soil infested with *Trichodorus*, as soon as the nematode comes in contact with young roots or root tips, it bends its head at approximately a right angle to the root surface, places its lip region against the cell wall, and punctures the wall with direct thrusts of the stylet. Once the stylet is inside the cell, a viscous substance is released through it into the cell, causing the cytoplasm of the plant cell to aggregate around the stylet tip. Part of the cytoplasm is then ingested by the nematode, after which it moves on to another cell within seconds or perhaps a few minutes from the beginning of feeding. Although a small opening up to 0.5 μm in diameter can be seen on the cell wall for many hours after the nematode has left the cell, no cell contents appear to be lost through the hole and

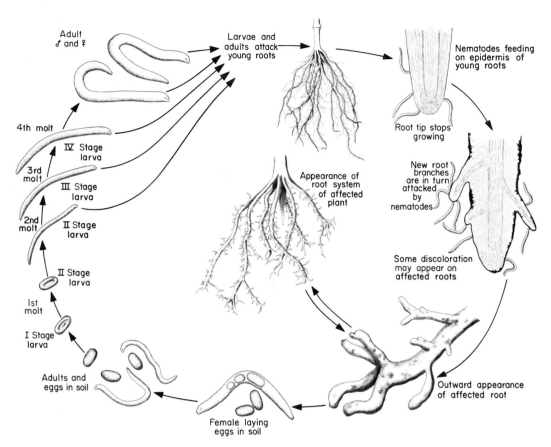

FIGURE 256.
Disease cycle of the stubby-root nematode *Trichodorus christii*.

gradually the aggregated cytoplasm is dispersed and the cell returns to normal.

All free larval stages and the adults can attack plants and feed on them. Feeding is restricted to the epidermal cells at or near the root tip on older roots, and the whole length of young succulent roots (Fig. 256).

Although one root tip may be attacked by many nematodes simultaneously or over a period of time, the mechanical damage caused by *Trichodorus* in feeding is slight and does not account for the gross changes on roots nor for the symptoms of the aboveground part of the plant. Parasitized roots show loss of meristematic activity at the root tip, have no definite root cap or region of elongation, and their region of mitosis is much smaller than that of healthy roots. Branch roots are more abundant and closer together in infected than in healthy roots. These effects seem to be the result of inhibitory or stimulatory action of substances secreted by nematodes into cells rather than of direct mechanical injury.

Control.   Stubby-root nematodes can be controlled through application of nematicides over the entire field. Methyl bromide–chloropicrin mixtures, Telone, DD, or EDB give good, but temporary, control of this nematode. Six to eight weeks after treatment stubby-root nematodes begin to reappear in the field and, if susceptible hosts are present, nematode populations build up rapidly. Slow-acting nematicides, such as Nemagon, retard or prevent the rapid buildup of nematodes, thus increasing the effectiveness of the treatment. Fallow or fallow and dry cultivation give fairly effective control of *Trichodorus*.

SELECTED REFERENCES

Allen, M. W. 1957. A review of the nematode genus *Trichodorus* with descriptions of ten new species. *Nematologica* **2**:32–62.

Chen, T. A., and W. F. Mai. 1965. The feeding of *Trichodorus christiei* on individually isolated corn root cells. *Phytopathology* **55**:128 (abstr.).

Rohde, R. A., and W. R. Jenkins. 1957. Host range of a species of *Trichodorus* and its host–parasite relationships on tomato. *Phytopathology* **47**:295–298.

Russell, C. C., and V. G. Perry. 1966. Parasitic habit of *Trichodorus christiei* on wheat. *Phytopathology* **56**:357–358.

Zuckerman, B. M. 1962. Parasitism and pathogenesis of the cultivated highbush blueberry by the stubby root nematode. *Phytopathology* **52**:1017–1019.

16
plant diseases caused by protozoa

Certain trypanosomatid flagellates, i.e., protozoa of the class Mastigophora, order Kinetoplastida, family Trypanosomatidae, have been known to parasitize plants for almost 70 years. That flagellates may be pathogenic to their host plants was suggested several times by the investigators of these parasites and rather good evidence was presented that some plant diseases are caused by flagellates. However, because these parasites could not be isolated in pure culture and could not be inoculated into healthy plants so that they could reproduce the disease, as Koch's postulates dictate, flagellates have not yet been fully accepted as plant pathogens. Yet, the pathogenicity of mycoplasmas and rickettsialike bacteria in plants is almost universally accepted although the same Koch's postulates are equally unfulfilled with these organisms as they are with the flagellates. Since the evidence supporting the pathogenicity of flagellates is no less than that available for mycoplasmas and rickettsialike bacteria it is reasonable to assume that at least some flagellates are considered capable of causing disease in plants and it is apparent that the role of flagellates, as well as the role of other protozoa, in plant pathology deserves more attention than it has received in the past.

The protozoa are mostly one-celled, microscopic animals, generally motile, and have typical nuclei. They may live alone or in colonies, may be free living, symbiotic, or parasitic. Some protozoa subsist on other organisms such as bacteria, yeasts, algae, and other protozoa, some saprophytically on dissolved substances in the surroundings, and some by photosynthesis as in plants. Protozoa move by flagella (class Mastigophora, the flagellates), by pseudopodia (class Rhizopoda, the amoebae), by cilia (class Ciliata, the ciliates), or by movements of the cell itself (class Sporozoa, a diverse group of parasitic protozoa).

Of the protozoa, apparently only the flagellates have been reported as associated with plant diseases so far, but there are no good reasons why the other classes might not be found in the future to also be parasitic on plants.

The Mastigophora, or flagellates, are characterized by one or more long slender flagella at some or all stages of their life cycle (Figs. 2, 3, 257, and 258). The flagella are used for locomotion and food capture, and perhaps as sense organs. The body of the flagellates usually has a definite long, oval, or spherical form that is maintained by a thin, flexible covering membrane or, in some groups, it may be armored. Flagellates generally reproduce by longitudinal fission (Figs. 3, 257C, and 258B). Although many flagellates are saprophytic and some contain plastids with colored pigments, including functional chlorophyll, others are parasites of man and various animals and some cause serious diseases. The best known flagellate pathogenic to humans is the blood parasite, *Trypanosoma*, the cause of sleeping sickness in Africa, which is transmitted by tsetse flies.

Flagellates were first found to be associated with plants in 1909, when Lafont reported that *Phytomonas davidi* parasitizes the latex-bearing cells—the laticifers—of the laticiferous plant *Euphorbia* (Euphorbiaceae). Since then several other species of *Phytomonas* have been reported from plants belonging to the families Asclepiadaceae (e.g., *P. elmassiani* on milkweed), Moraceae (e.g., *P. bancrofti* on a ficus species), Rubiaceae (e.g., *P. leptovasorum* on coffee), and species of unknown identity on coconut palm and on oil palm. All plant flagellates belong to the order Kinetoplastida, family Trypanosomatidae. The plant-infecting Phytomonads are apparently transmitted by insects but so far insect vectors are known only for *P. elmassiani*. Many of the investigators who studied the flagellates in laticiferous plants feel that although the flagellates parasitize the plants— since they live off their latex—the plants do not become diseased and, therefore, the flagellates are not pathogenic to these plants. According to some reports, however, symptoms apparently do develop in some flagellate-infected laticiferous plants, which would indicate that the flagellates are pathogenic to their hosts.

The nonlaticiferous hosts, coffee, coconut palm, and oil palm are apparently infected by pathogenic *Phytomonas* species and develop characteristic

---

FIGURE 257.
Electron micrographs of the trypanosomatid flagellate *Phytomonas* in the phloem of young inflorescences of coconut palms affected with hartrot. (A) Cross section of a differentiating vascular bundle in a palm that had early symptoms of the disease. Recently matured sieve elements are filled with flagellates; M, immature metaxylem; S, immature sieve elements (scale bar 10 μm). (B) Cross section of the phloem in a palm that had advanced symptoms of the disease, showing C, companion cell; F, fiber; P, phloem parenchyma cell; S, sieve elements free of flagellates (scale bar 5 μm). (C) Cross section of a flagellate undergoing longitudinal fission (scale bar 0.5 μm). (D) Longitudinal section of a sieve element filled with the flagellates. Arrows point to the DNA portion of kinetoplasts (scale bar 1 μm). (E) Similar to B but at a higher magnification; C, companion cell; P, parenchyma cell (scale bar 2 μm). (Photos courtesy M. V. Parthasarathy, from *Science* **192**:1346–1348. Copyright © 1976 by the American Association for the Advancement of Science.)

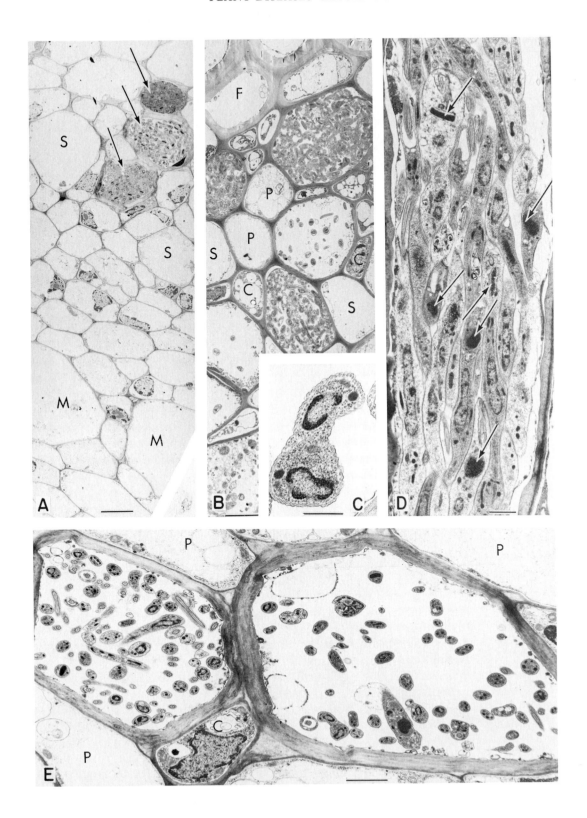

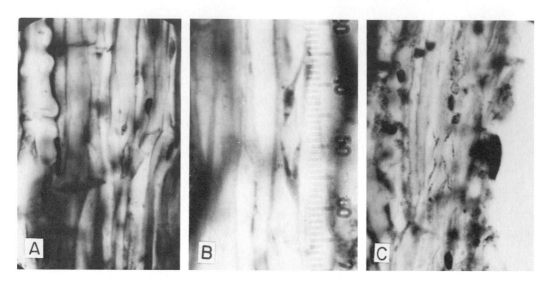

FIGURE 258.
Flagellates associated with the coffee wilt disease. (A) Single protozoon in vascular vessel of diseased *Coffea liberica*. (B) Flagellates in vessel of *C. liberica*, one of them in the process of division. (C) Long and thin flagellates in vessels of coffee tree showing advanced symptoms of the disease. (Photos courtesy J. H. van Emden.)

internal and external symptoms, and severe and economically important diseases. Flagellates apparently cause the phloem necrosis disease of coffee, the hartrot disease of coconut palm, and the marchitez sorpresiva (sudden wilt) disease of oil palms.

• *Phloem Necrosis of Coffee*

It occurs in Surinam, British Guiana, and probably Brazil, San Salvador, and Colombia. It affects trees of *Coffea liberica*. Infected trees show sparse yellowing and dropping of leaves and, as these increase gradually, only the young top leaves remain on the otherwise bare branches. As the roots begin to die back, the condition of the tree worsens and the tree dies (Fig. 259A). Sometimes, in the beginning of the dry season, trees wilt and die within 3 to 6 weeks (Fig. 259B). Internally, the roots and trunk of trees show multiple division of cambial cells and production of a zone of smaller and shorter phloem vessels of disorderly structure right next to the wood cylinder (Fig. 259C, D). At this stage the bark in the roots and in the trunk is firmly attached to the wood and cannot be separated from it.

The pathogen, *Phytomonas leptovasorum*, is a trypanosomatid flagellate. When symptoms first appear there are only a few, big (14 to 18 × 1.0 to 1.2 μm), spindle-shaped flagellates in the phloem (Fig. 258A, B). As multiple division of cambial cells and abnormal phoem production become apparent and many leaves turn yellow and fall, the flagellates are numerous, slender, and spindle shaped, 4 to 14 × 0.3 to 1.0 μm (Fig. 258C). A few shorter (2.0 to 3.0 μm) forms of the flagellate, called "leishmania forms," also appear in the oldest sieve tubes. When the

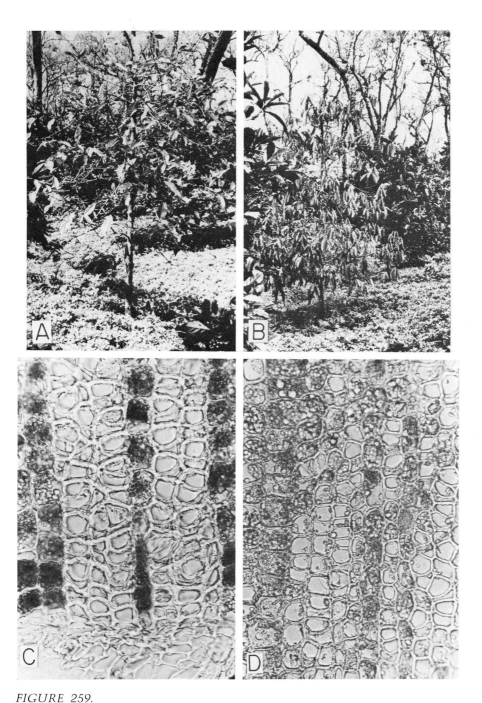

FIGURE 259.
Coffee wilt of *Coffea liberica* caused by the flagellate protozoon *Phytomonas leptovasorum*. (A) Affected tree during rainy season. Note loss of leaves and yellowing but no acute wilting. (B) Affected tree at the onset of the dry season. Note sudden wilting. (C) Cross section of abnormal phloem tissue from flagellate-affected and wilting coffee tree. (D) Cross section of healthy phloem tissue of coffee tree. (Photos courtesy J. H. van Emden.)

FIGURE 260.
"Hartrot" of coconut palms caused by flagellate protozoons. (A) Malayan Dwarf palms, 3 years old, suffering from "hartrot" disease. Note broken leaves and the collapsed spear. (B) Unopened inflorescence of Mayalan dwarf showing necrotic spike tops. It is inflorescence No. 1, and the first symptom of the disease. (C) Inflorescence showing necrotic spike tops. (D) Ceylonese dwarf-yellow, 4 years old, suffering from "hartrot" disease. Note nuts on the ground, one of the first symptoms. (Photos courtesy W. G. van Slobbe.)

multiple division develops into a multilayered sheath around the wood cylinder that extends from the roots up to 2 meters above the ground line and the tree is almost dead, there is a great abundance of small (3 to 4 × 0.1 to 0.2 μm), "spaghetti" flagellates only in the living tissues of the stem, while previously occupied cells are evacuated.

The flagellates can be traced from the roots upward into the trunk where they seem to migrate vertically in the phloem and laterally through the sieve plates into healthy sieve tubes. They also seem to move downward into unaffected roots. Flagellates could not be found in the tree outside areas with multiple division.

The disease can be transmitted through root grafts but not through green branch or leaf grafts. Following grafting of healthy trees with roots infected with flagellates, the flagellates can be observed in the previously healthy roots within a few weeks, the tree begins to develop external symptoms 4 to 5 months later, and it then dies shortly afterward. The disease spreads in the field from one tree to another, and healthy trees often become infected when transplanted in areas from which a diseased tree had been removed. No vector of the disease is known.

- *Hartrot of Coconut Palms*

Hartrot has been known in Surinam since 1906, sometimes under the names lethal yellowing, bronze-leaf wilt, Coronie wilt, and "unknown disease." Many of the symptoms of hartrot (Fig. 260, A–D) are similar to those caused by the lethal yellowing disease of coconut palms in the Caribbean, West Africa, and Florida, but the causes of the two diseases seem to be unrelated. Trees affected with lethal yellowing contain only mycoplasmas and no flagellates, while trees affected with hartrot contain only flagellates and no mycoplasmas.

Flagellates of the genus *Phytomonas* occur in mature sieve elements of young leaves and inflorescences of hartrot-affected coconut palms (Fig. 257). In advanced stages of the disease, 10 to 100 of the mature sieve elements contain flagellates and many of them are plugged with flagellates which are usually oriented longitudinally within the phloem. The flagellates measure 12 to 18 × 1.0 to 2.5 μm. The number and spread of the flagellates in sieve tubes increase proportionally with the development of the disease.

No vector and no means of transmission of the disease are known yet.

SELECTED REFERENCES

Emden, J. H. van. 1962. On flagellates associated with a wilt of *Coffea liberica. Meded. Landb. Hogesch. Opzoek Stns. Gent* **27**:776–784.

Harvey, R. B., and S. B. Lee. 1943. Flagellates of laticiferous plants. *Plant Physiol.* **18**:633–655.

Holmes, F. O. 1924. Herpetomonad flagellates in the latex of milkweed in Maryland. *Phytopathology* **14**:146–151.

Holmes, F. O. 1925. The relationship of *Herpetomonas elmassiani* to its plant and insect hosts. *Biol. Bull.* **49**:323–327.

Lafont, A. 1909. Sur la prèsence d'un parasite de la classe des flagellès dans le latex de l'*Euphorbia pilulifera. C. R. Soc. Biol.* **66**:1011–1013.

McGhee, R. B., and W. L. Hanson. 1964. Comparison of the life cycle of *Leptomonas oncopelti* and *Phytomonas elmassiani*. *J. Protozool.* **11**:555–562.

McGhee, R. B., and Ann H. McGhee. 1971. The relation of migration of *Oncopeltus fasciatus* to distribution of *Phytomonas elmassiani* in the eastern United States. *J. Protozool.* **18**:344–352.

Parthasarathy, M. V., W. G. vanSlobbe, and Carole Soudant. 1976. Trypanosomatid flagellate in the phloem of diseased coconut palms. *Science* **192**:1346–1348.

Stahel, G. 1933. Zur Kenntnis der Siebröhren-krankheit (Phloëmnekrose) des Kaffeebaumes in Surinam. III. *Phytopathol. Z.* **6**:335–357.

Thomas, D. L., R. E. McCoy, and A. F. Espinosa. 1977. Association of flagellated protozoa with marchitez sorpresiva (sudden wilt) disease of oil palms in Ecuador. *Proc. Am. Phytopathol. Soc.* **4**, No. 253, 137.

Vermeulen, H. 1963. A wilt of *Coffea liberica* in Surinam and its association with a flagellate, *Phytomonas leptovasorum*. *J. Protozool.* **10**:216–222.

Vermeulen, H. 1968. Investigations into the cause of the phloem necrosis disease of *Coffea liberica* in Surinam, South America. *Neth. J. Pl. Path.* **74**:202–218.

general references

The following are books available on general aspects of plant pathology and on diseases of specific crops.

Akai, S., and S. Ouchi (Eds.). 1971. "Morphological and Biochemical Events in Plant-Parasite Interaction." Proc. Univ. Hawaii Conf. (1970). Phytopathol. Soc. Japan, Tokyo. 415 p.

Alexopoulos, C. J. 1962. "Introductory Mycology," 2nd ed., 613 p. Wiley, New York.

Altman, J. 1966. "Phytopathological Techniques. Laboratory Manual," 259 p. Pruett Press, Boulder, Colorado.

Anderson, H. W. 1956. "Diseases of Fruit Crops," 501 p. McGraw-Hill, New York.

Anonymous. 1953. "Plant Diseases," U. S. Dept. Agr. Yearbook, 940 p. with 32 color plates. U. S. Dept. Agr., Washington, D. C.

Baker, K. F., and R. J. Cook. 1974. "Biological Control of Plant Pathogens," 433 p. W. H. Freeman, San Francisco, California.

Barnes, E. H. 1968. "Atlas and Manual of Plant Pathology," 325 p. Appleton, New York.

Bawden, F. C. 1964. "Plant Viruses and Virus Diseases," 4th ed., 361 p. Ronald Press, New York.

Boyce, J. S. 1961. "Forest Pathology," 3rd ed., 572 p. McGraw-Hill, New York.

Bruehl, G. W. (Ed.). 1975. "Biology and Control of Soil-borne Plant Pathogens," 216 p. Amer. Phytopathol. Soc., St. Paul, Minnesota.

Carefoot, G. L., and E. R. Sprott. 1967. "Famine on the Wind: Man's Battle Against Plant Disease," 231 p. Rand McNally, Chicago, Illinois.

Carter, W. 1973. "Insects in Relation to Plant Disease," 759 p. Wiley, New York.

Chester, K. S. 1950. "Nature and Prevention of Plant Diseases," 525 p. Blackiston, Philadelphia, Pennsylvania.

Christie, J. R. 1959. "Plant Nematodes, Their Bionomics and Control," 256 p. Fla. Agr. Expt. Sta., Gainesville, Florida.

Chupp, C., and A. F. Sherf. 1960. "Vegetable Diseases and Their Control," 693 p. Ronald Press, New York.

Corbett, M. K., and H. D. Sisler (Eds.). 1964. "Plant Virology," 527 p. Univ. of Florida Press, Gainesville, Florida.

Couch, H. B. 1973. "Diseases of Turf Grasses," 348 p. Krieger, Huntington, New York.

Day, P. R. 1974. "Genetics of Host-Parasite Interaction," 238 p. Freeman, San Francisco.

Day, P. R. (Ed.). 1977. "The Genetic Basis of Epidemics in Agriculture." *Annals N. Y. Acad. Sci.* **287**: 400 p.

669

Deverall, B. J. 1977. "Defense Mechanisms of Plants," 110 p. Cambridge Univ. Press, London and New York.

Dickson, J. G. 1956. "Diseases of Field Crops," 517 p. McGraw-Hill, New York.

Dowson, W. J. 1957. "Plant Diseases Due to Bacteria," 2nd ed., 232 p. Cambridge Univ. Press, London and New York.

Elliott, Charlotte. 1951. "Manual of Bacterial Plant Pathogens," 2nd ed., 186 p. Chronica Botanica, Waltham, Massachusetts.

Esau, K. 1968. "Viruses in Plant Hosts." 225 p. Univ. Wisconsin Press, Madison, Wisconsin.

Evans, E. 1968. "Plant Diseases and Their Chemical Control." Blackwell, Oxford.

Fawcett, H. S. 1936. "Citrus Diseases and Their Control," 2nd ed., 656 p. McGraw-Hill, New York.

Fischer, G. W., and C. S. Holton. 1957. "Biology and Control of the Smut Fungi," 622 p. Ronald Press, New York.

Forsberg, J. L. 1975. "Diseases of Ornamental Plants." Univ. Illinois Coll. Agr. Spec. Publ. No. 3 Rev., 220 p. Urbana, Illinois.

Frazier, N. W. (Ed.). 1970. "Virus Diseases of Small Fruits and Grapevines" (A Handbook), 290 p. Univ. California Press, Berkeley, California.

Fulton, J. P., D. A. Slack, N. D. Fulton, J. L. Dale, M. J. Goode, and G. E. Templeton. 1962. "Plant Pathology. Laboratory Manual," 95 p. Burgess, Minneapolis, Minnesota.

Garrett, S. D. 1956. "Biology of Root-Infecting Fungi," 293 p. Cambridge Univ. Press, London and New York.

Garrett, S. D. 1970. "Pathogenic Root-Infecting Fungi," 294 p. Cambridge Univ. Press, London and New York.

Gäumann, E. 1950. "Principles of Plant Infection," 543 p. Crosby Lockwood, London.

Goodman, R. Z., Z. Kiraly, and M. Zaitlin. 1967. "The Biochemistry and Physiology of Infectious Plant Diseases," 354 p. Van Nostrand-Reinhold, Princeton, New Jersey.

Gregory, P. H., and J. L. Monteith (Eds.). 1967. "Airborne Microbes." Cambridge Univ. Press, London and New York.

Hacskaylo, E. (Ed.). 1971. "Mycorrhizae," 255 p. U. S. Govt. Printing Office, Washington, D. C.

Heitefuss, R., and P. H. Williams (Eds.). 1976. "Physiological Plant Pathology." Encyclopedia of Plant Physiology, New Series, Vol. 4, 890 p. Springer-Verlag, New York.

Holton, C. S., G. W. Fischer, R. W. Fulton, Helen Hart, and S.E.A. McCallan (Eds.). 1959. "Plant Pathology. Problems and Progress, 1908–1958," 588 p. Univ. Wisconsin Press, Madison, Wisconsin.

Horsfall, J. G. 1956. "Principles of Fungicidal Action," 279 p. Chronica Botanica, Waltham, Massachusetts.

Horsfall, J. G., and E. B. Cowling (Eds.). 1977, 1978. "Plant Disease: An Advanced Treatise." Volumes I and II. Academic Press, New York.

Horsfall, J. G., and A. E. Dimond (Eds.). 1959–1960. "Plant Pathology. An Advanced Treatise," 3 vols. Academic Press, New York.

Jenkins, W. R., and D. P. Taylor. 1967. "Plant Nematology," 270 p. Van Nostrand-Reinhold, New York.

Kado, C. I., and H. O. Agrawal (Eds.). 1972. "Principles and Techniques in Plant Virology," 688 p. Van Nostrand-Reinhold, New York.

Kelman, A., et al. (Eds.). 1967. "Sourcebook of Laboratory Exercises in Plant Pathology," 388 p. Freeman, San Francisco, California.

Kenaga, C. B. 1974. "Principles of Phytopathology." 2nd ed., 402 p. Balt Publ., Lafayette, Indiana.

Kenaga, C. B., E. B. Williams, and R. J. Green. 1971. "Plant Disease Syllabus." Balt Publ., Lafayette, Indiana.

Klinkowski, M., E. Mühle, and E. Reinmuth. 1966. "Phytopathologie und Pflanzenschutz," 2 vols. Akademie-Verlag, Berlin.

Kuijt, J. 1970. "The Biology of Parasitic Flowering Plants." Univ. California Press, Berkeley, California.

Leach, J. G. 1940. "Insect Transmission of Plant Diseases," 615 p. McGraw-Hill, New York.

Maramorosch, K. (Ed.). 1969. "Viruses, Vectors and Vegetation," 666 p. Interscience, New York.

Maramorosch, K., and H. Koprowski. (Eds.). 1967–1975. "Methods in Virology," Vols. I–VI. Academic Press, New York.

Matthews, R.E.F. 1970. "Plant Virology," 778 p. Academic Press, New York.

Mirocha, C. J., and I. Uritani (Eds.). 1967. "The Dynamic Role of Molecular Constituents in Plant Parasite Interaction," 372 p. Bruce, St. Paul, Minnesota.

National Academy of Sciences. 1972. "Genetic Vulnerability of Major Crops," 307 p. Washington, D. C.

Nelson, R. R. (Ed.). 1973. "Breeding Plants for Disease Resistance. Concepts and Applications," 401 p. Penn. State Univ. Press, University Park, Pennsylvania.

Ou, S. H. 1972. "Rice Diseases," 368 p. Commonw. Mycol. Inst., Kew, Surrey, England.

Peace, T. R. 1962. "Pathology of Trees and Shrubs, with Special Reference to Britain," 753 p. Oxford Univ. Press, London and New York.

Peachy, J. E. (Ed.). 1969. "Nematodes of Tropical Crops," 355 p. Tech. Commun. Commonw. Bur. Helminth. No. 40. Farnham Royal, Bucks, England.

Pirone, P. P. 1978. "Diseases and Pests of Ornamental Plants," 5th ed., 600 p. Interscience, New York.

Plakidas, A. G. 1964. "Strawberry Diseases." Louisiana State Univ. Press, Baton Rouge, Louisiana.

Posnette, A. F. (Ed.). 1963. "Virus Diseases of Apples and Pears," 141 p. Tech. Commun. Bur. Hort., E. Malling, No. 30, Farnham Royal, Bucks, England.

Roberts, D. A., and C. W. Boothroyd. 1972. "Fundamentals of Plant Pathology," 402 p. Freeman, San Francisco, California.

Sauchelli, V. 1969. "Trace Elements in Agriculture." Van Nostrand-Reinhold, Princeton, New Jersey.

Sharvelle, E. G. 1969. "Chemical Control of Plant Diseases," 340 p. University Publ., College Station, Texas. 340 p.

Shurtleff, M. C. 1966. "How to Control Plant Diseases in Home and Garden," 2nd ed., 649 p. Iowa State Univ. Press, Ames, Iowa.

Shurtleff, M. C. (Ed.). 1973. "A Compendium of Corn Diseases," 64 p. Amer. Phytopathol. Soc., St. Paul, Minnesota.

Sinclair, J. B., and M. C. Shurtleff (Eds.). 1975. "Compendium of Soybean Diseases," 69 p. Amer. Phytopathol. Soc., St. Paul, Minnesota.

Smart, G. C., Jr., and V. G. Perry (Eds.). 1968. "Tropical Nematology," 153 p. Univ. Florida Press, Gainesville, Florida.

Smith, K. M. 1973. "A Textbook of Plant Virus Diseases," 3rd ed., 684 p. Academic Press, New York.

Sprague, R. 1950. "Diseases of Cereals and Grasses in North America," 538 p. Ronald Press, New York.

Stakman, E. C., and J. G. Harrar. 1957. "Principles of Plant Pathology," 581 p. Ronald Press, New York.

Stapp, C. 1961. "Bacterial Plant Pathogens" (Translated by A. Schoenfeld), 292 p. Oxford Univ. Press, London and New York.

Stevens, N. E., and R. B. Stevens. 1952. "Disease in Plants," 219 p. Chronica Botanica, Waltham, Massachusetts.

Stevens, R. B. 1974. "Plant Disease," 459 p. Ronald Press, New York.

Stover, R. H. 1972. "Banana, Plantain and Abaca Diseases," 316 p. Commonw. Mycol. Inst., Kew, Surrey, England.

Streets, R. B., Sr. 1969. "The Diagnosis of Plant Diseases." Coop. Ext. Service and Agric. Exp. Sta., Univ. Arizona Press, Tucson, Arizona.

Streets, R. B., Sr. 1969. "Diseases of the Cultivated Plants of the Southwest," 390 p. Univ. Arizona Press, Tucson, Arizona.

Strobel, G. A., and D. E. Mathre. 1970. "Outlines of Plant Pathology." 465 p. Van Nostrand-Reinhold, New York.

Tarr, S.A.J. 1972. "The Principles of Plant Pathology," 632 p. Winchester Press, New York.

Tattar, T. A. 1978. "Diseases of Shade Trees," 384 p. Academic Press, New York.

Thorne, G. 1961. "Principles of Nematology," 553 p. McGraw-Hill, New York.

U.S.D.A. 1976. "Virus Diseases and Noninfectious Disorders of Stone Fruits in North America," 433 p.

Agric. Handbook No. 437, U. S. Govt. Printing Office, Washington, D. C.

Van der Plank, J. E. 1963. "Plant Diseases: Epidemics and Control," 349 p. Academic Press, New York.

Van der Plank, J. E. 1968. "Disease Resistance in Plants," 206 p. Academic Press, New York.

Van der Plank, J. E. 1975. "Principles of Plant Infection," 216 p. Academic Press, New York.

Walker, J. C. 1952. "Diseases of Vegetable Crops." 529 p. McGraw-Hill, New York.

Walker, J. C. 1968. "Plant Pathology," 3rd ed., 819 p. McGraw-Hill, New York.

Wallace, T. 1961. "The Diagnosis of Mineral Deficiencies in Plants," 125 p. with 312 color plates. H.M.S.O., London.

Webber, G. F. 1973. "Bacterial and Fungal Diseases of Plants in the Tropics," 673 p. Univ. Florida Press, Gainesville, Florida.

Webster, J. M. (Ed.). 1972. "Economic Nematology," 563 p. Academic Press, New York.

Wellman, F. L. 1972. "Tropical American Plant Disease," 989 p. Scarecrow Press, Metuchen, New Jersey.

Westcott, Cynthia. 1960. "Plant Disease Handbook," 2nd ed., 825 p. Van Nostrand-Reinhold, Princeton, New Jersey.

Western, J. H. (Ed.). 1971. "Diseases of Crop Plants," 404 p. MacMillan Press, London.

Wheeler, B.E.J. 1969. "An Introduction to Plant Diseases." 374 p. Wiley, New York.

Wheeler, H. 1975. "Plant Pathogenesis," 106 p. Springer-Verlag, New York.

Wiese, M. V. (Ed.). 1977. "Compendium of Wheat Diseases," Amer. Phytopathol. Soc., St. Paul, Minnesota.

Wood, R.K.S. 1967. "Physiological Plant Pathology." Blackwell, Oxford.

PERIODICALS

Acta Phytopathologica
Advances in Botany
Advances in Virus Research
American Journal of Botany
Annales des Epiphytes
Annals of Applied Biology
Annals of the Phytopathological Society of Japan
Annual Review of Microbiology
Annual Review of Phytopathology
Annual Review of Plant Physiology
Biochimica et Biophysica Acta
Biological Abstracts
Botanical Gazette
Botanical Review
Boyce Thompson Institute, Contributions
Canadian Journal of Botany
Canadian Journal of Science
Canadian Plant Disease Survey
Hilgardia
Journal of Agricultural Research
Journal of Bacteriology
Journal of Cell Biology
Journal of Virology

Mycologia
Nature
Nematologica
Netherlands Journal of Plant Pathology
Physiological Plant Pathology
Phytochemistry
Phytopathologia Mediterranea
Phytopathologische Zeitschrift
Phytopathology
Plant Disease Reporter
Plant Pathology
Plant Physiology
Plant Protection Bulletin (FAO)
Proceedings, Helminthological Society of Washington
Review of Plant Pathology
Science
State Experiment Station Bulletins, Technical Bulletins, Memoirs, Annual Reports, etc.
U. S. Dept. Agr. Bulletins
Virology
Zeitschrift für Pflanzenkrankheiten

glossary*

Å (Ångstrom) — A unit of length equal to 1/10 millimicron (mµ) or 1/10,000 micron (µ).

Abscission layer — A zone of delicate, thin-walled cells surrounding a lesion on a leaf, the breakdown of which disjoins the affected area from the rest of the leaf.

Acervulus — A subepidermal, saucer-shaped, asexual fruiting body producing short conidiophores and conidia.

Actinomycetes — A group of microorganisms apparently intermediate between bacteria and fungi, and classified as either.

ADP (Adenosine diphosphate) — A compound which upon phosphorylation (addition of phosphate and energy) forms high energy bonds as ATP.

Adventitious roots — Roots that appear in an unusual place or position, e.g., on the stem.

Aeciospore — A binucleate rust spore produced in an aecium.

Aecium — A cup-shaped fruiting body of the rust fungi which produces aeciospores.

Aerobic — A microorganism that lives or a process that occurs in the presence of molecular oxygen.

Aflatoxin — A mycotoxin produced by *Aspergillus flavus* and other species of this fungus.

Agar — A gelatinlike material obtained from seaweed and used to prepare culture media on which microorganisms are grown and studied.

Agglutination — A serological test in which viruses or bacteria suspended in a liquid collect into clumps whenever the suspension is treated with antiserum containing antibodies specific against these viruses or bacteria.

Alkaloid — An organic compound with alkaline properties, and usually poisonous, produced by certain plants.

Alternate host — One of two kinds of plants on which a parasitic fungus

*NOTE
To make the plural of Latin words ending in:
-us (e.g., acervulus), change us to i (e.g., acervuli);
-um (e.g., aecium), change um to a (e.g., aecia);
-a (e.g., hypha), change a to ae (e.g., hyphae);
-is (e.g., tylosis), change is to es (e.g., tyloses).

673

(e.g., rust) must develop to complete its life cycle.

Amylase – Enzyme that breaks down starch.

Anaerobic – Relating to a microorganism that lives or a process that occurs in the absence of molecular oxygen.

Anastomosis – The union of a hypha or vessel with another resulting in intercommunication of their contents.

Antheridium – The male sexual organ found in some fungi.

Anthracnose – A leaf- or fruit-spot type of disease caused by fungi that produce their asexual spores in an acervulus.

Antibiotic – A chemical compound produced by one microorganism which inhibits or kills other microorganisms.

Antibody – A new or altered protein produced in a warm-blooded animal in reaction to an injected foreign antigen.

Antigen – Foreign proteins, and occasionally complex lipids and carbohydrates, which upon injection into an animal induce the production of antibodies.

Antiserum – The blood serum of a warm-blooded animal that contains antibodies.

Apothecium – An open cup- or saucer-shaped ascocarp of some ascomycetes.

Appressorium – The swollen tip of a hypha or germ tube that facilitates attachment and penetration of the host by the fungus.

Ascocarp – The fruiting body of ascomycetes bearing or containing asci.

Ascogenous hypha – Hyphae arising from the fertilized ascogonium and producing the asci.

Ascogonium – The female gametangium or sexual organ of ascomycetes.

Ascomycetes – A group of fungi producing their sexual spores, ascospores, within asci.

Ascospore – A sexually produced spore borne in an ascus.

Ascus – A saclike hypha containing ascospores (usually eight).

Asexual reproduction – Any type of reproduction not involving the union of gametes or meiosis.

ATP (Adenosine triphosphate) – A compound formed by phosphorylation of ADP and which stores and releases energy for the various cell functions.

Autoecious fungus – A parasitic fungus that can complete its entire life cycle on the same host.

Auxin – A plant growth-regulating substance controlling cell elongation.

Axillary bud – A bud formed in the upper angle between a twig or a periole and the stem.

Bacillus – A rod-shaped bacterium.

Bactericide – A chemical compound that kills bacteria.

Bacteriocins – Nonreplicating, bactericidal protein-containing substances produced by certain strains of bacteria and active against some other strains of the same or closely related species.

Bacteriophage – A virus that infects specific bacteria and usually kills them.

Bacteriostatic – A chemical or physical agent that prevents multiplication of bacteria without killing them.

Bacterium – A unicellular microscopic plant that lacks chlorophyll and multiplies by fission.

Base – An alkaline, usually nitrogenous organic compound; used particularly for the purine and pyrimidine moieties of the nucleic acids of cells and viruses.

Basidiomycetes – A group of fungi producing their sexual spores, basidiospores, on basidia.

Basidiospore – A sexually produced spore borne on a basidium.

Basidium – A club-shaped structure on which basidiospores are borne.

Biotype – A subgroup within a species usually characterized by the possession of a single or a few characters in common.

Blight – A disease characterized by general and rapid killing of leaves, flowers, and stems.

Blotch – A disease characterized by large, and irregular in shape, spots or blots on leaves, shoots, and stems.

Budding – A method of vegetative propagation of plants by implantation of buds from the mother plant onto a rootstock.

Callus – A mass of thin-walled cells, usually developed as the result of wounding or infection.

Cambium – A one- or two-cell-thick layer of persistently meristematic tissue that produces all secondary tissues and results in growth in diameter.

Canker – A necrotic, often sunken lesion on a stem, branch, or twig of a plant.

Capsid – The protein coat of viruses forming the closed shell or tube that contains the nucleic acid and consisting of protein subunits or capsomeres.

Capsomere – Also called a protein subunit; a small protein molecule that is the structural and chemical unit of the protein coat (capsid) of a virus.

Capsule – A relatively thick layer of mucopolysaccharides that surrounds some kinds of bacteria.

Carbohydrate – Foodstuffs composed of carbon, hydrogen, and oxygen (CH_2O) with the last two in a 2 to 1 ratio, as in water, H_2O.

Catalyst – A substance that accelerates a chemical reaction but is not used up in the reaction.

Cellulase – An enzyme that breaks down cellulose.

Cellulose – A polysaccharide composed of hundreds of glucose molecules linked in a chain and found in the plant cell walls.

Chemotherapy – Control of a plant disease with chemicals (chemo-therapeutants) that are absorbed and are translocated internally.

Chlamydospore – A thick-walled asexual spore formed by the modification of a cell of a fungus hypha.

Chlorosis – Yellowing of normally green tissue due to chlorophyll destruction or failure of chlorophyll formation.

Chronic symptoms – Symptoms that appear over a long period of time.

Circulative viruses – Viruses that are acquired by their vectors through their mouthparts, accumulate internally, then are passed through their tissues and introduced into plants again via the mouthparts of the vectors.

Cistron – The sequence of nucleotides within a certain area of a nucleic acid (DNA or RNA).

Cleistothecium – An entirely closed ascocarp.

Clone – The aggregate of individual organisms produced asexually from one sexually produced individual.

Coding – The process by which the sequence of nucleotides within a certain area of RNA determines the sequence of amino acids in the synthesis of the particular protein.

Codon – The coding unit, consisting of three adjacent nucleotides.

Concentric – Forming one circle around another with a common center.

Conjugation – A process of sexual reproduction involving the fusion of gametes morphologically similar.

Conidiophore – A specialized hypha on which one or more conidia are produced.

Conidium – An asexual fungus spore formed from the end of a conidiophore.

Coremium – An asexual fruiting body consisting of a cluster of erect hyphae bearing conidia.

Cork – An external, secondary tissue impermeable to water and gases. It is often formed in response to wounding or infection.

Cortex – The stem or root tissue between the epidermis and the phloem.

Cotyledon – The seed leaf; one in the monocotyledons, two in the dicotyledons.

Cross protection – The phenomenon in which plant tissues infected with one strain of a virus are protected from infection by other strains of the same virus.

Culture – To artificially grow microorganisms on a prepared food material; a colony of microorganisms artificially maintained on such food material.

Culture medium – The prepared food material on which microorganisms are cultured.

Cuticle – A membranous layer on outer wall of epidermal cells consisting primarily of wax and cutin.

Cutin – A waxy substance comprising the inner layer of the cuticle.

Cyst – An encysted zoospore (fungi); in nematodes, the carcass of dead adult females of the genus *Heterodera* which may contain eggs.

Cytokinins – A group of plant growth-regulating substances that regulate cell division.

Cytoplasm – All the living substance of a cell outside of the nucleus.

Damping-off – Destruction of seedlings near the soil line, resulting in the seedlings falling over on the ground.

Denatured protein – Protein whose properties have been altered by treatment with physical or chemical agents.

Density-gradient centrifugation – A method of centrifugation in which particles are separated in layers according to their density.

Detoxification – The inactivation or destruction of a toxin by alteration, binding, or breakdown of the toxic molecule.

Dieback – Progressive death of shoots, branches, and roots generally starting at the tip.

Dikaryotic – Mycelium or spores containing two sexually compatible nuclei per cell. Common in the basidiomycetes.

Disease – Any disturbance of a plant that interferes with its normal structure, function, or economic value.

Disease cycle – The chain of events involved in disease development, including the stages of development of the pathogen and the effect of the disease on the host.

Disinfectant – A physical or chemical agent that frees a plant, organ, or tissue from infection.

Disinfestant – An agent that kills or inactivates pathogens in the environment or on the surface of a plant or plant organ before infection takes place.

Dissemination – Transfer of inoculum from its source to healthy plants.

Dormant – Being in a state of reduced physiological activity.

Downy mildew – A plant disease in which the mycelium and spores of the fungus appear as a downy growth on the host surface; caused by fungi in the family Peronosporaceae.

Ectoparasite – A parasite feeding on a host from the exterior.

Egg – A female gamete. In nematodes, the first stage of the life cycle containing a zygote or a larva.

Enation – Tissue malformation or overgrowth induced by certain virus infections.

Endodermis – A layer of cells with thick walls and no intercellular spaces that surrounds the vascular tissues of the roots.

Endoparasite – A parasite which enters a host and feeds from within.

Enzyme – A protein produced by living cells that can catalyze a specific organic reaction.

Epidemic – A widespread and severe outbreak of a disease.

Epidermis – The superficial layer of cells occurring on all plant parts.

Epiphytically – Existing on the surface of a plant or plant organ without causing infection.

Epiphytotic – A widespread and destructive outbreak of a disease of plants.

Eradicant – A chemical substance that destroys a pathogen at its source.

Eradication – Control of plant disease by eliminating the pathogen after it is established or by eliminating the plants that carry the pathogen.

Etiolation – Yellowing of tissue and elongating of stems caused by reduced light or darkness.

Exudate – Liquid discharge from diseased or healthy plant tissue.

Fermentation – Oxidation of certain organic substances in the absence of molecular oxygen.

Fertilization – The sexual union of two protoplasts resulting in doubling of chromosome numbers.

Filamentous – Threadlike; filiform.

Fission – Transverse splitting in two of bacterial cells, asexual.

Flagellum — A whiplike structure projecting from a bacterium or zoospore and functioning as an organ of locomotion. Also called a cilium.

Flagging – The loss of rigidity and drooping of leaves and tender shoots preceding the wilting of a plant.

Fleck – A minute spot.

Free-living – Of a microorganism that lives freely, unattached; or a pathogen living in the soil, outside its host.

Fructification – Production of spores by fungi. Also, a fruiting body.

Fruiting body – A complex fungal structure containing spores.

Fumigant – A toxic gas or volatile substance that is used to disinfest certain areas from various pests.

Fumigation – The application of a fumigant for disinfestation of an area.

Fungicide – A compound toxic to fungi.

Fungistatic – A compound that prevents fungus growth without killing the fungus.

Fungus – An undifferentiated plant lacking chlorophyll and conductive tissues.

Gall – A swelling or overgrowth produced on a plant as a result of infection by certain pathogens.

Gametangium – A cell containing gametes or nuclei that act as gametes.

Gamete – A male or female reproductive cell or the nuclei within a gametangium.

Gel – A jellylike colloidal mass.

Gene – A material substance in the chromosome which determines or conditions one or more hereditary characters. The smallest functioning unit of the genetic material.

Genotype – The aggregate of genes in an organism.

Germ tube – The early growth of mycelium produced by a germinated fungus spore.

Giant cell – A multinucleate mass of protoplasm formed by coalescence of several adjacent plant cells. Also called a syncytium. Found in plants infected by certain nematodes.

Gibberellins – A group of plant growth-regulating substances with a variety of functions.

Glycolysis – The breakdown of glucose to pyruvic acid.

Grafting – A method of plant propagation by transplantation of a bud or a scion of a plant on another plant. Also, the joining of cut surfaces of two plants so as to form a living union.

Growth inhibitor – A natural substance that inhibits the growth of a plant.

Growth regulator – A natural substance that regulates the enlargement, division, or activation of plant cells.

Gum – Complex polysaccharidal substances formed by cells in reaction to wounding or infection.

Gummosis – Production of gum by or in a plant tissue.

Guttation – Exudation of water from plants, particularly along the leaf margin.

Habitat – The natural place of occurrence of an organism.

Haploid — A cell or an organism whose nuclei have a single complete set of chromosomes.

Hatching factor — A material produced by the roots of certain plants that is believed to increase the hatching of eggs of certain nematodes.

Haustorium — A projection of hyphae into host cells which acts as a penetration and absorbing organ.

Herbaceous plant — A higher plant that does not develop woody tissues.

Hermaphrodite — An individual bearing both functional male and female reproductive organs.

Heteroecious — Requiring two different kinds of hosts to complete its life cycle. Pertaining particularly to rust fungi.

Heterokaryosis — The condition in which a mycelium contains two genetically different nuclei per cell.

Heterothallic fungi — Fungi producing compatible male and female gametes on physiologically distinct mycelia.

Heterotrophic — Depending on an outside source for organic nutrients.

Homothallic fungus — A fungus producing compatible male and female gametes on the same mycelium.

Hormone — A growth regulator. Frequently referring particularly to auxins.

Host — A plant that is invaded by a parasite and from which the parasite obtains its nutrients.

Host range — The various kinds of host plants that may be attacked by a parasite.

Hyaline — Colorless, transparent.

Hybrid — The offspring of two individuals differing in one or more heritable characteristics.

Hybridization — The crossing of two individuals differing in one or more heritable characteristics.

Hydathodes — Structures with one or more openings that discharge water from the interior of the leaf to its surface.

Hydrolysis — The enzymatic breakdown of a compound through the addition of water.

Hyperplasia — A plant overgrowth due to increased cell division.

Hypersensitivity — Excessive sensitivity of plant tissues to certain pathogens. Affected cells are killed quickly, blocking the advance of obligate parasites.

Hypertrophy — A plant overgrowth due to abnormal cell enlargement.

Hypha — A single branch of a mycelium.

Immune — Exempt from infection by a given pathogen.

Immunity — The state of being immune.

Imperfect fungus — A fungus that is not known to produce sexual spores.

Imperfect stage — The part of the life cycle of a fungus in which no sexual spores are produced.

Incubation period — The period of time between penetration of a host by a pathogen and the first appearance of symptoms on the host.

Indexing — A procedure to determine whether a given plant is infected by a virus. It involves the transfer of a bud, scion, sap, etc. from one plant to one or more kinds of (indicator) plants that are sensitive to the virus.

Indicator — A plant that reacts to certain viruses or environmental factors with production of specific symptoms and is used for detection and identification of these factors.

Infection — The establishment of a parasite within a host plant.

Infectious disease — A disease that is caused by a pathogen which can spread from a diseased to a healthy plant.

Infested — Containing great numbers of insects, mites, nematodes, etc. as applied to an area or field. Also applied to a plant surface or soil contaminated with bacteria, fungi, etc.

Injury — Damage of a plant by an animal, physical, or chemical agent.

Inoculate — To bring a pathogen into contact with a host plant or plant organ.

Inoculation — The arrival or transfer of a pathogen onto a host.

Inoculum — The pathogen or its parts

that can cause disease. That portion of individual pathogens that are brought into contact with the host.

Integrated control — An approach that attempts to use all available methods of control of a disease or of all the diseases and pests of a crop plant for best control results but with the least cost and the least damage to the environment.

Intercalary — Formed along and within the mycelium — not at the hyphal tips.

Intercellular — Between cells.

Intracellular — Within or through the cells.

Invasion — The spread of a pathogen into the host.

In vitro — In culture. Outside the host.

In vivo — In the host.

Isolate — A single spore or culture and the subcultures derived from it. Also used to indicate collections of a pathogen made at different times.

Isolation — The separation of a pathogen from its host and its culture on a nutrient medium.

L-form bacteria — Bacteria that have, temporarily or permanently, lost the ability to produce a cell wall as a result of growth in the presence of antibiotics inhibiting cell wall synthesis.

Larva — The life stage of a nematode between the embryo and the adult; an immature nematode.

Latent infection — The state in which a host is infected with a pathogen but does not show any symptoms.

Latent virus — A virus that does not induce symptom development in its host.

Leaf spot — A self-limiting lesion on a leaf.

Lenticel — A structure of the bark, some fruits, etc. which permits the inward and outward passage of gases.

Lesion — A localized area of discolored, diseased tissue.

Life cycle — The stage or successive stages in the growth and development of an organism that occur between the appearance and reappear-ance of the same stage (e.g., spore) of the organism.

Lignin — A complex organic substance or group of substances that impregnates the cell walls of xylem vessels and certain other plant cells.

Lipase — An enzyme that breaks fats into glycerin and fatty acids.

Lipids — Substances whose molecules consist of glycerin and fatty acids and sometimes certain additional types of compounds.

Local lesion — A localized spot produced on a leaf upon mechanical inoculation with a virus.

Lumen — The cavity of the cell within the cell walls.

μ (micron) — A unit of length equal to 1/1000 of a millimeter.

mμ (millimicron) — A unit of length equal to 1/1000 of a micron.

mm (millimeter) — A unit of length equal to 1/10 of a centimeter (cm) or 0.03937 of an inch.

Macroscopic — Visible without the aid of a magnifying lense or a microscope.

Malignant — Used of a cell or tissue that divides and enlarges autonomously, i.e., its growth can no longer be controlled by the organism on which it is growing.

Masked symptoms — Virus-induced plant symptoms that are absent under certain environmental conditions but appear when the host is exposed to certain conditions of light and temperature.

Mechanical inoculation — Inoculation of a plant with a virus through transfer of sap from a virus-infected plant to a healthy plant.

Meristem — The undifferentiated tissue the cells of which can divide continually and differentiate into specialized tissues.

Mesophyll — The leaf parenchyma cells between epidermal layers.

Messenger RNA — A chain of ribonucleotides that codes for a specific protein.

Metabolism — The process by which cells or organisms utilize nutritive material to build living matter and structural components or to break down cellular material into simple substances to perform special functions.

Microscopic – Very small; can be seen only with the aid of a microscope.

Middle lamella – The cementing layer between adjacent cell walls; it generally consists of pectinaceous materials, except in woody tissues, where pectin is replaced by lignin.

Migratory – Migrating from plant to plant.

Mildew – A fungal disease of plants in which the mycelium and spores of the fungus are seen as a whitish growth on the host surface.

Mold – Any profuse or woolly fungus growth on damp or decaying matter or on surfaces of plant tissue.

Molt – The shedding or casting off of the cuticle.

Mosaic – Symptom of certain viral diseases of plants characterized by intermingled patches of normal and light green or yellowish color.

Mottle – An irregular pattern of indistinct light and dark areas.

Mummy – A dried, shriveled fruit.

Mutant – An individual possessing a new, heritable characteristic as a result of a mutation.

Mutation – An abrupt appearance of a new characteristic in an individual as the result of an accidental change in genes or chromosomes.

Mycelium – The hypha or mass of hyphae that make up the body of a fungus.

Mycoplasmas – Pleomorphic microorganisms that, like the bacteria, lack an organized and bounded nucleus but, unlike the bacteria, also lack a true cell wall and the ability to synthesize the substances to form a cell wall.

Mycoplasmalike organisms – Microorganisms found in the phloem and phloem parenchyma of diseased plants and assumed to be the causes of the disease; they resemble mycoplasmas in all respects except that they cannot yet be grown on artificial nutrient media.

Mycorrhiza – A symbiotic association of a fungus with the roots of a plant.

Mycotoxicoses – Diseases of animals and humans caused by consumption of feeds and foods invaded by fungi that produce mycotoxins.

Mycotoxins – Toxic substances produced by several fungi in infected seeds, feeds, or foods and capable of causing illnesses of varying severity and death to animals and humans that consume such substances.

Natural openings – Stomata, lenticels, hydathodes, and nectarthodes.

Necrotic – Dead and discolored.

Nectarthode – An opening at the base of a flower from which nectar exudes.

Nematicide – A chemical compound or physical agent that kills or inhibits nematodes.

Nematode – Generally microscopic, wormlike animals that live saprophytically in water or soil, or as parasites of plants and animals.

Noninfectious disease – A disease that is caused by an environmental factor, not by a pathogen.

Nucleic acid – An acidic substance containing pentose, phosphorus, and pyrimidine and purine bases. Nucleic acids determine the genetic properties of organisms.

Nucleolus – A dense protoplasmic body within the nucleus.

Nucleoprotein – Referring to viruses: consisting of nucleic acid and protein.

Nucleoside – The combination of a sugar and a base molecule in a nucleic acid.

Nucleotide – The phosphoric ester of a nucleoside. Nucleotides are the building blocks of DNA and RNA.

Nucleus – The dense protoplasmic body found in all cellular organisms and which is essential in all synthetic and developmental activities of a cell.

Obligate parasite – A parasite that in nature can grow and multiply only on living organisms.

Oogonium – The female gametangium

of some phycomycetes (Oomycetes) containing one or more gametes.

Oomycete – A fungus that produces oospores. An order of the phycomycetes.

Oospore – A sexual spore produced by the union of two morphologically different gametangia (oogonium and antheridium).

Osmosis – The diffusion of a solvent through a differentially permeable membrane.

Ostiole – A porelike opening in perithecia and pycnidia through which the spores escape from the fruiting body.

Ovary – The female reproductive structure that produces or contains the egg.

Oxidation – A chemical reaction in which oxygen combines with another substance or in which hydrogen atoms or electrons are removed from a substance.

Oxidative phosphorylation – The utilization of energy released by the oxidative reactions of respiration to form high-energy ATP bonds.

Ozone (O_3) – A highly reactive form of oxygen that in relatively high concentrations may injure plants.

Palisade parenchyma – Elongated cells found just beneath the upper epidermis of leaves and containing chloroplasts.

PAN (Peroxyacyl nitrates) – Air pollutants produced as by-products in the exhausts of internal combustion engines and which are injurious to plants.

Papillate – Bearing a papilla, i.e., a hump or swelling.

Paraphysis – A sterile hypha present in some fruiting bodies of fungi.

Parasexualism – A mechanism whereby recombination of hereditary properties is based on mitosis.

Parasite – An organism living on or in another living organism (host) and obtaining its food from the latter.

Parenchyma – A tissue composed of thin-walled cells which usually leave intercellular spaces between them.

Pathogen – An entity that can incite disease.

Pathogenicity – The relative capability of a pathogen to cause disease.

Pectin – A methylated polymer of galacturonic acid found in the middle lamella and the primary cell wall.

Pectinase – An enzyme that breaks down pectin.

Penetration – The initial invasion of a host by a pathogen.

Pentose pathway – Oxidation of glucose by elimination of one carbon atom as CO_2 and formation of five-carbon sugars.

Perfect stage – The sexual stage (e.g., fruiting bodies) in the life cycle of a fungus.

Pericycle – Tissue generally found in the root and bound externally by the endodermis and internally by the phloem.

Perithecium – The globular or flask-shaped ascocarp of the Pyrenomycetes, having an opening or pore (ostiole).

Phage – A virus that attacks bacteria; also called bacteriophage.

Phellogen – Also called cork cambium; it is cambium giving rise externally to cork and in some plants internally to phelloderm.

Phenolic – Applied to a compound that contains one or more phenolic rings.

Phenotype – The external visible appearance of an organism.

Phloem – Food-conducting tissue, consisting of sieve tubes, companion cells, phloem parenchyma, and fibers.

Photosynthesis – The process by which carbon dioxide and water are combined in the presence of light and chlorophyll to form carbohydrate.

Phycomycetes – A group of fungi whose mycelium has no cross walls.

Phytoalexin – A substance which inhibits the development of a fungus on hypersensitive tissue, formed mainly when host plant cells come in contact with the parasite.

Phytopathogenic – Term applicable to a microorganism that can incite disease in plants.

Phytotoxic – Toxic to plants.

Pistil – The central organ of flowers, typically consisting of ovary, style, and stigma.

Plasmalemma – The cytoplasmic membrane found on the outside of the protoplast adjacent to the cell wall.

Plasmid – A self-replicating piece of DNA that is stably inherited in an extrachromosomal state; generally not required for survival of the organism.

Plasmodesma (Plural = plasmodesmata) – A fine protoplasmic thread connecting two protoplasts and passing through the wall which separates the two protoplasts.

Plasmodium – A naked, slimy mass of protoplasm containing numerous nuclei.

Plasmolysis – The shrinking and separation of the cytoplasm from the cell wall due to removal of water from the protoplast.

Plerome – The plant tissues inside the cortex.

Polyhedron – A spheroidal particle or crystal with many plane faces.

Polysaccharide – A large organic molecule consisting of many units of a simple sugar.

Polysome (or polyribosome) – A cluster of ribosomes associated with a messenger RNA.

Precipitin – An antibody that causes precipitation of soluble antigens.

Primary infection – The first infection of a plant in the spring by the overwintering pathogen.

Primary inoculum – The overwintering pathogen or its spores that cause primary infections.

Proliferation – A rapid and repeated production of new cells, tissues, or organs.

Promycelium – The short hypha produced by the teliospore; the basidium.

Propagative virus – A virus that multiplies in its insect vector.

Protectant – A substance that protects an organism against infection by a pathogen.

Protein – A high-molecular-weight compound consisting of amino acids. It may be a structural protein or an enzyme.

Protein subunit – A small protein molecule that is the structural and chemical unit of the protein coat of a virus; a capsomere.

Protophloem – The conductive tissue of actively growing parts of the plant. Its sieve tubes function for a brief period and are replaced by metaphloem elements.

Protoplast – The organized living unit of a single cell; the cytoplasmic membrane and everything inside it.

Protoxylem – The conductive tissue that appears at the beginning of vascular differentiation and usually matures before the organ completes its elongation; it is followed by formation of metaxylem which matures after tissue elongation is finished.

Purification – The separation of virus particles in a pure form, free from cell components.

Pustule – Small blisterlike elevation of epidermis as spores emerge.

Pycnidium – An asexual, spherical, or flask-shaped fruiting body lined inside with conidiophores and producing conidia.

Pycniospore – Also called a spermatium. A spore produced in a pycnium.

Pycnium – Also called a spermagonium. A fruiting body of the rust fungi that produces small spores called pycniospores or spermatia which cannot infect plants but function as gametes or gametangia.

Quarantine – Control of import and export of plants to prevent spread of diseases and pests.

Race – A genetically and often geographically distinct mating group within a species; also a group of pathogens that infect a given set of plant varieties.

Receptive hypha – A specialized hypha protruding out of a pycnium and functioning as a female gamete or gametangium.

Reduction – Any chemical reaction involving the removal of oxygen from or the addition of hydrogen to a substance; it occurs with concomitant expenditure of energy.

Resistance – The ability of an organism to overcome, completely or in some degree, the effect of a pathogen or other damaging factor.

Resistant – Possessing qualities that hinder the development of a given pathogen.

Respiration – A series of chemical oxidations within the cell controlled and catalyzed by enzymes in which carbohydrate and fats are broken down, releasing energy to be used by the cell or organism in its various functions.

Resting spore – A sexual or other thick-walled spore of a fungus that is resistant to extremes in temperature and moisture and which often germinates only after a period of time from its formation.

Rhizoid – A short, thin hypha growing in a rootlike fashion toward the substrate.

Rhizosphere – The soil near a living root.

Ribosome – A subcellular particle involved in protein synthesis.

Rickettsiae – Microorganisms similar to bacteria in most respects but generally capable of multiplying only inside living host cells; parasitic or symbiotic.

Rickettsialike bacteria – Bacteria found in the xylem or phloem of diseased plants and assumed to be the causes of the disease; they are arthropod-transmitted, and most of them have not yet been grown on artificial nutrient media.

Ringspot – A circular area of chlorosis with a green center; a symptom of many virus diseases.

RNA (Ribonucleic acid) – A nucleic acid involved in protein synthesis; also, the only nucleic acid (genetic material) of many viruses.

RNase (Ribonuclease) – An enzyme that breaks down RNA.

Rosette – Short, bunchy habit of plant growth.

Rot – The softening, discoloration, and often disintegration of a succulent plant tissue as a result of fungal or bacterial infection.

Russet – Brownish roughened areas on skin of fruit as a result of cork formation.

Rust – A disease giving a "rusty" appearance to a plant and caused by one of the Uredinales (rust fungi).

Sanitation – The removal and burning of infected plant parts, decontamination of tools, equipment, hands, etc.

Saprophyte – An organism that uses dead organic material for food.

Scab – A roughened, crustlike diseased area on the surface of a plant organ. A disease in which such areas form.

Scion – A piece of twig or shoot inserted on another in grafting.

Sclerotium – A compact mass of hyphae with or without host tissue, usually with a darkened rind, and capable of surviving under unfavorable environmental conditions.

Scorch – "Burning" of leaf margins as a result of infection or unfavorable environmental conditions.

Scutellum – The single cotyledon of grass embryo.

Secondary infection – Any infection caused by inoculum produced as a result of a primary or a subsequent infection; an infection caused by secondary inoculum.

Secondary inoculum – Inoculum produced by infections that took place during the same growing season.

Sedentary – Staying in one place; stationary.

Septate – Having cross walls.

Septum – A cross wall (in a hypha or spore).

Serology – A method using the specificity of the antigen-antibody reaction for the detection and identification of antigenic substances and the organisms that carry them.

Serum – The watery portion of the blood remaining after coagulation.

Sexual – Participating in or produced as a result of a union of nuclei in which meiosis takes place.

Shock symptoms – The severe, often necrotic symptoms produced on the first new growth following infection with some viruses; also called acute symptoms.

Shot-hole – A symptom in which small diseased fragments of leaves fall off and leave small holes in their place.

Sieve plate – Perforated wall area between two phloem cells through which their protoplasts are connected.

Sieve tube – A series of phloem cells forming a long cellular tube through which food materials are transported.

Sign – The pathogen or its parts or products seen on a host plant.

Slime molds – Fungi of the class Myxomycetes; also, superficial diseases caused by these fungi on low-lying plants.

Smut – A disease caused by the smut fungi (Ustilaginales); it is characterized by masses of dark, powdery spores.

Sooty mold – A sooty coating on foliage and fruit formed by the dark hyphae of fungi that live in the honeydew secreted by insects such as aphids, mealybugs, scales, and whiteflies.

Sorus – A compact mass of spores or fruiting structure found especially in the rusts and smuts.

Spermagonium (or pycnium) – A fruiting body of the rust fungi in which the gametes or gametangia are produced.

Spermatheca – An enlarged portion of the female nematode reproductive system between the oviduct and uterus in which sperm is stored.

Spermatium (or pycniospore) – The male gamete or gametangium of the rust fungi.

Spiroplasmas – Pleomorphic, wall-less microorganisms that are present in the phloem of diseased plants. They are often helical in culture and are thought to be a kind of mycoplasma.

Sporangiophore – A specialized hypha bearing one or more sporangia.

Sporangiospore – Nonmotile, asexual spore borne in a sporangium.

Sporangium – A container or case of asexual spores.

Spore – The reproductive unit of fungi consisting of one or more cells; it is analogous to the seed of green plants.

Sporidium – The basidiospore of the smut fungi.

Sporodochium – A fruiting structure consisting of a cluster of conidiophores woven together on a mass of hyphae.

Sporophore – A hypha or fruiting structure bearing spores.

Sporulate – To produce spores.

Spur – A short twig on which much of the fruit of many trees is produced.

Starch – A polysaccharide consisting of glucose units; the principal food storage substance of plants.

Stele – The central cylinder, inside the cortex, of roots and stems of vascular plants.

Stem-pitting – A symptom of some viral diseases characterized by depressions on the stem of the plant.

Sterigma – A slender protruberance on a basidium that supports the basidiospore.

Sterile fungi – A group of fungi that are not known to produce any kind of spores.

Sterilization – The elimination of pathogens from soil by means of heat or chemicals.

Stolon – A hypha of some fungi that grows horizontally along the surface of the substrate.

Stoma (plural = stomata) – A minute, organized opening on the surface of leaves or stems through which gases pass.

Strain – The descendants of a single isolation in pure culture; an isolate. Also a group of similar isolates; a race. In plant viruses, a group of virus isolates having most of their antigens in common.

Stroma – A compact mycelial structure on or in which fructifications are usually formed.

Stylet – A long, slender, hollow feeding structure of nematodes and some insects.

Stylet-borne – A virus borne on the stylet of its vector; a noncirculative virus.

Substrate – The material or substance on which a microorganism feeds and develops. Also, a substance acted upon by an enzyme.

Succulent – A plant having tender, juicy, or watery tissues.

Suscept – Any plant that can be attacked by a given pathogen; a host plant.

Susceptible – Lacking the inherent ability to resist disease or attack by a given pathogen; nonimmune.

Susceptibility – The inability of a plant to resist the effect of a pathogen or other damaging factor.

Symbiosis – A mutually beneficial association of two different kinds of organisms.

Symptom – The external and internal reactions or alterations of a plant as a result of a disease.

Symptomless carrier – A plant which, although infected with a pathogen (usually a virus), produces no obvious symptoms.

Syncytium – A multinucleate mass of protoplasm surrounded by a common cell wall.

Synergism – The concurrent parasitism of a host by two pathogens in which the symptoms or other effects produced are of greater magnitude than the sum of the effects of each pathogen acting alone.

Systemic – Spreading internally throughout the plant body; said of a pathogen or a chemical.

Teliospore – The sexual, thick-walled resting spore of the rust and smut fungi.

Telium – The fruiting structure in which teliospores are produced.

Terminal oxidation – The oxidation of respiratory substrates and intermediates by the transfer of electrons (H+ ions) via various carriers to compounds (cytochromes) which are capable of yielding electrons to O_2, forming H_2O.

Tissue – A group of cells of similar structure which performs a special function.

Tolerance – The ability of a plant to sustain the effects of a disease without dying or suffering serious injury or crop loss. Also, the amount of toxic residue allowable in or on edible plant parts under the law.

Toxicity – The capacity of a compound to produce injury.

Toxin – A compound produced by a microorganism and being toxic to a plant or animal.

Transduction – The transfer of genetic material from one bacterium to another by means of a bacteriophage.

Transfer RNA (tRNA) – The RNA that moves amino acids to the ribosome to be placed in the order prescribed by the messenger RNA.

Transformation – The change in the DNA of a bacterium by absorption and incorporation of DNA fragments released by another bacterium. Also, the change of a normal to a malignant cell.

Translocation – Transfer of nutrients or virus through the plant.

Transmission – The transfer or spread of a virus or other pathogen from one plant to another.

Transpiration – The loss of water vapor from the surface of leaves.

Tumor – A malignant overgrowth of tissue or tissues.

Tylosis – An overgrowth of the protoplast of a parenchyma cell into an adjacent xylem vessel.

Uredium – The fruiting structure of the rust fungi in which uredospores are produced.

Uredospore – A binucleate, repeating spore of the rust fungi.

Variability – The property or ability of an organism to change its characteristics from one generation to the other.

Vascular — Term applied to a plant tissue or region consisting of conductive tissue; also, to a pathogen that grows primarily in the conductive tissues of a plant.

Vector — An animal able to transmit a pathogen.

Vegetative — Asexual; somatic.

Vesicle — A bubblelike structure produced by a zoosporangium and in which the zoospores are released or are differentiated.

Vessel — A xylem element or series of such elements whose function is to conduct water and mineral nutrients.

Virescent — A normally white or colored tissue that develops chloroplasts and becomes green.

Virion — A complete virus particle.

Viroids — Small, low-molecular-weight ribonucleic acids (RNA) that can infect plant cells, replicate themselves, and cause disease.

Virulence — The degree of pathogenicity of a given pathogen.

Virulent — Capable of causing a severe disease; strongly pathogenic.

Viruliferous — Said of a vector containing a virus and capable of transmitting it.

Virus — A submicroscopic obligate parasite consisting of nucleic acid and protein.

Xylem — A plant tissue consisting of tracheids, vessels, parenchyma cells, and fibers; wood.

Wilt — Loss of rigidity and drooping of plant parts generally caused by insufficient water in the plant.

Witches' broom — Broomlike growth or massed proliferation caused by the dense clustering of branches of woody plants.

Yellows — A plant disease characterized by yellowing and stunting of the host plant.

Zoosporangium — A sporangium which contains or produces zoospores.

Zoospore — A spore bearing flagella and capable of moving in water.

Zygospore — The sexual or resting spore of zygomycetes produced by the fusion of two morphologically similar gametangia.

Zygote — A diploid cell resulting from the union of two gametes.

index

F   3
G   4
H
I   5
J   6